NATURAL PRODUCTS AND MOLECULAR THERAPY

ANNALS OF THE NEW YORK ACADEMY OF SCIENCES
Volume 1056

NATURAL PRODUCTS AND MOLECULAR THERAPY

Edited by Girish J. Kotwal and Debomoy K. Lahiri

The New York Academy of Sciences
New York, New York
2005

∞ The paper used in this publication meets the minimum requirements of the American National Standard for Information Sciences—Permanence of Paper for Printed Library Materials, ANSI Z39.48-1984.

Library of Congress Cataloging-in-Publication Data has been applied for.

GYAT/PCP

Printed in the United States of America

ISBN 1-57331-594-X (cloth)

ISBN 1-57331-595-8 (paper)

ISSN 0077-8923

ANNALS OF THE NEW YORK ACADEMY OF SCIENCES

Volume 1056
November 2005

NATURAL PRODUCTS AND MOLECULAR THERAPY

Editors
GIRISH J. KOTWAL AND DEBOMOY K. LAHIRI

This volume is the result of the **First International Conference on Natural Products and Molecular Medicine**, which was held on January 13–15, 2005, at the University of Cape Town, Cape Town, South Africa.

CONTENTS

Part II. Structural Biology of Natural Proteins

Part III. Natural Enhancers of Health and Fertility

Part IV. Traditional African, Asian, and Mediterranean

Part V. Antiviral, Antimicrobial Live Vaccine and Plant Vaccines

Part VI. Cutting Edge Technologies and Delivery System

Part VII. Natural Anticancer Therapy

Part VIII. Neuroprotection

Part IX. Wound Healing and Laser Therapy

Preface

The First International Conference on Natural Products and Molecular Therapy was held in Cape Town, South Africa, from January 13–15, 2005. The aim of this international conference was to understand the protective and therapeutic roles of various natural products for the most significant diseases, such as AIDS, atherosclerosis, Alzheimer's disease, arthritis, avian flu, and cancer, that currently pose a great threat to global human well-being and health. The major goal of this meeting was to comprehend the cellular, molecular bases of action of different promising natural products and their applications to various diseases, including cancer, brain disorders, and infectious diseases. Another major goal was to discuss the scrutiny and evaluation of promising natural therapies and advance the need for rigorous scientific analysis of natural products, especially with regard to their short-term and longer-term toxicity and side effects. The participants of this conference also discussed how neuroprotection could be achieved by herbal extracts, plant products, and dietary supplements of hormones. The role of antioxidants and free radicals in health and disease was another major theme of discussion. The choice of South Africa as a venue was appropriate on the basis of its vast untapped natural products and the diversity of plant species that remain to be studied. In addition, South Africa has a first-world infrastructure when it comes to organizing meetings and scientific dialogues.

In addition to the mechanistic studies, this conference was able to maintain the focus on research findings involving botanical, nutritional, and pharmacological strategies for improving general health and slowing the normal aging process. To this end, the core of the meeting was the assembly of a group of scientists well-recognized for their molecular and mechanistic work in this area and its wide-range applications. These researchers presented their studies and, most importantly, provided a nucleus for the exchange of information and development of new ideas. The goal of this meeting, to bring together a critical mass of scientists who might not otherwise have the opportunity to discuss key issues in an informal and relaxed atmosphere, was met. This resulted in the development of new concepts concerning preventive strategy against human diseases and their mitigation, as discussed in the selected articles compiled in the present volume of the *Annals*. By this means, the meeting was successful in promoting the development of fresh research perspectives and therapeutic possibilities using the judicious application of natural products, either alone or in combination with the existing pharmacological agents and hormone, directed towards better health, body, and mind.

The major themes in this conference centered around the cellular, molecular, and structural aspects of natural products to understand their mode of action. Some of the topics that were covered were as follows: structural biology of therapeutic natural products, therapeutic and preventive aspects of nutrition, cellular and biochemical bases for African medicine, ayurvedic medicine, Chinese and other East Asian medicine; natural enhancement of fertility; natural antivirals against HIV and other viral infections; and antimicrobial agents. The themes also included the enhancement and neuroprotection of central nervous system function, natural therapy of can-

Ann. N.Y. Acad. Sci. 1056: xi–xiii (2005). © 2005 New York Academy of Sciences.
doi: 10.1196/annals.1352.41

cer, natural vaccines, enhancement of cardiovascular function, molecular bases for digestive tract natural therapy and natural products. A separate session was solely devoted to the issues related to the commercialization aspects of some of these approaches.

Some of the major highlights of this conference were the reports on the discovery of a plant extract with broad-spectrum antiviral activity against several viral agents including HIV, the anti-HIV activity of pomegranate juice, the anti-inflammtory activities of the vaccinia virus complement control protein, and the multiple activities of garlic and curcumin. In addition, the promise of melatonin in neuroprotection and brain disorders and the potential of dietary supplementation of melatonin to reverse the aging process were recognized. On the technique side, microarray techniques, molecular cloning, and state-of-the-art methods for prediction of long-term mutagenesis were other highlights of the conference.

The conference was sponsored by several local and international agencies: the Medical Research Council of South Africa, the Department of Research and Innovation, University of Cape Town, the Division of Medical Virology, Institute of Infectious Diseases and Molecular Medicine (IIDMM), University of Cape Town, K-Biotech, Pvt. Ltd., Merck of South Africa, Acorn Technologies, Set-point scientific, K-Biotech, Pvt. Ltd., Separation Scientific, Old Mutual Insurance, and Inqaba biotec, among others. The editors also thank members of the organizing committee, the session chairs, and Prof. Tony Mbewu, for their contribution to the success of the meeting.

The editors gratefully acknowledge the assistance of Mr. Amod Kulkarni in the preparation of the proceedings and in proposing a cover design. The editors acknowledge the valuable contribution of the following persons, who assisted the editors in the peer-review process of the science and/or text of the manuscripts:

Acknowledgments

John P. Atkinson (Washington University, St. Louis, MO, USA); Jason Bailey (Indiana University School of Medicine, Indianapolis, IN, USA); Paul Barlow (University of Edinburgh, Edinburgh, UK); Eric Block (University of Albany, Albany, NY, USA); Chris Boshoff (the Wolfson Institute for Biomedical Research, University College London, London, UK); Xufen Le Bourhis (Université des Sciences et technolgies de Lille Batiment Rennes, Cedex, France); Santy Daya (Rhodes University, Grahamstown, South Africa); Linda Dixon (Institute for Animal Health Pirbright Lab., Pirbright, Woking, Surrey, UK); Carl Groth (Karolinska Institute, Stockholm, Sweden); Y.K. Gupta (Director Industrial Toxicology research Centre, Lucknow, India); Lawrence A. Hunt (University of Louisville School of Medicine, Louisville, KY, USA); Stuart Isaacs (University of Pennsylvania, Philadelphia, PA, USA); Poonam Kakkar (Industrial Toxicology Research Centre, Lucknow, India); Anumantha Kanthasamy (Department of Biomedical Sciences, Iowa State University, IO, USA); Girish J. Kotwal (Division of Medical Virology, IIDMM, UCT, Cape Town, South Africa); Amod P. Kulkarni (Division of Medical Virology, IIDMM, UCT, Cape Town, South Africa); Debomoy K. Lahiri (Indiana University School of Medicine, Indianapolis, IN, USA); Samuel B. Lehrer (Tulane University, New Orleans, LA, USA); Bryan Maloney (Indiana University School of Medicine, IN, USA); Lionel Opie (Cape Heart Centre and the institute, Chris Bernard Building,

UCT, Medical School, UCT, Cape Town, South Africa); Timothy Skern (Division of Medical Virology, IIDMM, UCT, Cape Town, South Africa); Scott Smith (University of Louisville School of Medicine, Louisville, KY, USA); Edward Sturrock (Division of Medical Biochemistry, IIDMM, UCT, Cape Town, South Africa); Anna Lise Williamson (Division of Medical Virology, IIDMM, Cape Town, South Africa).

One of us (G.J.K.) thanks The Wellcome Trust of England for making it possible for him to work in South Africa and for research support through the funding received as a Senior International Wellcome Trust Fellow for Biomedical Sciences in South Africa, and the other (D.K.L.) thanks the National Institutes of Health for his research and Bryan Maloney for editorial assistance.

The abovementioned description is just a glimpse of what was presented in the conference and may fail to do justice to a wide range of the sessions and programs in what was a truly remarkable gathering. The consensus in the meeting was that natural products are still a viable approach in terms of their protective and therapeutic roles, which could translate into improving human health. Obviously, more research and funding are necessary to fully utilize their potentials against the most significant diseases. With proper scientific research and the implementation of adequate safeguards, various natural products can be used alone therapeutically or can be complemented with other currently used pharmacological interventions. On that positive note, we all await the next meeting being planned in the same venue in the near future.

—GIRISH JAYANT KOTWAL
—DEBOMOY KUMAR LAHIRI

Introductory Remarks

African Perspective on Natural Products and Molecular Medicine

PROF. A.D. MBEWU

President of the Medical Research Council of South Africa

INTRODUCTION

Natural products have been an exceptionally rich source of leads for commercially successful drugs in several major therapeutic categories.[1] Yet, until recently, research in this field was regarded as fringe and slightly odd. This landmark conference in the field of "natural products and molecular therapy" belies that notion, and comforts one in the belief that this is a valid field of scientific inquiry. The faculty of scientists gathered here, from Brazil to Japan, from Iceland to South Africa, has been truly remarkable, uniting institutes in both developing and developed countries. This gathering suggests that the 250,000 plant species worldwide, the millions of fungi, bacterial, and marine organisms, and the billions upon billions of natural molecules extant in the world could really contain solutions to the diseases that afflict mankind.

The breadth of scientific disciplines represented has been very broad, ranging from practitioners of alternative and complementary medicine to structural biologists studying computer-generated images of the molecular structure of natural medicines.

The determination by the organizing committee to showcase the union of the disciplines of natural medicine with those of molecular medicine in the pursuit of greater understanding of natural products was truly inspirational, a high-risk strategy that has proven very successful. We hope that this will not be the last of such gatheringsand feel honored that South Africa was chosen to host the first of such meetings. For us as South Africans, the study of indigenous knowledge systems is not simply scientific endeavor. It provides the opportunity to reclaim our scientific and sociocultural heritage, which was for so long stigmatized and discredited by the systems of colonialism and apartheid as primitive rituals and witchcraft.

Traditional knowledge systems are indeed much older than the 150-year-old field of allopathic medicine and draw on the rich heritage and knowledge of the earliest civilizations of the world in Central America, Africa, India, and China.

The South African Ministry of Health has developed the legislative framework to underpin the development of natural products for human health through the Medicines and Related Substances Act, which seeks to regularize the use of complementary, alternative, and African traditional medicines in South Africa in order to protect the public from bogus products, whilst avoiding the pitfall of putting such products in the same regulatory environment as pharmaceuticals drugs, whose testing and control are a very different matter.

Ann. N.Y. Acad. Sci. 1056: xv–xix (2005). © 2005 New York Academy of Sciences.
doi: 10.1196/annals.1352.042

Similarly, the Ministry of Health has ensured that traditional practitioners are not subsumed by the medical and allied health professionals, but rather are able to regulate their own affairs and training through implementation of the Traditional Health Practitioners Act. Our 200,000 traditional practitioners are consulted by over 80% of our population (including the majority of those living with HIV and AIDS) and will in the future increasingly ensure the proper accreditation and licensing of bona fide healers.

Our President, Thabo Mbeki, often talks of the human solidarity that is needed to solve the problems of global poverty, underdevelopment, and diseases of mass burden. Reviewing the list of presenters makes one aware of a global solidarity in the pursuit of greater knowledge concerning natural medicines, a solidarity rarely seen in the selfish, individualistic, and materialistic modern world, save in the global response to massive disasters such as the Asian tsunami.

I often describe the rationale for natural medicines as nature consisting of a gigantic biotechnology facility with hundreds of thousands of product lines, each developed over the course of hundreds of millions of years of experimentation, in millions of experimental reaction vessels called living organisms. These are not simply computer-generated structures, but novel chemical entities that have been tested in nature's bioassays over the course of millennia for useful biological and physiological activity. Those that have been proven useless have been discarded in the billion-year-long drug development pipeline through natural selection. Where else would one go, therefore, to seek out new cures or clues that might lead to novel therapies?

PHYTOCHEMISTRY

In fact, that is exactly what the pharmaceutical and biotechnology industries have done over the past 120 years. Organ transplantation, for example, was transformed by the introduction of immunosuppressant drugs that are of natural origin such as cyclosporin-A, sirolimus (rapamycin), tacrolimus (FK-506), gusperimus (15-deoxyspergualin), and mizoribine (Allison, 2000).

Similarly, treatment of heart failure was revolutionized by the introduction of ACE inhibitors, resulting in real mortality reductions following myocardial infarction and heart failure. ACE inhibitors were developed from research on the bradykinin potentiating peptides from the venom of the snake *Bothrops jaracusa.*

Natural products have and continue to have an enormous impact on the development of anticancer drugs such as paclitaxel (or taxol from the Pacific yew *Taxus brevifolia* (Cragg, 1999; Cragg and Newman, 1999), doxectaxel, irinotecan, etoposide, and teniposide. Some are natural products and some are semisynthetic derivatives of natural products. Cragg *et al.* (1997) reported that over 60% of new antibacterials and anticancer drugs approved between 1983 and 1994 were derived from natural products.

In neuropharmacology, natural products have given rise to anticholinesterase drugs for Alzheimer's disease. These include hupeazine A from *Huperzia serrata,* galanthamine (from the Caucasian snowdrop *Galanthus nivalis*), and rivastigmine, a synthetic analogue of the naturally occurring physostigmine.

RECENTLY LAUNCHED MEDICINES DERIVED FROM NATURAL PRODUCTS

The Scrip Report of 2001 refers to the *Pharmaprojects* database showing 143 products derived from natural origins recently launched at that time. They come from plant (30), bacterial (16), fungal (8), and animal sources (73).

The therapeutic categories of these agents include hormonal (47), metabolic (14), cardiovascular (13), cancer (11), inflammatory (10), skin (10), infectious (9), neuropharmacological (8), and imuunopsuppressant (5) categories.

SYNTHETIC AND SEMISYNTHETIC DERIVATIVES

Africa was the origin of phytochemistry as records of an Egyptian herbarium in the British Museum show. For many centuries, plants were used in their native or partially purified state. Only in the last 150 years was the modern pharmaceutical industry born with the extraction of active chemical moieties from the plants to produce lead compounds for preclinical and clinical testing. In most instances, a synthetic or semisynthetic product was subsequently produced using the chemical structure of the plant extract as the template.

The rationale for this approach was that it enabled production of stable synthetic analogues that could be manufactured in bulk quantities, to faithfully reproducible standards and often with augmented potencies.

Even more important, however, was the fact that synthetic analogues, unlike plant extracts, can be patented, thus protecting the revenue stream of the company commercially exploiting the discovery, for a good 20 years or more.

The *Pharmaprojects* database in 2001 listed 83 natural products that were in clinical trials or in the registration phase of development. Cancer was the largest therapeutic category, followed by metabolic disorders (diabetes, obesity), neuropharmacological indications, and infections.

In addition, the database in 2001 listed 30 products of semisynthetic origin, which were directly related to lead compounds that were natural products. The Scrip Report of 2001 lists these in detail.

Of interest was andrographolide from the Chinese medicinal plant *Andrographis paniculata* under investigation for its antiviral properties.

Calanolides and the structurally related inophyllums were isolated from plants in the genus *Calophyllum,* and were able to inhibit HIV-1 viral replication. They were isolated from a Malaysian tree (McKee, 1998) and could represent a novel class of non-nucleoside reverse transcriptase inhibitors. In 2001 they were entering Phase II clinical trials in a joint venture between an American biotech company and the state of Sarawak in Malaysia.

NATURAL PRODUCTS IN THEIR ORIGINAL STATE

Nevertheless, some natural products seem more efficacious in their native state. For example, one antituberculosis compound being developed in South Africa is a large molecule that seems too unstable for faithful production of a synthetic analogue.

The Medical Research Council (MRC), therefore, in its indigenous knowledge research, follows a twin-track approach of classical phytochemistry as well as the use of unprocessed natural products in clinical trials.

PATHOPHYSIOLOGY

Natural products, however, do not simply provide a source for novel lead compounds. They have the potential to be used in fundamental research, unraveling the etiology and pathophysiology of disease. Such greater understanding derived from investigating their mechanisms of action could be useful in developing interventions in different therapeutic classes.

Paclitaxel, for example, revealed much about the pathophysiology of cancer because of its novel molecular mechanism of action. It prevents cell division by stabilizing microtubules, leading to cell death. Once the mechanism was known, it was possible to screen for other compounds with similar actions.

The discovery of the combrestatins, extracted from the bark of a tree in Kwazulu Natal, South Africa, gave a stimulus to the science of angiostasis and the birth of an entirely new therapeutic class of anticancer compounds.

Similarly, the study of the physiological and molecular biological effects of natural products with immune-modulating properties could provide clues as to the correlates of immune protection in HIV disease and AIDS, thus assisting in HIV vaccine development.

CONCLUSION

South Africa contains 25,000 unique species of plants, one tenth of the world's known floral biodiversity. Apparently there are more plant species on Table Mountain than in the entire British Isles. In addition, it contains much unique biodiversity in fauna and potentially enormous untapped marine biodiversity off its shores.

Over 3,000 of these plants and animal products are used by the 200,000 traditional practitioners in South Africa, who are consulted by up to 80% of the population. These practices are hundreds, if not thousands, of years old and constitute what we call indigenous knowledge. It seems fitting, therefore, that the study of medicine that began over 5,000 years in Africa in the medicinal herbariums of Ancient Egypt should return to African shores.

The South African MRC has a very active natural products research program. In some instances, our scientists have extracted and characterized the active chemical moieties for possible development as novel drugs. In other cases, the approach has been to use the natural product in its native state and study its safety and efficacy.

We look forward, therefore, to the exciting developments ahead in this quest to use the methodologies of molecular medicine in the study of natural products. We are impatient to reap the knowledge that arises from your endeavors so that we can increasingly understand the workings of our own bodies, minds, and spirits and how all three interact with the environment and with our fellow human beings in society. The application of such knowledge, we are sure, will result in the promotion of health, prevention of sickness, and amelioration and cure of many of the diseases that afflict South Africans and human beings throughout the world. Through these endeavors we are confident that indigenous knowledge systems research will provide health, economic, and cultural benefits for both South Africa and the world

REFERENCES

1. NATURAL PRODUCT PHARMACEUTICALS. May 2001. A Diverse Approach to Drug Discovery. Scrip Reports. Alan Harvey. PJB Publications Ltd. London.
2. PHARMAPROJECTS. CD ROM. Feb. 2001. Richmond, PJB Publications Ltd. London.

Vaccinia Virus Complement Control Protein Diminishes Formation of Atherosclerotic Lesions

Complement Is Centrally Involved in Atherosclerotic Disease

PERLA THORBJORNSDOTTIR,[a] RAGNHILDUR KOLKA,[a]
EGGERT GUNNARSSON,[b] SLAVKO H. BAMBIR,[b]
GUÐMUNDUR THORGEIRSSON,[c] GIRISH J. KOTWAL,[d] AND
GUÐMUNDUR J. ARASON[a]

[a]*Department of Immunology, Institute of Laboratory Medicine, Landspitali University Hospital, 101 Reykjavik, Iceland*

[b]*Institute for Experimental Pathology, University of Iceland, 112 Reykjavik, Iceland*

[c]*Department of Medicine, Landspitali University Hospital, 101 Reykjavik, Iceland*

[d]*Division of Medical Virology, Institute of Infectious Diseases and Molecular Medicine, University of Cape Town Health Sciences Center, Cape Town 7925, South Africa*

ABSTRACT: Complement is known to be activated in atherosclerotic lesions, but the importance of this event in disease pathology is a matter of debate. Studies of rabbits fed a high-fat diet have indicated complement activation as a rate-limiting step, whereas results from genetically modified mouse strains (ApoE$^{-/-}$ or LDLR$^{-/-}$) have failed to support this finding. To resolve whether this reflects differences between species or between genetically driven and diet-induced disease, we studied the effect of a complement inhibitor, vaccinia virus complement control protein (VCP), on C57BL/6 mice, the background strain of ApoE$^{-/-}$ and LDLR$^{-/-}$ mice. Atherosclerosis was induced by a high-fat diet, and VCP (20 mg/kg) was injected once per week after the eighth week. Fatty streak development was monitored at 15 weeks by microscopic examination of oil red-O–stained sections from the root of the aorta. VCP injections led to significant (50%) reduction of lesion size ($P = 0.004$). Lesions were marked by gradual accumulation of lipids and macrophages but did not develop beyond the fatty streak stage. VCP activity disappeared from serum in 4 days, and the possibility therefore exists that a higher level of protection may be achieved by more frequent injections. We conclude that the development of fatty streaks in diet-induced atherosclerotic disease can be significantly retarded by prophylactic treatment with a complement inhibitor. These results support previous findings from complement-deficient rabbits and suggest that the pathogenesis of atherosclerosis in diet-induced disease differs from that induced by major defects in lipid metabolism.

Address for correspondence: Guðmundur J. Arason, Department of Immunology, Institute of Laboratory Medicine, Landspitali University Hospital, Hringbraut, 101 Reykjavik, Iceland. Voice: +354-543-5800; fax: +354-543-4828.
garason@landspitali.is

Ann. N.Y. Acad. Sci. 1056: 1–15 (2005). © 2005 New York Academy of Sciences.
doi: 10.1196/annals.1352.001

KEYWORDS: atherosclerosis; complement; vaccinia virus complement control protein; inflammation; mouse model; C57BL/6

INTRODUCTION

Atherosclerosis is the leading cause of death in all regions of the world except in sub-Saharan Africa, with prevalence values of <85% at age 50 and an overall global mortality rate of >40%.[1,2] It is a slow process that starts as a benign accumulation of low-density lipoproteins (LDLs) in the intima of large and medium-sized arteries in the first decade of life but leads to clinical problems primarily in the middle-aged and elderly.[3,4] Early lesions (fatty streaks) may recede, but clinical disease is diagnosed after further development into fibrotic plaques and complicated lesions. The most severe clinical condition is associated with lesion rupture, causing infarction of an artery supplying the heart (myocardial infarction, MI), the brain (stroke), or peripheral tissues (peripheral artery disease, PAD). Atherosclerosis is a chronic inflammatory disease[3–5] in which the presence of LDL particles in the vascular wall leads to recruitment of monocytes from the blood, their transformation into macrophages, and a dynamic but ultimately unsuccessful attempt to eliminate the LDL particles by phagocytosis. Both the innate and the adaptive immune system appear to contribute to the development of the lesions, and as in many other inflammatory diseases, activation of complement appears to mediate at least part of the tissue damage.[6–10]

Complement forms an important part of the innate immune system.[11,12] It consists of about 30 proteins, some of which act within a cascade-like reaction sequence, whereas others serve as control proteins or cellular receptors. The key components are present in the blood in precursor form and need to be activated. Complement can be activated by any of three pathways: (1) the antibody-dependent classic pathway (C1-C4-C2-C3), (2) the carbohydrate-dependent lectin pathway (MBL-C4-C2-C3), and (3) the alternative pathway (C3b-fB-C3), which is triggered directly by pathogen surfaces. Activated complement has many functions, including initiation of inflammation, recruitment of leukocytes, clearance of immune complexes, neutralization of pathogens, regulation of antibody responses, and cytolysis (the lytic pathway, C5b-C6-C7-C8-C9). The complement system is a very powerful mediator of inflammation, and complement activation generates proinflammatory peptides such as the anaphylatoxins C3a and C5a, which recruit and activate leukocytes, the cell-bound opsonins C4b and C3b, which facilitate phagocytosis of the target, and the membrane attack complex (MAC, C5b-9), which lyses target cells and may activate bystander cells to release proinflammatory mediators. Uncontrolled activation of complement and consequent host cell damage are prevented by a vast array of regulatory proteins, either circulating in plasma or expressed at the cell surface.

Considerable clinical and experimental evidence implicates complement in the pathogenesis of atherosclerosis.[7–10] Immunoglobulins, C3, C4, complement regulators, and terminal complexes C5b-9 have all been immunolocalized in human atherosclerotic plaques,[8,13,14] suggesting local activation of complement. Animal studies indicate that complement activation forms a link between LDL deposition and monocyte recruitment.[15] Deficiency in complement C6 has been shown to protect against diet-induced atherosclerosis in rabbits.[16,17] Two independent lines of evidence suggest a central involvement of C4 in disease pathology.[18–21] These results suggest that complement has a role in the development of lesions and are consistent with the notion that chronic activation of complement by modified LDLs leads to monocyte recruitment, foam cell formation, and lesion progression. By contrast, studies on mice with a combined deficiency in complement (C3 or C5) and a key element in lipid metabolism (LDL-R, ApoE, or both) reported that the development of atherosclerotic lesions was not severely affected.[22–24] This may reflect a difference between diet-induced and genetically driven disease, but further research is warranted to resolve this issue. We here report on a study employing the C57BL/6 mouse model and the use of a complement inhibitor to shed more light on the involvement of complement in the disease.

The tendency to develop atherosclerosis differs between animal species as well as strains. The C57BL/6 strain is very susceptible to diet-induced atherosclerosis; furthermore, lesions are restricted to the fatty streak stage,[25–27] which is convenient in this case because we wanted to study the initial stages in lesion development. Genetically modified mouse strains ($ApoE^{-/-}$, $LDLR^{-/-}$) that can form more extensive lesions are also available.[27] They are particularly useful in experiments requiring more accelerated disease and/or progression of lesions beyond the fatty streak stage. These mice do show lesions more similar to human atherosclerotic lesions, but in humans, the course of the disease is quite different as it is diet-induced and usually not because of genetic defects in major lipid transport proteins. The suitability of $ApoE^{-/-}$ and $LDLR^{-/-}$ mice for revealing the atherogenic effect of complement was recently questioned;[10] the atherogenic drive of these mice is much stronger than that of the human, and it is difficult to predict the effect of combined life-long deficiency of complement as well as key components in lipid transport. To test whether the discrepancies in previous animal studies are due to differences between species or between genetically driven and diet-induced disease, we used wild-type mice of the C57BL strain, the background strain of $ApoE^{-/-}$ and $LDLR^{-/-}$ mice,[22–24] and tested the importance of complement by using an inhibitor instead of gene knockout.

Complement inhibitors have not previously been used in models of atherosclerotic development. Various recombinant human complement inhibitors have been developed, as well as monoclonal antibodies, synthetic peptides, and peptidomimetics, which either block activation of a certain component, neutralize an activation fragment, or antagonize complement receptors.[28] Complement inhibitors have been used in reperfusion injury with very promising results, and therapy with complement inhibitors is approaching the clinic.[29–35] Vaccinia virus complement control protein (VCP) is a strong inhibitor of the classic, lectin, and alternative pathways of complement, acting on both C4 and C3.[36] Identified in 1988 as a product of the vaccinia virus, VCP is a 35-kDa soluble protein with structural[37] and functional[38] resemblance to human C4 binding protein (C4-BP) as well as other proteins of the regulators of complement activation (RCA) family. VCP binds to C4b, blocks the formation of the classic pathway C3 convertase, binds C3b, causes the accelerated decay of the classic pathway convertase, and blocks the conversion of C3 to C3b in both the classic and alternative pathways by promoting factor I cleavage of C3b.[39,40] Like its soluble mammalian RCA counterparts C4-BP and factor H, but unlike the surface-bound RCA molecules decay accelerating factor (DAF), membrane cofactor protein (MCP), and complement receptor 1 (CR1), VCP displays heparin-binding capabilities.[41] This suggests an *in vivo* role in connection with heparan sulfate proteoglycans lining the endothelial cell layer.[42,43] By blocking complement activation at multiple sites, VCP downregulates proinflammatory chemotactic factors (C3a, C4a, and C5a), resulting in reduced cellular influx and inflammation. The therapeutic potential of VCP has been extensively studied in several inflammatory diseases such as xenotransplantation,[44,45] Alzheimer's disease[46] and brain and spinal cord injury,[47–50] and the results make VCP very interesting as a potential therapeutic molecule in inflammatory conditions.[36]

The aim of the current study was to determine the role of complement in fatty streak formation in a mouse model and to examine whether the complement inhibitor VCP could hold promise as a possible prophylactic drug in atherosclerosis.

MATERIAL AND METHODS

Animals

Atherosclerosis was induced in female C57BL/6J mice by feeding them a high-fat diet.[25] The C57BL/6 mouse strain is very susceptible to diet-induced atherosclerosis.[26] Control mice were fed a chow diet. The mice were 5–6 weeks old at study initiation, weighing on average of 16–18 g.

Material

Female C57BL/6J mice were purchased from Jackson (Maine, USA), high-fat diet from ICN (Irvine, CA, USA), and chow diet from Special Diet Service (Witham, UK). VCP was produced in recombinant form in yeast (see below). Heparin columns (5 ml HiTrap™) were from Amersham (Uppsala, Sweden) and endotoxin-removing columns (1 ml Detoxi-Gel™) from Pierce (Rockford, IL, USA). Bovine serum albumen (BSA), Coomassie blue, orcein, and hematoxylin and eosin were from Sigma (St. Louis, MO, USA), Tris and H_2O_2 from Merck (Darmstadt, Germany), O.C.T.™ from Sakura Finetek (Zoeterwoude, Netherlands), glycergel mounting medium from Dako (Copenhagen, Denmark), diaminobenzidine from Pharmingen-BD (San Diego, CA, USA), and EZ complement kit from DiaMedix (Miami, FL, USA). Complement fixation test diluent (CFD) was from Flow Laboratories (Irvine, Scotland), oil red-O and HRP-conjugated goat anti-rat IgG from ICN, and monoclonal rat anti-mouse MOMA-2 from Serotec (Oxford, UK).

Production and Purification of VCP

The cloning of VCP in the *Picia pastoris* yeast expression system (Invitrogen) has previously been described.[41] Two milliliters of buffered minimal glycerol (BMG) were inoculated and grown overnight at 30°C. This starter culture was then used to inoculate 2 L of BMG, and the cultures were grown for 2 days at 30°C with vigorous shaking. The cells were harvested by centrifuging (3,500 rpm, 30 min), resuspended in 2 L of buffered minimal methanol (BMM) containing 4% methanol, and incubated for 2 days with vigorous shaking. The VCP-containing medium was then collected after centrifuging (12,000 rpm, 1 h), and the 2 L were concentrated to 50 ml using a 300-ml Amicon-stirred cell with a 10-kDa molecular weight cut-off

(Millipore, Billerica, MA, USA). Half the medium was then passed at a rate of 1 ml/min through three 5-ml heparin columns linked in a series, and after washing with 30 ml of 100 mM NaCl, the protein was eluted with 15 ml NaCl ranging from 250–550 mM. The fractions were visualized by SDS-PAGE (Invitrogen) with Coomassie blue staining. This was repeated for the second half of the medium, and VCP fractions were pooled, concentrated in a 50-ml Amicon stirred cell, and superconcentrated and desalted using several 2 ml centrifugal filters (Centricon, Millipore). Protein concentrations were assayed using a protein estimation kit (Bio-Rad, Hercules, CA, USA), and the specific activity of the samples was determined by hemolysis assay (see below). Purified VCP is very resistant to adverse conditions,[51] but for convenience it was transported as lyophilized product. Before use, it was restored in distilled water, purified by passage through a heparin column and endotoxin-removing gel, concentrated by freeze drying, and resolved to 2.2 mg/ml in 0.9% saline.

VCP Pharmacokinetics

The pharmacokinetics of injected VCP was followed in four mice injected i.p. with 10 mg/kg VCP. Blood was sampled from the tail vein at 1, 3, and 7 hours and at 1 and 4 days. The EZ Complement CH50 kit was used to measure VCP inhibition. The sensitized cells were concentrated 2 times by decanting half of the buffer and allowed to equilibrate to room temperature. Each test involved 75 μl of sensitized red blood cells, 15 μl of mouse serum, and 10 μl of CFD. Results were expressed as a percentage of a positive control (red blood cells lysed with H_20). One tube was used for spontaneous lysis. The tubes were incubated at 37°C for 1 hour and centrifuged at 150 g for 5 minutes. Absorbances were read at 405 nm.

Experiments

The study included three groups of mice: (1) 10 mice fed a high-fat diet and injected with VCP (study group), (2) 10 mice fed a high-fat diet and sham-injected with 0.9% saline (disease control group), and (3) 10 mice fed a chow diet and injected with saline (negative control group). Mice were injected in the tail vein with VCP or saline at weekly intervals from week 8 to week 15. At the end of the experiment, the mice were killed by cervical dislocation and the hearts excised. This experiment was designed to make optimal use of the available VCP; 15 weeks on atherogenic diet are sufficient for fatty streak development,[25] and this development occurs primarily after 7–8 weeks,[26] as confirmed in a preliminary experiment.

Histology

The normal structure of the heart was studied in longitudinal sections of whole formalin-fixed hearts stained with hematoxylin and eosin (HE), orcein, or Masson gold trichrome. For evaluation of atherosclerotic lesions, the lower half of the heart was discarded after being cut parallel to the atria. The upper half of the heart was mounted in O.C.T., quick frozen in liquid nitrogen, and kept at –70°C. Sections wereobtained according to the method of Paigen *et al.*[25] Briefly, the tissue blocks were trimmed on a cryotome (Reichert-Jung, Cambridge, UK) and then cut at 10 μm beginning with the lower portion of the heart. Sections were discarded until the three valve cusps were visible, and sections were then retained until the aorta was round and muscular and valve cusps no longer visible. Every sixth section was kept for immunohistochemical staining; the remaining sections were stained with oil red-O and counterstained with hematoxylin.

Immunohistochemistry

Cryostat sections were air-dried and fixed with acetone. Macrophages were visualized by rat anti-MOMA-2 (incubated overnight) after pretreatment for 10 minutes in 3% H_2O_2 in Tris (to block endogen peroxidase) and 20 minutes in a mixture of rat serum and 1% BSA in Tris (to avoid nonspecific antibody reaction). The anti-MOMA-2 antibody was detected with HRP-conjugated goat antibody against rat IgG (30 min) and visualized with diaminobenzidine (10 min). Hematoxylin was used for counterstaining.

Evaluation of Atherosclerotic Lesions

Evaluation of atherosclerotic lesions was confined to the 280-μm interval just beyond the aortic sinus and at the beginning of the aorta, and based on every fourth section as described by Paigen *et al.*[25] For objective analysis of the atherosclerotic progression, a photomicroscope attached to a digital camera (Leica, Bensheim, Germany) was used. The Leica Qwin program was used for computer analysis of images, and to minimize experimental error, lesion size was expressed as a percentage of the circumference of the aorta in the section examined. The sections were evaluated in a blinded fashion.

Statistical Analysis

Percent lesion size from the aortic root area (lesion size*100/aortic circumference) was compared among the three experimental groups using the Mann-Whitney rank sum test. Significance was set at $P < 0.05$.

RESULTS

Normal Structure of Mouse Heart

FIGURE 1 shows the normal anatomy of the mouse heart. The aorta is evident at the base of the right atrium (arrow). One of the three valve leaflets is visible (arrowhead). The valve is rich in collagen as seen with the Masson gold staining (FIG. 1a and b), but the aortic wall is mostly composed of fibrotic tissue, which stains prominently with orcein (FIG. 1c).

Atherosclerotic Lesions

After 15 weeks on atherogenic diet, cross-sections of the 280-μm interval between the aortic sinus and the beginning of the aorta (FIG. 2) showed atherosclerotic lesions composed of oil red-O–positive lipid deposits (FIG. 3). These lesions did not progress beyond the fatty streak stage during the course of the experiment. Lipid

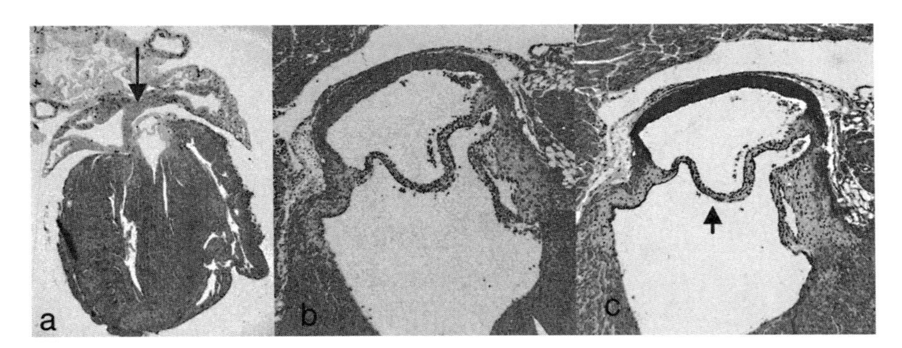

FIGURE 1. Structure of the mouse heart. (**a**) Longitudinal section of the whole heart with the aorta (*arrow*); (**b** and **c**) higher magnification of the aorta with valve leaflet (*arrowhead*). (**a** and **b**) Masson gold trichrome and (**c**) orcein.

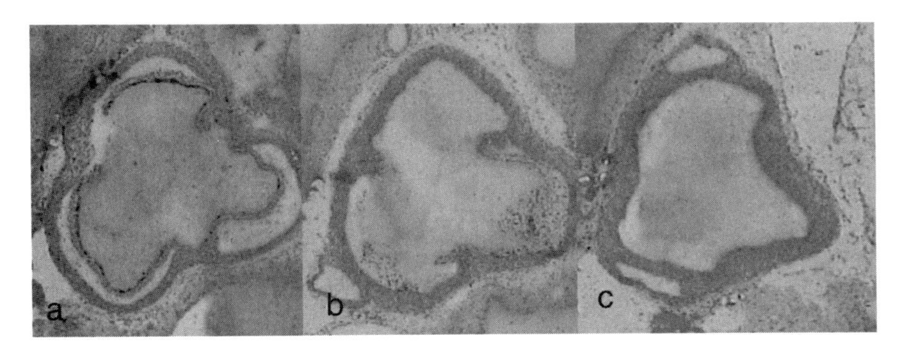

FIGURE 2. Cross-sections of the aorta, showing the beginning (**a**), middle (**b**), and end (**c**) of the 280-μm interval studied (oil-red O). All three valves are still visible in **a,** while only the cusps are showing in **b**; in **c**, they are disappearing and the aorta is almost round in shape with the wall becoming muscular.

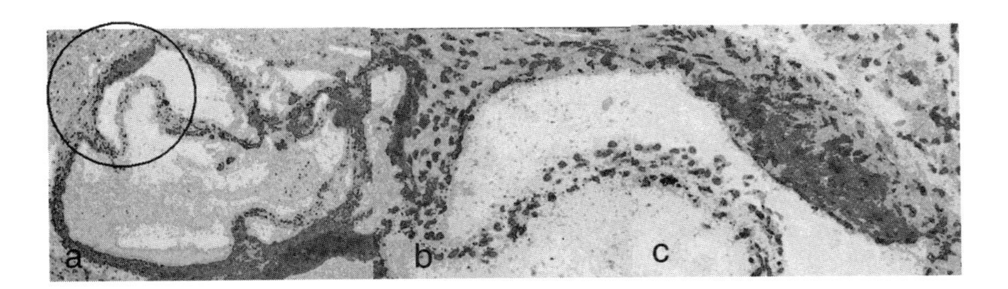

FIGURE 3. Fatty streaks in the aorta of a mouse fed a high-fat diet for 15 weeks (oil-red O); (**a**) overview and (**b-c**) magnification of the *circled area*. Fatty streaks are typically seen in the semilunar valve (**b**) and the aortic wall (**c**), where lipids are deposited in the *tunica intima* but even more prominently in the *tunica media*.

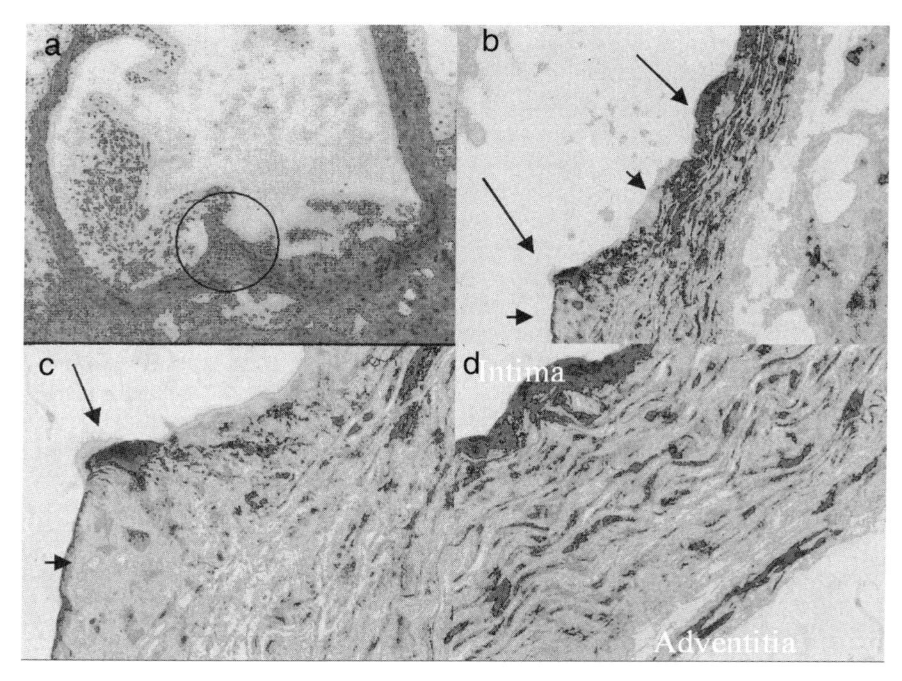

FIGURE 4. Atherosclerotic lesions in the aorta of a mouse fed a high-fat diet for 15 weeks. Sections stained in parallel with oil-red O (**a**) and immunohistochemistry (**b-c**) show the entry of macrophages (*arrows*) into the lesion encircled in **a**. Note the accumulation of foam cells in the intima, which leads to aortic lesions (*arrowhead*). (**d**) Positive immunohistochemical reaction of the intima and adventitia.

deposition was not confined to the intima but extended far into the media (FIG. 3c), as no internal elastic lamina separates these layers in this part of the aorta. Lesions were also commonly seen in the base of the semilunar valve (FIG. 3b). The presence of macrophages in fatty streaks was verified by immunohistochemical staining of sections cut in parallel with oil red-O–stained sections (FIG. 4).

Quantitative Analysis of Atherosclerotic Lesions

Injection of VCP led to a reduction in serum hemolytic activity from 50% to 40%; the effect, however, was delayed in mice injected i.p. with a peak at 24 hours. VCP activity had disappeared after 4 days. Evaluation of lesions in mice fed an atherogenic diet for 15 weeks and injected at weekly intervals in the last 7 weeks with either VCP or saline indicated a significant reduction in size in the VCP-injected mice compared to the saline-injected mice (FIG. 5). This was confirmed by computer-assisted evaluation of digital images captured by the Qwin program ($P = 0.004$; FIG. 6). No lesions were evident in control mice fed a normal diet (FIG. 5c). No differences were observed in food intake or weight between the groups (results not shown). Mortalities were similar in experimental and control groups (2 and 3 mice, respectively).

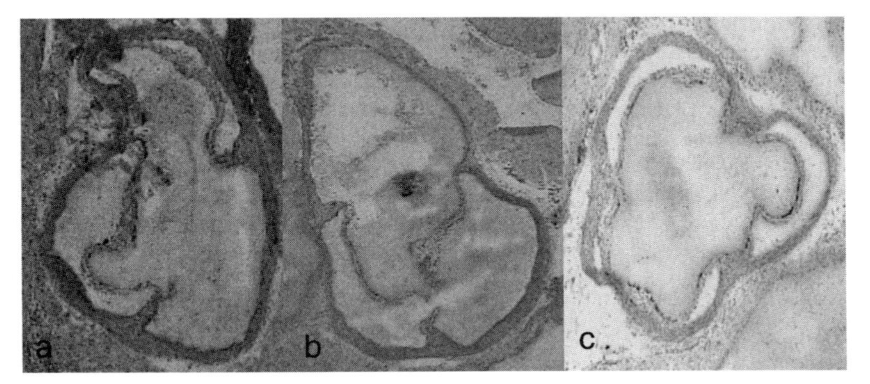

FIGURE 5. Development of fatty streaks in mice fed a high-fat diet for 15 weeks and injected with saline (**a**) or 20 mg/kg VCP (**b**) in weeks 8–15. Note the lipid plaques (*arrows*) in the *intima* (**b**) or *intima* and *media* (**a**). No lipid staining is evident in a mouse fed a chow diet and injected with saline (**c**). Oil-red O staining.

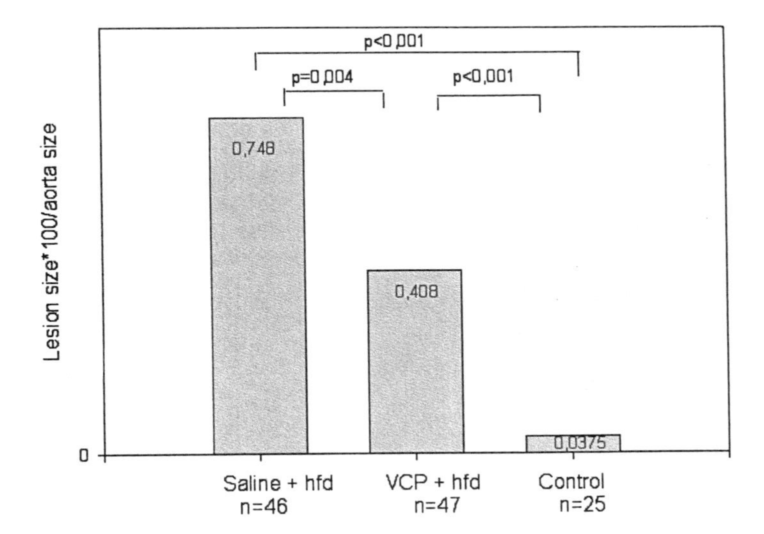

FIGURE 6. Percent lesion area in aortic sections from mice (**a**) fed a high-fat diet and injected with saline (n = 10), (**b**) fed a high-fat diet and injected with 20 mg/kg VCP (n = 10), and (**c**) control mice (n = 3); the number (n) of sections examined is indicated on the x-axis.

DISCUSSION

In this study we demonstrate, for the first time, that the development of fatty streaks in an animal model of diet-induced atherosclerotic disease can be significantly retarded by the injection of a complement inhibitor, VCP. Previous studies had demonstrated that (1) complement is deposited in atherosclerotic lesions from patients as well as experimental animals,[8,13,14] (2) activation of complement is the first sign of inflammation in the arterial wall in animal models of atherosclerosis, taking place concurrently with LDL deposition and before monocyte recruitment,[15] and (3) the concentration of terminal complement complex C5b-9, a marker of complement activation, increases in parallel with lesion progression.[13] An apparent controversy, however, existed over the relative importance of complement in this scenario because (4) in C6-deficient rabbits fed a high-fat diet, complement appeared to play an obligatory and rate-limiting role[16,17] but (5) in genetically modified mice (LDLR$^{-/-}$ and/or ApoE$^{-/-}$), deficiency of complement (C3 or C5) did not retard lesion progression.[22–24] Deficiency of C5 in ApoE$^{-/-}$ mice did not lead to any visible changes,[22] whereas deficiency of C3 in LDLR$^{-/-}$ or LDLR$^{-/-}$ ApoE$^{-/-}$ mice led to increased lesion size,[23,24] with increased build-up of LDL and macrophages and slower transition from fatty streaks to fibrotic plaques in the former mouse strain.[23] To account for the controversy between findings (3) and (4), it had been suggested that the process of atherogenesis might differ between animal species[9] or between genetically driven and diet-induced disease.[10] Our results strongly favor the latter supposition, as we used mice of the same strain (C57BL) as in previous experiments, only without genetic modifications. Our results thus support the view of complement as an obligatory step in diet-induced atherogenesis, although its importance appears to be abolished or reduced in cases in which the disease is driven by major defects in lipid metabolism.

After 15 weeks on an atherogenic diet, lipids and macrophages were abundant in lesions, but lesions did not progress beyond the fatty streak stage. This is consistent with previous observations on the C57BL/6 mouse model.[25–27] We did not stain specially for smooth muscle cells (SMCs), but normal histology showed no SMCs in the intimal layer of the aorta. The observations that lesions do not progress beyond the foam cell stage and that lesions are significantly smaller in VCP-injected mice support previous data suggesting that complement activation is important in the first stages of lesion formation of diet-induced atherosclerosis.[15–17] It should be pointed out that our experimental design did not explore the maximal protection attainable by VCP, as its inhibitory effect disappeared from serum less than 4 days after injection, but we had only sufficient VCP for injecting once per week. In addition, we had to confine our experiment to the period of maximal fatty streak formation in weeks 8–15, allowing for some lipid deposition before treatment. Due to its heparin-binding sites,[41–43] VCP may be sequestered in the body for periods exceeding its half-life in serum,[52] and this may explain the relatively high level of protection attained even by weekly injections. It is however likely that the 50% protection observed is a compromise of a much higher level of protection immediately after injection and no or lower protection towards the end of the week. Combined with the observation that total deficiency of C6 resulted in complete inhibition of fatty streak development in rabbits,[17] our results suggest that a much higher level of protection may be achieved with changes in the injection regimen. These results have strong implications from a practical as well as a scientific point of view, as they raise hope that VCP, or indeed other complement inhibitors, may lead to future development of drugs with prophylactic potential. Several studies have previously suggested a beneficial effect of VCP or other complement inhibitors in xenotransplantation,[44,45] Alzheimer's disease,[46] brain and spinal cord injury,[47–50] and reperfusion injury,[29–35] but the focus had been mainly on therapeutic use. From a wider point of view, our results shed new light on the role of complement in atherosclerosis, suggesting that it may be a rate-limiting

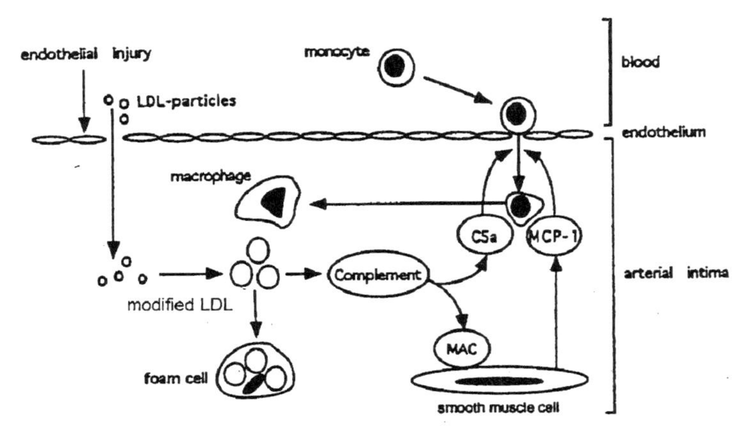

FIGURE 7. Schematic diagram of cellular interactions in atherogenesis induced by complement. (From: TORZEWSKI, J., D.E. BOWYER, J. WALTENBERGER & C. FITZSIMMONS. 1997. Processes in atherogenesis: complement activation. Atherosclerosis **132:** 131–138. Reprinted with permission from Elsevier.)

step in diet-induced disease, although in disease driven by defects in lipid metabolism, it may be redundant.

The current study provides no clue as to what in the atherosclerotic lesion activates complement, but several candidates have been proposed.[7–9] Complement may become activated directly by modified LDL present in the forming plaque (FIG. 7).[7,15,53,54] It could also become activated by immunoglobulins or C-reactive protein (CRP); IgM and IgG are retained in the atherosclerotic lesions[13,55] and can form immune complexes with oxLDL.[56] CRP deposition has also been confirmed in human atherosclerotic lesions and found to be consistent with the severity of the lesion.[13,57] CRP may modify the outcome by limiting the activation to the C3 level, as evidence has shown that it prevents the formation of MAC but favors opsonization.[58,59] The presence of C5b-9 (MAC) in the deeper layers of human lesions,[60] however, suggests that the effect of CRP may be confined to the upper layers where the activation takes place and that CRP and/or MAC inhibitors are less abundant in the deeper layers. Complement activation may modulate lesion development in various ways. Generation of anaphylatoxins may play a part in leukocyte recruitment.[7] Generation of C5b-9 within the arterial wall may injure vascular cells, triggering the release of growth factors and cytokines from endothelial cells, macrophages, and SMCs.[8] Of these cells, smooth muscle cells are the most likely target for C5b-9 formation because they are poorly protected by complement regulatory proteins.[61] Such an attack on SMCs with release of MCP-1 might explain the initial monocyte recruitment into the arterial wall.[7,62] It might, in fact, also explain SMC proliferation in the lesion, as sublytic attack of MAC has been shown to be mitogenic for SMCs.[63] Complement activation is associated with apoptosis. C5b-9 deposits in atherosclerotic lesions are localized not only on intact SMCs and on cell debris[64] but also on apoptotic cells,[14] indicating that activation of the complement system by apoptotic cells may contribute to lesion development. Complement activation may therefore play a role in both fatty streak formation and the chronic inflammatory processes involved in lesion progression.

In conclusion, our findings of 50% protection against fatty streak formation in wild-type C57BL mice by weekly injections of VCP combined with previous data[17] implicate complement activation as a rate-limiting step in lesion formation in diet-induced atherosclerotic disease and raise hope that disease progression may in the future be prevented or retarded by prophylactic use of complement inhibitors.

ACKNOWLEDGMENTS

This work was supported by The Science Fund of Landspitali University Hospital, The Memorial Fund of Helga Jónsdóttir and Sigurlidi Kristjánsson, and Grant 011330001 from the Icelandic Research Council. G.J.K. is currently a Senior International Wellcome Trust Fellow for Biomedical Sciences in South Africa. We thank Prof. Guðmundur Georgsson and Drs. Scott A. Smith and Purushottam Jha for useful advice, Margrét Jónsdóttir, Steinunn Árnadóttir, and Anna Guðrún Viðarsdóttir for help with histologic preparation, Tryggvi Eiríksson for freeze-drying the VCP, and Gudmundur Örn Gudmundsson for help with the Leica Qwin program.

REFERENCES

1. Bonow, R.O. *et al.* 2002. World Heart Day 2002: the international burden of cardiovascular disease: responding to the emerging global epidemic. Circulation **106:** 1602–1605.
2. Tuzcu, E.M. *et al.* 2001. High prevalence of coronary atherosclerosis in asymptomatic teenagers and young adults: evidence from intravascular ultrasound. Circulation **103:** 2705–2710.
3. Lusis, A.J. 2000. Atherosclerosis. Nature **407:** 233–241.
4. Glass, C.K. & J.L. Witztum. 2001. Atherosclerosis; the road ahead. Cell **104:** 503–516.
5. Libby, P. 2002. Inflammation in atherosclerosis. Nature **420:** 868–874.
6. Binder, C.J. *et al.* 2002. Innate and acquired immunity in atherogenesis. Nat. Med. **8:** 1218–1226.
7. Torzewski, J. *et al.* 1997. Processes in atherogenesis: complement activation. Atherosclerosis **132:** 131–138.
8. Niculescu, F. & H. Rus. 1999. Complement activation and atherosclerosis. Mol. Immunol. **36:** 949–955.
9. Oksjoki, R., P.T. Kovanen & M.O. Pentikainen. 2003. Role of complement activation in atherosclerosis. Curr. Opin. Lipidol. **14:** 477–482.
10. Bhakdi, S. *et al.* 2004. Beyond cholesterol: the enigma of atherosclerosis revisited. Thromb. Haemost. **91:** 639–645.
11. Walport, M.J. 2001. Complement. First of two parts. N. Engl. J. Med. **344:** 1058–1066.
12. Walport, M.J. 2001. Complement. Second of two parts. N. Engl. J. Med. **344:** 1140–1144.
13. Vlaicu, R. *et al.* 1985. Immunohistochemical localization of the terminal C5b–9 complement complex in human aortic fibrous plaque. Atherosclerosis **57:** 163–177.
14. Niculescu, F., T. Niculescu & H. Rus. 2004. C5b-9 terminal complement complex assembly on apoptotic cells in human arterial wall with atherosclerosis. Exp. Mol. Pathol. **76:** 17–23.
15. Seifert, P.S. *et al.* 1989. Prelesional complement activation in experimental atherosclerosis. Terminal C5b-9 complement deposition coincides with cholesterol accu-

mulation in the aortic intima of hypercholesterolemic rabbits. Lab. Invest. **60:** 747–754.

16. GEERTINGER, P. & H. SOERENSEN. 1977. On the reduced atherogenic effect of cholesterol feeding in rabbits with congenital complement (C6) deficiency. Artery **1:** 177–184.
17. SCHMIEDT, W. *et al.* 1998. Complement C6 deficiency protects against diet-induced atherosclerosis in rabbits. Arterioscler. Thromb. Vasc. Biol. **18:** 1790–1795.
18. KRAMER, J.T. *et al.* 1991. A marked drop in the incidence of the null allele of the B gene of the fourth component of complement (C4B*Q0) in elderly subjects: C4B*Q0 as a probable negative selection factor for survival. Hum. Genet. **86:** 595–598.
19. KRAMER, J.K. *et al.* 1994. C4B*Q0 allotype as risk factor for myocardial infarction. Br. Med. J. **309:** 313–314.
20. ARASON, G.J. *et al.* 2003. An age-associated decrease in the frequency of C4B*Q0 indicates that null alleles of complement may affect health or survival. Ann. N.Y. Acad. Sci. **1010:** 496–499.
21. ARASON, G.J. *et al.* 2005. Secret of the short life span: smoking and complement interact in dictating cardiovascular disease morbidity and survival. Science. Under evaluation.
22. PATEL, S. *et al.* 2001. ApoE(-/-) mice develop atherosclerosis in the absence of complement component C5. Biochem. Biophys. Res. Commun. **286:** 164–170.
23. BUONO, C. *et al.* 2002. Influence of C3 deficiency on atherosclerosis. Circulation **105:** 3025–3031.
24. PERSSON, L. *et al.* 2004. Lack of complement factor C3, but not factor B, increases hyperlipidemia and atherosclerosis in apolipoprotein E-/- low-density lipoprotein receptor-/- mice. Arterioscler. Thromb. Vasc. Biol. **24:** 1062–1067.
25. PAIGEN, B. *et al.* 1987. Quantitative assessment of atherosclerotic lesions in mice. Atherosclerosis **68:** 231–240.
26. PAIGEN, B. *et al.* 1990. Atherosclerosis susceptibility differences among progenitors of recombinant inbred strains of mice. Arteriosclerosis **10:** 316–323.
27. SMITH, J.D. 1998. Mouse models of atherosclerosis. Lab. Anim. Sci. **48:** 573–579.
28. MORGAN, B.P. & C.L. HARRIS. 2003. Complement therapeutics; history and current progress. Mol. Immunol. **40:** 159–170.
29. HORSTICK, G. *et al.* 2001. Application of C1-esterase inhibitor during reperfusion of ischemic myocardium: dose-related beneficial versus detrimental effects. Circulation **104:** 3125–3131.
30. LAZAR, H.L. *et al.* 1999. Total complement inhibition: an effective strategy to limit ischemic injury during coronary revascularization on cardiopulmonary bypass. Circulation **100:** 1438–1442.
31. BUERKE, M. *et al.* 2001. Novel small molecule inhibitor of C1s exerts cardioprotective effects in ischemia-reperfusion injury in rabbits. J. Immunol. **167:** 5375–5380.
32. ZACHAROWSKI, K. *et al.* 1999. Reduction of myocardial infarct size with sCR1sLe(x), an alternatively glycosylated form of human soluble complement receptor type 1 (sCR1), possessing sialyl Lewis x. Br. J. Pharmacol. **128:** 945–952.
33. VAKEVA, A.P. *et al.* 1998. Myocardial infarction and apoptosis after myocardial ischemia and reperfusion: role of the terminal complement components and inhibition by anti-C5 therapy. Circulation **97:** 2259–2267.
34. GRANGER, C.B. *et al.* 2003. COMMA Investigators. Pexelizumab, an anti-C5 complement antibody, as adjunctive therapy to primary percutaneous coronary intervention in acute myocardial infarction: the COMplement inhibition in myocardial infarction treated with angioplasty (COMMA) trial. Circulation **108:** 1184–1190.
35. MAHAFFEY, K.W. *et al.* 2003. COMPLY Investigators. Effect of pexelizumab, an anti-C5 complement antibody, as adjunctive therapy to fibrinolysis in acute myocardial infarction: the COMPlement inhibition in myocardial infarction treated with thromboLYtics (COMPLY) trial. Circulation **108:** 1176–1183.
36. JHA, P. & G.J. KOTWAL. 2003. Vaccinia complement control protein: multi-functional protein and a potential wonder drug. J. Biosci. **28:** 265–271.
37. KOTWAL, G.J. & B. MOSS. 1988. Vaccinia virus encodes a secretory polypeptide structurally related to complement control proteins. Nature **335:** 176–178.

38. KOTWAL, G.J. *et al.* 1990. Inhibition of the complement cascade by the major secretory protein of vaccinia virus. Science **250:** 827–830.
39. MCKENZIE, R. *et al.* 1992. Regulation of complement activity by vaccinia virus complement-control protein. J. Infect. Dis. **166:** 1245–1250.
40. SAHU, A. *et al.* 1998. Interaction of vaccinia virus complement control protein with human complement proteins: factor I-mediated degradation of C3b to iC3b1 inactivates the alternative complement pathway. J. Immunol. **160:** 5596–5604.
41. SMITH, S.A. *et al.* 2000. Conserved surface-exposed K/R-X-K/R motifs and net positive charge on poxvirus complement control proteins serve as putative heparin binding sites and contribute to inhibition of molecular interactions with human endothelial cells: a novel mechanism for evasion of host defense. J. Virol. **74:** 5659–5666.
42. MURTHY, K.H. *et al.* 2001. Crystal structure of a complement control protein that regulates both pathways of complement activation and binds heparan sulfate proteoglycans. Cell **104:** 301–311.
43. SMITH, S.A. *et al.* 2003. Mapping of regions within the vaccinia virus complement control protein involved in dose-dependent binding to key complement components and heparin using surface plasmon resonance. Biochim. Biophys. Acta **1650:** 30–39.
44. ANDERSON, J.B. *et al.* 2002. Vaccinia virus complement control protein ameliorates hyperacute xenorejection by inhibiting xenoantibody binding. Transplant. Proc. **34:** 3277–3281.
45. ANDERSON, J.B. *et al.* 2003. Vaccinia virus complement control protein inhibits hyperacute xenorejection in a guinea pig-to-rat heterotopic cervical cardiac xenograft model by blocking both xenoantibody binding and complement pathway activation. Transpl. Immunol. **11:** 129–135.
46. DALY, J., 4th & G.J. KOTWAL. 1989. Pro-inflammatory complement activation by the A beta peptide of Alzheimer's disease is biologically significant and can be blocked by vaccinia virus complement control protein. Neurobiol. Aging **19:** 619–627.
47. KEELING, K.L. *et al.* 2000. Local neutrophil influx following lateral fluid-percussion brain injury in rats is associated with accumulation of complement activation fragments of the third component (C3) of the complement system. J. Neuroimmunol. **105:** 20–30.
48. REYNOLDS, D.N. *et al.* 2003. Vaccinia virus complement control protein modulates inflammation following spinal cord injury. Ann. N.Y. Acad. Sci. **1010:** 534–539.
49. REYNOLDS, D.N. *et al.* 2004. Vaccinia virus complement control protein reduces inflammation and improves spinal cord integrity following spinal cord injury. Ann. N.Y. Acad. Sci. **1035:** 165–178.
50. KULKARNI, A.P. *et al.* 2004. Neuroprotection from complement-mediated inflammatory damage. Ann. N.Y. Acad. Sci. **1035:** 147–164.
51. SMITH, S.A. *et al.* 2002. Vaccinia virus complement control protein is monomeric, and retains structural and functional integrity after exposure to adverse conditions. Biochim. Biophys. Acta **1598:** 55–64.
52. JHA, P. *et al.* 2003. Prolonged retention of vaccinia virus complement control protein following IP injection: implications in blocking xenorejection. Transplant. Proc. **35:** 3160–3162.
53. SEIFERT, P.S. *et al.* 1990. Isolation and characterization of a complement-activating lipid extracted from human atherosclerotic lesions. J. Exp. Med. **172:** 547–557.
54. BHAKDI, S. *et al.* 1995. On the pathogenesis of atherosclerosis: enzymatic transformation of human low density lipoprotein to an atherogenic moiety. J. Exp. Med. **182:** 1959–1971.
55. VLAICU, R. *et al.* 1985. Quantitative determinations of immunoglobulins and complement components in human aortic atherosclerotic wall. Med. Interne **23:** 29–35.
56. YLA-HERTTUALA, S. *et al.* 1994. Rabbit and human atherosclerotic lesions contain IgG that recognizes epitopes of oxidized LDL. Arterioscler. Thromb. **14:** 32–40.
57. REYNOLDS, G.D. & R.P. VANCE. 1987. C-reactive protein immunohistochemical localization in normal and atherosclerotic human aortas. Arch. Pathol. Lab. Med. **111:** 265–269.

58. Jarva, H. *et al.* 1999. Regulation of complement activation by C-reactive protein: targeting the complement inhibitory activity of factor H by an interaction with short consensus repeat domains 7 and 8-11. J. Immunol. **163:** 3957–3962.
59. Bhakdi, S. *et al.* 2004. Possible protective role for C-reactive protein in atherogenesis: complement activation by modified lipoproteins halts before detrimental terminal sequence. Circulation **109:** 1870–1876.
60. Oksjoki, R. *et al.* 2003. Association between complement factor H and proteoglycans in early human coronary atherosclerotic lesions: implications for local regulation of complement activation. Arterioscler. Thromb. Vasc. Biol. **23:** 630–636.
61. Seifert, P.S. *et al.* 1992. CD59 (homologous restriction factor 20), a plasma membrane protein that protects against complement C5b-9 attack, in human atherosclerotic lesions. Atherosclerosis **96:** 135–145.
62. Torzewski, J. *et al.* 1996. Complement-induced release of monocyte chemotactic protein-1 from human smooth muscle cells. A possible initiating event in atherosclerotic lesion formation. Arterioscler. Thromb. Vasc. Biol. **16:** 673–677.
63. Niculescu, F. *et al.* 1999. Sublytic C5b-9 induces proliferation of human aortic smooth muscle cells: role of mitogen activated protein kinase and phosphatidylinositol 3-kinase. Atherosclerosis **142:** 47–56.
64. Rus, H.G. *et al.* 1986. Immunoelectron-microscopic localization of the terminal C5b-9 complement complex in human atherosclerotic fibrous plaque. Atherosclerosis **61:** 35–42.

Cytokine Therapy

ANTONY CUTLER AND FRANK BROMBACHER

University of Cape Town, Health Science Faculty, Institute for Infectious Disease and Molecular Medicine (IIDMM), Cape Town, South Africa

ABSTRACT: **Cytokines are a unique class of intercellular regulatory proteins that play a crucial role in initiating, maintaining, and regulating immunologic homeostatic and inflammatory processes. Indeed, measurement of cytokine profiles in patients provides a useful indication of disease status. Due to their multiple functions, including regulatory and effector cellular function in many diseases, these molecules, their receptors, and their signal transduction pathways are promising candidates for therapeutic interference. The therapeutic administration of cytokines, modulation of cytokine action, or at times gene therapy is being used for a wide range of infectious and autoimmune diseases, in immunocompromised patients with AIDS, and in neoplasia.**

KEYWORDS: **interleukin; monoclonal antibody; autovaccination; inflammation**

INTRODUCTION

Cytokines are not produced constitutively. Indeed, cytokines are usually produced upon stimulation and are secreted by a variety of cell types, and distinct cytokines often have redundant or overlapping functions. They are extremely potent biomolecules acting at a picomolar, sometimes femtomolar range and may have multiple (pleiotropic) actions on different target cells. Cytokine released by a cell may act in a short radius on its own (autocrine) or on other cells (paracrine), thus providing a means for cross-talk at the cellular level (FIG. 1).

Common structural features of interleukins make it possible to group cytokines within "families," and tremendous progress in the cloning of genes for cytokine receptors and their characterization has led to the recognition that cytokine receptors may be grouped into four principal families based on common structural features. Most of these receptors form heterodimers, but some form homodimers or heterotrimers. Interestingly, many of the multiple chain receptors form subfamilies with one chain shared by all members of the subfamily. Due to their multiple functions, including regulatory and effector function in many diseases, these molecules, their receptors, and their signal transduction pathways are promising candidates for therapeutic interference.

Advances in the understanding of the role of cytokines in immune and inflammatory disorders have led to the development of cytokine-based therapies. Therapies

Address for correspondence: Frank Brombacher, Institute for Infectious Disease and Molecular Medicine, University of Cape Town, S1. 27, Observatory 7925, Anzio Road, South Africa. Voice: +27-21-406-6616\6147; fax: +27-21-406-6029.

brombac@uctgsh1.uct.ac.za

Ann. N.Y. Acad. Sci. 1056: 16–29 (2005). © 2005 New York Academy of Sciences.
doi: 10.1196/annals.1352.002

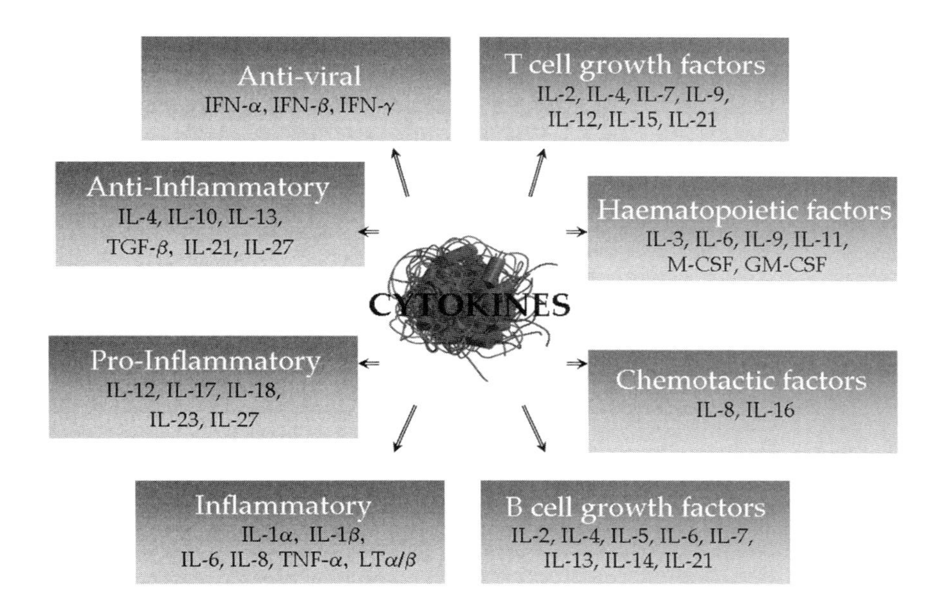

FIGURE 1. Functional roles of cytokines.

have been developed with the express aim to block/inhibit or restore the activity of specific cytokines. A variety of approaches have been employed to produce therapeutics with the aim of manipulating cytokine function. Cytokines have been cloned and produced in recombinant form; coupled to toxin, recombinant cytokine and receptor molecules have been fused to the Fc portion of human IgG1 or to albumin to stabilize and increase the serum half-life of the protein. Natural and synthetic antagonist proteins have been cloned/produced to interfere with ligand/receptor interaction. Cytokines delivered by gene therapy and antisense oligonucleotide treatment are also being assessed. Currently, the most utilized approach to cytokine therapy is that of blocking or neutralizing cytokine action with monoclonal antibodies (mAbs) (TABLE 1a and b). Mouse mAbs with specificity for human cytokines have been "humanized" by grafting the mouse CDR onto the Fc portion of the human IgG1. Fully human anticytokine mAbs have now been approved for clinical use (TABLE 1a).

Drugs that block inflammatory cytokines, such as tumor necrosis factor (TNF)-α, are among the most successful therapeutics approved for clinical use.[1–3] Of patients with rheumatoid arthritis (RA), 60–70% exhibit amelioration of disease upon treatment with anti-TNF-α. Fusion proteins of sTNF receptor[4] and interleukin (IL)-1Rα antagonist[5] are also routinely used in the treatment of RA. It has become apparent that TNF blockers are more efficient than therapies aimed at IL-1 antagonism or blocking. The scope of therapy using TNF-α and IL-1 blockers/antagonists is now evolving to target other inflammatory diseases. Targeting inflammatory cytokines has been a fruitful approach in treating inflammatory autoimmune disease. This success has led to the development of mAbs targeting a number of cytokines/cytokine

TABLE 1a. Cytokine therapies approved for clinical use and further clinical trial

Cytokine	Drug	Trade name/ code	Application	Clinical trial phase/ outcome	Company	Ref/ Link
IL-1Rα	Recombinant IL-1Rα antagonist	Kineret / Anakinra	Rheumatoid arthritis (RA)	Approved	Amgen	5
			Sepsis	Approved		
			Osteoarthritis	Phase II		
IL-2	Recombinant IL-2	Proleukin	Metastatic renal cell carcinoma	Approved	Chiron	6
			Metastatic melanoma	Approved		
			Non-Hodgkin's lymphoma	Phase II		
IL-2	Humanized IL-2Rα-chain blocking mAb	Daclizamab	Renal transplantation	Licensed	PDL/ Roche	
			Asthma	Phase II		
			GvHD	Phase III		[1]
			Multiple sclerosis	Phase II		38
			HIV	Phase I		[2]
			Psoriasis	Phase I/II		[3]
			Ulcerative uveitis	Halted		[4]
	Humanized IL-2Rα-chain blocking mAb	Simulect / Basiliximab	Renal transplantation	Licensed	Novartis	39
IL-11	Recombinant IL-11	Oprelvekin / Neumega	Chemotherapy-induced thrombocytopenia	Approved	Genetics Institute, Inc/ Wyeth	34
			Crohn's disease	Phase III		
			RA	Phase II		
			Psoriasis	Phase II		
			Colitis	Phase I		
TNF-α	Human anti-TNF-α mAb	Adalimumab/ Humira/ D2E7	RA	Approved	CAT/ Abbott	3
			Juvenile RA	Phase II		
			Ankylosing spondylitis	Phase III		
			Psoriatic arthritis	Phase III		
			Crohn's disease	Phase III		
			Chronic plaque psoriasis	Phase II		
TNF-α	Humanized anti-TNFα mAb	Remicade /Infliximab	RA	Approved	Centocor	1, 2
			Crohn's disease	Approved		
			Ulcerative colitis	Phase II		
			Psoriasis	Phase II		
			Psoriatic arthritis	Phase II		
			Ankylosing spondylitis	Phase II		
			Juvenile idiopathic arthritis	Phase II		
			Pediatric Crohn's disease	Phase II		

TABLE 1a. Cytokine therapies approved for clinical use and further clinical trial

Cytokine	Drug	Trade name/ code	Application	Clinical trial phase/ outcome	Company	Ref/ Link
TNF-α	Soluble p75 TNF receptor-Fc fusion	Etanercept/ Enbrel	RA	Approved	Amgen	4
			Juvenile RA	Approved		
			Ankylosing spondylitis	Approved		
			Psoriatic arthritis	Approved		
			Psoriasis	Approved		
			Severe plaque psoriasis	Approved		
IFN-γ	Bioengineered IFN-γ–1b	Actimmune	Chronic granulomatous disease	Approved	Intermune Pharma	12
			Osteoporosis	Approved		40
			Idiopathic pulmonary fibrosis	Phase III		
			Ovarian cancer	Phase III		
IFN-α	IFN-α–con-1	Infergen	Hepatitis C	Approved	Intermune Pharma	9
IFN-α	IFN-α–n3 leukocyte derived	Alferon-N	HPV gental warts	Approved	Hemi sperx Biopharma	10
			Hepatitis C	Phase II/III		
			West Nile virus	Phase II/III		
			HIV	Phase I/II		
IFN-α	Pegylated IFN-α–2a	Pegasys	Chronic hepatitis C	Approved	Roche	
			Hepatitis B	Phase III/ Filed		8
IFN-α	Recombinant IFN-α–2a	Roferon-A	Hairy cell leukemia	Approved	Roche	7
			Kaposi's sarcoma	Approved		
			Chronic myleloid leukemia	Approved		
			Chronic hepatitis C	Approved		
IFN-α	Recombinant IFN-α–2b	Intron-A	Hairy cell leukemia	Approved	Schering-Plough	7
			Kaposi's sarcoma	Approved		
			Chronic hepatitis B/ C	Approved		
			Malignant melanoma	Approved		
			Follicular lymphoma	Approved		
			Condylomata acuminata	Approved		
IFN-α	PEG recombinant IFN-α–2b	PEG Intron	Hepatitis C	Approved	Schering-Plough	
			Malignant melanoma	Phase III		
IFN-β	IFN-β–1a	Avonex	Relapsing multiple sclerosis	Approved	Biogen Idec	13
IFN-β	IFN-β–1a	Rebiferon	Relapsing multiple sclerosis	Approved	Serono	13
			Chronic hepatitis C	Phase III		

TABLE 1a. Cytokine therapies approved for clinical use and further clinical trial

Cytokine	Drug	Trade name/ code	Application	Clinical trial phase/ outcome	Company	Ref/ Link
IFN-β	IFN-β–1b	Betaseron	Early/relapsing multiple sclerosis	Approved	Berlex	13
GM-CSF	Recombinant GM-CSF	Leukine/Sar-gramostim	Leukemia	Approved	Schering-AG	
			Bone marrow/stem cell transplants	Approved		
			Crohn's disease	Phase III		

[1]http://www.clinicaltrials.gov/show/NCT00053976 (GvHD).
[2] http://www.clinicaltrials.gov/ct/show/NCT00080431?order=5.
[3] http://www.clinicaltrials.gov/ct/show/NCT00050661?order=10.
[4] http://www.pdl.com/applications/press_releases.cfm?newsId=241.

receptors, such as IL-6, IL-8, IL-18, and gamma interferon (IFN-γ) (TABLE 1b) in a variety of clinical conditions.

Another approach to cytokine therapy that has been applied successfully is treatment using recombinant cytokines. IL-2 has been used in cancer,[6,7] various derivatives of IFN-α in viral infection and cancer,[8–10] IL-11 in the treatment of post-chemotherapy induced thrombocytopenia,[11] IFN-γ in cancer and osteoporosis,[12] and IFN-β in multiple sclerosis[13] (TABLE 1a). Therapy using recombinant cytokines has also been unsuccessful in some cases. Recombinant IL-10 provided no beneficial therapy over side effects in a number of settings (TABLE 1b). However, this potent immunosuppressive cytokine may yet have some therapeutic application.[14] Recombinant IL-12 can also induce adverse side effects.[15]

AUTOVACCINATION AS A STRATEGY IN CYTOKINE THERAPY

Vaccination against autologous cytokines opens novel possibilities for therapy of chronic diseases resulting from excessive production of a particular factor. Tolerance against self-antigens can be overcome by physical association of foreign proteins to self-antigen. This concept was first validated for hormones such as bovine luteinizing hormone[16] and human chorionic gonadotropin[17] and subsequently extended to cytokines.[18–20] Recently, the Van Snick group chemically coupled mouse IL-9 and IL-12 to ovalbumin (Ova) that produced very immunogenic complexes, leading to complete inhibition of cytokine function in immunized mice.[21] Important for potential cytokine therapy was our finding, in collaboration with Van Snick's group, that IL-9–vaccinated mice showed increased resistance to cutaneous leishmaniasis,[22] and IL-12–vaccinated mice were protected against experimental autoimmune encephalomyelitis (EAE),[23] as summarized in TABLE 2. Similar results have also been reported after immunization with mouse TNF-α proteins fused to an Ova helper sequence[19] or with a DNA vaccine comprising a tetanus toxoid sequence inserted in IL-5 coding DNA.[24] The latter experiments showed clear inhibition of some pathologic hallmarks of arthritis and asthma, respectively.

TABLE 1b. Cytokine therapies in clinical trial

Cytokine	Drug	Trade name/ code	Application	Clinical trial phase/ outcome	Company	Ref/ link
IL-1	Blocking mAb	AMG-108	Osteoarthritis	Phase II	Amgen	[1]
IL-1	IL-1 type 1R / receptor accessory protein Fc fusion	IL-1 TRAP	RA	Phase II	Regeneron	[2]
IL-2	Adenoviral-IL-2	TG-1024	Renal cell carcinoma Soft tissue sarcoma Melanoma Head + neck cancer	Phase I/II Phase I/II Phase I/II Phase I/II	Transgene	[3]
IL-2	MVA-MUC1-IL-2	TG-4010	Kidney/prostate/ lung cancers	Phase II	Transgene	[4]
IL-2	MVA-HPV-IL2	TG-4001	Cerivcal cancer	Phase II	Transgene	[5]
IL-2	Humanized IL-2 receptor β-chain blocking mAb	MIKβ1	Large granular lymphocytic leukemia	Pre-phase I	NCI	[6]
IL-4/13	Recombinant IL-4Rα / IL-13Rα1 Fc fusion	IL-4/13 trap	HIV Asthma	Phase I Phase I	Regeneron	41
IL-4	IL-4 coupled to Pseudomonas exotoxin	IL-4PE NBI-3001	Malignant glioma Kidney/Lung cancers	Phase II Phase I	Neurocrine Bioscience	42
IL-4/13	IL-4Rα antagonist	Bay 16-9996	Asthma	Trial halted	Bayer	[7]
IL-4	Humanized anti-IL4 mAb	SB240683	Asthma	Trial halted Phase IIa	PDL/GSK	[8]
IL-5	Humanized anti-IL-5	SCH-55700	Asthma/Allergy	Phase II	Celltech/ Schering-Plough	43
IL-5	Humanized Anti-IL-5	Mepolizumab	Asthma Hypereosinophilic syndrome Atopic dermatitis	Phase II Phase III Phase II	GSK	[9] 44
IL-6	Recombinant IL-6	Atexakin	Peripheral neuropathy	Phase II/ halted?	Serono	

TABLE 1b. Cytokine therapies in clinical trial

Cytokine	Drug	Trade name/ code	Application	Clinical trial phase/ outcome	Company	Ref/ link
IL-6	Humanized anti-IL-6 receptor	MRA	Crohn's disease SLE RA Myeloma Systemic onset juvenile idiopathic arthritis	Phase II Phase I Phase III Phase I/II Phase II/III	Chugai Pharmaceuticals/ Roche	45
IL-8	Anti-IL-8 mAb	ABX-IL-8	RA Psoriasis COPD	Trials halted; no therapeutic value in Phase II	Abgenix	[10]
IL-8	IL-8 antagonist	656933	COPD	Phase I	GSK	
IL-9	Anti-IL-9 mAb	IL-9mAb	Asthma	Phase I	Genaera / Medimmune	
IL-10	Recombinant IL-10	Ilodecakin / Tenovil	Psoriasis Psoriatic arthritis Crohn's disease Bile duct diseases Biliary tract diseases Gallbladder diseases Pancreatitis Pancreatic diseases HIV RA Ulcerative colitis	Phase II/ halted Phase II/ halted Phase II/ halted Phase II/ halted Phase II/ halted Phase II/ halted Phase II/ halted Phase II/ halted Phase II/ halted Phase III/ halted Phase III	Schering-Plough	14
IL-12	Human anti-IL-12	ABT874/ J695	Crohn's disease Multiple sclerosis	Phase II Phase II	Abbott/CAT	[11]
IL-12	Recombinant IL-12	rIL-12	Asthma Non-Hodgkin's/ Hodgkin's lymphoma	Phase I/ halted Phase II		15, 46

TABLE 1b. Cytokine therapies in clinical trial

Cytokine	Drug	Trade name/ code	Application	Clinical trial phase/ outcome	Company	Ref/ link
IL-13	Human anti-IL-13	CAT 354	Asthma	Phase I	CAT	[13]
IL-13	IL-13 coupled to truncated Pseudomonas endotoxin A	IL-13-PE38QQR	Glioblastoma multiforme	Phase I/II	Neopharm	47
IL-13	Anti-IL-13 receptor α1	IL-13 receptor α 1	Asthma	Preclinical	Amrad/ Merck	[14]
IL-15	Humanized anti-IL-15	Humax IL-15 /AMG-714	RA	Phase II	Genmab/ Amgen	48
IL-18	Recombinant IL-18 binding protein	Tadeking-α	RA Psoriasis	Phase IIa Phase II	Serono	[15]
IL-18	Recombinant human IL-18 immuno-modulator	485232	Immunologically sensitive target melanoma + renal carcinoma Lymphoma	Phase II Phase I	GSK	[16]
IL-21	Recombinant IL-21	494C10	Metastatic melanoma Renal cell carcinoma	Phase I Phase I	NCI/ Zymogenetics	[17]
TNFα	Pegylated truncated soluble TNF p55 type I receptor	sTNFR1 / pegsunercept	RA	Phase II	Amgen	[18]
TGFβ1	Oligo-antisense	AP11014	Lung/colon/ prostate cancer	Preclinical	Antisense Pharma	
TGFβ1	Human anti-TGF-β1 mAb	CAT-192 / metelimumab	Systemic sclerosis Scleroderma	Phase I Phase II	CAT/Genzyme	[19]
TGF-β2	Human anti-TGFβ2 mAb	CAT-152 / Trabio	Glaucoma surgery	Phase II/III	CAT	[20]
TGF-β2	Oligo-antisense	-	Noncurable non-small cell lung cancer	Phase II	NovaRx	49

TABLE 1b. Cytokine therapies in clinical trial

Cytokine	Drug	Trade name/ code	Application	Clinical trial phase/ outcome	Company	Ref/ link
TGF-β2	Oligo-antisense	AP12009	High grade glioma pancreatic cancer malignant melanoma	Phase IIb Phase I Phase I	Antisense Pharma	[20]
TGF-β	Human anti-TGF-β	GC-1008	Idiopathic pulmonary fibrosis	Preclinical	CAT/Genzyme	
IFN-γ	Humanized anti-IFN-γ mAb	HuZAF/fontolizumab	Crohn's disease	Phase I/II	PDL	50
IFN-γ	Adenoviral IFN-γ	TG1042	Cutaneous B-cell lymphoma Cutaneous T-cell lymphoma	Phase I Phase I/II	Transgene	51

[1] http://www.amgen.com/rnd/pipeline.html
[2] http://www.hopkins-arthritis.som.jhmi.edu/edu/eular2004/ra-treatments-il1.html
[3] http://www.transgene.fr/us/product_pipeline/iframe_ad_il2.htm
[4] http://www.transgene.fr/us/page.php?fam=1&rub=3&iframe=product_pipeline/iframe_mva_muc1_il2.htm
[5] http://www.transgene.fr/us/page.php?fam=1&rub=2&iframe=product_pipeline/iframe_mva_hpv_il2.htm
[6] http://www.clinicaltrials.gov/ct/show/NCT00076180
[7] Preclinical Evaluation of Bay 16-9996 a Dual Il-4/Il-13 Receptor Antagonist N. Fitch, M. Morton, A. Bowden, M.-C. Shanafelt, A. Shanafelt, G. Wetzel, J. Moy Biotechnology Research, Bayer Corp, Berkeley, CA. Abstract J.A.C.I 107: 2 section 209
[8] http://www.pdl.com/applications/press_releases.cfm?newsId=68
[9] http://www.gsk.com/financial/product_pipeline/docs/pipeline.pdf
[10] http://www.abgenix.com/ProductDevelopment/?view=DevelopmentStrategy
[11] http://www.cambridgeantibody.com/html/technology_products/product_pipeline/abt874__treatment_for_autoimmune_
inflammatory_disorders
[12] http://www.nci.nih.gov/search/ViewClinicalTrials.aspx?cdrid=67489&version=patient&protocolsearchid=1344505
[13] http://www.cambridgeantibody.com/html/technology_products/product_pipeline/cat354
[14] http://www.amrad.com.au/Files/IL-13a1%20-%20Aug%202004.pdf
[15] www.serono.com/pipeline/pipeline.jsp?major=2 - 35k - 28 Dec 2004
[16] http://www.clinicaltrials.gov/ct/gui/show/NCT00085904
[17] http://www.zymogenetics.com/clinical/il-21.html
[18] http://www.clinicaltrials.gov/ct/show/NCT00037700?order=1
[19] http://www.clinicaltrials.gov/ct/show/NCT00043706
[20] http://www.cambridgeantibody.com/html/technology_products/product_pipeline/trabio treatment_for_scarring_following_glaucoma_surgery
[21] http://www.antisense-pharma.com/products/f_products.htm

Useful links: http://www.phrma.org/newmedicines/resources/2004-10-25.145.pdf
http://imgt.cines.fr/textes/IMGTrepertoire/GenesClinical/monoclonalantibodies/other.html

TABLE 2. Auto vaccination

Cytokine	Inhibition titer[a]	Disease	Effect
Interleukin-9	3–5 × 10^{-4}[b]	Leishmaniasis	Delayed[22]
Interleukin-12	Approx. 2 × 10^{-3}	EAE	Protection[23]

[a]Fifty percent inhibition of cytokine-dependent inhibition by sera dilution.
[b]Depending on the mouse strain used.

TARGETING IL-4/IL-13 IN ALLERGY AND ASTHMA

Allergies are reproducible reactions of the immune system against stimuli, which are tolerated by most people. Most allergies are mediated by Th2-dependent immune mechanisms. These are characterized by very fast responses towards the antigen, typically within minutes, with most or all of the Th2 cytokines (i.e., IL-4, IL-5, IL-9, and IL-13) possibly involved in the disease phenotype, therefore being potential targets for cytokine therapy against allergic diseases (FIG. 2). After initial contact with the allergen (sensitization phase), symptoms are felt within minutes (effector phase) in immediate type (type 1) allergies. This type includes common reactions such as hay fever, some food allergies, and allergies against house dust mites, animal hair, and insect stings, among others. Allergic asthma is a complex disorder characterized by local and systemic allergic inflammation associated with a chronic pulmonary eosinophilia and elevated serum IgE levels causing mast cell activation, airway remodeling, and reversible airway obstruction. Asthma symptoms, especially shortness of breath, are primarily related to airway obstruction, and death is almost invariably due to asphyxiation. Increased airway hyperresponsiveness (AHR) and mucus hypersecretion by goblet cells are two of the principal causes of airway obstruction observed in asthmatic patients. IL-4 is known to be the central promoter of Th2 development *in vivo* and *in vitro*. However, recent studies by us and others have challenged a sole *in vivo* role of IL-4 in Th2 differentiation, as in the absence of IL-4 or its receptor, Th2 differentiation was still present.[25] Despite these observations, the importance of IL-4 in allergic reactions has been confirmed in experimental murine models, mostly using OVA as the allergen. Indeed, in the absence of IL-4–mediated functions, either by *in vivo* blocking of IL-4 or its receptor or by genetically disrupting the gene, OVA-sensitized and challenged mice showed reduced AHR, eosinophilic inflammation, IgE production, and mastocytosis.[26–30] Much conflicting data have been gathered on the role of IL-5 in allergic asthma, despite its clear role as a differentiation factor for eosinophils,[31–33] with rather disappointing results in clinical studies in asthma targeting IL-5.[34] Monoclonal antibodies specific for IL-9 are now being used in phase I clinical trials in asthma. The possible role of IL-13 had remained elusive until 1998 when we and others demonstrated that IL-13 is a key factor in the asthmatic phenotype, able to directly induce goblet cell hyperplasia and AHR in mice[35,36] (FIG. 2). These results prompted discussions about IL-13 as a promising target for anti-asthmatic therapies, resulting in three drug leads now in clinical phase I trials (TABLE 1b). My group is currently involved in the establishment of several cell-type–specific knockout mouse models to further define the role of IL-4 and IL-13 in infectious and allergic diseases (TABLE 3).

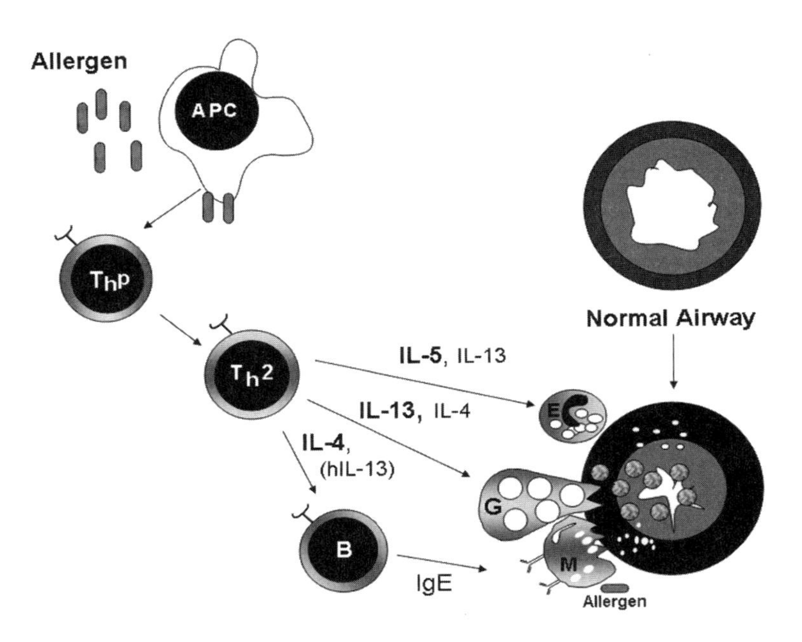

FIGURE 2. Proposed allergic asthma model. Role of Th2 cells, effector cells, and cytokine network in the pathogenicity of asthma. Thp, parental T helper cell; Th2, T helper 2 cell; B, B lymphocyte; E, eosinophil; G, goblet cell; M, mast cell; APC, antigen presenting cell.

TABLE 3. Interleukin-4Rα cell-specific knockout mice

Cell type	Strategy	Status
Macrophage/neutrophil Il-4Rα KO	$LysM^{Cre}T \times Il\text{-}4R\alpha^{flox/flox}$	Herbert *et al.* 2004
T-cell Il-4Rα KO	$Ick^{Cre}T \times IL\text{-}4R\alpha^{flox/flox}$	Ms in preparation
T-cell Il-4Rα KO	$CD4^{Cre}T \times IL\text{-}4R\alpha^{flox/flox}$	In characterization
Smooth muscle cell Il-4Rα KO	$SM\text{-}MHC^{Cre}T \times IL\text{-}4R\alpha^{flox/flox}$	In characterization
Endothelial cell Il-4Rα KO	$FABPi^{Cre} \times IL\text{-}4R\alpha^{flox/flox}$	In characterization
Goblet cell Il-4Rα KO		Breeding
T-cell Il-4Rα reconstitution	$hIL\text{-}4R\alpha T \times IL\text{-}4R\alpha^{-/-}$	Seki *et al.* 2004

FUTURE PERSPECTIVES

The therapeutic potential of targeting specific inflammatory cytokines with known function in RA has demonstrated the legitimacy of targeting cytokines as a means of ameliorating disease. Success in targeting cytokines in inflammation has opened a new approach to the treatment of inflammatory disease, and more mAbs specific for inflammatory cytokines such as IL-6 may be used in the clinic. Cytokine therapy has and will be successful when disease-causing mechanisms are relatively "simple" and cytokine driven. However, given the pleiotropic and redundant nature of cytokines, this will be hard to achieve in complex disease states. Perhaps a future direction in complex disorders may be the simultaneous inactivation/activation of multiple cytokines, such as simultaneous inactivation of IL-4, IL-5, IL-9, and IL- 13 in asthma. The use of interference RNA may achieve such a goal given the recent advances by Soutchek *et al.*[37] in delivery of iRNA.

ACKNOWLEDGMENTS

This work was supported in part by the Medical Research Council (MRC) and National Research Foundation (NRF) of South Africa. F.B. is a holder of a Wellcome Trust Research Senior Fellowship for Medical Science in South Africa (Grant 056708/Z/99).

REFERENCES

1. Andreakos, E. 2003. Targeting cytokines in autoimmunity: new approaches, new promise. Exp. Opin. Biol. Ther. **3:** 435–447.
2. Elliott, M.J. *et al.* 1994. Randomised double-blind comparison of chimeric monoclonal antibody to tumour necrosis factor alpha (cA2) versus placebo in rheumatoid arthritis. Lancet **344:** 1105–1110.
3. Scheinfeld, N. 2003. Adalimumab (HUMIRA): a review. J. Drugs Dermatol. **2:** 375–377.
4. Nanda, S. & J.M. Bathon. 2004. Etanercept: a clinical review of current and emerging indications. Exp. Opin. Pharmacother. **5:** 1175–1186.
5. Bresnihan, B. 2001. The prospect of treating rheumatoid arthritis with recombinant human interleukin-1 receptor antagonist. BioDrugs **15:** 87–97.
6. Dutcher, J. 2002. Current status of interleukin-2 therapy for metastatic renal cell carcinoma and metastatic melanoma. Oncology (Huntingt) **16:** 4–10.
7. Baron, E. & S. Narula. 1990. From cloning to a commercial realization: human alpha interferon. Crit. Rev. Biotechnol. **10:** 179–190.
8. Marcellin, P. *et al.* 2004. Peginterferon alfa-2a alone, lamivudine alone, and the two in combination in patients with HBeAg-negative chronic hepatitis B. N. Engl. J. Med. **351:** 1206–1217.
9. Melian, E.B. & G.L. Plosker. 2001. Interferon alfacon-1: a review of its pharmacology and therapeutic efficacy in the treatment of chronic hepatitis C. Drugs **61:** 1661–1691.
10. Syed, T.A. & O.A. Ahmadpour. 1998. Human leukocyte derived interferon-alpha in a hydrophilic gel for the treatment of intravaginal warts in women: a placebo-controlled, double-blind study. Int. J. STD AIDS **9:** 769–772.
11. Isaacs, C. *et al.* 1997. Randomized placebo-controlled study of recombinant human interleukin-11 to prevent chemotherapy-induced thrombocytopenia in patients with breast cancer receiving dose-intensive cyclophosphamide and doxorubicin. J. Clin. Oncol. **15:** 3368–3377.

12. The International Chronic Granulomatous Disease Cooperative Study Group. 1991. A controlled trial of interferon gamma to prevent infection in chronic granulomatous disease. N. Engl. J. Med. **324:** 509–516.
13. Stuart, W. H. *et al.* 2004. Selecting a disease-modifying agent as platform therapy in the long-term management of multiple sclerosis. Neurology **63:** S19–27.
14. Asadullah, K. *et al.* 2004. Interleukin-10: an important immunoregulatory cytokine with major impact on psoriasis. Curr. Drug Targets Inflamm. Allergy **3:** 185–192.
15. Bryan, S.A. *et al.* 2000. Effects of recombinant human interleukin-12 on eosinophils, airway hyper-responsiveness, and the late asthmatic response. Lancet **356:** 2149–2153.
16. Johnson, H.E. *et al.* 1988. Active immunization of heifers against luteinizing hormone-releasing hormone, human chorionic gonadotropin and bovine luteinizing hormone. J. Anim Sci. **66:** 719–726.
17. Talwar, G.P. *et al.* 1994. A vaccine that prevents pregnancy in women. Proc. Natl. Acad. Sci. USA **91:** 8532–8536.
18. Ciapponi, L. *et al.* 1997. Induction of interleukin-6 (IL-6) autoantibodies through vaccination with an engineered IL-6 receptor antagonist. Nat. Biotechnol. **15:** 997-1001.
19. Dalum, I. *et al.* 1999. Therapeutic antibodies elicited by immunization against TNF-alpha. Nat. Biotechnol. **17:** 666–669.
20. Gringeri, A. *et al.* 1996. Absence of clinical, virological, and immunological signs of progression in HIV-1-infected patients receiving active anti-interferon-alpha immunization: a 30-month follow-up report. J. Acquir. Immune Defic. Syndr. Hum. Retrovirol. **13:** 55–67.
21. Richard, M. *et al.* 2000. Anti-IL-9 vaccination prevents worm expulsion and blood eosinophilia in *Trichuris muris*-infected mice. Proc. Natl. Acad. Sci. USA **97:** 767–772.
22. Arendse, B., J. Van Snick & F. Brombacher. 2005. Interleukin-9 is a susceptibility factor in *Leishmania major* infection by promoting detrimental Th2/type 2 responses. J. Immunol. In press.
23. Uyttenhove, C. *et al.* 2004. Development of an anti-IL-12 p40 auto-vaccine: protection in experimental autoimmune encephalomyelitis at the expense of increased sensitivity to infection. Eur. J. Immunol. **34:** 3572–3581.
24. Hertz, M. *et al.* 2001. Active vaccination against IL-5 bypasses immunological tolerance and ameliorates experimental asthma. J. Immunol. **167:** 3792–3799.
25. Mohrs, M., C. Holscher & F. Brombacher. 2000. Interleukin-4 receptor alpha-deficient BALB/c mice show an unimpaired T helper 2 polarization in response to *Leishmania major* infection. Infect. Immun. **68:** 1773–1780.
26. Lukacs, N.W. *et al.* 1994. Interleukin-4–dependent pulmonary eosinophil infiltration in a murine model of asthma. Am. J. Respir. Cell Mol. Biol. **10:** 526–532.
27. Brusselle, G.G. *et al.* 1994. Attenuation of allergic airway inflammation in IL-4 deficient mice. Clin. Exp. Allergy **24:** 73–80.
28. Coyle, A.J. *et al.* 1995. Interleukin-4 is required for the induction of lung Th2 mucosal immunity. Am. J. Respir. Cell Mol. Biol. **13:** 54–59.
29. Gavett, S.H. *et al.* 1997. Interleukin-4 receptor blockade prevents airway responses induced by antigen challenge in mice. Am. J. Physiol. **272:** L253–261.
30. Grunewald, S.M. *et al.* 1998. An antagonistic IL-4 mutant prevents type I allergy in the mouse: inhibition of the IL-4/IL-13 receptor system completely abrogates humoral immune response to allergen and development of allergic symptoms in vivo. J. Immunol. **160:** 4004–4009.
31. Campbell, H.D. *et al.* 1987. Molecular cloning, nucleotide sequence, and expression of the gene encoding human eosinophil differentiation factor (interleukin 5). Proc. Natl. Acad. Sci. USA **84:** 6629–6633.
32. Kopf, M. *et al.* 1995. Immune responses of IL-4, IL-5, IL-6 deficient mice. Immunol. Rev. **148:** 45–69.
33. Kopf, M. *et al.* 1996. IL-5-deficient mice have a developmental defect in CD5+ B-1 cells and lack eosinophilia but have normal antibody and cytotoxic T cell responses. Immunity **4:** 15–24.
34. Ichinose, M. & P.J. Barnes. 2004. Cytokine-directed therapy in asthma. Curr. Drug Targets Inflamm. Allergy **3:** 263–269.

35. GRUNIG, G. *et al.* 1998. Requirement for IL-13 independently of IL-4 in experimental asthma. Science **282:** 2261–2263.
36. WILLS-KARP, M. *et al.* 1998. Interleukin-13: central mediator of allergic asthma. Science **282:** 2258–2261.
37. SOUTSCHEK, J. *et al.* 2004. Therapeutic silencing of an endogenous gene by systemic administration of modified siRNAs. Nature **432:** 173–178.
38. BIELEKOVA, B. *et al.* 2004. Humanized anti-CD25 (daclizumab) inhibits disease activity in multiple sclerosis patients failing to respond to interferon beta. Proc. Natl. Acad. Sci. USA **101:** 8705–8708.
39. NASHAN, B. *et al.* 1997. Randomised trial of basiliximab versus placebo for control of acute cellular rejection in renal allograft recipients. CHIB 201 International Study Group. Lancet **350:** 1193–1198.
40. STRIETER, R.M. *et al.* 2004. Effects of interferon-gamma 1b on biomarker expression in patients with idiopathic pulmonary fibrosis. Am. J. Respir. Crit Care Med. **170:** 133–40.
41. ECONOMIDES, A.N. *et al.* 2003. Cytokine traps: multi-component, high-affinity blockers of cytokine action. Nat. Med. **9:** 47–52.
42. WEBER, F. *et al.* 2003. Safety, tolerability, and tumor response of IL4-Pseudomonas exotoxin (NBI-3001) in patients with recurrent malignant glioma. J. Neurooncol. **64:** 125–137.
43. KIPS, J.C. *et al.* 2003. Effect of SCH55700, a humanized anti-human interleukin-5 antibody, in severe persistent asthma: a pilot study. Am. J. Respir. Crit Care Med. **167:** 1655–1659.
44. GARRETT, J.K. *et al.* 2004. Anti-interleukin-5 (mepolizumab) therapy for hypereosinophilic syndromes. J. Allergy Clin. Immunol. **113:** 115–119.
45. NISHIMOTO, N. & T. KISHIMOTO. 2004. Inhibition of IL-6 for the treatment of inflammatory diseases. Curr. Opin. Pharmacol. **4:** 386–391.
46. YOUNES, A. *et al.* 2004. Phase II clinical trial of interleukin-12 in patients with relapsed and refractory non-Hodgkin's lymphoma and Hodgkin's disease. Clin. Cancer Res. **10:** 5432–5438.
47. BARTH, S. 2001. hIL-13-PE38QQR. NeoPharm. Curr. Opin. Invest. Drugs **2:** 1309–1313.
48. MCINNES, I.B. & J.A. GRACIE. 2004. Interleukin-15: a new cytokine target for the treatment of inflammatory diseases. Curr. Opin. Pharmacol. **4:** 392–397.
49. FAKHRAI, H. *et al.* 1996. Eradication of established intracranial rat gliomas by transforming growth factor beta antisense gene therapy. Proc. Natl. Acad. Sci. USA **93:** 2909–2914.
50. SANDBORN, W.J. 2004. How future tumor necrosis factor antagonists and other compounds will meet the remaining challenges in Crohn's disease. Rev. Gastroenterol. Disord. **4** (Suppl 3)**:** S25–33.
51. LIU, M. *et al.* 2004. Gene-based vaccines and immunotherapeutics. Proc. Natl. Acad. Sci. USA **101** (Suppl 2)**:** 14567–14571.

Fighting Food Allergy

Current Approaches

NATALIE E. NIEUWENHUIZEN[a] AND ANDREAS L. LOPATA[b]

[a]*University of Cape Town, Faculty of Health Sciences, Institute of Infectious Disease and Molecular Medicine (IIDMM), Cape Town, South Africa*

[b]*Division of Immunology, Allergy Section (NHLS), IIDMM, Cape Town, South Africa*

ABSTRACT: Food allergy is defined as an adverse immunologic reaction to allergens present in food and is associated with symptoms ranging from gastrointestinal discomfort to anaphylactic shock and death. The increase in prevalence and potential fatality of disease has led to increased efforts to find effective therapies and prophylaxis. While specific immunotherapy (SIT) is effective in desensitization against inhalant allergens, it is unadvised against food allergy because of the high risk of adverse side effects. A review of the recent literature shows that various approaches have been taken to develop safer and more effective SIT regimens. Here we discuss the use of recombinant allergens, peptides, DNA vaccines, immunostimulatory DNA sequences, and other bacterial products in SIT. In addition, we review nonspecific therapies such as anti-IgE administration and cytokine therapy as well as natural therapies such as probiotics and Chinese herbal medications. In conclusion, anti-IgE treatment and SIT using hypoallergenic recombinant allergens in combination with Th1-inducing adjuvants appear the most promising approaches. New initiatives to increase our understanding of the pathophysiology and immunologic mechanisms of food allergy along with the molecular characterization of food allergens should pave the way towards safer and more effective ways of combating this debilitating and potentially life-threatening disease.

KEYWORDS: food allergy; recombinant allergens; cytokines; immunotherapy; natural therapy; murine models; immunostimulatory DNA; CpG; DNA vaccine

INTRODUCTION

The steady increase in allergic disease has intensified the need for successful therapeutic approaches. Up to 4% of adults and 8% of children are food allergic, with an estimated 30,000 food-induced anaphylactic events seen annually in the United States alone, 150 of them fatal. Fish, shellfish, soybean, wheat, egg, milk, peanuts, and tree nuts cause the majority of all allergic reactions to foods[1–4] with seafood allergy the most common in the United States[5] and nuts causing over 90% of fatalities.[6,7] Currently, strict avoidance of the allergenic foods is the only way to prevent

Address for correspondence: Andreas L. Lopata, IIDMM, Division of Immunology, University of Cape Town, Medical School, Observatory 7925, Cape Town, South Africa. Voice: +27-21-406-6033/6147; fax: +27-21-406-6029.
alopata@uctgsh1.uct.ac.za

**Ann. N.Y. Acad. Sci. 1056: 30–45 (2005). © 2005 New York Academy of Sciences.
doi: 10.1196/annals.1352.003**

reactions, but about 50% of affected individuals experience accidental exposure and reactions every 3–5 years, with hidden allergens in restaurant and processed food often to blame.[8–10]

Food allergy is defined as an adverse immunologic reaction to allergenic molecules present in foods, with reactions ranging from mild discomfort, urticaria, and allergic asthma to life-threatening anaphylactic shock.[2,3] The acute symptoms of allergy are usually due to the release of inflammatory mediators by mast cells and basophils, including histamine, platelet-activating factor, leukotrienes, mast cell proteases, and cytokines. These mediators are released when allergen binds to IgE antibody attached to FcεRI receptors on the cell surface, causing degranulation. Studies show a skewing towards a Th2 response in food allergy, with elevated interleukin (IL)-4, IL-5, and IL-13 levels, whereas tolerant individuals have higher levels of the Th1 cytokines interferon (IFN)-gamma and tumor necrosis factor (TNF)-alpha, and the regulatory cytokine IL-10.[11–15]

SPECIFIC IMMUNOTHERAPY

Specific immunotherapy to inhalant allergens has proven to be highly successful.[16,17] Injections of allergen over a long period of time leads to reduced sensitivity to allergen, which is correlated with a switch from a Th2 type response to a Th1 or regulatory type response, with raised levels of IFN-γ, the regulatory cytokine IL-10, and so-called "blocking" IgG antibodies.[18–22] It was recently shown that sublingual immunotherapy produces significantly fewer side effects than injections, with equivalent efficacy.[16,23,24] In addition, there are some indications that specific immunotherapy may abrogate symptoms due to oral allergy syndrome, a condition in which patients with pollen allergy experience reactions upon ingestion of food containing cross-reactive allergens.[25–29] It is estimated that up to 50% of patients with pollen allergy experience oral allergy syndrome. This so-called class II food allergy is generally mediated by less stable allergens (e.g., Bet vs 8 in peanuts)[30] and is drawing increased attention due to the steady rise in the prevalence of inhalant allergies. However, results of desensitization trials are inconclusive.[25–29] Cross-reactivity between related allergens (peanut and soybean) has been used in a murine model as a safe way to desensitize peanut-allergic mice.[31] Nevertheless, at this stage, specific immunotherapy against food allergy in humans is currently not viable because of the high risk of adverse reactions and limited efficacy.[7,32,33] Much research is therefore going into making immunotherapy safer and more efficacious by altering the various components: the antigens, the adjuvants, and the delivery systems.[34]

RECOMBINANT PROTEINS AND PEPTIDES

Currently, specific immunotherapy relies on crude or commercial allergen extracts that contain proteins and allergens in variable proportions, increasing the risk of unwanted side effects, as in patients treated for house dust-mite allergy who developed anaphylactic reactions to snails due to cross-reactivity of allergens.[35,36] The use of defined and purified recombinant allergens in immunotherapy would resolve these issues.[37–39] In addition, current recombinant technology allows the

TABLE 1. Specific immunotherapy in food allergy

Allergy	Treatment	Approach	Relevance	Molecular effects	Clinical effects	Ref.
Peanut	Peanut	Rush therapy (injection)	Human	Increased oral tolerance	Considerable systemic reactions	32
Peanut	Peanut	Rush therapy (injection)	Human	Increased oral tolerance	Considerable systemic reactions	33
Apple (oral allergy syndrome)	rBet v1 (Birch pollen)	SIT	Human	IgG4 upregulated; regulatory T cells not altered	Symptom reduction in food challenge (DBPCFC)	148
Apple (oral allergy syndrome)	Birch pollen	SIT (comparing subcutaneous with sub-lingual)	Human	Specific IgE reduced	No symptom reduction	149
Peanut	Modified Ara h 1-3	Allergen plus heat-killed *L. mono-cytogenes*; subcutaneous injection	Mouse	Th2 cyto-kines reduced; specific IgE reduced	Reduced anaphylactic reactions	150
Peanut	Modified Ara h 1-3	Allergen plus heat-killed *E. coli*; rectal sensitization	Mouse	Th2 cytokines and IgE reduced	Reduced symptoms	151
Peanut	Soya extract	Desensitization, intra-peritoneal, with soya extract	Mouse	Th2 cytokines and IgG1 reduced	Reduced anaphylactic reactions	31
Ovalbumin	Ovalbumin	Allergen plus heat-killed *Lactobacillus casei* strain Shirota (LcS)	Mouse	Th1 skewed response; specific IgE/IgG1 reduced	Reduced symptoms	152
Milk	Beta-lacto-globulin (BLG)	Carrageenan	Mouse	Reduced antigen-specific proliferative response	Reduced symptoms	153
Peanut	Peanut	Allergen plus heat-killed *Listeria mono-cytogenes*	Dog	Specific IgE reduced; Skin prick reactivity reduced	Reduced symptoms	66

construction of hypoallergenic derivatives of natural allergens, which should reduce the adverse effects of immunotherapy. Such allergens should not be able to activate cells via crosslinking of IgE antibodies, but should preserve T-cell epitopes (immunogenicity) and activate B cells to induce blocking IgG antibodies. Hypoallergenic derivatives of the major birch pollen allergen Bet v1 have been extensively studied. Recombinant Bet v1 (rBet v1) and even rBet v1 fragments induced blocking IgG antibodies in animal models and were even able to shift a type Th2 immune response towards a Th1 phenotype.[40] Recent patient trials with rBet v1 were successful.[38] A range of recombinant food allergens has been produced and successfully tested in animal models (TABLE 1).[41–44] However, no human immunotherapy trials have been performed with recombinant food allergens. Another approach has been to use synthetic peptides of allergen T-cell epitopes in specific therapy. This approach seemed to induce a regulatory T-cell response rather than a Th1 response and worked to some extent, but allergic reactions were reported during the therapy.[45,46]

ADJUVANTS

A variety of traditional adjuvants (e.g., aluminium hydroxide) have been used in immunotherapy without detailed understanding of their molecular mechanisms.[47] Now new adjuvants with stronger Th1-inducing properties are being investigated.

Immunostimulatory DNA

Bacterial CpG motifs, unmethylated dinucleotides flanked by specific sequences, are recognized by the mammalian immune system via toll-like receptor 9 (and possibly other pattern-recognition receptors) and trigger a Th1 response.[48–51] The same is true of synthetic oligodeoxynucleotides containing such CpG motifs (CpG-ODN).[50,52,53] Experiments in murine models of allergic asthma, rhino-sinusitis, and conjunctivitis show that administration of CpG-ODN prevents symptoms and reduces already established disease by reducing Th2 and IgE.[54–57] Allergen/CpG-ODN conjugates have been shown to be less allergenic and more immunogenic than native allergen.[58,59] The major allergen from ragweed, Amb a 1, linked to an immunostimulatory DNA sequence promoted Th1 cytokine expression and downregulated Th2 expression *in vitro*,[60] reversed established airway hyperreactivity in a murine model of asthma,[61,62] and yielded promising results in Phase II clinical trials.[63] Although no food allergen–CpG conjugates have been tested, a similar approach using Th1-inducing heat-killed bacteria has yielded good results in mice[64,65] and allergic dogs.[66]

Monophosphoryl Lipid A

Another bacterial derivative undergoing testing is 3-deacylated monophosphoryl lipid A (MPL), a detoxified form of lipid A derived from the lipopolysaccharide of *Salmonella minnesota* R595. Its effects as a Th1-inducing immunostimulatory adjuvant have been described in several publications.[67,68] Clinical trials using MPL as an adjuvant in immunotherapy for grass allergy demonstrated that MPL was efficacious, needing only four injections to induce antibody changes normally requiring lengthy injection schedules,[69] and that MPL induced strong blocking IgG responses.[70] It has not been tested in any food allergy vaccines.

DELIVERY SYSTEMS

Several routes of administration are under consideration for use in food allergy vaccines.

Microencapsulation

Oral administration of antigens usually leads to tolerance and has been effective in decreasing allergic sensitization to antibiotics and other medications.[71,72] Obviously, native food allergens cannot be administered in this way, but it may be possible with hypoallergenic or CpG-conjugated derivatives. Microencapsulation provides a promising way of delivering allergens without degradation in the stomach,[73] thereby inducing oral tolerance, and it has already been applied in clinical trials.[74]

DNA Vaccines

Plasmid DNA injected intramuscularly, intraperitoneally, or with a gene gun results in transcription and translation of encoded genes and elicits an antibody response in the host.[75–77] This method of immunization, known as a DNA vaccine, preferentially induces a Th1 immune response and suppression of IgE.[78,79] These effects appear to be mediated by both $CD8^+$ and $CD4^+$ cells,[77,80,81] and plasmid DNA requires immunostimulatory sequences (CpG) for optimal immunogenicity.[82–86] Several allergens have been tested in DNA vaccines in murine models, including Ara h 2 (peanut), bovine beta-lactoglobulin (cow's milk), Cry j 1 (Japanese cedar), phospholipase A2 (bee venom), Der f 11 and Der p 1 (dust mites), and Bet v 1 and Phl p 2 (grass).[80,83–85,87–90] Most studies found elicitation of a Th1 response and increased IL-10. Mice vaccinated against phospholipase A2 were protected against fatal anaphylaxis after allergen challenge,[83] whereas mice receiving an oral DNA vaccine containing the peanut allergen Ara h 2 orally[87] experienced significantly less severe and delayed allergic reactions upon subsequent sensitization and challenge. However, prophylactic effects, while promising, are not sufficient to aid patients who have existing food allergy. In mice pre-sensitized to phospholipase A, therapeutic gene vaccination prevented only 30% of mice from experiencing anaphylaxis.[83] However, DNA vaccination provides the option of co-delivering genes or adjuvant molecules with immunomodulatory properties together with the antigen sequence.[86] Vaccination of mice with a plasmid containing the cDNA for OVA fused to the cDNA of IL-18, a potent Th1 inducer, reversed established airway hyperreactivity, while a plasmid containing OVA alone had only a prophylactic effect.[91]

NONSPECIFIC IMMUNOTHERAPY

While specific immunotherapy aims to correct the underlying immunologic mechanisms leading to allergy, nonspecific therapy targeting Th2 cytokines or anti-IgE may help abrogate symptoms and could be useful in minimizing the side effects associated with conventional immunotherapy.

Anti-IgE Therapy

Blocking humanized anti-IgE antibody (Omalizumab/Xolair) has been successfully used in patients with moderate to severe allergic asthma (TABLE 2). TNX-901, another anti-IgE antibody, was tested in 84 peanut-allergic patients in clinical trials.[92] TNX-901 reduced serum IgE levels and successfully increased the sensitivity threshold to peanuts from an average of half a peanut to almost nine peanuts. While patients would still have to eliminate peanuts from the diet, TNX-901 therapy would ensure protection from accidental ingestion or exposure, which is the cause of most fatalities.[7,8,93] In addition, anti-IgE administered during specific immunotherapy for food may reduce the risk of anaphylaxis.[94,95] On other fronts, an inhibitor of the IgE receptor on mast cells, R112, is in Phase I/II trials for allergic rhinitis,[96] and preclinical experiments are being conducted on a human IgG-IgE Fc fusion protein that inhibits mast cells, basophils, and B cells[97,98] (TABLE 2). These drugs may also have application in food allergy.

Cytokines

Cytokines and anticytokine monoclonal antibodies have been tested in clinical trials in a number of diseases (Cutler & Brombacher, this issue), but to date they have not been tested against food allergies. However, cytokines have been targeted in other allergic diseases such as asthma and atopic dermatitis (TABLE 2), which have similar underlying mechanisms. IL-4, which signals via the IL-4R alpha, is known to be important for the generation of Th2 cytokines such as IL-5, IL-9, and IL-13[99] and IgE production,[100] although recent studies have found that a Th2 response occurred even in the absence of the IL-4R alpha in *Leishmania major* infection.[101] This redundancy perhaps explains the disappointing results in clinical trials using anti-IL-4 monoclonal antibodies to treat asthma (TABLE 2). However, trials of a recombinant human soluble IL-4R antagonist did show some clinical benefits in asthma.[102] It is now known that IL-13, which also binds to the IL-4Rα, is the central mediator of allergic asthma[103-105] and increases susceptibility to fatal anaphylaxis.[106] Furthermore, fatal anaphylaxis caused by food allergy in humans is often associated with a history of asthma,[7] and murine studies showed that helmith-induced protection against peanut allergy via IL-10 correlated with a fivefold decrease in IL-13.[107] Phase I trials using anti-IL-13 against asthma and development of anti-IL-13 receptor are currently underway.[108] Antibodies against other cytokines include anti-IL-5 for eosinophilia (phase II for asthma and atopic dermatitis),[109,110] anti-IL-9 (phase 1 for asthma),[111] and anti-eotaxin 1 (phase I/IIa for rhinitis)[112] (TABLE 2). A few cytokines with anti-allergic properties have also been tested in clinical trials. Murine models show that IL-10 is crucial for tolerance and successful immunotherapy[107,113–115] and that IFN-γ plays a role in oral tolerance.[116] Recombinant IL-10 has been produced but not tested against allergies, although it is in phase III trials for colitis,[117,118] (TABLE 2). Recombinant IFN-γ has shown promising results in phase II trials for dermatitis,[119, 120] with patients interestingly showing improvement in allergic rhinitis and conjunctivitis as well as skin symptoms.[119] Phase I trials of recombinant IL-12, a Th1-promoting cytokine, were halted due to inefficacy and adverse side effects.[121] In a murine model, oral administration of liposome-encapsulated rIL-12 had both preventative and therapeutic effects on peanut allergy.[122] It is likely that some kind of cytokine combination therapy would be required.

TABLE 2. Nonspecific immunotherapy in allergy

Target	Active principal	Generic name/ code	Company	Application	Status	Ref.
IgE	Humanized anti-IgE	Xolair (omalizumab)	Genentech, Novartis, Tanox	Peanut allergy Asthma	Phase II Approved	154 153
IgE	Humanized anti-IgE	TNX-901	Tanox, Genentech, Novartis	Peanut allergy	Phase II	92
FcεRI	FcεRI inhibitor	R112	Rigel	Allergic rhinits	Phase II	96
FcεRI and FcγRII	Human IgG-IgE Fc fusion protein	GE2	Biogen-Idec Inc.	Allergy	Preclinical	97, 98
IL-4	Humanized anti-IL-4 mAb	SB240683	PDL/GSK	Asthma	Phase II halted	156
IL-5	Humanized anti-IL-5	SCH-55700	Celltech/ Schering-Plough	Asthma / Allergy	Phase II	157
IL-5	Humanized anti-IL-5	Mepolizumab	GSK	Asthma Atopic dermatitis	Phase II Phase II	109, 110, 158
IL-9	Anti-IL-9 mAb	IL-9 mAb	Genaera / MedImmune	Asthma	Phase I	111
IL-4R	IL-4R antagonist	Bay 16-9996	Bayer	Asthma	Phase II halted	159
IL-4R	IL-4R antagonist	Altrakincept/ Nuvance	Immunex Corp.	Asthma	Phase II	102
IL-4 / IL-13	Recombinant IL-4Rα/ IL-13Rα/Fc fusion	IL-4/IL-13 trap	Regeneron	Asthma	Phase I	160, 161
IL-13	Anti-IL-13 human IgG4	CAT-354	Cambridge Antibody Technology	Asthma	Phase I	108
IL-13Rα	Anti-IL-13 receptor alpha 1	IL-13 receptor α 1	Amrad / Merck	Asthma	Preclinical	162
Eotaxin	Anti-eotaxin1	CAT213	Cambridge Antibody Technology	Rhinitis	Phase I/IIa	112
—	Recombinant IL-10	Ilodecakin / Tenovil	Schering-Plough	No trials in allergy Colitis	 Phase III	 117, 163
—	Recombinant IL-12	rIL-12	Genetic Institute	Asthma	Phase I halted	121
—	Recombinant IFN-γ	Actimmune	Intermune Pharma	Atopic dermatitis	Phase II	164

TABLE 3. Natural therapies in food allergy

Allergy	Model / application	Treatment	Relevance	Molecular effects	Clinical effects	Ref.
Peanut	Oral peanut sensitization using cholera toxin	Chinese herbal formula FAHF-1	Mouse	Reduced specific IgE, mast cell degranulation, histamine, Th2 cytokines	Abolished anaphylactic reactions	127
Peanut	Oral peanut sensitization using cholera toxin	Chinese herbal formula FAHF-2	Mouse	Reduced specific IgE, Th2 cytokines, enhanced IFN-γ	Abolished anaphylactic reactions	126
Cow's milk	Infants with cow's milk allergy and IgE-associated dermatitis (plus conventional treatment)	*Lactobacillus rhamnosus* GG or probiotic mix	Human	Increase in IFN-γ secretion	Alleviated skin symptoms	142
Various	Children with moderate to severe atopic dermatitis	*Lactobacillus rhamnosus* or *Lactobacillus reuteri*	Human	Decreased eosinophil cationic protein, decreased lactulose to mannitol ratio (stabilization of gut perme-ability); no change in cytokine production	Improvement in eczema, symptom scores, decreased gastrointestinal symptoms	143, 144
Various	Children with atopic dermatitis	*Lactobacillus rhamnosus* GG	Human	Elevated serum IL-10	—	138
—	Pregnant mothers, then infants for 6 months	*Lactobacillus* GG	Human	—	Frequency of atopy halved in infants at age 2 and 4	145, 165
—	Newborns	Probiotic *E.coli* strain	Human	—	Significant reduction in prevalence of allergies at ages 10 and 20	147
Cow's milk	Rat pups fed cow's milk	*Lactobacillus* GG	Rat	—	Restored aberrant gut permeability	139
Peanut	Oral peanut sensitization	Antibiotic treatment to remove intestinal flora	Mouse	Increased specific IgE, increased histamine and mast cell degranulation after challenge	Removing flora rendered mice susceptible to peanut-induced anaphylaxis. Reconstitution reversed the effect	140

NATURAL THERAPIES

A few authors have reviewed nonallopathic therapies for allergy, including herbal remedies, homeopathy, and acupuncture.[123–125] In prevention or therapy of food allergies, the most promising results have been obtained using Chinese herbal formulas or probiotics (TABLE 3). In murine models of peanut allergy, the formulas FAHF-1 and FAHF-2 completely blocked anaphylaxis and downregulated Th2 and IgE responses without any apparent toxicity.[126,127] The effects of probiotics, live microorganisms that are thought to exert beneficial effects on the host by improving gut microflora,[128,129] are more controversial.[130–135] It is thought that probiotic administration promotes a change in the local microflora and mucosal immune response and corrects aberrations in gut permeability.[134,136–139] Murine models found that toll-like receptor 4-dependent signals provided by macrobiotics inhibited the development of allergic responses to food,[140] whereas disruption in this microbiota enhanced the allergic response to OVA and to mold spores.[141] Clinical trials have demonstrated that probiotic therapy may help to alleviate atopic dermatitis and gastrointestinal symptoms caused by food allergy.[136,142–144] In addition, when probiotics were given to expectant mothers with a familial history of atopy as well as postnatally for 6 months to their infants, the frequency of atopic eczema in the treated infants was half that of the placebo group.[145,146] Another study found that administration of a probiotic *Escherichia coli* strain after birth resulted in a reduced incidence of allergies recorded at the ages of 10 and 20.[147] It is therefore possible that probiotics may have a role in the prevention of food allergy. However, such effects appear to be species-specific, and further research is needed.[132,133]

CONCLUSIONS

Several novel therapeutic and prophylactic therapies against food allergy are under investigation. To date, anti-IgE treatment and specific immunotherapy using hypoallergenic recombinant allergens together with Th1-inducing adjuvants such as CpG DNA or heat-killed bacteria appear the most promising. New initiatives to increase our understanding of the pathophysiology and immunologic mechanisms of food allergy along with the molecular characterization of food allergens should pave the way towards safer and more effective ways of combatting this debilitating and potentially life-threatening disease.

ACKNOWLEDGMENTS

This work was supported in part by a GlaxoSmithKline Research Award and the Medical Research Council (MRC) of South Africa.

REFERENCES

1. HELM, R.M. & A.W. BURKS. 2000. Mechanisms of food allergy. Curr. Opin. Immunol. **12:** 647–653.
2. SAMPSON, H.A. 2003. Food allergy. J. Allergy Clin. Immunol. **111:** S540–547.

3. SAMPSON, H.A. 2004. Update on food allergy. J. Allergy Clin. Immunol. **113:** 805–819.
4. SAMPSON, H.A. 2004. Food-induced anaphylaxis. Novartis Found Symp. **257:** 161–171.
5. SICHERER, S.H., A. MUNOZ-FURLONG & H.A. SAMPSON. 2004. Prevalence of seafood allergy in the United States determined by a random telephone survey. J. Allergy Clin. Immunol. **114:** 159–165.
6. SKOLNICK, H.S. *et al.* 2001. The natural history of peanut allergy. J. Allergy Clin. Immunol. **107:** 367–374.
7. BOCK, S.A., A. MUNOZ-FURLONG & H.A. SAMPSON. 2001. Fatalities due to anaphylactic reactions to foods. J. Allergy Clin. Immunol. **107:** 191–193.
8. PERRY, T.T. *et al.* 2004. Distribution of peanut allergen in the environment. J Allergy Clin. Immunol. **113:** 973–976.
9. MONERET-VAUTRIN, D.A. & G. KANNY. 2004. Update on threshold doses of food allergens: implications for patients and the food industry. Curr. Opin. Allergy Clin. Immunol. **4:** 215–219.
10. MORISSET, M. *et al.* 2003. Thresholds of clinical reactivity to milk, egg, peanut and sesame in immunoglobulin E-dependent allergies: evaluation by double-blind or single-blind placebo-controlled oral challenges. Clin. Exp. Allergy **33:** 1046–1051.
11. TURCANU, V., S.J. MALEKI & G. LACK. 2003. Characterization of lymphocyte responses to peanuts in normal children, peanut-allergic children, and allergic children who acquired tolerance to peanuts. J. Clin. Invest. **111:** 1065–1072.
12. SCHADE, R.P. *et al.* 2003. The cow's milk protein-specific T cell response in infancy and childhood. Clin. Exp. Allergy **33:** 725–730.
13. TIEMESSEN, M.M. *et al.* 2004. Cow's milk-specific T-cell reactivity of children with and without persistent cow's milk allergy: key role for IL-10. J. Allergy Clin. Immunol. **113:** 932–939.
14. ANDRE, F., J. PENE & C. ANDRE. 1996. Interleukin-4 and interferon-gamma production by peripheral blood mononuclear cells from food-allergic patients. Allergy **51:** 350–355.
15. NOMA, T. *et al.* 1996. Cytokine production in children outgrowing hen egg allergy. Clin. Exp. Allergy **26:** 1298–307.
16. WILSON, D.R., M.T. LIMA & S.R. DURHAM. 2005. Sublingual immunotherapy for allergic rhinitis: systematic review and meta-analysis. Allergy **60:** 4–12.
17. SCHMIDT-WEBER, C.B. & K. BLASER. 2004. Immunological mechanisms in specific immunotherapy. Springer Semin. Immunopathol. **25:** 377–390.
18. VARNEY, V.A. *et al.* 2003. Usefulness of specific immunotherapy in patients with severe perennial allergic rhinitis induced by house dust mite: a double-blind, randomized, placebo-controlled trial. Clin. Exp. Allergy **33:** 1076–1082.
19. DURHAM, S.R. *et al.* 1996. Grass pollen immunotherapy inhibits allergen-induced infiltration of $CD4^+$ T lymphocytes and eosinophils in the nasal mucosa and increases the number of cells expressing messenger RNA for interferon-gamma. J. Allergy Clin. Immunol. **97:** 1356–1365.
20. WACHHOLZ, P.A. *et al.* 2002. Grass pollen immunotherapy for hayfever is associated with increases in local nasal but not peripheral Th1:Th2 cytokine ratios. Immunology **105:** 56–62.
21. ROLLAND, J. & R. O'HEHIR. 1998. Immunotherapy of allergy: anergy, deletion, and immune deviation. Curr. Opin. Immunol. **10:** 640–645.
22. NASSER, S.M. *et al.* 2001. Interleukin-10 levels increase in cutaneous biopsies of patients undergoing wasp venom immunotherapy. Eur. J. Immunol. **31:** 3704–3713.
23. BUFE, A. *et al.* 2004. Efficacy of sublingual swallow immunotherapy in children with severe grass pollen allergic symptoms: a double-blind placebo-controlled study. Allergy **59:** 498–504.
24. PASSALACQUA, G. *et al.* 2004. Efficacy and safety of sublingual immunotherapy. Ann. Allergy Asthma Immunol. **93:** 3–12.
25. MA, S.H., S.H. SICHERER & A. NOWAK-WEGRZYN. 2003. A survey on the management of pollen-food allergy syndrome in allergy practices. J. Allergy Clin. Immunol. **112:** 784–788.
26. HANSEN, K.S. *et al.* 2004. Food allergy to apple and specific immunotherapy with birch pollen. Molec. Nutr. Food Res. **48:** 441–448.

27. Bolhaar, S.T.H.P. *et al.* 2004. Efficacy of birch-pollen immunotherapy on cross-reactive food allergy confirmed by skin tests and double-blind food challenges. Clin. Exp. Allergy **34:** 761–769.
28. Kelso, J.M. 2000. Pollen-food allergy syndrome. Clin. Exp. Allergy **30:** 905–907.
29. Bucher, X. *et al.* 2004. Effect of tree pollen specific, subcutaneous immunotherapy on the oral allergy syndrome to apple and hazelnut. Allergy **59:** 1272–1276.
30. Mittag, D. *et al.* 2004. Ara h 8, a Bet v 1-homologous allergen from peanut, is a major allergen in patients with combined birch pollen and peanut allergy. J. Allergy Clin. Immunol. **114:** 1410–1417.
31. Pons, L. *et al.* 2004. Soy immunotherapy for peanut-allergic mice: modulation of the peanut-allergic response. J. Allergy Clin. Immunol. **114:** 915–921.
32. Oppenheimer, J.J. *et al.* 1992. Treatment of peanut allergy with rush immunotherapy. J. Allergy Clin. Immunol. **90:** 256–262.
33. Nelson, H.S. *et al.* 1997. Treatment of anaphylactic sensitivity to peanuts by immunotherapy with injections of aqueous peanut extract. J. Allergy Clin. Immunol. **99:** 744–751.
34. Burks, W., S.B. Lehrer & G.A. Bannon. 2004. New approaches for treatment of peanut allergy: chances for a cure. Clin. Rev. Allergy Immunol. **27:** 191–196.
35. van Ree, R. *et al.* 1996. Possible induction of food allergy during mite immunotherapy. Allergy **51:** 108–113.
36. Pajno, G.B. *et al.* 2002. Harmful effect of immunotherapy in children with combined snail and mite allergy. J. Allergy Clin. Immunol. **109:** 627–629.
37. Valenta, R. 2002. Recombinant allergen-based concepts for diagnosis and therapy of Type I allergy. Allergy **57:** 66–67.
38. Niederberger, V. *et al.* 2004. Vaccination with genetically engineered allergens prevents progression of allergic disease. Proc. Natl. Acad. Sci. USA **101:** 14677–14682.
39. Bohle, B. & S. Vieths. 2004. Improving diagnostic tests for food allergy with recombinant allergens. Methods **32:** 292–299.
40. Vrtala, S. *et al.* 2000. T cell epitope-containing hypoallergenic recombinant fragments of the major birch pollen allergen, Bet v 1, induce blocking antibodies. J. Immunol. **165:** 6653–6659.
41. Renz, H. 1999. How can animal models lead to improved specific immunotherapy (SIT)? Allergy **54:** 39–40.
42. Herz, U. *et al.* 1996. The relevance of murine animal models to study the development of allergic bronchial asthma. Immunol. Cell Biol. **74:** 209–217.
43. Herz, U., H. Renz & U. Wiedermann. 2004. Animal models of type I allergy using recombinant allergens. Methods **32:** 271–280.
44. Helm, R.M. & A.W. Burks. 2002. Animal models of food allergy. Curr. Opin. Allergy Clin. Immunol. **2:** 541–546.
45. Haselden, B.M., A.B. Kay & M. Larche. 2000. Peptide-mediated immune responses in specific immunotherapy. Int. Arch. Allergy Immunol. **122:** 229–237.
46. Norman, P. S. 1996. Clinical experience with treatment of allergies with T cell epitope containing peptides. Adv. Exp. Med. Biol. **409:** 457–461.
47. Wilcock, L.K., J.N. Francis & S.R. Durham. 2004. Aluminium hydroxide down-regulates T helper 2 responses by allergen-stimulated human peripheral blood mononuclear cells. Clin. Exp. Allergy **34:** 1373–1378.
48. Bauer, M. *et al.* 2001. Bacterial CpG-DNA triggers activation and maturation of human CD11c^{-}, CD123^{+} dendritic cells. J. Immunol. **166:** 5000–5007.
49. Bauer, S. *et al.* 2001. Human TLR9 confers responsiveness to bacterial DNA via species-specific CpG motif recognition. Proc. Natl. Acad. Sci. USA **98:** 9237–9242.
50. Hartman, G., G.J. Weiner & A.M. Krieg. 1999. CpG DNA: a potent signal for growth, activation, and maturation of human dendritic cells. Proc. Natl. Acad. Sci. USA **96:** 9305–9310.
51. Stacey, K.J. *et al.* 2000. Macrophage activation by immunostimulatory DNA. Curr. Top. Microbiol. Immunol. **247:** 41–58.
52. Chu, R.S. *et al.* 1997. CpG oligodeoxynucleotides act as adjuvants that switch on T helper 1 (Th1) immunity. J. Exp. Med. **186:** 1623–1631.
53. Liang, H. *et al.* 1996. Activation of human B cells by phosphorothioate oligodeoxynucleotides. J. Clin. Invest. **98:** 1119–1129.

54. KLINE, J.N. *et al.* 1998. Modulation of airway inflammation by CpG oligodeoxynucleotides in a murine model of asthma. J. Immunol. **160:** 2555–2559.
55. KLINE, J.N. *et al.* 1999. CpG oligodeoxynucleotides do not require TH1 cytokines to prevent eosinophilic airway inflammation in a murine model of asthma. J. Allergy Clin. Immunol. **104:** 1258–1264.
56. SEREBRISKY, D. *et al.* 2000. CpG oligodeoxynucleotides can reverse Th2-associated allergic airway responses and alter the B7.1/B7.2 expression in a murine model of asthma. J. Immunol. **165:** 5906–5912.
57. MAGONE, M.T. *et al.* 2000. Systemic or mucosal administration of immunostimulatory DNA inhibits early and late phases of murine allergic conjunctivitis. Eur. J. Immunol. **30:** 1841–1850.
58. HORNER, A.A. *et al.* 2002. Immunostimulatory DNA-based therapeutics for experimental and clinical allergy. Allergy **57** (Suppl 72)**:** 24–29.
59. TIGHE, H. *et al.* 2000. Conjugation of immunostimulatory DNA to the short ragweed allergen Amb a 1 enhances its immunogenicity and reduces its allergenicity. J. Allergy Clin. Immunol. **106:** 124–134.
60. SIMONS, F.E. *et al.* 2004. Selective immune redirection in humans with ragweed allergy by injecting Amb a 1 linked to immunostimulatory DNA. J. Allergy Clin. Immunol. **113:** 1144–1151.
61. MARSHALL, J.D. *et al.* 2001. Immunostimulatory sequence DNA linked to the Amb a 1 allergen promotes T(H)1 cytokine expression while downregulating T(H)2 cytokine expression in PBMCs from human patients with ragweed allergy. J. Allergy Clin. Immunol. **108:** 191–197.
62. SANTELIZ, J.V. *et al.* 2002. Amb a 1-linked CpG oligodeoxynucleotides reverse established airway hyperresponsiveness in a murine model of asthma. J. Allergy Clin. Immunol. **109:** 455–462.
63. TULIC, M.K. *et al.* 2004. Amb a 1-immunostimulatory oligodeoxynucleotide conjugate immunotherapy decreases the nasal inflammatory response. J. Allergy Clin. Immunol. **113:** 235–241.
64. LI, X.M. *et al.* 2003. Persistent protective effect of heat-killed *Escherichia coli* producing "engineered," recombinant peanut proteins in a murine model of peanut allergy. J. Allergy Clin. Immunol. **112:** 159–167.
65. LI, X.M. *et al.* 2003. Engineered recombinant peanut protein and heat-killed *Listeria monocytogenes* coadministration protects against peanut-induced anaphylaxis in a murine model. J. Immunol. **170:** 3289–3295.
66. FRICK, O.L. *et al.* 2005. Allergen immunotherapy with heat-killed *Listeria monocytogenes* alleviates peanut and food-induced anaphylaxis in dogs. Allergy **60:** 243–250.
67. BALDRICK, P., D. RICHARDSON & A.W. WHEELER. 2001. Safety evaluation of a glutaraldehyde modified tyrosine adsorbed housedust mite extract containing monophosphoryl lipid A (MPL) adjuvant: a new allergy vaccine for dust mite allergy. Vaccine **20:** 737–743.
68. WHEELER, A.W., J.S. MARSHALL & J.T. ULRICH. 2001. A Th1-inducing adjuvant, MPL, enhances antibody profiles in experimental animals suggesting it has the potential to improve the efficacy of allergy vaccines. Int. Arch. Allergy Immunol. **126:** 135–139.
69. DRACHENBERG, K.J. *et al.* 2001. A well-tolerated grass pollen-specific allergy vaccine containing a novel adjuvant, monophosphoryl lipid A, reduces allergic symptoms after only four preseasonal injections. Allergy **56:** 498–505.
70. MOTHES, N. *et al.* 2003. Allergen-specific immunotherapy with a monophosphoryl lipid A-adjuvanted vaccine: reduced seasonally boosted immunoglobulin E production and inhibition of basophil histamine release by therapy-induced blocking antibodies. Clin. Exp. Allergy **33:** 1198–2008.
71. STEVENSON, D.D. 2000. Approach to the patient with a history of adverse reactions to aspirin or NSAIDs: diagnosis and treatment. Allergy Asthma Proc. **21:** 25–31.
72. STEVENSON, D.D. 2003. Aspirin desensitization in patients with AERD. Clin. Rev. Allergy Immunol. **24:** 159–168.
73. LITWIN, A. *et al.* 1996. Immunologic effects of encapsulated short ragweed extract: a potent new agent for oral immunotherapy. Ann. Allergy Asthma Immunol. **77:** 132–138.

74. TEPAS, E.C. *et al.* 2004. Clinical efficacy of microencapsulated timothy grass pollen extract in grass-allergic individuals. Ann. Allergy Asthma Immunol. **92:** 25–31.
75. TANG, D.C., M. DEVIT & S.A. JOHNSTON. 1992. Genetic immunization is a simple method for eliciting an immune response. Nature **356:** 152–154.
76. ULMER, J.B. *et al.* 1993. Heterologous protection against influenza by injection of DNA encoding a viral protein. Science **259:** 1745–1749.
77. HSU, C.H. *et al.* 1996. Immunoprophylaxis of allergen-induced immunoglobulin E synthesis and airway hyperresponsiveness *in vivo* by genetic immunization. Nat. Med. **2:** 540–544.
78. RAZ, E. *et al.* 1996. Preferential induction of a Th1 immune response and inhibition of specific IgE antibody formation by plasmid DNA immunization. Proc. Natl. Acad. Sci. USA **93:** 5141–5145.
79. YOSHIDA, A. *et al.* 2000. Advantage of gene gun-mediated over intramuscular inoculation of plasmid DNA vaccine in reproducible induction of specific immune responses. Vaccine **18:** 1725–1729.
80. PENG, H.J. *et al.* 2002. Induction of specific Th1 responses and suppression of IgE antibody formation by vaccination with plasmid DNA encoding Der f 11. Vaccine **20:** 1761–1768.
81. LEE, D.J. *et al.* 1997. Inhibition of IgE antibody formation by plasmid DNA immunization is mediated by both CD4+ and CD8+ T cells. Int. Arch. Allergy Immunol. **113:** 227–230.
82. SATO, Y. *et al.* 1996. Immunostimulatory DNA sequences necessary for effective intradermal gene immunization. Science **273:** 352–354.
83. JILEK, S. *et al.* 2001. Antigen-independent suppression of the allergic immune response to bee venom phospholipase A(2) by DNA vaccination in CBA/J mice. J. Immunol. **166:** 3612–3621.
84. HOCHREITER, R. *et al.* 2003. TH1-promoting DNA immunization against allergens modulates the ratio of IgG1/IgG2a but does not affect the anaphylactic potential of IgG1 antibodies: no evidence for the synthesis of nonanaphylactic IgG1. J. Allergy Clin. Immunol. **112:** 579–584.
85. ADEL-PATIENT, K. *et al.* 2001. Genetic immunisation with bovine beta-lactoglobulin cDNA induces a preventive and persistent inhibition of specific anti-BLG IgE response in mice. Int. Arch. Allergy Immunol. **126:** 59–67.
86. HARTL, A. *et al.* 2004. DNA vaccines for allergy treatment. Methods **32:** 328–339.
87. ROY, K. *et al.* 1999. Oral gene delivery with chitosan--DNA nanoparticles generates immunologic protection in a murine model of peanut allergy. Nat. Med. **5:** 387–391.
88. LUDWIG-PORTUGALL, I. *et al.* 2004. Prevention of long-term IgE antibody production by gene gun-mediated DNA vaccination. J. Allergy Clin. Immunol. **114:** 951–957.
89. TODA, M. *et al.* 2000. Inhibition of immunoglobulin E response to Japanese cedar pollen allergen (Cry j 1) in mice by DNA immunization: different outcomes dependent on the plasmid DNA inoculation method. Immunology **99:** 179–186.
90. KWON, S.S., N. KIM & T.J. YOO. 2001. The effect of vaccination with DNA encoding murine T-cell epitopes on the Der p 1 and 2 induced immunoglobulin E synthesis. Allergy **56:** 741–748.
91. MAECKER, H.T. *et al.* 2001. Vaccination with allergen-IL-18 fusion DNA protects against, and reverses established, airway hyperreactivity in a murine asthma model. J. Immunol. **166:** 959–965.
92. LEUNG, D.Y. *et al.* 2003. Effect of anti-IgE therapy in patients with peanut allergy. N. Engl. J. Med. **348:** 986–993.
93. WENSING, M. *et al.* 2002. The range of minimum provoking doses in hazelnut-allergic patients as determined by double-blind, placebo-controlled food challenges. Clin. Exp. Allergy **32:** 1757–1762.
94. HAMELMANN, E., C. ROLINCK-WERNINGHAUS & U. WAHN. 2003. Is there a role for anti-IgE in combination with specific allergen immunotherapy? Curr. Opin. Allergy Clin. Immunol. **3:** 501–510.
95. Kuehr, J. *et al.* 2002. Efficacy of combination treatment with anti-IgE plus specific immunotherapy in polysensitized children and adolescents with seasonal allergic rhinitis. J. Allergy Clin. Immunol. **109:** 274–280.

96. www.rigel.com/rigel/pr_1091415726.
97. ZHU, D. *et al.* 2002. A novel human immunoglobulin Fc gamma Fc epsilon bifunctional fusion protein inhibits Fc epsilon RI-mediated degranulation. Nat. Med. **8:** 518–521.
98. ZHANG, K. *et al.* 2004. Inhibition of allergen-specific IgE reactivity by a human Ig Fcgamma-Fcepsilon bifunctional fusion protein. J. Allergy Clin. Immunol. **114:** 321–327.
99. KOPF, M. *et al.* 1993. Disruption of the murine IL-4 gene blocks Th2 cytokine responses. Nature **362:** 245–248.
100. MAGNAN, A. *et al.* 2000. Relationships between natural T cells, atopy, IgE levels, and IL-4 production. Allergy **55:** 286–290.
101. MOHRS, M. *et al.* 1999. Differences between IL-4- and IL-4 receptor alpha-deficient mice in chronic leishmaniasis reveal a protective role for IL-13 receptor signaling. J. Immunol. **162:** 7302–7308.
102. BORISH, L.C. *et al.* 2001. Efficacy of soluble IL-4 receptor for the treatment of adults with asthma. J. Allergy Clin. Immunol. **107:** 963–970.
103. GRUNIG, G. *et al.* 1998. Requirement for IL-13 independently of IL-4 in experimental asthma. Science **282:** 2261–2263.
104. WALTER, D.M. *et al.* 2001. Critical role for IL-13 in the development of allergen-induced airway hyperreactivity. J. Immunol. **167:** 4668–4675.
105. WILLS-KARP, M. *et al.* 1998. Interleukin-13: central mediator of allergic asthma. Science **282:** 2258–2261.
106. FALLON, P.G. *et al.* 2001. IL-13 overexpression predisposes to anaphylaxis following antigen sensitization. J. Immunol. **166:** 2712–2716.
107. BASHIR, M.E. *et al.* 2002. An enteric helminth infection protects against an allergic response to dietary antigen. J. Immunol. **169:** 3284–3292.
108. www.bioportfolio.com/news/cat_8.htm.
109. LECKIE, M.J. *et al.* 2000. Effects of an interleukin-5 blocking monoclonal antibody on eosinophils, airway hyper-responsiveness, and the late asthmatic response. Lancet **356:** 2144–2148.
110. PHIPPS, S. *et al.* 2004. Intravenous anti-IL-5 monoclonal antibody reduces eosinophils and tenascin deposition in allergen-challenged human atopic skin. J. Invest. Dermatol. **122:** 1406–1412.
111. www.mediummune.com/pipeline/il9.asp.
112. http://www.cambridgeantibody.com/html/technology_products/product_pipeline/cat_213_treatment_for_allergies_including_asthma.
113. VISSERS, J.L. *et al.* 2004. Allergen immunotherapy induces a suppressive memory response mediated by IL-10 in a mouse asthma model. J. Allergy Clin. Immunol. **113:** 1204–1210.
114. FROSSARD, C.P. *et al.* 2004. Lymphocytes in Peyer patches regulate clinical tolerance in a murine model of food allergy. J. Allergy Clin. Immunol. **113:** 958–964.
115. AKBARI, O., R.H. DEKRUYFF & D.T. UMETSU. 2001. Pulmonary dendritic cells producing IL-10 mediate tolerance induced by respiratory exposure to antigen. Nat. Immunol. **2:** 725–731.
116. LEE, H.O. *et al.* 2000. Interferon gamma induction during oral tolerance reduces T-cell migration to sites of inflammation. Gastroenterology **119:** 129–138.
117. DUMONT, F.J. 2003. Therapeutic potential of IL-10 and its viral homologues: an update. Exp. Opin. Therapeutic Patents **13**: 1551–1577.
118. FRANCIS, J.N., S.J. TILL & S.R. DURHAM. 2003. Induction of IL-10+CD4$^+$CD25$^+$ T cells by grass pollen immunotherapy. J. Allergy Clin. Immunol. **111:** 1255–1261.
119. STEVENS, S.R. *et al.* 1998. Long-term effectiveness and safety of recombinant human interferon gamma therapy for atopic dermatitis despite unchanged serum IgE levels. Arch. Dermatol. **134:** 799–804.
120. NOH, G.W. & K.Y. LEE. 1998. Blood eosinophils and serum IgE as predictors for prognosis of interferon-gamma therapy in atopic dermatitis. Allergy **53:** 1202–1207.
121. BRYAN, S.A, *et al.* 2000. Effects of recombinant human interleukin-12 on eosinophils, airway hyper-responsiveness, and the late asthmatic response. Lancet **356:** 2149–2153.

122. LEE, S.Y. *et al.* 2001. Oral administration of IL-12 suppresses anaphylactic reactions in a murine model of peanut hypersensitivity. Clin. Immunol. **101:** 220–228.
123. ZIMENT, I. & D.P. TASHKIN. 2000. Alternative medicine for allergy and asthma. J. Allergy Clin. Immunol. **106:** 603–614.
124. NIGGEMANN, B. & C. GRUBER. 2003. Side-effects of complementary and alternative medicine. Allergy **58:** 707–716.
125. BIELORY, L. & K. LUPOLI. 1999. Herbal interventions in asthma and allergy. J. Asthma **36:** 1–65.
126. SRIVASTAVA, K.D. *et al.* 2005. The Chinese herbal medicine formula FAHF-2 completely blocks anaphylactic reactions in a murine model of peanut allergy. J. Allergy Clin. Immunol. **115:** 171–178.
127. LI, X.M. *et al.* 2001. Food allergy herbal formula-1 (FAHF-1) blocks peanut-induced anaphylaxis in a murine model. J. Allergy Clin. Immunol. **108:** 639–646.
128. FULLER, R. 1991. Probiotics in human medicine. Gut **32:** 439–442.
129. DUGAS, B. *et al.* 1999. Immunity and probiotics. Immunol. Today **20:** 387–390.
130. KIRJAVAINEN, P.V. & G.R. GIBSON. 1999. Healthy gut microflora and allergy: factors influencing development of the microbiota. Ann. Med. **31:** 288–292.
131. MATRICARDI, P.M. 2002. Probiotics against allergy: data, doubts, and perspectives. Allergy **57:** 185–187.
132. VANDERHOOF, J.A. & R.J. YOUNG. 2004. Current and potential uses of probiotics. Ann. Allergy Asthma Immunol. **93:** S33–S37.
133. PAGANELLI, R. *et al.* 2002. Probiotics and food-allergic diseases. Allergy **57** (Suppl 72)**:** 97–99.
134. O'Sullivan, G.C. *et al.* 2005. Probiotics: an emerging therapy. Curr. Pharm. Des. **11:** 3–10.
135. BROWN, A.C. & A. VALIERE. 2004. Probiotics and medical nutrition therapy. Nutr. Clin. Care **7:** 56–68.
136. MIRAGLIA DEL GIUDICE, M., JR., M.G. DE LUCA & C. CAPRISTO. 2002. Probiotics and atopic dermatitis. A new strategy in atopic dermatitis. Dig. Liver Dis. **34** (Suppl 2)**:** S68–S71.
137. SIMON, G.L. & S.L. GORBACH. 1984. Intestinal flora in health and disease. Gastroenterology **86:** 174–193.
138. PESSI, T. *et al.* 2000. Interleukin-10 generation in atopic children following oral *Lactobacillus rhamnosus* GG. Clin. Exp. Allergy **30:** 1804–1808.
139. PESSI, T. *et al.* 1998. Probiotics reinforce mucosal degradation of antigens in rats: implications for therapeutic use of probiotics. J. Nutr. **128:** 2313–2318.
140. BASHIR, M.E. *et al.* 2004. Toll-like receptor 4 signaling by intestinal microbes influences susceptibility to food allergy. J. Immunol. **172:** 6978–6987.
141. NOVERR, M.C. *et al.* 2005. Development of allergic airway disease in mice following antibiotic therapy and fungal microbiota increase: role of host genetics, antigen, and interleukin-13. Infect. Immun. **73:** 30–38.
142. POHJAVUORI, E. *et al.* 2004. Lactobacillus GG effect in increasing IFN-gamma production in infants with cow's milk allergy. J. Allergy Clin. Immunol. **114:** 131–136.
143. ROSENFELDT, V. *et al.* 2004. Effect of probiotics on gastrointestinal symptoms and small intestinal permeability in children with atopic dermatitis. J. Pediatr. **145:** 612–616.
144. ROSENFELDT, V. *et al.* 2003. Effect of probiotic Lactobacillus strains in children with atopic dermatitis. J. Allergy Clin. Immunol. **111:** 389–395.
145. KALLIOMAKI, M. *et al.* 2001. Probiotics in primary prevention of atopic disease: a randomised placebo-controlled trial. Lancet **357:** 1076–1079.
146. KALLIOMAKI, M. *et al.* 2003. Probiotics and prevention of atopic disease: 4-year follow-up of a randomised placebo-controlled trial. Lancet **361:** 1869–1871.
147. LODINOVA-ZADNIKOVA, R., B. CUKROWSKA & H. TLASKALOVA-HOGENOVA. 2003. Oral administration of probiotic *Escherichia coli* after birth reduces frequency of allergies and repeated infections later in life (after 10 and 20 years). Int. Arch. Allergy Immunol. **131:** 209–211.
148. BOLHAAR, S.T. *et al.* 2004. Efficacy of birch-pollen immunotherapy on cross-reactive food allergy confirmed by skin tests and double-blind food challenges. Clin. Exp. Allergy **34:** 761–769.

149. HANSEN, K.S. *et al.* 2004. Food allergy to apple and specific immunotherapy with birch pollen. Mol. Nutr. Food Res. **48:** 441–448.
150. LI, X.M. *et al.* 2003. Engineered recombinant peanut protein and heat-killed *Listeria monocytogenes* coadministration protects against peanut-induced anaphylaxis in a murine model. J. Immunol. **170:** 3289–3295.
151. LI, X.M. *et al.* 2003. Persistent protective effect of heat-killed *Escherichia coli* producing "engineered," recombinant peanut proteins in a murine model of peanut allergy. J. Allergy Clin. Immunol. **112:** 159–167.
152. SHIDA, K. *et al.* 2002. *Lactobacillus casei* strain Shirota suppresses serum immunoglobulin E and immunoglobulin G1 responses and systemic anaphylaxis in a food allergy model. Clin. Exp. Allergy **32:** 563–570.
153. FROSSARD, C.P., C. HAUSER & P.A. EIGENMANN. 2001. Oral carrageenan induces antigen-dependent oral tolerance: prevention of anaphylaxis and induction of lymphocyte anergy in a murine model of food allergy. Pediatr. Res. **49:** 417–422.
154. www.foodallergy.org/research/antiigetherapy.html.
155. WALKER, S. *et al.* 2004. Anti-IgE for chronic asthma in adults and children. The Cochrane Library, Issue 4.
156. www.pdl.com/applications/press_releases.cfm?newsId=68.
157. KIPS, J.C. *et al.* 2003. Effect of SCH55700, a humanized anti-human interleukin-5 antibody, in severe persistent asthma: a pilot study 164/rccm.200206-5250C. Am. J. Respir. Crit. Care Med. **167:** 1655–1659.
158. FLOOD-PAGE, P. *et al.* 2003. Anti-IL-5 treatment reduces deposition of ECM proteins in the bronchial subepithelial basement membrane of mild atopic asthmatics. J. Clin. Invest. **112:** 1029–1036.
159. FITCH, N. *et al.* 2001. Preclinical evaluation of BAY 16-9966 a dual IL-4/IL-13 receptor antagonist. J. Allergy Clin. Immunol. **107:** S61–S61.
160. ECONOMIDES, A.N. *et al.* 2003. Cytokine traps: multi-component, high-affinity blockers of cytokine action. Nat. Med. **9:** 47–52.
161. www.regn.com/products/product_candidate.asp?v_c_id=8.
162. www.amrad.com.au/Filesa1%20-%20Aug%202004.pdf.
163. ASADULLAH, K., W. STERRY & H.D. VOLK. 2003. Interleukin-10 therapy: review of a new approach. Pharmacol. Rev. **55:** 241–269.
164. CHANG, T.T. & S.R. STEVENS. 2002. Atopic dermatitis: the role of recombinant interferon-gamma therapy. Am. J. Clin. Dermatol. **3:** 175–183.
165. KALLIOMAKI, M. & E. ISOLAURI. 2003. Role of intestinal flora in the development of allergy. Curr. Opin. Allergy Clin. Immunol. **3:** 15–20.

Parasite Mitochondria as a Target of Chemotherapy

Inhibitory Effect of Licochalcone A on the *Plasmodium falciparum* Respiratory Chain

FUMIKA MI-ICHI,[a] HIROKO MIYADERA,[a] TAMAKI KOBAYASHI,[a] SHINZABURO TAKAMIYA,[b] SEIJI WAKI,[c] SUSUMU IWATA,[d] SHOJI SHIBATA,[e] AND KIYOSHI KITA[a]

[a]*Department of Biomedical Chemistry, Graduate School of Medicine, The University of Tokyo, Hongo, Bunkyo-ku, Tokyo 113-0033, Japan*

[b]*Department of Molecular and Cellular Parasitology, Juntendo University, School of Medicine, Hongo, Bunkyo-ku, Tokyo 113-8421, Japan*

[c]*Gunma Prefectural College of Health Sciences, Maebashi, Gunma 371, Japan*

[d]*Research Laboratory of Minophagen Pharmaceutical Co. Zama-shi, Kanagawa 228, Japan*

[e]*Shibata Laboratory of Natural Medicinal Materials, Minophagen Pharmaceutical Co. Yotsuya, Shinjyuku-ku, Tokyo 160, Japan*

ABSTRACT: Parasites have exploited unique energy metabolic pathways as adaptations to the natural host habitat. In fact, the respiratory systems of parasites typically show greater diversity in electron transfer pathways than do those of host animals. These unique aspects of parasite mitochondria and related enzymes may represent promising targets for chemotherapy. Natural products have been recognized as a source of the candidates of the specific inhibitors for such parasite respiratory chains. Chalcones was recently evaluated for its antimalarial activity *in vitro* and *in vivo*. However, its target is still unclear in malaria parasites. In this study, we investigated that licochalcone A inhibited the bc_1 complex (ubiquinol-cytochrome *c* reductase) as well as complex II (succinate ubiquinone reductase, SQR) of *Plasmodium falciparum* mitochondria. In particular, licochalcone A inhibits bc_1 complex activity at very low concentrations. Because the property of the *P. falciparum* bc_1 complex is different from that of the mammalian host, chalcones would be a promising candidate for a new antimalarial drug.

KEYWORDS: parasite; mitochondria; chemotherapy; plasmodium; succinate; ubiquinone; licochalcone

Address for correspondence: Kiyoshi Kita, Department of Biomedical Chemistry, Graduate School of Medicine, The University of Tokyo, Hongo, Bunkyo-ku, Tokyo 113-0033, Japan. Voice: +81-3-5841-3526; fax: +81-3-5841-3444.
kitak@m.u-tokyo.ac.jp

Ann. N.Y. Acad. Sci. 1056: 46–54 (2005). © 2005 New York Academy of Sciences.
doi: 10.1196/annals.1352.037

INTRODUCTION

Biological systems for energy metabolism are essential for the survival, continued growth, and reproduction of all living organisms, including parasites. A key energy-transducing mechanism in this regard is the aerobic respiratory chain, a pathway mediating electrogenic translocation of protons out of mitochondrial or bacterial membranes. This generates the proton motive force that drives ATP synthesis by F_0F_1-ATPase, a mechanism that has essentially remained unchanged from bacteria to human mitochondria. Parasites, however, have exploited unique energy metabolic pathways as adaptations to the natural host habitat. In fact, the respiratory systems of parasites typically show greater diversity in electron transfer pathways than those of host animals. These unique aspects of parasite mitochondria may represent promising targets for chemotherapy.[1]

Our recent findings suggested that antirespiratory drugs affect parasite survival. A novel compound, nafuredin, inhibits NADH-fumarate reductase activity of *Ascaris suum* mitochondria, a unique anaerobic electron transport system in helminth mitochondria.[2] Also in trypanosome, cyanide-insensitive oxidase (trypanosome alternative oxidase [TAO]) has been targeted for the development of the anti-trypanosomal drug ascofranone, because it does not exist in the host.[3] Thus, parasite mitochondria have become a focus of chemotherapy.

Malaria is one of the most serious infectious diseases in the developing world. Mortality associated with malaria is estimated at more than 1 million deaths per year and is mainly caused by the erythrocytic stage cells of *Plasmodium falciparum*.[4] Development of new antimalarial drugs is urgently required, because drug-resistant parasites are widespread.[5] Much research has been undertaken to explore the potential of new antimalarials and their targets. In *Plasmodium*, the differences between the mitochondria of malaria parasites and those of the host are also expected to be the target for chemotherapy.[6] The inhibition site of atovaquone, which is a recently developed antimalarial agent, is considered to be the ubiquinone oxidation site of cytochrome *b* in the bc_1 complex of the parasite mitochondria.[7,8] Atovaquone is effective against chloroquine-resistant strains and is already being used for therapy in epidemic regions in Africa and Thailand.[9] In addition, complex II was shown to be an essential component for parasite survival by antisense DNA analysis. Antisense DNA for Ip inhibited the growth of the malaria parasite.[10] These data clearly indicate that the bc_1 complex and complex II would be the target of chemotherapy for malaria.

As these enzymes have quinone-binding sites (FIG. 1), ubiquinone plays an important role in the respiratory chain of *Plasmodium* mitochondria.[6] It is a point of contact between pyrimidine metabolism and energy metabolism, both of which are essential metabolic systems for malaria parasites. By targeting this point, therefore, it may be possible to inhibit simultaneously these two important metabolic systems. Because many antimalarial agents contain a quinone structure, the enzymes mediating the electron transfer between the enzymes and ubiquinone are considered to be good targets.[11]

Chalcones, which are a promising group of flavonoids for antitumor-promoting activity, were recently evaluated for their antimalarial activity *in vitro* and *in vivo*.[12] These compounds were originally isolated from Xin-jiang licorice and have been shown to have various biological effects, such as antitumor, antibacterial, antituber-

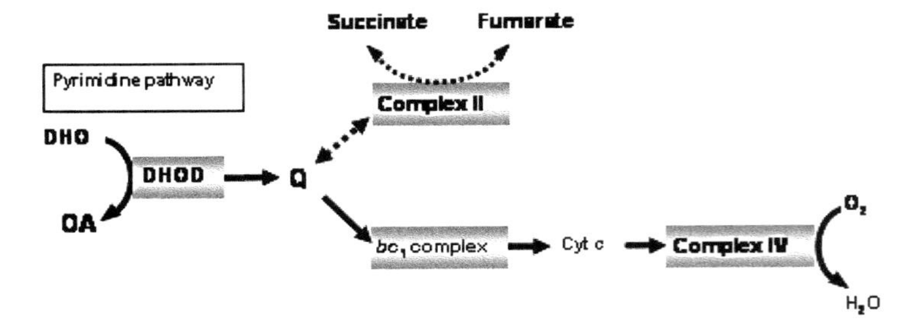

FIGURE 1. Respiratory chain of *P. falciparum* mitochondria.

culous, antiviral, antiprotozoal, and others.[13,14] However, their target and the inhibition mechanism were still unclear. This lack of molecular understanding of the target and mechanisms of action has impeded the development of efficient antimalarial drugs despite the many synthetic analogs, such as alkoxylated, hydroxylated, oxygenated, and quinolinyl chalcones, that have been synthesized.[15–18] A cysteine proteinase, falcipain, was evaluated as the target of chalcone analogs, but it was concluded in the same report that the antimalarial activity of chalcones was probably not due to the inhibition of falcipain and may follow a different mechanism because the antimalarial activity did not correlate well with the inhibition of the enzyme.[16]

In *Leishmania,* fumarate reductase (FRD) activity of complex II was characterized as the target of chalcone.[19] Generally, FRD activity is essential for the anaerobic respiratory chain of many parasites. Since complex II has a quinone-binding site, there is a possibility that complex II is a target also in *Plasmodium.* However, there has been no experimental evidence of the effect of chalcones on the respiratory chain of *Plasmodium,* because it was difficult to prepare the active mitochondria from *Plasmodium* spp. To solve this problem, we have established a protocol to prepare the active mitochondria from *P. falciparum,* showing high and reproducible respiratory enzyme activities.[10,20]

In this study, we showed that licochalcone A inhibited the bc_1 complex as well as complex II of *P. falciparum* mitochondria. The bc_1 complex was more sensitive to licochalcone A and its IC_{50} was 0.10 μM. Because the property of the *P. falciparum* bc_1 complex is different from that of mammals, chalcones should be a promising candidate for a new antimalarial drug.

MATERIAL AND METHODS

Culture

The *P. falciparum* parasite lines used were the Honduras-1. The parasites were cultured as described by Trager and Jensen in A(+) erythrocytes at 3% hematocrit in medium RPMI 1640 (Sigma), supplemented with 25 mM HEPES, 10 mM glucose, 0.32 mM hypoxanthine, and 10% (v/v) heat-inactivated human plasma, and in an atmosphere of 90% N_2, 5% O_2, and 5% CO_2. The growth medium was replaced daily.[21] The percentage parasitemia and stages were assessed daily by microscopic examination of thin blood smears stained with Giemsa.

Preparation of Mitochondria Fraction

Mitochondria were prepared as described previously.[20] The cells were harvested and treated with 0.075% (w/v) saponin in AIM (120 mM KCl, 20 mM NaCl, 10 mM Pipes, 1 mM $MgCl_2$, and 5 mM glucose; pH 6.7). After washing with AIM, the cells were disrupted with the N_2 cavitation method using cell disruption Bomb (4639, Parr, USA) at 1,200 psi for 20 min at 4°C in MSE buffer (225 mM mannitol, 75 mM sucrose, 0.1 mM EDTA, and 3 mM Tris HCl; pH 7.4) in the presence of 1 mM phenylmethylsulfonyl fluoride. Unbroken cells and cell debris were removed by centrifugation at 800× *g* for 5 min at 4°C. The mitochondrial fraction was then recovered as a precipitate by centrifugation at 23,000 × *g* for 20 min at 4°C.

Enzyme Assay

Succinate-ubiquinone reductase (SQR) activity and succinate dehydrogenase (SDH) activity were assayed as described previously.[20] Dihydroorotate dehydrogenase (DHOD) activity was measured by recording the absorbance change of dichlorophenol indophenol (DCIP) at 600 nm (Shimadzu spectrophotometer UV-3000). The reduction was started by the addition of 500 μM DHO to the reaction mixture containing 30 mM This-HCl (pH 8.0), 74 μg/ml DCIP, 100 μM UQ_2, and 2 mM KCN. DHO-cytochrome *c* reductase activity was measured by recording the absorbance change of cytochrome *c* at 550 nm, following the addition of 500 μM DHO to 30 mM This-HCl (pH 8.0), 20 μM cytochrome *c*, and 2 mM KCN. The effect of inhibitors to *Plasmodium* mitochondria was assayed in the presence or the absence of chalcone at 25°C. Licochalcone A was dissolved in DMSO and a final concentration of DMSO was fixed at 0.5% (v/v). Each reaction was initiated by the addition of DHO.

RESULTS

Inhibition of Complex II by Licochalcone A

Among the chalcones, licochalcone A (FIG. 2) was used in this study because it was demonstrated to exhibit potent antimalarial activity in the previous report.[12] First, we investigated the antimalarial activity of licochalcone A on *in vitro* culture of *P. falciparum*. The 50% inhibition concentration of growth (IC_{50} value) was 7.7 M, and this value was almost the same as that in the previous report (IC_{50} = 2 μM).

To investigate the effect of licochalcone A on complex II, we examined the effect of licochalcone A on SQR and SDH activity. The SQR activity of *Plasmodium* was sensitive to licochalcone A, and its IC_{50} value was 1.30 μM (TABLE 1). Rat mitochondrial SQR activity was also sensitive to licochalcone A, but its sensitivity was 10 times lower than that of *Plasmodium* complex II (TABLE 1). The SDH activity, on the other hand, was not affected even by 100 μM licochalcone A for both *Plasmodium* and rat liver (TABLE 1). The SDH activity is a partial reaction of SQR catalyzed by the electron transfer from succinate to the water-soluble electron acceptor. This

TABLE 1. Inhibitory effects of chalcones against *P. falciparum* and rat liver complex II

	Complex II activity (IC_{50})	
Species	SUC-UQ-DCIP	SUC-PMS-MTT
P. falciparum	1.30 μM	>100 μM
Rat	16.5 μM	>100 μM

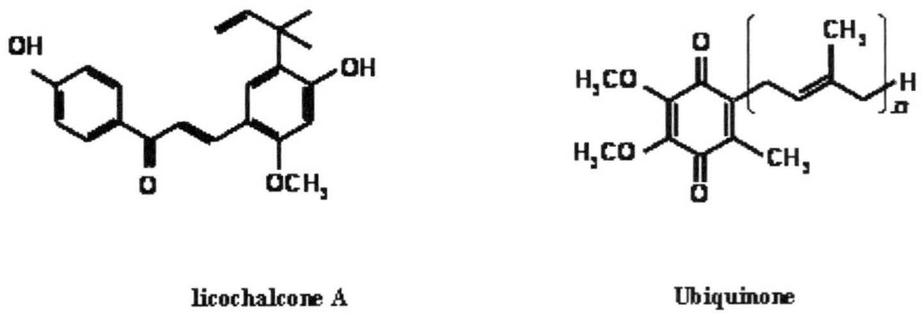

licochalcone A

Ubiquinone

FIGURE 2. Structure of licochalcone A and ubiquinone.

activity does not require the membrane anchor subunits, which are essential components for electron transfer to ubiquinone (UQ).[6] Therefore, this finding indicates that the licochalcone A binding site is localized at membrane anchor subunits.

Then, the mechanism of inhibition by licochalcone A was further studied with rat liver mitochondria, because specific activity of the *Plasmodium* enzyme was too low and mammalian complex II showed high enough activity for kinetic analysis. Double-reciprocal plots showed competitive inhibition of licochalcone A for UQ_2 (FIG. 3). This finding indicates that licochalcone A may block the electron transfer between the enzyme and UQ by binding to the UQ binding site.

Inhibition of the bc_1 Complex by Licochalcone A

From a study of the mechanism of complex II inhibition by licochalcone A, it was suggested that licochalcone A specifically inhibits the UQ binding site.Therefore, we examined the effect of licochalcone A on the enzymes possessing a UQ binding site such as the bc_1 complex and DHOD. To characterize the bc_1 complex, the ubiquinol-cytochrome *c* assay has generally been used.[22] However, we were unable to use this assay system because the activity of the *Plasmodium* bc_1 complex was too low to study detailed kinetics of the enzyme. Therefore, we measured the activity of DHO-cytochrome *c* reductase, which is much more reliable than the ubiquinol-cytochrome *c* assay. DHO-cytochrome *c* activity consists of two enzymes, DHOD and bc_1 complex (FIG. 4). Suitability of this system was confirmed by the specific and potent inhibitors of the bc_1 complex and DHOD, such as antimycin A and orotate, inhibiting DHO-cytochrome *c* activity (IC_{50} = 0.282 nM and 90 μM, respectively). To investigate the effect of licochalcone A on the bc_1 complex, the sensitivity of

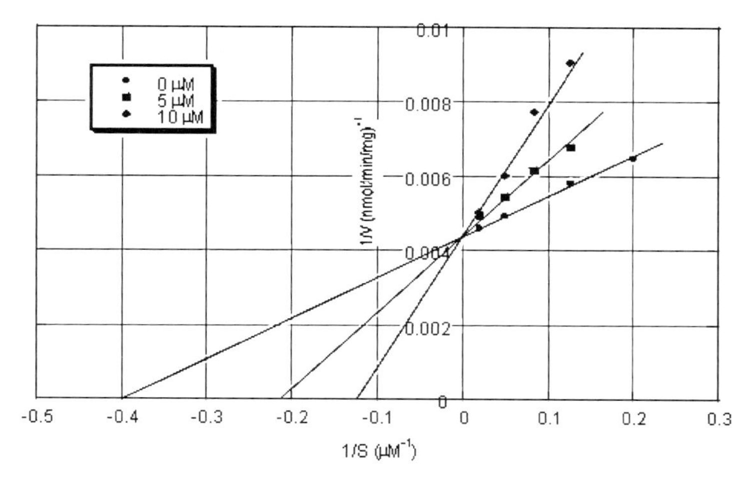

FIGURE 3. Double-reciprocal plots of licochalcone A inhibition of rat SQR activity. Inhibition of rat liver mitochondria by licochalcone A was measured in the presence of various concentrations of UQ_2. The assay was performed at 25°C in the presence 50 mM potassium phosphate buffer (pH 7.2), 74 μM DCIP, potassium succinate (10 mM), potassium cyanide (2 mM), and 10 μg of protein of rat liver mitochondria (298 nmol/min/mg). SQR activity was determined by the millimolar extinction coefficient for DCIP (21 $mM^{-1}cm^{-1}$ at 600 nm).

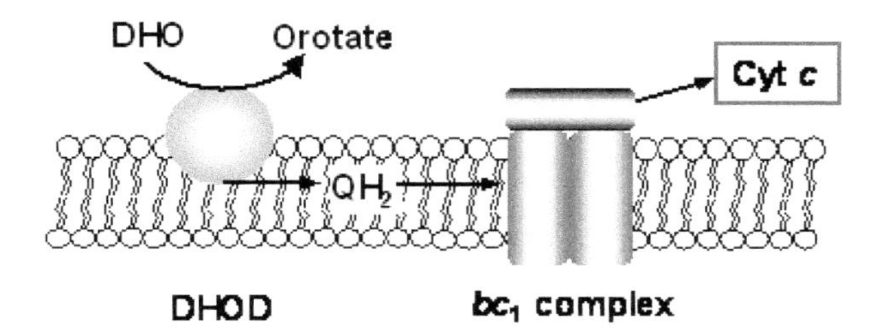

FIGURE 4. DHO-cytochrome *c* activity consists of two enzymes, DHOD and the bc_1 complex. DHO-cytochrome *c* reductase activity is determined by measuring the electron transfer from DHO to cytochrome *c* by monitoring the absorption change at 550 nm as described in the **Material and Methods** section.

TABLE 2. Mitochondrial enzyme activities of *P. falciparum* and inhibition by licochalcone A

Species	Activities	Enzyme	Specific activity (nmol/min/mg)	IC_{50}
P. falciparum	DHO-UQ-DCIP	DHOD[a]	7.10 ± 0.97	>100 μM
	DHO-cyt c	DHOD + bc_1 complex	2.56 ± 0.69	0.10 μM
Rat liver	DHO-UQ-DCIP	DHOD[a]	7.05 ± 0.62	1.42 μM
	DHO-cyt c	DHOD + bc_1 complex	10.8 ± 1.24	0.077 μM

[a]Dihydroorotate dehydrogenase.

DHO-cytochrome *c* activity was analyzed and compared with that of DHOD activity assayed using DCIP as an electron acceptor.

The DHO-cytochrome *c* activity of *P. falciparum* was sensitive to licochalcone A, and its IC_{50} was 0.10 μM (TABLE 2). The sensitivity was 10 times higher than that of complex II. DHOD, which is one of the components of DHO-cytochrome *c* activity, was not inhibited even by 100 μM licochalcone A, indicating that the target of licochalcone A is the bc_1 complex of *Plasmodium*. The bc_1 complex and DHOD of rat mitochondria were sensitive to licochalcone A and their IC_{50} was 0.077 and 1.42 μM (TABLE 2).

DISCUSSION

Licochalcone A is a characteristic chalcone of Xin-jiang licorice, which is the root of *Glycyrrhiza inflata* and its structure was established by one of the present authors as 3-a,a-dimethylallyl-4,4′-dihydroxy-6-methoxychalcone.[23] Licochalcone A has been reported to show anti-inflammatory activity, anti-tumor activity, and antimicrobial activity.[13] In this study, we showed that licochalcone A has multiple targets in *P. falciparum* mitochondria. Complex II and bc_1 complex were inhibited by licochalcone A at low concentrations (TABLE 1). Thus, energy metabolism is one of the indispensable systems for the survival of the parasite, and the enzymes of the energy-transducing pathway are promising targets of antiparasitic drugs as discussed previously.[1,24]

Considering the report of *Leishmania,*[19] it may be possible to speculate that the growth inhibition of *P. falciparum* by licochalcone A is due to the inhibition of complex II. The IC_{50} of *Plasmodium* SQR activity was very low, and its sensitivity was 10 times higher than that of rat mitochondrial complex II. The kinetics analysis of licochalcone A on rat SQR activity showed that licochalcone A was the competitive inhibiter of UQ. This finding indicates that the binding site of licochalcone A overlaps with the quinone-binding site. The important role of complex II in the erythrocytic stage of *Plasmodium* has been shown,[10,20,25] although more biochemical study is essential to understanding its physiological functions.

It should be noted that the *Plasodium* bc_1 complex was more sensitive to licochalcone A than was complex II. The bc_1 complex is an essential enzyme in the *Plasmodium* respiratory chain and a key molecule with regard to the chemotherapy of

Plasmodium. The bc_1 complex has two quinone binding sites, Qi site and Qo site. The Qo site of the *Plasmodium* bc_1 complex is the binding site of the antimalarial, atovaquone. Chalcone inhibits not only parasites, but also the bacterial bc_1 complex.[14] The low IC_{50} value (0.10 μM) of licochalcone A in the present study indicates that a major part of antimalarial activity may be explained by the inhibition of the *Plasmodium* bc_1 complex, although modification of the erythrocyte membrane has been suggested as a mechanism of the antiplasmodial activity of the compound.[26] To evaluate the effect of licochalcone A and its analogs on the *Plasmodium* bc_1 complex, the DHO-cytochrome *c* reductase assay system was useful. The ubiquinone- cytochrome *c* reductase assay was unable to perform because the artificial electron flow from ubiquinol to cytochrome *c* was very fast, and it could not be subtracted as a background. Since the DHO-cytochrome *c* reductase system is useful for the drug screening of the *Plasmodium* bc_1 complex inhibitor and the bc_1 complex has been suggested as a target of atovaquone and primaquine,[7,8,27] it would be interesting to investigate the effects of these compounds using this assay system.

In conclusion, we showed that licochalcone A inhibits the complex II and the bc_1 complex in the respiratory chain of *Plasmodium* mitochondria. As licochalcone A might have potential as a good lead compound, several analogs of licochalcone A have been analyzed for their efficacy. Interestingly, we found more effective derivatives, such as 3-hydroxy-3′-methylchalcone (3′Me-3C), on the parasite growth than licochalcone A among the compounds. Further study is now ongoing.

ACKNOWLEDGMENTS

This work was supported by a Grant-in-Aid for Scientific Research on Priority Areas from the Ministry of Education, Science, Culture and Sport of Japan (13226015 and13854011) as well as by a grant for Research on Emerging and Reemerging Infectious Diseases, and a grant for International Health Cooperation Research (15-C5) from the Ministry of Health, Labor, and Welfare of Japan. This study was also supported by a "Pilot Applied Research Project for the Industrial Use of Space" of the National Space Development Agency of Japan (NASDA) and Japan Space Utilization Promotion Center (JSUP).

REFERENCES

1. KITA, K., C. NIHEI & E. TOMITSUKA. 2003. Parasite mitochondria as a drug target: diversity and dynamic changes during the life cycle. Curr. Med. Chem. **10:** 1241–1253
2. OMURA, S., H. MIYADERA, H. UI, *et al.* 2001. An anthelmintic compound, nafuredin, shows selective inhibition of complex I in helminth mitochondria. Proc. Natl. Acad. Sci. USA **98:** 60–62.
3. NIHEI, C., Y. FUKAI & K. KITA. 2002. Trypanosome alternative oxidase as a target of chemotherapy. Biochim. Biophys. Acta **1587:** 234–239.
4. WHO World malaria situation in 1994. Weekly epidemiological record, 1997 WHO.
5. OLLIARIO, P. & P. TRIGG. 1995. Status of antimalarial drugs under development. Bull. WHO **73:** 565–571.
6. KITA, K., H. HIRAWAKE, H. MIYADERA, *et al.* 2002. Role of complex II in anaerobic respiration of the parasite mitochondria from *Ascaris suum* and *Plasmodium falciparum*. Biochim. Biophys. Acta **1553:**123–139.

7. Srivastava, I.K., H. Rottenberg & A.B. Vaidya. 1997. Atovaquone, a broad spectrum antiparasitic drug, collapses mitochondrial membrane potential in a malarial parasite. J. Biol. Chem. **272:** 3961–3966.
8. Srivastava, I.K., J.M. Morrisey, E. Darrouzet, *et al.* 1999. Resistance mutations reveal the atovaquone-binding domain of cytochrome *b* in malaria parasites. Mol Microbiol **33:** 704–711.
9. Looareesuwan, S., C. Viravan, H.K. Webster, *et al.* 1996. Clinical studies of atovaquone, alone or in combination with other antimalarial drugs, for treatment of acute uncomplicated malaria in Thailand. Am. J. Trop. Med. Hyg. **54:** 62–66.
10. Mi-ichi, F., S. Takeo, E. Takashima, *et al.* 2003. Unique properties of respiratory chain in *Plasmodium falciparum* mitochondria. Adv. Exp. Med. Biol. **531**: 117–133
11. Ellis, J.E. 1994. Coenzyme Q homologs in parasitic protozoa as targets of chemotherapeutic attack. Parasitol. Today. **10:** 296–301
12. Chen, M., T.G. Theander, S.B. Christensen, *et al.* 1994. licochalcone A, a new antimalarial agent, inhibits in vitro growth of the human malaria parasite *Plasmodium falciparum* and protects mice from *P. yoelii* infection. Antimicrob Agents Chemother. **38:** 1470–1475.
13. Shibata, S. 2000. A drug over the millennia: pharmacognosy, chemistry, and pharmacology of licorice. Yakugaku Zasshi **120:** 849–862.
14. Haraguchi, H., K. Tanimoto, Y. Tamura, *et al.* 1998. Mode of antibacterial action of retrochalcones from *Glycyrrhiza inflata*. Phytochemistry **48:** 125–129.
15. Liu, M., P. Wilairat & M.L. Go. 2001. Antimalarial alkoxylated and hydroxylated chalcones: structure-activity relationship analysis. J. Med. Chem. **44:** 4443–4452.
16. Domingue, J.N., J.E. Charris, G. Lobo, *et al.* 2001. Synthesis of quinolinyl chalcones and evaluation of their antimalarial activity. Eur. J. Med. Chem. **36:** 555–560.
17. Ram, V.J., A.S. Saxena, S. Srivastava & S. Chandra. 2000. Oxygenated chalcones and bischalcones as potential antimalarial agents. Bioorg. Med. Chem. Lett. **10:** 2159–2161.
18. Chen, M., B.S. Christensen, L. Zhai, *et al.* 1997. The novel oxygenated chalcone, 2,4-dimethoxy-4′-butoxychalcone, exhibits potent activity against human malaria parasite *Plasmodium falciparum in vitro* and rodent parasites *Plasmodium berghei* and *Plasmodium yoelii in vivo*. J. Infect. Dis. **176:** 1327–1333.
19. Chen, M., L. Zhai, S.B. Christensen, *et al.* 2001. Inhibition of fumarate reductase in *Leishmania major* and *L. donovani* by chalcones. Antimicrob. Agents Chemother. **45:** 2023–2029.
20. Takashima, E., S. Takamiya. F. Takeo, *et al.* 2001. Isolation of mitochondria from *Plasmodium falciparum* showing dihydroorotate dependent respiration. Parasitol. Int. **50:** 273–278.
21. Trager, W. & J.B. Jansen. 1976. Human malaria parasites in continuous culture. Science **193:** 673–675.
22. Krungkrai, J., S.R. Krungkrai, N. Suraveratum & P. Prapunwattana. 1997. Mitochondrial ubiquinol-cytochrome *c* reductase and cytochrome *c* oxidase: chemotherapeutic targets in malarial parasites. Biochem. Mol. Biol. Int. **42:** 1007–1014.
23. Saitoh, T. & S. Shibata. 1975. New type chalcones from licorice root. Tetrahed. Lett. **16:** 4461–4462
24. Sherman, I.W. 1979. Biochemistry of *Plasmodium* (malaria parasite). Microbiol. Rev. **43:** 453–495
25. Takeo, S., A. Kokaze, C.S. Ng, *et al.* 2000. Succinate dehydrogenase in *Plasmodium falciparum* mitochondria: molecular characterization of the SDHA and SDHB genes for the catalytic subunits, the flavoprotein (Fp) and iron-sulfur (Ip) subunits. Mol. Biochem. Parasitol. **107:** 191–205.
26. Ziegler, H.L., H.S. Hansen, D. Staerk, *et al.* 2004. The antiparasitic compound Licochalcone A is a potent echinocytogenic agent that modifies the erythrocyte membrane in the concentration range where antiplasmodial activity is observed. Antimicrob. Agents Chemother. **48:** 4067–4071.
27. Vaidya, A.B., M.S. Lashgari, L.G. Pologe & J. Morrisey. 1993. Structural features of *Plasmodium* cytochrome *b* that may underlie susceptibility to 8-aminoquinolines and hydroxynaphthoquinones. Mol. Biochem. Parasitol. **58:** 33–42.

Vaccinia Virus Complement Control Protein Ameliorates Collagen-Induced Arthritic Mice

PURUSHOTTAM JHA,[a] SCOTT A. SMITH,[a] DAVID E. JUSTUS,[a] AND GIRISH J. KOTWAL[a,b]

[a]*Departments of Microbiology and Immunology, University of Louisville School of Medicine, Louisville, Kentucky 40202, USA*

[b]*Division of Medical Virology, University of Cape Town Medical School, Observatory, Cape Town, South Africa*

ABSTRACT: The main objective of this study was to investigate the therapeutic efficiency of recombinant vaccinia virus complement control protein (rVCP) on collagen-induced arthritis (CIA) in DBA-1/J mice. Arthritis was induced in DBA-1J mice by injecting bovine collagen emulsified in complete Freunds adjuvant. We used rVCP to block complement activation and investigated its effect on different aspects of CIA including osteoclast formation and bone destruction. The osteoclast-like cells were detected using immunohistochemistry. Joint destruction was studied using X-ray of the intact knee joints. Cartilage destruction was monitored by staining the paraffin sections with toluidine blue. ELISA was used to measure the cytokine levels in the serum. Blocking complement activation in DBA/1J arthritic mice with rVCP resulted in significant inhibition of the clinical progression of the disease and reduction in joint destruction as revealed by X-ray analysis and toluidine blue staining of the joint sections. Inhibition of complement reduced the production of proinflammatory cytokines and the number of osteoclast-like cells in arthritic joints. In conclusion, blocking of complement in CIA by rVCP inhibits the inflammation and the formation of osteoclast-like cells and reduces cartilage destruction.

KEYWORDS: vaccinia virus; complement control; collagen-induced arthritis; bone erosion; osteoclasts; VCP

INTRODUCTION

Rheumatoid arthritis (RA) is a polyarticular inflammatory synovitis that affects up to 1% of the world's population. The hallmark of RA is progressive destruction of the joints, characterized by synovial hyperplasia and inflammation. Despite the recent development of numerous anti-inflammatory and disease-modifying drugs to limit pain and inflammation, new approaches are needed to inhibit the progression of this chronic and debilitating disease.[1] Collagen-induced arthritis (CIA) in mice is the most widely used animal model for studying RA. It produces clinical, histologi-

Address for correspondence: Girish J. Kotwal, Division of Medical Virology, University of Cape Town Medical School, Anzio Road, Observatory, Cape Town, South Africa. Voice: +27 21 406 6676 / +27 21 406 6126; fax: +27 21 406 6018.
gjkotw01@yahoo.com

Ann. N.Y. Acad. Sci. 1056: 55–68 (2005). © 2005 New York Academy of Sciences.
doi: 10.1196/annals.1352.004

cal, and immunological features similar to those of human RA. In CIA, both T- and B-cell responses play a role in disease pathogenesis after immunization with type II collagen. In recent years, numerous reports have implicated osteoclasts as an important cell in the effector phase of RA pathogenesis. Osteoclasts are derived from phagocytic precursors of the monocyte/macrophage lineage and are responsible for degradation of bone matrix. These synovial monocytes/macrophages under a suitable microenvironment differentiate into osteoclasts. Normal skeletal remodeling is tightly controlled by a balance between osteoclasts and osteoblasts.[2] The role of complement in the pathogenesis of CIA has already been demonstrated.[3–7] Complement activation products can cause damage to structures within the joint either directly via the membrane attack complex (MAC) or indirectly through C5a-mediated recruitment and activation of phagocytic cells. In addition, C5 deficiency has frequently been correlated with disease resistance[8,9] although this has not always been the case.[10] The role of the complement system has been studied in a variety of experimental models of arthritis in a number of different species.[11] Depletion of complement using cobra venom factor (CVF) in CIA, around the time of disease onset, delays the tissue responses until complement levels return to normal.[12] We hypothesize that complement activation in the synovium results not only in increased influx of monocytes/macrophages but also in their activation. This complement-mediated activation, in turn, results in increased osteoclastogenesis, which is the final mediator of joint destruction. In the current study, we used recombinant vaccinia virus complement control protein (rVCP) to block complement activation and investigated its therapeutic potential on different aspects of the disease including osteoclastogenesis and bone destruction. rVCP is a 35-kDa protein made up of 243 amino acid residues,[12] which was previously shown to inhibit the complement cascade.[13] Structurally it consists of four short consensus repeats (SCRs), which share amino acid identity to complement 4b binding protein (C4b-BP), membrane cofactor protein (MCP) and decay-accelerating factor (DAF). Apart from complement regulation, rVCP shares another common property with complement regulators MCP, DAF, and complement receptor type 1 (CR1), cell-surface association. rVCP cell-surface association is via its strong heparin-binding ability.[14] This as well as other properties of rVCP, its small size and physically robust characteristics,[15] makes it a good research tool for use as an effective blocker of the complement system.[16] The results of this study indicate that systemic rVCP treatment prevents bone damage, bone destruction, and reduces inflammation. The beneficial effect of rVCP was associated with reduced monocyte/macrophage influx and decreased formation of osteoclast-like cells. The effect was also associated with downregulation of interleukin-12 (IL-12), tumor necrosis factor-alpha (TNF-α), and IL-6. The results from the current work suggest that complement inhibitors have strong potential for use as therapeutic agents to prevent the progression of RA.

MATERIAL AND METHODS

Animals. DBA/1J mice were purchased from Jackson Laboratories, Bar Harbor, Massachusetts, USA. The animals were maintained in accordance with the guidelines established by the Institutional Animal Care and Use Committee (IACUC) at the University of Louisville Medical School. Institutional approval was obtained, and institutional guidelines regarding animal experimentation were followed.

Induction of Arthritis. Ten-week-old, male DBA/1J mice received a single intradermal injection of saline containing 200-g bovine type II collagen (BCII) emulsified in an equal volume of complete Freund's adjuvant (Chondrex, Redmond, Washington) at the base of the tail. Usually one to three paws becomes affected. Arthritis development was assessed by visual inspection three times a week by a protocol-blinded examiner. Clinical severity of arthritis was quantified according to a graded scale from 0 to 3 as follows: 0, normal; 1, detectable swelling in one joint of the paw; 2, swelling in more than one but not in all joints of the paw; and 3, severe swelling of the entire paw and/or ankylosis. Each of the four paws of the mouse was graded, and each mouse could achieve a maximum score of 12. A mean arthritic score value was calculated.

Efficacy of rVCP to Suppress CIA. DBA1/J mice were randomly divided into four groups (n = 5/group), and arthritis was elicited in all groups as indicated. To study the effect of VCP on CIA, animals received twice daily intraperitoneal (i.p.) injections of phosphate-buffered saline solution (PBS), 25 mg/kg of animal weight of rVCP, optimal dosage determined previously, [18] beginning either on days 9 (early intervention), 12 (delayed intervention), or 21 (established disease) after collagen immunization. Treatments were stopped on day 30 in animals that received rVCP from days 9 and 12. Treatment was continued until day 38 in mice receiving rVCP from day 21. All the animals were sacrificed on day 38. The experiment was repeated twice.

Histopathology. For histological assessment, mice were sacrificed, and the hind limbs were removed, fixed in 10% neutral-buffered formalin, and then decalcified in 5% formic acid and embedded in paraffin. Sections (5 μm) were prepared and stained with hematoxylin and eosin or toluidine blue (Sigma-Aldrich).

Immunohistochemistry. Immunohistochemistry was done as follows using a paraffin-embedded knee. The slides were deparaffinized and hydrated and were placed in 3% hydrogen peroxide for 5 minutes to quench endogenous peroxidase. The slides were rinsed and incubated for 10 minutes with EDTA to unmask the antigenic markers for rVCP. Again, the slides were rinsed and incubated for 20 minutes with serum from the same species as the origin of the secondary antibody. After rinsing, they were incubated for 60 minutes with chicken IgY anti-VCP. The slides were again rinsed and incubated for 60 minutes with goat anti-chicken IgG (Vector Laboratories). After rinsing, the slides were incubated for 30 minutes with Vectastain elite ABC reagent (Vector Laboratories, Burlingame, CA). A final rinsing was followed by incubation with diaminobenzidine tablets (Sigma, St. Louis, MO) for 5 minutes. The slides were counterstained with diluted hematoxylin (2 minutes) followed by dehydration (the reverse of the hydration step) and mounted with Permount (Fisher Scientific, Pittsburgh, PA).

Tartrate-Resistant Acid Phosphatase (TRAP) Staining. Fixed, paraffin-embedded tissues were sectioned at 5 μm and slides were dewaxed, hydrated to distilled water, placed in Target Retrieval Solution (DakoCytomation, Carpinteria, CA), and placed in a 75°C oven overnight for antigen retrieval. Slides were incubated for 1 hour in a working solution of mouse-Ig blocking reagent (MOM). Slides were washed in tris buffer and incubated for 5 minutes in a working solution of MOM diluent. Primary antibody clone 26E5, 1:50, NCL-TRAP from Novocastra (Vector Laboratories, Burlingame, CA) was used. After washing, all slides were incubated for 10 minutes in MOM biotinylated anti-mouse IgG reagent. Slides were washed and incubated for 5 minutes in Vectastain Elite ABC reagent. They were next washed in tris buffer and incubated in DAB (DakoCytomation) for 10 minutes. After washing, slides were counterstained with hematoxylin (Biomeda, Foster City, CA) for 1 minute.

Detection of Antibodies to C-II and Proinflammatory Cytokine in the Serum. Serum anticollagen antibody titers were measured using ELISA kits (Chondrex, Redmond, WA) as per the manufacturer's protocol. To examine whether rVCP affected humoral immunity to collagen, serum levels of anticollagen antibody were measured on days 21 and 33. Levels of IL-12, TNF-α, and IL-6 were determined using enzyme-linked immunosorbent assay (ELISA) (Biosource international Inc., Camarillo, CA) according to the protocols from the manufacturers. The detection limit for IL-12 was 2 pg/ml; that for IL-6 was 1 pg/ml. The detection limit for the TNF-α bioassay was 1 pg/ml.

Knee and Paws Radiology. Upon sacrifice, knees and paws were promptly removed and placed on ice. For radiologic analysis, samples were positioned over a radiographic cassette (Detector: Trophy RVG-ui, Marne-la-Vallée, France) to obtain a lateral view. A conventional X-ray source, GE 1000 AC generator (Milwaukee, WI) was used at exposure factors of 55 kV (peak) and 15 mA with a source-to-object distance of 54 cm and 2.7 mm Al equivalent filtration.

RESULTS

rVCP Is Delivered to Arthritic Joints When Given Intraperitoneally. Synovial tissue is the major site in the pathogenesis of RA as well as CIA. To study the presence of rVCP in the synovial tissue, arthritic mice were injected intraperitoneally with 25 mg/kg rVCP on day 24 (post immunization with collagen). This dosage was used on the basis of the results of pharmacokinetic studies (results not shown). Mice were sacrificed and the right knee was dissected. The knee was fixed and embedded in paraffin. Paraffin sections 5-μm thick were immunostained for rVCP using rabbit-anti-VCP. rVCP can be seen in the synovial tissue and at the site of pannus formation (FIG. 1).

rVCP Suppresses the Inflammation in Collagen-Induced Arthritis. We examined the anti-inflammatory effects of rVCP on type-II collagen-induced arthritis in DBA1/J mice. Animals were treated with rVCP and compared with those injected with PBS as the vehicle control. As shown (FIG. 2), rVCP completely inhibited type-II collagen–induced arthritis in mice when administered as early intervention (day 9) and delayed intervention (day 12). No sign of inflammation or edema was observed in rVCP-treated animals until well after stopping treatment. The difference in clinical scores between treated and untreated animals was maintained throughout the course of the experiment (FIG. 2). We also tested a recombinant protein that lacks SCR1 (rVCP~2,3,4). This protein lacks the ability to bind and inhibit complement but retains full heparin binding activity. No difference was observed in rVCP~2,3,4 treated arthritic mice as compared to untreated arthritic mice. However, there was a slight delay in onset (2 days) of disease in rVCP~2,3,4 treated mice (data not shown). Additional information gathered from this set of experiments verified the duration of the anti-inflammatory effect of rVCP on CIA. Once the treatment with rVCP was stopped (on day 30), inflammation could be seen again within 3 days. This

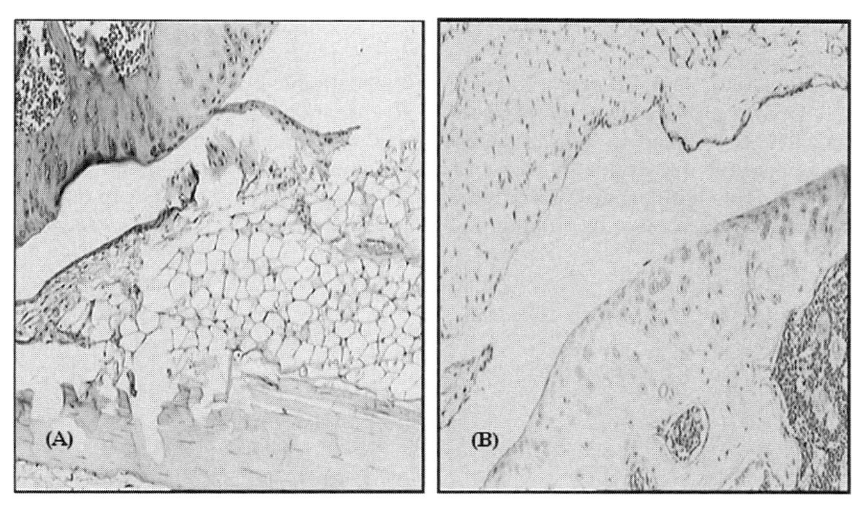

FIGURE 1. rVCP is delivered to arthritic joints when given intraperitoneally. DBA-1/J mice (with fully established arthritis) were injected with 25 mg/kg rVCP (**A**) or PBS (control **B**) i.p. and sacrificed 6 hours after injection. The hind knee was dissected and embedded in paraffin. Five-μm sections of the joints were stained with rabbit anti-rVCP (primary antibody) as described in Material and Methods. The dark stain in **A** shows the presence of rVCP in the inflamed synovial tissue.

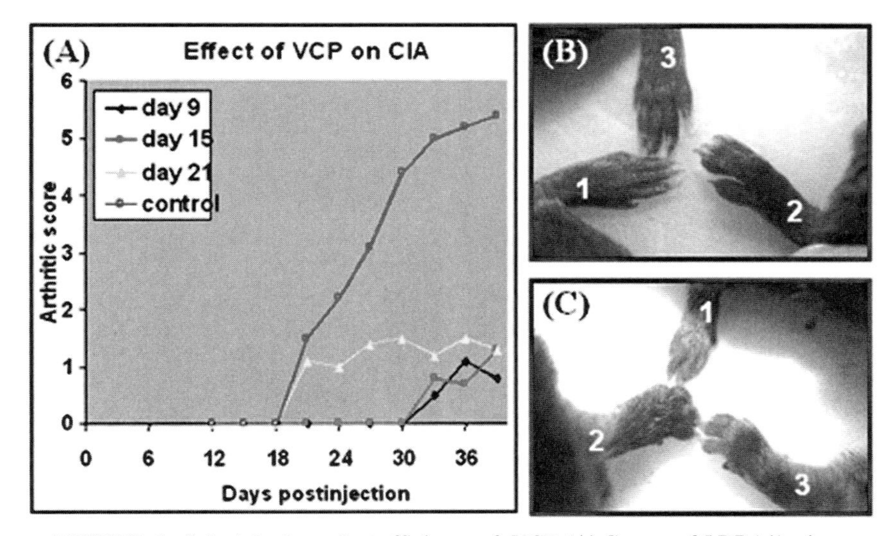

FIGURE 2. Schedule-dependent efficiency of rVCP. (**A**) Groups of 5 DBA/1 mice were immunized with bovine collagen type II in CFA. Thereafter, mice received rVCP, beginning at either day 9, day 15, or day 21 after initial collagen immunization. Data are group arithmetic means of 5 scores/group at each time point. (**B**) The hind paws of normal (**B1**), untreated arthritic (**B2**), and rVCP-treated arthritic mice (**B3**) at day 33. (**C**) The fore paws of mice of the same groups as in the top panel at day 38, 5 days after treatment with rVCP, were stopped. Note that inflammation and edema after rVCP treatment are stopped (**C3**).

indicates that complement inhibition has to be continuous to significantly inhibit the inflammatory progression.

Histopathological Changes. To assess histopathological changes in mice, five animals in each group were euthanized at day 30 after induction of CIA, and hematoxylin and eosin-stained sections of the knee joints were prepared and examined. PBS-treated control CIA groups showed marked inflammation of the synovium and subsynovial tissue with pannus formation eroding through cartilage and deep into subchondral bone. Cartilage destruction was diffuse and severe (FIG. 3). The knee joints

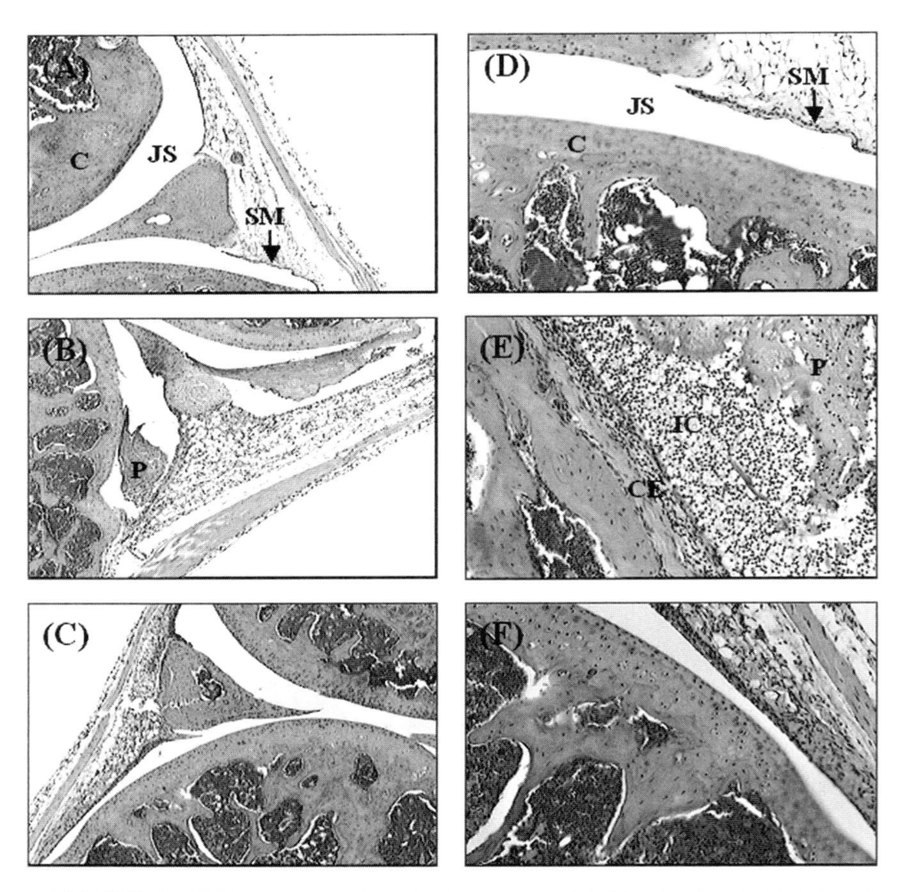

FIGURE 3. rVCP treatment reduces inflammatory cell infiltration in arthritic joints. At the end of the treatment (day 33), arthritic knees from the experimental and control groups were examined microscopically. (**A** and **D**) Normal nonarthritic knee joints showing intact synovial tissue with no signs of infiltrating inflammatory cells. (**B** and **E**) Arthritic knee joints from untreated arthritic mice, showing synovial hyperplasia and inflammatory cell infiltration. (**C** and **F**) No signs of synovial hyperplasia or inflammatory cell infiltration can be seen in joints of rVCP-treated arthritic mice. C, cartilage; CE, cartilage erosion; IC, inflammatory cells; JS, joint space; P, pannus; SM, synovial membrane.

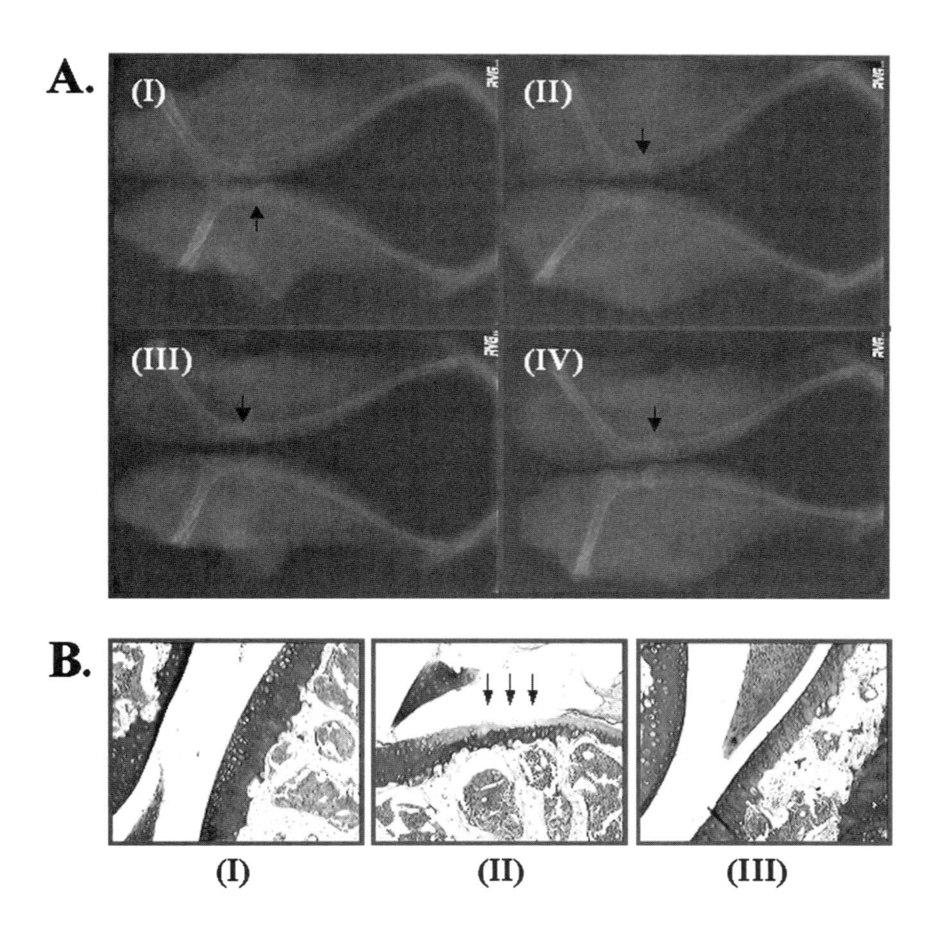

FIGURE 4. rVCP treatment reduces joint destruction in CIA. (**A**) Radiologic analysis of treated and untreated knee (**AI**) joints of normal (*top*) and arthritic (*bottom*) mice at day 33. Note enhanced joint damage and less compact bone structure (*arrow*) enhanced joint damage. (**AII**) Knee joints of arthritic mice treated with PBS (*top*) and rVCP (*bottom*) at day 33. Note the loss of joint space and diffused bone structure in arthritic mice treated with PBS (*arrow*). The rVCP-treated mice have more intact joint structure. (**AIII**) Knee joints of arthritic (*top*) and normal mice (*bottom*) at day 38. Joint destruction in knee of arthritic mice is indicated with *arrow*. (**AIV**) Knee joints of arthritic mice treated with PBS (*top*) and rVCP (*bottom*) at day 38. Note the prevention of joint damage and the compact bone structure of rVCP-treated arthritic mice (*bottom without arrow*). (**B**) Effects of rVCP treatment on cartilage erosion in established CIA. Toluidine blue staining of knee joint tissues from normal mice (**BI**), arthritic untreated mice (**BII**), or mice treated with rVCP (**BIII**). The erosion of cartilage (*arrows*) could be significantly inhibited in the knees of mice treated with rVCP. Objective lens magnification used was 40×.

and hind paws of mice with CIA revealed massive neutrophilic and mononuclear cell infiltration and loss of articular cartilage, typical of established CIA. In comparison with the PBS-treated controls, the severity of histopathologic changes was markedly reduced in mice given rVCP (FIG. 3). Joints from animals treated with rVCP were relatively normal in appearance, with well-preserved joint architecture and joint space.

rVCP Treatment Prevents Bone Erosion and Preserves Joint Integrity. Whole knee joints were radiologically studied for their degree of joint destruction. X-ray photographs of PBS- and rVCP-treated arthritic mice were taken at days 30 and 33 postimmunization with collagen II. In knee joints of the PBS-treated control group, marked zones of bone erosions in the femur and patella (arrows) were evident. Joint space was significantly reduced (FIG. 4A). Clear protection against joint destruction after rVCP treatment in established CIA was observed. Even at day 33, marked prevention of joint damage was still evident in the rVCP-treated group compared with the control group (FIG. 4AII and AIV). The joint space was found to be intact and comparable to that of the normal nonarthritic joints. Identical results were seen in paws from the same group of animals (data not shown).

One of the hallmarks of RA is the cartilage destruction mediated by enzymes and cytokines released from infiltrating inflammatory cells. From our findings it is clear that rVCP reduces this infiltration of inflammatory cells, and hence we expected that there would be less cartilage destruction in arthritic animals treated with rVCP versus control mice. The knee joints from normal and from both rVCP-treated and PBS-treated mice were stained with toluidine blue, which stains the proteoglycans of the cartilage in the joints. Joints from rVCP-treated mice retained their cartilage integrity, whereas those from PBS-treated mice showed a drastic loss of cartilage as demonstrated by loss of toluidine blue stain (FIG. 4B).

rVCP Reduces Osteoclast-Like Cells in the Joints. As previously noted, osteoclasts play a key role in bone resorption. To examine potential inhibition of osteoclastic activity after rVCP administration, tartrate-resistant acid phosphatase (TRAP) staining was performed on paraffin-embedded knee joint sections. TRAP activity is a characteristic phenotypic marker of osteoclasts and osteoclast precursors and is expressed in osteoclast-like cells in mice with well-established CIA. Mice of the PBS-treated control group revealed cartilage erosion (as seen with toluidine blue staining) and high numbers of $TRAP^+$ osteoclast-like cells in mice that had developed marked arthritis (FIG. 5A and B). However, all mice of the rVCP-treated group showed marked reduction in $TRAP^+$ osteoclast-like cells at sites of erosion in subchondral, trabecular, and cortical bone of the patella and femur/tibia region (FIG. 5C and D). rVCP reduces the production of anticollagen antibodies and proinflammatory cytokines in the joints. To evaluate whether or not treatment with rVCP affected immune responses to collagen, serum levels of anticollagen antibodies were measured (by ELISA) on days 21 and 33 (in arthritic mice that received rVCP treatment until day 33). Anticollagen antibodies were not detected in the sera of nonimmunized DBA control mice. Low levels of antibodies present on day 21 (day rVCP treatment was started) in the immunized mice were not different among the three treatment groups. Increased serum levels of anticollagen antibodies were observed between days 21 and 33 in the immunized mice treated with PBS alone. However, the changes in the serum level of anticollagen antibodies over 3 weeks were markedly reduced (13-fold reduction) after treatment with rVCP (FIG. 6A).

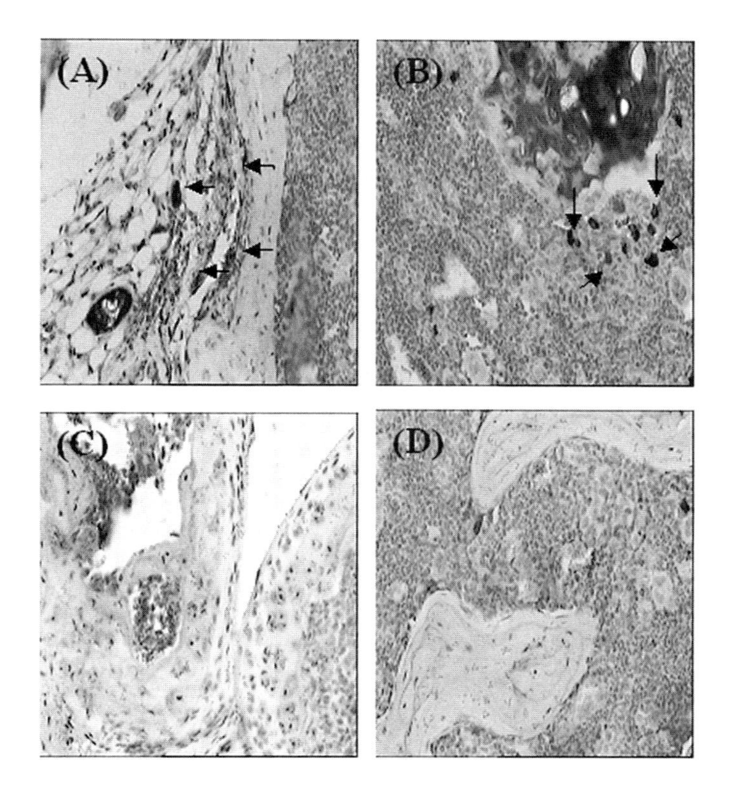

FIGURE 5. Immunohistological study for TRAP$^+$ cells. At day 33 of the disease, animals from treated and untreated groups were sacrificed and their right hind knees were processed for detection of TRAP$^+$ cells. Numerous TRAP$^+$-osteoclasts (*arrows*) were detected in both pannus (**A**) and subchondral (**B**) sections of bone of untreated arthritic joints as compared to rVCP-treated arthritic joints (**C** and **D**).

Since the inflammatory process in the synovium plays a major role in the development of arthritis, we next examined the change of proinflammatory cytokine expression in arthritic mice with or without rVCP treatment. Consistent with previous reports, high levels of IL-12, TNF-α, and IL-6 were observed in the sera from untreated arthritic mice. However, treatment with rVCP significantly inhibited the expression of these proinflammatory cytokines (FIG. 6B, C, and D). These findings suggest that rVCP treatment reduces arthritic inflammatory manifestations by downregulating the expression of proinflammatory cytokines in the joints.

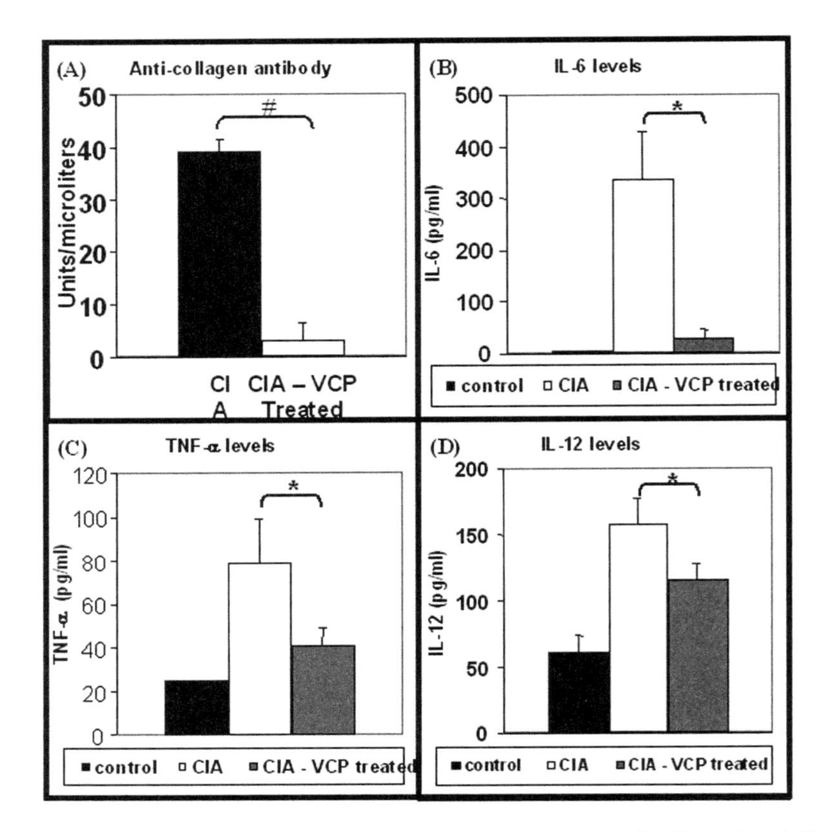

FIGURE 6. rVCP treatment causes reduction in anticollagen antibodies and proinflammatory cytokine levels in CIA. Concentrations of anti-CII antibody (**A**), IL-6 (**B**), TNF-α (**C**), and IL-12 (**D**) in serum samples, collected from the study groups, were assayed by ELISA as described in **Material and Methods**. Data are means ± SEM (n = 5/group). $^{\#}P$ <.005 rVCP versus PBS; $^{*}P$ <.009 rVCP versus PBS (Student's t test).

DISCUSSION

Rheumatoid arthritis is a systemic chronic autoimmune disease that lacks any unique clinical or laboratory feature to distinguish it from other joint diseases. Recently, it was reported that patients with RA could be stratified based on their gene expression profile in synovial tissue.[17] Activation of the complement cascade leads to the release of two very potent anaphylotoxins, C3a and C5a, and ultimately results in the formation of MAC. The role of complement activation in osteoclastogenesis has not been studied extensively, and it is not clear that the increased osteoclastogenesis in CIA is related to or independent of complement activation. As previously noted, osteoclasts are important cells of the effector phase in the pathogenesis of RA. If indeed the complement system is involved in osteoclastogenesis, then suppression of the complement system will not only reduce inflammation but also block bone destruction in RA.

Studies indicate that the majority of the complement present in RA as well as CIA joints is synthesized locally by synovial cells, infiltrating macrophages, and neutrophils. It appears that most of the damage caused by the complement system is done by locally synthesized complement. Therefore, it was very important to see if rVCP can be delivered to the synovial tissue of arthritic joints. The immunohistochemical analysis of joint sections revealed that rVCP, when injected i.p., can reach the synovial tissue and appears to attach to the synovial cells (FIG. 1). We believe that because of the strong heparin-binding activity, rVCP is retained at the site of injection and is released to the blood via the lymphatic system, which is the reason for its long retention time. We previously reported that rVCPs bind to the endothelium because of its heparin-binding ability.[18] This appears to be the case here. From a therapeutic viewpoint, this is advantageous, because rVCP can be retained for a longer period and hence is more effective.

Data from the current study clearly show that rVCP not only inhibits inflammation in CIA but also significantly inhibits bone and cartilage destruction. In the current study, rVCP treatment inhibits the clinical progression of CIA in all three treatment regimens examined, that is, early intervention, delayed intervention, and established disease. This indicates that the complement system is active, and it may play an important role throughout the disease process of CIA. Interestingly, no difference was observed in rVCP~2,3,4- (rVCP fragment that lacks complement-inhibiting activity but fully retains its ability to bind heparin) treated arthritic mice as compared to untreated arthritic mice. However, there was a slight delay in onset of disease in rVCP~2,3,4-treated mice (data not shown). This indicates that although the heparin-binding activity of rVCP alone is not significant in inhibiting inflammation and joint destruction, most likely this activity gives rVCP more ways to interfere with different aspects of the inflammatory process. One of the major effects of heparin-binding activity is prolonged retention time for rVCP *in vivo* as indicated by the previous pharmacokinetic studies.[18]

Histological analysis of knee and ankle joints indicates that rVCP treatment, starting at the onset of disease, reduces joint inflammation, cartilage damage, and loss of matrix proteoglycan and bone erosions. It had been shown earlier by other investigators that both pathways of complement activation are required for a normal antibody response to immunization with BCII in CIA.[19] Our data indicate that complement inhibition by VCP in mice reduces the production of anticollagen antibodies. The decrease in anticollagen antibody levels seen with rVCP treatment may be the result of a decrease in the activation of cells expressing C5a receptors or a decrease in the release of collagen antigens with a diminished B-cell response.

$TRAP^+$ osteoclast-like cells were observed in the erosive front in the knee joint of PBS-treated arthritic control mice showing severe joint destruction, demonstrating the active participation of osteoclast-like cells in joint destruction of CIA. A significant reduction in $TRAP^+$ osteoclast-like cells was observed in rVCP-treated versus untreated arthritic mice. This is an important observation because osteoclasts have been shown to be directly involved in bone resorption. Hematopoietic monocytes/macrophages are precursors of osteoclasts. Because rVCP reduces the number of these precursor cells in the joint space (data not shown), there is a smaller number of osteoclasts. Because of the reduction in C3a and C5a, the positive gradient of anaphylotoxins is disrupted and hence causes lower monocyte migrations. TNF-α, IL-6, and IL-12 appear to be the cytokines playing a central key role in immunopathogenesis of RA. The major source of TNF-α is the cells of the monocyte/macrophage lineage, with T cells, neutrophils, mast cells, and the endothelium also contributing to a lesser extent.[20] IL-6 plays an important role in the progression of arthritis[21] and is directly involved in bone resorption in an *in vivo* model of osteoporosis.[22] It has also been shown to be a powerful stimulator of osteoclast bone resorption, stimulating osteoclast development. Anti-IL-6 is inhibitory to osteoclast-like cell formation.[23] IL-12 is released by antigen-presenting cells such as monocytes/macrophages in response to bacterial products and immune signals. IL-12 plays an important role in the development of CIA. Like TNF-α, these cytokines are produced by infiltrating inflammatory cells. Since fewer monocytes/macrophages and neutrophils occurred in the rVCP-treated arthritic mice, we hypothesized that there should be a corresponding reduction in TNF-α, IL-6, and IL-12 levels in these mice. Indeed, ELISA results indicated this to be the case; we observed a marked reduction in levels of all three cytokines in rVCP-treated animals as compared to PBS-treated animals. It can be assumed that the reduction of TNF-α, IL-6, and IL-12 in sera may partly be due to a reduction of these inflammatory cells. In addition to the drop in infiltrating inflammatory cells, reduction in all of these proinflammatory cytokines in the rVCP-treated animals may explain the decrease in osteoclast-like cells and finally the diminished cartilage destruction. This indicates that the mechanism whereby the numbers of osteoclast-like cells are decreased by rVCP treatment is mediated by suppressing these important osteoclast activators. This is consistent with the reports that $C5aR^{-/-}$ mice express osteoprotegerin ligand (OPGL) at the baseline level, indicating the important role of C5a in osteoclastogenesis.[24] This study demonstrates that complement activation leads to increased osteoclastogenesis. Further studies are required to see if this is due to a direct or indirect effect of complement activation. By blocking complement activation, rVCP inhibits the clinical progression of CIA as well as the cellular parameters of the disease. The therapeutic effect of rVCP was seen not only in clinical signs, but also in histopathology and joint integrity (studied by X-ray analysis) in CIA. rVCP treatment resulted in reduction of proinflammatory cell infiltration, the cytokines responsible for inflammation and joint destruction in CIA.

CONCLUSION

The strong therapeutic efficacy of rVCP in the rodent model suggests that blocking complement activation can be considered a potential therapy for human rheumatoid arthritis.

ACKNOWLEDGMENTS

We would like to thank Sharon Lear of the Special Procedure Laboratory of the University of Louisville Department of Pathology for her assistance with tissue sectioning and immunohistochemistry. We would also like to thank Allan G. Farman of the University of Louisville School of Dentistry for helping with radiologic analysis.

This work was supported by funding from the Jewish Hospital Research Foundation, Louisville, Kentucky awarded to G.J.K. who is currently a senior Wellcome Trust International Fellow for biomedical science in South Africa.

REFERENCES

1. LEE, D.M. & M.E. WEINBLATT. 2001. Rheumatoid arthritis. Lancet **358:** 903–911.
2. JIMI, E., I. NAKAMURA, H. AMANO, *et al.* 1996. Osteoclast function is activated by osteoblastic cells through a mechanism involving cell-to-cell contact. Endocrinology **137:** 2187–2190.
3. DOHERTY, M., N. RICHARDS, J. HORNBY, *et al.* 1988. Relation between synovial fluid C3 degradation products and local joint inflammation in rheumatoid arthritis, osteoarthritis, and crystal associated arthropathy. Ann. Rheum. Dis. **47:** 190–197.
4. NYDEGGER, U.E., R.H. ZUBLER, R. GABAY, *et al.* 1977. Circulating complement breakdown products in patients with rheumatoid arthritis: correlation between plasma C3d, circulating immune complexes, and clinical activity. J. Clin. Invest. **59:** 862–868.
5. MALLYA, R.K., D. VERGANI, D.E. TEE, *et al.* 1982. Correlation in rheumatoid arthritis of concentrations of plasma C3d, serum rheumatoid factor, immune complexes and C-reactive protein with each other and with clinical features of disease activity. Clin. Exp. Immunol. **48:** 747–753.
6. JOSE, P.J., I.K. MOSS, R.N. MAINI, *et al.* 1990. Measurement of the chemotactic complement fragment C5a in rheumatoid synovial fluids by radioimmunoassay: role of C5a in the acute inflammatory phase. Ann. Rheum. Dis. **49:** 747–752.
7. NEUMANN, E., S.R. BARNUM, I.H. TARNER, *et al.* 2002. Local production of complement proteins in rheumatoid arthritis synovium. Arthritis Rheum. **46:** 934–945.
8. WANG, Y., S.A. ROLLINS, J.A. MADRI, *et al.* 1995. Anti-C5 monoclonal antibody therapy prevents collagen-induced arthritis and ameliorates established disease. Proc. Natl. Acad. Sci. USA **92:** 8955–8959.
9. WANG, Y., J. KRISTAN, L. HAO, *et al.* 2000. A role for complement in antibody-mediated inflammation: C5-deficient DBA/1 mice are resistant to collagen-induced arthritis. J. Immunol. **164:** 4340–4347.
10. ANDERSSON, M., T.J. GOLDSCHMIDT, E. MICHAELSSON, *et al.* 1999. T-cell receptor V beta haplotype and complement component C5 play no significant role for the resistance to collagen-induced arthritis in the SWR mouse. Immunology **73:** 191–196.
11. LINTON, S.M. & B.P. MORGAN. 1999. Complement activation and inhibition in experimental models of arthritis. Mol. Immunol. **36:** 905–914.
12. KOTWAL, G.J. & B. MOSS. 1988. Vaccinia virus encodes a secretory polypeptide structurally related to complement control proteins. Nature **335:** 176–178.
13. KOTWAL, G.J., S.N. ISAACS, R. MCKENZIE, *et al.* 1990. Vaccinia virus major secretory protein inhibits the classical complement cascade. Science **250:** 827–830.
14. SMITH, S.A., N.P. MULLIN, J. PARKINSON, *et al.* 2000. Conserved surface-exposed K/R-X-K/R motifs and net positive charge on poxvirus complement control proteins serve as putative heparin binding sites and contribute to inhibition of molecular interactions with human endothelial cells: a novel mechanism for evasion of host defense. J. Virol. **74:** 5659–5666.
15. SMITH, S.A., R. SREENIVASAN, G. KRISHNASAMY, *et al.* Mapping of regions within the vaccinia virus complement control protein involved in dose-dependent binding to key complement components and heparin using surface plasmon resonance. Biochim. Biophys. Acta **1650**: 30–39.
16. JHA, P. & G.J. KOTWAL. 2003. Vaccinia complement control protein: multi-functional protein and a potential wonder drug. J. Biosci. **28:** 265–271.
17. VAN DER POUW KRAAN, T.C., VAN, F.A. GAALEN, *et al.* 2003. Discovery of distinctive gene expression profiles in rheumatoid synovium using cDNA microarray technology: evidence for the existence of multiple pathways of tissue destruction and repair. Genes Immun. **4:** 187–196.

18. ANDERSON, J.B., S.A. SMITH, R. VAN WIJK, *et al.* 2003. Vaccinia virus complement control protein inhibits hyperacute xenorejection in a guinea pig-to-rat heterotopic cervical cardiac xenograft model by blocking both xenoantibody binding and complement pathway activation. Transpl. Immunol. **11:** 129–135.
19. JHA, P., S.A. SMITH, D.E. JUSTUS, *et al.* 2003. Prolonged retention of vaccinia virus complement control protein following IP injection: implications in blocking xenorejection. Transplant. Proc. **35:** 3160–3162.
20. HIETALA, M.A., I.M. JONSSON, A. TARKOWSKI, *et al.* 2002. Complement deficiency ameliorates collagen-induced arthritis in mice. J. Immunol. **169:** 454–459.
21. FELDMANN, M., F.M. BRENNAN & R.N. MAINI. 1996. Rheumatoid arthritis. Cell **85:** 307–310.
22. NOWELL, M.A., P.J. RICHARDS, S. HORIUCHI, *et al.* 2003. Soluble IL-6 receptor governs IL-6 activity in experimental arthritis: blockade of arthritis severity by soluble glycoprotein 130. J. Immunol. **171:** 3202–3209.
23. JILKA, R.L., R.S. WEINSTEIN, T. BELLIDO, *et al.* 1999. Increased bone formation by prevention of osteoblast apoptosis with parathyroid hormone. J. Clin. Invest. **104:** 439–446.
24. SUZUKI, Y., F. NISHIKAKU, M. NAKATUKA, *et al.* 1998. Osteoclast-like cells in murine collagen induced arthritis. J. Rheumatol. **25:** 1154–1160.
25. GRANT, E.P., D. PICARELLA, T. BURWELL, *et al.* 2002. Essential role for the C5a receptor in regulating the effector phase of synovial infiltration and joint destruction in experimental arthritis. J. Exp. Med. **196:** 1461–1471.

The Vaccinia Virus N1L Protein Influences Cytokine Secretion *in Vitro* after Infection

ZHOUNING ZHANG,[a] MELISSA-ROSE ABRAHAMS,[b]
LAWRENCE A. HUNT,[b] JILL SUTTLES,[a] WILLIAM MARSHALL,[c]
DEBOMOY K. LAHIRI,[d] AND GIRISH J. KOTWAL[a–c]

[a]*Department of Microbiology and Immunology, University of Louisville, School of Medicine, Louisville, Kentucky 40202, USA*

[b]*Division of Medical Virology, University of Cape Town Medical School, IIDMM, Observatory 7925, Cape Town, South Africa*

[c]*Department of Medicine, University of Massachusetts School of Medicine, Worchester, Massachusetts 01605, USA*

[d]*Laboratory of Molecular Genetics, Department of Psychiatry, Indiana University School of Medicine, Indianapolis, Indiana 46202, USA*

ABSTRACT: The vaccinia virus N1L ORF encodes a protein that enhances virulence and replication of the virus by an unknown mechanism. It has been studied for its ability to enhance viral replication and dissemination in the brain and more recently has been linked to an immunomodulatory role in which it inhibits the activation of cytokine transcription activators in Toll-like receptor signaling pathways after pathogen recognition. The effect of N1L on the release of cytokines from human primary monocytes was investigated. Secretion of the proinflammatory, antiviral cytokines TNF-α, IL-1β, IFN-α, IFN-β, and the anti-inflammatory cytokine IL-10 was found to be inhibited by the presence of the N1L protein.

KEYWORDS: N1L ORF; cytokines; NF-κB; TNF; IFN-β

INTRODUCTION

The vaccinia virus encodes various proteins that are not essential for their growth in cell culture and are used to escape host immune surveillance and establish a suitable habitat for virus propagation.[1,2] Some of these proteins are secreted, and the term *virokine* has been coined to describe viral-encoded proteins that are secreted from infected cells.[3] Several classes of virokines exist, covering a wide range of functions including inhibition of complement activity, mimicking of host cytokines and cytokine receptors, and mimicking of host chemokines.[2]

Among the identified nonessential viral proteins, some share no significant homology with any other known host proteins, such as the N1L protein, a highly

Address for correspondence: Girish J. Kotwal, Division of Medical Virology, IIDMM, Faculty of Health Sciences, University of Cape Town Medical School, Anzio Road, Observatory 7925, Cape Town, South Africa. Voice: +27-21-406-6676; fax: +27-21-406-6018.
gjkotw01@yahoo.com

Ann. N.Y. Acad. Sci. 1056: 69–86 (2005).
doi: 10.1196/annals.1352.005

conserved gene encoded at the left end of the vaccinia virus genome. The N1L open reading frame (ORF) encodes a 13.8-kDa protein. A strain of the Western Reserve (WR) vaccinia virus, known as vGK5, has an insertionally inactivated N1L ORF and was shown to display attenuation in mouse models when administered intranasally, intradermally, and intracranially.[4–6]

Attenuation was particularly pronounced after the intracranial inoculation of mice with vGK5, where the log LD_{50} increased from a value of 0.72 for the parental WR stain to 4.95 for vGK5, and virulence was reduced by a factor of 4 logs.[7] The presence of the N1L protein seemed to enhance not only the virus replication but also the spread of the virus in brain tissue, and this protein was subsequently concluded to be among the strongest determinants of vaccinia virus virulence.[6] The energy reserves of the brain are small[8] and are likely insufficient to support the rapid replication of vaccinia virus. N1L, therefore, was investigated for its influence on energy levels in the brain during viral replication. It was discovered that the N1L ORF product may influence utilization of the energy molecule adenosine triphosphate (ATP) *in vivo*. In addition, bioinformatics studies conducted on the N1L amino acid sequence allowed for identification of a putative adenine binding site and a protein kinase catalytic domain.[9]

A database search of homologous proteins revealed that N1L has limited similarity with the vaccinia virus A52R gene product at the amino acid level.[10] The A52R protein has been demonstrated to inhibit Toll, interleukin (IL)-1β, and IL-18–mediated signaling of NF-κB.[11,12] Activation of Toll-like receptors (TLRs) in response to various pathogens leads to distinct expression patterns of cytokines and costimulatory molecules important for the initiation of innate and adaptive immune responses.[13]

The effect of the N1L protein on TLR signaling was analyzed by expression of N1L in cells capable of signaling via different TLRs. Human embryonic kidney (HEK) cells were cotransfected with the N1L gene and other components of the TLR pathway. It was found that signaling via TNF-α, lymphotoxins, and TLRs was inhibited by N1L, which also inhibited downstream activation of NF-κB. These results suggested that the N1L protein functioned as an immunoregulator.[10] This model may not represent the physiological condition where N1L is expressed in virus-infected cells because all experiments were conducted in transfected cell lines in which the N1L was cotransfected and expressed in the absence of all other vaccinia virus proteins. Furthermore, the downstream effects of N1L after the NF-κB activation, such as cytokine production, were not investigated in these experiments.

In this study, we investigate the influence of the N1L protein on the release and level of cytokines in vaccinia virus–infected human primary monocytes. Monocytes confronted with infectious microorganisms express various cytokines to alarm the immune system[14] and have been demonstrated by *in vitro* experiments to be the most susceptible cells to vaccinia virus infection.[15]

MATERIAL AND METHODS

Production of Rabbit Polyclonal Antiserum against N1L Protein

A predicted antigenic index of the Western Reserve (WR) vaccinia virus N1L gene sequence (accession no. AF451287) was generated by MacVector (Kodak). A

17–amino acid peptide, WRNDNDQTYYNDNFKKY, located near the N-terminus of the N1L ORF was found to have the highest index score. For the production of antiserum against the N1L protein, synthesis of this peptide and production of N1L antiserum were done by Washington Biotechnology (USA) using standardized immunization protocols. A cysteine residue was added to the peptide N-terminus to facilitate conjugation with the carrier protein keyhole limpet hemocyanin (KLH). A New Zealand White rabbit was injected with 50–200 μg of KLH-coupled peptide emulsified in complete Freund's adjuvant (CFA) or incomplete Freund's adjuvant (IFA) per immunization, and serum was collected every 2 weeks.

Overexpression of Recombinant N1L in **Pichia pastoris**

The vaccinia virus N1L protein was expressed from the yeast *Pichia pastoris* using the *Pichia* Expression Kit (Invitrogen, Carlsbad, CA). The N1L ORF was polymerase chain reaction (PCR)-amplified from WR vaccinia virus template DNA (obtained from Dr. B. Moss, NIH) using the P_{N1L} forward primer 5′ TCT C TC GA G AAA AGA GAG GCT GAA GCT ATG AGG ACT CTA CTT ATT AGA 3' (containing a *Xho*I site) and P_{N1L} reverse primer 5′ GC G GCC G CT TTT TCA CCA TAT AGA TCA ATC 3′ (containing a *Not*I site). The *Xho*I and *Not*I restriction enzyme sites were incorporated to facilitate cloning of the N1L ORF into the *P. pastoris* pPIC9 cloning vector, which contains the corresponding sites. Cloning of inserts between these two sites results in disruption of the secretion signal sequence encoded by the vector; thus, the P_{N1L} forward primer was designed to include a secretion signal sequence. N1L PCR product was incorporated into the pPCR2.1 holding vector using a TA cloning kit (Invitrogen, Carlsbad, CA) and transformed into One Shot TOP10 chemically competent *Escherichia coli* cells (Invitrogen, Carlsbad, CA). Positive *E. coli* clones were identified after growth on ampicillin-containing medium and plasmid DNA was isolated and purified using a QIAprep Miniprep plasmid purification kit (Qiagen). The presence of the N1L ORF was confirmed by PCR amplification. Recombinant pPCR2.1 DNA was digested with *Xho*I and *Not*I restriction enzymes and the fragment containing the N1L ORF was purified. The pPIC9 yeast cloning vector was likewise digested with *Xho*I and *Not*I and T4 ligase (Invitrogen, Carlsbad, CA) was used to ligate the digested vector and N1L DNA. Recombinant pPIC9 plasmid DNA was isolated and purified from positive *E. coli* clones selected for on ampicillin-containing medium after transformation into TOP10 *E. coli* cells (Invitrogen, Carlsbad, CA). Successful incorporation of the N1L ORF into pPIC9 was confirmed by restriction enzyme digestion of purified recombinant vector DNA with *Xho*I and *Not*I and PCR amplification using the P_{N1L} primers and sequencing primers specific for the *P. pastoris* alcohol oxidase (AOX) gene primer sites flanking the N1L insert DNA.

Purified recombinant pPIC9 vector DNA (3 μg) was linearized using various restriction enzymes and transformed into chemically competent *P. pastoris* GS115 cells. Yeast clones transformed with the vector were identified on the basis of their ability to grow on histidine-deficient medium. *P. pastoris* genomic DNA was isolated from colonies produced on the selective media. The in-frame integration of the N1L ORF as well as the presence of an intact signal peptide was confirmed by PCR and sequencing using the P_{N1L} primers and AOX sequencing primers.

For large-scale expression of the recombinant N1L protein, a single recombinant *P. pastoris* colony was inoculated into 5–10 ml of buffered glycerol complex medium (BMGY; 1% yeast extract, 2% peptone, 100 mM potassium phosphate, pH 6.0, 1.34% yeast nitrogen base, 4×10^{-5}% biotin, 1% glycerol) and incubated at 30°C in a shaker incubator (200–300 rpm) until the exponential phase of growth was reached (OD_{600} = 2.0–6.0). The culture then was used to inoculate 1 L of BMGY medium which was likewise incubated until the exponential phase of growth was reached. Cells were harvested by centrifugation at 1,500–4,500 rpm for 5–10 min at room temperature, and cell pellets were resuspended in buffered methanol-complex medium (BMMY; 1% yeast extract, 2% peptone, 100 mM potassium phosphate, pH 6.0, 1.34% yeast nitrogen base, 4×10^{-5}% biotin, 0.5% methanol) medium to a final OD_{600} of 1.0 and incubated at 30°C with shaking. Expression was induced by the addition of methanol to the culture medium every 24 h until optimal time of induction was reached. Methanol concentrations of 0.5%, 1%, 2%, 3%, 4%, and 5% were tested to determine at which concentration optimal protein production occurred. Cells were once again pelleted at 3,000 rpm, and the supernatant was collected and cooled to 4°C.

Culture supernatants were concentrated by ammonium sulfate precipitation using 4.01 M saturated NH_2SO_4 at a 40% final concentration. Precipitated protein was further concentrated by ultrafiltration in an Amicon stir-cell centrifugal device with a 40-kDa cutoff membrane followed by ultrafiltration in an Amicon stir-cell with a 10-kDa cutoff membrane and finally size exclusion column chromatography in a Superdex75 gel filtration column (Amersham Biosciences Corp., Piscataway, NJ). Alternatively, yeast expression supernatants were collected by centrifugation at 10,000 rpm for 45 min at 4°C and concentrated by 2 spin cycles in a YM-10 Centriprep ultrafiltration centrifugal device (Millipore) according to the manufacturer's instructions.

Immunodetection of the Recombinant N1L Protein

Proteins contained in recombinant *P. pastoris* yeast expression supernatants were separated by SDS-PAGE on NuPage Novex precast 4–12% Bis-Tris gels and electrophoretically transferred by Western blotting at 300 mA for 2 h to a nitrocellulose membrane (Invitrogen, Carlsbad, CA). The membrane was blocked in 5% blocking solution (0.1% Tween-20, 5% [w/v] nonfat milk powder in phosphate-buffered saline [PBS]) for 1 h at room temperature. After extensive rinsing in PBS-T (0.1% Tween-20 in PBS), the membrane was incubated in N1L rabbit polyclonal antiserum (diluted 1: 1,000 in PBS) for 1 h at room temperature. The membrane was once again rinsed and incubated in horseradish peroxidase–conjugated anti-rabbit IgG (diluted 1:2,000 in 5% blocking solution) for 1 h at room temperature. After rinsing, bound IgG was detected using the enhanced chemiluminescence (ECL) detection system (Amersham).

In addition, proteins in *P. pastoris* expression supernatants were transferred onto an Immobilon-P PVDF (polyvibylidene difluoride) nitrocellulose membrane (Millipore) using the Minifold II Slot-Blot System (Schleicher & Schuell Bioscience Inc., Keene, NH). The membrane was processed as previously described using a 1:10,000 dilution of N1L polyclonal antiserum and 1:100,000 dilution of anti-rabbit IgG antibody.

Culturing of Vaccinia Virus Stocks

The parental wild-type vaccinia virus strain Western Reserve (WR) and the mutant strain vGK5 were propagated in HeLa cells (human epithelial carcinoma cell line, American Type Culture Collection [ATCC] no. CCL2). Viral titers were determined by plaque assay in BS-C-1 cells (African green monkey kidney epithelial cell line, ATCC no. CCL26) and plaque-forming units (PFUs) per inoculum volume or per cell were calculated according to the method described by Lorenz and Bogel.[16]

Isolation of Human Primary Monocytes from Human Peripheral Blood

Human blood samples were obtained from healthy donors in the Department of Medicine's Kidney Disease Program, University of Louisville, and a standard monocyte isolation protocol was followed with modifications.[17,18] Cell suspensions were dispensed to 24-well plates (Costar) at a density of 5×10^6 cells/ml. Cells were allowed to adhere to plate surfaces for 2 h, and nonadherent cells were removed by rinsing wells with prewarmed Dulbecco's PBS plus 2% fetal bovine serum (FBS; Invitrogen, Carlsbad, CA). Fresh RPMI-1640 (Hyclone) supplemented with 5% heat-inactivated FBS, 0.01 M HEPES, and 50 μg/ml gentamicin and designated R5 medium was added and cells were incubated overnight in preparation for experimental use.

Immunostaining of Vaccinia Virus–Infected Human Monocytes

Supernatants of vaccinia virus–infected human monocyte monolayers were removed at 24 h after infection and washed three times with PBS. Cells were fixed with 3% paraformaldehyde for 30 min, washed once again with PBS, and permeabilized with 0.01% Triton X-100 for 5 min. Cells were then incubated with anti-N1L rabbit IgG serum added to a final dilution of 1:1,000 for 30 min at 4°C followed by extensive rinsing with PBS. FITC-conjugated anti-rabbit IgG (Molecular Probes) was added, and the mixture was incubated at 4°C for 1 hour. Cells were washed an additional three times with PBS, and samples were examined under a Carl Zeiss Axiovert 200 fluorescence microscope. Images were captured using Carl Zeiss AxioVision version 4.2 software.

Cytokine Assays

An analysis of proinflammatory cytokine production by vaccinia virus–infected monocytes was performed using a cytometric bead array (CBA) human inflammation kit (BD Biosciences). The production of cytokines IL-8, IL-1β, IL-6, IL-10, TNF-α, and IL-12p70 was analyzed. Human monocyte monolayers prepared in 24- or 96-well plates were infected with vaccinia virus WR or vGK5 at a concentration of 1.0 PFU/cell or mock infected. After 2 h of absorption, inoculum from each well was removed, and cells were rinsed with warm Dulbecco's PBS containing 2% FBS. Thereafter, 1 ml/well of R5 medium was added for 24-well plates or 100 μl/well for 96-well plates. Media from each well were collected at various time points and stored at –80°C. For the CBA assay, serial dilutions of reconstituted human inflammation standards were prepared, and collected cell culture supernatants were diluted 1:50 in assay diluent. Human inflammation capture bead suspensions were prepared by mixing 5 μl of each test bead with 25 μl of mixed beads and vortexing vigorously to ensure thorough mixing. A volume of 25 μl of human inflammation PE detection reagent was added to each bead suspension followed by mixing. A 25-μl volume of standard dilution or test sample was added and mixtures were incubated at room temperature for 3 h away from direct light. After incubation, 1 ml of wash buffer was added to each assay mix followed by centrifugation at 200 *g* for 5 min. Supernatants were carefully aspirated and discarded. A 500-μl volume of Wash buffer was used to resuspend each bead pellet. Resulting suspensions were analyzed using a FACS-Calibur flow cytometer and data obtained were analyzed by BD CBA software.

Interferon Assays

Human interferon (IFN)-α and IFN-β ELISA kits (PBL Biomedical Laboratories, Piscataway, NJ) were used for the measurement of IFN-α and -β production in the supernatants of vaccinia virus–infected monocytes. Results were analyzed using an E-max Precision microplate reader (Molecular Devices Corp., Sunnyvale, CA). Human monocyte monolayers were infected with vaccinia virus WR or vGK5 at a concentration of 1.0 PFU/cell or mock infected. A 1:50 and 1:100 dilution of each supernatant was prepared using a dilution buffer. Thereafter, 100-μl aliquots of each diluted and undiluted sample as well as of serially diluted standards were loaded onto a 96-well plate and incubated at room temperature for 1 h. Wells then were rinsed twice with a final wash solution. A 100-μl volume of antibody solution was added to each well, followed by incubation for 1 h. Wells were rinsed three times and incubated with horseradish peroxidase–conjugated antibody for 1 h Wells were once again rinsed and 100 μl of TMB substrate solution was added to each well followed by incubation for 15 min in the dark. Stop solution (200 μl) was added per well. Absorbance readings were taken at 450 nm using an E-max Precision microplate reader (Molecular Devices) within 5 min of adding the stop solution. The interferon titers were determined by comparison with graphic data obtained for the serially diluted standards.

RESULTS

Overexpression and Immunodetection of the Recombinant N1L Protein

A polyclonal antiserum against the WR vaccinia virus N1L protein was raised up in New Zealand White rabbits. To test its ability to detect the N1L protein, we expressed recombinant N1L in the *P. pastoris* yeast GS115 strain. Optimal expression of recombinant N1L was achieved when inducing with methanol at a final concentration of 2% (FIG. 1). Methanol concentrations exceeding 2% caused sharp decreases in N1L protein production as well as in cell mass (data not shown). The rabbit polyclonal antiserum detected a protein in yeast expression supernatants after methanol induction, which was not detected in preinduction supernatants (FIG. 2A). Western blot analysis showed this protein to be approximately 13 kDa (FIG. 2B), thereby confirming the specificity of the antiserum against the N1L protein.

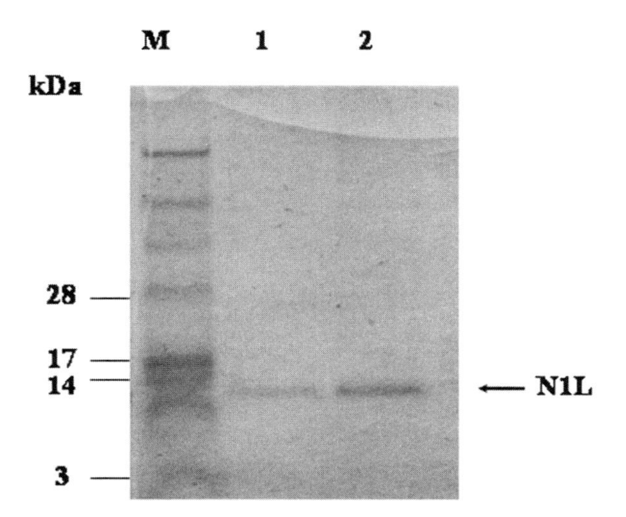

FIGURE 1. Recombinant N1L protein expression in *P. pastoris* GS115 yeast transformants. Yeast colonies transformed with the pPIC9 vector carrying the N1L gene were cultivated in BMGY medium. Protein expression was induced in methanol-containing medium and by the addition of 100% methanol to a 2% final concentration every 24 h. Expression was analyzed by 4–12% SDS-PAGE. (*lane M*) SeeBlue Plus2 Pre-stained protein marker (Invitrogen); (*lane 1*) expression 24 h after induction; (*lane 2*) expression 48 h after induction.

TABLE 1. Vaccinia virus replication in human monocytes[a]

	Virus titers (Log PRU/ml)					
	2 h		24 h		48 h	
Donor	WR	vGK5	WR	vGK5	WR	vGK5
1	3.53	3.48	3.5	3.36	3.58	3.33
2	3.7	3.72	3.72	3.68	3.6	3.57
3	3.0	2.8	2.88	2.82	2.78	2.8

[a]Fresh human monocytes were isolated from different healthy donors and infected with vaccinia virus WR or vGK5 at a multiplicity of 0.1 PFU/cell. Cells were collected at 2, 24, and 48 h after infection, and viral titers were determined by plaque assay in BSC-1 cells.

Vaccinia Virus Infection of Human Peripheral Blood Monocytes

To investigate the involvement of the N1L protein in the production of cytokines in vaccinia virus–infected cells, we infected human peripheral blood monocytes from healthy donors with vaccinia virus WR or the mutant strain vGK5 at a multiplicity of 0.1 PFU/cell. Viral titers determined for 24 and 48 h after infection are summarized in TABLE 1. As represented by the viral titers measured at 2 h after infection, both wild-type (WR) and mutant (vGK5) viruses were able to attach to

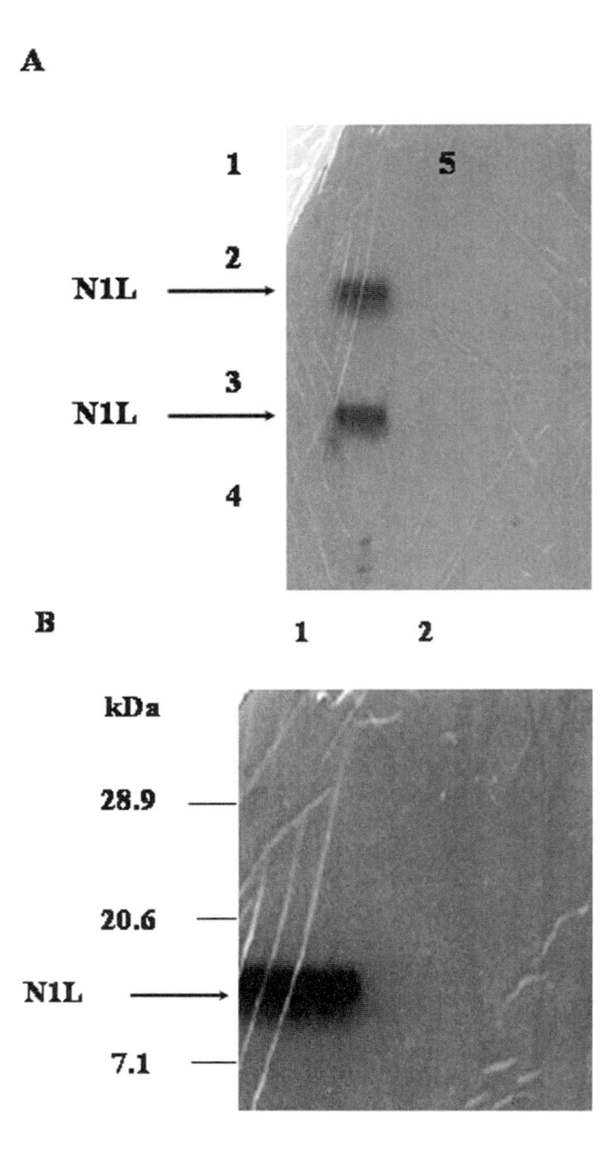

FIGURE 2. Immunodetection of the recombinant N1L protein. Proteins expressedfrom *P. pastoris* GS115 transformants were transferred onto nitrocellulose membranes and probed with rabbit polyclonal N1L antiserum for 1 h at room temperature. Antibody-bound protein was incubated with anti-rabbit IgG peroxidase-linked secondary antibody and detected by chemiluminescence. (**A**) Slot-blot. Bands correspond to: 1, preinduction expression supernatant; 2, 96 h postinduction expression supernatant; 3–4, retentate and filtrate from YM-10 Centriprep-concentrated 96 h after induction supernatant (17× concentrated); 5, recombinant VCP (vaccinia virus complement control protein) expressed in *P. pastoris* (negative control). (**B**) Western blot. (*Lane 1*) 24 h postinduction expression supernatant (10× concentrated); (*lane 2*) preinduction expression supernatant.

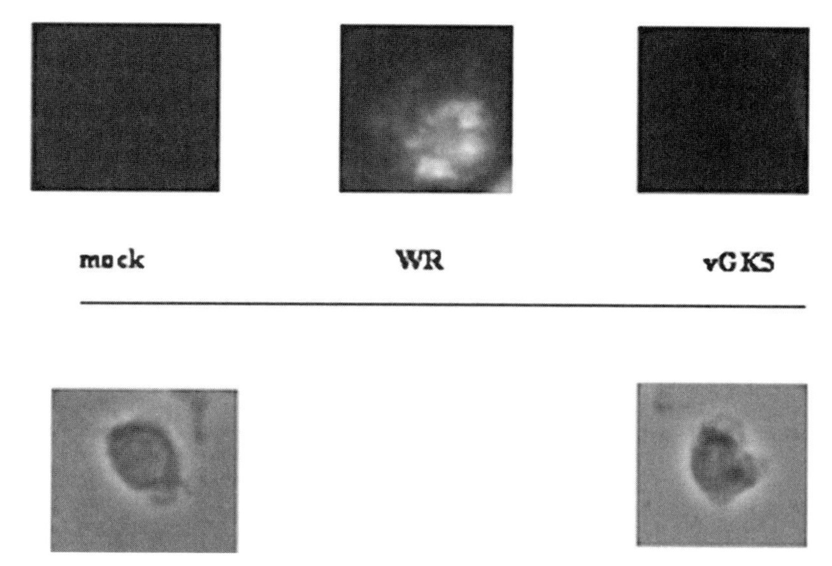

FIGURE 3. Immunostaining of vaccinia virus–infected human monocytes. Monocytes were infected with vaccinia virus WR or vGK5 at a multiplicity of 1.0 PFU/cell. Cells were examined at 24 h after infection under a fluorescence microscope (60× magnification). *Bottom panel* represents phase-contrast images of *upper panel* fluorescence images.

monocytes. No considerable difference in viral titers between the WR and vGK5 strains was observed. Viral titers at 24 and 48 h after infection were comparable to those observed at 2 h after infection. This indicated that no virus progeny was produced after the attachment and possible entry of virus particles into the monocytes. This result confers with that reported for vaccinia virus–infected human peripheral blood leukocytes in which viral replication was abortive, indicating that the virus was able to enter cells but did not complete its productive replication cycle and generate new infectious virus particles.[19]

Expression of the N1L Protein in Vaccinia Virus–Infected Human Monocytes

The N1L gene contains a transcriptional promoter that functions at both early and late times and was expressed throughout the infection cycle in BSC-1 cells, in which virus was able to replicate. It was of interest to test whether the N1L gene would be expressed in monocytes which do not support complete virus replication. Expression of the N1L gene was examined by immunofluorescence studies in vaccinia virus–infected monocytes. Cell monolayers were infected with 1.0 PFU/cell of vaccinia virus WR or vGK5 and expression was assayed at 24 h after infection. N1L protein was detected in the cytoplasm of WR-infected monocytes but not in vGK-infected or mock-infected cells (FIG. 3). This was consistent with earlier reports in which the virus underwent abortive replication in infected dendritic cells, but the early viral-encoded intracellular 6B6 protein was still expressed.[14]

Type I IFN Secretion in Vaccinia Virus–Infected Human Monocytes

Type I IFNs are produced in response to TLR signaling upon induction by virus, bacteria components, and synthetic compounds such as dsRNA.[20] It was of interest to examine how the N1L-deleted vGK5 strain affected the production of type I IFNs, IFN-α and IFN-β, compared with the wild-type WR strain.

Human primary monocytes were infected with vaccinia virus WR or vGK5, and the supernatant was collected at various time points and assayed for IFN secretion. Some cells were found to express relatively high levels of IFN even in the mock-infected condition and did not respond to virus infection (data not shown). When infected at a multiplicity of 0.1 and 0.5 PFU/cell, respectively, monocytes failed to

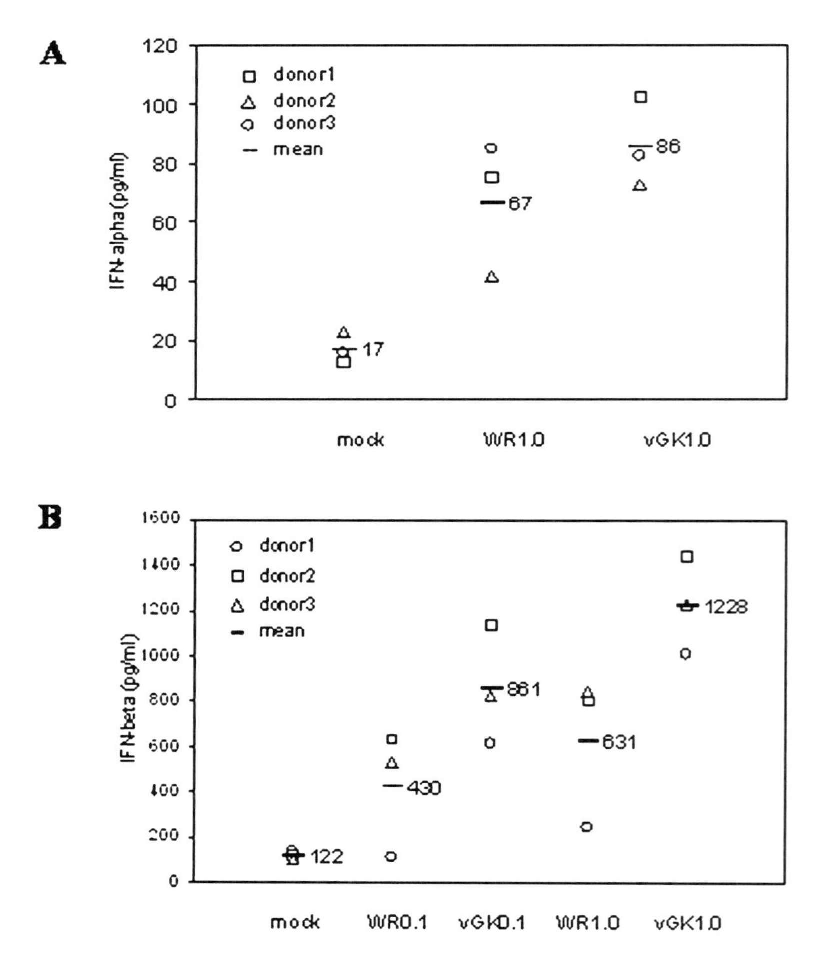

FIGURE 4. IFN-α and IFN-β secretion from vaccinia virus–infected human monocytes. (**A**) IFN-α secretion from monocytes 48 h after infection. (**B**) IFN-β secretion from monocytes at 24 h after infection. IFN concentrations were measured by ELISA.

produce significant amounts of IFN-α in the first 24 h of infection (data not shown). However, when infected at a multiplicity of 1.0 PFU/cell, the IFN-α concentration at 48 h after infection increased. vGK5-infected monocytes released slightly higher levels of IFN-α than did WR-infected monocytes. The difference, however, was not significant and variation was observed from donor to donor (FIG. 4A).

Preliminary data suggested that IFN-β production was detected as early as 8 h after infection and accumulated thereafter (data not shown). FIGURE 4B shows IFN-β secretion from WR and vGK5-infected monocytes. Infection at 0.1 PFU/cell was capable of inducing detectable levels of IFN-β in the medium, and IFN-β secretion was increased when infecting with 1.0 PFU/cell. Although variability was observed between monocytes from different donors, IFN-β production by vGK5-infected monocytes was consistently higher than production by WR-infected monocytes. Therefore, the expression of the N1L gene in the vaccinia virus WR strain appeared to inhibit secretion of IFN-β from monocytes.

IL-10 Secretion from Vaccinia Virus–Infected Human Monocytes

FIGURE 5 illustrates the secretion of IL-10 from vaccinia virus–infected monocytes at 24 h after infection. IL-10 secretion was both dose dependent and time dependent. Low levels of IL-10 were detected at 8 h after infection (data not shown). vGK5-infected monocytes produced much higher levels of IL-10 than those produced for WR-infected cells at both infection doses. The absolute concentration of IL-10 varied from donor to donor. A trend was, however, observed in which IL-10 expression in vaccinia virus–infected human monocytes appeared to be enhanced in the absence of the N1L gene.

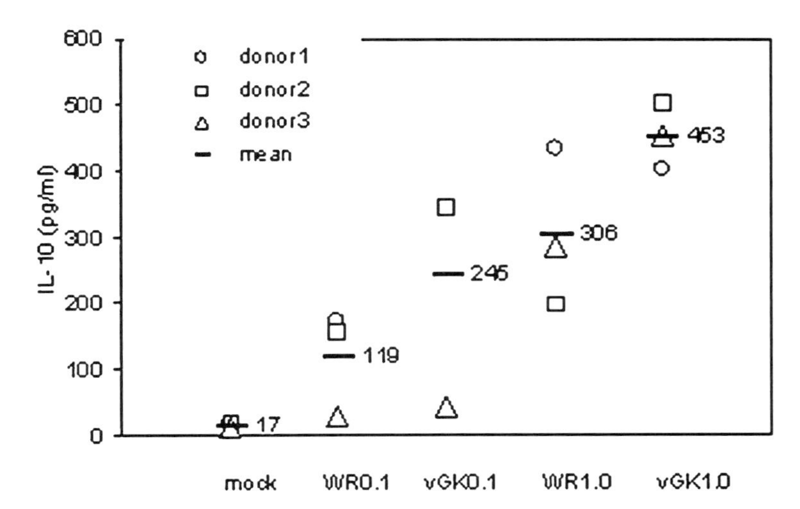

FIGURE 5. Comparison of IL-10 secretion from vaccinia virus WR and vGK5-infected human monocytes. Cells were infected at the same multiplicity of wild-type WR or mutant vGK5 vaccinia virus (0.1 PFU/cell or 1.0 PFU/cell). Growth medium served as a mock-infected control. IL-10 secretion from monocytes was monitored at 24 h after infection.

Proinflammatory Cytokine Secretion from Vaccinia Virus–Infected Human Monocytes

Monocytes and macrophages express a wide selection of cytokines in response to vaccinia virus infection. These include proinflammatory and anti-inflammatory cytokines, some of which have direct antiviral activity. Several proinflammatory cytokines, including IL-6, IL-8, IL-12, IL-1β, and TNF-α, were considered to determine whether their production differed in vaccinia virus WR and vGK5-infected monocytes.

IL-12 expression in vaccinia virus–infected monocytes was of particular interest because it has opposite effects in Th1 and Th2 cell differentiation compared with cytokine IL-10. Infection with the wild-type WR strain of vaccinia virus failed to induce considerable IL-12 secretion from monocytes, with IL-12 levels barely above

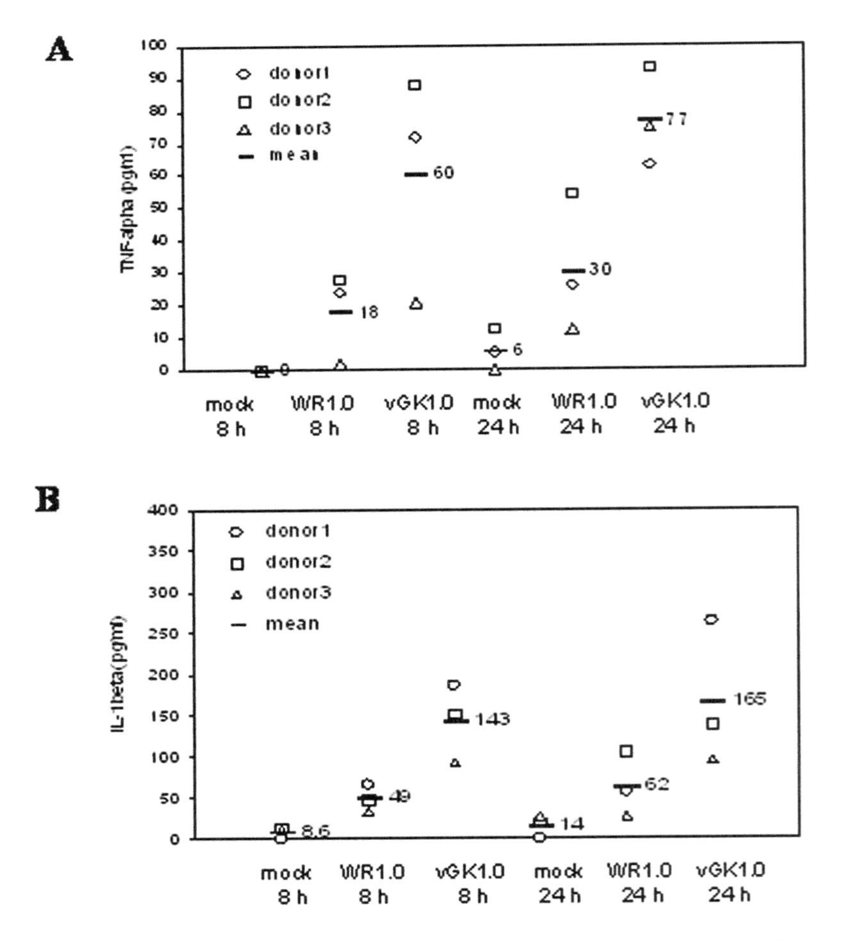

FIGURE 6. TNF-α (**A**) and IL-1β (**B**) secretion from vaccinia virus–infected human monocytes.

the detection limit (data not shown). The same was observed for vGK5-infected cells (data not shown), suggesting that monocytes are not efficiently induced to secrete IL-12 upon infection with vaccinia virus.

TNF-α production was detected in vaccinia virus–infected monocytes at 8 h after infection, and levels continued to increase over the remaining infection period (FIG. 6A). vGK5-infected monocytes produced higher levels of TNF-α than did WR-infected cells at both 8 and 24 h after virus infection. Monocytes from different donors displayed different responses, some consistently expressing higher levels of TNF-α than others. This observation suggested that the N1L gene product may inhibit induction and secretion of TNF-α after vaccinia virus infection.

Human monocytes produce both forms of IL-1, IL-lα, and IL-1β, yet IL-1β contributes the majority of IL-l activity.[21] Therefore, secretion of IL-1β from vaccinia virus–infected monocytes was examined. FIGURE 6B shows the results for IL-1β secretion in the supernatants of WR and vGK5-infected human monocytes. As with other cytokines, it was observed that monocytes from different donors varied in the amount of IL-1β produced in response to vaccinia virus WR infection. At 8 h after infection with WR, monocytes produced detectable amounts of IL-1β. vGK5-infected cells produced much higher levels of IL-1β compared with WR-infected cells at both 8 and 24 h after infection. Thus, the absence of the N1L gene product increased the IL-1β secretion in response to vaccinia virus infection.

Infection with vaccinia virus increased IL-6 expression in monocytes, yet variability in responses between different donors was great, and no consistent pattern in the production of this cytokine was observed after infection with either of the strains used (data not shown).

IL-8 was the most abundant of all the cytokines induced by vaccinia virus infection. Low doses of WR or vGK5 vaccinia virus were able to induce production of IL-8 in concentrations up to 25 ng/ml. The absolute concentration of IL-8 varied from donor to donor. When infected at 0.1 PFU/cell or 1.0 PFU/ml, vGK5-infected monocytes produced higher levels of IL-8 than did WR-infected cells, yet the difference was not significant (data not shown).

DISCUSSION

Recently, the vaccinia virus N1L protein was identified as an immunoregulator that inhibits signaling to transcription activator NF-κB by TNF-α as well as IL-1R and TLR4 signaling cascades.[10] All observations about the function of the N1L protein were based on studies in which the N1L gene was overexpressed in transfected cell lines, a scenario that may not mimic its physiologic condition. In this study, the role of the N1L protein in relation to cytokine production was investigated in vaccinia virus–infected human monocytes using viral strains that were identical except for the presence or absence of the N1L gene.

Absolute cytokine quantities were not comparable among individual donors from which monocytes were obtained. This interindividual variation in cytokine production by purified monocytes and PBMCs has been reported previously.[22] Variability between individuals in response to vaccinia virus infection was also observed. Although the vaccinia virus did not replicate in monocyte cultures (TABLE 1), the N1L protein was successfully expressed (FIG. 1). The inefficient replication of the vaccinia virus in monocytes remains to be explained. The question of how many virus particles actually enter the cells and how many only attach to the cell membrane is still to be answered. An engineered virus that expresses green fluorescent protein (GFP) may be helpful in answering this question. It was suggested that vaccinia virus replicates only in mitogen-stimulated cell populations and that active cell proliferation is required for virus replication.[15] Although isolated monocytes did not support virus replication, monocytes and macrophages did support replication of poxviruses in an *in vivo* model.[23]

Vaccinia virus was cytopathic to some cells, but monocytes seemed resistant to cell death during viral infection. Viability of the isolated monocytes and virus-treated monocytes was examined by Trypan blue exclusion. Within infection multiplicity of 0.01 PFU/cell to 5 PFU/cell, no significant decrease was observed between vaccinia virus–infected cells and mock-infected cells (data not shown). This indicated that vaccinia virus infection had no adverse effect on monocyte cell viability within the dose range used here.

Vaccinia virus infection results in the expression of IFN-α/β, which are best known for their antiviral properties and effects on innate and adaptive immune responses.[24] In variola (smallpox) virus–infected macaques, the earliest consistent transcriptional response was a drastic increase in transcripts of IFN-associated genes, and virus infection quickly became fatal in two animals that had minimal IFN-associated gene induction.[25] It has been shown that IFN-α/β knock-out mice demonstrated enhanced susceptibility to vaccinia virus infection.[26] Vaccinia virus infection induced IFN-β expression in monocytes, even at a low multiplicity of infection (0.1 PFU/ml). vGK5-induced cells expressed consistently higher levels of IFN-β compared with WR-infected cells (FIG. 4B). Thus, N1L seemed to inhibit IFN-β release. In contrast, infection with a low dose of WR or vGK5 failed to induce measurable IFN-α secretion from monocytes, and expression of this cytokine was delayed in comparison to IFN-β expression. Marginal IFN-α release reached its maximum at 48 h after infection and the inhibitory effect of N1L was not as prominent. Cells infected with WR did, however, produce less IFN-α than did cells infected with vGK5. Thus, both IFN-α and IFN-β were induced by vaccinia virus infection, yet they were not expressed with the same kinetics or at the same levels, and the inhibitory effects of N1L on their expression were not the same. During viral infection, transcription of IFN genes is induced after activation of IFN regulatory factors (IRF) 3 and 7.[27] IRF3 is constitutively expressed, whereas IRF7 expression, which is required for induction of IFN-α expression, requires transcription activation.[24] This may contribute to the delayed secretion of IFN-α in monocytes upon viral infection. IRF3 associates with the IFN-β promoter in the cell nucleus and drives the expression of IFN-β.[28] In cotransfected cells, N1L was shown to inhibit IRF3 activation when stimulated by dsRNA.[10] Therefore, lower levels of IFN-β in WR-infected cells may be caused by the N1L protein decreasing IFN-β expression through the IRF-3 pathway.

IL-10 is one of the most important anti-inflammatory cytokines secreted by monocytes when infected with pathogens. Its anti-inflammatory effects include the shutdown of stimulated Th1 cells, the inhibition of inflammatory cell migration, the inhibition of maturation of dendritic cells (DCs), and the suppression of proinflammatory molecule production by recruited macrophages.[29,30] To counteract the function of IL-10, several viruses have evolved IL-10 homologs.[31,32] When infected by vaccinia virus, monocytes were able to express IL-10 at 24 h after infection in a dose-dependent manner. vGK5 infection elicited slightly higher IL-10 expression than WR. The molecular mechanism of IL-10 expression is still not fully understood.[33,34] Although N1L inhibition of IL-10 secretion was not significant, based on the broad effects of IL-10 on immune cells, the decreased IL-10 expression by WR-infected cells may have profound effects on the immune response. IL-10 knockout mice showed marked enhancement of vaccinia virus clearance, suggesting that this cytokine naturally suppresses the host response to vaccinia virus infection.[35]

In contrast, IL-12 has been shown to play a critical role in the clearance of vaccinia virus infection, and IL-12-knockout mice demonstrated greater susceptibility to infection. When infected with vaccinia virus WR, monocytes failed to secrete measurable amounts of IL-12 within 48 h of infection. This was observed even at a multiplicity of 1.0 PFU/cell. It seemed that monocytes were not efficient in IL-12 secretion upon vaccinia virus infection. It was reported that when monocytes were stimulated by parapoxvirus ovis (PPVO), a member of a different genus other than vaccinia and variola viruses, IL-12 secretion was not detectable by ELISA.[36] Thus, when stimulated by vaccinia virus infection, IL-10 seemed to be the dominant cytokine produced by monocytes.

A drastic increase in IL-8 and IL-6 concentration was observed in sera of monkeys lethally inoculated with variola virus.[23] Similarly, infection with vaccinia virus led to secretion of IL-6 and IL-8 in monocytes, with IL-8 being the most abundant inflammatory cytokine produced. No significant differences in IL-8 response between vGK5 and WR-infected monocytes were observed, and N1L appeared to have little effect on IL-8 expression. Cells from different donors showed marked variation in IL-6 response to virus infection, and no conclusion regarding the effects of the N1L protein on this cytokine could be reached.

TNFs and TNF receptors participate in many aspects of host antiviral responses, including inhibition of virus replication and selective killing of virus-infected cells.[37,38] Furthermore, TNF-α is a proinflammatory cytokine involved in the recruitment of immune cells to the sites of infection. *In vivo,* a recombinant vaccinia virus overexpressing TNF-α was highly attenuated.[39] Poxviruses encode several TNF receptor homologs, such as crmB, C, D, and E, which prevent TNF-α from binding to its natural receptors, thus preventing TNFs from exerting their function.[40] Endogenous TNF-α and IL-1β levels in mock-infected monocytes were negligible but were quickly released after vaccinia virus stimulation. Detectable amounts of cytokine were secreted 8 h after infection. vGK5-infected monocytes produced higher levels of TNF-α and IL-1β compared with WR-infected cells, suggesting that the presence of N1L gene had an inhibitory effect on the release of both these cytokines. It was reported that monocytes/macrophages had enhanced the gene transcription and subsequent release of TNF-α, IL-1β and IL-6, and type I IFNs in response to influenza A virus infection. There was an accumulation of cytokine mRNA due to prolonged mRNA stability and an augmented gene transcription. Activation of transcription factors such as NF-κB was involved in activation of cytokine mRNA transcription.[40]

Studies using the lipopolysaccharide (LPS) inhibitor bactericidal/permeability-increasing protein BPI, a product of neutrophil primary granules that binds to LPS, rendering the BPI–LPS complex unable to activate monocytes,[41] were conducted to exclude the possibility of LPS contamination in the vaccinia virus stocks or monocyte cultures. These confirmed that cytokine secretion was caused by the infection of virus and not the contamination with LPS (data not shown).

In summary, vaccinia virus infection stimulated secretion of proinflammatory cytokines from human primary monocytes, including IL-1, IL-6, IL-8, and TNF-α. The N1L protein inhibited the release of TNF-α and IL-1β but had no significant effects on IL-6 and IL-8 release. The N1L protein had an inhibitory effect on the release of IL-10, IFN-α, and IFN-β.

Study of the N1L protein remains limited and many questions still need to be answered. Considering the strong regulatory effects of N1L on virus virulence and the inhibitory effects on NF-κB signaling pathway, it is of great interest to examine its effects on the host immune response *in vivo*. The specific function and mechanism of this protein will provide valuable information regarding the interaction between vaccinia virus and the host immune system and will justify the possible use of the attenuated recombinant virus vGK5 as a safer smallpox vaccine candidate because of its reportedly maintained immunogenicity.[4] In addition, virally expressed immunomodulators have shown exquisite specificity and potency and may be good candidates for the development of new drugs that could target diseases characterized by excessive inflammation and hyperactive immune reactions.[42,43]

REFERENCES

1. Kotwal, G.J. 1997. Microorganisms and their interaction with the immune system. J. Leukoc. Biol. **62:** 415–429.
2. Smith, S.A. & G.J. Kotwal. 2001. Virokines: novel immunomodulatory agents. Exp. Opin. Biol. Ther. **1:** 343–357.
3. Kotwal, G.J. 1999. Virokines: mediators of virus-host interaction and future immunomodulators in medicine. Arch. Immunol. Ther. Exp. (Warsz.) **47:** 135–138.
4. Kotwal, G.J., A.W. Hugin & B. Moss. 1989. Mapping and insertional mutagenesis of a vaccinia virus gene encoding a 13,800 Da secreted protein. Virology **171:** 579–587.
5. Bartlett, N., J.A. Symons, D.C. Tscharke & G.L. Smith. 2002. The vaccinia virus N1L protein is an intracellular homodimer that promotes virulence. J. Gen. Virol. **83:** 1965–1976.
6. Billings, B., S.A. Smith, Z. Zhang, *et al.* 2004. Lack of N1L gene expression results in a significant decrease of vaccinia virus replication in mouse brain. Ann. N.Y. Acad. Sci. **1030:** 297–302.
7. Billings, B. 2001. MSc Thesis. *In vivo* characterization of a neurovirulence factor encoded by vaccinia virus. Department of Microbiology and Immunology, University of Louisville, Louisville, KY.
8. Ames, A., III. 2000. CNS energy metabolism is related to function. Brain Res. Rev. **34:** 42–68.
9. Abrahams, M.H., Z. Zhang, S. Chien, *et al.* 2005. The vaccinia virus N1L ORF may encode a multifunctional protein possibly targeting different kinases, one of which influences ATP levels *in vivo*. Ann. N.Y. Acad. Sci. In press.
10. DiPerna, G., J. Stack, A.G. Bowie, *et al.* 2004. Poxvirus protein N1L targets the I-B Kinase complex, inhibits signalling to NF-κB by the tumor necrosis factor superfamily of receptors, and inhibits NF-κB and IRF3 signalling by Toll-like receptors J. Biol. Chem. **279:** 36570–36578.
11. Bowie, A., E. Kiss-Toth, J.A. Symons, *et al.* 2000. A46R and A52R from vaccinia virus are antagonists of host IL-1 and toll-like receptor signaling. Proc. Natl. Acad. Sci. USA **97:** 10162–10167.
12. Harte, M.T, I.R. Haga, G. Maloney, *et al.* 2003. The poxvirus protein A52R targets Toll-like receptor signaling complexes to suppress host defense. J. Exp. Med. **197:** 343–351.

13. Sato, A. & A. Iwasaki. 2004. Induction of antiviral immunity requires Toll-like receptor signaling in both stromal and dendritic cell compartments. Proc. Natl. Acad. Sci. USA **101:** 16274–16279.
14. Engelmayer, J., M. Larsson, M. Subklewe, *et al.* 1999. Vaccinia virus inhibits the maturation of human dendritic cells: a novel mechanism of immune evasion. J. Immunol. **163:** 6762–6768.
15. Miller, G. & J.F. Enders. 1968. Vaccinia virus replication and cytopathic effect in cultures in phytohemagglutinin-treated human peripheral blood leukocytes. J. Virol. **2:** 787–792.
16. Lorenz, R.J. & K. Bogel. 1973. Laboratory techniques in rabies: methods of calculation. Monogr. Ser. World Health Organ. **23:** 321–335.
17. Welsh, C.T., J.T. Summersgill & R.D. Miller. 2004. Increases in c-Jun N-terminal kinase/stress-activated protein kinase and p38 activity in monocyte-derived macrophages following the uptake of *Legionella pneumophila.* Infect. Immun. **72:** 1512–1518.
18. Mukundan, L., G.A. Bishop, K.Z. Head, *et al.* 2005. TNF receptor-associated factor 6 is an essential mediator of CD40-activated proinflammatory pathways in monocytes and macrophages. J. Immunol. **174:** 1081–1090.
19. Sanchez-Puig, J.M., L. Sanchez, G. Roy & R. Blasco. 2004. Susceptibility of different leukocyte cell types to vaccinia virus infection. Virol. J. **1:** 10–16.
20. Hertzog, P.J., L.A. O'Neill, J.A. Hamilton. 2003. The interferon in TLR signaling: more than just antiviral? Trends Immunol. **24:** 534–539.
21. Kern, J.A., R.J. Lamb, J.C. Reed, *et al.* 1988. Dexamethasone inhibition of interleukin 1 beta production by human monocytes. Post-transcriptional mechanisms. J. Clin. Invest. **81:** 237–244.
22. Yaqoob, P., E.A. Newsholme & P.C. Calder. 1999. Comparison of cytokine production in cultures of whole human blood and purified mononuclear cells. Cytokine **11:** 600–605.
23. Jahrling, P.B., L.E. Hensley, M.J. Martinez, *et al.* 2004. Exploring the potential of variola virus infection of cynomolgus macaques as a model for human smallpox. Proc. Natl. Acad. of Sci. USA **101:** 15196–15200.
24. Katze, M.G., Y. He & M Gale Jr. 2002. Viruses and interferon: a fight for supremacy. Nat. Rev. Immunol. **2:** 675–687.
25. Rubins, K.H., L.E. Hensley, P.B. Jahrling, *et al.* 2004. The host response to smallpox: analysis of the gene expression program in peripheral blood cells in a nonhuman primate model. Proc. Natl. Acad. Sci. USA **101:** 15190–15195.
26. Deonarain, R., A. Alcami, M. Alexiou, *et al.* 2000. Impaired antiviral response and alpha/beta interferon induction in mice lacking beta interferon. J. Virol. **74:** 3404–3409.
27. Taniguchi, T. & A. Takaoka. 2002. The interferon-/ system in antiviral responses: a multimodal machinery of gene regulation by the IRF family of transcription factors. Curr. Opin. Immunol. **14:** 111–116.
28. Juang, Y.T., W. Lowther, M. Kellum, *et al.* 1998. Primary activation of interferon A and interferon B gene transcription by interferon regulatory factor 3. Proc. Natl. Acad. Sci. USA **95:** 9837–9842.
29. Fiorentino, D.F., A. Zlotnik, T.R. Mosmann, *et al.* 1991. IL-10 inhibits cytokine production by activated macrophages. J. Immunol. **147:** 3815–3822.
30. Ralph, P., I. Nakoinz, A. Sampson-Johannes, *et al.* 1992. IL-10, T lymphocyte inhibitor of human blood cell production of IL-1 and tumor necrosis factor. J. Immunol. **148:** 808–814.
31. Rode, H-J., W. Janssen, A. Rosen-Wolff, *et al.* 1993. The genome of equine herpesvirus type 2 harbors an interleukin-10 (IL-10)-like gene. Virus Genes **7:** 111–116.
32. Kotenko, S.V., S. Saccani, L.S. Izotova, *et al.* 2000. Human cytomegalovirus harbors its own unique IL-10 homolog (cmvIL-10). Proc. Natl. Acad. Sci. USA **97:** 1695–1700.
33. Slezak, K., K. Guzik & H. Rokita. 2000. Regulation of interleukin 12 and interleukin 10 expression in vaccinia virus-infected human monocytes and U-937 cell line. Cytokine **12:** 900–908.

34. MOORE, K.W., R. DE WAAL MALEFYT, R.L. COFFMAN & A. O'GARRA. 2001. Interleukin-10 and the interleukin-10 receptor. Ann. Rev. Immunol. **19:** 683–765.
35. ZEH, H.J. & D.L. BARTLETT. 2002. Development of a replication-selective, oncolytic poxvirus for the treatment of human cancers. Cancer Gene Ther. **9:** 1001–1012.
36. FRIEBE, A., A. SIEGLING, S. FRIEDERICHS, *et al.* 2004. Immunomodulatory effects of inactivated parapoxvirus ovis (ORF virus) on human peripheral immune cells: induction of cytokine secretion in monocytes and Th1-like cells. J. Virol. **78:** 9400–9411.
37. MESTAN, J., W. DIGEL, S. MITTNACHT, *et al.* 1986. Antiviral effects of recombinant tumour necrosis factor *in vitro*. Nature **323:** 816–819.
38. WONG, G.H., J.F. KROWKA, D.P. STITES & D.V. GOEDDEL. 1988. *In vitro* anti-human immunodeficiency virus activities of tumor necrosis factor-alpha and interferon-gamma. J. Immunol. **140:** 120–124.
39. SAMBHI, S.K., M.R. KOHONEN-CORISH & I.A. RAMSHAW. 1991. Local production of tumor necrosis factor encoded by recombinant vaccinia virus is effective in controlling viral replication in vivo. Proc. Natl. Acad. Sci. USA **88:** 4025–4029.
40. READING, P.C., A. KHANNA & G.L. SMITH. 2002. Vaccinia virus CrmE encodes a soluble and cell surface tumor necrosis factor receptor that contributes to virus virulence. Virology **292:** 285–298.
41. HUBACEK, J.A., J. PITHA, Z. SKODOVÁ, V. *et al.* 2002. Polymorphisms in the lipopolysaccharide-binding protein and bactericidal/permeability-increasing protein in patients with myocardial infarction. Clin. Chem. Lab. Med. **40:** 1097–1100.
42. MOSS, B. 1996. Genetically engineered poxviruses for recombinant gene expression, vaccination, and safety. Proc. Natl. Acad. Sci. USA **93:** 11341–11348.
43. LUCAS, A. & G. MCFADDEN. 2004. Secreted immunomodulatory viral proteins as novel biotherapeutics. J. Immunol. **173:** 4765–4774.

The Vaccinia Virus N1L ORF May Encode a Multifunctional Protein Possibly Targeting Different Kinases, One of Which Influences ATP Levels *in Vivo*

MELISSA-ROSE ABRAHAMS,[a] ZHOUNING ZHANG,[b] SUFAN CHIEN,[b]
TIM SKERNS,[a,c] AND GIRISH J. KOTWAL[a,b]

[a]*Division of Medical Virology, University of Cape Town Medical School, IIDMM, Observatory 7925, Cape Town, South Africa*

[b]*Department of Microbiology and Immunology and Surgery, University of Louisville School of Medicine, Louisville, Kentucky 40202, USA*

[c]*Max F. Perutz Laboratories, Department of Medical Biochemistry, Medical University of Vienna, Vienna, Austria*

ABSTRACT: As the single-most potent virulence factor of the vaccinia virus, the 13.8-kDa protein enhances viral replication in the brain by an unknown mechanism. Due to the high energy demands of the brain and the at times inadequate energy supply and small energy reserves to support physiologic activity, the ability of this organ to support energy requirements for replication of a virus is unlikely. We investigated the possible role of the 13.8-kDa protein in the enhancement of adenosine triphosphate (ATP) utilization in the brain to sustain viral replication. *In vitro* and *in vivo* monitoring and comparison of ATP levels in mouse brain tissue infected with a wild-type vaccinia virus or a 13.8-kDa deletion strain (vGK5) revealed differences in ATP utilization and a significant difference in ATP levels *in vivo* after 5 days of infection. Because of poor replication of the wild-type Lister vaccinia virus in the brain, a role for the 13.8-kDa protein in the modulation of ATP levels to support viral replication in the brain could not be conclusively implicated. Evaluation of the amino acid sequence and predicted secondary structure of the 13.8-kDa protein and sequence and structural homologs thereof provided evidence of putative dimerization and adenine binding sites and a possible kinase-related function for this protein.

KEYWORDS: 13.8-kDa protein; ATP; vaccinia virus; neurovirulence factor; pathogenesis

Address for correspondence: Girish J. Kotwal, Division of Medical Virology, IIDMM, Faculty of Health Sciences, University of Cape Town Medical School, Anzio Road, Observatory-7925, Cape Town, South Africa. Voice: +27-(21)-406-6676; fax: +27-(21)-406-6018.
gjkotw01@yahoo.com

**Ann. N.Y. Acad. Sci. 1056: 87–99 (2005). © 2005 New York Academy of Sciences.
doi: 10.1196/annals.1352.006**

INTRODUCTION

The vaccinia virus 13.8-kDa protein is one of numerous major proteins, known as virokines, produced in vaccinia virus-infected cells.[1] This protein is a nonglycosylated, alpha-helical homodimeric[2] product of the N1L open reading frame encoded at the left end of the vaccinia virus genome.[3] It is highly conserved between vaccinia virus strains as well as poxvirus strains and is not essential for growth in tissue culture.[2] This protein has been localized within infected cells[2] and was successfully expressed in and secreted from yeast cells (unpublished data), yet its structure to date has not been determined.

A strain of vaccinia virus with an insertionally inactivated N1L gene, known as vGK5, was found to be attenuated after infection of mice via intradermal and intranasal routes.[2,3] *In vivo* studies using mice intracranially inoculated with the neurovirulent Western Reserve (WR) strain of vaccinia virus or the vGK5 strain revealed that viral replication in the brain in the absence of the 13.8-kDa protein was greatly impaired.[4] Therefore, a role for this protein in facilitating viral dissemination in the brain was established, yet a specific function remains to be elucidated.

Because a large portion of the characterized virokines produced by the vaccinia virus are active in helping the virus evade the host's immune system,[5] a possible immunomodulatory role for the 13.8-kDa protein was investigated and implicated by Diperna *et al.*[6] who found that the N1L gene product inhibited an intracellular signaling pathway in the innate immune response by targeting the I-κB kinase complex.

Comparison of the predicted three-dimensional structure of the 13.8-kDa protein to a database of known protein structures has provided a range of proteins[2,3,6] of limited homology including those belonging to the protein family, which includes the energy-converting enzyme adenlyate kinase.[4] Adenylate kinase catalyzes the interconversion of the energy molecule adenosine triphosphate (ATP) to its hydrolyzed intermediates adenosine diphosphate (ADP) and adenosine monophosphate (AMP).[7]

In this study, we used bioinformatics to further investigate the structural properties of the 13.8-kDa protein to identify possible functional moieties and considered specifically the relation between vaccinia virus replication and the brain. Replication of the vaccinia virus and poxviruses generally occurs largely independently of the host cell[8,9] as the virus genome is large enough to encode enzymes that function in genome transcription and replication.[10] Many of these enzymes have adenosine triphosphatase (ATPase) activities or the reactions that they catalyze require ATP specifically as an energy source and therefore the virus alters the host cell's metabolic network to fulfill its energy requirements.[11,12] As the organ in the body with the highest energy requirement, the brain encounters periods of increased physiologic activity during which the demand for energy is not met by the supply thereof.[13] The low energy reserves of the brain contribute to this problem.[13] It therefore seems unlikely that replication of the vaccinia virus could be supported within this organ. We hypothesized that the vaccinia virus 13.8-kDa protein enhances viral replication in the brain by influencing the utilization of ATP in an environment in which the supply of energy to support viral replication is limiting.

MATERIAL AND METHODS

Culturing of Vaccinia Virus Stocks and Virus Titration

The South African (SA) vaccine Lister strain of vaccinia virus was acquired for the purpose of utilizing a safer, less virulent strain than the previously used WR vaccinia virus. The Lister strain and recombinant vGK5 strain (derived from wild-type WR vaccinia virus) were grown on the chorioallantoic membranes of 9-day-old embryonated hens' eggs according to the method described by Joklik[14] with modifications[15] and purified by centrifugation through a sucrose cushion.

The concentrations of virus stocks were determined by infecting confluent BSC-1 (African Green Monkey kidney epithelial cells, ATCC# CCL26) cell monolayers with serial dilutions of virus suspension followed by incubation of infected cultures in a CO_2 incubator at 37°C (5% CO_2) until plaque formation. Plaques were counted after staining of cells in a 10% crystal violet solution (Merck) (10% formaldehyde, 10% crystal violet, and 80% ddH_2O), and the number of plaque-forming units (PFUs) per milliliter was calculated.

Sequencing of the South African Vaccine Lister Vaccinia Virus N1L Gene

To ensure that the SA vaccine Lister strain of vaccinia virus encoded an identical N1L gene to that of the previously studied WR strain, viral DNA was isolated by SDS lysis buffer (50 mM Tris-aminomethane, 700 mM NaCl, 10 mM EDTA, 1% SDS, pH 9.5) treatment of 100 µl of virus stock followed by phenol:chloroform:isoamyl (25:24:1) extraction and DNA precipitation in 95% ethanol and 3 M NaAc. The N1L open reading frame (ORF) was amplified by polymerase chain reaction (PCR) using the primers 5′ GAA TTC ATG AGG ACT CTA CTT ATT AG 3′ and 5′ GCG GCC GCT TAT TTT TCA CCA TAT AG 3′, which were complementary to the termini of the WR vaccinia virus strain N1L gene (accession no. AF451287). The PCR product was purified using the GFX PCR DNA and Gel Band Purification Kit (Amersham) and cloned using the pGEM-T easy cloning vector system (Promega) and sequenced (Stellenbosch University Sequencing Unit). Sequence results were compared to sequences in the Genbank database using the NCBI nucleotide-nucleotide BLAST (blastn) analysis system (http://www.ncbi.nlm.nih.gov/BLAST/).

N1L Amino Acid Sequence Homology Determination

An NCBI protein-protein BLAST (blastp) (www.ncbi.nlm.nih.gov/BLAST) analysis of the Lister N1L amino acid sequence (accession no. AAL57606) was carried out to identify homologous protein sequences from the protein data bank (pdb) server. Protein sequences displaying the highest percentage identities were analyzed and used to identify possible protein motifs.

Secondary Structure Prediction of the 13.8-kDa Protein

The Lister vaccinia virus N1L amino acid sequence (accession no. AAL57606) was submitted to the PSIPRED (http://bioinf.cs.ucl.ac.uk/psipred/) server for prediction of secondary protein structure. The sequence was likewise submitted to the 3D-

PSSM (www.sbg.bio.ic.ac.uk/~3dpssm/) web server to identify structural homologs to the predicted secondary structure of the 13.8-kDa protein.

In Vitro *ATP Level Analysis*

Mouse neuroblastoma SK-N-SH cell monolayers were transfected with the WR or recombinant vGK5 strain of vaccinia virus at a concentration of no more than 0.1 PFUs/cell to avoid high lethality of cells resulting in leakage of cell contents and degradation of ATP by ATPases. At 2, 6, 24, and 48 hours postinfection, cell monolayers were washed twice in phosphate-buffered saline solution (PBS) and scraped after treatment with 0.5 ml of ice-cold 0.1 M perchloric acid. Cells were centrifuged at 10,000 rpm for 10 minutes at 4°C, and supernatants were neutralized with 1 M KOH and once again centrifuged to sediment precipitates. ATP levels in the resulting supernatants were assayed using high performance liquid chromatography (HPLC).[16,17] Absorbance of samples at 254 nm was monitored, and ATP concentrations were determined according to the standard nucleotide calibration curve. The total protein concentration in neuroblastoma cell pellets was determined using a BCA protein assay kit (Pierce), and ATP concentrations were expressed as μmoles per gram of protein. As a control, ATP levels were monitored in the same manner in BSC-1 cells infected with 0.1 PFUs/ml of WR or vGK5 vaccinia virus.

In Vivo *ATP Level Analysis*

Groups of 3- to 4-week-old male and female BALB/c mice (Animal Unit, University of Cape Town) were intracranially inoculated with the SA vaccine Lister vaccinia virus (wild-type) or the recombinant vGK5 virus (knockout) or mock-infected with 25 μl of physiologic saline solution, respectively. Each mouse was anesthetized by intraperitoneal administration of a ketamine/xylazine cocktail (Animal Unit, University of Cape Town) followed by swabbing of its head with 70% ethanol and intracranial inoculation with virus stock in 25 μl of physiologic saline solution (0.85%) using a 1-cc tuberculin syringe with a 29-gauge needle (Cliniscience).

In an initial study conducted over 48 hours, three mice were mock-infected with saline, six mice were infected with 100 PFUs of the vGK5 strain, and six mice with a larger dose (1,000 PFUs) of the Lister strain to compensate for an anticipated later onset of virus replication as observed in infected BSC-1 cells (data not shown). Three mice from each of the wild-type and knockout-infected groups and two mock-infected mice were sacrificed at 24 hours postinfection and the remaining mice at 48 hours postinfection. In an expanded study, mice were infected with equal concentrations (100 PFUs) of Lister or vGK5 vaccinia virus and three mice from each of the wild-type, knockout, and mock-infected groups were sacrificed at 48, 72, 96, and 120 hours postinfection, respectively.

Brains were harvested immediately after sacrifice and frozen in liquid nitrogen. ATP extraction from mouse brain tissue and determination of ATP concentration using a luciferin/luciferase bioluminescence system was modified by the method described by Khan.[18] Brains were thawed on ice and homogenized in ddH_2O using an Ultra-turax T8 tissue homogenizer (IKA Labortechnik). Homogenates were treated with an equal volume of pre-cooled 20% perchloric acid ($HClO_4$) and centrifuged at 4,500 rpm for 10 minutes at 4°C in a Haraeus Multifuge 3 L-R benchtop centrifuge (Kendro Laboratory Products). A 500-µl aliquot of each supernatant was neutralized with 200 µl of 2.5 M KOH and centrifuged at 4,500 rpm for 5 minutes at 4°C. Supernatants were diluted 1:40 in 0.1 M Tris-EDTA buffer, pH 7.75 (100 mM Tris; 2 mM EDTA). A volume of 100 µl of each extract was loaded onto a sterile MultiScreen Opaque 96-well plate (Millipore), and ATP levels were measured in a Luminoskan Ascent or Veritas microplate luminometer (Turner Biosystems) after the addition of 50 µl of ENLITEN rLuciferin/Luciferase reagent (Promega) per well. ATP concentrations were calculated from a linear double log fitting ATP standard curve[19] formulated from light readings of serial dilutions of adenosine 5′ triphosphate (ATP) disodium salt (Sigma-Aldrich). The total protein content in 50 µl aliquots of each vaccinia virus-infected or mock-infected mouse brain tissue homogenate was estimated using a BioRad protein estimation kit with bovine serum albumin (BSA) standards. ATP concentrations were correlated with protein mass and presented as µmoles per milliliter of extract per milligram of protein.

RESULTS

Sequencing of the South African Vaccine Lister Vaccinia Virus N1L Gene

To ensure its suitability as a wild-type strain expressing an identical 13.8-kDa protein to that of the WR strain used in previous studies relating to this protein, the SA vaccine Lister vaccinia virus N1L gene was sequenced. An NCBI nucleotide-nucleotide BLAST (blastn) analysis of the resulting sequence revealed a 100% sequence homology to the N1L ORF of the WR vaccinia virus genome (data not shown).

N1L Amino Acid Sequence Homology Determination

A 24% sequence identity to the unactivated apo insulin-like growth factor-1 receptor kinase domain (IGFRK-0P)[20] (pdb 1P4OB) was obtained from a protein database search of homologous amino acid sequences to the Lister N1L ORF. This kinase domain contains functionally important loop regions which include a glycine-rich phosphate-binding loop, a catalytic loop, and an activation loop that contains the principal phosphorylation sites for activation of the protein. The kinase is able to bind ATP upon activation. The N1 and N6 nitrogen atoms of adenine bind to the side chains of residue 107 (glutamic acid) and residue 109 (methionine) of IGFRK-0P, respectively, and the β and γ phosphate groups of ATP bind to two residues in the glycine-rich phosphate-binding loop.[21]

An alignment of the two sequences allowed for identification of putative adenine binding sites (main chains of IGFRK-0P residues 107 and 109) and a catalytic domain (IGFRK-0P residues 162-171) within the N1L amino acid sequence (FIG. 1). However, an essential amino acid (histidine), situated at the start of the IGFRK-0P catalytic domain, was lacking in the N1L protein sequence, and no glycine-rich region for binding of phosphate groups was observed.

```
Lister N1L   MLLDELVDDGDVCTLIKNMRMTLSDGPL-----LDRLNQPVNNIEDAKRMIAISAKVARD 82
IGFRK-0P     LVIMELMTRGDLKSYLRSLRPAMANNPVLAPPSLSKMIQMAGEIADGMAYLNANKFVHRD 163
             ... **.  **. . .. .* ... .*.     * .. *  . * *    .     * **

Lister N1L   IGERSEIRWEESFTI 97
IGFRK-0P     LAARN-CMVAEDFTV 177
             .  *      * **.
```

FIGURE 1. Clustal multiple sequence alignment of the Lister vaccinia virus N1L and unactivated apo insulin-like growth factor-1 receptor kinase domain (IGFRK-0P) chain B amino acid sequences. The identity was generated by performing a blast search of the pdb database with the N1L sequence. The identity is 24% and the similarity 77%. Identical amino acids are indicated by an *asterisk* and conserved amino acids by a *single point*. The catalytic domain (residues 162-171) and adenine binding regions (residues 107 and 109) of IGFRK-0P are underlined.

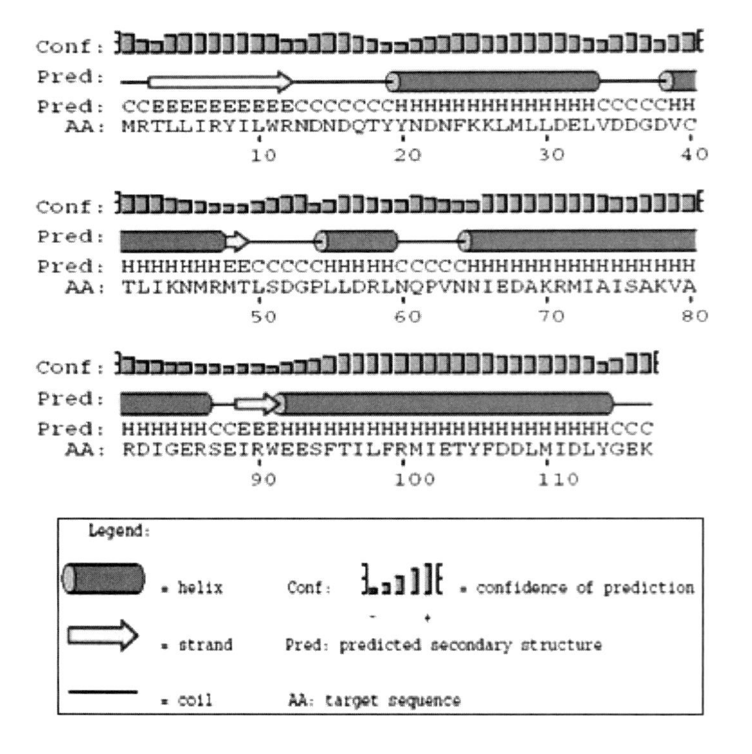

FIGURE 2. Predicted secondary structure of the vaccinia virus 13.8-kDa protein obtained from submission of the N1L amino acid sequence to the PSIPRED web server.

```
              1                                                 50
vaccini_PSS   CCHHHHHHEE ECCCCCCCHH HHHHHHHHHH ..HHHHHCCC CEEEEECCEE
vaccini_Seq   MRTLLIRYIL WRNDNDQTYY NDDFKKLMLL ..DELVDDGD VCTLIKNMRM
-----------                  +-+--- ++--++L+L+   -+++-+++  ++-I+N+++
dlaoa_2_Seq   .......... ....TLEELM KLSPEELLLR WANFHLENSG .WQKINNFSA
dlaoa_2_SS    .......... ....CHHHHH HCCHHHHHHH HHHHHHHHCC .CCCCCCCCC
CORE          .......... ....000000 0000000220 1400030000 .000500300

              51                                               100
vaccini_PSS   EECCC..... .......... ...CCCCCCC CCCCCCCCHH HHHHHHHHHH
vaccini_Seq   TLSDG..... .......... ...PLLDRLN QPVNNIEDAK RMIAISAKVA
-----------   -++D+                    ++-+-+N ++++N++D++ +-+++--++A
dlaoa_2_Seq   DIKDSKAYFH LLNQIAPKGQ KEGEPRIDIN MSGFNETDDL KRAESMLQQA
dlaoa_2_SS    CCCCCHHHHH HHHHHCCCCC CCCCCCCCCC CCCCCCCCHH HHHHHHHHHH
CORE          0600505320 4700141000 0000002050 0000000000 0060065006

              101                                137
vaccini_PSS   HHHCCCCEEE HHHHHHHHHH HHHHHHHHHH HHHCCCC
vaccini_Seq   RDIGERSEIR WEESFTILFR MIETYFDDLM IDLYGEK
-----------   +++G-R+-++ -++-++---+ ++-+++--L+ -
dlaoa_2_Seq   DKLGCRQFVT PADVVSGNPK LNLAFVANLF N......
dlaoa_2_SS    HHHCCCCCCC HHHHHHCCHH HHHHHHHHHH C......
CORE          0050300160 1009101000 1110641030 0......
```

FIGURE 3. Multiple sequence alignment of the amino acid sequences and secondary structures of the vaccinia virus 13.8-kDa protein and the N-terminal actin-crosslinking domain of fimbrin obtained from a sequence homology search conducted by the 3D-PSSM server. An identity of 10% was determined. Alpha helices are represented by the letter H, extended beta strands by the letter E, and coils by the letter C. The core number issued for each residue is a measure of the burial and the number of contacts made by that residue, where a value of 9 indicates very buried with many contacts and 0 is assigned to a residue that is not buried and makes few contacts.

Secondary Structure Prediction of the 13.8-kDa Protein

A secondary structure prediction of the vaccinia virus N1L gene revealed a predominantly alpha-helical structure with a higher confidence of prediction obtained for the protein C-terminus (FIG. 2). The N-terminal actin-crosslinking domain of fimbrin (pdb 1aoa), belonging to the alpha protein family, was found to be of the highest homology (10% identity) to the predicted secondary structure of the 13.8-kDa protein(FIG. 3). This agrees with earlier findings by Bartlett *et al.*[2]

In Vitro *ATP Level Analysis*

ATP levels in BSC-1 cells and mouse neuroblastoma cells infected with the WR (wild-type) or vGK5 (knockout) strain of vaccinia virus were measured over a 48-hour period. Cultures were infected at a maximum of 0.1 PFU/ml and ATP concentrations determined by HPLC. No significant differences in ATP levels between mock and virus-infected BSC-1 cell cultures were observed, and ATP levels remained relatively consistent over the period of infection (FIG. 4). By contrast, ATP levels were seen to fluctuate in mock- and virus-infected neuroblastoma cells. FIGURE 5 shows a rapid drop in the concentration of ATP in neuroblastoma cells after

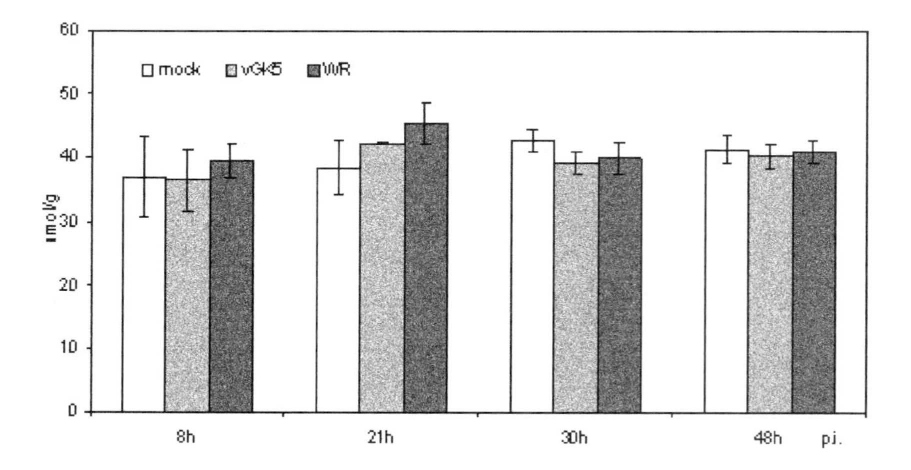

FIGURE 4. ATP levels in WR and vGK5 vaccinia virus-infected BSC-1 cells infected at 0.1 PFUs/cell. The concentration of ATP extracted at various timepoints was measured by HPLC. Data are presented as means ± standard deviation.

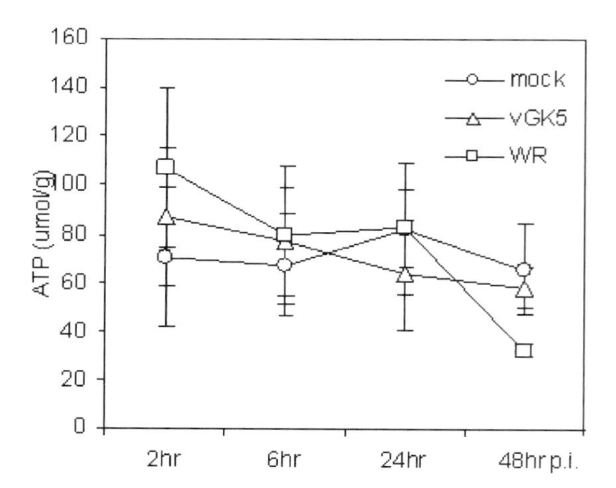

FIGURE 5. ATP level changes in WR and vGK5 vaccinia virus-infected neuroblasto-ıa cells. ATP was harvested at various time periods after infection, and ATP concentrations ere measured by an HPLC method. Concentrations expressed as μmoles per gram of pro-ein are presented as mean ± standard deviation (p.i., postinfection).

2 hours of infection with the wild-type virus, after which the concentration remained constant and finally decreased once again over the final 24 hours. This pattern of ATP utilization was not seen with knockout or mock-infected cells. In subsequent repeat assays, fluctuations in ATP levels during the course of infection as well as between different experiments accompanied by high standard deviations were observed for mock- and virus-infected cell cultures (data not shown). Variability in neuroblastoma cell response to vaccinia virus infection was evident.

In Vivo *ATP Level Analysis*

Three- to 4-week-old BALB/c mice were intracranially infected with the Lister (wild-type) or vGK5 (knockout) vaccinia virus strain, and changes in brain ATP levels were monitored over a period of 120 hours. ATP levels in both wild-type and knockout-infected mouse brains remained relatively unchanged after 48 hours of infection, and no significant differences in ATP levels between the two experimental groups were observed (FIG. 6A). Fluctuations in ATP levels were observed for knockout and mock-infected mice from 48 hours postinfection onwards (FIG. 6B), whereas ATP levels in wild-type–infected brains displayed minimal change over the 5-day infection period and appeared to be maintained from 72 hours postinfection onwards (FIG. 6B). A significant difference (*P* value 0.01613) in ATP concentrations between wild-type and knockout-infected mouse brains was observed at 120 hours postinfection (FIG. 6B).

DISCUSSION

The specific function of the vaccinia virus 13.8-kDa protein and the mechanism that it employs to enhance virus replication in the brain remains unclear. However, the high energy demand and small energy reserves of this organ suggest a role for this protein in influencing the energy utilization of the virus for replication. This may prove a worthwhile avenue of investigation. To establish whether a relation exists between the vaccinia virus 13.8-kDa protein and ATP levels or utilization in the brain, *in vitro* and *in vivo* studies were conducted in vaccinia virus-infected mouse brain tissue.

Following a direct comparison of ATP levels at various timepoints postinfection in neuroblastoma cell cultures infected with a wild-type vaccinia virus and cultures infected with a strain not expressing the 13.8-kDa protein, no significant differences were observed, and variability in ATP levels between different experiments occurred. This may be due to numerous reasons including that the assay system used was not sufficiently sensitive. The established cell line used may not have efficiently mimicked *in vivo* conditions, and a primary cell line may be required instead. No significant difference in replication of the wild-type and the knockout strains was detected in the neuroblastoma cell line (data not shown), which may account for the lack of significant differences in ATP levels. Analysis of the pattern of changes in ATP levels in this *in vivo* model, however, may provide an indication of the virus' utilization of this nucleoside. The drop in ATP levels at 2 and 24 hours postinfection observed for WR vaccinia virus-infected neuroblastoma cells (FIG. 5) possibly represents the consumption of energy by the virus for functions such as viral transcrip-

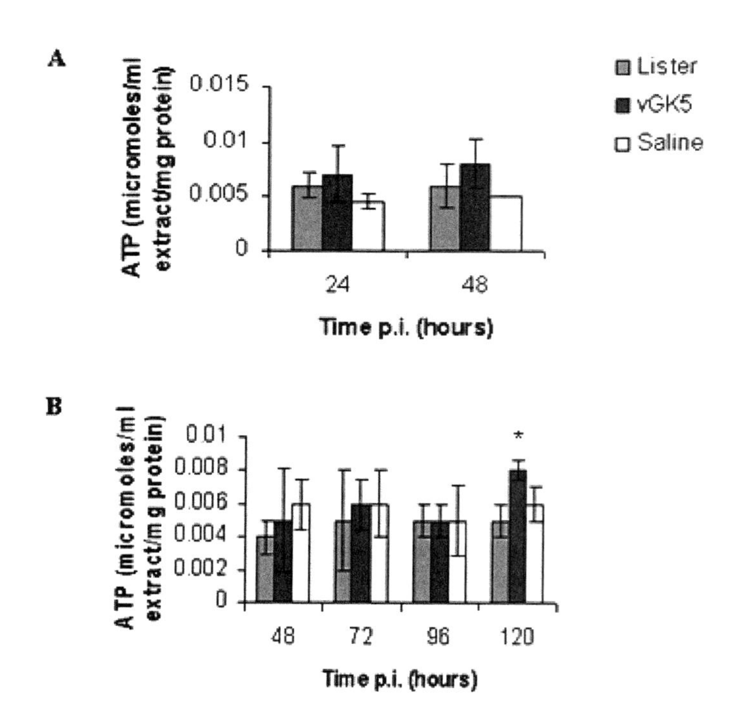

FIGURE 6. Changes in brain ATP levels in Lister vaccinia virus, vGK5 virus, and mock-infected BALB/c mice over a 48-hour (**A**) and 120-hour (**B**) infection period. ATP concentrations were measured using a luciferin/luciferase bioluminescence assay and are presented as means ± standard deviation (p.i., postinfection; **P* value 0.01613).

tion and replication. The low ATP concentration at 48 hours postinfection likely indicates ultimate depletion of the energy supply. Subsequently, no correlation was established between the 13.8-kDa protein and ATP levels in this *in vitro* system.

In *in vivo* experiments, the concentration of ATP in the brain in vGK5-infected mice proved to be significantly higher than that of mice infected with the wild-type Lister vaccinia virus at 5 days postinfection (FIG. 6B). This was an anticipated difference, as a nonreplicating strain would be expected to consume less energy. When analyzing the pattern of energy utilization in vaccinia virus-infected mice, the maintained ATP level seen in Lister vaccinia virus-infected mouse brain tissue from 72 hours postinfection onwards (FIG. 6B) may indicate that *in vivo* the 13.8-kDa protein plays a role in "balancing" ATP levels or maintaining a constant turnover of ATP to support viral replication. Any ATP level changes in mock-infected mice were speculated to be in response to the injury incurred as a result of the intracranial inoculation. The same could be proposed for ATP level changes in knockout-infected mice.

The Lister vaccinia virus, however, has been reported to display a different cell tropism from that of the WR vaccinia virus as well as a reduced capacity to replicate in the brain due to its dermotropic properties.[22,23] When taking this into account along with the attenuation associated with its use as a vaccine strain, the possibility of any differences in ATP utilization owing to differences in replicative ability rather than lack of the 13.8-kDa protein cannot be ruled out. Measurement of viral titers in wild-type and vGK5-infected brain tissue homogenates over the period of infection revealed that the Lister vaccinia virus replicated poorly in the brain with only slight increases in viral titer after 4 days of infection, and titers did not exceed those of the vGK5-infected brains at any point during infection (data not shown). In addition, it was noted that even when infecting with a 10-times higher concentration of the Lister strain, no significant differences in brain ATP levels between wild-type and knockout-infected mice were observed (FIG. 6A). We therefore conclude that ATP levels and ATP utilization in the brain may be influenced in the presence of the 13.8-kDa protein, yet further *in vivo* studies using the neurovirulent WR vaccinia virus as a more suited wild-type strain are required to allow for improved evaluation of an ATP-related function for the 13.8-kDa protein.

An adenylate kinase activity assay using purified 13.8-kDa protein found that this protein displayed no adenylate kinase activity (unpublished data). This, however, does not rule out the possibility of a kinase-related function. A comparison of the N1L amino acid sequence with that of the unactivated apo insulin-like growth factor-1 receptor kinase domain (IGFRK-0P) allowed for identification of a possible protein kinase domain within the 13.8-kDa protein, which if present would be rendered inactive due to the absence of an essential histidine residue. The insulin-like growth factor-1 receptor is a member of the transmembrane receptor tyrosine kinase superfamily,[20] which binds ATP in an activated state and is thought to play a role in the suppression of apoptosis.[24] The presence of a possible adenine-binding motif matching that of IGFRK-0P may indicate the ability of the 13.8-kDa protein to bind this nucleotide, but not ATP due to the lack of a putative phosphate binding region. Studies to investigate whether this viral protein binds adenine or adenine derivatives may be required to support theories regarding an ATP or protein kinase-related function.

Based on the predicted secondary structure of the 13.8-kDa protein, the C-terminal α-helix is proposed to be a dimerization domain. The homodimeric properties of this protein expressed intracellularly have been described by Bartlett *et al.*[2] who also suggested its predominantly alpha-helical structure. Structural homologies to the N-terminal actin-crosslinking domain of fimbrin were limited, and the significance of the identity between the two proteins is unknown as numerous other predominantly alpha-helical proteins displayed varying homologies to the 13.8-kDa predicted protein structure. Future studies on the binding and functional motifs and structure of the 13.8-kDa protein using tools such as X-ray crystallography and NMR will offer much insight into the function of this protein. Additional studies on the immunomodulatory properties of the 13.8-kDa protein, linking it with inhibition of the production of antiviral and proinflammatory cytokines, are to be published.

ACKNOWLEDGMENTS

We are grateful to Dr. Wolfie Katz of the Biovac Institute for providing the SA vaccine Lister strain of the vaccinia virus, Rodney Lucas for assistance in performing the animal work, and Dr. S. Prince and Dr. M.F. Essop for the use of their luminometers. M.A. is a recipient of Medical Research Council (MRC), Poliomyelitis Research Foundation, and UCT scholarships. G.J.K. is a Senior Wellcome Trust International Fellow for Biomedical Science in South Africa.

REFERENCES

1. Kotwal, G.J. & B. Moss. 1988. Vaccinia virus encodes a secretary polypeptide structurally related to complement control proteins. Nature **355:** 176–178.
2. Bartlett, N., J.A. Symons, D.C. Tscharke & G.L. Smith. 2002. The vaccinia virus N1L protein is an intracellular homodimer that promotes virulence. J. Gen. Virol. **83:** 1965–1976.
3. Kotwal, G.J., A.W. Hügen & B. Moss. 1989. Mapping and insertional mutagenesis of a vaccinia virus gene encoding a 13,800-Da secreted protein. Virology **171:** 579–587.
4. Billings, B.B., S.A. Smith, Z. Zhang, *et al.* 2004. Lack of N1L gene expression results in a significant decrease of vaccinia virus replication in mouse brain. Ann. N.Y. Acad. Sci. **1030:** 297–302.
5. Smith, S.A. & G.J. Kotwal. 2001. Virokines: novel immunomodulatory agents. Expert Opin. Biol. Ther. **1:** 343–357.
6. DiPerna, G., J. Stack, A.G. Bowie, *et al.* 2004. Poxvirus protein N1L targets the I-B Kinase complex, inhibits signalling to NF-B by the tumor necrosis factor superfamily of receptors, and inhibits NF-B and IRF3 signalling by Toll-like receptors J. Biol. Chem. **279:** 36570–36578.
7. Stryer, L. 1995. Metabolism: basic concepts and design. *In* Biochemistry. 4th Ed. :442–462. W.H. Freeman & Co. New York.
8. Broyles, S.S. 2003. Vaccinia virus transcription. J. Gen. Virol. **84:** 2293–2303.
9. Fenger, T.W. 1984. Replication of DNA viruses. *In* Textbook of Human Virology. R.B. Belshe, Ed. :49–78. PSG Publishing Co., Inc. Littleton, MA.
10. Baxby, D. 1984. Poxviruses. *In* Textbook of Human Virology. R.B. Belshe, Ed. :929–950. PSG Publishing Co., Inc. Littleton, MA.
11. Gershowitz, A., R.F. Boone & B. Moss. 1978. Multiple roles for ATP in the synthesis and processing of mRNA by vaccinia virus: specific inhibitory effects of adenosine (, -Imido) triphosphate. J. Virol. **27:** 399–408.
12. Moss, B. 2001. Poxviridae and their replication. *In* Virology. D.M. Knipe & P.M. Howley, Eds. :2849–2883. Lippincott–Williams & Wilkins. Philadelphia.
13. Ames, A., III. 2000. CNS energy metabolism is related to function. Brain Res. Rev. **34:** 42–68.
14. Joklik, W.K. 1962. The purification of four strains of poxvirus. Virology **18:** 9–18.
15. Stannard, L.M., D. Marias, D. Kow & K.R. Dumbell. 1998. Evidence for incomplete replication of a penguin poxvirus in cells of mammalian origin. J. Gen. Virol. **79:** 1637–1646.
16. Notari, L., M. Isodoro, C. Cugnoli & A. Morelli. 2001. Adenylate kinase activity in rod outer segments of bovine retina. Biochim. Biophys. Acta **1504:** 438–443.
17. Pissarek, M., R. Reihjardt, C. Reichelt, *et al.* 1999. Rapid assay for one-run determination of purine and pyrimidine nucleotide contents in neocortical slices and cell cultures. Brain Res. Protocols **4:** 314–321.
18. Khan, H. 2003. Bioluminometric assay of ATP in mouse brain: determinant factors for enhanced test sensitivity. J. Biosci. **28:** 379–382.
19. Santos, D.A., A.I. Salgado & R.A. Cunha. 2003. ATP is released from nerve terminals and from activated muscle fibres on stimulation of the rat phrenic nerve. Neurosci. Lett. **338:** 225–228.
20. Munshi, S., D.L. Hall, M. Kornienko, *et al.* 2003. Structure of apo, unactivated insulin-like growth factor-1 receptor kinase at 1.5 A resolution. Acta Crystallogr. Sect. D Biol. Crystallogr. **59:** 1725–1730.

21. PAUTSCH, A., A. ZOEPHEL, H. AHORN, *et al.* 2001. Crystal structure of bisphosphorylated IGF-1 receptor kinase: insight into domain movements upon kinase activation. Structure **9:** 955–965.
22. BERANEK, C.F., R. SHÄFER, L. BOLOGA & N. HERSCHKOWITZ. 1982. Viral tropisms in mouse brain cell cultures. Med. Microbiol. Immunol. **170:** 201–207.
23. SOEKAWA, M., C. MORITA, R. MORIGUCHI & M. NAKAMURA. 1974. Neurovirulence of vaccinia viruses for mice. Zentralbl. Bakteriol. Hyg. J. Abt. Orig. A. **226:** 434–442.
24. BASERGA, R. 1999. The IGF–1 receptor in cancer research. Exp. Cell Res. **253:** 1–6.

Curcumin Inhibits the Classical and the Alternate Pathways of Complement Activation

AMOD P. KULKARNI, YOHANNES T. GHEBREMARIAM, AND GIRISH J. KOTWAL

Division of Medical Virology, IIDMM, Faculty of Health Sceinces, University of Cape Town, Medical School, Cape Town 7925, South Africa

ABSTRACT: Curcumin (Cur), the golden yellow phenolic compound in turmeric, is well studied for its medicinal properties. In the current investigation, Cur dissolved using sodium hydroxide solution (CurNa) was tested for *in vitro* complement inhibitory activity and compared with rosmarinic acid (RA) and quercetin (Qur) dissolved using sodium hydroxide (RANa and QurNa, respectively) and the vaccinia virus complement control protein (VCP). The comparative study indicated that CurNa inhibited the classical complement pathway dose dependently (IC_{50} = 404 μM). CurNa was more active than RANa, but less active than QurNa. VCP was about 2,212, 2,786, and 4,520 times more active than QurNa, CurNa, and RANa, respectively. Further study revealed that CurNa dose dependently inhibited zymosan-induced activation of the alternate pathway of complement activation.

KEYWORDS: curcumin; complement; classical pathway (CP); alternate pathway (AP); rosmarinic acid; VCP; quercetin; hemolysis; Bb; C3b

INTRODUCTION

Turmeric (*Curcuma longa,* family Zingiberaceae) is described as a healing (heals wounds; is an anti-inflammatory), reducing (reduces cholesterol), complexion-improving, and antidermatitis herb in Ayurveda.[1] Cur, along with other curcuminoids and volatile oil, is the potential active ingredient of this herb.[2] A vast number of publications regarding the anti-inflammatory and anticancer activity of Cur can be found, and it is described as yellow gold for its medicinal value. Cur is beneficial in wound healing and accelerates wound repair in mice exposed to γ-radiation. Its antioxidant and radioprotective activities might be responsible for the observed protective effect.[3] Cur and its derivatives inhibit the cyclooxygenases (COXs) to a greater extent than does indomethacin, a currently prescribed mixed COX inhibitor. This activity, along with free radical scavenging and other roles, makes it a strong chemopreventive agent.[4] Cur inhibits cytokine-mediated I-κB phosphorylation, degradation and I-κB kinase activity and thereby inhibits nuclear factor kappa-B (NF-κB) activation, which is involved in transcriptional regulation of critical pro-

Address for correspondence: Girish J. Kotwal, Division of Medical Virology, IIDMM, Faculty of Health Sciences, Medical School, University of Cape Town, Anzio Road, Observatory, Cape Town - 7925, South Africa. Voice: (27)-(21) 406-6676; fax: (27)-(21) 406 6018.
gjkotw01@yahoo.com

Ann. N.Y. Acad. Sci. 1056: 100–112 (2005).
doi: 10.1196/annals.1352.007

inflammatory molecules.[5] It also inhibits extracellular signal-regulated kinase (ERK) activation. All of these activities, that is, ERK, NF-κB, and COX inhibition by Cur, could be responsible for its anticancer activity.[5,6]

The anti-inflammatory activity of Cur is not restricted to a single factor. Cur treatment decreased the levels of interleukin-1β (IL-1β), C-reactive protein, and haptoglobin involved in inflammation.[7] It inhibited NF-κB activation and proliferation of tumor cells with potency greater than currently used nonsteroidal anti-inflammatory agents (NSAIDs), and it can be regarded as a more potent anti-inflammatory and antiproliferative agent than most of the currently prescribed NSAIDs.[8] Thus, Cur, a well-established anti-inflammatory agent, inhibits most of the proinflammatory molecules involved in the process of inflammation; however, its effect on the complement system is unknown.

Complement plays a damaging role in neuroinflammatory disorders. Alzheimer's disease (AD),[9] multiple sclerosis,[10,11] traumatic brain injury,[12] and spinal cord injury[13] are disorders of the brain in which activated complement components mediate inflammation either by direct actions or by activating other proinflammatory molecules. The complement components are involved in the chemotaxis of immune cells to the site of inflammation.[14] The complement system is involved in release of proinflammatory molecules such as cytokines, free radicals, and arachidonic acid from the immune cells.[14–16] Thus, for a drug to be effective in the treatment of neuroinflammatory disorders, it must be able to regulate the complement system effectively.

Although inflammation plays an important role in the etiology of neuroinflammatory disorders, and complement components are among the major proinflammatory molecules involved in inflammation, no complement inhibitory agent is available on the market. The main targets of NSAIDs used for treating inflammatory disorders are cyclooxygenases. However, COXs play a beneficial role, and controversies exist on the usefulness of COX inhibitors in neuroinflammatory disorders.[17]

Vaccinia virus complement control protein (VCP) is a small protein (28 kDa) secreted by vaccinia virus. It regulates both the complement pathway (CP) and the alternate pathway (AP).[18,19] Positive preclinical outcomes of VCP in rodent models of spinal cord injury[20] and traumatic brain injury[21] and a possible beneficial role in treating AD[22] suggest the possibility of using complement regulatory molecules in the treatment of neuroinflammatory disorders in which upregulated complement components play a devastating role.

Cur, RA, and Qur are polyhydroxy phenolic compounds structurally related to each other, RA being structurally closer to Cur (FIG. 1). RA and Qur are known to inhibit the complement system, but the ability of Cur to inhibit the complement is yet to be explored. RA inhibits C5 convertase;[23,24] however, it does not inhibit the C3 convertase, and its concentration required to inhibit C5 convertase is very high. It shows covalent attachment to the thioester containing the α-chain of the nascent C3b component of the complement and thereby prevents both CP and AP activation.[23] Qur is also known to inhibit both the AP and CP.[25] Cur has never been tested or reported to influence complement activity. Therefore, we decided to investigate its impact on the complement system as it is structurally similar to the aforementioned compounds. Also, its complement inhibitory activity was compared to that of Qur, RA, and VCP in the current investigation. Its effect on zymosan-mediated activation of the AP was also studied.

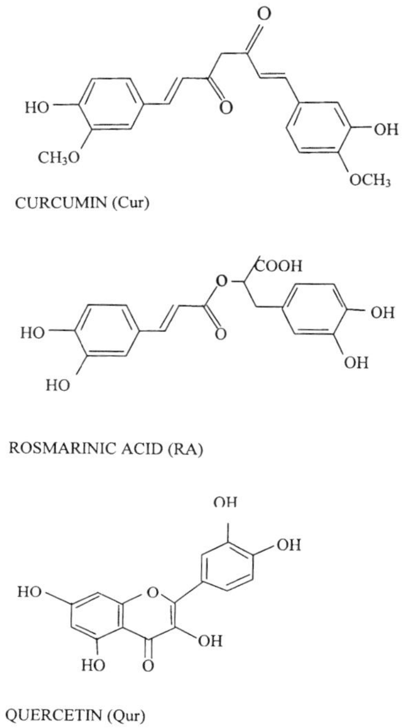

CURCUMIN (Cur)

ROSMARINIC ACID (RA)

QUERCETIN (Qur)

FIGURE 1. Cur, RA, and Qur are structurally related polyhydroxy phenolic compounds. (Structures are based on Sigma-Aldrich website: www.sigmaaldrich.com).

MATERIAL AND METHODS

Cur, RA, and Qur (dihydrate) were purchased from Sigma. All the other reagents used were of analytical grade. Normal human serum (NHS) collected from a healthy donor served as a source of complement in the assay. The sensitized sheep red blood cells (ssRBCs) were purchased from Diamedix Corporation (Miami, FL, USA). The Bb fragment enzyme immunoassay kit for the effect on the AP was from Quidel Corporation, and zymosan (Sigma) was a kind gift from Dr. Kevin Dennehy.

Preparation of Reagents

Stock solutions: Cur, RA, and Qur stock solutions were prepared in phosphate-buffered saline solution(PBS; pH 7.2) using either methanol or sodium hydroxide solution and stored at −20°C prior to the experiment. At the time of experiment, the drugs were diluted in PBS (pH 7.2) to obtain various concentrations (800, 640, 320, 160, and 80 μM) in the final assay. VCP 1, 2, or 3 μg was diluted to 20 μl using PBS, pH 7.2. Serum was diluted (1:20) using gelatin veronal-buffered saline. ssRBCs were also diluted in the same buffer.

Hemolysis Assay

Twenty microliters of the drug solutions (diluted to required concentrations using PBS, pH 7.2) were added to 75 μl of ssRBCs. Furthermore, 5 μl of serum (1:20

dilution) was added to the mixture at 0°C. The mixture was incubated for an hour at 37°C. The reaction was stopped after 1 hour, and the eppendorf tubes were subjected to centrifugation at 700 rpm for 30 seconds. As the drug solutions were colored, the assay was modified. In normal hemolysis assay, the supernatant is analyzed directly for estimation of the percentage hemolysis.[26] Here, the supernatant was discarded, the cell pellet was washed with 100 µl of gelatin veronal-buffered saline, and hemolyzed using 100 µl of double-distilled water (ddH_2O) and analyzed at 405 nm using an ELISA microplate reader (Anthos-labtec, Salzburg, Austria). This was done to determine the percentage of cells that were not lysed during the assay. From the value of absorbance, the percentage inhibition was calculated. All experiments were repeated three times ($n = 3$).

Effect of CurNa on the AP

Quidel enzymatic immunoassay based on the quantitative estimation of Bb was used to study the effect of CurNa on the AP. (Please refer to the article by Ghebremariam *et al.* in this volume of the *Annals* for details regarding principles, methods, and references.) In short, 10 µl of different dilutions of CurNa, positive and negative controls, were incubated for an hour at 37°C with normal human serum (undiluted, 10 µl) and/or heat-inactivated serum in the presence of 1.25 µl of zymosan solution (10 mg/ml) pre-sonicated before use. Ten microliters of this solution was then diluted with specimen diluent (Quidel Corp.) to 100 µl and incubated for 30 minutes with mouse anti-Bb monoclonal antibody in pre-washed eight-well microplate assay strips (Quidel Corp.). The amount of Bb (in the sample and controls) was detected by adding 50 µl of horseradish peroxidase (HRP)–conjugated goat anti-human Bb antibody and developed by 1/20 diluted chromogenic substrate. The color developed was quantitated using microplate reader at 405 nm. From this percentage, inhibition of the AP by CurNa at different concentrations was determined and plotted on a log scale to determine the IC_{50} value of CurNa.

Statistical Analysis

One-way analysis of variance (ANOVA) was used for statistical analysis of the data, and means were compared using Student's t test to evaluate the statistical significance of the results.

RESULTS

CurNa inhibited NHS-mediated hemolysis in the modified hemolysis assay (as just described) in a dose-dependent manner. There was a statistically significant rise in the complement inhibitory activity of CurNa with the dose, as indicated by decreased hemolysis with the increased dose of CurNa. Surprisingly, between 320 and 640 µM, there was a sudden rise in the complement inhibitory activity of CurNa, and yet at 800 µM there was a statistically significant drop in activity. QurNa also showed a rise followed by a sudden drop in activity at 320- and 800-µM concentrations, respectively. CurNa was less active than RANa at lower concentrations. Its activity was significantly lower than that of RANa and QurNa at 80 µM. At 160 µM,

RANa inhibited complement to a greater extent than did CurNa and QurNa, but at all the other concentrations, CurNa and QurNa showed better activity than did RANa. CurNa showed dose-dependent inhibition of the CP, but in case of RANa, although there was an increase in activity with the dose, it was not statistically significant (FIG. 2). The complement inhibitory activity of CurNa was found to be better than that of Cur in methanolic solution. Similar observations were true for RANa. In the case of the methanolic solution of Qur, the results could not be interpreted because of its precipitation during the assay (results not shown). The IC_{50} value of VCP was found to be 0.145 μM (FIG. 3A). The IC_{50} values of CurNa, QurNa, and RANa were found to be about 320, 404, and 679 μM, respectively (FIG. 3B). Thus, CurNa's activity was between that of QurNa and RANa. However, at 160 and 640 μM, no significant difference was noted between the activities of CurNa and QurNa (FIG. 2). Comparison of these active constituents of herbal origin with VCP revealed that it was 2,212-, 2,786-, and 4,520-fold more potent than that of QurNa, CurNa, and RANa, respectively (FIG. 4). CurNa was also found to inhibit the zymosan-induced activation of the AP (FIG. 5). The IC_{50} value of curcumin was found to be 630 μM. The effect was dose dependent, and there was a decrease in the complement inhibitory activity beyond 1,000 μM.

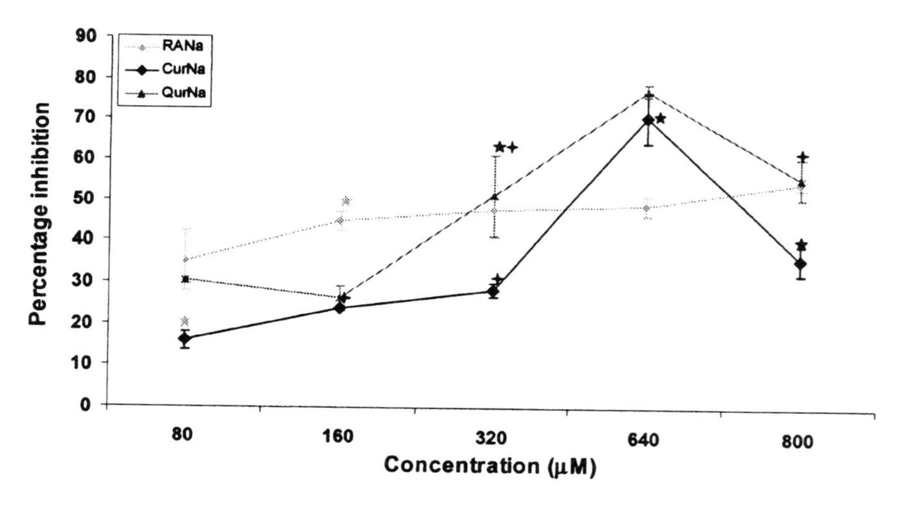

FIGURE 2. Hemolysis assay indicating complement inhibition by CurNa, QurNa, and RANa at different concentrations. The Y axis represents the percentage inhibition by the different drugs. The X axis represents concentrations of CurNa, RANa, and QurNa. All the experiments were repeated three times ($n = 3$).

✦Statistically significant change in activity ($P = .05$) as compared to the previous observation of the same drug.

★Statistically highly significant change in activity ($P = .01$) as compared to the previous observation of the same drug.

★✦Statistically very highly significant change in activity ($P = .001$) as compared to the previous observation of the same drug.

⁂ Statistically significant and/or highly significant difference between the activities ($P = .05$ and/or .01) of two different drugs at the same concentration.)

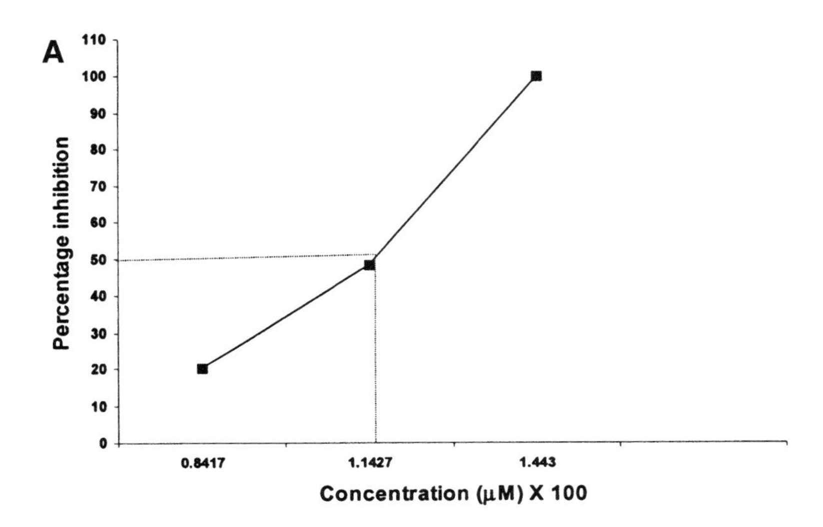

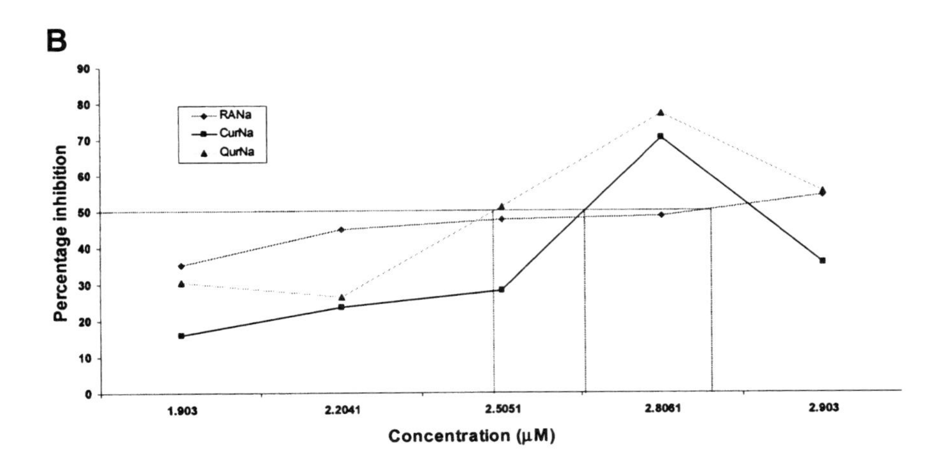

FIGURE 3. Determination of IC_{50} values of VCP (**A**) and the active ingredients of herbal origin (**B**). IC_{50} values of different drugs are determined on a log scale. The X axis in each figure represents the drug concentration in micromolars (μM) and the percentage inhibition of hemolysis is shown on the Y axis. In VCP (**A**), as the log values were falling on a negative side of the graph, the 100 × log scale is used (i.e., concentration = 100 × the actual concentration used in the experiments).

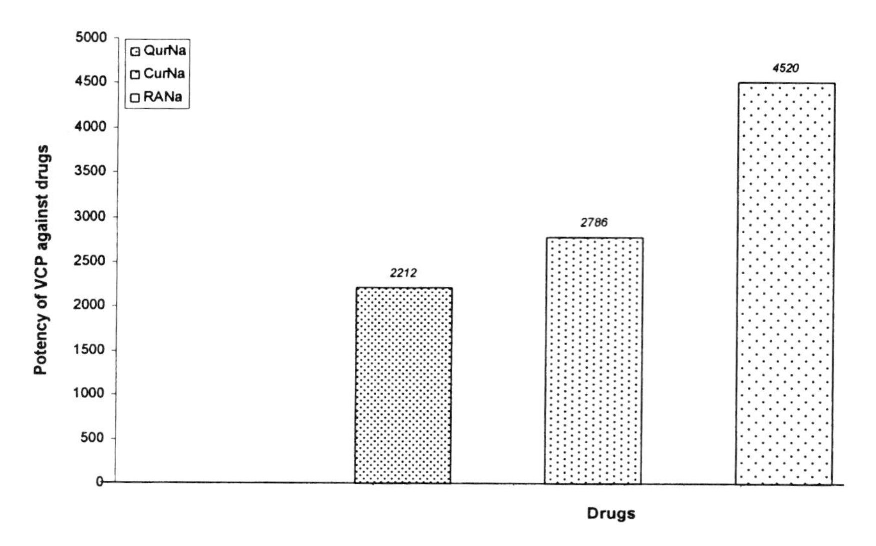

FIGURE 4. Comparison of the complement inhibitory activity of QurNa, CurNa, and RANa with VCP. On the Y axis, potency of VCP against different drugs, and on the X axis the three different drugs are shown.

DISCUSSION

Failure of some of the currently prescribed COX inhibitors and other anti-inflammatory agents in offering significant neuroprotection in neuroinflammatory disorders such as AD[27–29] suggests the need for the development of new anti-inflammatory agents. As described previously, neuroinflammatory disorders are associated with the complement components.[9–13] VCP is a complement regulatory molecule of viral origin and inhibits both the CP and AP.[18,19] Due to its ability to regulate the activated complement components, VCP was found to be beneficial in rodents' models for spinal cord injury and traumatic brain injury.[20,21] Positive preclinical outcome of VCP-like molecules suggests that regulation of the activated complement components could be effective in treating neuroinflammatory disorders. Tight regulation of the complement components is required because of the susceptibility of neurons to the activated complement components, as they express low levels of complement regulatory molecules. Natural ingredients of herbal origin offer significant potential in drug development. Cur, an active constituent of turmeric, is one of those compounds that is known for its medicinal value. The anti-inflammatory activity of Cur is well documented in the scientific literature, as mentioned previously. The compound is known for its ability to reduce the formation of Aβ plaques,[30] Aβ-induced cognitive deficits,[31] cancer,[4,6] and inflammatory activities.[3,7,32] As discussed previously, Cur is a novel compound that blocks or inactivates several pro-inflammatory molecules and/or signaling pathways involved in inflammation. The compound is under clinical investigation for the treatment of several disorders. The

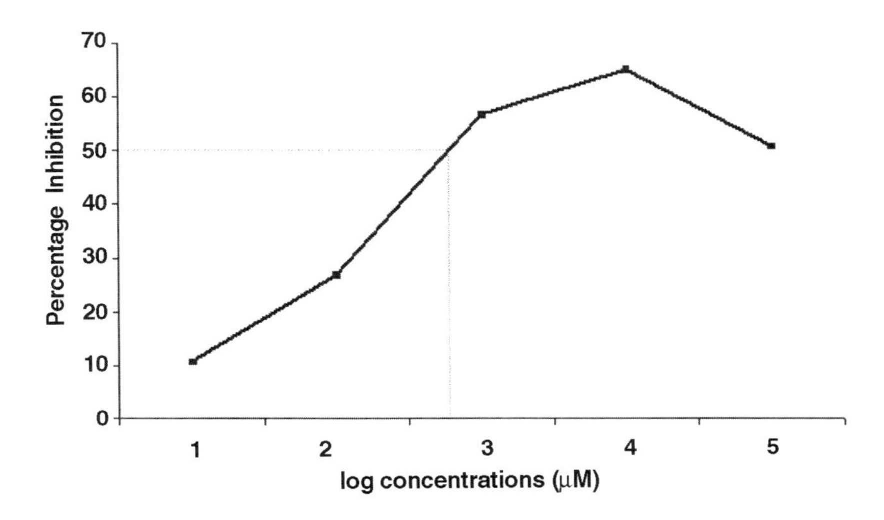

FIGURE 5. Effect of CurNa on the AP. Determination of the IC_{50} value. The percentage inhibition of the AP by CurNa is shown on the Y axis and different concentrations of CurNa used in the assay are shown on the X axis.

complement inhibitory activity, elucidated in this paper, has added a new dimension to the therapeutic potential of Cur.

Thus, Cur can be considered to prevent inflammation by its direct actions on several proinflammatory molecules including the inhibition of the complement components and the activation of the various proinflammatory molecules involved in the inflammation. In previous studies, the sodium salt of Cur was found to have a better anti-inflammatory profile than that of the original compound and phenylbutazone. Also, the observed anti-inflammatory activity of the sodium salt of Cur in the same experiment was not due to inhibition of prostaglandin synthesis or the release of steroids from the adrenal cortex.[32] The increased anti-inflammatory activity of sodium salt of Cur observed by previous researchers can be attributed to its better complement inhibitory activity. Our study revealed that CurNa inhibited the CP to a better extent than did the methanolic solutions. The reason for the better complement inhibitory activity of the alkaline solutions of these compounds (better than that of the methanolic solutions) can be attributed to their better solubility. The sudden rise and drop in the activities of CurNa and QurNa, as observed in the assay, can be explained on the basis of the cooperative effect. In an earlier study, inhibition of hemolysis by VCP was compared to that of the human C4b binding protein (hC4b-BP).[33]Inhibition by VCP was found to be linear, whereas that of hC4b-BP was a sigmoidal response indicating a cooperative effect. Similar results have now been observed here with Cur and Qur, which show an effect comparable to that of human C4b-BP, suggesting that the mode of inhibition of Cur and Qur is a cooperative effect.

In previous studies, RA was found to be more active than Qur,[25] but our findings demonstrated that QurNa is more active than RANa. If the result with RANa in the current investigation is correlated with the findings of previous researchers, it can be concluded that RA is a more potent inhibitor of complement than RANa, and QurNa is more effective than Qur. Hence, on the one hand, making sodium salt of Qur improves its activity and, on the other hand, it decreases the activity of RANa. The improved complement inhibitory activity of QurNa might be due to its better solubility, and the decreased activity of RANa might be due to the decreased stability of RA in alkaline solutions.

The complement inhibitory activity of all these compounds is several-fold less than that of VCP ($IC_{50} = 0.145$ μM). VCP was found to be about 2,212, 2,786, and 4,520 times more active than QurNa, CurNa, and RANa, respectively (FIG. 4). Thus, to use Cur in neuroinflammatory disorders, in which complement plays a major role, its dose should be several-fold higher than that of VCP. The several-fold activity of VCP might be due to its significantly higher molecular weight and/or greater size than that of Qur, Cur, and RA.

CurNa inhibits not only the CP, but also the zymosan-mediated activation of the AP at about a twofold higher concentration, that is, 630 μM (FIG. 5). Interestingly, in contradiction with previous studies,[25] in this assay, RaNa or QurNa did not show activity on the AP at 640 μM (results not shown). The Quidel Bb fragment enzyme immunoassay used in the current investigation is based on quantitative estimation of Bb, a cleavage product of factor B, in human serum. Bb factor is an index of complement activation by the AP. [34] In the current investigation, CurNa dose-dependently inhibited the formation of Bb. This dose-dependent inhibition of complement activation can be attributed to reduction in the formation of Bb by CurNa through different mechanisms. Bb is a fragment of factor B involved in the activation of the AP. [34] Formation of Bb from factor B is a two-step process, which requires Mg^{2+} ions and factor D. [34] As discussed later in this article, formation of the phenoxide ion might be involved in the action of Cur on the complement system. Due to its negative charge, it may chelate the Mg^{2+} ions and thereby prevent the cleavage of factor B. However, its inhibitory action cannot be solely attributed to this chelation effect, as Mg^{2+} ions are not an absolute requirement for the formation of the AP convertases.[35] Also, it may inhibit factor D (which is involved in the cleavage of factor B[34]), properdin (which stabilizes the alternate pathway convertases [34]), and/or may have some direct inhibitory action on the alternate pathway convertases. Furthermore, insight into its mode of action on individual complement components involved in the AP would be an interesting area of research.

Thus, like VCP, Cur also inhibits both pathways of complement activation. Although these two compounds differ in chemical structure and size, but still show regulatory actions on the complement system, they can be regarded as antineuroinflammatory compounds. It would be interesting at this stage to compare VCP with Cur. Both have their own advantages and disadvantages. VCP is a highly potent complement regulatory agent, but its larger size may pose a problem of bioavailability to the brain. Attempts are being made to deliver VCP to the brain bypassing and/or avoiding the blood brain barrier (BBB). Also, more potent analogues of VCP with reduced size and improved complement inhibitory ability might prove an interesting area for research. However, Cur can cross the BBB due to its small size, and it reduces *in vivo* plaque formation.[30] The safety and toxicity studies indicate that it is relatively safe even at higher doses, although it may cause gastric irritation in humans at very high doses.[36] Thus, it can be used even at high dose to regulate inappropriately activated complement components from playing a damaging role in neuroinflammatory disorders.

The biological activities of Cur can be at least partly explained on the basis of its complement inhibitory ability, as 30 different complement proteins are involved in a wide range of biologic activities. The complement components and receptors are involved in the activation of NF-κB, which plays an important role in inflammatory disorders such as AD.[37,38] The complement components not only are associated with NF-κB, but also mediate release of other proinflammatory mediators such as IL-6,[39] TNF-α, and eicosanoids and thereby mediate recruitment of immune cells to the site of injury.[40] Cur is known to have direct inhibitory effects on NF-κB and subsequent proinflammatory events.[41] The complement inhibitory activity of Cur, as observed in the current investigation, might be responsible for the indirect inhibition of complement-mediated activation of NF-κB and subsequent proinflammatory events. All of these direct and indirect actions of Cur on proinflammatory mediators, particularly on the complement system, make it a strong anti-inflammatory agent for the treatment of neuroinflammatory disorders.

The current investigation confirms the complement inhibitory activity of Cur, but further insight into its mode of action is necessary. VCP inhibits C3b through its positively charged modules.[42] Cur may inhibit the complement components by a mechanism different from VCP. Cur contains two phenolic groups (FIG. 1). Phenols are acidic in nature and readily form phenoxides in the presence of a base such as sodium hydroxide.[43] Cur may form a highly reactive, negatively charged phenoxide ion with sodium hydroxide in the same way. It is likely that it may react with the complement system in a manner similar to that of the bisphenolic compounds,[44] which bind and inhibit C1 components of the complement. The presence of negatively charged molecules such as enolate ions and phenoxide ions suggests that it may bind and inhibit the complement system by binding to the C1 component of complement activation. The nucleophilic nature of enolate ions as well as the phenoxide group also strongly suggests that Cur, like other phenolic compounds (e.g., dopamine, 5-hydroxytryptamine),[45] may react with the thioester bond of C3b and thereby inhibit the complement system. The possible interaction of negatively charged or acidic moieties of CurNa with basic amino acids such as lysine, arginine, and histidine of the complement components cannot be overlooked. Cur shows keto-enol tautomerism.[46] Like other β-dicarbonyl compounds,[47] the presence of the β-dicarbonyl group in the structure of curcumin further enhances the stability of the enolate ion. Enolate or enol of Cur is involved in the intermolecular and intramolecular hydrogen bonding.[46] Hence, hydrogen bonding might also be involved in binding of curcumin to complement components. As just discussed, there is an increase in the complement inhibitory activity of CurNa up to 640 μM. The increase in its intermolecular hydrogen bonding with increasing concentration can be attributed to the observed effect. As the concentration of CurNa increases beyond 640 μM, the intramolecular hydrogen bonding of CurNa may predominate over the intermolecular hydrogen bonding, and this might be responsible for a decrease in the activity of CurNa at concentrations beyond 640 μM.

Being a polyhydroxy phenolic compound and structurally related to RA, Cur may bind to the C3b component of the complement system and thereby inhibit the CP and

AP. Thus, its ability to bind C3b and other complement components must be detected. Results from previous studies suggest that glycosides of flavonoids are more potent inhibitors of the complement system than the corresponding flavonoids.[48,49] The glucosides have the advantage of greater solubility and better stability than those of the alkaline salt of Cur. Hence, the glucosides of Cur prepared by a simple and convenient method[50] need to be screened to discover more potent complement inhibitory analogues of Cur with better ability to inhibit the complement system.

In summary, this study reveals the complement inhibitory activity of CurNa. CurNa shows dose-dependent inhibition of the CP and AP, although it inhibits the CP at a relatively lower concentration. CurNa is a moderate inhibitor of the complement system and is a thousandfold less active than VCP. Further insight into the mechanism of its complement inhibitory activity is necessary. Attempts should be made to prepare more potent complement inhibitory analogues of Cur.

ACKNOWLEDGMENTS

G.J.K. is currently a Senior International Wellcome Trust Fellow for Biomedical Sciences in South Africa. A.P.K. and Y.T.G. are the recipients of the Poliomyelitis Research Foundation, Senior Entrance Merit and International Students' Scholarships at the University of Cape Town. We thank Dr. Kevin Dennehy for generously providing the zymosan.

REFERENCES

1. http://www.aarogya.com/Complementary/Ayurveda/commonherbs.asp
2. Anna Carolina, C.M.M. *et al.* 2003. Extraction of essential oil and pigments from *Curcuma longa* [L.] by steam distillation and extraction with volatile solvents. J. Agric. Food Chem. **51:** 6802–6807.
3. Jagetia, G.C. & G.K. Rajanikant. 2004. Role of curcumin, a naturally occurring phenolic compound of turmeric in accelerating the repair of excision wound, in mice whole-body exposed to various doses of γ-radiation. J. Surg. Res. **120:** 127–138.
4. Gafner, S. *et al.* 2004. Biologic evaluation of curcumin and structural derivatives in cancer chemoprevention model systems. Phytochemistry **65:** 2849–2859.
5. Jobin, C. *et al.* 1999. Curcumin blocks cytokine-mediated NF-κB activation and proinflammatory gene expression by inhibiting inhibitory factor i-kb kinase activity. J. Immunol. **163:** 3474–3483.
6. Chun, K.S. *et al.* 2003. Curcumin inhibits phorbol ester-induced expression of cyclooxygenase-2 in mouse skin through suppression of extracellular signal-regulated kinase activity and NF-kappaB activation. Carcinogenesis **24:** 1515–1524.
7. Banerjee, M. *et al.* 2003. Modulation of inflammatory mediators by ibuprofen and curcumin treatment during chronic inflammation in rat. Immunopharmacol. Immunotoxicol. **25:** 213–224.
8. Takada, Y. *et al.* 2004. Nonsteroidal anti-inflammatory agents differ in their ability to suppress NF-kappaB activation, inhibition of expression of cyclooxygenase-2 and cyclinD1, and abrogation of tumor cell proliferation. Oncogene **23:** 9247–9258.
9. Tacnet-Delorme, P. *et al.* 2001. β-amyloid fibrils activate the C1 complex of complement under physiological conditions: evidence for a binding site for Aβ on the C1q globular regions. J. Immunol. **167:** 6374–6381.
10. Sanders, M.E. *et al.* 1986. Activated terminal complement in cerebrospinal fluid in Guillain-Barre syndrome and multiple sclerosis (abstr). J. Immunol. **136:** 4456–4459.

11. SELLEBJERG, F. *et al.* 1998. Intrathecal activation of the complement system and disability in multiple sclerosis. J. Neurol Sci. **157:** 168–174.
12. SEWELL, D.L. *et al.* 2004. Complement C3 and C5 play critical roles in traumatic brain cryoinjury: blocking effects on neutrophil extravasation by C5a receptor antagonist. J. Neuroimmunol. **155:** 55–63.
13. ANDERSON, A.J. *et al.* 2004. Activation of complement pathways after contusion-induced spinal cord injury. J. Neurotrauma **21:** 1831–1846.
14. SORURI, A. *et al.* 2003. Anaphylatoxin C5a induces monocyte recruitment and differentiation into dendritic cells by TNF-alpha and prostaglandin E2-dependent mechanisms. J. Immunol. **171:** 2631–2636.
15. TAKANO, T. & A.V. CYBULSKY. 2000. Complement C5b-9-mediated arachidonic acid metabolism in glomerular epithelial cells: role of cyclooxygenase-1 and -2. Am. J. Pathol. **156:** 2091–2101.
16. FISCHER, W.H., M.A. JAGELS & T.E. HUGLI. 1999. Regulation of IL-6 synthesis in human peripheral blood mononuclear cells by C3a and C3a (desArg). J. Immunol. **162:** 453–459.
17. MINGHETTI, L. 2004. Cyclooxygenase-2 (COX-2) in inflammatory and degenerative brain diseases. J. Neuropathol. Exp. Neurol. **63:** 901–910.
18. KOTWAL, G.J. & B. MOSS. 1988. Vaccinia virus encodes a secretory polypeptide structurally related to complement control proteins. Nature **335:** 176–178.
19. MCKENZIE, R. *et al.* 1992. Regulation of complement activity by vaccinia virus complement-control protein. J. Infect. Dis. **166:** 1245–1250.
20. REYNOLDS, D.N. *et al.* 2004. Vaccinia virus complement control protein reduces inflammation and improves spinal cord integrity following spinal cord injury. Ann. N.Y. Acad. Sci. **1035:** 165–178.
21. HICKS, R.H. *et al.* 2002. Vaccinia virus complement control protein enhances functional recovery after traumatic brain injury. J. Neurotrauma **19:** 705–714.
22. KOTWAL, G.J., D.K. LAHIRI & R. HICKS. 2002. Potential intervention by vaccinia virus complement control protein of the signals contributing to the progression of central nervous system injury to Alzheimer's disease. Ann. N.Y. Acad. Sci. **973:** 317–322.
23. SAHU, A., N. RAWAL & M.K. PANGBURN. 1999. Inhibition of complement by covalent attachment of rosmarinic acid to activated C3b. Biochem. Pharmacol. **57:**1439–1446.
24. PEAKE, P.W. *et al.* 1991. The inhibitory effect of rosmarinic acid on complement involves the C5 convertase. Int. J. Immunopharmacol. **13:** 853–857.
25. CIMANGA, K. *et al.*1995. In vitro anticomplementary activity of constituents from *Morinda morindoides*. J. Nat. Prod. **58:** 372–378.
26. KOTWAL, G.J. *et al.* Inhibition of the complement cascade by the major secretary protein of vaccinia virus. Science **250:** 827–830.
27. AISEN, P. S. *et al.* 2003. Effects of rofecoxib or naproxen vs placebo on Alzheimer disease progression: a randomized controlled trial. JAMA **289:** 2819–2826.
28. REINES, S.A. *et al.* 2004. Rofecoxib: no effect on Alzheimer's disease in a 1-year, randomized, blinded, controlled study. Neurology **62:** 66–71.
29. SCHARF, S. *et al.* 1999. A double-blind, placebo-controlled trial of diclofenac/misoprostol in Alzheimer's disease. Neurology **53:** 197–201.
30. YANG, F. *et al.* 2005. Curcumin inhibits formation of amyloid {beta} oligomers and fibrils, binds plaques, and reduces amyloid *in vivo*. J. Biol. Chem. **280:** 5892–5901.
31. FRAUTSCHY, S.A. *et al.* 2001. Phenolic anti-inflammatory antioxidant reversal of A beta-induced cognitive deficits and neuropathology. Neurobiol. Aging **22:** 993–1005.
32. MUKHOPADHYAY, A. *et al.* 1982. Anti-inflammatory and irritant activities of curcumin analogues in rats (abstr). Ag. Actions **12:** 508–515.
33. KOTWAL, G.J. 1994. Purification of virokines using ultrafiltration. Am. Biotechnol. Lab. **12:** 76–77.
34. http://www.quidel.com
35. PRYZDIAL, E.L. & D.E. ISENMAN. 1986. A reexamination of the role of magnesium in the human alternative pathway of complement. Mol. Immunol. **23:** 87–96.
36. CHAINANI-WU, N.J. 2003. Safety and anti-inflammatory activity of curcumin: a component of tumeric (*Curcuma longa).*Alt. Complement Med. **9:** 161–168.

37. MAZZEO, A. *et al.* 2004. Immunolocalization and activation of transcription factor nuclear factor kappa B in dysimmune neuropathies and familial amyloidotic polyneuropathy (abstr). Arch. Neurol. **61:** 1097–1102.
38. MEDVEDEV, A.E. *et al.* 1998. Involvement of CD14 and complement receptors CR3 and CR4 in Nuclear Factor-κB activation and TNF production induced by lipopolysaccharide and group b streptococcal cell walls. J. Immunol. **160:** 4535–4542.
39. O'BARR, S. & N.R. COOPER. 2000. The C5a complement activation peptide increases IL-1b and IL-6 release from amyloid-b primed human monocytes: implications for Alzheimer's disease. J. Neuroimmunol. **109:** 87–94.
40. SORURI, A. *et al.* 2003. Anaphylatoxin C5a induces monocyte recruitment and differentiation into dendritic cells by TNF-α and prostaglandin E2-dependent mechanisms. J. Immunol. **171:** 2631–2636.
41. SINGH, S. & B.B. AGGARWAL. 1995. Activation of transcription Factor NF-κB is suppressed by curcumin (diferulolylmethane). J. Biol. Chem. **270:** 24995–25000.
42. MURTHY, K.H. *et al.* 2001. Crystal structure of a complement control protein that regulates both pathways of complement activation and binds heparan sulfate proteoglycans. Cell **104:** 301–311.
43. WADE, L.G. 1995. Structure and synthesis of alcohol. *In* Organic Chemistry. B.M. Cappuccio, Ed. :402. Prentice Hall, Inc. A Simon & Schuster Company. New Jersey.
44. BUREEVA, S. *et al.* 2005. Inhibition of classical pathway of complement activation with negative charged derivatives of bisphenol A and bisphenol disulphates. Bioorg. Med. Chem. **13:** 1045–1052.
45. SAHU, A. & M.K. PANGBURN. 1996. Investigation of mechanism-based inhibitors of complement targeting the activated thioester of human C3. Biochem. Pharmacol. **51:** 797–804.
46. TONNESEN, H.H., J. KARLSEN & A. MOSTAD. 1982. Structural studies of curcuminoids. I. The crystal structure of curcumin. Acta Chem. Scand. B. **36:** 475–479.
47. WADE, L.G. 1995. Additions and condensations of enols and enolate ions. *In* Organic Chemistry. B.M. Cappuccio, Ed. :1079. Prentice Hall, Inc. A Simon & Schuster Company. New Jersey.
48. MIN, B.S. *et al.* 2003. Anti-complement activity of constituents from the stem-bark of *Juglans mandshurica*. Biol. Pharm. Bull. **26:** 1042–1044.
49. PIERONI, A. *et al.* 2000. Studies on anti-complementary activity of extracts and isolated flavones from *Ligustrum vulgare* and *Phillyrea latifolia* leaves (Oleaceae). J. Ethnopharmacol. **70:** 213–217.
50. KAMINAGA, Y. *et al.* 2003. Production of unnatural glucosides of curcumin with drastically enhanced water solubility by cell suspension cultures of *Catharanthus roseus.* FEBS Lett. **555:** 311–316.

Humanized Recombinant Vaccinia Virus Complement Control Protein (hrVCP) with Three Amino Acid Changes, H98Y, E102K, and E120K Creating an Additional Putative Heparin Binding Site, Is 100-fold More Active Than rVCP in Blocking Both Classical and Alternative Complement Pathways

YOHANNES T. GHEBREMARIAM, ODUTAYO O. ODUNUGA, KRISTEN JANSE, AND GIRISH J. KOTWAL

Division of Medical Virology, Institute of Infectious Diseases and Molecular Medicine, Faculty of Health Sciences, University of Cape Town, Observatory 7925, Cape Town, South Africa

ABSTRACT: Vaccinia virus complement control protein (VCP) is able to modulate the host complement system by regulating both pathways of complement activation. Efficient downregulation of complement activation depends on the ability of the regulatory protein to effectively bind the activated third (C3b) and fourth (C4b) complement components. Based on native crystallographic structure, molecular modeling, and sequence alignment with other *Orthopoxviral* complement control proteins (CCPs) and their host homologs, putative sites have been found on VCP as contact points for C3b/C4b. Here, we report that using site-directed mutagenesis, modified proteins have been generated. In addition, we report that the generated modified proteins with postulated contact point substitutions have shown greater ability to regulate both the classical and the alternative pathways of complement activation than the recombinant Western Reserve VCP, with one modified protein showing nearly 100-fold more potency in regulating both complement activation pathways independently. The augmented *in vitro* inhibitory activity of the modified protein together with the newly created putative heparin binding site suggests its promising potential as a competent therapeutic agent in modulating various complement-mediated ailments, for example, traumatic brain injury, Alzheimer's disease, rheumatoid arthritis, multiple organ dysfunction syndrome, reperfusion injury, and xenorejection.

KEYWORDS: site-directed mutagenesis; rVCP; humanized rVCPs; 100-fold potency; complement; heparin

Address for correspondence: Girish J. Kotwal, Division of Medical Virology, IIDMM, Faculty of Health Sciences, University of Cape Town, Observatory 7925, Cape Town, South Africa. Voice: +27-21-406-6676; fax: +27-21-406-6018.
gjkotw01@yahoo.com

Ann. N.Y. Acad. Sci. 1056: 113–122 (2005). © 2005 New York Academy of Sciences.
doi: 10.1196/annals.1352.024

Poxviruses, cytoplasmic DNA viruses belonging to the family of poxviridae, encode a wide variety of proteins to evade the consequence of host defense and successfully challenge the host, ranging from small animals to humans.[1] The repertoire of encoded proteins engage in blocking the host complement, cytokine, and/or chemokine activity.[2,3] The vaccinia virus complement control protein (VCP) is secreted by vaccinia virus to block the complement system and protect the virus from neutralization. VCP, a member of the regulators of complement activation (RCA), shares a structural similarity to the human complement C4b binding protein (C4b-BP). Functionally, it is similar to other cellular complement regulatory proteins such as factor H (fH), soluble complement receptor type-1 (sCR1, CD35), decay-accelerating factor (DAF, CD55), membrane cofactor protein (MCP, CD46),[4,5] and other poxviral homologs, for example, the cowpox virus-encoded inflammation modulatory protein (IMP)[2] and the variola virus-encoded serine protease inhibitor of complement enzymes (SPICE).[6] VCP inhibits both the alternative and the classical pathways of complement activation because of its ability to bind complement components C3b and C4b and by facilitating the factor I–mediated degradation of these active components[7] while retaining its property to bind heparin.[4] Its ability to bind C4b might also contribute toward the inhibition of the lectin pathway of complement activation, because this pathway is known to associate with the classical pathway of complement activation.[8]

Efficient downmodulation of complement activation depends on the ability of the complement control protein to effectively bind C3b/C4b complement components. Sequence alignment of VCP with SPICE has revealed only 11 amino acid differences, yet SPICE was shown to possess almost 100-fold more potency in inactivating human C3b,[6] suggesting that some of the substitutions are exceptionally important in C3b interaction. Interestingly, 45% of the amino acid differences reside in the second module of the proteins (SCR-2),[9] suggesting a significant contribution of this module in SPICE's enhanced activity in modulating the complement system.

Moreover, native crystallographic structure[5] and computer modeling exercises[10] have disclosed the sites on VCP that putatively interact with C3b/C4b in its native and heparin-bound conformation.[10]

Based on these and other relevant observations, we used site-specific mutagenesis studies to enhance the activity of the yeast cell–expressed recombinant VCP (rVCP). We successfully generated modified proteins with deduced contact point substitution. Upon hemolysis assay–based functional comparison of the classical pathway, it was shown that the biologic activity of rVCP was greatly enhanced, as high as 100-fold. Moreover, the modified proteins have also been shown to regulate the zymosan-induced alternative pathway of complement activation more effectively than the rVCP, with one protein showing as high as 110-fold increased potency. This study reveals that a significantly greater potency can be achieved by making a small number of specific amino acid substitutions in the rVCP amino acid sequence.

MATERIAL AND METHODS

Cloning of rVCP/hrVCP

Genomic DNA was isolated from the Western Reserve (WR) strain of vaccinia virus, and a degenerative set of primers was designed to amplify the open reading frame of VCP while introducing 5′ *Eco*RI and 3′ *Not*I sites. The mutagenesis experiments were based on the native crystal structure,[5] molecular modeling,[10] and sequence alignment with other complement control proteins in the databank.

The amplicons were cloned into the expression vector pPIC9, downstream of an alcohol oxidase promoter, and GS115 *Pichia pastoris* yeast cells were transfected using the spheroplasting technique (Invitrogen Life Technologies). The cells putative to express pPIC9 integrated inserts were screened by polymerase chain reaction using the AOX-1 universal primers (Invitrogen Life Technologies), and the recombinant clones were used for subsequent expression studies.

Expression of rVCP/hrVCP

Expression of the recombinant proteins was performed by inoculating single colonies into 400 ml of buffered minimal glycerol medium (BMGY) at 30°C at 200 rpm until the OD_{600} reached between 2 and 6. The cells were harvested and resuspended in 100 ml of buffered minimal methanol medium (BMMY) containing 4% methanol[5] under similar conditions for 96 hours, and the samples were analyzed with SDS-PAGE .

Samples expressing a band at the size of the standard VCP were filter sterilized (Adcock Ingram filter units), diluted in binding buffer (50 mM sodium acetate, pH 5.5), and passed through a 5-ml HiTrap heparin column (Millipore) pre-equilibrated with the same binding buffer. The protein then was eluted with a linear NaCl gradient of 0 to 100% at a flow rate of 1 ml/min. The fractions containing pure rVCP/hrVCP were tested for biologic activity using hemolysis assay[11] for the classical pathway and using Bb fragment enzyme immunoassay for the alternative complement pathway.

Regulation of the Classical Complement Pathway

This assay is based on antigen–antibody complex and subsequent complement activation and quantitates the classical complement pathway and terminal complement components.[12] The activities of rVCP and the humanized recombinant VCPs (hrVCPs) were determined by testing their ability to inhibit complement-mediated lysis of sensitized sheep red blood cells (ssRBCs)[11](Diamedix Corporation). Samples containing equal amounts of the different proteins were tested at different concentrations (FIG. 2A) and their activities were compared (FIG. 2A, B). Standard human serum was used as a source of complement and was activated by incubating at 37°C with the IgG-sensitized sheep red cells. After 1-h incubation, the samples were centrifuged at 7,000 rpm for 30 seconds, and the supernatants were analyzed for the extent of lysis by measuring absorbance at 405 nm.

Regulation of the Alternative Complement Pathway

This enzymatic immunoassay selectively quantitates the degree of the alternative complement pathway activation.[13] This assay measures the amount of the generated proteolytic enzyme Bb, a cleavage product of the alternative pathway-specific factor B zymogen,[13] activated in the presence of C3b, and the alternative pathway enzyme factor D.[7] Although, the kit seems to have been developed for the assessment of naturally activated alternative complement components because of microbial infections or autoimmunity, we modified the procedure by preactivating the alternative pathway in standard human serum using the yeast cell–derived zymosan (Sigma-Aldrich).

Undiluted standard human serum was first mixed with different amounts of rVCP, ^{rVCP}E108K or ^{rVCP}H98Y,E102K,E120K and then with 12.5 μg of presonicated (2 × 30 s) zymosan and incubated at 37°C for 1 h.[14] The samples were centrifuged at 3,700 rpm for 2 min to remove any pelleted zymosan particles. Serum sample alone and serum with equivalent amounts of zymosan in the absence of rVCP/hrVCPs were used as negative and positive controls, respectively. All the samples were diluted to a final concentration of 1:10 with complement specimen diluent before use. The samples (100 μl each) were added into prewashed eight-well microplate assay strips immobilized with mouse anti-human Bb monoclonal antibody (Quidel Corporation) and incubated for 30 min at room temperature. The bound Bb was detected by adding 50 μl of horseradish peroxidase–conjugated goat anti-human Bb antibody and developed by adding the 1:20 diluted chromogenic substrate.[13] The concentration of Bb fragment was measured at 405 nm and the potency of rVCP was compared with the hrVCPs (FIG. 3).

RESULTS

Based on X-ray crystallographic structure,[5] computer modeling exercises,[10] and sequence alignment with its variola virus homolog, SPICE, and its host homologs CR1 and MCP, putative C3b/C4b binding sites have been mapped on rVCP. Using site-directed mutagenesis studies, we generated the single mutant ^{rVCP}E108K to make the residue identical to that of SPICE and the triple mutant

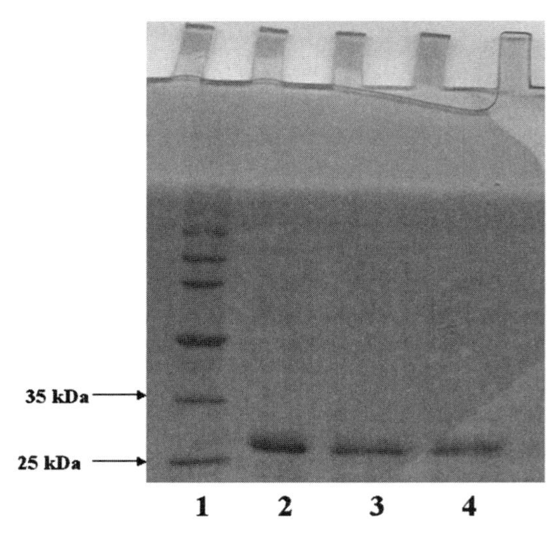

FIGURE 1. Coomassie blue–stained SDS-PAGE (12%) analysis of recombinant VCPs. (*lane 1*) Molecular weight marker; (*lane 2*) rVCP; (*lane 3*) ^{hrVCP}E108K; and (*lane 4*) ^{hrVCP}H98Y,E102K,E120K.

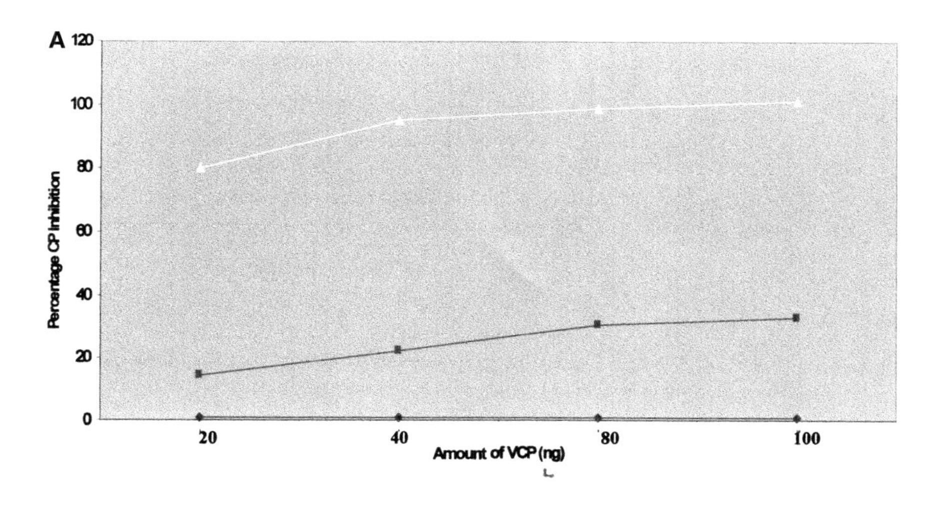

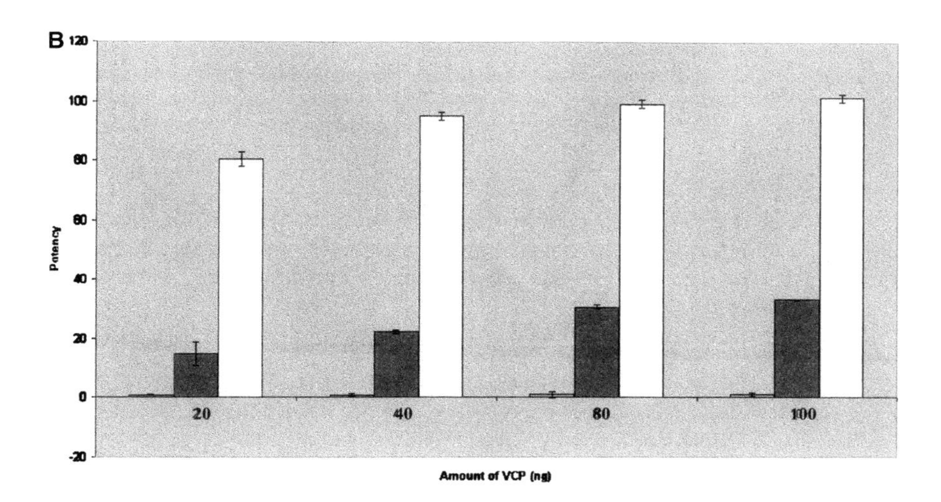

FIGURE 2. (**A**) Inhibition of complement-mediated red cell lysis by rVCP/hrVCPs. The coordinates represent the percentage of inhibition by rVCP (*diamonds*), ^{hrVCP}E108K (*squares*), or ^{hrVCP}H98Y,E102K,E120K (*triangles*) at the specified amounts. Experiments were performed in duplicates. (**B**) Comparison of the potency of rVCP with that of ^{hrVCP}E108K and ^{hrVCP}H98Y,E102K,E120K, respectively, at the indicated amounts. ^{hrVCP}E108K showing ~25-fold increased potency (ranging from 19.7 to 33.7 times) and ^{hrVCP}H98Y,E102K,E120K displaying ~100-fold increased activity (ranging from 91.8 to 116 times) than rVCP in regulating the classical pathway of complement activation. Experiments were performed in duplicate and the error bars represent standard deviation.

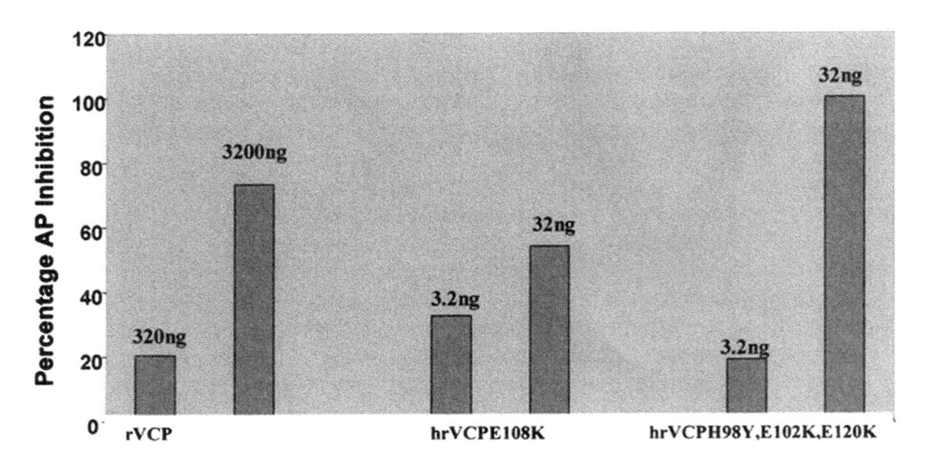

FIGURE 3. Comparison of the potency of rVCP with that of ^{hrVCP}E108K and ^{hrVCP}H98Y,E102K,E120K. ^{hrVCP}E108K (IC_{50} value of 32 ng) showing ~55-fold increased activity and ^{hrVCP}H98Y,E102K,E120K (IC_{50} value of 16 ng) showing 100-fold increased potency than rVCP (IC_{50} value of 1760 ng) in inhibiting the alternative pathway of complement activation.

^{rVCP}H98Y,E102K,E120K based on the sequence alignment with SPICE and MCP (H98Y), modeling studies (E102K), and alignment with CR1 and SPICE (E120K). The recombinant proteins were purified and analyzed with SDS-PAGE as shown in FIGURE 1. The purity of the proteins was comparable, the band intensities were measured, and the concentrations were estimated using the BioRad microplate assay protocol (BioRad).

The biologic activity of the different recombinant proteins then was compared for both the classical and the alternative complement pathways in a series of dilutions. Compared with the rVCP, the single mutant ^{rVCP}E108K showed a 25-fold increase in inhibiting the classical pathway (FIG. 2B) and a nearly 55-fold increase in inhibiting the alternative complement pathway with an IC_{50} value of 32 ng (FIG. 3). The triple mutant ^{rVCP}H98Y,E102K,E120K revealed almost 100-fold increased potency in inhibiting both the classical complement pathway with an IC_{50} value of 50 ng and the zymosan-induced alternative complement pathway with an IC_{50} value of 16 ng compared with the rVCP with an IC_{50} value of 1,760 ng in inhibiting the latter pathway (FIG. 3). This suggests that the deduced amino acids truly interact with C3b/C4b and enable the humanized rVCPs (hrVCPs) to effectively block complement activation. The values of the inhibitory concentrations at 50% (IC_{50}) suggest that the triple mutant ^{rVCP}H98Y,E102K,E120K possesses approximately threefold more potency in inhibiting the alternative pathway than the classical pathway. None of the amino acid substitutions is in the previously described heparin-binding region of rVCP.[5,15] Furthermore, the purification studies indicate that the heparin binding affinity is intact.

In fact, the residue change at position 102 (E102K) has created an additional putative heparin binding site of the K-X-K type[4] and has increased the number of putative heparin binding sites from one[4] to two in this module (SCR-2) and from four[4] to five in the entire protein. Therefore, the generated recombinant protein may have rendered *in vivo* biologic advantages such as controlling the aggregation of inflammatory cells such as neutrophils and natural killer (NK) cells[16] and increasingly maintaining *in vivo* retention in the heparan sulfate granules of endothelial cells.[17]

DISCUSSION

The ever-increasing advance in molecular biology and proteomics has helped researchers to understand and manipulate several human and non-human proteins. A significant identity between human and some orthopoxviral proteins has been noted for more than a decade.[18] These proteins, known as complement control proteins (CCPs), together with their mammalian homologs belong to the family of regulators of complement activation (RCA). However, there are considerable differences in terms of their biologic activity,[6,9,10,19] and, more interestingly, there is host discrimination between some of the well-studied poxviral complement control proteins such as VCP and SPICE.[6] Despite their sharing of more than 95% amino acid identity,[9,19] the remaining 11 amino acid differences seem to have created a preferential advantage for SPICE in inactivating the human C3b approximately 100-fold more efficiently.[6] However, Sfyroera *et al.*[9] have recently reported that SPICE is nearly 1,000-fold more active than VCP in inhibiting the lipopolysaccharide-induced activation of the alternative complement pathway by measuring bound human C3b in an ELISA-based assay. It was reasonable to assume that not all the 11 amino acid differences between VCP and SPICE are equally crucial in conferring SPICE's potency, because some of them are not surface-exposed/less exposed than the others.[5,6,10] It has been shown previously that the reversion of the amino acid tyrosine (Tyr) at position 98 in MCP has attenuated its C4b cofactor activity.[20]

Recently, the substitution of Tyr in VCP at this position has been shown to increase the complement regulatory activity by 28-fold, making it a good target for mutagenesis studies to replace the corresponding amino acid His-98 in VCP toward its MCP and SPICE homologs. Rosengard *et al.*[6] also demonstrated that the amino acid residues (94-103) in SPICE and VCP are important in C3b/C4b interaction. Moreover, a recent modeling study has revealed that His-98 and Glu-102 are surface-exposed and therefore expected to create a redundant interface for C3b/C4b interaction.[10]

The amino acid residue Glu-108 in VCP is also visualized as surface-exposed and is anticipated to contact C3b/C4b. Moreover, Sfyroera *et al.*[9] have reported a 17-fold increased potency in modulating the alternative pathway of complement activation by substituting the amino acid at this position to its SPICE analog (Glu-108-Lys), rendering it an important residue for modification. In addition, the amino acid Glu-120 in VCP is also expected to interact with both C3b and C4b,[10] and Lys-120 has substituted this residue in both CR1 and SPICE. Moreover, amino acid residue substitution at this position has been shown to affect C4b cofactor activity in MCP.[19] The generation of the mutant protein with the substitution of the single amino acid at this position to its SPICE analog (VCP-E120K) has been shown to immensely enhance the ability of the modified protein to regulate the alternative pathway of complement activation as high as 87-fold. It therefore was based on these evidences and inferences that we generated the single ^{rVCP}E108K and the triple ^{rVCP}H98Y,E102K,E120K mutants that possess between 25- and 100-fold more *in vitro* potency than the rVCP, respectively. This enhancement in VCP's activity is achieved most likely because, as suggested recently,[9] the one and the three amino acid substitutions increased the overall basic amino acid content of the modified proteins and reduced the acidic (negative) electrostatic potential at the second module (SCR-2). The *in vivo* role of these modified proteins remains to be elucidated. However, the augmented hemolysis assay–based activity suggests their promising potential in modulating complement-mediated inflammatory conditions. Especially those that are predominantly driven by the classical pathway of complement activation as in the pig-to-baboon organ xenotransplantation (extensively reviewed elsewhere in the book). The significantly increased ability to regulate the zymosan-induced alternative pathway of complement activation promises a candid potential in attenuating complement-mediated ailments predominantly driven by the alternative complement pathway such as the guinea pig-to-rat cardiac xenotransplantation,[21] renal reperfusion injury,[22] and zymosan-induced multiple organ dysfunction syndrome (MODS).[23,24] Moreover, the overall increase in the number of putative heparin binding sites might have biologically privileged the triple ^{rVCP}H98Y,E102K,E120K mutant. The *in vitro* ability of VCP to inhibit zymosan-induced complement activation was first demonstrated by Mahesh *et al.*[24] We feel that it is important to establish an animal model with MODS triggered by zymosan to further understand the principles of the alternative pathway-driven multiple organ failure and therapeutically interfere using such effectively modified complement control proteins.

In conclusion, although a similar potency to that of SPICE was achieved with only three amino acid substitutions, the substitution Lys-102-Glu does not exist in SPICE, making it difficult to conclude that only the three amino acids are exceptionally critical for achieving the 100-fold increased potency in SPICE. Therefore, we recommend the generation of a double mutant ^{rVCP}H98Y,E120K or a triple mutant ^{rVCP}H98Y,E108K,E120K to specifically evaluate the role of these residues in SPICE.

ACKNOWLEDGMENTS

This work was supported by the Poliomyelitis Research Foundation and the University of Cape Town (Y.T.G.). O.O.O. was a recipient of the Claude Harris Leon Foundation, the National Research Foundation (NRF, South Africa), and the University of Cape Town, Faculty of Health Sciences postdoctoral fellowships. G.J.K. is currently a Senior International Wellcome Trust Fellow for biomedical sciences in South Africa. We thank Dieter Blaas for the rVCP/hrVCPs surface potential drawings which were done using the program "GRASP." We also thank Dr. Kevin Dennehy for generously providing the zymosan.

REFERENCES

1. Smith S.A. & G.J. Kotwal. 2002. Immune response to poxvirus infections in various animals. Crit. Rev. Microbiol. **28:** 149–185.

2. KOTWAL, G.J. 2000. Poxviral mimicry of complement and chemokine system components: what is the end game? Immunol. Today **21:** 242–248
3. ALCAMÍ, A., J.A. SYMONS & G.L. SMITH. 2000. The vaccinia virus soluble alpha/beta interferon (IFN) receptor binds to the cell surface and protects cells from the antiviral effects of IFN. J. Virol. **74:** 11230–11239.
4. SMITH, S.A., N.P. MULLIN, J. PARKINSON, *et al.* 2000. Conserved surface-exposed K/R-X-K/R motifs and net positive charge on poxvirus complement control proteins serve as a putative heparin binding sites and contribute to inhibition of molecular interactions with human endothelial cells: a novel mechanism for evasion of host defense. J. Virol. **74:** 5659–5666.
5. MURTHY, K.H.M., S.A. SMITH, V.K. GANESH, *et al.* 2001. Crystal structure of a complement control protein that regulates both pathways of complement activation and binds heparan sulfate proteoglycans. Cell **104:** 301–311.
6. ROSENGARD, A.M., Y. LIU, Z. NIE & R. JIMENEZ. 2002. Variola virus immune evasion design: expression of a highly efficient inhibitor of human complement. Proc. Natl. Acad. Sci. USA **99:** 8808–8813.
7. KOTWAL, G.J. 1997. Microogranisms and their interaction with the immune system. J. Leukoc. Biol. **62:** 415–429.
8. SUANKRATAY, C., C. MOLD, Y. ZHANG, *et al.* 1999. Mechanism of complement-dependent haemolysis via the lectin pathway: role of the complement regulatory proteins. Clin. Exp. Immunol. **117:** 442–448.
9. SFYROERA, G., M. KATRAGADDA, D. MORIKIS, *et al.* 2005. Electrostatic modeling predicts the activities of orthopoxvirus complement control proteins. J. Immunol. **174:** 2143–2151.
10. VANNAKAMBADI, G.K., S.A. SMITH, G.J. KOTWAL & K.H.M. MURTHY. 2004. Structure of vaccinia complement protein in complex with heparin and potential implications for complement regulation. Proc. Natl. Acad. Sci. USA **101:** 8924–8929.
11. KOTWAL, G.J., S.N. ISAACS, R. MCKENZIE, *et al.* 1990. Inhibition of the complement cascade by the major secretory protein of vaccinia virus. Science **250:** 827–829.
12. DIAMEDIX CORPORATION. 2003. EZ complement CH50 test for in vitro diagnostic use. A subsidiary of IVAX Diagnostics, Inc. Miami, FL.
13. QUIDEL CORPORATION. 2004. An enzymatic immunoassay for the quantitation of Bb fragment of Factor B of the alternative complement pathway. San Diego, CA.
14. KAZATCHKINE, M., D.T. FEARON, J.E. SILBERT & K.F. AUSTEN. 1979. Surface-associated heparin inhibits zymosan-induced activation of the human alternative complement pathway by augmenting the regulatory action of the control proteins on particle-bound C3b. J. Exp. Med. **150:** 1202–1215.
15. SMITH, S.A., R. SREENIVASAN, G. KRISHNASAMY, *et al.* 2003. Mapping of regions within the vaccinia virus complement control protein involved in dose-dependent binding to key complement components and heparin using surface plasmon resonance. Biochim. Biophys. Acta **1650:** 30–39.
16. AL-MOHANNA, F., R. PARHAR & G.J. KOTWAL. 2001. Vaccinia virus complement control protein is capable of protecting xenoendothelial cells from antibody binding and killing by human complement and cytotoxic cells. Transplantation **71:** 796–801.
17. JHA, P., S.A. SMITH, D.E. JUSTUS & G.J. KOTWAL. 2003. Prolonged retention of vaccinia virus complement control protein following IP injection: implications in blocking xenorejection. Transplant. Proc. **35:** 3160–3162.
18. KOTWAL, G.J. & B. MOSS. 1988. Vaccinia virus encodes a secretory polypeptide structurally related to complement control proteins. Nature **335:** 176-178.
19. LACHMAN, P.J. 2002. Microbial subversion of the immune response. Proc. Natl. Acad. Sci. USA **99:** 8461–8462.
20. LISZEWSKI, M.K., M. LEUNG, W. CUI, *et al.* 2000. Dissecting sites important for complement regulatory activity in membrane cofactor protein MCP (CD46). J. Biol. Chem. **275:** 37692–37701.
21. ANDERSON, J.B., S.A. SMITH, R. VAN WIJK, *et al.* 2003. Vaccinia virus complement control protein inhibits hyperacute xenorejection in a guinea pig-to-rat heterotopic cervical cardiac xenograft model by blocking both xenoantibody binding and complement pathway activation. Transplant. Immunol. **11:** 129–135.

22. STAHL, G.L., Y. XU, L. HAO, *et al.* 2003. Role for the alternative complement pathway in ischemia/reperfusion injury. Am. J. Pathol. **162:** 449–455.
23. MAHESH, J., C.G. MILLER, J.C. PEYTON, *et al.* 1997. Elucidation of the mechanisms of action of MIP-1 alpha and C5a in regulating the lethality of Zymosan-induced mods. 4th International Congress on the Immune Consequences of Trauma, Shock and Sepsis. Munich, Germany.
24. MAHESH, J., J. DALY, W.G. CHEADLE & G.J. KOTWAL. 1999. Elucidation of the early events contributing to zymosan-induced multiple organ dysfunction syndrome using MIP-1α, C3 knockout, and C5-deficient mice. Shock **12:** 340–349.

Intervention Strategies and Agents Mediating the Prevention of Xenorejection

YOHANNES T. GHEBREMARIAM,[a] SCOTT A. SMITH,[a,b] J.B. ANDERSON,[b]
D. KAHN,[c] AND GIRISH J. KOTWAL[a,b]

[a]*Division of Medical Virology, IIDMM, University of Cape Town 7925, HSC, Cape Town, South Africa*

[b]*Department of Microbiology and Immunology, University of Louisville School of Medicine, Louisville, Kentucky 40202, USA*

[c]*Division of General Surgery, Department of Surgery, University of Cape Town, GSH, HSC, Cape Town, South Africa*

ABSTRACT: Xenotransplantation, the transplantation of cells, tissues, and/or organs across species, has proven to be an enormous challenge, resulting in only limited achievements over the last century. Unlike allotransplantation, the immunologic barriers involved in xenotransplant rejection are aggressive and usually occur within minutes in a hyperacute fashion. The use of organs from phylogenetically related concordant species may not be practical. Discordant xenotransplantation is characterized by hyperacute graft rejection, and to use nonprimate discordant organs for human benefit will require manipulation of the taxonomic differences. The hyperacute rejection process is primarily due to the attachment of preformed xenoreactive antibodies to the donor vascular endothelium, which results in hyperactivation of the complement system beyond the control of the natural complement regulatory proteins. Understanding the complex and diverse immune components involved in hyperacute, acute, and accelerated rejections has resulted in the development of different hematologic and molecular strategies. Plasmapheresis has been used to remove xenoantibodies, and xenoperfusion techniques are used to create a suitable and familiar environment for the xenograft. Various molecular approaches, such as the development of transgenic animals expressing human complement regulatory proteins such as CD59 or decay accelerating factor (DAF), to downregulate complement activation or the production of pigs lacking the xenoreactive antigen by knockout of the Galα-1,3-galactosyl transferase gene have also been attempted. A combination of these techniques together with the administration of soluble complement inhibitors such as the vaccinia virus complement control protein (VCP) may well contribute to prolong graft survival. However, various issues including the possible emergence of new viral infections have confounded the topic of xenotransplantation. Here the different modulatory approaches and agents mediating interventions in xenorejection are discussed.

KEYWORDS: xenorejection; complement; hyperacute rejection; organ transplantation; VCP; inflammation; modulation

Address for correspondence: Girish J. Kotwal, Division of Medical Virology, IIDMM, University of Cape Town 7925, HSC, Cape Town, South Africa. Voice: +27-21-406-6676; fax: +27-21-406-6018.
gjkotw01@yahoo.com

Ann. N.Y. Acad. Sci. 1056: 123–143 (2005).
doi: 10.1196/annals.1352.028

INTRODUCTION

The advent of clinical transplantation has resulted in a radical transformation in the field of medicine. The transplantation of vital human organs such as the heart, liver, and kidneys, made possible largely due to the discovery of an immunosuppressive agent, cyclosporine, from a natural source, is now a routine procedure in many major centers around the world. However, the supply of living related and cadaveric organs is inadequate to meet the demand. For example, in the United States alone, up to 45,000 patients seek heart transplants every year; however, only 2,000 hearts are available for allotransplantation.[1] This severe organ shortage is unfortunately increasing by approximately 10–15% each year.[2] Conventional therapies are unable to significantly support the failing organ(s), and the cost of care for patients on waiting lists is generally unaffordable. As a result, many patients die while on waiting lists.[3] The use of non-human cells, tissues, and organs (xenotransplants) could well be an alternative solution to overcome this shortage.[4–6]

Cellular xenotransplantation has attained considerable success preclinically and has been able to reach into different phases of clinical trials in an endeavor to treat patients with different primary cell disorders.[7–10] An attempt has been made to reverse hyperglycemia by using pig isletlike cell clusters in patients with chronic type I diabetes mellitus.[7] This study elucidated the safety of grafting and the survival of insulin-producing piglet cells posttransplantation.[7] In addition, there is histological indication for the survival of fetal pig neural cells beyond 7 months after transplantation in a patient with Parkinson's disease.[11] Similarly, tissue xenotransplantation has also been evaluated, and despite the outcomes, it proceeded into various stages of trials. The transplantation of skin, cornea, and bone marrow has been attempted experimentally and/or clinically.[12–15] Baboon bone marrow transplantation was attempted in AIDS patients in the hope of producing chimeric cells resistant to HIV infection.[12,13]

Despite the feasibility of cellular and tissue xenotransplantation, repeated failures and potential risks of solid organ xenotransplantation have created audible concern and demanded a more careful approach to its undertaking.[16,17] The effort of several researchers has revived the field of xenotransplantation as an alternative life support. Since Ernest Unger of the University of Berlin first attempted it in 1910,[18] the transplantation of non-human primate organs into humans has been performed sparingly. For example, the surgical team led by James Hardy of the University of Mississippi performed the world's first heart xenotransplantation using a chimpanzee as the donor.[19] Thirteen years later, in 1977, two cardiac xenotransplants were performed in Cape Town, South Africa, using baboon and chimpanzee donor hearts accompanied by high doses of immunosuppressive drugs.[20] Subsequently, different researchers attempted to control short-term xenorejection using the conventional immunosuppressive therapy currently used for human allografts.

By definition, however, the transplantation across such phylogenetically related concordant species does not result in hyperacute rejection.[21] The routine use of non-human primate organs for human transplantation is not a viable option, as it may not be feasible to obtain an adequate number of organs from non-human primates, as they are endangered, have single births, and have a long gestation period. Moreover, there is reluctance to use non-human primate organs for religious and/or ethical reasons.[22] Therefore, the use of non-primate discordant species as potential organ donors has been explored.

Different researchers actively working in the field have focused on pigs as the most suitable secondary organ source.[23,24] This is mainly because of similarities in organ morphology, large litter size, and short gestation period, which can easily overcome the large shortfall of organ donors. In addition, fewer objections would be raised by animal rights groups. For example, in the United States alone, over 95 million pigs are slaughtered each year for food purposes;[24] however, comparatively fewer numbers of pigs would be used for transplantation.

The transplantation of organs from species taxonomically distant to humans, such as swine, results in an immediate fulminate rejection process known as hyperacute rejection (HAR). HAR occurs within minutes to hours, depending on the type of organ being transplanted, in contrast to cell-mediated human allograft rejection, which takes place progressively from within a few days.[3,25] However, if HAR is diverted successfully, the graft experiences another phase of rejection within days, known as delayed xenograft rejection (DXR) by some and as acute vascular rejection (AVR) by others. If graft survival can be prolonged beyond DXR, it is speculated that there could be a xenorejection process equivalent to T-cell–mediated allograft rejection and ultimately some form of chronic rejection. A brief description of each of the experienced or anticipated types of rejection follows.

TYPES AND MECHANISMS OF XENOREJECTION

Xenotransplantation is defined as the transplantation of cells, tissues, and/or organs across species. Solid organ xenotransplantation is broadly categorized as follows: (1) concordant xenotransplantation is transplantation between closely related species such as hamster-to-rat, wolf-to-dog, and apes- or old-world monkeys-to-humans, which does not manifest in a hyperacute rejection,[26,27] and (2) discordant xenotransplantation is transplantation between distantly related species such as the guinea pig-to-rat, pig-to-dog, pig-to-baboon, and pig-to-human, which ends in hyperacute rejection (HAR.)[28–30]

Hyperacute Rejection (HAR)

The epicentral cause of HAR is hyperactivation of the complement system either due to the attachment of preformed xenoreactive natural antibodies (XNAs) to the donor endothelial cells, with subsequent activation of the antibody-dependent component of the complement pathway, ordue to failure of the complement regulatory proteins to adequately inhibit the spontaneous activation of the alternative complement pathway. The predominant driving force seems to vary depending on the species of the donor and the recipient,[31] that is, whether the recipients have preformed xenoantibodies and whether the complement regulatory proteins on the graft endothelium fail to control the recipient's heterologous complement.[29] For example, in the guinea pig-to-rat vascular graft, the alternative pathway of complement activation seems to predominate, depending on the strain type of the recipient rat.[32–34] In this case, the membrane inhibitor of reactive lysis, CD59, which is commonly found on the surface of endothelial cells, fails to significantly suppress spontaneous activation of the alternative pathway. The classical pathway initiates preformed or induced antibody-dependent hyperacute rejection as in the mouse-to-sensitized rat and pig-to-

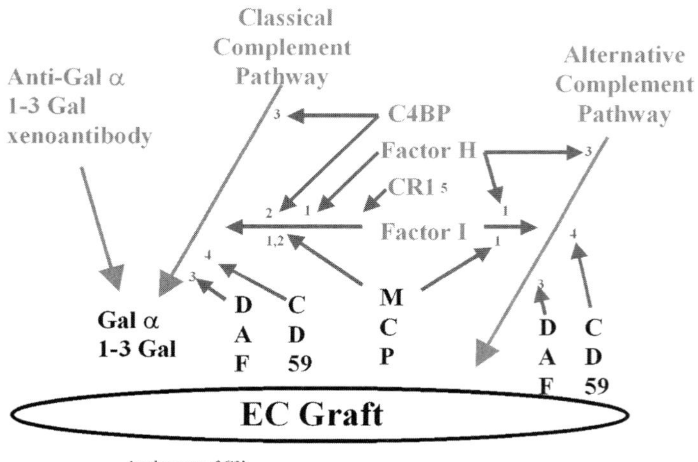

FIGURE 1. Mediators of hyperacute rejection. Interaction between membrane-bound complement regulatory proteins (MCP, CD 59, and DAF) and soluble-complement regulatory proteins (C4BP, factor H, CR 1, and Factor I) with complement activation. Anti-Galα–1,3 Gal antibodies attach to Galα-1,3 Gal epitopes on the graft endothelial cells (ECs) with subsequent activation of complement via the classical pathway, or spontaneous activation of the alternative pathways goes unchecked by the heterologous membrane-bound complement regulatory proteins. Which pathway predominates depends on the species involved.

primate xenotransplants. In pig-to-primate vascular grafts, the primates have preformed natural xenoreactive antibodies (xenoantibodies) to the pig grafted tissue, which drives the classical pathway of complement activation. In the mouse-to-sensitized rat vascular graft, the rat has induced antibodies to mouse tissues, which also drives the classical pathway. The different mediators in HAR are depicted in FIGURE 1. The degree to which the graft is rejected is equivalent to the amount of antibodies, effective complement regulatory proteins, and/or xenoantigens that are present.

During the process of evolution the terminal galactosyl transferase gene sequence was truncated in humans and old world monkeys, and therefore these species produce preformed xenoreactive natural antibodies (XNAs) that bind to the transplanted donor endothelium expressing the Gal α 1-3 gal antigenic moiety during transplantation with a non-primate organ.[23] In fact, these xenoantibodies are probably produced during early childhood in response to certain microorganisms that colonize the colon.[35,36] This xenoantibody binding causes activation of the transplanted "resting" vascular endothelial cells followed by complement attachment to the Fc receptors of the xenoantibodies, resulting in complement activation and HAR. The

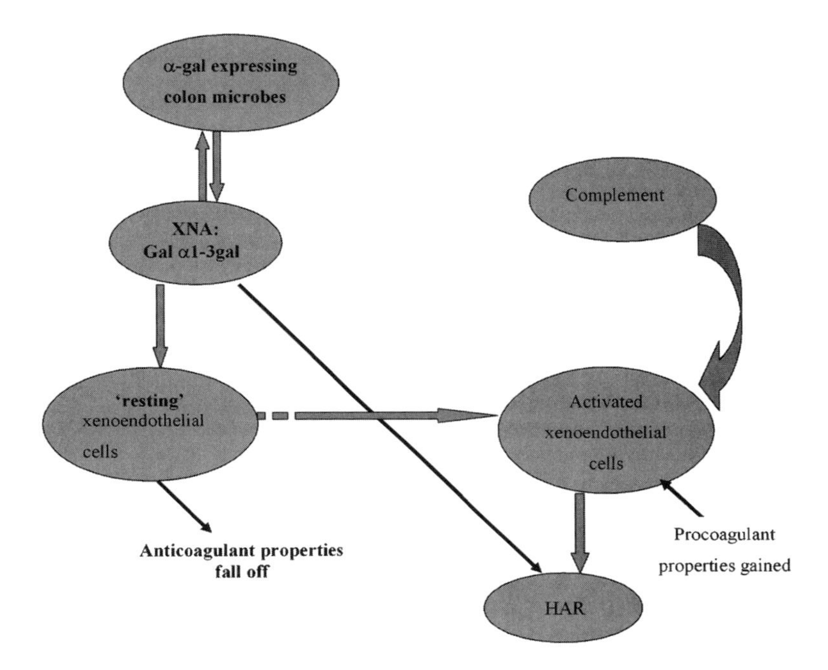

FIGURE 2. Proposed mechanism in hyperacute rejection (HAR). Elucidation of the role of xenoantibodies and complement in triggering hyperacute rejection.

proposed mechanism for HAR is shown in FIGURE 2. All or most of the XNAs, which constitutes approximately 1% of circulating immunoglobulins in humans, apes, and old-world monkeys, are directed against the Gal α1-3 gal antigenic moiety.[37,38]

The level of expression of the α-galactosyl epitope is proportional to the severity of graft rejection. For example, its expression level appears in millions of copies per cell in pigs.[38,39] Hence, when pig organs are transplanted into humans, these XNAs react with the terminal Gal α1-3 gal epitopes on the pig endothelium. Although the anti-gal α1-3 gal antibodies seem to predominate the HAR response, other additional prominent antibodies may play a critical role if the α-Gal antibodies are removed.

Besides XNA and complement, the other pivotal component in HAR is the vascular endothelial cell. The titer level of the recipient's XNAs, the multiplicity of the donor's antigenic expression, and the effectiveness of the recipient's complement regulatory proteins dictate whether the endothelial cells are driven to activation or cell cytotoxicity.[3] The contribution of the endothelial cell in the HAR response is probably linked to the species involved, which delineates the roles of the XNAs and complement. In aggressive rejections lasting only a few minutes, endothelial cells do not have the time to become fully activated. However, in slower rejections lasting several minutes to hours and beyond, the endothelial cell probably plays a more actively significant role as it has time to be activated.

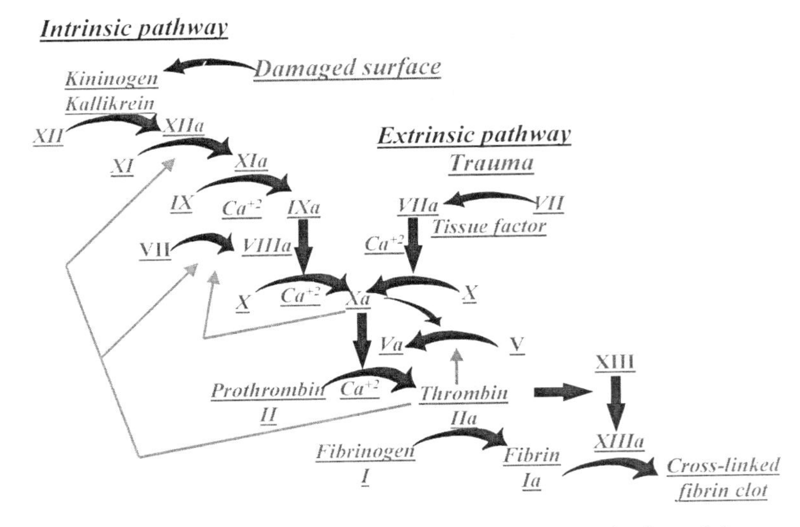

FIGURE 3. Coagulation cascade. The *block arrows* represent the flow of the coagulation cascade and the *line arrows* represent the amplification feedback loops. The intrinsic pathway is now believed to be mainly an amplification loop of the extrinsic pathway.

In the subsequent phases of rejection (after HAR) that occur day(s) after transplantation, the endothelial cells have ample time to be fully activated and can become an even greater contributor to the rejection process.

The contribution that the endothelial cell can potentially make is linked to its structural and/or functional alterations. Upon activation the cell expresses various components that increase vascular permeability. Moreover, the cell's anticoagulant properties decrease and procoagulant properties are gained. FIGURE 3 portrays the sequence of events in the normal coagulation cascade, and FIGURE 4 shows the quiescent endothelial cell's effect on the coagulation cascade.

Two important anticoagulant molecules, thrombomodulin and heparan sulfate proteoglycan, are expressed on the surface of such "resting" endothelial cells. Thrombomodulin binds thrombin and accelerates the conversion of protein C (PC) to activated PC (APC). Thus, in conjugation with protein S, it splits factors VIIIa and Va, resulting in reduced thrombosis as annotated below (FIG. 5).

Heparan sulfate proteoglycans (HSPs) foster the integrity of blood vessels by securing the attachment of plasma proteins on the surface of "dormant" endothelial cells. Heparan sulfate also mediates the activation of antithrombin III to discourage thrombin formation (FIG. 6). Moreover, HSPs protect endothelial cells from the toxic effects of free radicals by promoting the activation of scavenging enzymes.[40]

Activated endothelial cells express platelet-activating factor (PAF), tissue factor (TF), and plasminogen activator inhibitor (PAI-1), which all promote coagulation (FIG. 7). PAF and TF activate factors V and VII, respectively, and PAI-1 inhibits tissue plasminogen activator (TPA). The transcription of thrombomodulin gets downregulated, and factors VIIIa and Va are delayed. Heparan sulfate is cleaved and lost from the surface of the endothelial verge. A large portion of the heparan sulfate (5%

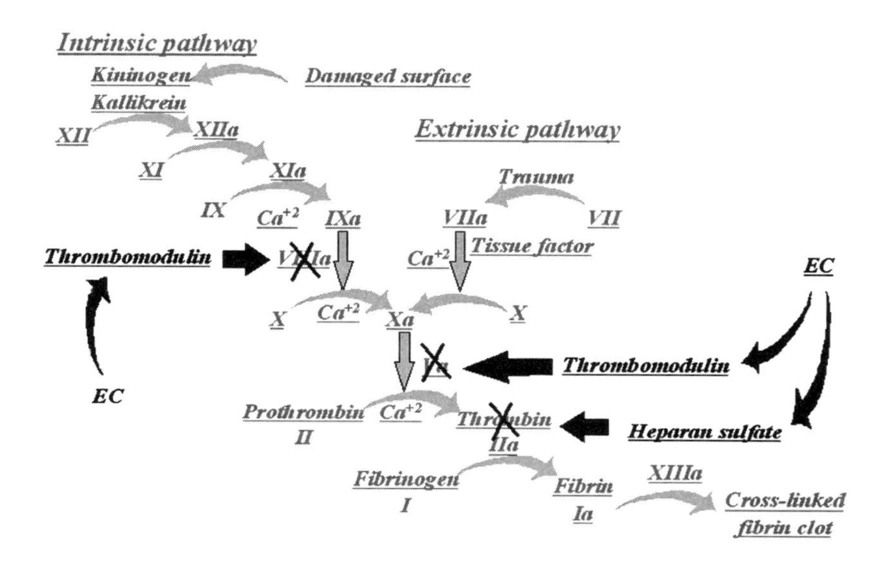

FIGURE 4. Quiescent endothelial cells (ECs) control of the coagulation cascade promoting anticoagulation. Illustration of the proteins secreted by quiescent endothelial cells, which directly affect the coagulation cascade to promote an anticoagulation state.

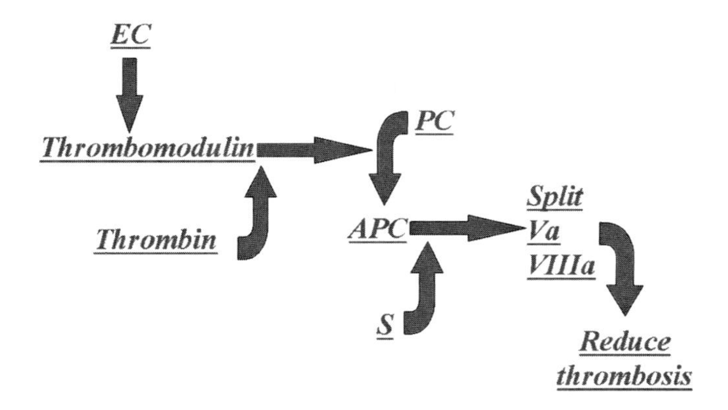

FIGURE 5. Quiescent endothelial cell (EC) control of coagulation via thrombomodulin and its mechanism of action to promote anticoagulation.

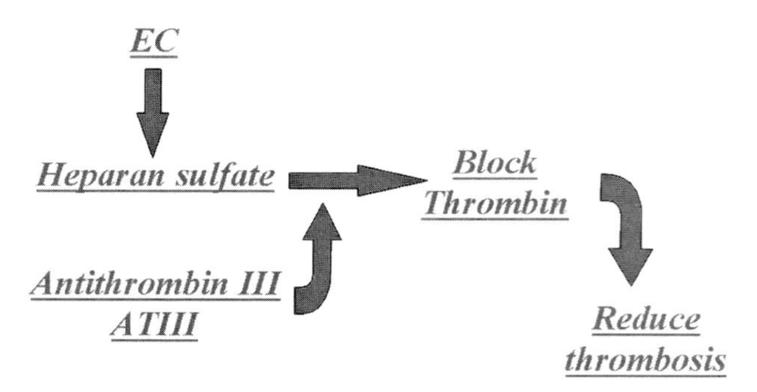

FIGURE 6. Quiescent endothelial cell (EC) control of coagulation via heparan sulfate and its mechanism of action to promote anticoagulation.

within 4 min and 50% within 30 min) is lost from the endothelial cells upon exposure to anti-endothelial antibodies and complement.[41,42] The loss of heparan sulfate appears to depend on the generation of C5a.[43]

Depositions of C5b-7 complexes on endothelial cells cause the cells to transiently retract from each other.[44] The gaps formed between the endothelial cells result in exposure of tissue factor in the subendothelium to factor VII in the blood, resulting in inhibition of the coagulation cascade. In addition, the subendothelium may be exposed to platelets and Von Willebrand factor (vWF), resulting in platelet activation. The activated platelets then release their prothrombotic substances, leading to stable clot formation. The formation of intracellular gaps is only temporary. The gaps are reversed by the addition of C8 and C9 into the C5b-7 complex during the formation of the membrane attack complex (MAC). MAC may also promote coagulation by inducing endothelial cell vesiculation and gathering the prothrombinase complex. Complement components C3 and C5 also cause vasoconstriction and act as chemotactic factors for inflammatory cell influx. However, in the case of swift rejection, the chemotaxic effect is probably not fully realized. When rejection occurs within minutes, structural changes rather than metabolic alterations are the more likely mechanisms of coagulation initiation.

An additional consideration is that all of the coagulation regulatory proteins mentioned may vary in their efficacy depending on the species involved.[45] Some paradoxic molecular interactions may also occur, such as porcine endothelium's ability to activate prothrombin directly, which has been shown *in vitro.*[46,47] Moreover, platelets can be directly activated by porcine vWF, which is expressed on the activated endothelium.[48]

After XNA attachment to the surface of endothelial cells, activation of host complement and coagulation cascades lead to interstitial hemorrhage, intravascular coagulation, and ischemic necrosis.[49] The histopathologic hallmarks of HAR (FIG. 2) consist of a disrupted vascular endothelium, with massive interstitial edema and hemorrhage.[50,51]

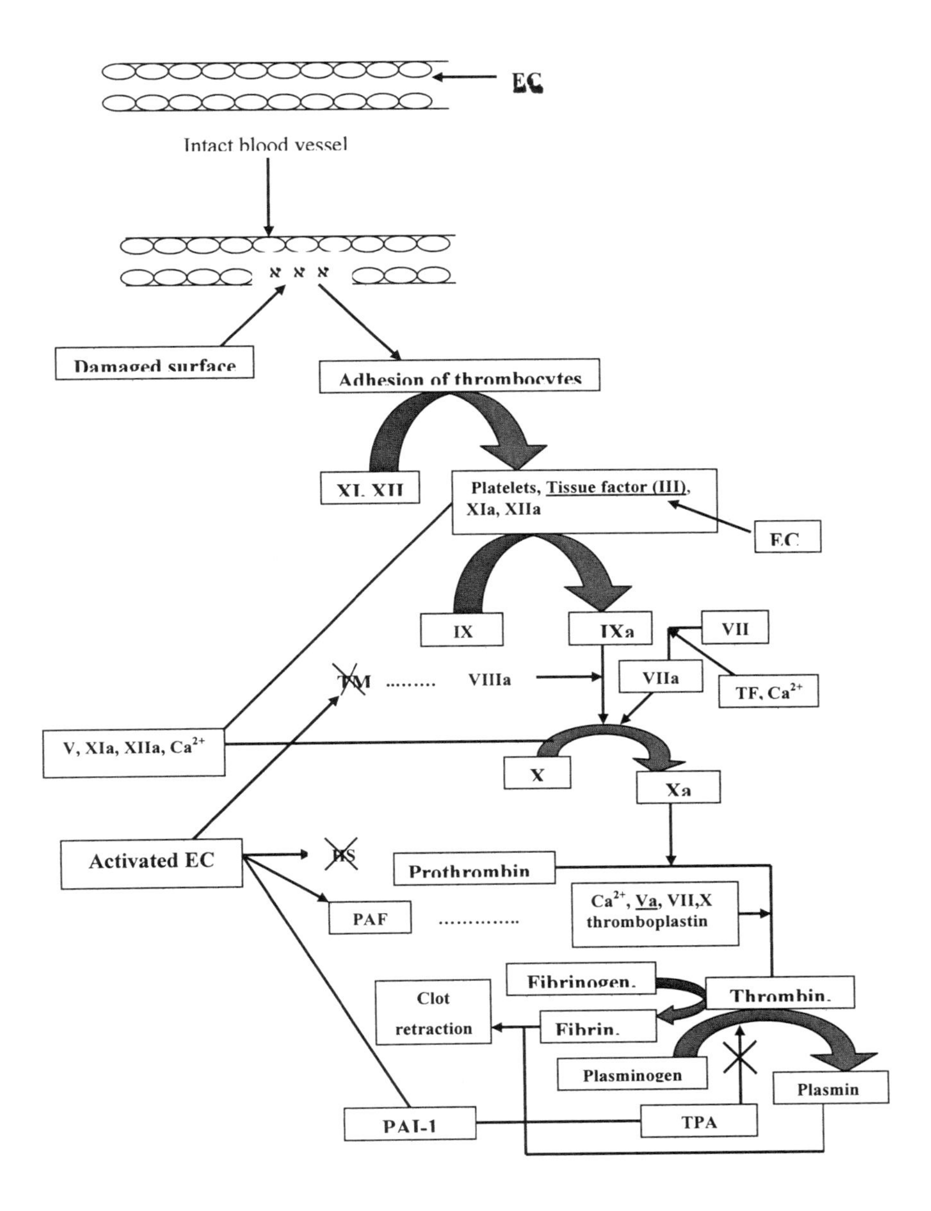

FIGURE 7. Activated endothelial cell (EC) enhances coagulation. Proteins that fall off during EC activation and their effect on the coagulation cascade.

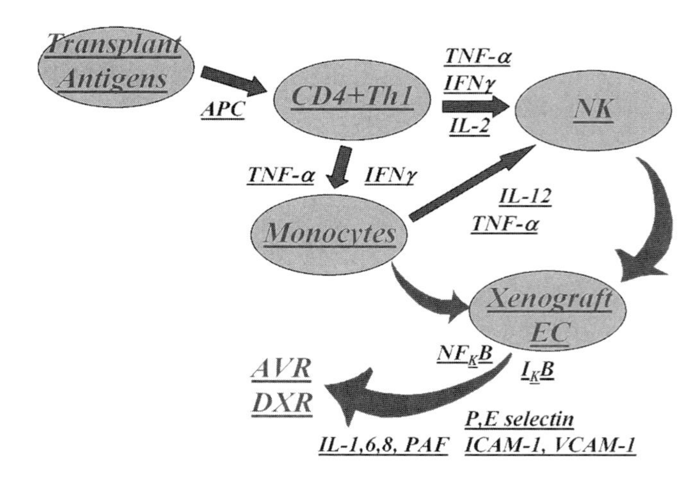

FIGURE 8. Sequence of events in delayed xenograft rejection. Types of cells involved in delayed xenograft rejection.

Delayed Xenograft Rejection

If HAR is averted,[52,53] current evidence shows that a delayed form of rejection occurs within a few days,[54] which leads to more gradual graft rejection. This process has been named acute vascular rejection (AVR) by some[55] and delayed xenograft rejection (DXR) by others.[54] In this chapter, we use the latter name. FIGURE 8 shows the proposed mechanism of DXR. The exact mechanism of DXR remains unclear.

DXR involves activated monocytes and natural killer (NK) cells.[56] Endothelial cells become activated, and the consequence of this activation plus the lethal products of the activated monocytes and NK cells promotes DXR.[57] The inability of various complement regulatory molecules to function across species also exacerbates the inflammatory response against the xenograft.[58] The myriad of histopathologic hallmarks of DXR varies from those seen in HAR to those revealing the presence of cytokines and infiltrating cell types such as NK cells and macrophages.[52,59,60] DXR still remains a formidable obstacle that has not been successfully overcome.

Acute Cellular Rejection

Because of early graft destruction due to HAR or DXR, evaluation of the eventual cellular response has not been studied in detail. However, FIGURE 9 depicts the proposed mechanism of acute cellular rejection (ACR).

Preliminary data in pig-to-human or non-human primate xenotransplantation indicates that human T cells directly recognize porcine xenoantigen on endothelial and dendritic cells. Vigorous cell proliferation and cytotoxic secretions then follow, and the cellular rejection process is anticipated to be at least equivalent to that of an allograft.[55] The impact of the current allograft oriented immunosuppressive drugs on ACR, provided that HAR and DXR can be overcome, remains to be seen.

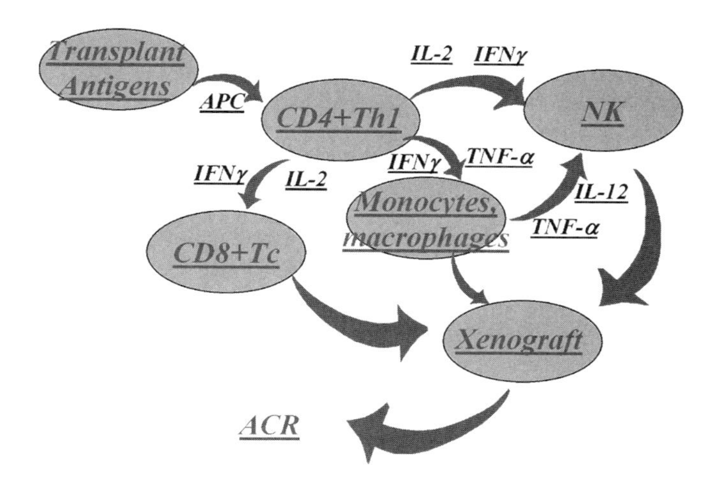

FIGURE 9. Sequence of events in acute cellular rejection. Types of cells involved in acute cellular rejection.

Chronic Rejection

Since discordant xenografts have never survived the early rejection processes, nothing specific is known about chronic rejection in discordant xenotransplantation. However, some investigators have speculated that the chronic rejection (CR) of xenografts may well be more unprovoked than that of allografts.[61] In allografts it is a vasculopathy that occurs in the vessels, leading to intimate proliferation and finally to lamina occlusion with ischemia or infarction in parts of the transplanted organ. Its cause is believed to be immune related.

Accelerated Rejection

Accelerated rejection (AR) involves the same cells and mechanisms as those in HAR except that there are no preformed antibodies. This type of rejection would occur instead of HAR in concordant xenografts such as mouse-to-naïve rat or hamster-to-naïve rat xenografts. In such cases, the antibodies against the endothelial cells of the xenograft are generated as with any exposure to an antigenic stimulus. After the production of antibodies (within a few days), the rejection process proceeds as in HAR. Attachment of the antibodies to the grafted cells is followed by complement activation, ending in graft failure. FIGURE 10 shows the mechanism of AR. AR differs from HAR not only in the time of rejection, but also in the predominant class of antibodies involved. In AR the antibodies are considered to be of the IgG class, whereas in HAR they are predominantly of the IgM isotype. However, this might vary depending on the type of organ being transplanted.

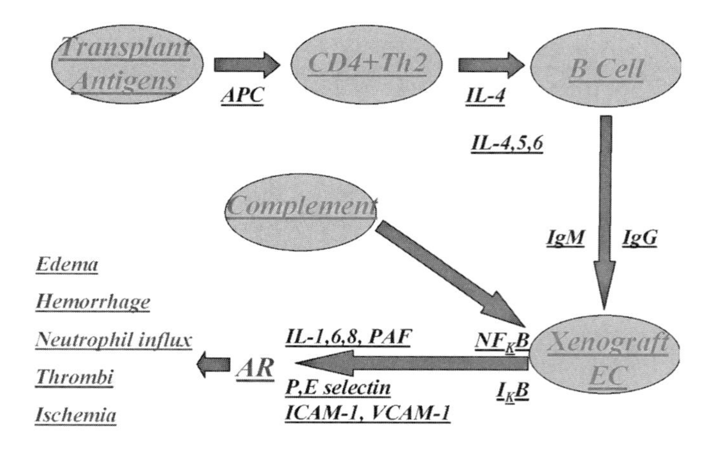

FIGURE 10. Sequence of events in accelerated rejection. Illustration of the types of cells involved in accelerated rejection.

APPROACHES TO OVERCOMING XENOREJECTION

A number of techniques have been applied to manipulate XNA, complement, and/or endothelial cell activation. Some of the techniques have been used independently, whereas others are used in combination to prolong the survival of xenografts. Various hematological, molecular, and therapeutic strategies that have been attempted in the history of organ xenotransplantation are discussed below.

Hematological Approaches

Based on advancing knowledge in the field of immunohematology, kaleidoscopic approaches have been formulated to regulate hyperacute graft rejection. Exchange of the recipient's plasma with albumin, fresh frozen plasma, or other volume replacement fluids has been successfully used to reduce the xenoantibody titers from the recipient's blood.[62] Although this technique has been proven to prolong graft survival, it is nonspecific and hence depletes many biologically important immunoglobulins, which may compromise the recipient's residual defense mechanism against microbial infections. Moreover, due to exhaustion of important plasma proteins such as PC and ATIII (refer to FIGS. 5 and 6 for their specific role), the coagulation cascade is deregulated.[62] However, this technique has resulted in a 23-fold prolonged survival of a discordant graft in a guinea pig-to-rat cardiac xenotransplantation.[63] Similarly, depletion of high antibody titers has been obtained using an affinity column adsorption technique.[62] The latter approach is more specific towards the depletion of the xenoantibodies. Today, electrophoresis is used to fractionate plasma proteins and return the vital components to the recipient's circulation.

Even though XNAs are the predominant antibodies involved in HAR, the importance of donor versus recipient ABO blood group compatibility has been suggested to avoid the possible interaction of some reactive antibodies.[1,64] Therefore, as in allotransplantation, it is advisable to perform blood type compatibility and cross-match tests as criteria before xenotransplantation. Other immunohematological options have been tried. For example, Cooper *el al.*[4] and Asano *et al.*[65] have performed extensive studies aimed at generating immunological tolerance by either transplanting pig bone marrow or infusing pig hematopoietic cells to baboons. However, they both observed a gradual loss of infused/transplanted pig cells, possibly due to the phagocytic action of macrophages.[4] Moreover, both groups have also performed splenectomy prior to xenografting experiments to minimize possible antibody rebounding from the recipient's spleen.[4,65] Interestingly, the Asano group[65] has achieved a survival period of more than 300 days, with one animal surviving more than 680 days, in a rhesus monkey-to-baboon cardiac xenotransplant following the combination of treatment with immunosuppressive drugs, total lymphoid irradiation, and infusion of pig bone marrow cells.

The xenoperfusion (*ex vivo* perfusion) approach has also been attempted to remove XNAs from human blood by perfusion through a pig kidney or liver for 3 hours. A significant reduction (absorption) of IgM and IgG isotype xenoantibodies was obtained.[66] As just discussed, the IgM isotype is mainly involved in HAR, whereas IgG predominates in AR.

Molecular Approaches

The paradigm shift in molecular biology over the last few years has been a cornerstone in transplantation research. It has been possible to humanize pig organs by applying genetic engineering tools. Generation of transgenic pigs expressing the human complement regulatory proteins such as CD59[67] or decay-accelerating factor (DAF) (CD55)[68–70] has been achieved successfully and has a dramatic effect on the survival of xenografts transplanted from such modified pig organs.[6,71,72]

Since HAR, primarily driven by galα-1,3-gal antibodies, is the main barrier in xenotransplantation, an increasing amount of molecular research is now focusing on this phase of rejection. Bracy and colleagues[36] have applied genetic engineering-based gene therapy strategies to prevent the production of xenoantibodies in mice, and Lai *et al.*[23] have extrapolated this technology to produce heterozygous galα-1,3-gal knockout pigs.

Therapeutic Approaches

The other approach to abrogate hyperactivation of the complement system is the use of membrane-anchored proteins such as the membrane cofactor protein (MCP), DAF, and CD59 or the use of soluble proteins such as cobra venom factor (CVF), soluble complement receptor 1 (sCR1), and the vaccinia virus complement control protein (VCP).

Transgenic mice expressing DAF and MCP have been developed to attenuate complement activation with promising results.[21] Furthermore, McCurry *et al.*[73] generated transgenic pigs expressing CD59 and DAF on the surface of endothelial cells by applying the passive translocation expression system from engineered red cells.

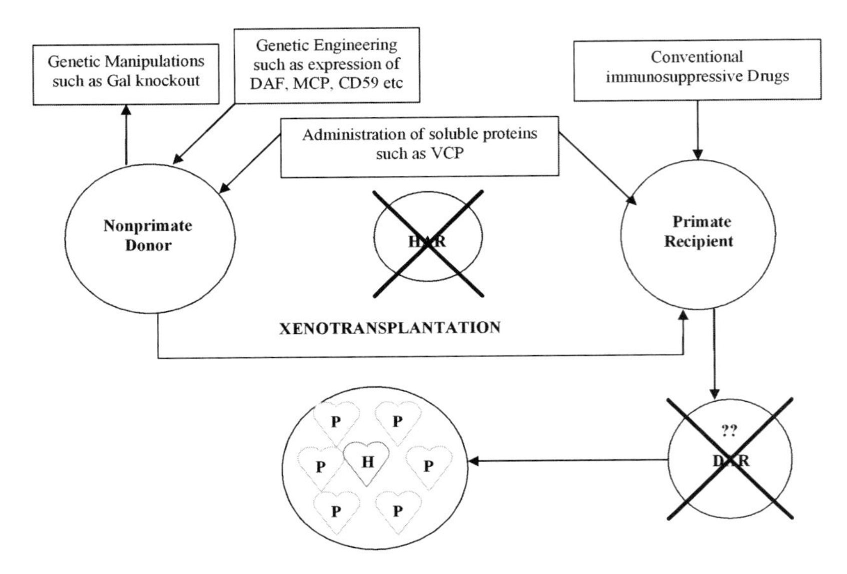

FIGURE 11. Hypothetical approach to prolonging the survival of xenograft. H = human; P = pig.

Their comparative cardiac xenotransplant study has reported the survival of a transgenic pig heart for up to 25 hours compared to the 60–90-minute survival of an unmodified heart. Different researchers have continued to show the therapeutic potential of DAF.[6,71,72] Schmoeckel *et al.*[74] have concluded that these DAF transgenics make an organ from a discordant species function as an organ from a concordant species (FIG. 11). However, the degree to which DAF regulates hyperactivation of the complement system is not sufficient for its use in clinical transplantation.[25,75,76] Moreover, other studies have shown that the alternative pathway of complement activation can still be activated despite the presence of membrane-bound DAF and MCP.[78]

The other concern with using natural human complement receptors such as DAF and MCP is that they are used by viruses and other microorganisms as routes of entry into cells.[78,79] For example, DAF (CD55) is used as a receptor by the enteric cytopathic human orphan viruses, a family of viruses known to cause myocarditis, and even protects HIV viral particles from complement-mediated attack.[78] MCP (CD46) is a putative receptor for the measles virus.[61] This amplifies the challenging issue of xenozonotic disease(s) as discussed below. To effectively and safely modulate complement and eliminate hyperacute rejection, other potent complement inhibitory proteins are being investigated.

The soluble complement receptor type 1 (sCR1) is a 120-kDa soluble complement inhibitory protein. This protein is highly effective in inhibiting both the classical and the alternative pathways of complement activation[28] and has been shown to prolong the survival of different xenografts.[80,81] However, because sCR1 lacks the ability to be retained *in vivo* for an extended period of time, it needs to be administered continuously.[28] This limited bioavailability, and its bulky conformation, which increases the likelihood of it inducing a complex immune response when administered chronically, may affect its chances of proceeding to the clinic.

Cobra venom factor (CVF) is another glycoprotein (149 kDa) that has been shown to have an extremely potent complement inhibitory activity.[52,82] However, it mainly targets the C3/C5 convertase of the alternative pathway, but it lacks any activity against preformed XNAs. Therefore, its application may only be restricted to certain models. Despite its potency to eliminate HAR, the excessive aggregation of inflammatory cells, such as neutrophils, in xenografts after treatment with CVF also suggests its limited potential. Moreover, because it is a foreign protein, it can stimulate a strong immune response,[53] and there is a considerable risk of contamination with other toxic venom enzymes such as the phospholipase A2,[53] a myotoxin that can lead to muscular degeneration.

The other soluble complement regulatory protein that has shown promising potential in xenotransplantation research is the vaccinia virus complement control protein (VCP). VCP (35 kDa) was isolated in 1988[83] from vaccinia virus, a cytoplasmic DNA virus belonging to the family of poxviridae. It was the first microbial protein to have a postulated role in modulation of the complement system and evasion of host defense.[83–85] It inhibits both the classical and the alternative pathways of complement activation by binding to complement components C3b and C4b and facilitating the factor I-mediated cleavage of these active components. This activity inhibits the release of chemotactic factors (C3a, C4a, and C5a), macrophage influx, and proinflammatory cytokines, which directly or indirectly contributes to rapid xenorejection. VCP has been shown to inhibit xenoantibody attachment to ECs by binding to anti-galα-1,3-gal antibodies and regulates complement activation both *in vitro*[86,87] and *in vivo*[31,88,89] in a dose-dependent manner. Previously, Anderson *et al.*[88] have shown that VCP can significantly reduce cardiac tissue damage in the mouse-to-sensitized rat and guinea pig-to-rat[31] models. Moreover, the ability of VCP to inhibit the complement system in baboons has already been established,[90] and its role in nonprimate-to-primate xenotransplantation remains to be evaluated. The other promising feature of VCP in xenotransplantation research is its ability to simultaneously bind complement,[91] while bound to heparan sulfate and its uptake and prolonged retention in the heparin granules of mast cells,[92] which gets released upon degranulation caused by complement activation. This reduces the need for continuous infusion and monitoring, unlike the sCR1, of the recipient. This heparin-binding property, which most of the other complement regulatory proteins lack, empower VCP with the ability to inhibit aggregation of inflammatory cells, such as neutrophils and NK cells, which are believed to play a key role in DXR.[82] These combined abilities, its molecular mimicry to host proteins, and its small size give VCP a preferential advantage as a complement regulatory protein with a potential to abolish HAR and early cellular graft rejection.

We believe that solid organ xenotransplantation may well be considered at least as a bridging life support via the combined approaches shown in FIGURE 11. These collective approaches may help to maintain the degree of xenorejection as low as that seen in allograft rejection, and thus increase the availability of donor organs. However, the issue of biosafety (cross-species infection) that is hampering the progress towards the clinical application needs to be resolved.

BIOSAFETY CONCERNS

Because pigs are considered as potential organ donors,[23,24,93] the greatest concern is the likelihood of cross-infection by some porcine viruses, particularly the porcine endogenous retroviruses (PERVs), which have been shown to infect some human cell lines *in vitro*.[94–97] However, no evidence of PERV infection was perceived in lymphocytic cells and in the tissue biopsies of remote organs following pig-to-cynomolgus monkey renal xenografting for up to 41 weeks posttransplantation.[93] Moreover, in the investigative study conducted by Specke and colleagues,[97] the *in vivo* inoculation of small immunocompromised animals and triple immunosuppressive drug-treated non-human primates, with different PERVs, did not cause an infection during the 12-week follow-up period, as established by western blotting and polymerase chain reaction (PCR) analysis.[97] Here, it is noteworthy to mention the long term retrospective study conducted by Paradis *et al.*[98] to screen the presence of PERV in patients treated with diverse living pig tissues. The study confirmed the absence of PERV RNA in all 160 patients.[98] Several other studies have also reported the absence of PERV infection following xenografting from modified or unmodified pigs.[96,99-101] Recently, Irgang *et al.*[102] also confirmed the absence of infection in patients treated with PERV-excluding membrane-processed porcine hepatic cells. In addition, highly conserved epitopes among the diverse retroviral genome have been mapped, and antibodies directed to these epitopes were shown to neutralize PERV infectivity. This can be considered a significant achievement to minimize the biosafety concerns and progress towards the design of effective vaccine.[103]

CONCLUDING REMARKS

Although a number of techniques have been attempted to overcome the immunological barriers involved in xenorejection, none of them has been successful in eliminating XNA's rapid return to the circulation. Therefore, we suggest the need to consider integrating the various approaches outlined in the chapter, as each approach may contribute to bypass the multiple and subsequent phases of rejection and hopefully facilitate the success of clinical xenotransplantation.

ACKNOWLEDGMENTS

The Poliomyelitis Research Foundation (PRF) and the University of Cape Town support Y.T.G. The Jewish Hospital Research Foundation of Louisville is gratefully acknowledged for the research support on xenotransplantation in G.J.K.'s lab. G.J.K. is currently a Senior International Wellcome Trust Fellow in Biomedical Sciences in South Africa.

REFERENCES

1. MICHLER, R.E. 1996. Xenotransplantation: risks, clinical potential, and future prospects. Emerging Infect. Dis. **2:** 64–70.
2. COOPER, D.K.C. 1993. Xenografting: how great is the clinical need? Xenotransplantation **1:** 25–26.

3. Platt, J.L. *et al.* 1990. Transplantation of discordant xenografts: a review of progress. Immunol. Today **11:** 450–457.
4. Cooper, D.K.C., B. Gollackner, C. Knosalla & K. Teranishi. 2002. Xenotransplantation – how far have we come? Transplant. Immunol. **9:** 251–256.
5. Wang, X.M., G. Chen, S. Chen, *et al.* 2001. A novel model of pig-to-monkey kidney xenotransplantation. Transplant. Proc. **33:** 3859.
6. Goddard, M.J., J.J. Dunning, J. Horsley, *et al.* 2002. Histopathology of Cardiac xenograft rejection in the pig-to-baboon model. J. Heart Lung Transplant. **21:** 474–484.
7. Groth, C.G., O. Korsgren, A. Tibell, *et al.* 1994. Transplantation of porcine fetal pancreas to diabetic patients. Lancet **344:** 1402–1404.
8. Aebischer, P., M. Schluep, N. Déglon, *et al.* 1996. Intrathecal delivery of CNTF using encapsulated genetically modified xenogeneic cells in amyotrophic lateral sclerosis patients. Nat. Med. **2:** 696–699.
9. Ram, Z. *et al.* 1997. Therapy of malignant brain tumors by intratumoral implantation of retroviral vector-producing cells. Nat. Med. **3:** 1354-1361.
10. Aebischer, P., A.F. Hottinger & N. Déglon. 1999. Cellular xenotransplantation. Nat. Med. **5:** 852.
11. Deacon, T., J. Schumacher, J. Dinsmore, *et al.* 1997. Histological evidence of fetal pig neural cell survival after transplantation into a patient with Parkinson's disease. Nat. Med. **3:** 350–353.
12. Lehrman, S. 1995. AIDS patient given baboon bone marrow. Nature **378:** 756.
13. Relf, M.V. 1996. Xenotransplantation of baboon bone marrow cells: a historical review of the protocols as a possible treatment modality for HIV/AIDS. J. Assoc. Nurses AIDS Care **7:** 27–35.
14. Ohno, K., L.R. Nelson, K. Mitooka & W.M. Bourne. 2002. Transplantation of cryopreserved human corneas in a xenograft model. Cryobiology **44:** 142–149.
15. Nickoloff, B.J. & F.O. Nestle. 2004. Recent insights into the immunopathogenesis of psoriasis provide new therapeutic opportunities. J. Clin. Invest. **113:** 1664–1675.
16. US guidelines on xenotransplantation.1999. Nat. Med. **5:** 465.
17. Butler, D. 1999. Europe is urged to hold back on xenotransplant clinical trials. Nature **397:** 281–282.
18. Reemtsma, K. 1991. Xenotransplantation: a brief history of clinical experiences: 1900–1965: early attempts at renal xenografting. *In* Xenotransplantation: The Transplantation of Organs and Tissues Between Species. D.K.C. Cooper, E. Kemp, K. Reemtsma & D.J.G. White, Eds. :9–22. Springer-Verlag. Berlin, Heidelberg.
19. Cooper, D.K.C. & Y. Ye. 1991. Experience with clinical heart xenotransplantation. *In* Xenotransplantation: The Transplantation of Organs and Tissues Between Species. D.K.C. Cooper, E. Kemp, K. Reemtsma & D.J.G. White, Eds. :541–557. Springer-Verlag. Berlin, Heidelberg.
20. Barnard, C.N., A. Wolpowitz & J.G. Losman. 1977. Heterotopic cardiac transplantation with a xenograft for assistance of the left heart in cardiogenic shock after cardiopulmonary bypass. S. Afr. Med. J. **52:** 1035–1038.
21. Bach, F.H. & H. Auchincloss. 1995. Transplantation Immunology. New York. Wiley-Liss. Xi, 409.
22. The Right Reverend Lord John Habgood, A.G. Spagnolo, E. Sgreccia & A.S. Daar. 1997. Religious views on organ and tissue donation. *In* Organ and Tissue Donation for Transplantation. Jeremy R. Chapman, Mark Deierhoi & Celia Wight, Eds. :23–33. Arnold, a member of the Hodder Headline Group.
23. Lai, L., D. Kolber-Simonds, K.-W. Park, *et al.* 2002. Production of α-1,3-Galactosyltransferase knockout pigs by nuclear transfer cloning. Science **295:** 1089–1092.
24. Prather, R.S., R.J. Hawley, R.J. Carter, *et al.* 2003. Transgenic swine for biomedicine and agriculture. Theriogenology **59:** 115–123.
25. van Denderen, B.J. *et al.* 1997. Combination of decay-accelerating factor expression and alpha 1,3-galactosyltransferase knockout affords added protection from human complement-mediated injury. Transplantation **64:** 882–888.
26. Galili, U. *et al.* 1988. Man, apes, and Old World monkeys differ from other mammals in the expression of alpha-galactosyl epitopes on nucleated cells. J. Biol. Chem. **263:** 17755–17762.

27. GALILI, U. 1998. Anti-Gal antibody prevents xenotransplantation. Sci. Med. **5:** 28–37.
28. RYAN, U.S. 1995. Complement inhibitory therapeutics and xenotransplantation. Nat. Med. **1:** 967–968.
29. COOPER, D.K.C. 1991. Xenotransplantation: The Transplantation of Organs and Tissues between Species. Berlin, New York. Springer-Verlag.
30. LUO, Y. *et al.* 1998. Comparative histopathology of hepatic allografts and xenografts in the nonhuman primate. Xenotransplantation **5:** 197–206.
31. ANDERSON, J.B., S.A. SMITH, R. VAN WIJK, *et al.* 2003. Vaccinia virus complement control protein inhibits hyperacute xenorejection in a guinea pig-to-rat heterotopic cervical cardiac xenograft model by blocking both xenoantibody binding and complement pathway activation. Transplant Immunol. **11:,** 129–135.
32. MIYAGAWA, S. *et al.* 1988. The mechanism of discordant xenograft rejection. Transplantation **46:** 825–830.
33. GAMBIEZ, L., E. SALAME, C. CHEREAU, *et al.* 1992. Natural IgM play a major role in hyperacute rejection of discordant heart xenografts. Transplant. Proc. **24:** 441–442.
34. GAMBIEZ, L., E. SALAME, C. CHEREAU, *et al.* 1992. The role of natural IgM in the hyperacute rejection of discordant heart xenografts. Transplantation **54:** 577–583.
35. GALILI, U. *et al.* 1988. Interaction between human natural anti-alpha-galactosyl immunoglobulin G and bacteria of the human flora. Infect. Immun. **56:** 1730–1737.
36. BRACY, J.L., D.H. SACHS & J. IACOMINI. 1998. Inhibition of xenoreactive natural antibody production by retroviral gene therapy. Science **281:** 1845–1847.
37. GALILI, U. *et al.* 1984. A unique natural human IgG antibody with anti-alpha-galactosyl specificity. J. Exp. Med. **160:** 1519–1531.
38. GALILI, U. *et al.* 1987. Evolutionary relationship between the natural anti-Gal antibody and the Gal alpha 1-3Gal epitope in primates. Proc. Natl. Acad. Sci. USA **84:** 1369–1373.
39. GALILI, U., C.R. GREGORY & R.E. MORRIS. 1995. Contribution of anti-Gal to primate and human IgG binding to porcine endothelial cells. Transplantation **60:** 210–213.
40. PLATT, J.L. & F.H. BACH. 1991. Mechanisms of tissue injury in hyperacute xenograft rejection: a model of hyperacute xenograft rejection. *In* Xenotransplantation: The Transplantation of Organs and Tissues between Species. D.K.C. Cooper, E. Kemp, K. Reemtsma & D.J.G. White, Eds. :69–79. Springer-Verlag. Berlin, Heidelberg.
41. PLATT, J.L. *et al.* 1990. Release of heparan sulfate from endothelial cells. Implications for pathogenesis of hyperacute rejection. J. Exp. Med. **171:** 1363–1368.
42. IHRCKE, N.S. & J.L. PLATT. 1996. Shedding of heparan sulfate proteoglycan by stimulated endothelial cells: evidence for proteolysis of cell-surface molecules. J. Cell Physiol. **168:** 625–637.
43. PLATT, J.L. *et al.* 1991. The role of C5a and antibody in the release of heparan sulfate from endothelial cells. Eur. J. Immunol. **21:** 2887–2890.
44. SAADI, S. & J.L. PLATT. 1995. Transient perturbation of endothelial integrity induced by natural antibodies and complement. J. Exp. Med. **181:** 21–31.
45. JANSON, T.L., H. STORMORKEN & H. PRYDZ. 1984. Species specificity of tissue thromboplastin. Haemostasis **14:** 440–444.
46. JURD, K.M., R.V. GIBBS & B.J. HUNT. 1996. Activation of human prothrombin by porcine aortic endothelial cells – a potential barrier to pig to human xenotransplantation. Blood Coagul. Fibrinolysis **7:** 336–343.
47. JURD, K.M., R.V. GIBBS & B.J. HUNT. 1996. Activation of human prothrombin by porcine endothelium. Transplant. Proc. **28:** 610.
48. PARETI, F.I. *et al.* 1992. Interaction of porcine von Willebrand factor with the platelet glycoproteins Ib and IIb/IIIa complex. Br. J. Haematol. **82:** 81–86.
49. HAISCH, C.E. *et al.*1990. The vascular endothelial cell is central to xenogeneic immune reactivity. Surgery **108:** 306–311.
50. Rose, A.G. *et al.* 1991. Histopathology of hyperacute rejection of the heart: experimental and clinical observations in allografts and xenografts. J. Heart Lung Transplant. **10:** 223–234.
51. ROSE, A.G. & D.K.C. COOPER. 1996. A histopathologic grading system of hyperacute (humoral, antibody-mediated) cardiac xenograft and allograft rejection. J. Heart Lung Transplant. **15:** 804–817.

52. KOBAYASHI, T. *et al.*1997. Delayed xenograft rejection of pig-to-baboon cardiac transplants after cobra venom factor therapy. Transplantation **64:** 1255–1261.
53. LEVENTHAL, J.R., A.P. DALMASSO, J.W. CROMWELL, *et al.* 1993. Prolongation of cardiac xenograft survival by depletion of complement. Transplantation **55:** 857-866.
54. BACH, F.H., S.C. ROBSON, H. WINKER, *et al.* 1995. Barriers to xenotransplantation. Nat. Med. **1:** 869-873.
55. COOPER, D.K.C. & J.L. PLATT. 1997. Xenotransplantation: The Transplantation of Organs and Tissues between species. 2nd Ed. Berlin, New York. Springer. Xxxiv, 854.
56. BACH, F.H., S.C. ROBSON, C. FERRAN, *et al.* 1995. Xenotransplantation: endothelial cell activation and beyond. Transplant. Proc. **27:** 77–79.
57. BACH, F.H. *et al.* 1996. Delayed xenograft rejection. Immunol. Today **17:** 379–384.
58. BACH, F.H. *et al.* 1997. Modification of vascular responses in xenotransplantation: inflammation and apoptosis. Nat. Med. **3:** 944–948.
59. KOZLOWSKI, T. *et al.*1999. Porcine kidney and heart transplantation in baboons undergoing a tolerance induction regimen and antibody adsorption. Transplantation **67:** 18–30.
60. HANCOCK, W.W. & F.H. Bach. 1994. The immunology of discordant xenotransplantation. Xenotransplantation **2:** 68–77.
61. SEOW, J. 2003. Clinical xenotransplantation. The Lancet **362:** 1421–1422.
62. BACH, F.H., J.L. PLATT & D.K.C. COOPER. 1991. Accommodation: the role of natural antibody and complement in discordant xenograft rejection: methods of removal of preformed natural antibodies. *In* Xenotransplantation: The Transplantation of Organs and Tissues Between Species. D.K.C.E. Kemp, K. Reemtsma & D.J.G. White, Eds. :81–99. Springer-Verlag. Berlin, Heidelberg.
63. REDING, R., S. ff. H. DAVIES, D.J.G. WHITE, *et al.* 1989. Effect of plasma exchange on guinea pig-to-rat heart xenografts. Transplant. Proc. **21:** 534–536.
64. MAJADO, M.J. *et al.* 2002. ABO system and blood crossmatch study in baboons: importance of designing a primate blood bank for orthotopic pig-to-baboon liver xenotransplantation. Transplant. Proc. **34:** 327–328.
65. ASANO, M., S.R. Gundry, H. Izutani, *et al.* 2003. Baboons undergoing orthotopic concordant cardiac xenotransplantation surviving more than 300 days: effect of immunosuppressive regimen. J. Thoracic Cardiovasc. Surg. **125:** 60–70.
66. TUSO, P.J., D.V. CARMER, C.YASUNAGA, *et al.* 1993. Removal of natural human xenoantibodies to pig vascular endothelium by perfusion of blood through pig kidneys and livers. Transplantation **55:** 1375–1378.
67. DIAMOND, L.E. *et al.* 1996. Characterization of transgenic pigs expressing functionally active CD59 on cardiac endothelium. Transplantation **61:** 121–129.
68. LANGFORD, G.A., N. YANNOUTSOS, E. COZZI, *et al.* 1994. Production of pigs transgenic for human decay accelerating factor. Transplant. Proc. **26:** 1400–1401.
69. Cozzi, E. & D.J.G. WHITE. 1995 The generation of transgenic pigs as potential organ donors for humans. Nat. Med. **1:** 964–966.
70. ROSENGARD, A.M., N.R.B. CARRY, G.A. Langford, *et al.* 1995. Tissue expression of human complement inhibitor, decay-accelerating factor, in transgenic pigs. Transplantation **59:** 1325–1333.
71. BRANDL, U. *et al.* 2003. 25 days of baboons after orthotopic xenotransplantation of hDAF transgenic pig hearts using a moderate immunosuppression (Abstr). J. Heart Lung Transplant. **22:** S100–S101.
72. RAMIREZ, P. *et al.* 2002. Transgenic pig-to-baboon liver xenotransplantation: clinical, biochemical, and immunologic pattern of delayed acute vascular rejection. Transplant. Proc. **34:** 319–320.
73. MCCURRY, K.R. *et al.* 1995. Human complement regulatory proteins protect swine-to-primate cardiac xenografts from humoral injury. Nat. Med. **1:** 423–427.
74. SCHMOECKEL, M. *et al.* 1997. Transgenic human decay accelerating factor makes normal pigs function as a concordant species. J. Heart Lung Transplant. **16:** 758–764.
75. KOIKE, C. *et al.* 1996. Establishment of a human DAF/HRF20 double transgenic mouse line is not sufficient to suppress hyperacute rejection. Surg. Today **26:** 993–998.
76. MAKRIDES, S.C. 1998. Therapeutic inhibition of the complement system. Pharmacol. Rev. **50:** 59–87.

77. KRAUS, D., M.E. MEDOF & C. MOLD.1998. Complement recognition of alternative pathway activators by decay-accelerating factor and factor H. Infect. Immun. **66:** 399–405.
78. KOTWAL, G.J. 1997. Microogranisms and their interaction with the immune system. J. Leuk. Biol. **62:** 415–429.
79. WANG, G. *et al.* 1998. Functional differences among multiple isoforms of guinea pig decay-accelerating factor. J. Immunol. **160:** 3014–3022.
80. XIA, W. *et al.* 1992. Prolongation of guinea pig cardiac xenograft survival in rats by soluble human complement receptor type 1. Transplant. Proc. **24:** 479–480.
81. PRUITT, S.K. *et al.* 1994. The effect of soluble complement receptor type 1 on hyperacute rejection of porcine xenografts. Transplantation **57:** 363–370.
82. SUN, Q-Y., G. CHEN, H. GUO, *et al.* 2003. Prolonged cardiac xenograft survival in guinea pig-to-rat model by a highly active cobra venom factor. Toxicon **42:** 257–262.
83. KOTWAL, G.J. & B. MOSS. 1988. Vaccinia virus encodes a secretory polypeptide structurally related to complement control proteins. Nature **335:** 176–178.
84. KOTWAL, G.J., S.N. ISAACS, R. MCKENZIE, *et al.* 1990. Inhibition of the complement cascade by the major secretory protein of vaccinia virus. Science **250:** 827–829.
85. ISAACS, S.N., G.J. KOTWAL & B. MOSS. 1992. Vaccinia virus complement-control protein prevents antibody-dependent complement-enhanced neutralization of infectivity and contributes to virulence. Proc. Natl. Acad. Sci. USA **89:** 628–632.
86. AL-MOHANNA, F., R. PARHAR & G.J. KOTWAL. 2001. Vaccinia virus complement control protein is capable of protecting xenoendothelial cells from antibody binding and killing by human complement and cytotoxic cells. Transplantation **71:** 796–801.
87. MURTHY, K.H.M., S.A. SMITH,V.K. GANESH, *et al.* 2001. Crystal structure of a complement control protein that regulates both pathways of complement activation and binds heparan sulfate proteoglycans. Cell **104:** 301–311.
88. ANDERSON, J.B., S.A. SMITH, R. VAN WIJK, *et al.* 2002. Vaccinia virus complement control protein ameliorates hyperacute xenorejection by inhibiting xenoantibody binding. Transplant. Proc. **34:** 3277–3281.
89. ANDERSON, J.B., S.A. SMITH & G.J. KOTWAL. 2002. Vaccinia virus complement control protein inhibits hyperacute xenorejection. Transplant. Proc. **34:** 1083–1085.
90. KAHN, D., S.A. SMITH & G.J. KOTWAL. 2003. Dose-dependent inhibition of complement in baboons by vaccinia virus complement control protein: implications in xenotransplantation. Transplant. Proc. **35:** 1606–1608.
91. SMITH, S.A. *et al.* 2000. Conserved surface-exposed K/R-X-K/R motifs and net positive charge on poxvirus complement control proteins serve as a putative heparin binding sites and contribute to inhibition of molecular interactions with human endothelial cells: a novel mechanism for evasion of host defense. J. Virol. **74:** 5659–5666.
92. JHA, P., S.A. SMITH, D.E. JUSTUS & G.J. KOTWAL. 2003. Prolonged retention of vaccinia virus complement control protein following ip injection: implications in blocking xenorejection. Transplant. Proc. **35:** 3160–3162.
93. WINKLER, M.E., U. MARTIN, M. LOSS, *et al.* 2000. Porcine endogenous retrovirus is not transmitted in a discordant porcine-to-cynomolgus xenokidney transplantation model with long-term survival of organ recipients. Transplant. Proc. **32:** 1162.
94. ALLAN, J.S. 1996. Xenotransplantation at a crossroads: prevention versus progress. Nat. Med. **2:** 18–21.
95. BACH, F.H. *et al.* 1998. Uncertainty in xenotransplantation: Individual benefit versus collective risk. Nat. Med. **4:** 141–144.
96. HENEINE, W., A. TIBELL, G.J. SWITZER, *et al.* 1998. No evidence of infection with porcine endogenous retrovirus in recipients of porcine islet-cell xenografts. Lancet **352:** 695–699.
97. SPECKE, V., H.-J. SCHUURMAN, R. PLESKER, *et al.* 2002. Virus safety in xenotransplantation: first exploratory in vivo studies in small laboratory animals and non-human primates. Transplant. Immunol. **9:** 281–288.
98. PARADIS, K. *et al.* 1999. Search for cross-species transmission of porcine endogenous retrovirus in patients treated with living pig tissue. Science **285:** 1236–1241.

99. DINSMORE, J.H., C. MANHART, R. RAINERI, *et al.* 2000. No evidence for infection of human cells with porcine endogenous retrovirus (PERV) after exposure to porcine fetal neuronal cells. Transplantation **70:** 1382–1389.
100. SWITZER, W.M. *et al.* 2001. Lack of cross-species transmission of porcine endogenous retrovirus infection to nonhuman primate recipients of porcine cells, tissues, or organs. Transplantation **71:** 959–965.
101. KUDUS, R., J.F. PATZER II, R. LOPEZ, *et al.* 2002. Clinical and laboratory evaluation of the safety of a bioartificial liver assist device for potential transmission of porcine endogenous retrovirus. Transplantation **73:** 420–429.
102. IRGANG,, M. *et al.* 2003. Porcine endogenous retroviruses: no infection in patients treated with a bioreactor based on porcine liver cells. J. Clin. Virol. **28:** 141–154.
103. FIEBIG, U., O. STEPHAN, R. KURTH & J. DENNER. 2003. Neutralizing antibodies against conserved domains of p15E of porcine endogenous retroviruses: basis for a vaccine for xenotransplantation? Virology **307:** 406–413.

Hemolytic Uremic Syndrome

An Example of Insufficient Complement Regulation on Self-Tissue

JOHN P. ATKINSON, M. KATHRYN LISZEWSKI, ANNA RICHARDS, DAVID KAVANAGH, AND ELIZABETH A. MOULTON

Department of Medicine/Rheumatology Division, Washington University School of Medicine, St. Louis, Missouri 63110, USA

ABSTRACT: Hemolytic uremic syndrome (HUS) is a triad of microangiopathic hemolytic anemia, thrombocytopenia, and acute renal failure. HUS is classified as either diarrhea associated, most commonly caused by infection with *Escherichia coli* O157, or the less common atypical HUS (aHUS), which may be familial or sporadic. Approximately 50% of patients with aHUS have mutations in one of the complement control proteins: factor H, factor I, or membrane cofactor protein (MCP). These proteins regulate complement activation through cofactor activity, the inactivation of C3b by limited proteolytic cleavage, a desirable event in the fluid phase (no target) or on healthy self-tissue (wrong target). Complement activation follows the endothelial cell injury that characterizes HUS. This disease represents a model of what takes place when inappropriate complement activation occurs on self-tissues due to the presence of mutated complement regulatory proteins. Screening for mutations in factor H, factor I, or MCP is expensive and time consuming. One approach is to perform antigenic screening for factor H and factor I deficiency and to look for low levels of MCP (CD46) expression by flow cytometry. Complement regulatory protein deficiency impacts treatment decisions as patients with aHUS have a recurrence rate in renal transplants of ~50%, whereas those with factor H mutations have an even higher risk (~80%). By contrast, MCP deficiency can be corrected in part by a renal allograft. However, caution in the use of live-related donations is needed because of the high rates of incomplete penetrance of the described mutations.

KEYWORDS: hemolytic uremic syndrome (HUS); complement activation; complement regulation; cofactor activity; renal transplantation

ATYPICAL HEMOLYTIC UREMIC SYNDROME

Hemolytic uremic syndrome (HUS) is a clinical triad of microangiopathic hemolytic anemia, thrombocytopenia, and acute renal failure.[1] The major pathologic feature is fibrin and platelet-rich microthrombi in kidney vessels.[2–4] The syndrome

Address for correspondence: John P. Atkinson, Washington University School of Medicine, Department of Medicine/Rheumatology Division, 660 South Euclid Avenue, Campus Box 8045, St. Louis, MO 63110, USA. Voice: 314-362-8391; fax: 314-362-1366.
jatkinso@im.wustl.edu

Ann. N.Y. Acad. Sci. 1056: 144–152 (2005).
doi: 10.1196/annals.1352.032

is most commonly associated with a preceding diarrheal illness, usually caused by infection with a Shiga toxin-producing *Escherichia coli* O157:H7.[5,6] Atypical hemolytic uremic syndrome (aHUS) has been classified as either sporadic or familial (if more than one family member is affected). However, this designation may not be meaningful, as in most cases an unaffected parent will carry the disease-causing mutation in the "so-called" sporadic cases. aHUS is characterized by incomplete penetrance in many of the reported studies, pointing to the involvement of additional genetic and/or environmental factors in the phenotypic expression of the disease (reviewed in Refs. 1 and 7–9). The disease has a penetrance of ~50%, with most individuals being affected before age 30. aHUS also tends to be recurrent, and loss of renal function occurs in most patients. Dialysis successfully treats the acute renal failure phase of aHUS. The disease commonly recurs in renal allografts.[1,7–9]

PREDISPOSITION TO aHUS SECONDARY TO COMPLEMENT REGULATORY PROTEIN DEFICIENCY

Approximately 50% of patients with aHUS have a mutation in one of three genes that each encodes a complement regulator (TABLE 1). The three proteins are (1) factor H, a plasma protein synthesized in the liver that binds C3b and serves as a cofactor for serine protease factor I to cleave C3b, (2) membrane cofactor protein (MCP or CD46), a widely expressed transmembrane glycoprotein that binds C3b and C4b and is a cofactor for their cleavage by factor I, and (3) factor I, a plasma serine protease synthesized in the liver that cleaves C3b and C4b but only in association with a cofactor protein such as factor H or MCP. In several series of aHUS patients, 20–30% had a factor H mutation,[10–17] 10–20% an MCP mutation,[18–20] and 10–20% a factor I mutation.[21,22] A few cases of combined mutation involving these three proteins have been reported,[20,23] as have three patients with autoantibodies to factor H.[24]

TABLE 1. Abnormalities of complement regulators in aHUS (CA, cofactor activity; DAA, decay accelerating activity)

Regulator	Size (kDa)	Site of synthesis (and location)	Function	Heterozygous mutations in aHUS (%)
Factor H (FH)	150,000	Liver (plasma protein)	Binds C3b, CA, and DAA for convertases containing C3	20–30
MCP (CD-46)	55,000/65,000 (two size variants due to alternative splicing)	Most cells (transmembrane protein)	Binds C3b and C4b, CA for both of these ligands	10–20
Factor I (FI)	88,000 (two chains of 50,000 and 38,000)	Liver (plasma protein)	Serine protease requires a cofactor protein to cleave C3b or C4b	10–20

In most cases, the affected individual is heterozygous for a mutation in one of these three genes, possessing one wild-type gene and protein product and one mutated gene and protein product. One parent who usually has not developed aHUS carries the mutated gene. Other children of these families carrying the mutation have approximately a 50% chance of developing the disease. Most of the afflicted individuals develop the disease before 10 years of age, but initial presentations postpartum and between ages 20 and 40 have been described.

Complement regulator protein deficiency predisposes to aHUS, whereas the illness is probably triggered by an environmental factor. In diarrheal HUS, the vero toxin is the most commonly identified etiologic agent. aHUS is rarely preceded by diarrhea. The "toxin equivalent" in aHUS is not known, the most common presentation being one without an obvious infection or other precipitating event. However, in a recent pediatric series,[25] serious infections, especially with streptococcal pneumonia, were seen in ~30% of cases. Other associations include pregnancy and the postpartum period; drugs such as the oral contraceptive pill, cyclosporin A, and mitomycin C; malignancies; and connective tissue diseases. The usual patient presents with a short history of a nonspecific illness, but laboratory testing reveals hemolytic anemia with a microangiopathic blood smear, thrombocytopenia, and acute renal failure. Enteropathic HUS patients typically recover,[26,27] whereas ~50% of patients with aHUS require dialysis long-term. Thus, aHUS tends to both recur and have a much worse renal outcome.

CRITICAL ROLE OF COFACTOR ACTIVITY IN CONTROLLING COMPLEMENT ACTIVATION

Three proteins regulate complement via cofactor activity (CA) (FIGS. 1 and 2). CA refers to inactivation of the pivotal C3b activation fragment by limited proteo-

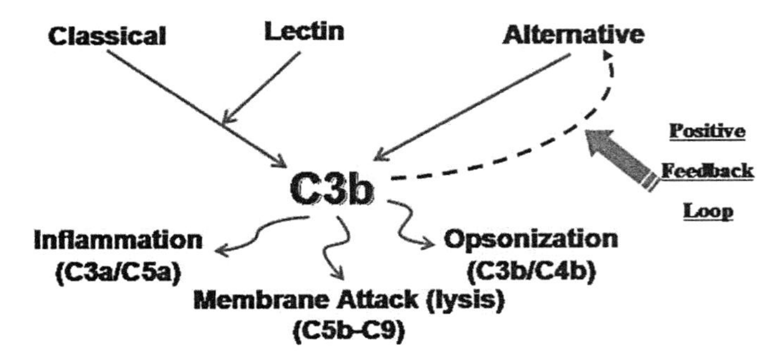

FIGURE 1. Complement pathways. C3b is deposited on a target by immune complexes that engage the classical pathway, lectins such as mannose-binding protein that bind to sugars (lectin pathway), or the continuous low-grade turnover of the alternative pathway. The deposited C3b serves as a nidus for amplification of C3b deposition via the positive feedback loop activity of the alternative pathway. Proper control of this amplification process is critical to prevent injury to host cells. Presence of antibodies or lectins on a membrane relatively lacking in complement regulators is conducive to excessive activation on self-tissue. (Reprinted with permission from Atkinson.[32])

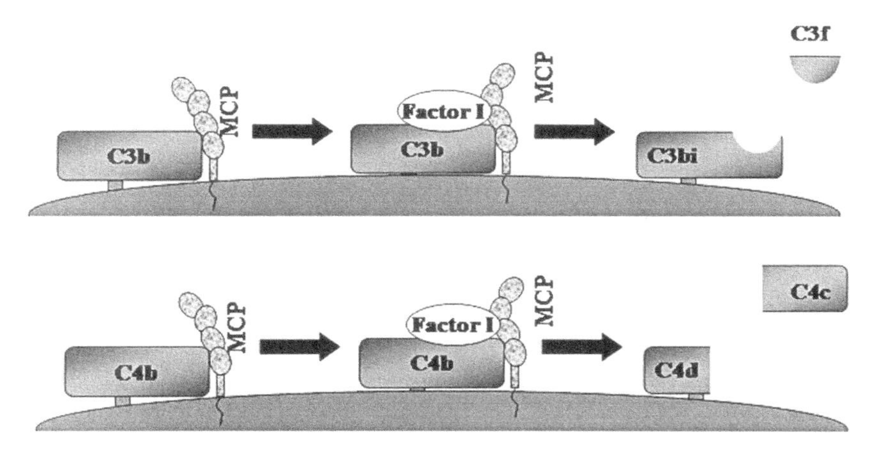

FIGURE 2. Cofactor activity. C3b or C4b attached to a membrane (as shown in this diagram) or plasma is bound by a cofactor protein. In the example in this figure, the membrane regulator MCP is the cofactor protein. Upon MCP binding, the serine protease factor I is now able to cleave C3b or C4b. This limited cleavage of the α-chain inactivates the protein so that no further complement activation can occur. Factor H has this same cofactor activity for C3b.

lytic cleavage, a highly desirable event in the fluid phase (no target) or on healthy self-tissue (wrong target). This proteolytic cleavage reaction involves the alpha chain of C3b and requires both a C3b binding protein (cofactor protein) and the serine protease factor I. For cleavage of C3b, factor H and MCP are cofactor proteins. In the case of C4b, MCP serves as a cofactor protein on the cell surface, but in plasma, C4b binding protein is the major fluid phase cofactor protein for C4b.

One can envision the following pathological sequence of events in HUS. First, an initiating or triggering event must induce endothelial cell injury and activation. In typical or diarrheal HUS, the Shiga toxin of *E. coli* O157:H7 serves this role. Following endothelial cell injury, complement activation occurs. Altered cell membranes may be recognized by natural antibodies and lectins. Limited complement activation ensues that normally would be desirable as it facilitates membrane repair. If the toxin kills the cell, then the underlying basement membrane is exposed to plasma. Lacking a cellular cover, this tissue now requires protection from complement attack as extracellular matrix and basement membranes lack intrinsic regulators.

If the individual has inherited a mutation in one of the complement regulators, then excessive complement activation occurs on injured endothelial cell or exposed basement membrane (FIG. 3). This initiates a procoagulant state.[7] For example, (1) clustered C3b may serve as ligand for complement receptors on phagocytic cells, (2) the liberated anaphylatoxins C3a and C5a can signal through their receptors on an endothelial cell to upregulate procoagulant proteins and to attract inflammatory cells, and (3) membrane attack complex may induce signaling events, sublytic membrane perturbation, or outright cell lysis.

Once the basement membrane is exposed, it must be protected from complement attack. In this regard, MCP is a membrane protein so it cannot protect the basement

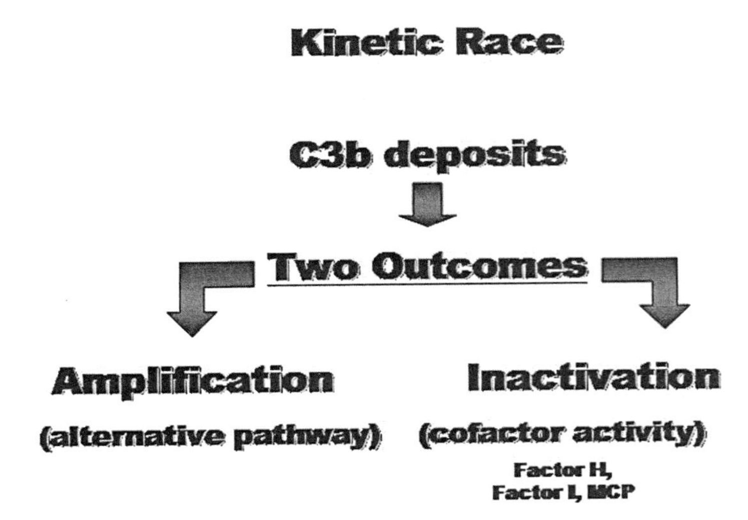

FIGURE 3. Outcom of C3b deposition on a biologic surface. C3b is deposited throgh pathways illustrated in FIGURE 1. On microbes, massive amplification is the desired result. On healthy self-tissue, inactivation is the desired result. On damaged or modified self-tissue, a delicate balance is presumably in place to allow for both repair and recovery and to limit the immune response. If regulators such as factor H or MCP are deficient, excessive activation occurs, leading to further cell damage.

membrane. However, factor H, a plasma protein, can attach to the glycosaminoglycans and polyanions on basement membranes through its heparin binding sites.[7,28–31] Once bound, factor H functions like a membrane protein. The location of factor H mutation in aHUS supports this line of reasoning since most mutations are not in the C3b binding/CA site but instead are in regions responsible for heparin binding. This observation implies that factor H protects acellular areas of tissue.

Several important implications and conclusions can be derived from the points discussed above. First, since a deficiency in either factor H or MCP predisposes to aHUS, both must be required to protect host tissue. A likely scenario is that MCP patrols cell surfaces, whereas factor H protects the plasma and the acellular tissue matrix. Second, CA is critical to proper control of complement activation. Decay accelerating activity, which is the disassociation of the multicomponent C3 and C5 convertases, is not an adequate control mechanism. Decay accelerating factor (DAF or CD55) is normal in aHUS (Ref. 20 and unpublished data). DAF is an effective early inhibitor to block the complement cascade, but this activity of dissociation merely delays, not permanently destroys, the amplification potential of C3b. For example, the alternative pathway C3 convertase C3bBb is decayed by DAF. This transiently destroys the enzyme because the decayed catalytic Bb cannot rebind; however, another factor B may now bind to the same C3b and form a new enzyme. Such an amplification system would eventually overwhelm DAF. In contrast, CA cleaves C3b, and the resulting fragment can no longer bind factor B. The sequence of events in aHUS, therefore, may be endothelial cell damage→excessive complement activation→ procoagulant endothelial cell microenvironment→formation of fibrin and platelet microthrombi→renal damage. A similar scenario likely occurs in typical HUS, but in this case, Shiga toxin alone damages the endothelium so severely that even normal levels of complement inhibitors cannot control the reaction to the injury (in most patients, though, repair and subsequent recovery of renal function do occur).

COMPLEMENT ACTIVATION ON HOST CELLS AND TISSUES

HUS represents the interaction between cellular injury and the complement system. We propose that it is a model system for what occurs during damage to self-tissue.[32] Important clinical processes featuring such diverse situations as cellular toxins, ischemia-reperfusion injury, autoantibodies/immune complexes, and apoptosis may be related in this manner. In each case, there is some type of cellular injury in the setting of endogenous complement inhibitors. This should be contrasted to complement activation on microbes, which needs to be as rapid and robust as possible (at least sufficient to mediate pathogen entrapment and death). On healthy self-tissue, complement activation is not desirable; however, on injured yet viable self-tissue, a "controlled" amount of complement activation may facilitate healing and wound repair. On apoptotic and necrotic cells, more extensive complement activation may be desirable to safely dispose of intracellular debris with several purposes in mind, such as removing the "garbage" to facilitate repair while minimizing an immune response.

Special situations whereby complement activation on modified self-tissue has been co-opted to serve important biologic functions are also likely to exist. For example, complement activation occurs after the acrosomal reaction on viable spermatozoa, probably to facilitate egg-sperm interactions.[33]

DIAGNOSTIC AND THERAPEUTIC IMPLICATIONS

In aHUS, mutational screening should be considered because 50% of patients will have a factor H, MCP, or factor I mutation. Currently, this type of genetic testing is expensive and time consuming. Assessing protein levels of these three proteins can be helpful. Factor H and factor I levels can be monitored by antigenic assays. MCP may most easily be screened by flow cytometry of human peripheral blood cells. If the level of any one of these three regulators is reduced by 50% or greater, this substantially enhances the likelihood of mutation in the corresponding gene. Normal levels, however, do not rule out a mutation because about one third of patients with a mutation express normal levels of a dysfunctional protein. Unfortunately, functional assays are not yet widely available for screening of these three proteins, but they are available in laboratories specializing in the complement system and in a few commercial laboratories.

Currently, one screening approach is to assess factor H, factor I, and MCP levels. If low, genetic analysis should be undertaken. If normal but the suspicion remains high (familial HUS), then functional screening could be ordered. C3, C4, factor B, and total hemolytic or whole complement assays have not proven helpful in screening patients because they are usually normal. DAF (Ref. 20 and unpublished data), factor B,[34] and complement receptor 1 (CR1)[19] have not been linked to aHUS. CD59 has not been studied. More subtle defects, such as combined partial deficiencies or haplotype associations, may account for some of the remaining 50% of patients who have no defect in these three complement regulators.[14,20,35]

The presence of a complement regulatory protein deficiency has an impact on treatment decisions. In general, patients with aHUS have a high recurrence rate (~50%) in the transplanted kidney.[7] Patients with a deficiency of factor H have an even greater (~80%) chance of recurrence.[7] Factor I deficiency would be predicted to have recurrence rates similar to those of factor H. Because factor H and factor I are synthesized in the liver, a renal transplant will not address the deficiency. Additionally, patients with factor H or factor I would also be more susceptible to acute humoral and chronic vascular graft rejection because of their decreased ability to inhibit complement activation. By contrast, MCP deficiency can be corrected (at least in the kidney) by a renal transplant with a graft that expresses normal MCP levels. Therefore, a family member carrying a mutation must not be the donor. In the three reported cases of transplantation in MCP deficiency, aHUS has not recurred.[18]

SUMMARY

Mutations in complement regulatory proteins factor H, factor I, and MCP have now been recognized in ~50% of cases of aHUS. All of these mutations impair CA regulation. We advise the analysis of patients with aHUS for mutations before renal transplantation.

REFERENCES

1. Richards, A., J. Goodship & T.H. Goodship. 2002. The genetics and pathogenesis of haemolytic uraemic syndrome and thrombocytopenic purpura. Curr. Opin. Nephrol. Hypertens. **11:** 431–436.
2. Richardson, S.E., M.A. Karmali, L.E. Becker, *et al.* 1988. The histopathology of the hemolytic uremic syndrome associated with verocytotoxin-producing *Escherichia coli* infections. Human Pathol. **19:** 1102–1108.
3. Inward, C.D., A.J. Howie, M.M. Fitzpatrick, *et al.* 1997. Renal histopathology in fatal cases of diarrhoea-associated haemolytic uraemic syndrome. Pediatr. Nephrol. **11:** 556–559.
4. Taylor, C.M., C. Chua, J.J. Howie, *et al.* 2004. Clinico-pathological findings in diarrhoea-negative haemolytic uraemic syndrome. Pediatr. Nephrol. **19:** 419–425.
5. Karmali, M.A., M. Petric, C. Lim, *et al.* 1985. The association between idiopathic hemolytic uremic syndrome and infection by verotoxin-producing *Escherichia coli*. J. Infect. Dis. **151:** 775–781.
6. Karmali, M., B. Steele, M. Petric, *et al.* 1983. Sporadic cases of haemolytic uraemic syndrome associated with faecal cytotoxin and cytotoxin-producing *Escherichia coli* in stools. Lancet **1:** 619–620.
7. Goodship, T.H.J., V. Fremeaux-Bacchi & J.P. Atkinson. 2005. Genetic testing in atypical HUS and the role of membrane cofactor protein (MCP; CD46) and factor I. *In* Progress in Inflammation Research: Complement and Kidney Disease. Springer. In press.
8. Zipfel, P.F., H.P.H. Neumann & M. Jozsi. 2003. Genetic screening in haemolytic uraemic syndrome. Curr. Opin. Nephrol. Hypertens. **12:** 653–657.

9. ZIPFEL, P.F., S. HEINEN & M. JOZSI. 2006. Complement and diseases: defective alternative pathway control results in kidney and eye disease. Molec. Immunol. **43:** 97–106.
10. BUDDLES, M., R. DONNE, A. RICHARDS, *et al.* 2000. Complement factor H gene mutation associated with autosomal recessive atypical hemolytic uraemic syndrome. Am. J. Human Genet. **66:** 1721–1722.
11. WARWICKER, P., T.H.J. GOODSHIP, R.L. DONNE, *et al.* 1998. Genetic studies into inherited and sporadic hemolytic uremic syndrome. Kidney Int. **53:** 836–844.
12. RICHARDS, A., M.R. BUDDLES, R.L. DONNE, *et al.* 2001. Factor H mutations in hemolytic uremic syndrome cluster in exons 18-20, a domain important for host cell recognition. Am. J. Human Genet. **68:** 485–490.
13. CAPRIOLI, J., P. BETTINAGLIO, P.F. ZIPFEL, *et al.* 2001. The molecular basis of familial hemolytic uremic syndrome: mutation analysis of factor H gene reveals a hot spot in short consensus repeat 20. J. Am. Soc. Nephrol. **12:** 297–307.
14. CAPRIOLI, J., F. CASTELLETTI, S. BUCCHIONI, *et al.* 2003. Complement factor H mutations and gene polymorphisms in haemolytic uraemic syndrome: the C-257T, the A2089G and the G2881T polymorphisms are strongly associated with the disease. Human Molec. Genet. **12:** 3385–3395.
15. DRAGON-DUREY, M.-A., V. FREMEAUX-BACCHI, C. LOIRAT, *et al.* 2004. Heterozygous and homozygous factor H deficiencies associated with hemolytic uremic syndrome or membranoproliferative glomerulonephritis: report and genetic analysis of 16 cases. J. Am. Soc. Nephrol. **15:** 787–795.
16. PEREZ-CABALLERO, D., C. GONZALEZ-RUBIO, M.E. GALLARDO, *et al.* 2001. Clustering of missense mutations in the C-terminal region of factor H in atypical hemolytic uremic syndrome. Am. J. Human Genet. **68:** 478–484.
17. NEUMANN, H.P.H., H. SALZMANN, B. BOHNERT-IWAN, *et al.* 2003. Haemolytic uraemic syndrome and mutations of the factor H gene: a registry based study of German speaking countries. J. Med. Genet. **40:** 676–681.
18. RICHARDS, A., E.J. KEMP, M.K. LISZEWSKI, *et al.* 2003. Mutations in human complement regulator, membrane cofactor protein (CD46), predispose to development of familial hemolytic uremic syndrome. Proc. Natl. Acad. Sci. USA **100:** 12966–12971.
19. NORIS, M., S. BRIOSCHI, J. CAPRIOLI, *et al.* 2003. Familial haemolytic uraemic syndrome and an MCP mutation [see comment]. Lancet **362:** 1542–1547.
20. ESPARZA-GORDILLO, J., E. GOICOECHEA DE JORGE, A. BUIL, *et al.* 2005. Predisposition to atypical hemolytic uremic syndrome involves the concurrence of different susceptibility alleles in the regulators of complement activation gene cluster in 1q32. Human Molec. Genet. **14:** 703–712.
21. KAVANAGH, D., E.J. KEMP, E. MAYLAND, *et al.* 2005. Mutations in complement factor I predispose to the development of atypical hemolytic uremic syndrome. J. Am. Soc. Nephrol. **16:** 2150–2155.
22. FREMEAUX-BACCHI, V., M.-A. DRAGON-DUREY, J. BLOUIN, *et al.* 2004. Complement factor I: a susceptibility gene for atypical haemolytic uraemic syndrome. J. Med. Genet. **41:** 1–5.
23. CASTELLETTI, F., S. BUCCHIONI, S. BRIOSCHI, *et al.* 2005. Combined mutations in factor H (CFH) and membrane cofactor protein (MCP) in hemolytic uremic syndrome (HUS) (abstr). Molec. Immunol. **43:** 166.
24. DRAGON-DUREY, M.-A., C. LOIRAT, S. CLOAREC, *et al.* 2005. Anti-factor H autoantibodies associated with atypical hemolytic uremic syndrome. J. Am. Soc. Nephrol. **16:** 555–563.
25. CONSTANTINESCU, A.R., M. BITZAN, L.S. WEISS, *et al.* 2004. Non-enteropathic hemolytic uremic syndrome: causes and short-term course. Am. J. Kidney Dis. **43:** 976–982.
26. GARG, A.X., W.F. CLARK, M. SALVADORI, *et al.* 2005. Microalbuminuria three years after recovery from *Escherichia coli* O157 hemoltyic uremic syndrome due to municipal water contamination. Kidney Int. **67:** 1476–1482.
27. GARG, A.X., R.S. SURI, N. BARROWMAN, *et al.* 2003. Long-term renal prognosis of diarrhea-associated hemolytic uremic syndrome: a systematic review, meta-analysis, and meta-regression [see comment]. J. Am. Med. Assoc. **290:** 1360–1370.

28. Jozsi, M., T. Manuelin, S. Heinen, *et al.* 2004. Attachment of the soluble complement regulator factor H to cell and tissue surfaces: relevance for pathology. Histol. Histopathol. **19:** 251–258.
29. Manuelin, T., J. Hellwage, S. Meri, *et al.* 2003. Mutations in factor H reduce binding affinity to C3b and heparin and surface attachment to endothelial cells in haemolytic uraemic syndrome. J. Clin. Invest. **111:** 1181–1190.
30. Perkins, S.J. & T.H. Goodship. 2002. Molecular modelling of the C-terminal domains of factor H of human complement: a correlation between haemolytic uraemic syndrome and a predicted heparin binding site. J. Molec. Biol. **316:** 217–224.
31. Sanchez-Corral, P., D. Perez-Caballero, O. Huarte, *et al.* 2002. Structural and functional characterization of factor H mutations associated with atypical haemolytic uraemic syndrome. Am. J. Human Genet. **71:** 1285–1295.
32. Atkinson, J.P. 2003. Complement system on the attack in autoimmunity. J. Clin. Invest. **112:** 1639–1641.
33. Riley-Vargas, R.C., S. Lanzendorf & J.P. Atkinson. 2005. Targeted and restricted complement activation on acrosome-reacted spermatozoa. J. Clin. Invest. **115:** 1241–1249.
34. Kavanagh, D., E.J. Kemp, A. Richards, *et al.* 2005. Does complement factor B have a role in the pathogenesis of atypical HUS? Molec. Immunol. In press.
35. Fremeaux-Bacchi, V., E.J. Kemp, J. Goodship, *et al.* 2005. Development of atypical HUS is influenced by susceptibility factors in factor H and membrane co-factor protein-evidence from two independent cohorts. J. Med. Genet. In press.

Oligomeric Structure of Nitrilases

Effect of Mutating Interfacial Residues on Activity

B.T. SEWELL,[a] R.N. THUKU,[a] X. ZHANG,[b] AND M.J. BENEDIK[b]

[a]*Electron Microscope Unit, IIDMM, University of Cape Town, Cape Town, South Africa*

[b]*Department of Biology, Texas A&M University, College Station, Texas 77843-3258, USA*

ABSTRACT: Nitrilases are important industrial enzymes that convert nitriles into their corresponding acids or, occasionally, amides. Atomic resolution structures of four members of the nitrilase superfamily have been determined, but these differ from microbial nitrilases in that they do not form typical large homo-oligomeric complexes. At least two nitrilases, the cyanide dihydratases from *Pseudomonas stutzeri* AK61 and *Bacillus pumilus* C1, form unusual spiral structures of 14 and 18 subunits, respectively. Evidence suggests that the formation of the spiral structure is essential for activity. Sequence analysis reveals that the nitrilases differ from the nonspiral-forming homologs by two insertions of between 12 and 14 amino acids and a C-terminal extension of up to 35 amino acids. The insertions are positioned at an intermolecular interface in the spiral and probably contribute to its formation. The other interfaces responsible for the formation and/or stabilization of the spirals can also be identified. Comparative structure modeling enables identification of the residues involved in these interacting surfaces, which are remote from the active site. Mutation of these interacting residues usually leads to loss of activity. The effect of the mutations on activity in most cases can be rationalized in terms of a possible effect on spiral formation.

KEYWORDS: nitrilase; oligomeric structure; mutations; structure-activity; interfacial residues

INTRODUCTION

The nitrilases (E.C.3.5.5.1) are industrial enzymes that are being used to manufacture the biologically active enantiomers such as: (R)-mandelic acid, (S)-phenyllactic acid, and (R)-3-hydroxy-4-cyano-butyric acid, which is a key intermediate in the synthesis of the blockbuster drug Lipitor® (Pfizer Inc.).[1,2] They convert nitriles to the corresponding acids and ammonia. We have determined the low resolution structures of two cyanide-degrading members of this family from *Pseudomonas stutzeri* ($CynD_{stu}$)[3] and *Bacillus pumilus* ($CynD_{pum}$)[4] and have found that they are defined-length spiral structures having 14 and 18 subunits, respectively, under conditions of optimum activity (pH 7–8). Members of the nitrilase superfamily are mod-

Address for correspondence: B.T. Sewell, Electron Microscope Unit, University of Cape Town, 7701 Cape Town, South Africa.
sewell@uctvms.uct.ac.za

**Ann. N.Y. Acad. Sci. 1056: 153–159 (2005). © 2005 New York Academy of Sciences.
doi: 10.1196/annals.1352.025**

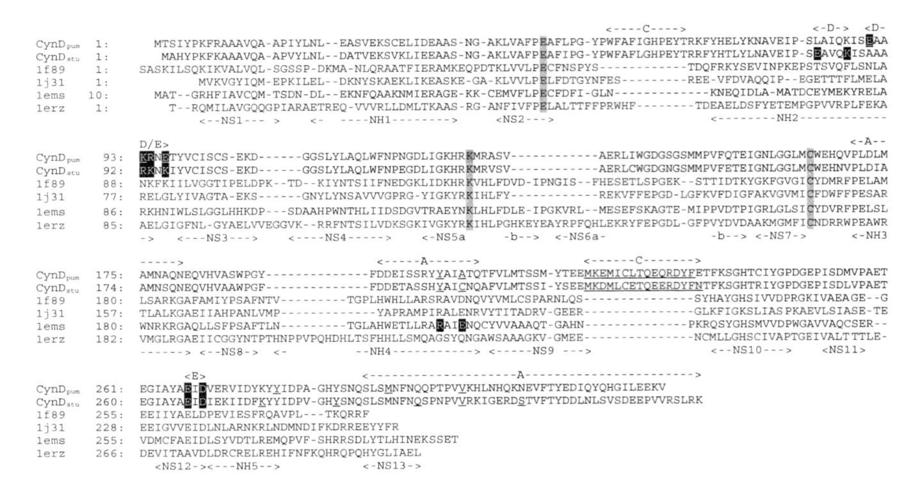

FIGURE 1. Alignment of the sequences of the cyanide dihydratases from *B. pumilus* C1 ($CynD_{pum}$) and *P. stutzeri* AK61 ($CynD_{stu}$) with the sequences of the four homologs for which the atomic structures have been determined. Pairwise alignment of 1f89, 1j31, 1ems, and 1erz was done with ALIGN,[12] and the CynD sequences were aligned with GenTHREADER.[13] The secondary structural elements referring to 1ems are indicated in the *bottom line* and use the notation of Pace *et al.*[6] The approximate regions of the interacting surfaces A, C, D, and E are indicated in the *top line*. Charged residues that may be involved in these interactions are *white* on a *black background*. The residues mutated as indicated in TABLE 1 are *underlined*. The conserved active site residues are *black* on a *grey background*.

erately ubiquitous and are believed to demonstrate structural homology despite varying sequence conservation and differing substrates.[5]

The atomic structures of four distant homologs have been determined (1ems,[6] 1erz,[7] 1j31,[8] and 1f89[9]). All the structures have twofold symmetry, which conserves the αββααββα fold comprising a dimer of the 35–40-kDa protein (FIG. 1). The nitrilases for which the atomic structure has been solved exist as dimers or tetramers, whereas the microbial nitrilases are found as high molecular weight homo-oligomers. Modeling based on these structures has enabled us to interpret the low-resolution maps and has, in particular, enabled us to identify the interfaces that lead to spiral oligomer formation and to postulate which residues are involved in the interactions across the interfaces (FIG. 2). The location of the interfacial regions in the oligomeric structure of $CynD_{stu}$ is shown in FIGURE 3.

Two independent lines of evidence suggest that oligomer formation is essential for activity. In Rhodococci, the nitrilases in several cases have been isolated as inactive dimers. Nagasawa *et al.*[10] have shown that these dimers form active decamers in the presence of the substrate benzonitrile. In the case of $CynD_{pum}$, the protein exists as an active octadecamer at neutral pH; however, we have demonstrated the formation of long helical fibers at pH 5.4. Whereas most cyanide-degrading enzymes decrease monotonically in activity as a function of pH below the optimum, the onset of fiber formation corresponds to a small increase in activity for $CynD_{pum}$, consis-

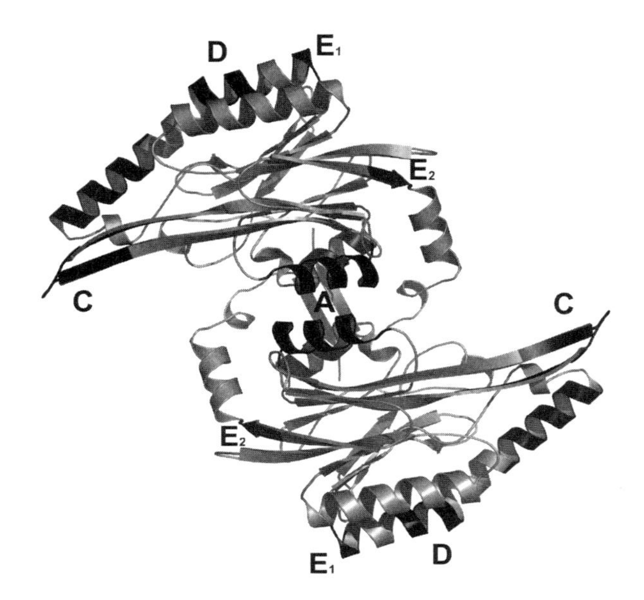

FIGURE 2. A ribbon diagram depicting a model of a dimer formed by $CynD_{stu}$. The model was made using MODELLER[14] and was based on the alignment with 1ems shown in FIGURE 1. The locations of residues predicted to be involved in interfacial interactions are indicated in *black*. The A surface comprises the α-helices 170-179 and 192-205 as well as β-sheet 289-296. The remainder of the C-terminal extension, also thought to contribute to the A surface, is not modeled. The residues participating in the D surface are 82-87. The E surface comprises two components, E_1 (92-96) and E_2 (266-268). The C surface comprises two insertions. Residues 63-74 were constrained in the modeling to be an α-helix, and residues 216-233 were constrained to be a β-sheet with a bend at 224-230. This configuration suggests that the C surface contains a four-stranded β-sheet with two strands made up by extensions of NS9 and NS10, being contributed symmetrically by each molecule.

tent with a model whereby the terminal monomers of the short spirals become activated as they participate in the extended fibers.[11]

In this study we further explore the dependence of the activity of these enzymes on quaternary structure by modification of the residues at the interfaces that lead to spiral formation in such a way that the interface would be damaged.

RESULTS AND DISCUSSION

Creation and Expression of Mutant Nitrilases

Recombinant clones of the cyanide-degrading nitrilase genes were created in pET26b and, when introduced into *Escherichia coli* BL21(DE3) for expression, produce abundant active enzyme.[4,11] The plasmid p2784 carries the *P. stutzeri* nitrilase $CynD_{stu}$ and p2890 carries the *B. pumilus* nitrilase $CynD_{pum}$. These plasmids were

TABLE 1. Effect of mutating interfacial residues on activity

Mutant	Surface	Change and location	Activity
B. pumilus			
1. Delta 303	A	Vgtg->stop	Full activity
2. Delta 293	A	Matg->stop	Partial activity
3. Delta 279	A	Ytat->stop	Inactive
4. Y201D/A204D	A	Ytat->Dgac, Agcg->Dgac	Inactive
5. Delta 219-233	C	MKEMICLTQEQRDYF was deleted. 235 Egaa->Naac	Inactive
6. 90	D	EAAKRNE->AAARKNK	Full activity
P. stutzeri			
7. Delta 310	A	Sagt->stop	Inactive
8. Delta 302	A	Vgtg->stop	Inactive
9. Delta 296	A	Qcag->stop	Inactive
10. Delta 285	A	Ytat->stop	Inactive
11. Delta 276	A	Kaaa->stop	Inactive
12. Y200D/C203D	A	Ytac->Dgac, Ctgc->Dgac	Inactive
13. Delta 220-234	C	MKDMLCETQEERDYF deleted	Inactive
Hybrids			
14. Pum – Stu	A	Residues 1-286 from *B. pumilus*, 287-end from *P. stutzeri*	Full activity
15. Stu – Pum	A	Residues 1-286 from *P. stutzeri,* 287-end from *B. pumilus*	Inactive

the backbone of subsequent site-directed mutations, created using the Stratagene Quickchange method, and are listed in TABLE 1. The mutations were confirmed by DNA sequencing, and those mutants that remained inactive were further analyzed by SDS-PAGE or Western blotting to demonstrate that in all cases protein was made and the deficiency in activity was not due to rapid turnover. Methods for expression and measurement of cyanide-degrading activity were as previously described.[4,11]

Analysis of Interfaces in the Terminating Spirals of $CynD_{pum}$ and $CynD_{stu}$ — the C Surface

The spirals formed by $CynD_{pum}$ and $CynD_{stu}$ have a global twofold axis, which coincides with the twofold axis of the dimer formed by the alpha helices labeled NH3 and NH4 and the beta sheet NS13 in the NitFhit (1ems) structure.[6] We have previously called this the A surface. Since the C-terminus of the Nit element of Nit-Fhit participates in forming this interface, we speculate that a component of the much larger carboxy-terminal extension (relative to Nit) is involved in this interface

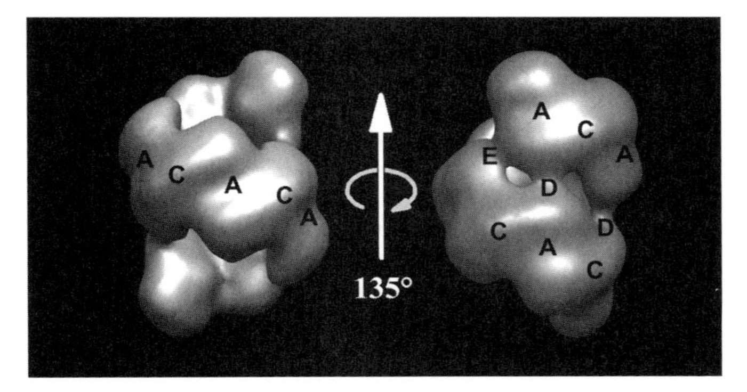

FIGURE 3. Location of the interfacial regions in the low resolution map of the terminating spiral of $CynD_{stu}$. For further information see Sewell *et al.*[3]

in an analogous manner. A region of density on the inner surface of the spiral, coincident with the dyad axis, is available to accommodate some of these residues. Spiral extension is made possible by a further interaction, previously called the C surface by us.[3] This surface is a further symmetric interaction between the subunits. We have speculated that the residues involved in this interaction are the insertions at the amino terminal end of NH2 (12–13 residues) and at the bend between NS9 and NS10 (13–15 residues). Both of these insertions are located in such a way that they could plausibly fit empty density in the spiral structures. The dyadic symmetry of these interactions is broken, in the case of the $CynD_{stu}$ spiral or the terminating $CynD_{pum}$ spiral at neutral pH, by the interactions (E surface, described below) that lead to spiral termination, thereby reducing the symmetry operator to a pseudo-dyad. Damage to the C surface, as exemplified by mutations 5 and 13 (TABLE 1), render the enzyme inactive, as would be predicted. In these mutants, the insertion between NS9 and NS10 is excised, thus reducing the length of the β-sheet to that found in the non-spiral-forming homologs.

D Surface

In the $CynD_{stu}$ structure we noted four sites of interaction across the groove (FIG. 3).[3] These sites of interaction are of two different types. The first type is located on the pseudo-dyad axis relating the C surface, but on the other side of the spiral. Modeling suggests that the interactions involve residues near the carboxy-terminal end of the NH2 helix. We will name this interaction the D surface. In the case of $CynD_{stu}$ the residues 82EAVQK87 are appropriately located in our homology model to form a pair of pseudo-symmetric salt bridges at this point of interaction. The possibility of a salt bridge at this point does not exist in $CynD_{pum}$ where the corresponding sequence is 83LAIQK88, but rather there is a possibility of forming a pair of salt bridges at 90EAAKRNE97. Conversely, the $CynD_{stu}$ sequence at this region is 89AAARKNK96, which could not form a pair of symmetrically related salt bridges.

If our model-relating structure to activity is correct, then this raises the possibility that there are different interactions maintaining the helical structure in each case. In particular, we suggest that these differences are located in the D surface, and the 90EAAKRNE97 stretch represents a potentially important difference in the sequences of $CynD_{stu}$ and $CynD_{pum}$. Removal of this putative D-surface sequence by mutation 6 (TABLE 1) has no measurable effect on activity. This indicates that the $CynD_{pum}$ spiral is stable without these interactions. The stability could be due to compensating interactions at the A surface to be discussed.

The other interaction across the groove leads to termination of the spiral in the case of $CynD_{stu}$ by closing the gap that would otherwise be available for a further subunit to be added. This is caused by asymmetric interactions suggested by modeling to be located in NH2, on the one side, involving residues 92RKNK96 and in NS12, involving the highly conserved residues 266EID268, on the other. These two structural elements interact to form the E surface.

A Surface

The A-surface region comprising the interactions of helices NH3 and NH4 interacting across the dyad axis is well conserved in the four crystallographically determined homologs. The excellent fit of a dimer model built around this interface to thenegative stain electron microscopic density provides some evidence that this feature is common to the cyanide dihydratases. In 1ems, residues R211 and E214 (corresponding to residues Y200 and C203 in $CynD_{stu}$) interact across the twofold axis to form a pair of salt bridges. It was therefore considered that damage to this surface would occur if these two residues were replaced by residues of similar charge. This is indeed the case, as shown by mutation of these residues (4 and 12 of TABLE 1), leading to inactivation of the enzyme. The mutations, which introduced like charges, resulted in loss of activity in both the *B. pumilus* and *P. stutzeri* enzymes.

It is likely that the A surface is further comprised of a component of the C-terminal "extension" as well. A series of C-terminal truncations were generated in both enzymes; the effect, however, was different for each of the enzymes. Even very small changes affected the activity of $CynD_{stu}$, but $CynD_{pum}$ proved to be more robust, tolerating truncations back to residue 293 before activity was lost (1-3, 7-11 of TABLE 1). This robustness is consistent with the effect reported for the D-surface mutation 6, and indeed, interactions at the D surface may stabilize the spiral having a truncated C-terminal tail. This is further demonstrated with mutants 14 and 15, which are hybrid proteins carrying the N-terminal region from one enzyme and the C-terminal tail from the other. Swapping the C-terminal domains between these enzymes shows that $CynD_{pum}$ functions regardless of which C-terminus it carries, but $CynD_{stu}$ cannot function with the $CynD_{pum}$ C-terminus.

CONCLUSIONS

We have identified a number of interacting ion pairs that are likely to contribute to the formation and stabilization of the spiral. Modification of residues distant from the active site usually does not influence the activity of the enzyme. However, in the case of the spiral-forming nitrilases, interactions occur between subunits, which we postulate impinge on the activity of the enzyme. We have sought to systematically disrupt the interactions at the interfaces and have observed the effect on the activity of the enzymes. Systematic disruption of the interfaces generally deactivates the enzyme, which is evidence in support of our postulate. In addition, we have shown that different interactions lead to spiral stabilization despite the similarities between $CynD_{pum}$ and $CynD_{stu}$, imposing a different requirement for the highly variable C-terminal domain.

ACKNOWLEDGMENTS

We gratefully acknowledge the Robert A. Welch Foundation and the Gulf Coast Hazardous Substance Research Center (#069UHH0789 to M.J.B.) and the Wellcome Trust (to B.T.S.) for support of this project. R.N.T. is grateful for support from the Carnegie Corporation of New York and CSIR Bio/Chemtek.

REFERENCES

1. BANERJEE, A., R. SHARMA & U.C. BANERJEE. 2002. The nitrile-degrading enzymes: current status and future prospects. Appl. Microbiol. Biotech. **60:** 33–44.
2. O'REILLY, C. & P.D. TURNER. 2003. The nitrilase family of CN hydrolyzing enzymes: a comparative study. J. Appl. Microbiol. **95:** 1161–1174.
3. SEWELL, B.T., M. BERMAN, P.R. MEYERS, *et al.* 2003. The cyanide degrading nitrilase from *Pseudomonas stutzeri* AK61 is a two-fold symmetric, 14-subunit spiral. Structure **11:** 1413–1422.
4. JANDHYALA, D., M.N. BERMAN, P.R. MEYERS, *et al.* 2003. CynD, the cyanide dihydratase from *Bacillus pumilus*: Gene cloning and structural studies. Appl. Environ. Microbiol. **69:** 4794–4805.
5. PACE, H.C. & C. BRENNER. 2001. The nitrilase superfamily: classification, structure and function. Genome Biology **2:** reviews 0001.1–0001.9.
6. PACE, H.C., S.C. HODAWADEKAR, A. DRAGANESCU, *et al.* 2000. Crystal structure of the worm NitFhit Rosetta Stone protein reveals a Nit tetramer binding two Fhit dimers. Curr. Biol. **10:** 907–917.
7. NAKAI, T., T. HASGAWA, E. YAMASHITA, *et al.* 2000. Crystal structure of *N*-carbamyl-D-amino acid amidohydrolase with a novel catalytic framework common to amidohydrolases. Structure **8:** 729–737.
8. SAKAI, N., Y. TAJIKA, M. YAO, *et al.* Crystal structure of the hypothetical protein Ph0642 from *Pyrococcus horikoshii* and structure based prediction of enzymatic reaction. RCSB Protein Databank (1j31).
9. KUMARAN, D., S. ESWARAMOORTHY, S.E. GERCHMAN, *et al.* 2003. Crystal structure of a putative CN hydrolase from yeast. Proteins: Struct. Funct. Genet. **52:** 283–291.
10. NAGASAWA, T., M. WIESER, T. NAKAMURA, *et al.* 2000. Nitrilase of *Rhodococcus rhodochrous* J1: conversion into the active form by subunit association. Eur. J. Biochem. **267:** 138–144.
11. JANDHYALA, D.M., R.C. WILLSON, B.T. SEWELL & M.J. BENEDIK. 2005. Analysis of three microbial cyanide degrading enzymes. Appl. Microbiol. Biotech. **68:** 327–335.
12. COHEN, G.E. 1997. ALIGN: a program to superimpose protein coordinates, accounting for insertions and deletions. J. Appl. Cryst. **30:** 1160–1161.
13. MCGUFFIN L.J. & D.T. JONES. 2003. Improvement of the GenTHREADER method for genomic fold recognition. Bioinformatics **19:** 874–881.
14. SALI, A. & T.L. BLUNDELL. 1993. Comparative protein modelling by satisfaction of spatial restraints. J. Mol. Biol. **234:** 779–815.

Development of Domain-Selective Angiotensin I-Converting Enzyme Inhibitors

PIERRE REDELINGHUYS, ALOYSIUS T. NCHINDA, AND EDWARD D. STURROCK

Division of Medical Biochemistry, Institute of Infectious Disease and Molecular Medicine, University of Cape Town, Cape Town, South Africa

ABSTRACT: Somatic angiotensin-converting enzyme (ACE) is an essential component of the renin-angiotensin system and consequently plays a key role in blood pressure and electrolyte homeostasis. Thus, ACE inhibitors are widely used in the treatment of cardiovascular disease, causing a decrease in the production of angiotensin II and an increase in the circulating vasodilator bradykinin. The ectodomain of ACE consists of two parts (N and C domains), each bearing an active site that differs in substrate and inhibitor specificity. Advances in the elucidation of the functional roles of these two domains and an expanded view of the renin-angiotensin system underscore the need for the next generation of domain-selective inhibitors with improved pharmacologic profiles. Moreover, recent breakthroughs in determining the crystal structure of testis ACE (identical to the C domain) and its homologue ACE2 provide new mechanistic insights into the interactions of ACE inhibitors and substrates with active site pockets. This review summarizes the structural basis and recent synthetic chemistry approaches to the development of novel domain-selective inhibitors.

KEYWORDS: angiotensin I-converting enzyme; inhibitors; domain selectivity; drug design

INTRODUCTION

Hypertension has emerged as one of the greatest public health challenges of the twenty-first century, affecting an estimated 26% of the world's adult population.[1] This figure is expected to increase to 29% by 2025, the greatest increase occurring in economically developing countries. The emergence of hypertension as a major risk factor in cardiovascular and kidney disease[2] has necessitated the development of novel therapeutic approaches that have centered upon the renin-angiotensin system.

Angiotensin-converting enzyme (ACE), a member of the M2 gluzincin family of metallopeptidases,[3,4] occupies a central position in the renin-angiotensin system, where it is a key regulator of blood pressure, fluid, and electrolyte homeostasis.[5] The

Address for correspondence: Edward D. Sturrock, Division of Medical Biochemistry, Institute of Infectious Disease and Molecular Medicine, University of Cape Town, Cape Town, South Africa. Voice: +27-21-4066312; fax: +27-21-4066470.
sturrock@curie.uct.ac.za

Ann. N.Y. Acad. Sci. 1056: 160–175 (2005). © 2005 New York Academy of Sciences.
doi: 10.1196/annals.1352.035

larger somatic isoform of ACE (sACE) is a 1277 amino acid, 150–180-kDa[6] type I transmembrane glycoprotein that is expressed in a variety of tissues including vascular endothelial cells, intestinal brush border cells, and renal proximal tubule epithelial cells.[7–9] It is shed into the systemic circulation as a soluble ectodomain *via* cleavage at the Arg^{1203}–Ser^{1204} bond of the juxtamembrane stalk by a zinc metalloproteinase.[9] sACE comprises a C-terminal cytosolic tail, a hydrophobic membrane-anchoring domain, a juxtamembrane stalk, and an ectodomain consisting of two parts (C and N domains). The C and N domains, resulting from internal duplication of an ancestral gene, display a high level of homology.[7] Both domains contain a characteristic HEMGH zinc-coordinating motif, crucial for the catalytic activity of sACE.[10,11]

Alternate transcription from an internal promoter within the sACE gene has given rise to a smaller 701 amino acid, the 90–110-kDa testicular isoform of the C domain, which is restricted to male germinal cells. Testis ACE (tACE) lacks an N-terminal domain and consequently possesses only one active site per molecule.[5] tACE is expressed in male spermatozoa under hormonal control[12] and plays an important role in fertilization.[13] Recently, it was shown to cleave glycosylphosphatidylinositol (GPI)-anchored sperm proteins by a mechanism independent of is peptidyl dipeptidase activity, triggering sperm cell capacitation.[14]

Both somatic and testis ACE are heavily *N*-glycosylated with carbohydrates such as *N*-acetylglucosamine, fucose, mannose, glucose, and sialic acid, constituting 30% of their molecular weights.[15] sACE contains 17 potential *N*-linked glycosylation site,[16] whereas tACE contains 7 such sites in addition to a unique 36-residue N-terminus, which is heavily *O*-glycosylated.[17] sACE displays dual substrate specificity, acting both as an exo- and endopeptidase. For substrates like substance P and luteinizing hormone-releasing hormone (LHRH), where the C-termini are amidated, sACE acts not only as a dipeptidyl carboxypeptidase but also as an endopeptidase. Naqvi *et al.*[18] recently characterized the molecular basis of the exopeptidase activity of the sACE C-domain. This involves interactions between the substrate C-terminal P_2' side-chain and the S_2' pocket of the C domain as well as carboxylate-docking interactions with residues Lys^{1087} and Tyr^{1096}. These interactions are thought to stabilize the ground state, restricting the registration of substrates with a C-terminal carboxylate, limiting their processing to the cleavage of a C-terminal dipeptide. Other ACE substrates include acetyl-Ser-Asp-Lys-Pro and neurotensin.[18,19] With respect to its role as a regulator of cardiovascular homeostasis, the principal physiologic substrates of ACE are angiotensin I and bradykinin. Acting as a dipeptidyl carboxypeptidase, ACE mediates the hydrolysis of the decapeptide angiotensin I to the active vasopressor octapeptide angiotensin II via removal of a C-terminal dipeptide, His-Leu. The vasodilator bradykinin is inactivated via sequential hydrolysis of two carboxy-terminal dipeptides at Pro^7-Phe^8 and Phe^5-Ser^6.[15,20]

Both the N and C domains of membrane-bound sACE are responsible for the inactivating hydrolysis of bradykinin. However, in the soluble, circulating state, angiotensin I hydrolysis is also affected by the N domain, although it is the membrane-bound form of sACE that is primarily responsible for cardiovascular homeostasis.[21–23]

The interaction of angiotensin II with its cognate endothelial angiotensin type-1 (AT_1) receptor results in vasoconstriction, aldosterone and vasopressin release, reno tubular sodium resorption, and decreased renal blood flow. The net effect, alongside inactivation of the vasodilator bradykinin, is an elevation in blood pressure. Therefore, increased ACE activity is intimately linked to hypertension.

Despite the more than 60% sequence identity between the C- and N domains of sACE, each domain (both of which harbor an active site) demonstrates distinct substrate specificities. In addition to the differential hydrolysis of angiotensin I and bradykinin (discussed previously), the N domain preferentially hydrolyzes the hemoregulatory peptide *N*-acetyl-Ser-Asp-Lys-Pro (AcSDKP) as well as angiotensin 1-7.[24] Recently, the N domain was found to be responsible for the degradation of Alzheimer amyloid β-peptide, inhibiting its aggregation and cytotoxicity.[25] With respect to substrate hydrolysis, the C domain demonstrates a significantly greater chloride dependence than does the N domain, whereas the N domain is thermally more stable than the C domain.[26,27]

EARLY DEVELOPMENT OF ACE INHIBITORS

In 1977, captopril (**1,** FIG. 1), the first nonpeptidic ACE inhibitor, was developed.[28–30] Its development was initiated by the discovery of bradykinin-potentiating peptides isolated from the venom of the Brazilian snake *Bothrops jararaca.*[31–33] Some bradykinin-potentiating peptides have been shown to display domain-specific ACE inhibition.[34]

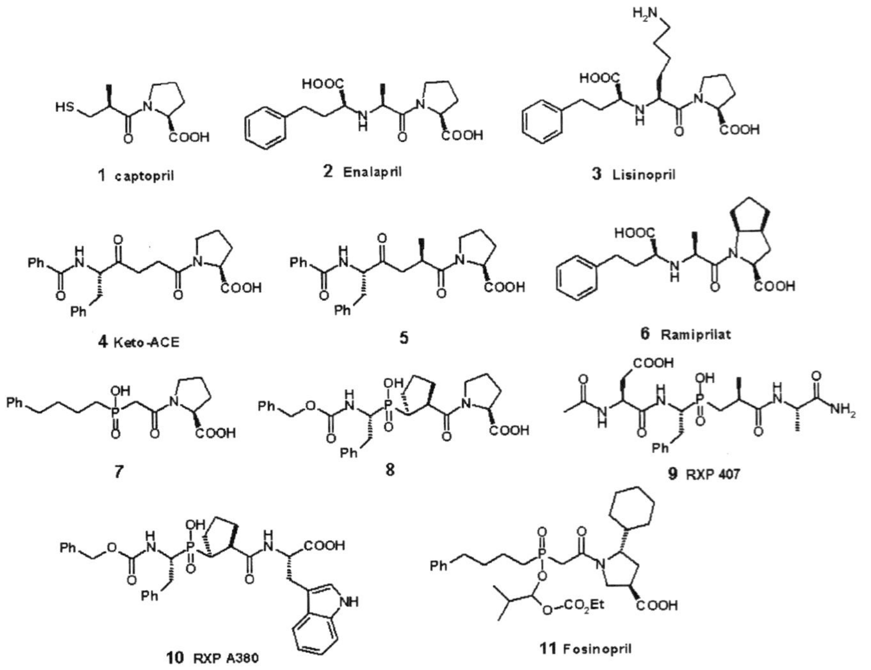

1 captopril
2 Enalapril
3 Lisinopril
4 Keto-ACE
5
6 Ramiprilat
7
8
9 RXP 407
10 RXP A380
11 Fosinopril

FIGURE 1. Some commonly known ACE inhibitors with different zinc-binding functionalities.

Captopril, together with other orally active ACE inhibitors such as enalapril and lisinopril (**2** and **3**, FIG. 1), are used extensively in the successful treatment and management of hypertension, congestive heart failure, myocardial infarction, and diabetic nephropathies.[24] However, these ACE inhibitors were designed without the knowledge of the three-dimensional structure of ACE.

The design of captopril was based on the structure of bovine pancreas carboxypeptidase A, a zinc-dependent carboxypeptidase thought to have similar catalytic mechanisms to those of ACE.[28–30] Lisinopril and enalapril were developed based on their inhibition of thermolysin.[35,36] These ACE inhibitors bind to somatic ACE at a 1:1 stoichiometry, indicating binding to only one of the two active sites.[12,27]

The lack of domain specificity of these ACE inhibitors (lisinopril being only slightly more C-domain specific) may be the underlying cause of their adverse side effects. Because both the C and N domains are involved in bradykinin hydrolysis, nonspecific inhibition of both domains may be linked to the phenomena of persistent cough (5–20% of patients) and potentially life-threatening angioedema (0.1–0.5% of patients).[37–40] This has been associated with systemic bradykinin accumulation as a result of the inappropriate suppression of N-domain bradykinin hydrolysis.

SYNTHESIS AND SELECTIVITY OF ACE INHIBITORS

Currently available ACE inhibitors include natural products and synthetic peptides belonging to the following classes of reactive compounds: sulfhydryls (captopril **1**); ketones (keto-ACE **4** and ketomethylene tripeptides **5**); carboxylates (lisinopril **3**); hydroxamic acids; and silanediols.

Although different approaches have been used in the synthesis of specific classes of inhibitor, they rely primarily on the interaction between a strong chelating group and the zinc-binding functionality of ACE. Because inhibitor binding is governed by the strength of this interaction, a drawback of this approach has been poor compound selectivity.[41] This has necessitated the development of approaches that include interactions with specific ACE substrate-binding pocket residues.

The recently elucidated structure of a testis ACE-lisinopril complex[42] as well as subsequent structures of ACE-inhibitor complexes[43,44] has provided valuable insights into the molecular basis of the specific interactions between the ACE substrate-binding pockets and ACE inhibitors (FIG. 2).

Significant interactions with the inhibitor lisinopril **3** occur via residues occupying the S_1, S_1' and S_2' pockets of the enzyme, highlighting the potential importance of these pockets in determining the domain selectivity of ACE inhibitors. The S_1' pocket of *t*ACE is shown to be very deep, and previous structure-activity relationship data have shown that the S_1' pocket can tolerate large hydrophobic P_1' side chains. This tendency is exhibited by most ACE inhibitors irrespective of their zinc-binding groups.

Moreover, the lysyl amine forms a salt bridge with D^{377} of tACE, which is replaced by a glutamine in the N domain (FIG. 2B), and an E162D substitution also occurs in the S_1' subsite. Thus, extension of P_1' substituents into the S_1' pocket of ACE may provide a means of developing ACE inhibitors with specific C and N domain selectivities. The stereochemistry of the P_1' substituents is very important in the determination of ACE inhibitory potency. The P_1' substituents with *S*-stereo-

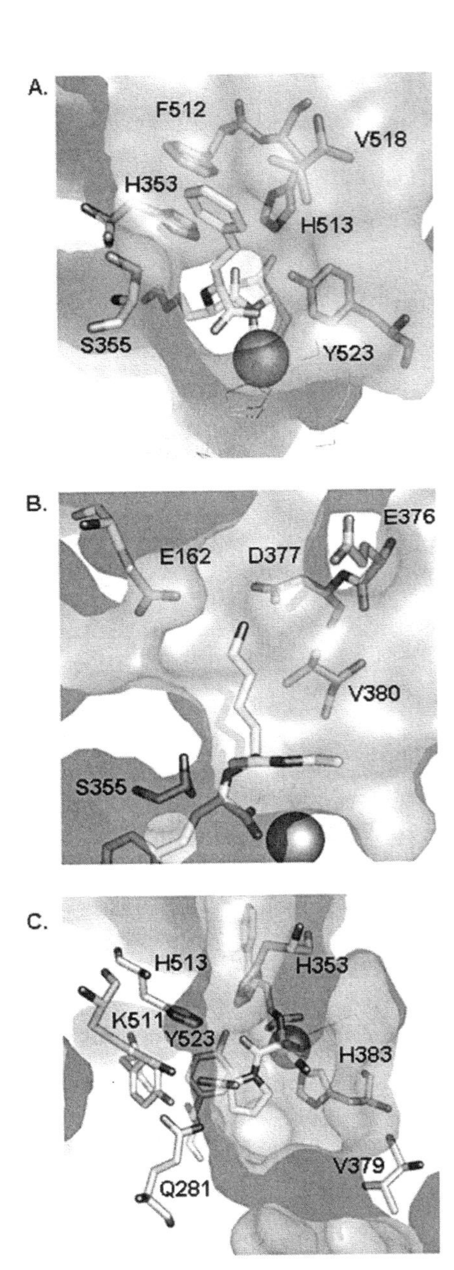

FIGURE 2. S_1 (A), S_1' (B) and S_2' (C) subsites of tACE and their interactions with lisinopril. The catalytic zinc atom is shown.

chemistry have been shown to possess greater potency than their *R* counterparts. The P_2' group of lisinopril fits into a relatively large S_2' pocket (FIG. 2C). This interaction with S_2' subsite residues such as K^{511} and Q^{281} via the hydrophobic prolyl ring[45] increases the ACE potency by at least 25-fold. Most available ACE inhibitors incorporate a proline or a tryptophan moiety at the P_2' position. The C-domain selective nature of a bulky P_2' residue may arise from the hydrophobic interactions with Val^{379}, which is replaced by a serine in the N domain.

Based on interactions with the zinc-binding functionality and surrounding subsite residues, two classes of compounds have been designed and developed: (1) those in which the zinc-binding group is flanked on both sides by amino acid residues (Type I and II inhibitors), and (2) those in which the amino acid residues are present on the right-hand side of the zinc-binding groups (Type III inhibitors) (FIGS. 3 and 4). A variety of different ACE inhibitors have been synthesized using this approach, incorporating a range of different zinc-binding functionalities.[46]

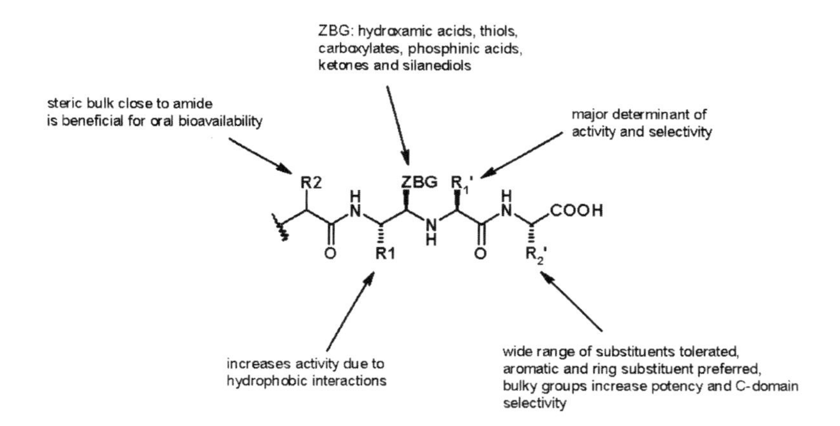

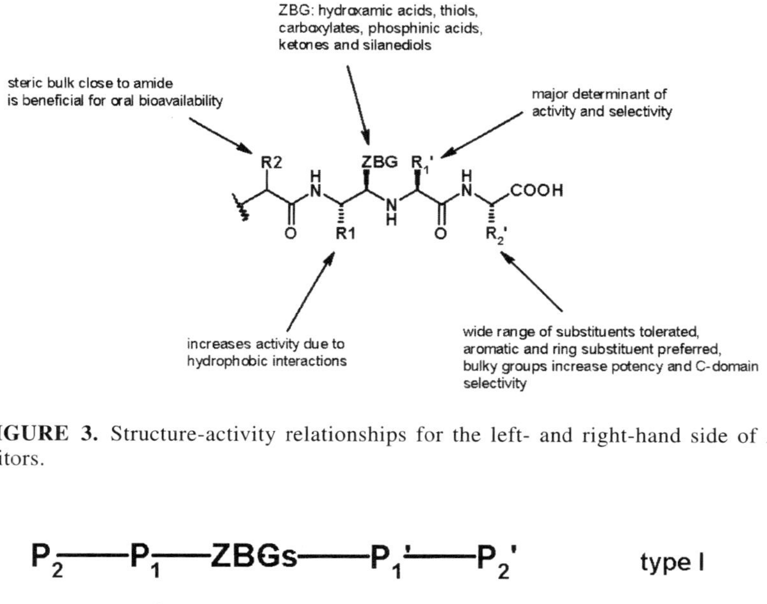

FIGURE 3. Structure-activity relationships for the left- and right-hand side of ACE inhibitors.

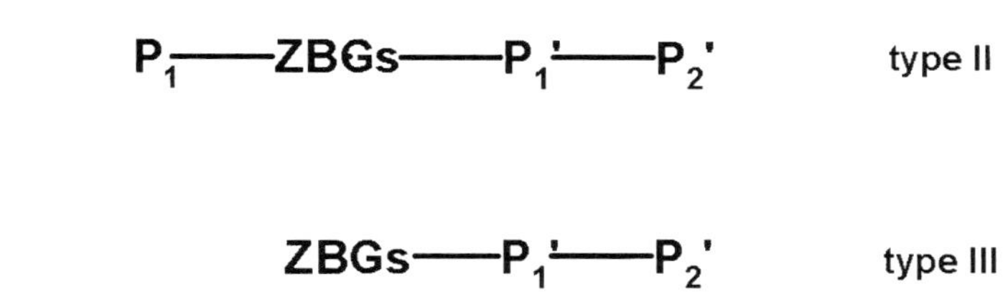

P_2—P_1—ZBGs—P_1'—P_2' type I

P_1—ZBGs—P_1'—P_2' type II

ZBGs—P_1'—P_2' type III

FIGURE 4. Schematic representation of the three major types of ACE inhibitors

Synthesis of left- or right-hand side inhibitors such as hydroxamates (to be discussed) and thiolates has permitted probing of either the primed or the unprimed side of the active site. Here, potency is primarily reliant upon the strength of the zinc-chelating group, but selectivity is compromised. By contrast, compounds that mimic the peptide substrate sequence to the left- and right-hand sides of the scissile bond and incorporate a weaker zinc-binding group, such as a phosphinate, exhibit more domain-selective inhibition.[41]

Phosphinic Acid Derivatives

Phosphinic peptide chemistry has been used to develop inhibitors that interact with both primed and unprimed sides of the active site. One of the early phosphinic acid ACE inhibitors developed in the 1980s was [hydroxyl-(4-phenylbutyl)phosphinyl]acetyl-*L*-proline (**7**).[47] This inhibitor had an IC_{50} value of 180 nM. The insertion of a methylene spacer between the phosphinic acid zinc-binding group and the carbon bearing the P_1' substituents results in increased potency (compound **8**) (FIG. 1). Furthermore, despite weaker phosphinic acid zinc-binding functionality, C- and N-domain selectivity has been greatly enhanced by various modifications of the P_1 and P_1' inhibitor residues. Using this approach, Dive and co-workers[41,48,49] recently reported the synthesis of two domain-specific phosphinic peptide ACE inhibitors, RXP 407 (**9**) and RXP A380 (**10**) (FIG. 1). RXP 407 is more selective for the N domain (K_i = 12 nM), whereas RXP A380 is approximately 1,000-fold more C-domain selective (K_i = 3 nM). RXP 407, characterized by the unusual incorporation of an aspartate residue in the P_2 position, was obtained using a solid-phase peptide synthesis approach, whereas the synthesis of RXP A380 was a classical synthetic chemistry approach similar to the right-hand to left-hand methodology as for type I ACE inhibitors. The N-domain selectivity of RXP407 has been attributed to an interaction between its acidic P_2 aspartate and an S_2 subsite arginine as well as its C-terminal carboxamido group. In the C domain, this S_2 subsite residue is replaced by a glutamate. The C-domain selectivity of RXPA380 can be attributed to the trans-amide geometry of the P_1'-P_2' residues required for an effective hydrogen bonding interaction between enzyme and inhibitor. Docking of RXP A380 at the active site showed that the pseudo-proline in the S_1' pocket imposes a particular orientation to the P_2' tryptophan residue, thus improving the C-domain selectivity. Given their *in vivo* stability, these domain-selective phosphinic acid derivatives have been useful in probing the functional roles of the N and C domains, confirming the role of both domains in bradykinin hydrolysis.[21,50,51] Thus, domain-specific inhibition of the C domain may be necessary and sufficient for the treatment of hypertension and phosphinic acid derivatives, potential lead compounds in the development of therapeutic drugs. Fosinopril (**11,** FIG. 1), belonging to the same class of inhibitors, is currently in clinical use for the treatment of hypertension.[41]

Phe-Ala-Pro and Ketomethylene Analogues

Phe-Ala-Pro and ketomethylene analogues (compounds **2–6**, FIG. 1) are type I and II ACE inhibitors, respectively (FIG. 4). The synthesis of type II inhibitors proceeds by synthesizing the building block P_1-ZBGs (the P_1 residues incorporating the zinc-binding group) followed by successive coupling of the P_1' and P_2' residues. A

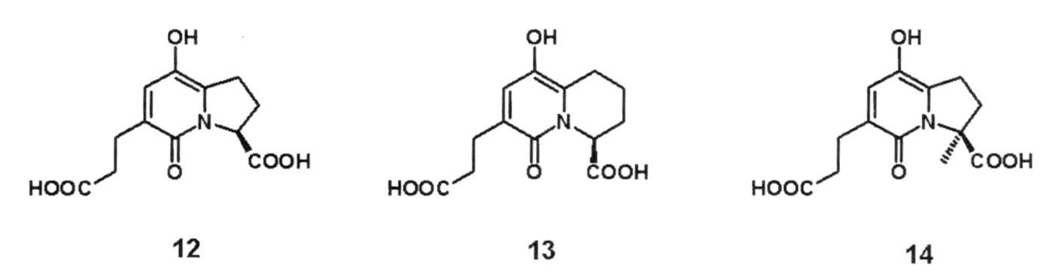

FIGURE 5. Nonpeptidic bicyclic ACE inhibitors.

similar approach is followed in the synthesis of type I ketomethylene analogues such as keto ACE, except that the P_2 residue is coupled with the P_1 residue. Exploiting their P_1' and P_2' residues, Phe-Ala-Pro analogues, such as enalaprilat (**2**) and ramiprilat (**6**), were some of the first commercially available ACE inhibitors for the treatment of hypertension. These inhibitors show IC_{50} values in a similar nanomolar range as those of lisinopril and captopril.

In general, peptides and peptidic compounds are usually unsuitable as drug candidates, possessing undesirable physical properties such as poor solubility, susceptibility to degradative enzymes, and poor oral bioavailability. To reduce the peptidic nature of ketomethylene inhibitors the P_1' and P_2' substituents may be cyclized to form a lactam, where there is a correlation between the inhibitory potency and the ring size.[52–54] Thus, the bicyclic dicarboxylic acids (**12** and **13**) and the nonepimerizable 3-methyl analogue (**14**, FIG. 5) were found to inhibit ACE in the low nanomolar range, similar to those of the commercially available inhibitors captopril and enalapril.

The synthetic chemistry of these ACE inhibitors was reviewed by De Lima.[46] Keto-ACE (**4**, FIG. 1), originally described in 1980, has emerged as a potential lead compound for C-domain–specific ACE inhibitors, with a 40–50-fold greater specificity for this domain compared with the N domain.[55,56] Keto-ACE and its analogues, which were found to inhibit ACE in the nanomolar range, contain a ketomethylene isostere replacement at the scissile bond that is believed to mimic the tetrahedral transition state of the proteolytic reaction.[57]

Keto-ACE, a tripeptide analogue of Phe-Gly-Pro, contains a bulky P_1 group and a P_2 benzyl ring that might confer C-domain selectivity.[58] With the availability of the three-dimensional C-domain (tACE) structure, the interaction of keto-ACE with the active site pockets of ACE was investigated. In this respect, synthesis of keto-ACE analogues with Trp or Phe at the P_2' position led to a marked increase in C-domain selectivity (unpublished data), highlighting the importance of the C-domain S_2' pocket. Inhibitory potency may further be enhanced by the incorporation of hydrophobic substituents such as a phenyl group at the P_1' position. In this instance, the stereochemistry of P_1' substituents is important. P_1' substituents with *S*-stereochemistry have been shown to possess greater inhibitory potency than have their *R* counterparts.[59]

Iterative docking experiments performed on a keto-ACE analogue synthesized in our research group (unpublished work) illustrated that the inhibitor makes protein-ligand contacts with the S_1, S_2, S_1' and S_2' residues of the ACE active site. The orientation of its benzyl ring permits a stacking interaction with the aromatic side chain of F^{391}. In the N domain, this residue is replaced by a less favorable Y^{369} at the S_2 subsite, explaining in part the C-domain selectivity of keto-ACE. Additional contacts include those between the ketone group and the catalytic zinc atom, an interaction believed to be one of the most significant forces holding the complex together. Moreover, the peptide backbone of the inhibitor makes hydrogen bonding contacts with the main chain of the protein.

Silanediols

One approach to protease inhibition that has proven very successful is the incorporation of a nonhydrolyzable isostere of the tetrahedral intermediate of amide hydrolysis (structure **15,** FIG. 6). Ketone or aldehyde hydrates (**16**) are examples that effectively mimic the tetrahedral intermediate, but such structures are often reactive and form undesirable covalent bonds with other nucleophilic species *in vivo.*[60,61] Other molecular structures that mimic the "geminal diol motif," such as **17** or **18,** have also been incorporated into peptide derivatives and shown to serve as effective isosteres. The phosphorous geminal diol (**17**) is very similar to the phosphinic acid zinc-binding group functionality present in the phosphinic acid class of ACE inhibitors.[47,48] Silicon forms a dialkylsilanediol compound (structure **18**) that is a stable isostere, providing the diol component that is sufficiently hindered to prevent the formation of a siloxane polymer. In fact, it has been demonstrated that silanediol-based dipeptide analogues are potent inhibitors of metalloproteases and aspartic proteases.

Synthesis of the first silanediol analogues of type I ACE inhibitors (FIG. 7) was reported by Sieburth and co-workers.[59,62,63] The synthetic approaches to these silanediols used classical synthetic chemistry methodology resembling the right- to left-hand side coupling approach described for the ketomethylene tripeptide analogues. Because silanediols are more stable than carbon diols, they are likely to exhibit a longer half-life. The silanediol (**19**) was found to be approximately fourfold

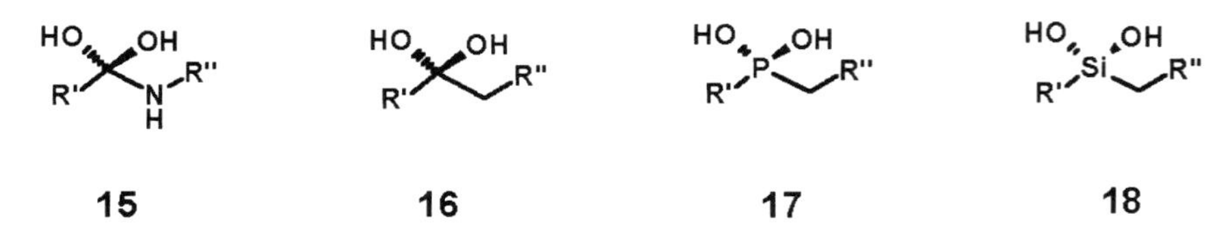

FIGURE 6. Schematic representation of peptide bond hydrolysis isosteres.

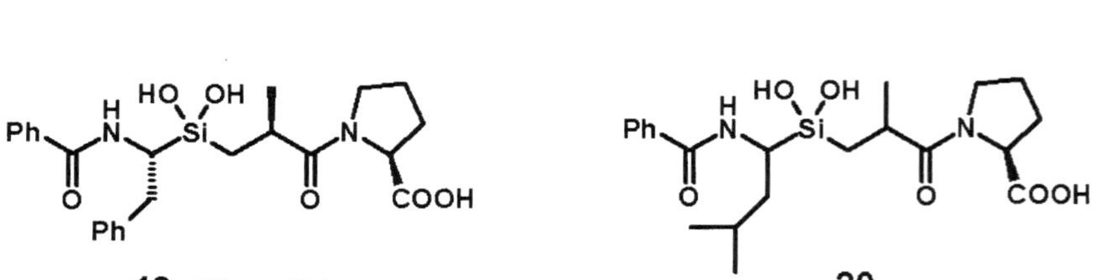

19 Silanediols

20

FIGURE 7. Silanediol ACE inhibitors with different P_1 substituents.

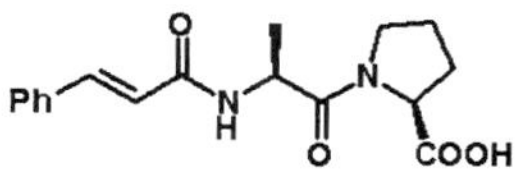

21 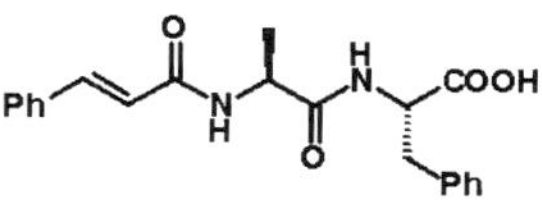 22 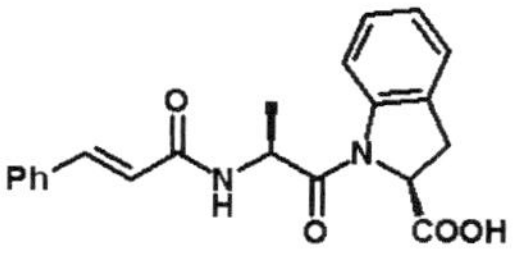 23

FIGURE 8. Irreversible ACE inhibitors comprising α,β-unsaturated carbonyls.

less potent than the ketone analogue (**5**)[59] due to the weaker zinc chelation of silanediols as compared with ketones.

Compound **20** is a weaker ACE inhibitor than compound **19**, presumably due to the *i*-butyl group of compound **20** compared to the benzyl group. Moreover, the introduction of a hydrophobic methyl phenyl provides a modest increase in potency over an analogue with a *tert*-butyl group at P_1, suggesting that there is improved recognition with the S_1 pocket.

Irreversible ACE Inhibitors

In principle, irreversible inhibitors should be devoid of the inherent drawbacks associated with the classical reactive warhead groups, that is, lack of specificity, excessive reactivity, and instability. Park Choo *et al.*[64] reported a novel class of ACE inhibitors comprised of α,β-unsaturated carbonyls (FIG. 8, compounds **21–23**). These compounds had IC_{50} values of 0.23, 2.0, and 3.19 mM, respectively.

Although the IC_{50} values of these compounds were significantly higher than those reported for most ACE inhibitors, the rationale was that the α,β-unsaturated moiety might react with a catalytic nucleophile at the active site to afford a Michael-type adduct, rendering the enzyme inactive. Because of this covalent bond formation between the enzyme and the inhibitor, this class of inhibitors has the advantages of increased half-life, oral bioavailability, and increased resistance to hydrolytic degradation.The tenfold difference in inhibitory potential between compounds **22** and **23,** on the one hand, and **21,** on the other, might be due to the extra π-π interaction between the electron-rich phenyl or benzene ring and the active site S_2' residues that reduce the covalent strength of the Michael-type adduct, thereby rendering the enzyme more active.

Captopril Analogues

Because the S_2' subsites of both the C and N domains are relatively large and can accommodate various linear and cyclic side chains, the selectivity of ACE inhibitors might depend on interactions with the S_2' pockets.[48,49,65] Before the solution of the ACE crystal structure,[42,43] many ACE inhibitors were designed with bulky P_2' residues. Captopril (**1**) is one of the simplest ACE inhibitors, comprising a P_1' and P_2' residue and a thiol zinc-binding group. Hanessian and co-workers[66] reported the synthesis of modified captopril analogues (**24–26**) in which the P_2' proline residue of captopril had been modified by a conformationally constrained heterocycle (FIG. 9). These analogues showed increased ACE inhibition. Synthesis of these con-

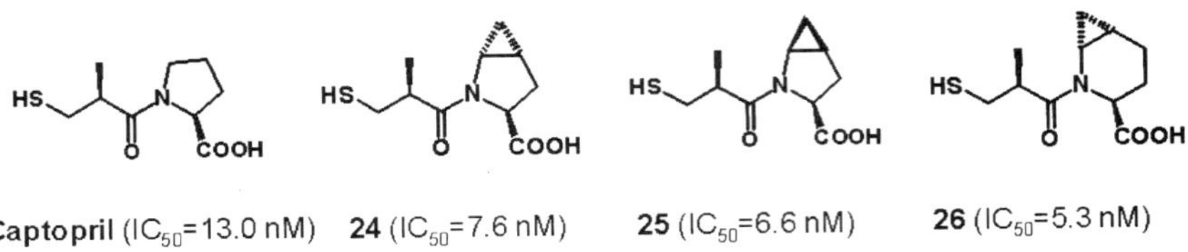

Captopril (IC_{50}=13.0 nM) **24** (IC_{50}=7.6 nM) **25** (IC_{50}=6.6 nM) **26** (IC_{50}=5.3 nM)

FIGURE 9. A series of captopril analogues with different P_2' substituents.

27 **28**

FIGURE 10. Hydroxamates.

formationally constrained captopril derivatives follows the type III synthetic approach in which the P_2' proline moieties are synthesized first and then coupled with the ZBGs-P_1' portion.

Hydroxamates

ACE inhibitors of the general formula (**27**, FIG. 10), incorporating a hydroxamic acid zinc-binding group, an N-alkylated amide, and a 1,2-cyclohexanedicarboxylic acid moiety, show remarkable specificity and potency *in vitro* and *in vivo*.[67,68] A methyl- or ethyl-substituted amide in the P_1' position resulted in a 250-fold decrease in IC_{50}. Hydroxamates were far better inhibitors than the corresponding carboxylic and sulfhydrylic analogues (CONHOH<<SH<<COOH, for IC_{50} values). However, this preference for a zinc-ligand group, although applicable with these non-amino acid compounds, is not necessarily the case with all ACE inhibitors. The synthetic route used by Turbanti *et al.*[68] employs the condensation of the zinc-binding derivative with a cyclomethylenedicarboxylic anhydride or with the corresponding acid in the presence of a carbodiimide.

Thus, this approach is amenable to the coupling of various P_1' and P_2' residues that make more selective contacts with amino acids in the active site of the C- or N domain. The stereochemistry of carboxylate and hydroxamic acid zinc-binding functionalities also plays a role in the potency of the inhibitor, with the carboxylates and hydroxamates bearing the *S*-stereochemistry more potent than those with the *R*-stereochemistry.

CONCLUSION

A number of compounds have been identified that exhibit a domain preference for ACE based on specific interactions with the active-site pockets. Because the C and N domains display about 90% identity at their active site, targeting specific residues that are not conserved in both domains, might aid in the discovery of potent and domain-selective ACE inhibitors. This review discusses new insights to consider during the design of domain-selective ACE inhibitors.

The compounds might serve as attractive chemical entities for drug discovery research in the field hypertension and heart-related diseases. However, caution is needed when using structural information in a predictive manner in the design of ACE inhibitors, because conformational changes may have occurred in the active site upon binding of the inhibitor in the ACE crystal structure.

The clinical benefits of ACE inhibitors are undeniable, and their use in the treatment of conditions extending beyond that of hypertension is growing. In this regard, ACE inhibitors have been shown to be beneficial in treating hypertension accompanied by the cardiometabolic syndrome and type II diabetes mellitus.[69] ACE inhibitors have been shown to retard myocardial remodeling and contractile dysfunction leading to heart failure,[70] improve endothelial function,[71] and reduce atherogenesis.[72] Recently, the role of ACE as a signal transduction molecule was investigated. Here the ACE inhibitors ramiprilat and perindoprilat together with bradykinin were shown to enhance phosphorylation of the ACE cytoplasmic tail in endothelial cells. This "outside-in signaling" pathway results in an increase in ACE expression.[73] Furthermore, ACE inhibitors are associated with increased expression of cyclooxygenase-2 (COX-2), resulting in increased production of prostaglandin E2 and prostacyclin.[74]

The burgeoning use of ACE inhibitors, especially in the treatment of conditions comorbid with hypertension, underscores the importance of understanding at a structural level the nature of their interactions with the ACE active site. Moreover, in improving the efficacy of ACE inhibitors and reducing the prevalence of adverse side effects, the need exists for the development of drugs that can specifically distinguish between the physiologically diverse C and N domains of this enzyme.

ACKNOWLEDGMENTS

This work was supported by the Wellcome Trust, U.K. (SIRF 070060) and the National Research Foundation, South Africa.

REFERENCES

1. Kearney, P.M., M. Whelton, K. Reynolds, *et al.* 2005. Global burden of hypertension: analysis of worldwide data. Lancet **365:** 217–223.
2. Carretero, O.A. & S. Oparil. 2000. Essential hypertension. Part I: definition and etiology. Circulation **101:** 329–335.
3. Rawlings, N.D., D.P. Tolle & A.J. Barrett. 2004. Evolutionary families of peptidase inhibitors. Biochem. J. **378:** 705–716.
4. Rawlings, N.D., D.P. Tolle & A.J. Barrett. 2004. MEROPS: the peptidase database. Nucleic Acids Res. **32:** D160–D164.
5. Ehlers, M.R.W. & J.F. Riordan. 1989. Angiotensin-converting enzyme: new concepts concepts concerning its biological role. Biochemistry **28:** 5311–5318.
6. Corvol, P., T.A. Williams & F. Soubrier. 1995. Peptidyl dipeptidase A: angiotensin I-converting enzyme. Methods Enzymol. **248:** 283–305. Academic Press.

7. SOUBRIER, F., F. ALHENC-GELAS, C. HUBERT, *et al.* 1988. Two putative active centers in human angiotensin I-converting enzyme revealed by molecular cloning. Proc. Natl. Acad. Sci. USA **85:** 9386–9390.
8. SOUBRIER, F., C. HUBERT, P. TESTUT, *et al.* 1993. Molecular biology of the angiotensin I-converting enzyme: I. Biochemistry and structure of the gene. J. Hypertens. **11:** 471–476.
9. HOOPER, N.M., E.H. KARRAN & A.J. TURNER. 1997. Membrane protein secretases. Biochem. J. **321:** 265–279.
10. WEI, L., F. ALHENC-GELAS, P. CORVOL & E. CLAUSER. 1991. The two homologous domains of human angiotensin I-converting enzyme are both catalytically active. J. Biol. Chem. **266:** 9002–9008.
11. JASPARD, E., L. WEI & F. ALHENC-GELAS. 1993. Differences in the properties and enzymatic specificities of the two active sites of angiotensin I-converting enzyme (kininase II). J. Biol. Chem. **268:** 9496–9503.
12. EHLERS, M.R.W., Y.-N.P. CHEN & J.F. RIORDAN. 1991. Purification and Characterization of recombinant human testis angiotensin-Converting Enzyme Expressed in Chinese hamster ovary cells. Protein Exp. Purif. **2**: 1–9.
13. HAGAMAN, J.R., J.S. MOYER, E.S. BACHMAN, *et al.* 1998. Angiotensin-converting enzyme and male fertility. Proc. Natl. Acad. Sci. USA **95:** 2552–2557.
14. KONDOH, G., H. TOJO, Y. NAKATANI, *et al.* 2005. Angiotensin-converting enzyme is a GPI-anchored protein releasing factor crucial for fertilization. Nat. Med. **11:** 160–166.
15. EHLERS, M.R.W. & J.F. RIORDAN. 1990. Angiotensin-Converting Enzyme: Biochemistry and Molecular Biology. Hypertension: Pathophysiology, Diagnosis, and Management Chapt. **76:** 1217–1231. Raven Press. New York.
16. RIPKA, J.E., J.W. RYAN, F.A. VALIDO, *et al.* 1993. *N*-glycosylation of forms of angiotensin converting enzyme from four mammalian species. Biochem. Biophys. Res. Commun. **196:** 503–508.
17. YU, X.C., E.D. STURROCK, Z. WU, *et al.* 1997. Identification of *N*-linked glycosylation sites in human testis angiotensin-converting enzyme and expression of an active deglycosylated form. J. Biol. Chem. **272:** 3511–3519.
18. NAQVI, N., K. LIU, R.M. GRAHAM & A. HUSAIN. 2005. Molecular basis of exopeptidase activity in the C-terminal domain of human angiotensin I-converting enzyme. J. Biol. Chem. **280:** 6669–6675.
19. CORVOL, P., M. EYRIES & F. SOUBRIER. 2004. *In* Handbook of Proteolytic Enzymes. A.J. Barrett, N.D. Rawlings & J.F. Woessner, Eds. :332–346. Academic Press Inc. New York.
20. BAUDIN, B. 2002. New aspects on angiotensin-converting enzyme: from gene to disease. Clin. Chem. Lab. Med. **40:** 256–265.
21. VAN ESCH, J.H., B. TOM, V. DIVE, *et al.* 2005. Selective angiotensin-converting enzyme C-domain inhibition is sufficient to prevent angiotensin I-induced vasoconstriction. Hypertension **45:** 120–125.
22. ESTHER, C.R., E.M. MARINO, T.E. HOWARD, *et al.* 1997. The critical role of tissue angiotensin-converting enzyme as revealed by gene targeting in mice. J. Clin. Invest. **99:** 2375–2385.
23. FUCHS, S., H.D. XIAO, J.M. COLE, *et al.* 2004. Role of the *N*-terminal catalytic domain of angiotensin-converting enzyme investigated by targeted inactivation in mice. J. Biol. Chem **279:** 15946–15953.
24. FERNANDEZ, J.H., M.A. HAYASHI, A.C. CAMARGO & G. NESHICH. 2003. Structural basis of the lisinopril-binding specificity in N- and C-domains of human somatic ACE. Biochem. Biophys. Res. Commun. **308:** 219–226.
25. OBA, R., A. IGARASHI, M. KAMATA, *et al.* 2005. The *N*-terminal active centre of human angiotensin-converting enzyme degrades Alzheimer amyloid beta-peptide. Eur. J. Neurosci. **21:** 733–740.
26. VORONOV, S., N. ZUEVA, V. ORLOV, *et al.* 2002. Temperature-induced selective death of the C-domain within angiotensin-converting enzyme molecule. FEBS Lett. **522:** 77–82.

27. ANDUJAR-SANCHEZ, M., A. CAMARA-ARTIGAS & V. JARA-PEREZ. 2004. A calorimetric study of the binding of lisinopril, enalaprilat and captopril to angiotensin-converting enzyme. Biophys. Chem. **111:** 83–189.
28. CUSHMAN, D.W., H.S. CHEUNG, E.F. SABO & M.A. ONDETTI. 1977. Design of potent competitive inhibitors of angiotensin-converting enzyme. Carboxyalkanoyl and mercaptoalkanoyl amino acids. Biochemistry **16:** 5484–5491.
29. ONDETTI, M.A., B. RUBIN & D.W. CUSHMAN. 1977. Design of specific inhibitors of angiotensin-converting enzyme: new class of orally active antihypertensive agents. Science **196:** 441–444.
30. CUSHMAN, D.W. & M.A. ONDETTI. 1999. Design of angiotensin converting enzyme inhibitors. Nat. Med. **5:** 1110–1113.
31. FERREIRA, S.H. & M. ROCHA E SILVA. 1965. Potentiation of bradykinin and eledoisin by BPF (bradykinin potentiating factor) from *Bothrops jararaca* venom. Experientia **21:** 347–349.
32. FERREIRA, S.H., D.C. BARTELT & L.J. GREENE. 1970. Isolation of bradykinin-potentiating peptides from *Bothrops jararaca* venom. Biochemistry **9:** 2583-2593.
33. FERREIRA, S.H., L.H. GREENE, V.A. ALABASTER, *et al.* 1970. Activity of various fractions of bradykinin potentiating factor against angiotensin I converting enzyme. Nature **225:** 379–380.
34. HAYASHI, M.A. & A.C. CAMARGO. 2005. The bradykinin-potentiating peptides from venom gland and brain of *Bothrops jararaca* contain highly site specific inhibitors of the somatic angiotensin-converting enzyme. Toxicon **45:** 1163–1170.
35. PATCHETT, A. A., E. HARRIS, E. W. TRISTRAM, *et al.* 1980. A new class of angiotensin-converting enzyme inhibitors. Nature **288:** 280–283.
36. PATCHETT, A.A. & E.H. CORDES. 1985. The design and properties of N-carboxyalkyl-dipeptide inhibitors of angiotensin-converting enzyme. Adv. Enzymol. Relat. Areas Mol. Biol. **57:** 1–84.
37. BICKET, D.P. 2002. Using ACE inhibitors appropriately. Am. Fam. Physician **66:** 461–468.
38. DICKSTEIN, K. & J. KJEKSHUS. 2002. Effects of losartan and captopril on mortality and morbidity in high-risk patients after acute myocardial infarction: the OPTIMAAL randomised trial. Optimal trial in myocardial infarction with angiotensin ii antagonist losartan. Lancet **360:** 752–760.
39. ADAM, A., M. CUGNO, G. MOLINARO, *et al.* 2002. Aminopeptidase P in individuals with a history of angio-oedema on ACE inhibitors. Lancet **359:** 2088–2089.
40. MORIMOTO, T., T.K. GANDHI, J.M. FISKIO, *et al.* 2004. An evaluation of risk factors for adverse drug events associated with angiotensin-converting enzyme inhibitors. J. Eval. Clin. Pract. **10:** 499–509.
41. DIVE, V., D. GEORGIADIS, M. MATZIARI, *et al.* 2004. Phosphinic peptides as zinc metalloproteinase inhibitors. Cell Mol. Life Sci. **61:** 2010–2019.
42. NATESH, R., S.L. SCHWAGER, E.D. STURROCK & K.R. ACHARYA. 2003. Crystal structure of the human angiotensin-converting enzyme-lisinopril complex. Nature **421:** 551–554.
43. KIM, H.M., D.R. SHIN, O.J. YOO, *et al.* 2003. Crystal structure of Drosophila angiotensin I-converting enzyme bound to captopril and lisinopril. FEBS Lett. **538:** 65–70.
44. NATESH, R., S.L. SCHWAGER, H.R. EVANS, *et al.* 2004. Structural details on the binding of antihypertensive drugs captopril and enalaprilat to human testicular angiotensin I-converting enzyme. Biochemistry **43:** 8718–8724.
45. TZAKOS, A.G. & I.P. GEROTHANASSIS. 2005. Domain-selective ligand-binding modes and atomic level pharmacophore refinement in angiotensin I converting enzyme (ACE) inhibitors. Chembiochemistry **6:** 1089–1103.
46. DE LIMA, D.P. 1999. Synthesis of angiotensin-converting enzyme (ACE) inhibitors: An important class of antihypertensive drugs. Quimica Nova **22:** 375–381.
47. KRAPCHO, J., C. TURK, D. W. CUSHMAN, *et al.* 1988. Angiotensin-converting enzyme inhibitors. Mercaptan, carboxyalkyl dipeptide, and phosphinic acid inhibitors incorporating 4-substituted prolines. J. Med. Chem. **31:** 1148–1160.
48. DIVE, V., J. COTTON, A. YIOTAKIS, *et al.* 1999. RXP 407, a phosphinic peptide, is a potent inhibitor of angiotensin I converting enzyme able to differentiate between its two active sites. Proc. Natl. Acad. Sci. USA **96:** 4330–4335.

49. GEORGIADIS, D., P. CUNIASSE, J. COTTON, *et al.* 2004. Structural determinants of RXPA380, a potent and highly selective inhibitor of the angiotensin-converting enzyme C-domain. Biochemistry **43:** 8048–8054.
50. JUNOT, C., M.F. GONZALES, E. EZAN, *et al.* 2001. RXP 407, a selective inhibitor of the N-domain of angiotensin I-converting enzyme, blocks in vivo the degradation of hemoregulatory peptide acetyl-Ser-Asp-Lys-Pro with no effect on angiotensin I hydrolysis. J. Pharmacol. Exp. Ther. **297:** 606–611.
51. GEORGIADIS, D., F. BEAU, B. CZARNY, *et al.* 2003. Roles of the two active sites of somatic angiotensin-converting enzyme in the cleavage of angiotensin I and bradykinin: insights from selective inhibitors. Circ Res. **93:** 148–154.
52. CLIVE, D.L.J. & D.M. COLTART. 1998. Synthesis of the angiotensin-converting enzyme inhibitors (+/-)-A58365. Tetrahedron Lett.**39:** 2519–2522.
53. CLIVE, D.L.J., H. YANG & R.Z. LEWANCZUK. 2001. Synthesis and in vitro activity of a non-epimerizable analogue of the angiotensin-converting enzyme inhibitors A58365A. C. R. Acad. Sci. Paris. Chimie **4:** 505–512.
54. REICHELT, A., S.K. BUR & S.F. MARTINS. 2002. Applications of vinylogous Mannich reactions: Total synthesis of the angiotensin-converting enzyme inhibitors (-)-A58365A. Tetrahedron **58:** 6323–6328.
55. ALMQUIST, R.G., W.R. CHAO, M.E. ELLIS & H.L. JOHNSON. 1980. Synthesis and biological activity of a ketomethylene analogue of a tripeptide inhibitor of angiotensin converting enzyme. J. Med. Chem. **23:** 1392–1398.
56. DEDDISH, P.A., B. MARCIC, H.L. JACKMAN, *et al.* 1998. N-domain-specific substrate and C-domain inhibitors of angiotensin-converting enzyme: angiotensin-(1-7) and keto-ACE. Hypertension **31:** 912–917.
57. MEYER, R.F., E.D. NICOLAIDES, F.J. TINNEY, *et al.* 1981. Novel synthesis of (S)-1-[5-(benzoylamino)-1,4-dioxo-6-phenylhexyl]-L-proline and analogues: potent angiotensin converting enzyme inhibitors. J. Med. Chem. **24:** 964–969.
58. ACHARYA, K.R., E.D. STURROCK, J.F. RIORDAN & M.R. EHLERS. 2003. Ace revisited: a new target for structure-based drug design. Nat. Rev. Drug Discov. **2:** 891–902.
59. KIM, J., G. HEWITT, P. CARROLL & S.M. SIEBURTH. 2005. Silanediol inhibitors of angiotensin-converting enzyme: synthesis and evaluation of four diastereomers of Phe[Si]Ala dipeptide analogues. J. Org. Chem. **70:** 5781–5789.
60. SANDERSON, P.E. 1999. Small, noncovalent serine protease inhibitors. Med. Res. Rev. **19:** 179–197.
61. BURSAVICH, M.G. & D.H. RICH. 2002. Designing non-peptide peptidomimetics in the 21st century: inhibitors targeting conformational ensembles. J. Med. Chem. **45:** 541–558.
62. MUTAHI, M., T. NITTOLI, L. GUO & S.M. SIEBURTH. 2002. Silicon-based metalloprotease inhibitors: synthesis and evaluation of silanol and silanediol peptide analogues as inhibitors of angiotensin-converting enzyme. J. Am. Chem. Soc. **124:** 7363–7375.
63. KIM, J. & S.M. SIEBURTH. 2004. Silanediol peptidomimetics. Evaluation of four diastereomeric ACE inhibitors. Bioorg. Med. Chem. Lett. **14:** 2853–2856.
64. CHOO, H.-Y. P. ., K.-H. PEAK, J. PARK, *et al.* 2000. Design and synthesis of α,β-unsaturated carbonyl compounds as potential ACE inhibitors. Eur. J. Med. Chem. **35:** 643–648.
65. BERSANETTI, P.A., M.C. ANDRADE, D.E. CASARINI, *et al.* 2004. Positional-scanning combinatorial libraries of fluorescence resonance energy transfer peptides for defining substrate specificity of the angiotensin I-converting enzyme and development of selective C-domain substrates. Biochemistry **43:** 15729–15736.
66. HANESSIAN, S., U. REINHOLD, M. SAULNIER & S. CLARIDGE. 1998. Probing the importance of spacial and conformational domains in captopril analogues for angiotensin-converting enzyme activity. Bioorg. Med. Chem. Lett. **8:** 2123–2128.
67. SUBISSI, A., M CRISCUOLI, G. SARDELLI, *et al.* 1992. Pharmacology of idrapril: a new class of angiotensin converting enzyme inhibitors. J. Cardiovasc. Pharmacol. **20:** 139–146.
68. TURBANTI, L., G. CERBAI, C. DI BUGNO, *et al.* 1993. 1,2-Cyclomethylenecarboxylic monoamide hydroxamic derivatives. A novel class of non-amino acid angiotensin converting enzyme inhibitors. J. Med. Chem. **36:** 699–707.

69. McFarlane, S.I., A. Kumar & J.R. Sowers. 2003. Mechanisms by which angiotensin-converting enzyme inhibitors prevent diabetes and cardiovascular disease. Am. J. Cardiol. **91:** 30H–37H.
70. Satoh, S., Y. Ueda, N. Suematsu, *et al.* 2003. Beneficial effects of angiotensin-converting enzyme inhibition on sarcoplasmic reticulum function in the failing heart of the Dahl rat. Circ. J. **67:** 705–711.
71. Hamdi, H. K. & R. Castellon. 2003. ACE inhibition actively promotes cell survival by altering gene expression. Biochem. Biophys. Res. Commun. **310:** 1227–1235.
72. Scribner, A. W., J. Loscalzo & C. Napoli. 2003. The effect of angiotensin-converting enzyme inhibition on endothelial function and oxidant stress. Eur. J. Pharmacol. **482:** 95–99.
73. Kohlstedt, K., R.P. Brandes, W. Muller-Esterl, *et al.* 2004. Angiotensin-converting enzyme is involved in outside-in signaling in endothelial cells. Circ. Res. **94:** 60–67.
74. Kohlstedt, K., R. Busse & I. Fleming. 2005. Signaling via the angiotensin–converting enzyme enhances the expression of cyclooxygenase-2 in endothelial cells. Hypertension **45:** 126–132.

Opportunities for New Therapies Based on the Natural Regulators of Complement Activation

EVE BROOK,[a] ANDREW P. HERBERT,[a] HUW T. JENKINS,[a]
DINESH C. SOARES,[a,b] AND PAUL N. BARLOW[a,b]

[a]*The Edinburgh Biomolecular NMR Unit and*
[b]*Institute of Structural and Molecular Biology, Schools of Chemistry and Biological Sciences, University of Edinburgh, Edinburgh, UK*

ABSTRACT: While the complement system is an essential component of immunity, shutting down all or part of it could be beneficial in a wide range of clinical situations. Designer, small-molecule, protease inhibitors and antagonists of protein-protein interactions are under development, while an approach based on a humanized monoclonal antibody to the C5 component works effectively against the later stages of complement activation and is close to completing clinical trials. The cobra venom factor depletes plasma of essential complement components, and a humanized (nonimmunogenic) version is being sought. Perhaps the most promising approach to comprehensive complement down-regulation, however, is the exploitation of innate regulators of complement activation, with two products in clinical trials. The potential for more efficacious complement blockers of this kind is growing because of better targeting, but a deeper knowledge at the atomic level of mechanisms of action of these regulators is needed to underpin a rational approach to design of still more potent complement inhibitors.

KEYWORDS: complement system; therapeutics; regulators of complement activation; structural biology; disease

INTRODUCTION

The debilitating symptoms of many autoimmune, degenerative, and iatrogenic human and animal diseases stem from direct or indirect tissue damage caused by proteins of the complement system. In this review, we consider therapeutic strategies for downregulating unwanted or excessive complement activation, emphasizing the exploitation of the natural regulators of the complement activation (RCA) family.[1] Interestingly, members of the poxvirus family have hijacked the gene for a mammalian complement regulator and evolved it into a potent defensive mechanism for the virus,[2] thus establishing a precedent for this strategy.

Without access to the myriad possibilities afforded by millions of years of evolution, we must rely on detailed knowledge of structure-function relationships among

Address for correspondence: Professor Paul N. Barlow, Joseph Black Chemistry Building, University of Edinburgh, West Mains Road, Edinburgh EH9 3JR. Voice: +44 (0) 131 650 4727; fax: +44 (0) 131 650 7055.
Paul.Barlow@ed.ac.uk

**Ann. N.Y. Acad. Sci. 1056: 176–188 (2005). © 2005 New York Academy of Sciences.
doi: 10.1196/annals.1352.033**

the RCAs to perform a comparable feat. Before discussing anticomplement therapies, we provide background on: the complement system; complement and disease; and regulation of complement at a structural level.

BACKGROUND

Complement

The complement system is a set of ~30 plasma and cell-attached proteins that act in an orchestrated way to rid the body of invasive organisms and other unwanted particles.[3,4] The alternative pathway of complement activation provides a first line of defense because it is permanently "switched on" at a low level but kept in check by regulators that act only on self-surfaces. The presence of any unprotected surface induces rapid escalation of the system. The complement system may additionally be activated by sugars on bacterial surfaces (lectin pathway) or by antibodies bound to polyvalent antigens (classical pathway). All three pathways result in formation of bimolecular convertase enzymes that drive a cascade of proteolytic events, providing a basis for amplification.

In the alternative pathway (FIG. 1), the protein C3 undergoes hydrolysis upon encounter with nucleophiles found, for example, on biologic surfaces, to yield C3(H_2O). C3(H_2O) associates with a second protein, factor B. In this context, factor B is cleaved to form Bb, and the resultant C3(H_2O).Bb complex is a proteolytically active entity that is normally surface-associated. It cleaves C3 into C3a and C3b. C3b, like C3(H_2O), associates with factor B, forming the alternative pathway convertase, C3b.Bb. This bimolecular enzyme converts more molecules of C3 to C3b, thus resulting in the formation of extra convertases and the creation of a positive feedback loop. The lectin and classical pathways (FIG. 1) entail cleavage of C4, a homologue of C3, resulting in formation of C4b. C4b binds to surfaces and forms a convertase through association with C2, creating the bimolecular, proteolytically active, complex C4b.2a. Like the alternative pathway convertase, the lectin/classical pathway convertase thus formed cleaves C3 to C3b and C3a. The C3b product is available for formation of more alternative pathway convertase complexes (i.e., C3b.Bb). Both alternative and classical pathway convertases can associate with further molecules of C3b to form trimolecular complexes, C3b.Bb.C3b and C4b.2a.C3b, which have C5 convertase activity, i.e., they cleave C5 to C5b and C5a. Surface-bound C5b may subsequently act as a nucleation site for assembly of a multiprotein membrane attack complex (MAC) that penetrates cell membranes, causing leakage and cell death. A further consequence of complement activation, which is independent of MAC formation, is deposition of multiple molecules of C3b and C4b on the target surface. This is called opsonization and provides the basis for immune clearance, an important process by which C3b/C4b-coated particles are cleared from the bloodstream.[5] Furthermore, the cleavage products of C3b, C3d, and C3dg are ligands for other receptors, able to trigger signaling pathways that bring about, for example, enhanced levels of antibody production by B-cells.[6] Finally, the small fragments, C3a, C4a, and C5a, cleaved from C3, C4, and C5 during complement activation, are the anaphylotoxins, chemotactic peptides that mediate a series of proinflammatory phenomena.

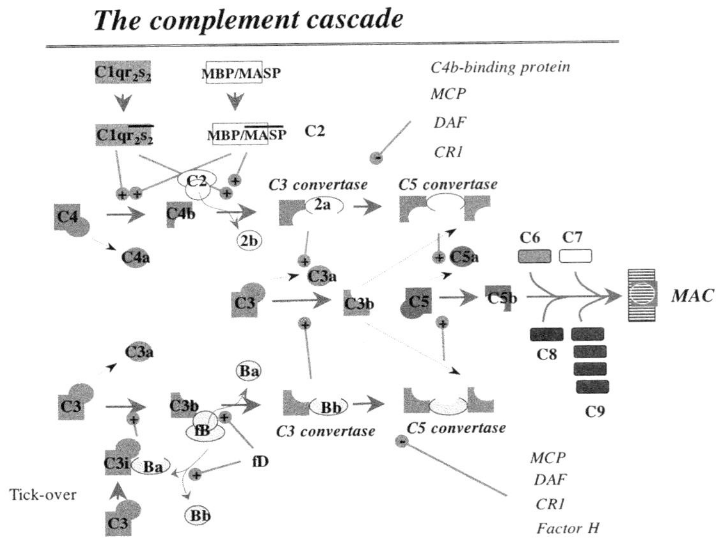

FIGURE 1. Schematic diagram to summarize the complement system. The pathways run from left to right. The classical and lectin pathways are triggered by binding of antibodies to complement component C1 or specific sugars to mannose binding protein/mannose binding protein associated serine protease (MBP/MASP). The alternative pathway is always in "tick over" mode due to the possibility of C3 undergoing conversion to C3(H_2O) = C3i. The RCAs are C4b binding protein, MCP, DAF, CR1, and fH. All act in subtly different ways upon the convertases. See text for a description of pathway and explanation of abbreviations.

Thus, activation of complement disrupts the membrane of the target cell, removes foreign particles to the liver for destruction, enhances the adaptive immune response, and, via anaphylotoxins, recruits lymphocytes and macrophages to the site of infection.

Regulation of Complement Activation

Not only is the complement system effective in immune defense, but it also has potential for damage to self. It is therefore vital that complement activation is both tightly regulated and capable of being rapidly triggered. These properties are achieved largely as a result of assembly and decay of the aforementioned C3 convertase enzymes, C3b.Bb and C4b.2a, which occupy pivotal positions within the pathway. It is the positive feedback loop involving C3b production and convertase formation that provides complement with the ability to act in a rapid (explosive) fashion. The bimolecular convertases (C3b.Bb and C4b.2a) have a half-life in the order of minutes. Once they have decayed into their components, they cannot reassemble; for example, C3b must associate with a fresh molecule of intact factor B to produce a new convertase complex. Consequently, there is the potential for a complement response to die out when the stimulus to activation is destroyed and/or subjected to immune clearance. In addition, a critical set of proteins is dedicated to regulation of the complement cascade via interactions with the convertases. These are known as the regulators of complement activation (RCA) family and represent the main focus of this paper.

The RCA proteins have two principal activities in terms of their action on the convertases. Some of the RCAs accelerate the decay of the convertases, reducing their half-life to seconds. This is decay accelerating activity (DAA). The mechanism is unknown. Some RCAs act as cofactors for the protease factor I; factor I cleaves C3b and C4b to generate C3dg and C4dg (cofactor activity, CA), thus further reducing potential for convertase formation. The five human RCAs have overlapping functional profiles. Thus, decay accelerating factor (DAF, CD55), a GPI-linked RCA, possesses only DAA, whereas the type 1 membrane glycoprotein, membrane cofactor protein (MCP, CD46), exhibits solely CA. However, factor H (fH) is a soluble protein with sites for both DAA and CA, but acts exclusively upon the alternative pathway convertase. C4b-binding protein (C4BP) performs a parallel role to factor H, but acts predominantly upon the CP convertase. Finally, complement receptor type 1 (CR1, CD35) is unique in having both DAA and CA and acting on both the AP and the CP convertases. Molecules of at least one of DAF, MCP, and CR1 are expressed on the surfaces of every cell in the body where they play a protective role. The soluble regulators, C4BP and factor H, are particularly important for preventing activation at self-surfaces that lack membranes because such surfaces are unlikely to possess the cell-bound regulators DAF, MCP, and CR1. C4BP[7] and factor H[8] have polyanion binding sites that enable them to bind to glycosaminoglycans and sialic acid, markers that are absent from many bacterial cells.

The RCA family is exceptional in that each is composed predominantly of multiple examples of a single type of domain or module known as the CCP module (also sometimes termed the short consensus repeat[9] or sushi domain) (FIG. 2). These CCP modules (51-67 amino acid residues in length) have a consensus sequence including four cysteines that participate in a 1-3, 2-4 pattern of disulfide linkages and a tryptophan that contributes to a compact hydrophobic core. The modules are connected together by short linkers of between three and eight residues (most commonly four). Apart from the consensus residues there is wide variation in sequence, and the family of CCP modules (that includes many noncomplement proteins) has been divided into clusters on the basis of similarity.[10] The soluble RCA, fH, is made up entirely from 20 CCP modules; DAF is the smallest human RCA, with four CCP modules at its N-terminus followed by a Pro/Ser/Thr-rich stalk and a GPI-anchor. Membrane cofactor protein also has four CCP modules and a Pro/Ser/Thr-rich stalk, but these are followed by a transmembrane region and a small intracellular domain; CR1 is the largest of the single-chain RCAs with 30 CCP modules in its most common variant followed by a transmembrane region and intracellular domain. Finally, C4BP is the only polymeric RCA; in humans, each of its seven α-chains contains eight CCP modules, whereas its single β-chain has three CCP modules.

Virtually all RCA-mediated protein-protein interactions occur via binding sites extending over several adjacent CCP modules. Thus, in CR1[11] there are three functional regions, each extending over three modules, within the extracellular domain, one copy of "site 1" and two near-identical copies of "site 2." Site 1 resides within the N-terminal, membrane-distal, three CCP modules of CR1. It has convertase DAA along with the ability to bind C3b and C4b. Each copy of site 2 has high affinity for C3b and lower affinity for C4b, and it has cofactor activity.

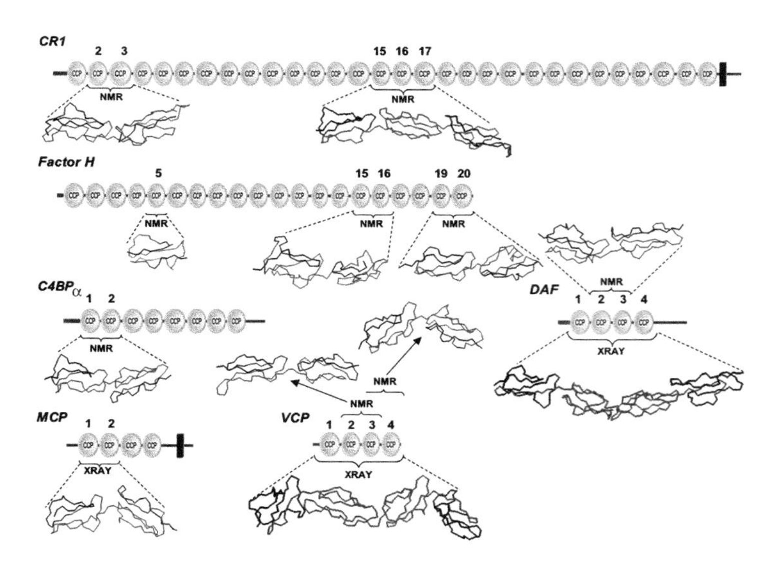

FIGURE 2. Structural information available for the RCA family and the viral mimic, VCP. Where they are known, 3D structures of modules are shown as backbone traces, and their relative positions within the protein are shown diagrammatically. Note that CR1~1,2, fH~19,20, and C4BPα~1,2 are unpublished structures.

Good progress has been made in determining the 3D structures of the RCAs in recent years.[1,12,13] Progress was founded upon recombinant production of fragments of RCA proteins containing up to four CCP modules (FIG. 2). X-ray diffraction has delivered the more precise structural information for each individual module. NMR, used to determine structures in solution rather than when they are packed in the crystal, is less liable to artefacts arising from crystal-packing forces or the use of high concentrations of precipitants, both of which could influence the orientation and flexibility of intermodular arrangements.[14] In each CCP module, β-strands and β-strand-like extended regions predominate (FIG. 3).[15] These align approximately with the long axis of the oblate module, wrap around the hydrophobic core, and run antiparallel to one another.[16,17] There are five such extended regions running "up-down-up-down-up," so that N- and C-termini lie at opposite poles. Variations occur between modules in the extent of β-strand formation and the pattern of H-bonds between strands. The extended regions/β-strands are linked by loops and turns of varying lengths that in general lie close to termini of the module and may contact adjoining modules. These loops and turns are sites of sequence variation, deletions, and insertions and are generally quite flexible.[18]

Neighboring modules are joined in an elongated, end-to-end arrangement with only a small area of contact between modules (with the exception of the extensive interactions between the N-terminal two CCP modules of complement receptor type 2).[19] While fragments containing CCP modules appear rigid according to crystallography, NMR reveals that varying degrees of flexibility exist in solution.[20] Both structural techniques suggest that intermodular angles (average ones in the case of

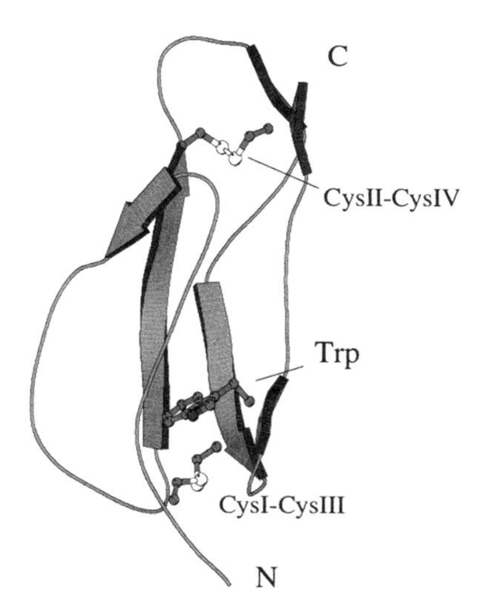

FIGURE 3. A representation of the 3D structure of a typical CCP module, the sixteenth from factor H. Disulfides and the highly conserved tryptophan are indicated.

NMR-derived structures) are not conserved between module pairs, even when the modules involved are of similar sequence and have similar functions. This means that reliably predicting intermodular angles on the basis of homology with known structures is impossible, a serious limitation given that most binding sites on RCAs span several modules. Intriguingly, accumulating evidence suggests that most contact residues of RCAs lie within, or close to, the junctions between neighboring modules. This corresponds with intuition in that: (1) junctions form clefts into which ridges on the partner protein could bind, and (2) these are sites of chemical and structural diversity because they are composed of residues from poorly conserved loop and linker regions. Given the flexibility between modules, it follows that there is also potential for intermodular reorientation upon ligand binding. This complicates greatly the interpretation of mutagenesis data and the search for common motifs or mechanisms across the family of proteins, despite overlapping functional profiles. As a consequence, despite many years of effort and large bodies of structural and mutagenesis information, we still have no clear molecular mechanism either for decay acceleration or cofactor activity. This makes it hard to rationally design new RCA-based ligands.

As a consequence of its potential for inflicting damage on invaders, a large range of complement evasion strategies have evolved among microorganisms.[21] Strikingly, the most widespread anticomplement measure involves hijacking of RCAs; many bacteria express proteins that act as virulence factors through their ability to sequester factor H or C4PB. Another approach is to "steal" an RCA gene and adapt it for the purposes of the invader; thus, pox viruses produce a protein that closely mimics the mammalian RCAs in both sequence and function.[22]

Complement and Disease

Despite the actions of the RCA protein family and other inhibitory proteins, such as CD59, that act elsewhere within the complement system, there remain numerous situations in which complement-mediated damage to self-tissues is pathologic. A particularly graphic example of complement's destructive power is provided by the events that follow xenotransplantation. Naturally occurring xenoreactive antibodies trigger a complement-mediated "hyperacute rejection," resulting in decimation of the xenograft within minutes. This phenomenon remains one of the major hurdles to a potentially revolutionary strategy for overcoming the chronic shortage of donor-organs.

Complement has also been implicated in the debilitating symptoms of a very long list of clinical conditions.[23,24] Varying amounts of evidence exist for complement involvement in the following autoimmune diseases: rheumatoid arthritis, systemic lupus erythematosus, most types of glomerulonephritis, autoimmune myocarditis, multiple sclerosis, myasthenia gravis, antiphospholipid syndrome, noncardiac myositis, type I diabetes mellitus, and asthma. There is also evidence for an important role of complement in degenerative diseases, such as Alzheimer's disease, and in reperfusion following ischemia during myocardial infarction and stroke. Inappropriate complement activation can also be triggered by burns and by treatments such as cardiopulmonary bypass and hemodialysis. One in five people, according to some estimates, will have a complement-mediated symptom at some stage in their lives. Considerable effort has been invested in the search for safe and effective agents to use in the therapeutic downregulation of the complement system.

INHIBITING COMPLEMENT

Several candidate anticomplement drugs have made it as far as clinical trials. Only brief mention is made here of the majority of contenders, because the focus of this article is the natural regulators (and their engineered derivatives).

Small Molecule Inhibitors

Small molecule inhibitors were reviewed recently.[25] A range of anticomplement protease inhibitors were reported. Structure-based drug design techniques have yielded an effective inhibitor of C1s (a proteolytic subunit of C1)[26] and factor D (converts factor 2 to 2a and factor B to Bb). The major difficulty with this approach is achieving specificity of action. An alternative strategy is to target C3 with ligands that block its interactions with partner proteins. An excellent example of this approach is compstatin, a cyclic peptide derivative that is the product of rational design and structure (inhibitory potential studies).[27] The exact mechanism whereby compstatin acts is unknown. It is, however, a good convertase inhibitor, and its efficacy in this respect is manifested in several potential clinical applications. It is beneficial in animal models of cardiac surgery and cardiopulmonary surgery; it also shows promise in *ex vivo* models of xenotransplantation.

The 74-amino acid C5a fragment, produced upon cleavage of C5 by the C5 convertases, is an anaphylatoxin. This very potent proinflammatory peptide is one of the strongest chemotactic substances known. Acting via specialized receptors on its surfaces, it recruits neutrophils and macrophages to a site of complement activation. The potential for antagonizing its receptor is obvious, and this has been explored successfully using novel, small cyclic molecules derived from the C-terminus of C5a. Administration of such compounds to rats or mice by oral and other routes exhibited a broad range of anti-inflammatory activities.[28] This work demonstrates the therapeutic potential of anticomplement therapies.

Antibody-Based Therapies

An alternative approach is to block C5 cleavage production by derivatives of antibodies to C5. Two antibody-derived molecules, eculizumab and pexelizumab,[29,30] are at an advanced stage of development for cardiopulmonary bypass,[31] myocardial infarction, and stroke as well as chronic conditions such as rheumatoid arthritis, dermatomyosis, and pemphigus. For example, a study showed that patients with the rare, occasionally fatal, blood disease, paroxysmal nocturnal hemoglobinuria, who received eculizumab maintained statistically significant reductions in red blood cell destruction and blood transfusions.[32]

Cobra Venom Factor

The cobra venom factor (CVF) protein is a close relative of C3. Following introduction into the blood stream, it combines with factor B, forming a C3 convertase, CVF.Bb, which has a half-life of several hours (rather than minutes), and during this time it can consume all of the available C3. Thus, it effectively decomplements the organism for up to 48 hours.[33] The primary scientific use of CVF has been to demonstrate the role played by complement in many of the diseases just mentioned. There is also interest in "humanizing" recombinant CVF to render it less immunogenic and more suitable for clinical purposes.[34] Based on a comparison of cobra C3 with its own CVF, a handful of residues in CVF have been identified as being important for the ability of CVF to form a nonlabile interaction with factor B/Bb. The introduction of these critical residues into the human C3 sequence could produce a powerful means of depleting complement proteins.

EXPLOITING THE RCAs

A possible advantage of therapy based on inhibition at the level of C5 is that it leaves intact earlier stages of complement, including the generation of opsonins and immune clearance. Therefore, a patient taking anti-C5 therapy would not be entirely immunocompromised. If, however, the goal is to suppress all the principal outcomes of complement activation–immune clearance, the generation of all three anaphylatoxins, and the formation of the membrane attack complex–then the regulators of complement activation represent an excellent starting point.

An early, thoroughly investigated example is soluble CR1 (sCR1) or TP10.[35] This recombinant protein contains all 30 of its CCP modules encompassing functional site 1 and two copies of site 2. Nanomolar concentrations of TP10 can downregulate C3 and C5 convertases in both alternative and classical pathways of complement via decay acceleration and cofactor activities. Thereby, it diminishes production of opsonins, anaphylatoxins, and membrane attack complexes. In an exciting initial report,[36] TP10 reduced infarction size by 44% in a rat model of reperfusion injury of ischemic myocardium. Fifteen years later, the efficacy of TP10 in ischemia-reperfusion injury associated with cardiopulmonary bypass, the most promising application so far, is still the subject of ongoing investigation. A placebo-controlled, double-blind study involving 564 high-risk patients[37] showed that TP10 significantly inhibited complement activity immediately after cardiopulmonary bypass, and this inhibition persisted for several days. Although TP10 was clinically ineffective when considered over the entire population of patients, it did significantly decrease the incidence of mortality and myocardial infarction in male patients. On the basis of this ray of hope, a new Phase II clinical trial involving female patients has been launched with results expected in early 2006 (http://www.avantimmune.com). In a much smaller (n = 15), related, phase I/II trial, the possible use of TP10 in infants undergoing cardiopulmonary bypass was investigated and no adverse events were attributable to TP10.

Another possible application of TP10 is in xenotransplantation. In pig-to-primate heart transplantation, a survival time of up to 7 days (compared to 30–40 minutes for the control) under continuous treatment with TP10 was reported.[38] However, TP10 was apparently less successful in xenotransplantation when administered in conjunction with the use of transgenic donor pigs expressing human DAF,[39,40] probably reflecting the multilayered complexity of the immune response and our lack of detailed knowledge of how complement is regulated in various tissues.

The very large size (~190 kDa) of TP10, expense of production, and complexity mean it is unlikely to find wide application in humans. A smaller, more elegant solution might lie in a molecule termed APT070.[41] This therapeutic compound consists only of the N-terminal three CCP modules of CR1 followed by a positively charged peptide sequence (known as an "addressin") and a myristoyl group. The peptide and fatty acid moieties help to direct the CCP module component to the phospholipid bilayer and hold it there; their addition significantly enhances the complement inhibitory properties of the triple CCP module fragment alone. In some respects, therefore, APT070 resembles GPI-anchored DAF. Indeed, the three CCP modules encompass functional "site 1" of CR1 that is the main locus for the decay accelerating activity of CR1 and also has a binding site for C3b/C4b. It does not have cofactor activity. APT070 prevents complement-mediated tissue damage in animal models of rheumatoid arthritis, renal transplant reperfusion injury, and vascular shock.[41] Most recently,[42] it has been investigated as a possible therapy in Guillain-Barré syndrome. Despite these promising results, the company that currently possesses the rights to APT070 (Inflazyme Pharmaceuticals Ltd., Richmond, British Columbia, Canada; http://www.inflazyme.com) is no longer pursuing its development.

Building on the idea that engineered complement regulators are effective anti-complement agents, the possible uses of hybrids with antibodies have been explored.[23] In some cases, the antibody Fab arms are attached to the complement regulator in the hope that they will target the attached regulator to a specific site. Alternatively, regulator molecules may be attached to the Fc portion of an antibody, thus they form the equivalents to the Fab arms, and the Fc helps to extend the plasma half-life. Some fusion proteins act as prodrugs in that the full inhibitory potential of the attached regulator may only be revealed by proteolysis, thus providing an additional opportunity for targeting specific sites.

Finally, mention must be made of vaccinia complement protein (VCP) and its relative from the smallpox (variola) virus, smallpox inhibitor of complement enzymes (SPICE). Neither VCP nor SPICE is glycosylated; both consist entirely of four CCPs. Of all the RCAs, VCP and SPICE, therefore, have the simplest composition. VCP is also the best-characterized RCA from a structural standpoint, initially thanks to NMR studies[43,44] and latterly to X-ray crystallography.[45] Furthermore, 3D structures have been determined of VCP in complex with two ligands, heparin[46] and suramin.[47] Unfortunately, a commensurate body of site-directed mutagenesis data is not available. Nonetheless, domain deletion experiments[48] demonstrate that all four CCP modules are required, whereas module 4 contains the main heparin-binding site. Also of interest is a functional and sequence comparison with SPICE.[49] Neither VCP nor SPICE possess potent decay accelerating activity, but they both have cofactor activity for factor I-catalyzed proteolysis of C3b and C4b. SPICE is a better cofactor than VCP, and the cofactor activity of SPICE is ~25–50% that of sCR1. Thus, the putative, single functional site in SPICE (*cf* three functional sites in CR1) is comparable in potency to each of the three functional sites in sCR1. There are 11 amino acid residue differences between SPICE and VCP, all located in CCPs 2, 3, and 4. None of the amino acid residues that differ between the two proteins is near intermodular junctions, and all are surface exposed, implying no significant structural differences between the two. Clearly, some or all of the 11 substitutions in SPICE compared to VCP are contact residues, either for C3b/C4b or factor I, and they warrant further inspection. It is striking that VCP can bind heparin and simultaneously inhibit complement activation via its DAA and CA, implying that it has the ability to protect surfaces.[45] SPICE binds heparin equally well. These properties have encouraged the hope that VCP (and SPICE) could be used therapeutically in humans despite their probable immunogenicity. In support of this, the Kotwal group has shown the possible benefits of VCP in numerous animal models of disease.[50]

FUTURE OF ANTICOMPLEMENT THERAPY

Within the small-molecule category of anticomplement drug candidates, the C3 antagonist compstatin and the C5a receptor antagonists developed by Promics Ltd. (http://www.promics.com.au) appear to be the most likely to reach the clinic. Small-molecule antagonists and inhibitors are generally regarded as the preferred lead compounds for commercial drug development in terms of absorption and permeation in biologic systems and ease and reliability of production. Therefore, even moderate success of such products in the marketplace would stimulate new interest from "big pharma" and scaled-up efforts at designing a second generation of such ligands.

Turning to biopharmaceuticals, humanized monoclonal antibodies such as the ones against C5a under development by Alexion Pharmaceuticals (Cheshire, CT, USA) represent a now tried-and-tested technology of great promise; there are hundreds of antibodies in clinical trials and dozens have been approved. Approval for one of their products in paroxysmal nocturnal hemoglobinuria appears imminent, and this molecule might well win the race to become the first anticomplement drug on the market and certainly the first to be used to treat a chronic condition.

Engineered RCA proteins (or viral proteins used without modification) may have a greater therapeutic potential as a consequence of their efficacy in blocking all the major outcomes of a complement activation event and their specificity. This is illustrated to some extent by the relative success of sCR1, although it looks as if this agent will only be used in limited and specific circumstances where it can be physically delivered to where it is needed, as for example in cardiac surgery. The ~190-kDa sCR1 was not rationally designed; because the entire extracellular portion of CR1 was used, no knowledge of structure or function was applied. A significant advance was the truncation of sCR1 to three CCP modules and the attachment of a targeting entity to create APT070. Although this product is not currently being developed for the clinic, it illustrates the benefits of rational design based on knowledge of the location of functional sites in CR1 and on understanding of the improved effectiveness of surface-associated complement regulators. APT070 additionally exemplifies the advantages of a targeting strategy, as also evident from the improved efficacy of RCA-antibody fusions.

That SPICE (and, to a lesser extent, VCP) can encapsulate much of the (so-far identified) activity of sCR1 in just four CCP modules suggests that a simple truncation, as was done in the creation of APT070, does not go far enough. Indeed, a strategy not yet well explored is the use of knowledge-based site-directed mutagenesis as a means of tailoring the activities of the CCP modules within the regulatory portions of hybrid molecules such as TP10 and APT070. This lack of progress is probably attributable to the dearth of knowledge regarding the mechanism at the atomic level despite a wealth of structural information. The results of mutagenesis, however, although they are not easy to interpret for reasons described earlier, are tantalizing. For example, single amino acid residue swaps between functional sites 1 and 2 of CR1 were sufficient to delete the cofactor activity in site 2 and introduce it into site 1.[11] Numerous other mutagenesis studies support the suggestion that significant decreases and increases in inhibitory activity may be brought about with a few judiciously selected amino acid substitutions. Because there appear to be different sites (in CR1, for example) for cofactor activity and decay accelerating activity and non-identity of C3b- and C4b-interaction sites (for example, in DAF), it is also possible to tailor an RCA to selectively inhibit a specific step in the complement cascade. The fusion of much more powerful and selective complement regulatory components with the elegant solutions for targeting and extending plasma half-lives just described could yield a new generation of powerful complement inhibitors able to act at much lower plasma concentrations than the present ones. Currently, however, there are far too many gaps in our knowledge of how RCAs work at the atomic level, and this is holding back intelligent design of highly potent and specific inhibitors.

REFERENCES

1. KIRKITADZE, M.D. & P.N. BARLOW. 2001. Structure and flexibility of the multiple domain proteins that regulate complement activation. Immunol. Rev. **180:** 146–161.
2. KOTWAL, G.J. & B. MOSS. 1988. Vaccinia virus encodes a secretory polypeptide structurally related to complement control proteins. Nature **335:** 176–178.
3. WALPORT, M.J. 2001. Complement. First of two parts. N. Engl. J. Med. **344:** 1058–1066.
4. WALPORT, M.J. 2001. Complement. Second of two parts. N. Engl. J. Med. **344:** 1140–1144.
5. ROSS, G.D. & M.E. MEDOF. 1985. Membrane complement receptors specific for bound fragments of C3. Adv. Immunol. **37:** 217–267.

6. DEMPSEY, P.W. *et al.* 1996. C3d of complement as a molecular adjuvant: bridging innate and acquired immunity. Science **271:** 348–350.
7. BLOM, A.M. *et al.* 1999. A cluster of positively charged amino acids in the C4BP alpha-chain is crucial for C4b binding and factor I cofactor function. J. Biol. Chem. **274:** 19237–19245.
8. PANGBURN, M.K., M.A. ATKINSON & S. MERI. 1991. Localization of the heparin-binding site on complement factor H. J. Biol. Chem. **266:** 16847–16853.
9. REID, K.B. & A.J. DAY. 1989. Structure-function relationships of the complement components. Immunol. Today **10:** 177–180.
10. SOARES, D.C. *et al.* 2005. Large-scale modelling as a route to multiple surface comparisons of the CCP module family. Protein Eng. Des. Sel. **18:** 379–388.
11. KRYCH-GOLDBERG, M. & J.P. ATKINSON. 2001. Structure-function relationships of complement receptor type 1. Immunol. Rev. **180:** 112–122.
12. SMITH, B.O. *et al.* 2002. Structure of the C3b binding site of CR1 (CD35), the immune adherence receptor. Cell **108:** 769–780.
13. LUKACIK, P. *et al.* 2004. Complement regulation at the molecular level: the structure of decay-accelerating factor. Proc. Natl. Acad. Sci. USA **101:** 1279–1284.
14. HAMMEL, M. *et al.* 2002. Solution structure of human and bovine beta(2)-glycoprotein I revealed by small-angle X-ray scattering. J. Mol. Biol. **321:** 85–97.
15. BARLOW, P.N. *et al.* 1991. Secondary structure of a complement control protein module by two-dimensional 1H NMR. Biochemistry **30:** 997–1004.
16. NORMAN, D.G. *et al.* 1991. Three-dimensional structure of a complement control protein module in solution. J. Mol. Biol. **219:** 717–725.
17. BARLOW, P.N. *et al.* 1992. Solution structure of the fifth repeat of factor H: a second example of the complement control protein module. Biochemistry **31:** 3626–3634.
18. O'LEARY, J.M. *et al.* 2004. Backbone dynamics of complement control protein (CCP) modules reveals mobility in binding surfaces. Protein Sci. **13:** 1238–1250.
19. SZAKONYI, G. *et al.* 2001. Structure of complement receptor 2 in complex with its C3d ligand. Science **292:** 1725–1728.
20. SOARES, D.C. & P.N. BARLOW. 2005. Complement control protein modules in the regulators of complement activation. *In* Structural Biology of the Complement System. D. Morikis & J.D. Lambris, Eds. :19–62. Taylor and Francis. Boca Raton. Florida.
21. LINDAHL, G., U. SJOBRING & E. JOHNSSON. 2000. Human complement regulators: a major target for pathogenic microorganisms. Curr. Opin. Immunol. **12:** 44–51.
22. CIULLA, E. *et al.* 2005. Evolutionary history of orthopoxvirus proteins similar to human complement regulators. Gene **355:** 40–47.
23. MORGAN, B.P. & C.L. HARRIS. 2003. Complement therapeutics; history and current progress. Mol. Immunol. **40:** 159–170.
24. MIZUNO, M. & B.P. MORGAN. 2004. The possibilities and pitfalls for anti-complement therapies in inflammatory diseases. Curr. Drug Targets Inflamm. Allergy **3:** 87–96.
25. HOLLAND, M.C., D. MORIKIS & J.D. LAMBRIS. 2004. Synthetic small-molecule complement inhibitors. Curr. Opin. Invest. Drugs **5:** 1164–1173.
26. BUERKE, M. *et al.* 2001. Novel small molecule inhibitor of C1s exerts cardioprotective effects in ischemia-reperfusion injury in rabbits. J. Immunol. **167:** 5375–5380.
27. SOULIKA, A.M. *et al.* 2003. Studies of structure-activity relations of complement inhibitor compstatin. J. Immunol. **171:** 1881–1890.
28. WOODRUFF, T.M. *et al.* 2005. Increased potency of a novel complement factor 5a receptor antagonist in a rat model of inflammatory bowel disease. J. Pharmacol. Exp. Ther. **314:** 811–817.
29. WHISS, P.A. 2002. Pexelizumab alexion. Curr. Opin. Invest. Drugs **3:** 870–877.
30. KAPLAN, M. 2002. Eculizumab (alexion). Curr. Opin. Invest. Drugs **3:** 1017–1023.
31. FLEISIG, A.J. & E.D. VERRIER. 2005. Pexelizumab–a C5 complement inhibitor for use in both acute myocardial infarction and cardiac surgery with cardiopulmonary bypass. Expert Opin. Biol. Ther. **5:** 833–839.
32. HILLMEN, P. *et al.* 2004. Effect of eculizumab on hemolysis and transfusion requirements in patients with paroxysmal nocturnal hemoglobinuria. N. Engl. J. Med. **350:** 552–559.

33. VOGEL, C.W. *et al.* 1996. Structure and function of cobra venom factor, the complement-activating protein in cobra venom. Adv. Exp. Med. Biol. **391:** 97–114.
34. KOLLN, J. *et al.* 2004. Functional analysis of cobra venom factor/human C3 chimeras transiently expressed in mammalian cells. Mol. Immunol. **41:** 19–28.
35. RIOUX, P. 2001. TP-10 (AVANT Immunotherapeutics). Curr. Opin. Invest. Drugs **2:** 364–371.
36. WEISMAN, H.F. *et al.* 1990. Soluble human complement receptor type 1: *in vivo* inhibitor of complement suppressing post-ischemic myocardial inflammation and necrosis. Science **249:** 146–151.
37. LAZAR, H.L. *et al.* 2004. Soluble human complement receptor 1 limits ischemic damage in cardiac surgery patients at high risk requiring cardiopulmonary bypass. Circulation **110:** 11274–1129.
38. PRUITT, S.K. *et al.* 1997. Effect of continuous complement inhibition using soluble complement receptor type 1 on survival of pig-to-primate cardiac xenografts. Transplantation **63:** 900–902.
39. AZIMZADEH, A. *et al.* 2003. Hyperacute lung rejection in the pig-to-human model. 2. Synergy between soluble and membrane complement inhibition. Xenotransplantation **10:** 120–131.
40. LAM, T.T. *et al.* 2005. The effect of soluble complement receptor type 1 on acute humoral xenograft rejection in hDAF-transgenic pig-to-primate life-supporting kidney xenografts. Xenotransplantation **12:** 20–29.
41. SMITH, R.A. 2002. Targeting anticomplement agents. Biochem. Soc. Trans. **30:** 1037–1041.
42. HALSTEAD, S.K. *et al.* 2005. Complement inhibition abrogates nerve terminal injury in Miller Fisher syndrome. Ann. Neurol. **58:** 203–210.
43. WILES, A.P. *et al.* 1997. NMR studies of a viral protein that mimics the regulators of complement activation. J. Mol. Biol. **272:** 253–265.
44. HENDERSON, C.E. *et al.* 2001. Solution structure and dynamics of the central CCP module pair of a poxvirus complement control protein. J. Mol. Biol. **307:** 323–339.
45. MURTHY, K.H. *et al.* 2001. Crystal structure of a complement control protein that regulates both pathways of complement activation and binds heparan sulfate proteoglycans. Cell **104:** 301–311.
46. GANESH, V.K. *et al.* 2004. Structure of vaccinia complement protein in complex with heparin and potential implications for complement regulation. Proc. Natl. Acad. Sci. USA **101:** 8924–8929.
47. GANESH, V.K. *et al.* 2005. Structural basis for antagonism by suramin of heparin binding to vaccinia complement protein. Biochemistry **44:** 10757–10765.
48. SMITH, S.A. *et al.* 2003. Mapping of regions within the vaccinia virus complement control protein involved in dose-dependent binding to key complement components and heparin using surface plasmon resonance. Biochim. Biophys. Acta **1650:** 30–39.
49. ROSENGARD, A.M. *et al.* 2002. Variola virus immune evasion design: expression of a highly efficient inhibitor of human complement. Proc. Natl. Acad. Sci. USA **99:** 8808–8813.
50. JHA, P. & G.J. KOTWAL. 2003. Vaccinia complement control protein: multi-functional protein and a potential wonder drug. J. Biosci. **28:** 265–271.

Arthropod-Derived Protein EV131 Inhibits Histamine Action and Allergic Asthma

WYNNE WESTON-DAVIES,[a] ISABELLE COUILLIN,[b,c] SILVIA SCHNYDER,[b,d] BRUNO SCHNYDER,[b,d] RENE MOSER,[d] OLGA LISSINA,[e] GUIDO C. PAESEN,[e] PATRICIA NUTTALL,[e] AND BERNHARD RYFFEL[b]

[a]*Evolutec Ltd., Reading, United Kingdom*

[b]*CNRS, Institute Transgenose, Orleans, France*

[c]*Key-Obs S. A., Orleans, France*

[d]*IBR Inc., Matzingen, Switzerland, and Biomedical Research Foundation, Matzingen, Switzerland*

[e]*Centre for Ecology and Hydrology, Oxford, United Kingdom*

ABSTRACT: Histamine is an important mediator of allergic responses. Arthropods express several biologically active proteins in their saliva, which may allow a prolonged blood meal on the host. Proteins identified and expressed include histamine, serotonin, tryptase, and complement binding proteins. We review here data that scavenging of endogenous histamine by the histamine-binding protein EV131 has a profound inhibitory effect on allergic asthma. Aerosol administration of EV131 prevented airway hyperreactivity and abrogated peribronchial inflammation, eosinophil recruitment, mucus hypersecretion, and IL-4 and IL-5 secretion. Saturation with histamine abrogated the inhibitory effect of EV131 on bronchial hyperreactivity. The data suggest that histamine plays a role in allergies and that scavenging of histamine by EV131 may represent a novel therapeutic strategy in the treatment of allergic diseases.

KEYWORDS: endotoxin; adult respiratory distress syndrome; histamine scavenging; mice

INTRODUCTION

Since its discovery by Sir Henry Dale in 1911, histamine has been a focus of interest for pharmacologists, clinicians, and immunologists. It took several decades to identify the receptors for histamine, and only in 1994 was the fourth mammalian histamine receptor discovered.[1] It was another 7 years before it was fully characterized.[2] Only since the discovery of the H_2, H_3, and H_4 receptors has the true complexity of histamine's interactions within the organism become apparent. H_1 receptors are associated with allergic responses, but even here there are major differences in the actions that histamine induces. It is responsible for relaxation of vascu-

Address for correspondence: B. Ryffel, 3B rue de La Ferollerie, F-45071 Orleans, France. Voice: +33-238-25-54-38; fax: +33-238-25-79-79.
bryffel@cnrs-orleans.fr

Ann. N.Y. Acad. Sci. 1056: 189–196 (2005). © 2005 New York Academy of Sciences.
doi: 10.1196/annals.1352.009

lar smooth muscle but contraction of bronchial smooth muscle even though both are apparently mediated through identical receptors. H_2 receptors originally were thought to control only the secretion of gastric acid by the parietal cells, a logical enough function, given that the evolutionary role of histamine is thought to be protection against parasites, and the stomach constitutes a major line of defense against the external environment. Subsequently, they were found to have a major regulatory function by providing negative feedback to a wide variety of different cells activated by H_1 stimulation.[3–6] H_3 receptors have, as yet, poorly understood functions in the central nervous system, where histamine acts as a neurotransmitter while the precise function of H_4 receptors has only just begun to be explored.

HISTAMINE AND ASTHMA

At the time it was first recognized that asthma might have an allergic basis, only the H_1 receptor was known. It was a logical step to see whether the small molecule-blocking H_1 receptor, which had already proved effective in other forms of allergy such as hay fever, would work in asthma. *In vitro* these receptors appeared to block the allergen-induced contraction of bronchial smooth muscle and even to reverse it in selected animal species such as the guinea pig. In human asthmatic subjects, they were found to be effective in blocking the bronchoconstricting effects of inhaled histamine but had little effect on inhaled methacholine or allergen.[7] Not surprisingly therefore, with the possible exception of a few agents such as ketotifen that also stabilized the mast cell, they proved to be ineffective in treating clinical asthma.[8]

White[9] defined the cardinal features of asthma and other allergic diseases that were capable of inhibition by the use of corticosteroids. Using sensitized normal and mast cell–deficient mice, Nagai *et al.*[10] showed that airway hyperreactivity was mast cell dependent but that, although drugs such as ketotifen reduced IL-5 production and eosinophil infiltration, they had no effect on airway hyperreactivity. The inference was that histamine was not an important mediator in airway hyperreactivity but that other mast cell products might be. Despite the fact that it was known that histamine is released in great quantity by mast cells and basophils during an asthmatic attack, interest shifted to the other mast cell products such as the cysteinyl leukotrienes.

Therefore, histamine appears to be only one of many inflammatory mediators at work in asthma, and blocking its activity is unlikely to make very much difference as long as the other pathways remain intact. The emphasis became focused on the longer term inflammatory processes involved in the airway remodeling that characterizes chronic asthma. Inflammatory cells, particularly eosinophils, were known to migrate to the affected airways and to take part in chronic inflammatory processes leading to fibrosis and permanent narrowing and loss of elasticity of the small air passages.

Whatever the role of histamine, it was not thought to be directly involved in the recruitment of these cells despite the fact that Clark had shown in 1975 that histamine was chemotactic for human eosinophils.[11,12] Unlike chemokines such as IL-8 that can attach to vascular endothelial cells and set up an effective concentration gradient to guide granulocytes to their site of action, histamine is rapidly and permanently removed from the circulation after its release. It was difficult therefore to see how it could possibly exert such a remote effect as recruitment of eosinophils from the bone marrow.

The attention on histamine was again attracted by the publication of a fourth histamine receptor.[13–16] It is widely distributed in tissues as diverse as vascular endothelium, bone marrow, eosinophils, neutrophils, and bronchial smooth muscle. It was quickly established that neither H_1 nor H_2 receptor antagonists could block it and that only the H_3 (and H_4) receptor blocker thioperadine had a substantial inhibitory effect. The precise function of the fourth receptor for leukocyte trafficking and airway inflammation is still a mystery.

BIOLOGICALLY ACTIVE PROTEINS FROM ARTHROPODS: THE LIPOCALIN PROTEIN SUPERFAMILY

A new and powerful research tool became available with the discovery of histamine-binding proteins in the saliva of arthropods/ticks. Ticks are blood-sucking arthropods that parasitize animals and may spend a long time, up to 14 days if undisturbed, feeding at the same site of the host. More than 25 years ago, parasitologists speculated that ticks must inject into the host substances that overcome the normal allergic and immunological responses, and this led to the discovery of histamine-blocking activity in tick saliva.[17] With the advent of modern biotechnology techniques such as polymerase chain reaction, gene libraries, and recombinant peptide production, it has become possible to identify and replicate some of these molecules. One of the first to be isolated was a molecule that bound histamine with an affinity an order of magnitude greater than that of the mammalian H_1 receptor.[18]

This first molecule, now known as rEV131, was followed by the discovery of a series of others that bound both histamine and serotonin or acted as a substrate for other inflammatory mediators such as tryptase.[19] rEV131 together with several other ligand-binding molecules discovered in tick saliva is a member of the lipocalin superfamily of proteins.[20] Lipocalins are ubiquitous throughout the animal and plant kingdoms and usually act as carrier molecules for substances such as vitamins and pheromones.[21] They are extremely robust molecules and rEV131 has proved to be thermostable in solution for periods in excess of 15 months at room temperature.

THE HISTAMINE-BINDING PROTEIN, INHIBITION OF ALLERGIC ASTHMA

Although scavenging histamine is the only known action of rEV131, the end results of this process seem to far exceed most experts' predictions of what such a molecule might achieve. The most puzzling result was probably its effect in a mouse asthma model in which, as well as reducing airway hyperreactivity to the level achieved by a potent corticosteroid, it substantially reduced eosinophil counts and mucus production.[24]

Upon antigen challenge, the mice developed a robust bronchial hyperreactivity (BHR) in response to aerosolized methacholine at 24 h (FIG. 1). Administration of EV131 by aerosol before Ova challenge inhibited methacholine-induced bronchoconstriction (70% of controls; $P < 0.01$). The vehicle alone, saline, had no effect on BHR. The effect of EV131 on BHR was matched by the inhibition induced by the

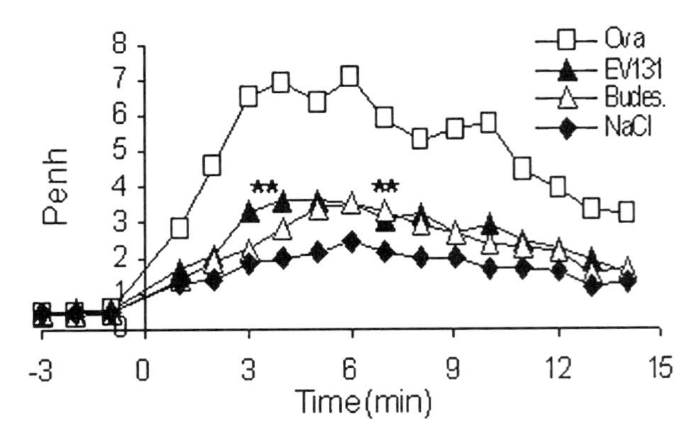

FIGURE 1. The histamine scavenger EV131 inhibits Ova-induced BHR in mice sensitized with antigen. BP2 mice were immunized at days 0 and 7 with ovalbumin (Ova). The mice were given 50 μL NaCl (saline) or EV131 (340 μg in NaCl) intratracheally and challenged with 50 μL of NaCl or Ova 1 h later. BHR to methacholine was measured 24 h after challenge by the noninvasive enhanced pause (Penh) and expressed as a function of time. The results represent the mean for two independent experiments (n = 8 mice per group). Inhibition by EV131 and by budesonide is significant at **P <0.01.

glucocorticosteroid budesonide, used as a control inhibitor (70%, P <0.01). Saturation with histamine abrogated the inhibitory effect of EV131 on bronchial hyperreactivity (data not shown). Therefore, *in situ* neutralization of histamine by EV131 acting as a soluble receptor with high-affinity binding for histamine had a profound effect on BHR, and we therefore asked whether the recruitment of eosinophils during the allergic asthma was influenced.

Antigen challenge caused a substantial recruitment of inflammatory cells into the BAL fluid at 72 h (P <0.05, FIG. 2A). Budesonide had a slightly more pronounced effect (P <0.01). Only a few eosinophils and neutrophils were found in the BAL fluid of saline-challenged mice. In contrast, the antigen challenge resulted in a significant increase of eosinophil counts in the BAL fluid (P <0.01, FIG. 2B). EV131 administration before the antigen challenge largely prevented the recruitment of eosinophils (P <0.01, FIG. 2B). Furthermore, we demonstrated that the prevention of eosinophil recruitment was caused by histamine scavenging because histamine-presaturated EV131 was ineffective (data not shown).

Finally, administration of EV131 reduced significantly the peribronchial eosinophilia, mucus hypersecretion, and hyperplasia of bronchial smooth muscles (FIG. 3). Therefore, complete *in vivo* neutralization of histamine with the high-affinity histamine-binding protein EV131 inhibited the inflammatory cell recruitment and suppressed the characteristic allergic inflammation of the airways.

How such wide-ranging anti-inflammatory effects as the recruitment of eosinophils from the bone marrow could happen at such a distance from the site of mast cell degranulation remained a mystery for some time. A possible explanation was the recent finding that histamine induces the release of the leukocyte attractant factor IL-16, inducing the recruitment of inflammatory cells, essentially granulocytes.[22] Fur-

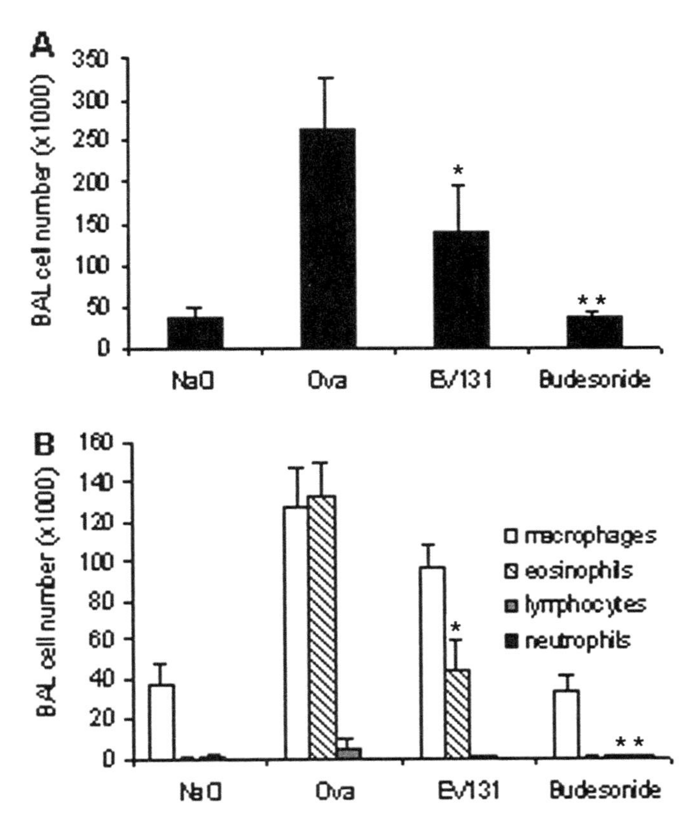

FIGURE 2. EV131 inhibits Ova-induced eosinophil recruitment in BAL fluid. BAL fluid was collected at 72 h after Ova challenge, and cytospin preparations were prepared: total cell counts (**A**) and differential cell counts in BAL (**B**). The results are pooled from two independent experiments; mean values and SEM are given (n = 8 mice per group). Differences are significant $*P < 0.05$ and $**P < 0.01$, respectively. Reproduced with permission from Ref. 24.

thermore, Gantner *et al.* discovered that this activity is mediated through both H_1 and H_4 receptors. Therefore, inhibition of IL-16 by histamine scavenging may be a partial explanation of how rEV131 works.

Evolution and the tick have found a better way of counteracting the downstream inflammatory effects of histamine than blocking an unknown number of receptors. Furthermore, the inhibitory effect of rEV131 on neutrophil infiltration in endotoxin-induced lung inflammation suggests a potential application in adult respiratory distress syndrome (see accompanying chapter in this volume by Ryffel *et al.*). In contrast, corticosteroids have little effect in this model.[23] Because there are no currently effective medications for adult respiratory distress syndrome, a condition that carries a mortality of at least 40% in humans, this could be a potentially valuable therapeutic discovery.

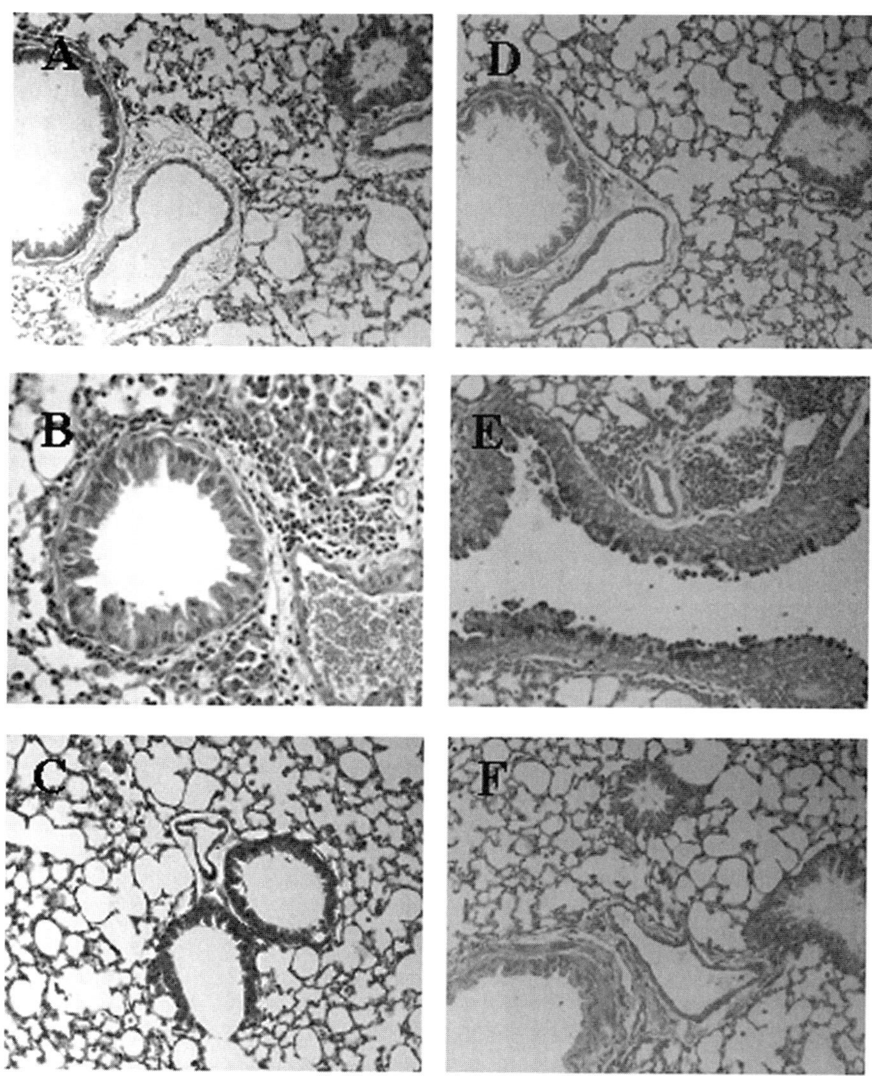

FIGURE 3. EV131 inhibits inflammation and mucus hypersecretion in Ova-challenged mice. Inflammatory changes of lungs of saline (**A**) and Ova- (**B**) challenged mice: significant peribronchial and perivascular eosinophil infiltration in mice challenged with Ova. EV131 inhibits peribronchial and perivascular eosinophil infiltration (**C**). Hematoxylin and eosin staining. EV131 inhibits mucus secretion of lungs in Ova-challenged mice. Mucus secretion changes in lungs of saline- (**D**) and Ova- (**E**) challenged mice: significant mucus secretion in mice challenged with Ova (*arrows* mark mucus secreted by epithelial cells). EV131 inhibits mucus secretion in mice challenged with Ova (**F**). PAS mucus staining. Intermediate power micrographs are shown in **A**, **C**, **D**, **E**, and **F** (magnification, ×200), and a higher magnification shows eosinophil infiltration in **B** (magnification, ×400). These micrographs are representatives of three independent experiments with four mice per group. Reproduced with permission from Ref. 24.

CONCLUSION

Assuming that rEV131 does nothing other than potently bind and thereby neutralize histamine, it is clear from results that have now been confirmed by repeated experiments that histamine must be a more important inflammatory mediator than has previously been recognized. It was the failure of H_1 and H_2 blocking agents to mitigate late-stage inflammation in conditions such as asthma that led researchers to conclude that histamine was not in itself an important mediator. It is the recent discovery of H_4 receptors and the use of selective blocking agents in appropriate models that have led to a reappraisal of the role of histamine. However, rEV131 provides the only practical means of denying histamine to all its receptors and it is providing new insights into the role of histamine in a range of conditions. In particular, it seems to have long-term effects on inflammatory cell trafficking and tissue remodeling as well as more immediate effects on vascular and bronchial smooth muscle. In many cases, the chronic effects are likely secondary to the influence of histamine on cytokine release from tissues such as vascular endothelium and T lymphocytes. Because these, in turn, appear to be mediated by a combination of different receptors, including H_4, and possibly others yet to be discovered, the possibility of controlling them by the use of small molecule blocking agents would appear to be limited. Evolution appears to have chosen irreversible binding of the ligand as a more efficient method of intervention.

It is hoped that rEV131 and its molecular relatives will become interesting therapeutic agents. Clinical trials in humans have already commenced, but even at this stage rEV131 is contributing to a complete rethinking of the importance of histamine in the etiology of asthma and other allergic conditions. Possibly it will do the same for the many autoimmune disorders, such as rheumatoid arthritis and inflammatory bowel disease, in which the mast cell has long been thought to play a part but in which histamine receptor blockers have produced disappointing results.

REFERENCES

1. Raible, D.G., T. Lenahan, Y. Fayvilevich, *et al.* 1994. Pharmacologic characterization of a novel histamine receptor on human eosinophils. Am. J. Respir. Crit. Care Med. **149:** 1506–1511.
2. Hough, L.B. 2001. Genomics meets histamine receptors: new subtypes, new receptors. Mol. Pharmacol. **593:** 415–419.
3. Lapin, D. & K. Whaley. 1980. Effects of histamine on monocyte complement production. I. Inhibition of C2 production mediated by its action on H2 receptors. Clin. Exp. Immunol. **41:** 497–504.
4. Francis, G.S. 1988. Modulation of peripheral sympathetic nerve transmission. J. Am. Coll. Cardiol. **12:** 250–254.
5. Nogrady, S.G. & C. Bevan. 1981. H2 receptor blockade and bronchial hyperreactivity to histamine in asthma. Thorax **36:** 268–271.
6. Jutel, M., S. Klunker, M. Akdis, *et al.* 2001. Histamine upregulates Th1 and downregulates Th2 responses due to different patterns of surface histamine 1 and 2 receptor expression. Int. Arch. Allergy Immunol. **124:** 190–192.
7. Nogrady, S.G. & C. Bevan. 1978. Inhaled antihistamines—bronchodilatation and effects on histamine- and methacholine-induced bronchoconstriction. Thorax **33:** 700–704.
8. Holgate, S.T. & J.P. Finnerty. 1989. Antihistamines in asthma. J. Allergy Clin. Immunol. **83:** 537–547.

9. WHITE, M.V. 1990. The role of histamine in allergic diseases. J Allergy Clin. Immunol. **86:** 599–605.
10. NAGAI, H., S. YAMAGUCHI, Y. MAEDA & H. TANAKA. 1996. Role of mast cells, eosinophils and IL-5 in the development of airways hyperresponsiveness in sensitized mice. Clin. Exp. Allergy **26:** 618–620.
11. CLARK, R.A., J.I. GALLIN & A.P. KAPLAN. 1975. The selective eosinophil chemotactic activity of histamine. J. Exp. Med. **142:** 1462–1476.
12. CLARK, R.A., J.A. SANDLER, J.I. GALLIN & A.P. KAPLAN. 1977. Histamine modulation of eosinophil migration. J. Immunol. **118:** 137–145.
13. ODA, T., N. MORIKAWA, Y. SAITO, *et al.* 2000. Molecular cloning and characterization of a novel type of histamine receptor preferentially expressed in leukocytes. J. Biol. Chem. **275:** 36781–36786.
14. NGUYEN, T., D.A. SHAPIRO, S.R. GEORGE, *et al.* 2001. Discovery of a novel member of the histamine receptor family. Mol. Pharmacol. **59:** 427–433.
15. ZHU, Y., D. MICHALOVICH, H.L. WU, *et al.* 2001. Cloning, expression and pharmacological characterization of a novel human histamine receptor. Mol. Pharmacol. **59:** 434–441.
16. LIU, C., X.J. MA, X. JIANG, *et al.* 2001. Cloning and pharmacological characterization of a fourth histamine receptor (H_4) expressed in bone marrow. Mol. Pharmacol. **59:** 420–426.
17. CHINNERY, W.A. & E. AYITEY-SMITH. 1977. Histamine blocking agent in the salivary gland homogenate of the tick *Rhipicephalus sanguineus sanguineus*. Nature **265:** 366–367.
18. PAESEN, G.C., P.L. ADAMS, K. HARLOS, *et al.* 1999. Tick histamine binding proteins: isolation, cloning, and three dimensional structure. Mol. Cell **3:** 661–671.
19. SANGAMNATDEJ, S., G.C. PAESEN, M. SLOVAK & P.A. NUTTALL. 2002. A high affinity serotonin- and histamine-binding lipocalin from tick saliva. Insect Mol. Biol. **11:** 79–86.
20. PAESEN, G.C., P.L. ADAMS, P.A. NUTTALL & D.L. STUART. 2000. Tick histamine-binding proteins: lipocalins with a second binding cavity. Biochim. Biophy. Acta **1482:** 92–101.
21. FLOWER, D.R. 1996. The lipocalin protein family: structure and function. Biochem. J. **318:** 1–14.
22. GANTNER, F., K. SAKAI, M.W. TUSCHE, *et al.* 2002. Histamine h(4) and h(2) receptors control histamine-induced interleukin-16 release from human CD8(+) T cells. J. Pharmacol. Exp. Ther. **303:** 300–307.
23. LEFORT, J., L. MOTREFF & B.B. VARGAFTIG. 2001. Airway administration of *Escherichia coli* endotoxin to mice induces glucocorticosteroid-resistant bronchoconstriction and vasopermeation. Am. J. Respir. Cell Mol. Biol. **24:** 345–351.
24. COUILLIN, I., I. MAILLET, B.B. VARGAFTIG, *et al.* 2004. Arthropod-derived histamine-binding protein prevents murine allergic asthma J. Immunol. **173:** 3281–3286.

Histamine Scavenging Attenuates Endotoxin-Induced Acute Lung Injury

BERNHARD RYFFEL,[a] ISABELLE COUILLIN,[a,c] ISABELLE MAILLET,[a,c] BRUNO SCHNYDER,[a] GUIDO C. PAESEN,[b] PATRICIA NUTTALL,[b] AND WYNNE WESTON-DAVIES[b,d]

[a]*CNRS Institute Transgenose, IEM, Orleans, France, and IIDMM, UCT, Cape Town, South Africa*

[b]*Centre for Ecology and Hydrology, Oxford, England*

[c]*Key-Obs S. A., Orleans, France*

[d]*Evolutec Ltd., Green Park, England*

ABSTRACT: Histamine is an important mediator of early and late inflammatory responses. Here we asked whether scavenging of endogenous histamine by the arthropod-derived histamine binding protein EV131 diminishes acute respiratory distress syndrome (ARDS) induced by inhaled endotoxin. We demonstrate that EV131 (360 μg given intranasally) reduced endotoxin-induced bronchoconstriction and recruitment of neutrophils. Furthermore, EV131 administration diminished TNF-α and protein leak in the bronchoalveolar lavage fluid. The data suggest that histamine attenuates endotoxin-induced bronchoconstriction and neutrophil recruitment. Therefore, scavenging of histamine by EV131 may represent a novel therapeutic strategy in ARDS.

KEYWORDS: endotoxin; acute respiratory distress syndrome (ARDS); histamine scavenging; mice

INTRODUCTION

The arthropod-derived histamine scavenging protein EV131, which binds histamine with an affinity 10 to 100 times greater than that of the mammalian H_1 receptor,[1] was investigated in an animal model of acute respiratory distress syndrome (ARDS). This recombinant protein, derived from a native protein found in the saliva of the female *Rhipicephalus appendiculatus* tick, belongs to a group of lipocalin-like molecules that, because of their capacity to sequester histamine, have been given the name histacalins.[2] We asked whether EV131 would inhibit pathologies mediated by histamine. In an allergic asthma model, EV131 given prior to antigen challenge of ovalbumin-immunized mice reduced airway hyperreactivity by 70% and abrogated peribronchial inflammation, pulmonary eosinophilia, mucus hypersecretion, and interleukin (IL)-4 secretion.[3] The inhibitory effect of EV131 on airway hyperreac-

Address for correspondence. B. Ryffel, 3B rue de La Ferollerie, F-45071 Orleans, France. Voice: +33 238 25 54 38; fax: +33 238 25 79 79.
bryffel@cnrs-orleans.fr

**Ann. N.Y. Acad. Sci. 1056: 197–205 (2005). © 2005 New York Academy of Sciences.
doi: 10.1196/annals.1352.034**

tivity was comparable to that of glucocorticosteroids.[3] These results demonstrate that histamine is a critical mediator of chronic allergic inflammation in the mouse model.

Because similarities exist between allergic asthma and ARDS, and neutrophils express histamine receptors,[4] we asked whether EV131 would influence endotoxin (LPS)-induced ARDS in mice. We established a model for ARDS in which neutrophil infiltration, capillary protein leakage, airways bronchoconstriction, and hyperreactivity are induced by the inhalation of *Escherichia coli* endotoxin (055:B5, Sigma) as previously described.[5] We report here that EV131 given prior to endotoxin administration reduces bronchoconstriction, neutrophil recruitment, vascular leak, and tumor necrosis factor (TNF) secretion in the lung.

METHODS

Mice

C57BL/6 mice were bred in our specific pathogen-free animal facility at CNRS. For experiments, adult animals (6–8 weeks old) of 23–25 g body weight were kept in isolated ventilated cages. All protocols complied with the French Government's ethical and animal experiment regulations.

Experimental Protocol of Acute Respiratory Distress Syndrome

Mice were given a low dose of ketamine/xylazine by the i.v. route. Lipopolysaccharide (LPS) (*E. coli*, serotype O111:B4; from Sigma, St Louis, MO) in a volume of 40μL was applied intranasally. EV131 was applied 1 h before LPS administration by the intranasal route 90, 180, and 360 μg/mouse (4–18 mg/kg) and budesonide (a glucocorticosteroid acting as a positive control, 375 μg/mouse (19 mg/kg). Three and 24 h after stimulation the mice were sacrificed by a high dose of ketamine/xylazine. Via a tracheal cannula, the lungs were lavaged 4 times with 0.5 ml of ice-cold phosphate-buffered saline (PBS) solution. The BAL fluid was analyzed for cell composition and cytokine quantifications as described before.[3] After bronchoalveolar lavage, the lung was perfused via heart puncture with ISOTON® II acid-free balanced electrolyte solution (Beckman Coulter, Krefeld, Germany). Half of the lung was stored at −20°C for MPO assay and the other half was fixed in 4% buffered formaldehyde overnight for histologic analysis. Experiments were performed at least twice. Groups of four to five animals were used.

Airways Resistance

The airways resistance was evaluated by whole-body plethysmography as described before.[6] Bronchoconstriction was investigated over a period of 3 h after LPS application (1 μg LPS i.n. per mouse). Unrestrained conscious mice were placed in whole-body plethysmography chambers (Buxco Electronic, Sharon, CO, USA). Mean airway bronchoconstriction was estimated by the enhanced respiratory pause (PenH) index. PenH can be conceptualized as the phase shift of the thoracic flow and the nasal flow curves; increased phase shift correlates with increased respiratory system resistance. PenH is calculated by the formula PenH = (Te/RT−1) × PEF/PIF, where Te is expiratory time, RT is relaxation time, PEF is peak expiratory flow, and PIF is peak inspiratory flow.

Microscopy and Myeloperoxidase Activity (MPO) in Lungs

One perfused lung lobe was fixed in 4% buffered formaldehyde for standard microscopic analysis: 3 μm thick sections were stained with hematoxylin and eosin (H&E). Neutrophil recruitment in the lung parenchyma was assessed by counting 20 microscopic lung sections at high power fields (HPF) from five mice per group.

Lung tissue MPO activity was evaluated as described previously.[5] In brief, the right heart ventricle was perfused with 10 ml saline to flush out the vascular contents, and lungs were frozen at −20°C until use. Lungs were homogenized for 30 s in a glass homogenizer at 4°C in 1 ml PBS and centrifuged (10,000 rpm, 10 min at 4°C), and the supernatant was discarded. The pellets were resuspended in 1 ml PBS containing 0.5% hexadecyltrimethyl ammonium bromide (HTAB) and 5 mM ethylenediamine tetraacetic acid (EDTA). Following centrifugation, 50 μl supernatant fluids were placed in test tubes with 200 μl PBS-HTAB-EDTA, 2 ml Hanks' balanced salt solution (HBSS), 100 μl of *o*-dianisidine dihydrochloride (1.25 mg/ml), and 100 μl H_2O_2 0.05%. After 15 min of incubation at 37°C in an agitator, the reaction was stopped with 100 μl NaN_3 1%. MPO activity was determined as absorbance at 460 nm against medium.

Dosage of TNF and Total Protein in BAL Fluid

For these determinations, the BAL fluid was obtained at 3 h. TNF in the BAL fluid was evaluated by enzyme-linked immunosorbent assay (ELISA) following the manufacturer's instructions (R&D). For protein determination, Bradford stain was added to the supernatant as described by the manufacturer (Bio-Rad, Ivry sur Seine, France) using albumin as standard, and absorbance was measured at 595 nm (Uvikon spectrophotometer, Kontron, Zurich, Switzerland).

Statistical Analysis

Data are presented as the means and standard deviation (SD) indicated by error bars. Statistical significance was determined by Student's *t* test for comparisons of two groups. *P* values of <0.05 were considered statistically significant.

RESULTS

EV131 Reduces Endotoxin-Induced Bronchoconstriction

We first established a dose-response effect of intranasal endotoxin (0.1–100 μg) that induced a substantial bronchoconstriction within 90 min, lasting to about 180 min (data not shown). A dose of 1 μg endotoxin, which induced a significant bronchoconstriction within 90 min, was selected for all experiments (FIG. 1). Then we administered EV131 at 360 μg intranasally 1 h before endotoxin, which inhibited endotoxin-induced bronchoconstriction expressed as PenH by about 75% (FIG. 1A,

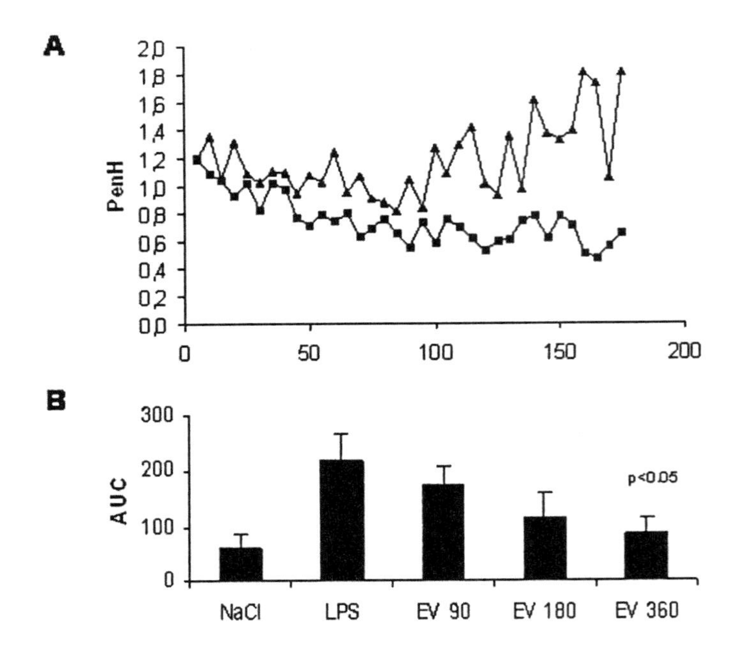

FIGURE 1. Endotoxin (LPS)-induced bronchoconstriction and inhibition by EV131. LPS was given at 1 μg by the intranasal route 1 h after NaCl (*triangles*) or EV131 (*squares*). (**A**) EV131 at 360 μg reduces endotoxin-induced bronchoconstriction, mean PenH values from four mice per group (results are representative for two independent experiments). (**B**) Dose-response effect expressed of AUC of PenH values from 90–180 min; mean values and SD are given ($n = 5$, $P < 0.05$).

$P < 0.05$). The dose-response effect of EV131-mediated inhibition of endotoxin-induced bronchoconstriction was further investigated. The mean PenH expressed as AUC (90–180 min) was inhibited by 50% in the range of 180 μg EV131, but was significant only at 360 μg (FIG. 1, $P < 0.05$). The glucocorticosteroid budesonide had a comparable effect (data not shown).

These data suggest that endogenous histamine has a physiologic role in bronchoconstriction induced by endotoxin, and hence neutralization of histamine by EV131 could ameliorate ARDS.

Histamine Scavenging by EV131 Reduces Neutrophil Recruitment into BAL Fluid

Intranasal administration of endotoxin causes a significant recruitment of total cell and especially neutrophils in the BAL fluid. We recovered about 3.54×10^5 leukocytes in the BAL 24 h after endotoxin treatment. Administration of EV131 (360 μg) reduced the total cell count in the BAL fluid (FIG. 2A). Differential cell counts in the BAL fluid showed that macrophage levels were identical (data not shown), whereas neutrophil recruitment was significantly inhibited by the high dose of

A

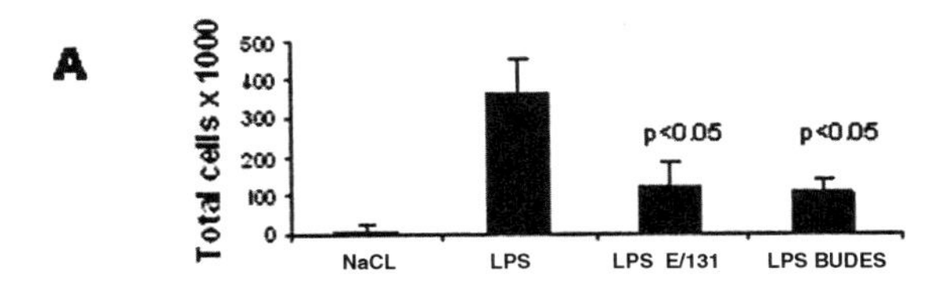

B

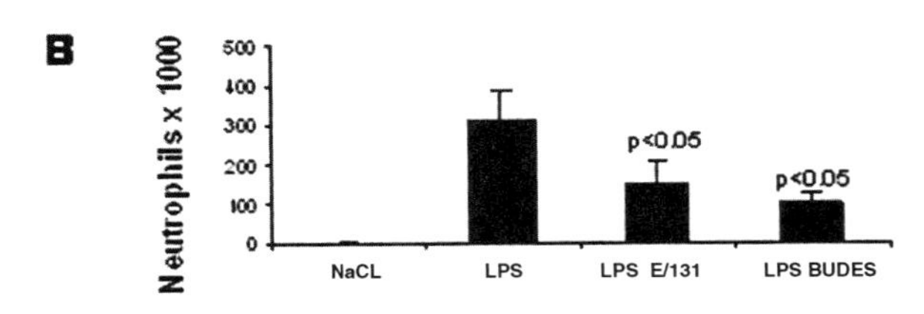

FIGURE 2. EV131 inhibits endotoxin-induced neutrophil recruitment in BAL fluid. EV131 (360 μg) or Budenoside (375 μg) was given 1 h prior to LPS at 1 μg by the intranasal route. BAL was obtained at 24 h, cells were counted, and cytospin preparations were analyzed. (**A**) Total cell count. (**B**) Neutrophil counts in BAL fluid. Mean values and SD are given ($n = 5$, $P < 0.05$).

EV131 (FIG. 2B, $P < 0.05$). Budesonide (375 μg) induced a comparable reduction in total cell and neutrophil counts. These results indicate that neutralization of endogenous histamine reduces endotoxin-induced neutrophil recruitment in the BAL fluid. We therefore asked whether the proinflammatory cytokine response is reduced by histamine neutralization.

Reduction of Endotoxin-Induced TNF-α and Protein in BAL Fluid by EV131

Because intranasal endotoxin administration induces production of proinflammatory cytokines such as TNF-α in the BAL fluid of C57BL/6, we asked whether EV131 pre-treatment could attenuate TNF-α production. We established that TNF-α secretion is maximal at 3 h after endotoxin instillation, but still present after 24 h (data not shown). Furthermore, we show a dose-dependent trend for a reduction in TNF-α levels in the BAL fluid after administration of EV131, reaching significance at the highest dose (FIG. 3A, $P < 0.05$). Endotoxin causes epithelial-endothelial damage and hence vascular protein leak. Therefore, we tested a potential inhibitory effect of EV131 on protein levels in the BAL. In fact, 360 μg EV131 reduced protein levels in the BAL by about 50% as compared to the endotoxin control (FIG. 3B, $P < 0.05$). These data suggest that scavenging of histamine reduces the production of TNF-α and diminishes the vascular damage.

Inhibition of the Recruitment of Neutrophils into the Lung

We then asked whether the recruitment of neutrophils into the lung parenchyma was altered by the administration of the histamine scavenging protein. For this we

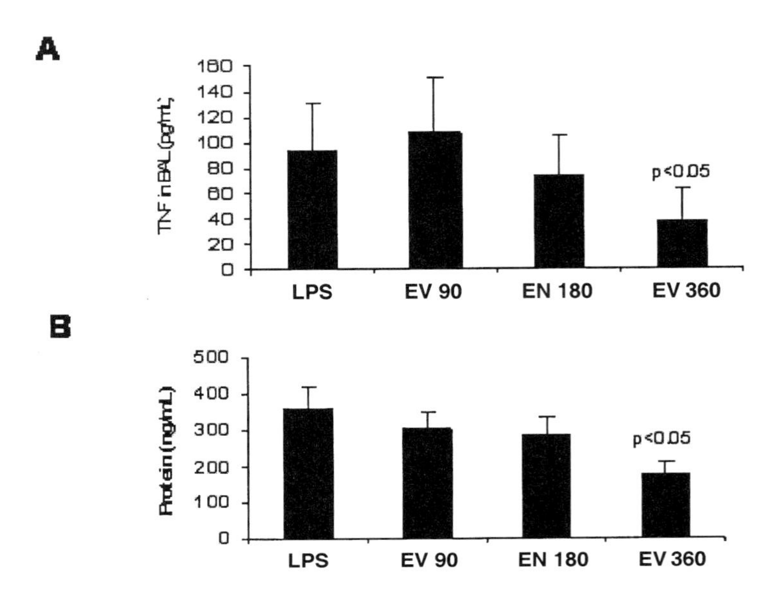

FIGURE 3. EV131 reduces LPS-induced TNF and protein levels in the BAL fluid. LPS was given at 1 μg by the intranasal route 1 h after EV131 at 90, 180, and 360 μg or saline. BAL was performed at 3 h. (**A**) TNF-α levels in the BAL fluid were quantified by ELISA; (**B**) total protein assessed by BioRad assay. Mean values and SD are given (n = 5, P <0.05).

performed histologic analysis of lung sections. Lungs from mice exposed to endotoxin showed peribronchial cellular infiltrates with neutrophils and peribronchial edema (FIG. 4A). The histamine scavenger EV131 inhibited neutrophil recruitment in the alveolar septae, peribronchial area, and alveoli. To quantify the neutrophils in the lung parenchyma, the number of neutrophils was assessed in high power field views of lung sections. Neutrophil counts were slightly reduced in lung sections after EV131 administration (FIG. 4B, P = 0.74). Myeloperoxidase activity of the fresh lung was then used as another parameter of lung neutrophil recruitment. EV131 significantly inhibited MPO activity (FIG. 4C, P <0.05). The data therefore reinforce the finding that histamine scavenging reduces endotoxin-induced neutrophil recruitment in the lung parenchyma and into BAL fluid.

DISCUSSION

Using a mouse model of ARDS, we show here that the neutralization of histamine by the histamine scavenger protein EV131 attenuates endotoxin-induced bronchoconstriction, TNF secretion, vascular leak and the recruitment of neutrophils in the lung and BAL, and local tissue damage. This appears to be the first report that histamine scavenging inhibits endotoxin induced lung pathology.

Histamine, first identified as a potent vasoactive amine, is now recognised for its multiple regulatory activities in the respiratory, digestive, immune and central ner-

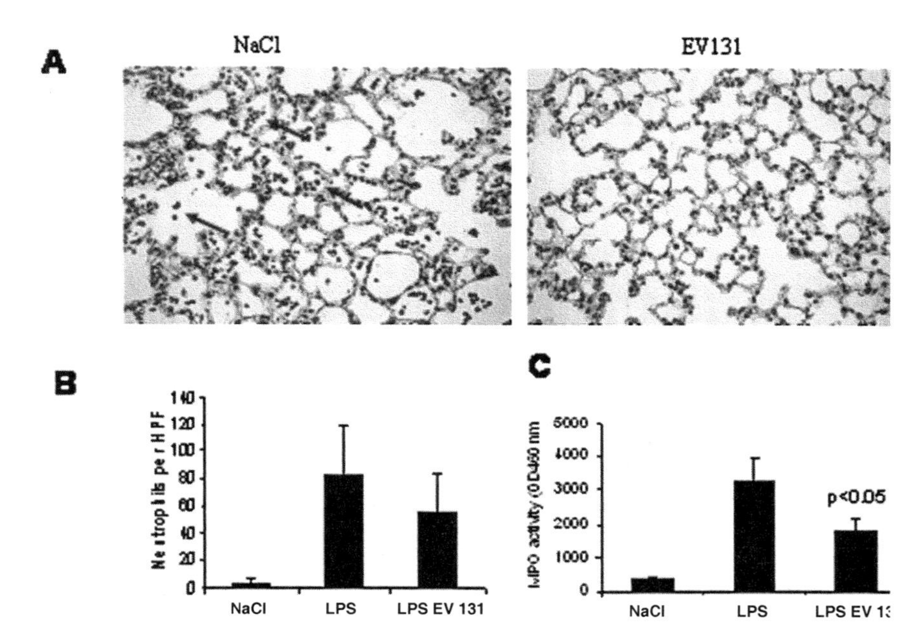

FIGURE 4. EV131 inhibits neutrophil recruitment in lung tissues upon endotoxin administration. (**A**) Endotoxin induced a distinct recruitment of neutrophils in the alveolar space inhibited by EV131. Lung sections were stained with H&E. (**B**) Quantification of neutrophils in 20 HPF per lung section ($P = 0.74$) and (**C**) by the MPO assay as described in the **Methods** section. Mean values and standard deviations ($n = 5$, $P < 0.05$).

vous systems.[4,7–11] There are at least two sources of histamine. The first is preformed histamine from mast cells and basophils which release histamine from intracellular granules in response to antigen mediated cross-linking of IgE receptors. In lungs, mast cells are present in bronchial walls near blood vessels and smooth muscle cells, and in the bronchial lumen.[12] The second source is de novo formation via the induction of histidine decarboxylase (HDC). The histamine newly formed via HDC induction is released without storage from neutrophils, blood platelets, dendritic cells [13], and T cells [11], in various organs or tissues such as liver, lung, spleen and bone marrow.

Histamine was one of the first inflammatory mediators of allergic asthma recognizd in human and guinea pig models.[14] Subsequent studies demonstrated that histamine modulates a variety of immune responses, including the production of inflammatory cytokines.[4,15] Indeed, a large elevation in HDC activity was seen in liver, lung, spleen, or bone marrow in response to inflammatory stimuli including bacterial products such LPS and peptidoglycans and to inflammatory cytokines such as IL-1a, IL-1b, TNF-α, IL-18, and IL-12.[16–20] These organs play important roles in innate immunity. Thus, the histamine that is formed in these organs via HDC induction may be involved in the processes implicated in innate immune responses.

Histamine exerts its effects through four receptor subtypes: HR1 and HR2, both expressed on lymphoid and nonlymphoid cells; HR3, mainly expressed in the brain[4]; and HR4, expressed in leukocytes.[21] It was recently shown that histamine, via HR4, can induce the release of IL-16 (a powerful granulocyte attractant) by $CD8^+$ cells.[22] In view of the critical role of histamine signaling in the inflammatory response, the beneficial effects of histamine scavenging by EV131 on the endotoxin-induced pulmonary inflammation modeling is not unexpected. In conclusion, we demonstrate that histamine scavenging by EV131 reduced endotoxin-induced acute inflammation of the lung, and therefore histamine blockade might represent a novel approach to the treatment or prevention of clinical ARDS.

REFERENCES

1. PAESEN, G.C. *et al.* 1999. Tick histamine-binding proteins: isolation, cloning, and three-dimensional structure. Mol. Cell. **3:** 661–671.
2. PAESEN, G.C. *et al.* 2000. Tick histamine-binding proteins: lipocalins with a second binding cavity. Biochim Biophys. Acta **1482:** 92–101.
3. COUILLIN, I, *et al.* 2004. Arthropod-derived histamine-binding protein prevents murine allergic asthma. J. Immunol. **173:** 3281–3286.
4. SCHNEIDER, E. *et al.* 2002. Trends in histamine research: new functions during immune responses and hematopoiesis. Trends Immunol. **23:** 255–263.
5. LEFORT, J., L. MOTREFF & B. B. VARGAFTIG. 2001. Airway administration of *Escherichia coli* endotoxin to mice induces glucocorticosteroid-resistant bronchoconstriction and vasopermeation. Am. J. Respir. Cell Mol. Biol. **24:** 345–351.
6. SCHNYDER-CANDRIAN, S. *et al.* 2005. Dual effects of p38 mapk on TNF-dependent bronchoconstriction and TNF-independent neutrophil recruitment in lipopolysaccharide-induced acute respiratory distress syndrome. J. Immunol. **175:** 262–269.
7. CARON, G. *et al.* 2001. Histamine polarizes human dendritic cells into Th2 cell-promoting effector dendritic cells. J. Immunol. **167:** 3682–3686.
8. CARON, G., *et al.* 2001. Histamine induces CD86 expression and chemokine production by human immature dendritic cells. J. Immunol. **166:** 6000–6006.
9. JUTEL, M. *et al.* 2002. Immune regulation by histamine. Curr. Opin. Immunol. **14:** 735–740.
10. JUTEL, M. *et al.* 2001. Histamine regulates T-cell and antibody responses by differential expression of H1 and H2 receptors. Nature **42413:** 420–425.
11. KUBO, Y. & K. NAKANO. 1999. Regulation of histamine synthesis in mouse $CD4^+$ and $CD8^+$ T lymphocytes. Inflamm. Res. **48:** 149–153.
12. METCALFE, D.D., D. BARAM & Y.A. MEKORI. 1997. Mast cells. Physiol. Rev. **77:** 1033–1079.
13. SZEBERENYI, J.B. *et al.* 2001. Intracellular histamine content increases during in vitro dendritic cell differentiation. Inflamm. Res. **50**(Suppl 2)**:** S112–113.
14. TOHDA, Y. *et al.* 1998. Roles of histamine receptor in a guinea pig asthma model. Int. J. Immunopharmacol. **20:** 565–571.
15. AKDIS, C.A. & K. BLASER. 2003. Histamine in the immune regulation of allergic inflammation. J. Allergy Clin. Immunol. **112:** 15–22.
16. ANDO, T. *et al.* 1994. Stimulation of the synthesis of histamine and putrescine in mice by a peptidoglycan of gram-positive bacteria. Microbiol. Immunol. **38:** 209–215.
17. ENDO, Y. 1982. Simultaneous induction of histidine and ornithine decarboxylases and changes in their product amines following the injection of Escherichia coli lipopolysaccharide into mice. Biochem. Pharmacol. **31:** 1643–1647.
18. ENDO, Y. 1983. Induction of histidine decarboxylase in mouse tissues by mitogens in vivo. Biochem. Pharmacol. **32:** 3835–3838.
19. WU, X. *et al.* 2004. Histamine production via mast cell-independent induction of histidine decarboxylase in response to lipopolysaccharide and interleukin-1. Int. Immunopharmacol. **4:** 513–520.

20. YAMAGUCHI, K., K. MOTEGI & Y. ENDO. 2000. Induction of histidine decarboxylase, the histamine-forming enzyme, in mice by interleukin-12. Toxicology **156:** 57–65.
21. ODA, T. *et al.* 2000. Molecular cloning and characterization of a novel type of histamine receptor preferentially expressed in leukocytes. J. Biol. Chem. **275:** 36781–36786.
22. GANTNER, F. *et al.* 2002. Histamine h(4) and h(2) receptors control histamine-induced interleukin-16 release from human CD8(+) T cells. J. Pharmacol. Exp. Ther. **303:** 300–307.

Curcumin: Getting Back to the Roots

SHISHIR SHISHODIA, GAUTAM SETHI, AND BHARAT B. AGGARWAL

Cytokine Research Laboratory, Department of Experimental Therapeutics, The University of Texas M. D. Anderson Cancer Center, Houston, Texas 77030, USA

ABSTRACT: The use of turmeric, derived from the root of the plant *Curcuma longa,* for treatment of different inflammatory diseases has been described in Ayurveda and in traditional Chinese medicine for thousands of years. The active component of turmeric responsible for this activity, curcumin, was identified almost two centuries ago. Modern science has revealed that curcumin mediates its effects by modulation of several important molecular targets, including transcription factors (e.g., NF-κB, AP-1, Egr-1, β-catenin, and PPAR-γ), enzymes (e.g., COX2, 5-LOX, iNOS, and hemeoxygenase-1), cell cycle proteins (e.g., cyclin D1 and p21), cytokines (e.g., TNF, IL-1, IL-6, and chemokines), receptors (e.g., EGFR and HER2), and cell surface adhesion molecules. Because it can modulate the expression of these targets, curcumin is now being used to treat cancer, arthritis, diabetes, Crohn's disease, cardiovascular diseases, osteoporosis, Alzheimer's disease, psoriasis, and other pathologies. Interestingly, 6-gingerol, a natural analog of curcumin derived from the root of ginger (*Zingiber officinalis*), exhibits a biologic activity profile similar to that of curcumin. The efficacy, pharmacologic safety, and cost effectiveness of curcuminoids prompt us to "get back to our roots."

KEYWORDS: **curcumin; antioxidant; anti-tumor; curcumin analogs**

INTRODUCTION

The turmeric (*Curcuma longa*) plant, a perennial herb belonging to the ginger family, is cultivated extensively in south and southeast tropical Asia. The rhizome of this plant is also referred to as the "root" and is the most useful part of the plant for culinary and medicinal purposes. The most active component of turmeric is curcumin, which makes up 2–5% of the spice. The characteristic yellow color of turmeric is due to the curcuminoids, first isolated by Vogel in 1842. Curcumin is an orange-yellow crystalline powder practically insoluble in water. The structure of curcumin ($C_{21}H_{20}O_6$) was first described in 1910 by Lampe and Milobedeska and shown to be diferuloylmethane.[1]

Turmeric is used as a dietary spice, coloring agent in foods and textiles, and a treatment for a wide variety of ailments. It is widely used in traditional Indian medicine to cure biliary disorders, anorexia, cough, diabetic wounds, hepatic disorders, rheumatism, and sinusitis. Turmeric paste in slaked lime is a popular home remedy

Address for correspondence: Bharat B. Aggarwal, Cytokine Research Laboratory, Department of Experimental Therapeutics, The University of Texas M. D. Anderson Cancer Center, Box 143, 1515 Holcombe Boulevard, Houston, TX 77030. Voice: 713-792-3503/6459; fax: 713-794-1613. aggarwal@mdanderson.org

Ann. N.Y. Acad. Sci. 1056: 206–217 (2005). © 2005 New York Academy of Sciences.
doi: 10.1196/annals.1352.010

for the treatment of inflammation and wounds. For centuries, curcumin has been consumed as a dietary spice at doses up to 100 mg/day. Recent phase I clinical trials indicate that human beings can tolerate a dose as high as 8 g/day with no side effects.[2] The focus of this review is to describe the effect of curcumin in various diseases.

DISEASE TARGETS OF CURCUMIN

Ancient texts of Indian medicine describe the use of curcumin for a wide variety of inflammatory diseases including sprains and swellings caused by injury, wound healing, and abdominal problems.[3] Texts on traditional medicine in China describe the uses of curcumin for the treatment of diseases that are associated with abdominal pain. There are over 1,500 citations in Medline relating to the biologic effect of curcumin. Perhaps most of the activities associated with curcumin are based on its ability to suppress inflammation. Curcumin has been shown to be effective in acute as well as chronic models of inflammation.

Antiinflammatory and Antioxidant Properties. Several studies have shown that curcumin is a potent antioxidant (FIG. 1). In fact, curcumin has been found to be at least 10 times more active as an antioxidant than even vitamin E.[4] Curcumin prevents the oxidation of hemoglobin and inhibits lipid peroxidation (for references see Ref. 1). The antioxidant activity of curcumin could be mediated through antioxidant enzymes such as superoxide dismutase, catalase, and glutathione peroxidase. Curcumin has been shown to serve as a Michael acceptor reacting with glutathione and thioredoxin 1.[5] Reaction of curcumin with these agents reduces intracellular GSH in

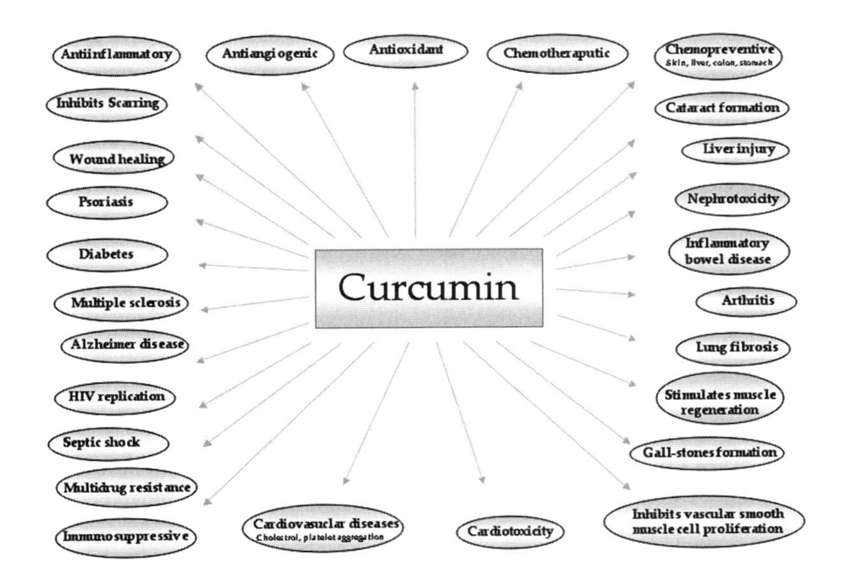

FIGURE 1. Disease targets of curcumin.

the cells. The suppression of lipid peroxidation by curcumin could lead to suppression of inflammation.

Anticancer Properties. The anticancer potential of curcumin in various systems was recently reviewed.[1] Curcumin has been shown to block transformation, tumor initiation, tumor promotion, invasion, angiogenesis, and metastasis. *In vivo,* curcumin suppresses carcinogenesis of the skin, forestomach, colon, and liver in mice. Curcumin also suppresses mammary carcinogenesis. Curcumin has been shown to inhibit the proliferation of a wide variety of tumor cells, including B-cell and T-cell leukemia, colon carcinoma, epidermoid carcinoma, and various breast carcinoma cells.

Cardioprotective Effects. Curcumin has been effective against atherosclerosis and myocardial infarction.[6] The proliferation of peripheral blood mononuclear cells (PBMCs) and vascular smooth muscle cells (VSMCs), which are hallmarks of atherosclerosis, is inhibited by curcumin. Curcumin prevents the oxidation of low-density lipoproteins (LDLs), inhibits platelet aggregation, and reduces the incidence of myocardial infarction.

Skin Diseases. Curcumin has been shown to be effective against different skin diseases including skin carcinogenesis, psoriasis,[7] scleroderma,[8] and dermatitis. Numerous reports suggest that curcumin accelerates wound healing. In addition, curcumin also prevents the formation of scars and plays a role in muscle regeneration following trauma.[6]

Diabetes. In type II diabetes, administration of curcumin reduced the blood sugar, hemoglobin, and glycosylated hemoglobin levels significantly in an alloxan-induced diabetic rat model. Diabetic rats maintained on a curcumin diet for 8 weeks excreted less albumin, urea, creatinine, and inorganic phosphorus. Dietary curcumin also partially reversed the abnormalities in plasma albumin, urea, creatine, and inorganic phosphorus in diabetic animals.[6]

Rheumatoid Arthritis. Curcumin has also been shown to possess antirheumatic and antiarthritic effects, most likely through the downregulation of COX2, tumor necrosis factor (TNF), and other inflammatory cytokines.[6]

Multiple Sclerosis. Multiple sclerosis is characterized by the destruction of oligodendrocytes and myelin sheath in the CNS. Curcumin inhibits experimental allergic encephalomyelitis by blocking interleukin (IL)-12 signaling in T cells, suggesting it would be effective in the treatment of multiple sclerosis.[6]

Alzheimer's Disease. Curcumin can suppress oxidative damage, inflammation, cognitive deficits, and amyloid accumulation in Alzheimer's disease.[9]

Inflammatory Bowel Disease. Ukil *et al.*[10] recently investigated the protective effects of curcumin on inflammatory bowel disease induced in a mouse model. Pretreatment of mice with curcumin for 10 days significantly ameliorated the appearance of diarrhea and the disruption of the colonic architecture.

Cystic Fibrosis. Cystic fibrosis, the most common lethal hereditary disease in the white population, is caused by mutations in the cystic fibrosis transmembrane conductance regulator gene. In a recent report, Egan *et al.*[11] demonstrated that curcumin corrected the cystic fibrosis defects in DeltaF508 CF mice.

Others. Curcumin was found to be a potent and selective inhibitor of human immunodeficiency virus (HIV-1) long-terminal repeat-directed gene expression, which governs the transcription of type 1 HIV-1 provirus. It has also been shown to prevent cataractogenesis in an *in vitro* rat model. Treatment with curcumin also prevented experimental alcoholic liver disease. Curcumin has a protective effect on cyclophosphamide-induced early lung injury. Nephrotoxicity, a problem observed in patients who are administered chemotherapeutic agents, can be prevented with curcumin.[6]

MOLECULAR TARGETS OF CURCUMIN

Various studies have shown that curcumin modulates numerous targets (FIG. 2). These include the growth factors, growth factor receptors, transcription factors, cytokines, enzymes, and genes regulating apoptosis.

Cytokines and Growth Factors. Numerous growth factors have been implicated in the growth and promotion of tumors. Curcumin has been shown to downregulate the expression of several cytokines including TNF, IL-6, IL-8, IL-12, and fibroblast growth factor-2.[6]

Receptors. Curcumin has been shown to downregulate both epithelial growth factor receptor (EGFR) and HER2/neu receptors. It also modulates androgen receptors.[6]

Transcription Factors. Curcumin may also operate through suppression of various transcription factors including NF-κB, STAT3, Egr-1, AP-1, PPAR-γ, and beta catenin activation.[6] These transcription factors play an essential role in various diseases. The constitutively active form of NF-κB has been reported in a wide variety of cancers. NF-κB is required for the expression of genes involved in cell proliferation, cell invasion, metastasis, angiogenesis, and resistance to chemotherapy. Bharti *et al.*[12] demonstrated that curcumin inhibited IL-6–induced STAT3 phosphorylation

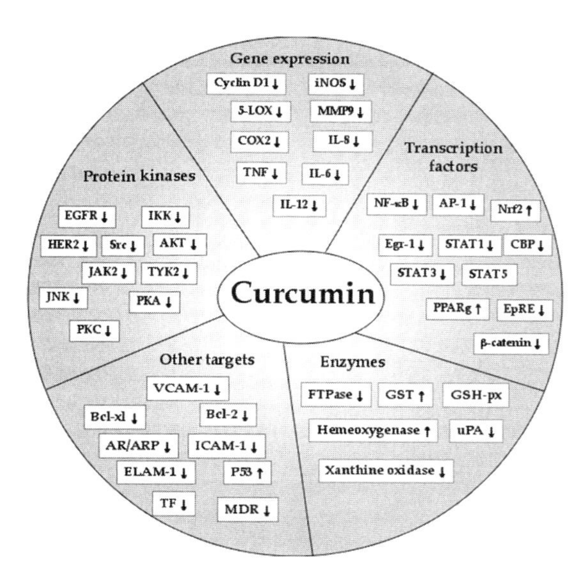

FIGURE 2. Molecular targets of curcumin.

and consequent STAT3 nuclear translocation. Activation of PPAR-γ inhibits the proliferation of nonadipocytes. Xu *et al.*[13] demonstrated that curcumin dramatically induced the gene expression of PPAR-γ and activated PPAR-γ. AP-1, another transcription factor that has been closely linked with proliferation and transformation of tumor cells, has been shown to be suppressed by curcumin. Studies also suggest that curcumin has a potential therapeutic effect on prostate cancer cells through downregulation of AR and AR-related cofactors.[6]

Proinflammatory Enzymes. Curcumin has been shown to suppress the expression of COX2, 5-LOX, and iNOS, most likely through the downregulation of NF-κB activation.[6]

Protein Kinases. Curcumin suppresses a number of protein kinases including mitogen-activated protein kinases, JNK, PKA, PKC, src tyrosine kinase, phosphorylase kinase, IκBα kinase, JAK kinase, and the growth factor receptor protein tyrosine kinases.[6]

Cell Cycle. Curcumin modulates cell-cycle–related gene expression. Specifically, curcumin induced G0/G1 and/or G2/M phase cell cycle arrest, upregulated CDKIs, p21WAF1/CIP1, p27KIP1, and p53, and slightly downregulated cyclin B1 and cdc2. We found that curcumin can indeed downregulate cyclin D1 expression[14–16] at the transcriptional and posttranscriptional levels.

Adhesion Molecules. Curcumin inhibits inflammation by blocking the adhesion of monocytes to endothelial cells by inhibiting the activation of the cell adhesion molecules ICAM-1, VCAM-1, and ELAM-1.[6]

Antiapoptotic Proteins. Curcumin induces apoptosis by inducing cytochrome *c* release, Bid cleavage, and caspase 9 and 3 activation and by downregulating the antiapoptotic proteins Bcl-2 and $BclX_L$.[1]

Multidrug Resistance. Multidrug resistance is associated with decreased drug accumulation in tumor cells due to increased drug efflux. Curcumin downregulates drug resistance by inhibiting the expression of the mdr gene, which is responsible for this phenomenon.[6]

LESSONS LEARNED FROM SYNTHETIC ANALOGS OF CURCUMIN

To elucidate which portion of the molecule is critical for the activity, a large number of structural analogs of curcumin have been synthesized (FIG. 3A). Some analogs are more active than native curcumin, whereas others are less active[17–32] (TABLE 1). It was found that the phenolic analogs were more active than the nonphenolic analogs.[33] The highest antioxidant activity was obtained when the phenolic group was sterically hindered by the introduction of two methyl groups at the ortho position. The phenolic group is essential for free radical scavenging activity, and the presence of the methoxy group further increases the activity.[34] Curcumin shows both antioxidant and pro-oxidant effects. Ahsan *et al.*[35] have shown that both of these effects are determined by the same structural moieties of the curcuminoids.

Dinkova-Kostova and Talalay[30] showed that the presence of hydroxyl groups at the ortho-position on the aromatic rings and the beta-diketone functionality were required for high potency in inducing Phase 2 detoxification enzymes. Curcumin is a noncompetitive inhibitor of rat liver microsomal delta 5 desaturase and delta 6 desaturase. Kawashima *et al.*[36] have shown that only half the structure is essential

TABLE 1. Relative potency of curcumin and its synthetic analogs

Effects	References
Analogs more potent than curcumin	
THC: lipid peroxidation under aqueous condition by pulse radiolysis technique	17
HC: preventing nitrite-induced oxidation of haemoglobin	18
NaC: carrageenin-induced rat hind paw edema	19
HMBME: inhibition of prostate cancer	20
BJC005, CHC011, and CHC007: formation of Fos-/Jun- DNA complex	21
Tocopheryl curcumin: inhibiting Tat transactivation of HIV-LTR	22
4,4′-DAC : histamine blocking activity	23
Copper chelates of 2-hydroxynapthyl curcumin: antitumor activity	24
Hydrazinocurcumin: BAECs proliferation	25
o-hydroxy substituted analog: inhibiting alcohol and PUFA induced oxidative stress	26
Di-O-glycinoyl curcumin and 2′-deoxy-2'-curcuminyl uridine: antiviral activity	27
Pyrazole and isoxazole analogs: Cox-2 inhibitory activity	28
1,7-bis-(2-hydroxy-4-methoxyphenyl)-1,6-heptadiene-3,5-dione): AL activity	29
Salicylcurcuminoid: antioxidant	
Analogs less potent than curcumin	
THC: lipid peroxidation under aerated condition by pulse radiolysis technique	17
THC: TPA-induced mouse ear edema and skin carcinogenesis	30
Analogs as potent as curcumin	
5-hydroxy-1,7-diphenyl-1,4,6-heptatriene-3-one: Scavenge hydroxyl radicals	31
Manganese complexes of curcumin and diacetylcurcumin: Scavenge hydroxyl radicals	32

ABBREVIATIONS: THC, tetrahydrocurcumin; NaC, sodium curcuminate; HMBME, 4-hydroxy-3-methoxybenzoic acid methyl ester; DAC, diacetylcurcumin; BAEC, bovine aortic endothelial cells; PUFA, thermally oxidized sunflower oil; Cox-2, cyclooxygenase-2; AL, anti-leishmanial.

for desaturase inhibition. A 3-hydroxy group of the aromatic ring is essential for the inhibition, and a free carboxyl group at the end opposite the aromatic ring interferes with the inhibitory effect.

Simon *et al.*[37] found that the presence of the diketone moiety in the curcumin molecule seems to be essential for its ability to inhibit the proliferation of MCF-7 human breast tumor cells. The aromatic enone and dienone analogs of curcumin were demonstrated to have potent antiangiogenic properties in an *in vitro* SVR assay.[38]

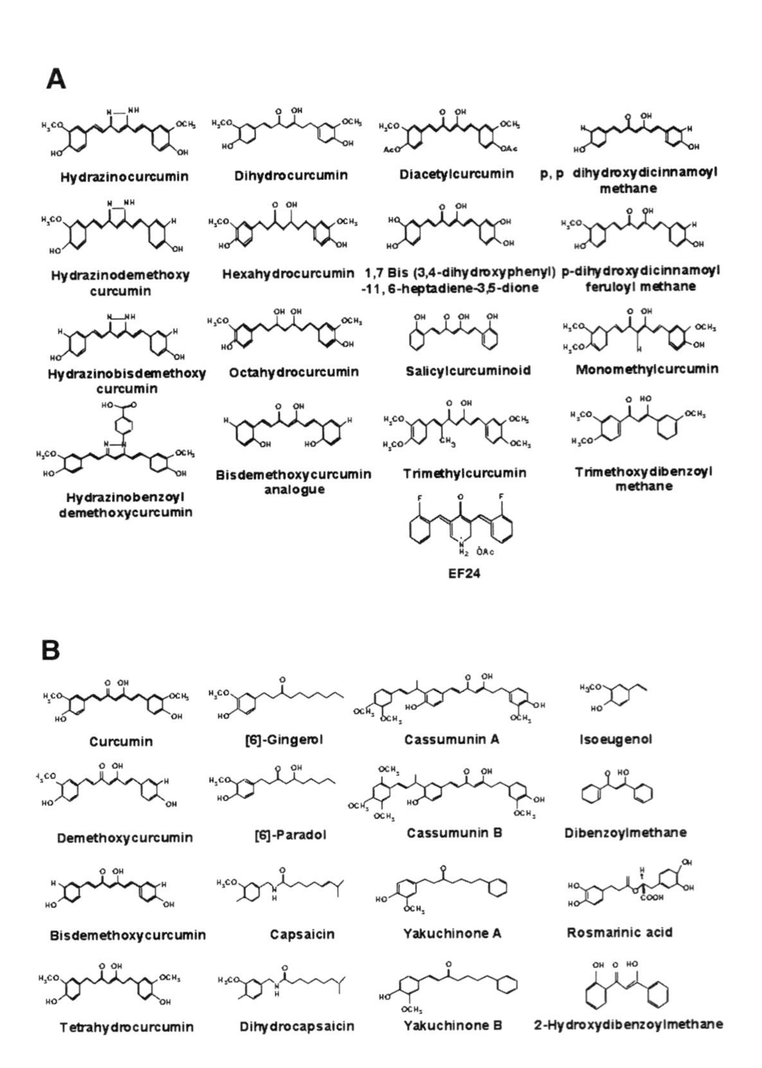

FIGURE 3. Structure of various analogs of curcumin. Hydrazinocurcumin,[25] hydrazinodemethoxycurcumin,[25] hydrazinobisdemethoxycurcumin,[25] hydrazinobenzoyldemethoxycurcumin,[25] dihydrocurcumin,[50] hexahydrocurcumin,[51] octahydrocurcumin,[51] bisdemethoxycurcumin,[52] diacetylcurcumin,[51] salicylcurcuminoid,[30] monomethylcurcumin,[53] trimethylcurcumin,[53] and 7–bis(3,-4-dihydroxyphenyl)11,6-heptadiene-3,5-dione.[53]

TABLE 2. Sources and site of action of natural analogs of curcumin[a]

Analogs	Source	Target	Ref.
6-Gingerol	Ginger (*Zingiber officinale* Roscoe)	TNF, NF-κB, AP-1, COX2, ODC, iNOS, p38MAPK, antifungal	39
8-Gingerol	Ginger (*Zingiber officinale* Roscoe)		39
6-Paradol	Ginger (*Zingiber officinale* Roscoe)	Caspase activation	40
Shogoal	Ginger (*Zingiber officinale* Roscoe)	*Helicobacter pylori*	41
Cassumunin A&B	Ginger (*Zingiber cassumunar*)	Antioxidant	42
Diarylheptanoids	Ginger (*Zingiber spp.*)	PGE2 and LT	43
Dibenzoylmethane	Licorice (*Glycyrrhiza echinata*)	COX2, LOX, HIF, VEGF	43
Galanals A&B	Zingiber (*Zingiber mioga* Roscoe)	Caspase 3, bcl2	44
Garcinol	Kokum (*Garcinia indica*)	NF-κB, COX-2, iNOS, HAT	45
Isoeugenol	Cloves (*Eugenia caryophyllus*)	NF-κB, antioxidant β-amyloid	46
Yakuchinone A&B	Galanga (*Alpinia officinarum*)	PG synthetase, COX2, iNOS, NF-kB, insecticidal, adhesion molecules, TNF, AP-1, 5-HETE	47

[a]For structure of these analogs, see FIGURE 3.

NATURAL ANALOGS OF CURCUMIN

Natural curcumin contains three major curcuminoids, namely, curcumin, demethoxycurcumin, and bisdemethoxycurcumin (FIG. 3B). Several analogs of curcumin have been identified from other plant sources. These include 6- and 8-gingerol, 6-paradol, cassumunin, galanals, diarylheptanoids, yakuchinones, isoeugenol, and dibenzoylmethane. Like curcumin, gingerol, paradol, cassumunin, shogaol, and diarylheptanoids are also derived from the roots of the plant (TABLE 2).[39–47] Although most of these analogs exhibit activities very similar to those of curcumin, whether they are more potent or less potent than curcumin has not been established. Yakuchinones[48] have been shown to be more potent inhibitors of 5-HETE production than curcumin. Synthetic cassumunins also show stronger protective activity than curcumin against oxidative cell death induced by hydrogen peroxide.[42] Garcinol is more potent than curcumin in inhibiting tumor cells.[45] The anticancer potential of galanals, however, is comparable to that of curcumin.[44] Curcumin has been shown to be more cytotoxic than isoeugenol, bis-eugenol, and eugenol.[49]

CONCLUSION

The medicinal properties of curcumin and its analogs have been known to mankind for ages. Modern science has now provided a scientific basis to the numerous reports of the medicinal effects of these most inexpensive, yet pharmacologically safe, polyphenols. Extensive research in the last few years has indicated that most diseases are caused by the dysregulation of multiple signaling pathways, thus casting doubt on how effective monotherapy against single targets will prove to be. Curcumin and its analogs have been found to attack multiple targets, which provides the basis for their effectiveness in so many different diseases. Although most of the NSAIDS are now either withdrawn or survive with black box warnings, curcumin is one that is not known to show any adverse effects, even at doses as high as 8 g a day. Thus, a trip back to our "roots" to explore the "roots" of *Curcuma longa* as a source for better treatments will certainly prove productive. As Hippocrates said almost 25 centuries ago, "let food be thy medicine and medicine be thy food."

ACKNOWLEDGMENTS

We would like to thank Walter Pagel for a careful review of the manuscript. Dr. Aggarwal is a Ransom Horne Jr. Distinguished Professor of Cancer Research. This work was supported in part by the Odyssey Program and the Theodore N. Law Award for Scientific Achievement at The University of Texas M. D. Anderson Cancer Center (to S.S.).

REFERENCES

1. Aggarwal, B.B., A. Kumar & A.C. Bharti. 2003. Anticancer potential of curcumin: preclinical and clinical studies. Anticancer Res. **23:** 363–398.
2. Cheng, A.L., C.H. Hsu, J.K. Lin, *et al.* 2001. Phase I clinical trial of curcumin, a chemopreventive agent, in patients with high-risk or pre-malignant lesions. Anticancer Res. **21:** 2895–2900.
3. Ammon, H.P. & M.A. Wahl. 1991. Pharmacology of *Curcuma longa*. Planta Med. **57:** 1–7.
4. Khopde, S.M., K.I. Priyadarsini, P. Venkatesan, *et al.* 1999. Free radical scavenging ability and antioxidant efficiency of curcumin and its substituted analogue. Biophys. Chem. **80:** 85–91.
5. Adams, B.K., J. Cai, J. Armstrong, *et al.* 2005. EF24, a novel synthetic curcumin analog, induces apoptosis in cancer cells via a redox-dependent mechanism. Anticancer Drugs **16:** 263–275.
6. Aggarwal, B.B., A. Kumar, M.S. Aggarwal, *et al.* 2005. Curcumin derived from turmeric (*Curcuma longa*): a spice for all seasons. *In* Phytochemicals in Cancer Chemoprevention. P.D. Debasis Bagchi & H.G. Preuss, Eds. :349–387. CRC Press. New York.
7. Heng, M.C., M.K. Song, J. Harker, *et al.* 2000. Drug-induced suppression of phosphorylase kinase activity correlates with resolution of psoriasis as assessed by clinical, histological and immunohistochemical parameters. Br. J. Dermatol. **143:** 937–949.
8. Tourkina, E., P. Gooz, J.C. Oates, *et al.* 2004. Curcumin-induced apoptosis in scleroderma lung fibroblasts: role of protein kinase cepsilon. Am. J. Respir. Cell Mol. Biol. **31:** 28–35.

9. YANG, F., G.P. LIM, A.N. BEGUM, *et al.* 2005. Curcumin inhibits formation of amyloid beta oligomers and fibrils, binds plaques, and reduces amyloid *in vivo*. J. Biol. Chem. **280:** 5892–5901.
10. UKIL, A., S. MAITY, S. KARMAKAR, *et al.* 2003. Curcumin, the major component of food flavour turmeric, reduces mucosal injury in trinitrobenzene sulphonic acid-induced colitis. Br. J. Pharmacol. **139:** 209–218.
11. EGAN, M.E., M. PEARSON, S.A. WEINER, *et al.* 2004. Curcumin, a major constituent of turmeric, corrects cystic fibrosis defects. Science **304:** 600–602.
12. BHARTI, A.C., N. DONATO & B.B. AGGARWAL. 2003. Curcumin (diferuloylmethane) inhibits constitutive and IL-6-inducible STAT3 phosphorylation in human multiple myeloma cells. J. Immunol. **171:** 3863–3871.
13. XU, J., Y. FU & A. CHEN. 2003. Activation of peroxisome proliferator-activated receptor-gamma contributes to the inhibitory effects of curcumin on rat hepatic stellate cell growth. Am. J. Physiol. Gastrointest. Liver Physiol. **285:** G20–30.
14. BHARTI, A.C., N. DONATO, S. SINGH, *et al.* 2003. Curcumin (diferuloylmethane) down-regulates the constitutive activation of nuclear factor-kappa B and IkappaBalpha kinase in human multiple myeloma cells, leading to suppression of proliferation and induction of apoptosis. Blood **101:** 1053–1062.
15. MUKHOPADHYAY, A., C. BUESO-RAMOS, D. CHATTERJEE, *et al.* 2001. Curcumin down-regulates cell survival mechanisms in human prostate cancer cell lines. Oncogene **20:** 7597–7609.
16. MUKHOPADHYAY, A., S. BANERJEE, L.J. STAFFORD, *et al.* 2002. Curcumin-induced suppression of cell proliferation correlates with donregulation of cyclin D1 expression and CDK4-mediated retinoblastoma protein phosphorylation. Oncogene **21:** 8852–8862.
17. KHOPDE, S.M., K.I. PRIYADARSINI, S.N. GUHA, *et al.* 2000. Inhibition of radiation–induced lipid peroxidation by tetrahydrocurcumin: possible mechanisms by pulse radiolysis. Biosci. Biotechnol. Biochem. **64:** 503–509.
18. VENKATESAN, P., M.K. UNNIKRISHNAN, S.M. KUMAR, *et al.* 2003. Effect of curcumin analogues on oxidation of haemoglobin and lysis of erythrocytes. Curr. Sci. **84:** 74–78.
19. RAO, T.S., N. BASU & H.H. SIDDIQUI. 1982. Anti-inflammatory activity of curcumin analogues. Indian J. Med. Res. **77:** 574–578.
20. KUMAR, A.P., G.E. GARCIA, R. GHOSH, *et al.* 2003. 4-Hydroxy-3-methoxybenzoic acid methyl ester: a curcumin derivative targets Akt/NF kappa B cell survival signaling pathway: potential for prostate cancer management. Neoplasia **5:** 255–266.
21. HAHM, E.R., G. CHEON, J. LEE, *et al.* 2002. New and known symmetrical curcumin derivatives inhibit the formation of Fos-Jun-DNA complex. Cancer Lett. **184:** 89–96.
22. BARTHELEMY, S., L. VERGNES, M. MOYNIER, *et al.* 1998. Curcumin and curcumin derivatives inhibit Tat-mediated transactivation of type 1 human immunodeficiency virus long terminal repeat. Res. Virol. **149:** 43–52.
23. DOUGLAS, D.E. 1993. 4,4′-Diacetyl curcumin: *in-vitro* histamine-blocking activity. J. Pharm. Pharmacol. **45:** 766.
24. JOHN, V.D., G. KUTTAN & K. KRISHNANKUTTY. 2002. Anti-tumour studies of metal chelates of synthetic curcuminoids. J. Exp. Clin. Cancer Res. **21:** 219–224.
25. SHIM, J.S., D.H. KIM, H.J. JUNG, *et al.* 2002. Hydrazinocurcumin, a novel synthetic curcumin derivative, is a potent inhibitor of endothelial cell proliferation. Bioorg. Med. Chem. **10:** 2439–2444.
26. RUKKUMANI, R., K. ARUNA, P.S. VARMA, *et al.* 2004. Comparative effects of curcumin and an analog of curcumin on alcohol and PUFA induced oxidative stress. J. Pharm. Pharm. Sci. **7:** 274–283.
27. MISHRA, S., S. TRIPATHI & K. MISRA. 2002. Synthesis of a novel anticancer prodrug designed to target telomerase sequence. Nucleic Acids Res. (Suppl.) **2:** 277–278.
28. SELVAM, C., S.M. JACHAK, R. THILAGAVATHI, *et al.* 2005. Design, synthesis, biological evaluation and molecular docking of curcumin analogues as antioxidant, cyclooxygenase inhibitory and anti-inflammatory agents. Bioorg. Med. Chem. Lett. **15:** 1793–1797.
29. GOMES DDE, C., L.V. ALEGRIO, M.E. DE LIMA, *et al.* 2002. Synthetic derivatives of curcumin and their activity against Leishmania amazonensis. Arzneimittelforschung **52:** 120–124.

30. DINKOVA-KOSTOVA, A.T. & P. TALALAY. 1999. Relation of structure of curcumin analogs to their potencies as inducers of Phase 2 detoxification enzymes. Carcinogenesis **20:** 911–914.
31. TONNESEN, H.H. & J.V. GREENHILL. 1992. Studies on curcumin and curcuminoids XXII: curcumin as a reducing agent and as a radical scavenger. Int. J. Pharm. **87:** 79–87.
32. VAJRAGUPTA, O., P. BOONCHOONG & L.J. BERLINER. 2004. Manganese complexes of curcumin analogues: evaluation of hydroxyl radical scavenging ability, superoxide dismutase activity and stability towards hydrolysis. Free Radic. Res. **38:** 303–314.
33. VENKATESAN, P. & M.N. RAO. 2000. Structure-activity relationships for the inhibition of lipid peroxidation and the scavenging of free radicals by synthetic symmetrical curcumin analogues. J. Pharm. Pharmacol. **52:** 1123–1128.
34. SREEJAYAN, N. & M.N. RAO. 1996. Free radical scavenging activity of curcuminoids. Arzneimittelforschung **46:** 169–171.
35. AHSAN, H., N. PARVEEN, N.U. KHAN, *et al.* 1999. Pro-oxidant, anti-oxidant and cleavage activities on DNA of curcumin and its derivatives demethoxycurcumin and bisdemethoxycurcumin. Chem. Biol. Interact. **121:** 161–175.
36. KAWASHIMA, H., K. AKIMOTO, S. JAREONKITMONGKOL, *et al.* 1996. Inhibition of rat liver microsomal desaturases by curcumin and related compounds. Biosci. Biotechnol. Biochem. **60:** 108–110.
37. SIMON, A., D.P. ALLAIS, J.L. DUROUX, *et al.* 1998. Inhibitory effect of curcuminoids on MCF-7 cell proliferation and structure-activity relationships. Cancer Lett. **129:** 111–116.
38. ROBINSON, T.P., T. EHLERS, I.R. HUBBARD, *et al.* 2003. Design, synthesis, and biological evaluation of angiogenesis inhibitors: aromatic enone and dienone analogues of curcumin. Bioorg. Med. Chem. Lett. **13:** 115–117.
39. KIM, S.O., J.K. KUNDU, Y.K. SHIN, *et al.* 2005. [6]-Gingerol inhibits COX-2 expression by blocking the activation of p38 MAP kinase and NF-kappaB in phorbol ester-stimulated mouse skin. Oncogene **24:** 2558–2567.
40. KEUM, Y.S., J. KIM, K.H. LEE, *et al.* 2002. Induction of apoptosis and caspase-3 activation by chemopreventive [6]-paradol and structurally related compounds in KB cells. Cancer Lett. **177:** 41–47.
41. MAHADY, G.B., S.L. PENDLAND, G.S. YUN, *et al.* 2003. Ginger (*Zingiber officinale* Roscoe) and the gingerols inhibit the growth of Cag A+ strains of *Helicobacter pylori*. Anticancer Res. **23:** 3699–3702.
42. MASUDA, T., H. MATSUMURA, Y. OYAMA, *et al.* 1998. Synthesis of (+/–)-cassumunins A and B, new curcuminoid antioxidants having protective activity of the living cell against oxidative damage. J. Nat. Prod. **61:** 609–613.
43. HONG, J., M. BOSE, J. JU, *et al.* 2004. Modulation of arachidonic acid metabolism by curcumin and related beta-diketone derivatives: effects on cytosolic phospholipase A(2), cyclooxygenases and 5-lipoxygenase. Carcinogenesis **25:** 1671–1679.
44. MIYOSHI, N., Y. NAKAMURA, Y. UEDA, *et al.* 2003. Dietary ginger constituents, galanals A and B, are potent apoptosis inducers in human T lymphoma Jurkat cells. Cancer Lett. **199:** 113–119.
45. PAN, M.H., W.L. CHANG, S.Y. LIN-SHIAU, *et al.* 2001. Induction of apoptosis by garcinol and curcumin through cytochrome c release and activation of caspases in human leukemia HL-60 cells. J. Agric. Food Chem. **49:** 1464–1474.
46. CHAINY, G.B., S.K. MANNA, M.M. CHATURVEDI, *et al.* 2000. Anethole blocks both early and late cellular responses transduced by tumor necrosis factor: effect on NF-kappaB, AP-1, JNK, MAPKK and apoptosis. Oncogene **19:** 2943–2950.
47. CHUN, K.S., J.Y. KANG, O.H. KIM, *et al.* 2002. Effects of yakuchinone A and yakuchinone B on the phorbol ester-induced expression of COX-2 and iNOS and activation of NF-kappaB in mouse skin. J. Environ. Pathol. Toxicol. Oncol. **21:** 131–139.
48. FLYNN, D.L., M.F. RAFFERTY & A.M. BOCTOR. 1986. Inhibition of 5-hydroxy-eicosatetraenoic acid (5-HETE) formation in intact human neutrophils by naturally-occurring diarylheptanoids: inhibitory activities of curcuminoids and yakuchinones. Prostaglandins Leukot. Med. **22:** 357–360.

49. FUJISAWA, S., T. ATSUMI, M. ISHIHARA, *et al.* 2004. Cytotoxicity, ROS-generation activity and radical-scavenging activity of curcumin and related compounds. Anticancer Res. **24:** 563–569.
50. PAN, M.H., T.M. HUANG & J.K. LIN. 1999. Biotransformation of curcumin through reduction and glucuronidation in mice. Drug Metab. Dispos. **27:** 486–494.
51. ISHIDA, J., H. OHTSU, Y. TACHIBANA, *et al.* 2002. Antitumor agents. Part 214: synthesis and evaluation of curcumin analogues as cytotoxic agents. Bioorg. Med. Chem. **10:** 3481–3487.
52. SRIVIVASAN, A., V.P. MENON, V. PERIASWAMY, *et al.* 2003. Protection of pancreatic beta-cell by the potential antioxidant bis-o-hydroxycinnamoyl methane, analogue of natural curcuminoid in experimental diabetes. J. Pharm. Pharm. Sci. **6:** 327–333.
53. OHTSU, H., Z. XIAO, J. ISHIDA, *et al.* 2002. Antitumor agents. 217. Curcumin analogues as novel androgen receptor antagonists with potential as anti-prostate cancer agents. J. Med. Chem. **45:** 5037–5042.

Transcription Factor NF-κB

A Sensor for Smoke and Stress Signals

KWANG SEOK AHN AND BHARAT B. AGGARWAL

Cytokine Research Laboratory, Department of Experimental Therapeutics, The University of Texas M. D. Anderson Cancer Center, Houston, Texas 77030, USA

ABSTRACT: **Nuclear factor-kappa B (NF-κB) is a transcription factor that resides in the cytoplasm of every cell and translocates to the nucleus when activated. Its activation is induced by a wide variety of agents including stress, cigarette smoke, viruses, bacteria, inflammatory stimuli, cytokines, free radicals, carcinogens, tumor promoters, and endotoxins. On activation, NF-κB regulates the expression of almost 400 different genes, which include enzymes (e.g., COX-2, 5-LOX, and iNOS), cytokines (such as TNF, IL-1, IL-6, IL-8, and chemokines), adhesion molecules, cell cycle regulatory molecules, viral proteins, and angiogenic factors. The constitutive activation of NF-κB has been linked with a wide variety of human diseases, including asthma, atherosclerosis, AIDS, rheumatoid arthritis, diabetes, osteoporosis, Alzheimer's disease, and cancer. Several agents are known to suppress NF-κB activation, including Th2 cytokines (IL-4, IL-13, and IL-10), interferons, endocrine hormones (LH, HCG, MSH, and GH), phytochemicals, corticosteroids, and immunosuppressive agents. Because of the strong link of NF-κB with different stress signals, it has been called a "smoke-sensor" of the body.**

KEYWORDS: **NF-κB; stress; smoke; gene expression; cancer**

WHAT IS NF-κB?

Nuclear transcription factor κB (NF-κB) was identified by David Baltimore in 1986 as a factor in the nucleus that binds the promoter of the kappa chain of immunoglobulins in B cells.[1] NF-κB has since been shown to be present in the cytoplasm of every cell type in its inactive state and is conserved in animals all the way from Drosophila to man. Five different mammalian NF-κB family members have been identified and cloned: NF-κB1 (p50/p105), NF-κB2 (p52/p100), RelA(p65), RelB, and c-Rel. All family members share a highly conserved Rel homology domain (RHD; ~300 aa) responsible for DNA binding, a dimerization domain, and the ability to interact with IκBs, the intracellular inhibitor for NF-κB. Two different NF-κB activation pathways have been identified, a canonical pathway initiated by NF-κB1 (p50/p105) and a noncanonical pathway initiated by NF-κB2

Address for correspondence: Bharat B. Aggarwal, Cytokine Research Laboratory, Department of Experimental Therapeutics, The University of Texas M. D. Anderson Cancer Center, Box 143, 1515 Holcombe Boulevard, Houston, TX 77030. Voice: 713-792-3503/6459; fax: 713-794-1613. aggarwal@mdanderson.org

Ann. N.Y. Acad. Sci. 1056: 218–233 (2005).
doi: 10.1196/annals.1352.026

(p52/p100). Before the NF-κB complex is translocated into the nucleus, NF-κB1 and NF-κB2 are cleaved to the active p50 and p52 subunits, respectively.

In resting cells, NF-κB, consisting of p50 and RelA, is sequestered in the cytoplasm in an inactive form through its association with one of several inhibitory molecules, including IκB-α, IκB-β, IκB-γ, p105, and p100, among which IκB-α is the most abundant. In response to environmental stimuli, including cytokine/chemokines, viral and bacterial pathogens, and stress-inducing agents, inactive NF-κB/IκB complex is activated by phosphorylation on two conserved serine (S) residues within their N-terminal domain of IκB proteins. Phosphorylation of these conserved S residues in response to stimulators leads to the immediate polyubiquitination of IκB proteins by the SCF-β-TrCP complex (FIG. 1). This modification subsequently targets IκB proteins for rapid degradation by the 26S proteasome.

Activation of the NF-κB signaling cascade results in complete degradation of IκB, allowing the translocation of NF-κB to the nucleus, where it induces transcription. Activated NF-κB binds to specific DNA sequences in target genes, designated as κB-elements, and regulates transcription of over 400 genes involved in immunoregulation, growth regulation, inflammation, carcinogenesis, and apoptosis.

WHAT ACTIVATES NF-κB?

Extensive research in the last two decades has shown that a large number of stimuli can activate NF-κB (TABLE 1). These include bacteria and fungi, bacterial and fungal products, viruses and viral proteins, inflammatory cytokines, parasites, mitogens, physiological stress, physical stress, oxidative stress, environmental and occupational particles, heavy metals, intracellular stresses, viral or bacterial products, UV light, X-rays, gamma radiation, chemotherapeutic agents, carcinogens, cigarette smoke, hydrogen peroxide, colony-stimulating factors, mechanical stress, psychological fear, Th1 cytokines, hypoxia and hyperoxia, chemotherapeutic agents, endotoxins, and tumor promoters. The diversity of the stimuli that can stimulate NF-κB activation suggests that it can be used as a "smoke-detector" or "stress-sensor."

The mechanisms by which these diverse stimuli activate NF-κB are not identical. Perhaps the best understood of these pathways is the tumor necrosis factor (TNF)-induced NF-κB activation pathway (FIG. 1). The sequential recruitment of TNFR, TRADD, TRAF2, RIP, and IKK leads to TNF-induced NF-κB activation.[2] Recent work from our laboratory has implicated ras,[3] syk,[4] and β-GSK[5] in TNF-induced NF-κB activation. Others have implicated AKT,[6] MEK3,[7] and FAK.[8] TNF-induced NF-κB activation is mediated through the production of reactive oxygen species as SOD[9] and γ-GCS[10] inhibited the activation. Numerous studies have indicated that NF-κB activated by several agents, however, differs from that of TNF.[11,12] For example, we have shown that NF-κB activated by pervanadate[13,14] and hydrogen peroxide[12] differs from that activated by TNF. Others have shown that activation of NF-κB by hypoxia,[15] UV,[16] γ-radiation,[17] X-rays,[18] ds RNA,[19] erythropoietin,[20] and hepatitis C virus[21] differs significantly from that activated by TNF. Although activation of NF-κB by most agents requires the activation of IκBα kinase (IKK), activation of NF-κB by UV, X-ray, hypoxia, pervanadate, erythropoietin, H_2O_2, and hepatitis C virus (NS5A) has been shown to be IKK-independent.

Table. 1 A list of inducers of NF-κB

Bacterial & Fungi	Staphylococcus	THANK	**Environmental Hazards**	Maleylated Bovine Serum	Activin A	Uremic toxins
Anaplasma	enterotoxin A and B	TNFα	Arsenic	Albumin	Adenosine	Very Low density Lipoproteins
phagocytophilum	(super antigen)	TNFβ	3,3',4,4'-tetrachlorobiphenyl	Modified (Oxidized)LDL	Adrenomedullin	(VLDL)
Angiostrongylus	Toxic Shock Syndrome	TWEAK (TNF-like weak	(PCB77)	Neurotrophin Receptor	AILb-A (from Aeginetia indica)	Violacein
cantonensis	Toxin 1	inducer of apoptotis)	Benzo[a]pyrene diol epoxide	proteolytic fragments	Albumin	Vitamin D_3
Bacteroides forsythus	Wogonin (Scutellaria		Chromium	Non-amyloid beta component	Allergin	
Bartonella henselae	baicalensis)	**Physiological (Stress)**	Cigarette smoke	of Alzheimer's disease	Alloxan	**Chemical Agents**
Bordetella pertussis		**Conditions**	Cigarette smoke condensate		Amino acid analogs	2-Deoxyglucose
Chlamydia pneumoniae	**Viruses**	Acute lung	Cobalt	**Overexpressed Proteins**	Anaphylatoxin C3a	Adriamycin
Ehrlichia chaffeensis	Adenovirus	injury/respiratory distress	Crocidolite asbestos fibers	CFTR	Anaphylatoxin C5a	Alumnum
EPEC,	Cytomegalovirus	syndrome	Dicamba (herbicide,	Erythropoietin-Receptor	Angiotensin II	Anisomycin
enterophathogenic E.	Epstein-Barr Virus (EBV)	Adhesion	peroxisome proliferator)	Hematopoietic progenitor	Antiphospholipid antibodies	Benzyl isothiocyanate
coli	Hepatitis B Virus	Angina pectoris	Diesel exhaust particles	kinase 1	Areca nut extract	Bisperoxovanadium (bpV)
Fusobacterium	Herpes Virus Saimiri	Antiphospholipid	Gliadin (from wheat)	Ig heavy chain	Baicalin (plant compound)	phosphotyrosyl phosphatase
nucleatum	Human Herpesvirus 6	antibodies	House dust mite	MHC Class I	Basic calcium phosphate crystals	inhibitors
Gardnerella vaginalis	HIV-1	Appendicitis	Dust particles		Bronchoalveolar lavage fluid	Brefeldin A
Helicobacter pylori	Herpex Simplex Virus -1	Asthma	Fear-potentiated startle	**Receptor Ligands**	Bradykinin	Cadmium
Lactobacilli	HTLV-I	Butter and Walnut diet	response	Antigen (IgM-Ligand)	Brusatol	Calchicine
Listeria monocytogenes	Influenza Virus	Cecal ligatin and puncture	Iron	BAFF (B cell-activating factor)	beta-carotene	Calcium Ionophores
Mycoplasma fermentans	Measles Virus	(mouse)	Lead	Beta-D-glucan ligand	Catalase	Calyculin A
Mycobacteria	Molony Murine Leukemia	Crohn's disease/ulcerative	Lead chromate	CD11b/CD18-Ligand	C2-Ceramide (N-acetyl-sphingosine)	Ceramide-beta-galactose
tuberculosis	Virus	colitis	Manganese	(Complement)	Cerulein	2-chloroethyl ethyl sulfide (mustard
Neisseria gonorrhoeae	mRNA (in vitro	Corneal epidemic	Nickel	CD28-Ligand (B7-1)	Chelidonium majus extract	analog)
Neisseria meningiditis	transcribed)	keratoconjunctivitis	Noise	CD2-Ligand	Collagen lattice	Cobalt chloride
Porphyromonas	Newcastle disease virus	Coronary artery by-pass	Oily fly ash	CD35-Ligand (Complement)	Collagen Type I	Con A
gingivalis	Respiratory Syncytial	Depolarization	Silica Particles	CD3-Ligand	Cryptdins	Cycloheximide
Prevotella intermedia	Virus	Hemorrhage	Wood smoke	CD40-Ligand	Cysteinyl leukotrines	Cyclopiazonic Acid
Pseudomonas	Rhinovirus	Hypercholesterolemia	Zymosan (yeast cell wall	CD43-Ligand	Deoxycholic acid (bile acid)	Diquat
aerogenosa	Sendai paramyxovirus	Hyperglycemia	product)	CD4-Ligand (gp120)	Des-Arg10-kallidin (B1 receptor	2,4-dinitrofluorobenzene
Rhodococcus equi	Sindbis Virus	Hyperhomocysteinemia		CD66a-Ligand	agonist)	Ethanol
Rickettsia rickettsii	Vaccinia Virus Akara	Hyperosmotic Shock	**Therapeutically used drugs**	M3 Cholinergic receptor	1,25-dihydroxycholecalciferol	Ferrocene
Salmonella dublin	West Nile Flavavirus	Hyperoxia	ABR-25757 (oxo-quinoline-	agonist	Double-stranded polynucleotides	Forskolin
Salmonella typhimurium		Ischemia (transient, focal)	3-carboxamide)	Fc-2a-Receptor-Ligand (IgG2a)	f-Met-Leu-Phe	Gadolinium chloride
Shigella flexneri	**Viral products**	Ischemic preconditioning	2-(1-adamantylamino)-6-	Flt-1-Ligand	Fibrinogen	Glass fibers
Staphyllococcus aureus	Adenovirus 5: E1A	Liver Regeneration	methylpyridine (AdAMP)	Heat shock protein 60 (HSP60)	Free fatty acids	HDAC inhibitors (sodium butyrate
Streptococci (group A)	Adenovirus: E3/19K	Human labor (childbirth)	1-b-D-Arabinofuranosyl-	Ly6A/E-Ligand	Heat shock protein 25 (Hspb1)	and trichostatin A)
Streptococcus (group B)	African Swine Fever	Mechanical Ventilation (in	cytosine (ara-C))	N-CAM	Heat shock protein 60 (HSP 60)	Linoleic acid
Streptomyces	Virus: IAP	vitro)	Anthralin	PGG-Glucan (Betafectin)	Hemoglobin	L-NMA
californicus (fungus)	Antibody to Dengue Virus	Muscle disuse	Azidothymidine (AZT)	Sphingosine 1-phosphate	Homocysteine	Lysophosphatidic acid
Trichomonas vaginalis	nonstructural protein 1	Muscular Dystrophy (type	Baicalein	Trail-receptor-1-Lignad (Trail)	Hyaluronan	Malondialdehyde
Ureaplasma urealyticum	CMV: ie1	2A)	Bleomycin	Trail-receptor-2-Ligand (Trail)	12(R)-Hydroxyeicosatrienoic acid	MDMA ("Ecstacy")
Yersinia enterocolitica	Double-stranded RNA	Neuronal firing	Bryostatin-1	Trail-receptor-4-Ligand (Trail)	6-hydroxydopamine	MEN 17055 (disaccharide

Bacterial or Fungal Products
- Apicularen A
- CpG
- Cytolysin (Vibrio vulnificus)
- Diphosphoryl lipid A (Rhodobacter sphaeroides)
- Enterotoxin (Bateroides fragilis)
- Exotoxin B
- Fimbria protein ATTLE (P gingivalis)
- Fumonisin B1 (Fusarium verticillioides)
- G(Anh) M Tetra (E coli)
- Glycosylphosphatidylinositols (Plasmodium falciparum)
- Lipoprotein LpK (Mycobacterioum leprae)
- Lipoteichoic acid (Listeria)
- Lipopolysaccharide (LPS)
- Membrane lipoproteins (Mycoplasma fermentans)
- Membrane lipoproteins (Mycoplasma penetrans)
- Muramyl Peptides
- Mycobacterium lipoarabinomannan
- PlcA (Phospholipase) (Listeria)
- PlcB (Phospholipase) (Listeria)
- Porins (Gram negative bacteria)
- Porin 1B (Gonoccocus)
- EBV: EBNA-2
- EBV: LMP1
- HBV: HBx
- HBV: LHBs
- HBV: MHBst
- HCV: Core protein
- Herpes Saimiri: HVS13
- Herpes Saimiri:StpC
- HIV-1: gp160
- HIV-1: Nef
- HIV-1: p9 (9 aa peptide)
- HIV-1: Tat
- HTLV-I: Tax
- HTLV-II: Tax
- Influenza Virus: Hemagglutinin
- Parvovirus B19: NSI
- SV40: small T-antigen

Eukaryotic parasites
- Leishmania (lipophosphglycan)
- Phospholipomannan (C. albicans)
- Pneumocystis
- Theileria parva
- Trypanoplasma borreli

Cytokines and Cytokine Receptors
- CD30
- Salivary cystatins (SA1 and SA2)
- IL-1
- IL-2
- IL-12
- IL-15
- IL-17
- IL-18
- LIF
- p43
- Pentraxin-3
- S100B
- Overventilation (perfused lungs)
- Pancreatitis
- Proteinuria
- Reoxygenation
- Rheumatoid arthritis
- Senescence (keratinocytes)
- Shear Stress
- Neuoronal trimethyltin injury
- Uni-axial cyclic cell stretching
- T-cell selection

Physical Stress
- Bile duct ligation
- Cyclic mechanical muscle strain
- Exercise
- Gamma Radiation
- Heavy ion irradiation
- Laminar shear stress
- PPME Photosensitization
- Ultraviolet irradiation (UV-A, B, C)
- Mechanical lung ventilation
- Obesity
- Wounding combined with HeNe irradiation
- Wounding combined with thermal irradiation

Oxidative Stress
- Butyl Peroxide
- Cerulein
- Glutathione
- Hydrogen Peroxide
- Ozone
- Peroxynitrite
- Pervanadate
- Reoxygenation
- Bucillamine metabolite SA 981
- Camptothecin
- Celecoxib
- Ciprofibrate
- Cisplatin
- Cycloprodigiosin
- Dacarbazine
- Daio-Orengedeokuto
- Daunomycin
- Daunorubicin
- Diazoxide
- 5,6-dimethylxanthenone-4-acetic acid
- Doxorubicin
- Etoposide
- Flavone-8-acetic acid
- Haloperidol
- Kunbi-Boshin-Hangam-Tang
- Lithium
- Methamphetamine
- Mitoxantrone
- Norepinephrine
- Olitipraz
- Phenobarbital
- Protocatechuic acid (from herb radix Salviae miltiorrhizae)
- SN38 (metabolite of CPT-11)
- Tamoxifen
- Taxol (Paclitaxel)
- Vinblastine
- Vincristine
- WR1065

Modified Proteins
- Advanced glycated end products (AGEs)
- Amyloid Protein Fragment (bA4)
- Anti-PR3
- Glycosylated oxyhaemaglobin

Apoptotic Mediators
- Anti-Fas/Apo-1
- Poly(ADP) Ribose Polymerase (PARP)
- Trail

Mitogens, growth factors and hormones
- Bone morphogenic protein 2
- Bone morphogenic protein 4
- Cortical Releasing Hormone
- Epidermal Growth Factor
- Folicle Stimulating Hormone
- Gastrin
- GMCSF
- Hepatocyte Growth Factor
- Human Growth Hormone
- Insulin
- Insulin-like growth factor 1
- Lysophosphatidic acid
- M-CSF
- Mullerian Inhibiting Substance
- Nerve Growth Factor
- Pigment epithelium-derived factor (PEDF)
- Platelet Activating Factor (PAF)
- Platelet-Derived Growth Factor
- Plant steroids (diosgenin, hecogenin, tigogenin)
- Prostatin
- All-trans retinoic acid
- RET/PTC3 Fusion oncoprotein
- S100B
- Serum
- Sulphatide (L-selectin crosslinker)
- TGF-alpha
- TGF-beta2

Physiological Mediators
- Acrp30/adiponectin
- hPepT1 (apical di-/tripeptide transporter
- Kaianic acid (Kainate)
- Leukotriene B4
- L-Glutamate
- Long-term potentiation (LTP)
- Lysophosphatidylcholine (LysoPC)
- Mixed meal ingestion (hi glucose)
- Monosodium urate crystals
- Neuormelanin
- Neutrophil elastase
- Nitric oxide
- NS-398 (high dose)
- Oleic acid
- Osteopontin
- PAF (platelet activating factor)
- Palmitate
- PCSC (polysaccharide from Poria cocos)
- Phellinus linteus proteoglycan
- Platelet type arachidonate 12-lipoxygenase
- Polysaccharides of Poria cocos
- Potassium
- Prolactin N-terminal fragment (16K PRL)
- Proteolysis-inducing factor (PIF)
- Regulatory RNA
- Rev-erbalpha
- S100B
- Saturated fatty acids
- Sleep deprivation
- St John's Wort (hyperforin)
- Streptozotocin
- Substance P
- Tauroursodeoxycholic acid (TUDCA)
- T-cell costimulatory receptor 4-1BB
- Thioredoxin
- Thrombin
- Titanium and copper implants
- Trypsin (SLIGRL)
- Tuberous sclerosis complex
- anthracycline)
- Monensin
- N-methyl-D-aspartate
- Mycophenolic acid
- Nafenopin
- Nickel sulfate
- Nicotine
- N-nitrosomorphine
- Nocodazol
- Okadaic Acid
- Peplomycin
- PHA
- Phorbol ester
- Phosphodiester CpG DNAs
- Podophyllotoxin
- Prostratin (a phorbol ester)
- Pyrogallol
- Quercetin (high concentrations)
- Quinolinic acid
- Saflower polysaccharides
- Sanglifehrin A
- Staurosporine
- Thapsigargin
- Transglutaminase 2
- Tunicamycin
- Vinblastine
- WF10WY-14 643 (peroxisome proliferator)

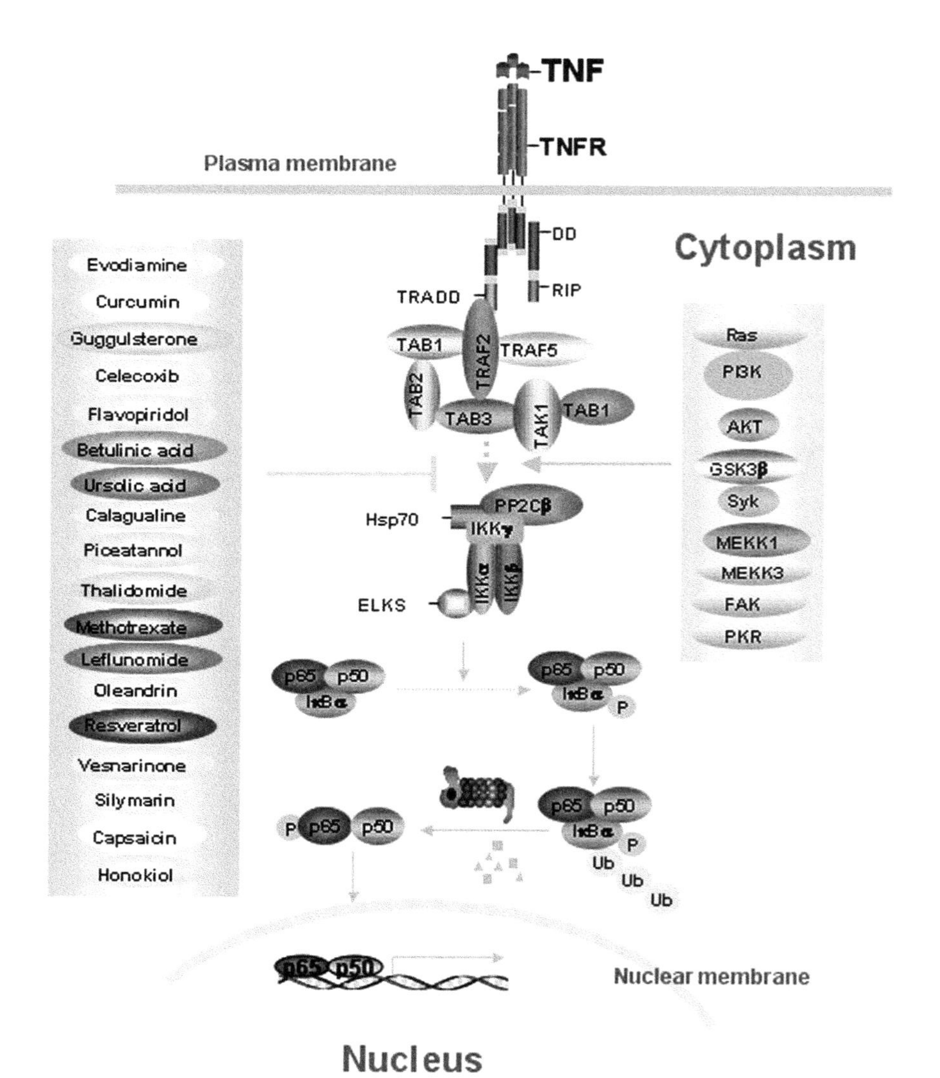

FIGURE 1. Schematic pathway for TNF-induced NF-κB activation and its inhibition by various natural products.

WHAT GENES ARE REGULATED BY NF-κB?

Although initially identified in kappa chain of immunoglobulin, the NF-κB binding sequences have now been identified in over 400 different genes (TABLE 2). These include inflammatory cytokines (e.g., TNF, IL-1, IL-6, and chemokines), adhesion molecules, inflammatory enzymes (e.g., COX-2, 5-LOX), viral proteins, telomerase, angiogenesis proteins (VEGF), antiapoptotic proteins, and cell cycle–regulatory

Table. 2 A list of target genes of NF-κB-regulated

Cytokines/Chemokines	**Immunoreceptors**	**Cell adhesion molecules**	Bradikinin B1-receptor	
CCL5*	B7.1	E-selectin	Amiloride-sensitive sodium channel	**Enzymes**
CCL15/Leukotactin	BRL-1*	Endoglin	A1 adenosine receptor	Liver alcohol dehydrogenase
CCL22	CCR5	Fibronectin		Collagenase 1
CCL28	CCR7	ICAM-1*	**Regulators of apoptosis**	Glutathione S-transferase
CINC-1*	CD137	MadCAM-1*	TRAF-1*	Hyaluronan synthase
CXCL 11*	CD154	P-selectin, tenascin-C	TRAF-2*	H+-K+ATPase α2
Eotaxin	CD40	VCAM-1*	IEX-1L*	Lysozyme
Fractalkine	CD40 ligand	DC-SIGN*	IAPs*	Matrix metalloproteinaase-9
Gro a-g	CD48		Fas-ligand	GD3-synthase
Gro-1	CD83	**Acute phase response proteins**	CD95 (Fas)	Gelatinase B
ICOS*	Fc epsilon receptor II		c-FLIP	PIM-1
IFN-g	IL-2 receptor α-chain	Angiotensinogen	Nr13	PKCδ
IL-1α	Immunoglobulin Cγ1	β-defensin-2	Caspase-11	Phospholipase C δ1
IL-1β	IgG γ4	C4b binding protein	Bcl-2	Serpin 2A
IL-1 receptor antagonist	Immunoglobulin epsilon heavy chain	Complement factor B	Bcl-x_L	Transglutaminase
IL-2	Immunoglobulin k light chain	Complement factor C4	Bfl1/A1	TIRT*
IL-6	Invariant chain II	C-reactive protein		
IL-8	MHC class I (H-2Kb)	Lipopolysaccharide binding protein	**Viruses**	**Miscellaneous**
IL-9	MHC Class I HLA-B7	Pentraxin PTX3	Adenovirus (E3 region)	α-1 acid glycoprotein
IL-10	β_2-microglobulin	SAA1 and SAA2*	Avian leukosis virus	Apolipoprotein C III
IL-11	Nod2	Tissue factor-1	Bovine leukemia virus	AMH*
IL-12 (p40)	Polymeric Ig receptor	Urokinase-type plasminogen activator	Cytomegalovirus	Cyclin D1
IL-13	T-cell receptor β chain		Epstein-Barr virus (Wp promoter)	Factor VIII
IL-15				

β-Interferon
IP-10*
KC*
ENA-78 (CXCL5)
GCP-2 (CXCL6)
Lymphotoxin α
Lymphotoxin β
MCP-1/JE*
MIP-1α,β*
MIP-2
MIP-3α
mob-1
Neutrophil activating peptide-78
RANTES*
TCA3*
TNFα
TNFβ
TRAIL*
TFF3*
T-cell receptor/CD3γ
p80 TNF-receptor
Complement B
Complement component 3
Complement receptor 2
Proteasome subunit LMP2
Peptide transporter TAP1
Tapasin

Growth Factors

Bone morphogenic protein-2
Granulocyte colony stimulating factor
Granulocyte macrophage colony stimulating factor
Erythropoietin, macrophage colony stimulating factor (M-CSF)
Neurokinin-1 receptor
Hepatocyte growth factor
Platelet-derived growth factor B chain
Proenkephalin
Vascular endothelial growth factor

Stress-response genes

Angiotensin II
Cytochrome p450 gene
COX-2*
Ferritin H chain
12-Lipoxygenase
iNOS*
Mn SOD*
NQO1*
Phospholipase A2

Cell surface receptors

RAGE- receptor for advanced glycation end products
Platelet activator receptor-1
Neuropeptide Y Y1-receptor
Mu-opioid receptor
Mdr1*
Lox-1*
Gal1 receptor
CD69
Hepatitis B virus (pregenomic promoter)
HIV-1
HSV*
JC virus
Human papillomavirus type 16
SIV*
SV-40*

Transcription/ growth control factors

A20
Androgen receptor
c-myb
c-myc
IRF-1*
IRF-2
IRF-4
IRF-7
Rel/NF-κB proteins (p52/p100, p50/p105, c-Rel, and RelB)
IκB proteins (IκBα, IκBβ, Bcl-3, JunB, Stat5a, WT1, p53, Ras)
Gadd45β
Galectin 3
Epsilon-globin
HMG-14*
K3 keratin
Laminin B2 chain
Mts1
MUC-2
Perforin
Pregnancy-specific glycoprotein mCGM3
Prostate-specific antigen
S100A6 (calcyclin)
Syndecan-4
Vimentin
Wilm's tumor suppressor gene
a1-antitrypsin,

---

*CCL5, C-C chemokine ligand 5; CINC-1, cytokine-induced neutrophil chemoattractant-1; CXCL 11, CXC chemokine ligand 11; ICOS, inducible co-stimulator; IP-10, IFN-gamma-inducible protein 10; KC, kupffer cells; MCP-1, monocyte chemoattractant protein-1; MIP, macrophage inflammatory protein; RANTES, regulated upon activation, normal T-cell expressed and secreted; TCA3, T cell activation; TRAIL, tumor necrosis factor-related apoptosis-inducing ligand; TFF3, trefoil factor 3; ICAM-1, intercellular adhesion molecule-1; MadCAM-1, mucosal addressin cell adhesion molecule; VCAM-1, vascular cell adhesion molecule; DC-SIGN, dendritic cell surface C-type lectin; SAA, serum amyloid A proteins; COX-2, cyclooxygenase-2; iNOS, inducible nitric oxide synthase; Mn SOD, superoxide dismutase; NQO1, NAD(P)H quinone oxidoreductase 1; Mdr1, Multiple drug resistance mediator 1; Lox-1, lectin-like oxidized low-density lipoprotein receptor-1; TRAF, TNF-receptor associated factor; IEX-1L, immediate early response factor-1; IAPs, Inhibitor of apoptosis; HSV, Herpes simplex virus; SIV, Simian immunodeficiency virus; SV-40, Simian virus 40; IRF, Interferon regulatory factor; TIRT, Telomerase catalytic subunit; AMH, Anti-mullerian hormone; HMG-14, High mobility group 14.

genes. Besides NF-κB, other transcription factors may modulate the expression of these genes. Microarray analysis has added even more genes to the list of those regulated by NF-κB.[22,23]

WHICH DISEASES ARE LINKED TO NF-κB ACTIVATION?

Constitutive NF-κB activation has now been shown to contribute to the pathogenesis of a large number of diseases (TABLE 3). These include cancer, diabetes, allergy, rheumatoid arthritis, Crohn's disease, cardiovascular diseases, atherosclerosis, Alzheimer's disease, muscular dystrophy, cardiac hypertrophy, catabolic disorders, hypercholesterolemia, ischemia/reperfusion, angina pectoris, acid-induced lung injury disease, renal disease, gut diseases, skin diseases, incontinentia pigmenti, appendicitis, pancreatitis, peritonitis, sepsis, silica-induced disease, sleep apnea, autoimmunity, lupus erythematosus, psychosocial stress diseases, neuropathological diseases, familial amyloid polyneuropathy, Parkinson's disease, Huntington's disease, and retinal disease. NF-κB activation has also been linked with the human aging process.

A constitutive NF-κB has been detected in most tumor cell types including esophageal cancer, laryngeal cancer, pharyngeal cancer, renal cancer, colon cancer, head and neck squamous carcinoma, lung cancer, bladder cancer, acute myelogenous leukemia, non-Hodgkin's lymphoma, B-cell lymphoma, adult T-cell leukemia, T-cell lymphoma, mantle cell lymphoma, multiple myeloma, acute lymphoblastic leukemia, cervical cancer, nasopharyngeal carcinoma, melanoma, thyroid cancer, liver cancer, breast cancer, ovarian cancer, and prostate cancer.[24,25] NF-κB can mediate transformation, proliferation, invasion, and angiogenesis of tumor cells. Mutated ras found in several tumors has been shown to activate NF-κB. Chemoresistance and radioresistance have also been linked to NF-κB activation. The *p*-glycoprotein linked to drug-resistance is also regulated by NF-κB. Similarily, COX-2 overexpressed in most tumors is also regulated by NF-κB. Cyclin D1, overexpressed by most tumors and required for G_1 to S transition, is also regulated by NF-κB. Similarily, VEGF and adhesion molecules required for angiogenesis and metastasis are also regulated by NF-κB.

Many inflammatory genes relevant to the pathogenesis of atherosclerosis are regulated by NF-κB, the activated form of which is present in atherosclerotic plaques. NF-κB has been shown to be activated in atherosclerosis and myocarditis, in association with angina, during transplant rejection, after ischemia/reperfusion, in congestive heart failure, in dilated cardiomyopathy, after ischemic and pharmacological preconditioning, in heat shock, in burn trauma, and in hypertrophy of isolated cardiomyocytes.

Bronchial asthma is one of the most common chronic diseases in modern society and yet, despite the availability of highly effective drugs, there is increasing evidence to suggest that its incidence is increasing. The pathogenesis of asthma involves persistent expression of a broad array of genes, which contain the κB site for NF-κB within their promoters, suggesting that NF-κB plays a pivotal role in the initiation and perpetuation of allergic inflammation.

Several reports suggest that amyloid β peptide can activate NF-κB in neurons, indicating a plausible mechanism by which amyloid may act during the pathogenesis

Table. 3 A list of NF-κB-mediated diseases

Ageing	Acid-induced lung inury disease (COPD)	Silica-induced
Headaches	Renal Disease	Sleep apnoea
Pain	Leptospiriosis renal disease	AIDS (HIV-1)
Cardiac hypertrophy	Gut Diseases	Autoimmunity
Muscular hystrophy (type 2A)	Skin Diseaes	Lupus
Catabolic disorders	Incontinentia pigmenti	Psychosocial stress diseases
Diabetes, Type 1	Asthma	Neuropathological diseases
Diabetes, Type 2	Arthritis	Familial amyloidotic polyneuropathy, inflamm neuropathy
Hypercholesterolemia	Crohns disease	Parkinson disease
Atherosclerosis	Ocular allergy	Alzheimers disease
Heart disease	Appendicitis	Huntington's disease
Chronic heart failure	Pancreatitis	Retinal disease
Ischemia/reperfusion	Periodonitis	Cancer
Angina pectoris	Inflammatory bowel disease	
Pulmonary disease	Sepsis	

Table. 4 A list of inhibitors of NF-κB*

Cytokine & Hormones	Aged garlic extract (allicin)	Nordihydroguaiaritic acid	Glucorticoid-induced leucine zipper	Compound 26**
Interleukin-4[+]	Anetholdithiolthione	Oleandrin+	protein	Cycloepoxydon
Interleukin-10	Anethole+	Orthophenanthroline	γ-glutamylcysteine synthetase[+]	Cyclolinteinone
Interleukin-11	Apocynin	Parthenolide	Heat shock protein 72	Cycloprodigiosin
Interleukin-13[+]	Apple juice	PDTC**	HSCO**	hycrochloride Dehydroxymethylepo
Growth hormone	Astaxanthin	Phenolic antioxidants (Hydroquinone	Losartin	xyquinomicin
HBEGEF**	Baicalein	and tert-butyl hydroquinone)	MnSOD**[+]	Diamide[+]
hCG**	Benidipine	Phenolic antioxidants**	NDPP1 (CARD protein)	Diarylheptanoid 7-(4'-hydroxy-3'-
Luteinizing hormone[+]	Betulinic acid[+]	Phenylarsine oxide (PAO, tyrosine	NF-2 protein	methoxyphenyl)-1-phenylhept-4-en-
α-MSH**	bis-eugenol	phosphatase inhibitor)	NLS cell permeable peptides	3-one 3-ditriazine)
Somatomammotropin	Butylated hydroxyanisole		p202a**	Dimethylfumarate
VEGF**	Caffeic Acid Phenethyl Ester (3,4-	**Phytochemicals**	Pioglitazone (PPARγ ligand)	Dioxin[+]
Estrogen	dihydroxycinnamic acid, CAPE)	Piceatannol[+]	Pituitary adenylate cyclase-activating	Disulfiram
Glucocorticoids	Caffeic Acid Phenethyl Ester [+]	PMC (2,2,5,7,8-pentamethyl-6-	polypeptide	E-73 (cycloheximide analog)
PG-15-deoxy-Δ(12,14)-PGJ(2)**	Calagualine[+]	hydroxychromane)	Protein-bound polysaccharide	Ecabet sodium
Prostaglandin A1	Capsaicin[+]	PMC**	PTEN	Epoxyquinone A monomer Fibrates
Prostaglandin E2	Carnosol	Polysaccharides	Suppressors of cytokine signaling-1	Erythromycin
	Carvedilol	Pyrrolinedithiocarbamate (PDTC)	Triglyceride-rich lipoproteins	Fosfomycin
Antiinflammatory agents	Catalposide	Quercetin	Vasoactive intestinal peptide	Flunixin meglumine
Acetaminophen	Catechol Derivatives	Quercetin (low concentrations)	ZAS3 protein**	Gangliosides
Aspirin (sodium salicylate)	Cepharanthine	Red wine		Gabexate mesilate
Flurbiprofen	Conophylline	Redox factor 1	**Stress**	Geldanamycin
Ibuprofen	Curcumin[+]	Ref-1 (redox factor 1)	Carbon monoxide	Glimepiride
Leflunamide metabolite**[+]	Dehydroepiandrosterone	Resiniferatoxin[+]	Electrical stimulation of vagus nerve	Glucosamine sulfate
Sulindac	DHEA-sulfate	Resveratrol[+]	Hypothermia	Herbimycin A

TABLE 4 — *continued.*

	Dibenzylbutyrolactone lignans	Rg(3) (ginseng derivative)	Metals**	Hydroquinone
Cell-signaling	Diethyldithiocarbamate	Rg(3), a ginseng derivative	Nitric Oxide	4-Hydroxynonenal
inhibitors	Diferoxamine	Rocaglamides	Saline (low Na^{+} istonic)	Hypochlorite
Atrovastat**	Dihydrolipoic Acid	Rotenone	Hyperosmolarity	Hypoethyl starch
D609**	Dilazep +	Rotenone		Isomallotochromanol
LY294002**	Dilazep + fenofibric acid	S-allyl-cysteine (SAC, garlic	**Vitamins**	Isomallotochromene
Quinadril**	Dimethyldithiocarbamates	compound)	BTEE**	Jesterone dimer
RO31-8220**	Dimethylsulfoxide	Sanguinarine+	Vitamin C	Kamebakaurin
SB203580**	Disulfiram	Saucerneol D and E	Vitamin D	Lactoferrin
SC236**	Ebselen	Sauchinone	Vitamin E	LDL (Extensively oxidized)
Sphondin	EGTA**	Sauchinone	Nitrosylcobalamin**	Leptomycin B
TNP-470**	Emodin^{+}	Silibinin^{+}		Mevinolin, 5'-methylthioadenosine
U0126**	Ent-kaurane diterpenoids	Silymarin+	**Virus derivatives**	Monochloramine
	Epigallocatechin-3-gallate	Tempol	Core Protein of Hepatitis C virus^{+}	MX781
IKK inhibitors	EPC-K1 (phosphodiester compound	Tepoxaline	E1A	Nafamostat mesilate
AS602868	of vitamin E and vitamin C)	Tepoxaline (5-(4-chlorophenyl)-N-	HIV-1 Vpu protein	N-ethyl-maleimide
BAY-117082**	Epigallocatechin-3-gallate (green tea	hydroxy-(4-methoxyphenyl) -N-	IκB-like proteins	Nicotine
BAY-117083**	polyphenols)	methyl-1H-pyrazole-3-propanamide)	K1 protein	Omega 3 fatty acids
BMS-345541	Epoxyquinol	Tert-butyl hydroquinone	Kaposi's sarcoma-associated herpesvirus	Pervanadate^{+}
DTD**	Epoxyquinol A	Tranilast	Pertussis toxin binding protein	Petrosaspongiolide M
E3330**	Erbstatin^{+}	Triptolide (PG490)	SspH1 and IpaH9.8**	Phenethylisothiocyanate
LF15-0195**	Ergolide	Uncaria tomentosa	YopJ**	Phenylarsine oxide^{+}
MOL 294**	Ergothioneine	Ursolic acid^{+}		Phenyl-N-tert-butylnitrone
PS1142	Ethyl Pyruvate	Vitamin C	**Synthetic compounds**	Phosphorylation
	Ethylene Glycol Tetraacetic Acid	Vitamin E derivatives	AS602868	Phytic acid
Protease/ Proteasome	Eugenol	Yakuchinone A and B	Decoy oligonucleotides**	Pranlukast
inhibitors	Fenofibric acid	Yakuchinone A and B	DTD**	Psychosine
ALLnL	Flavenoids (Crategus)		E3330**	Pyrithione
APNE	Flavopiridol^{+}	**Plant extracts**	Hydroquinone	Raxofelast
APNE**$^{+}$	Fluorochalcones	Apple	Macrolide antibiotics	Rebamipide
Boronic Acid Peptide	Gamma-glutamylcysteine synthetase	Aged garlic	MOL 294**	Rhein
BTEE	Ganoderma lucidum polysaccharides	Black raspberry	Pentoxifylline	Ribavirin
Cyclosporin A	Garcinol (from extract of Garcinia	Blueberry		Rifamides
DCIC**	indica fruit rind)	Ganoderma lucidum	**Others**	Rifampicin
Deoxyspergualin	Genistein^{+}	Ginkgo biloba	Adenosine^{+}	Rolipram

TABLE 4 — *continued.*

DFP**	Ginkgo biloba extract	Ochna macrocalyx bark	6-aminoquinazoline derivatives	Ro106-9920
FK506**	Glutathione	PC-SPES (8 herb mixture)	6(5H)-phenanthridinone + benzamide	Sanggenon C
Lactacystine, b-lactone	Glycyrrhizin	Phyllanthus amarus	7-amino-4-methylcoumarin	Serotonin derivative**
LLM**	Guaianolides	Qingkailing	15-Deoxyspergualin	Siah2**
MG101**	Hematein	Shuanghuanglian	ADP ribosylation inhibitors**	SLPI**
MG115**	Hypericin	Stinging nettle	Amentoflavone	Statins
MG132**	IRFI 042 (Vitamin E-like compound)	Tanacetum larvatum	Amrinone	Staurosporine
Pefabloc	Iron tetrakis	Uncaria tomentosum	Anandamide	Sulfasalazine
Peptide Aldehydes	Isoeugenol	Fungal gliotoxin	Anti-thrombin III	Surfactant protein A
PS-341**	KT-90**		APC0576**	Survanta
TLCK**	L-cysteine	**Polypeptides and enzymes**	Artemisinin	T-614**
TPCK**	Lacidipine		Astragaloside IV	Taurine + niacine
Ubiquitin Ligase Inhibitors	Lazaroids	Angiopoietin-1	Atorvastatin	Tetrathiomolybdate
Z-LLL	Lovastatin	Atrial Natriuretic Peptide	Aucubin	THI 52**
Z-LLnV	Lupeol	AvrA protein (Salmonella)	Azidothymidine	Thalidomide[+]
	Luteolin	β-amyloid protein	Benfotiamine (thiamine derivative)	Thiopental
Anti-oxidants	Magnolol	β-catenin	Bisphenol A	Triflusal
β-lapachone[+]	Manassantins A and B	Bovine serum albumin	o,o'-bismyristoyl thiamine disulfide	Tyrphostin AG-126
α-lipoic acid	Manganese superoxide dismutase	CaMKK**	Cacospongionolide B	Wedelolactone
α-tocopherol	Melatonin	Complement protein C5a	Capsiate	Wogonin
15-deoxyspergualine	Mesalamine	Cytochalasin D	Caprofin	
23-hydroxyursolic acid	N-acetyl-L-cysteine	D-amino acid peptide	Chitosan	
α-torphryl acetate	Nacyselyn	DQ 65-79**	Chromene derivatives	
α-torphryl succinate	Nordihydroguaiaritic acid		Clarithromycin	

*For most references see http://people.bu.edu/gilmore/nf-kb/lab/index.html

** ADP ribosylation inhibitors, nicotinamide and 3-aminobenzamide; APC0576, 5-(((5)-2,2-dimethylcyclopropanecarbonyl)amino)-2-(4-(((5)-2,2-dimethylcyclopropanecarbonyl)amino)-pheoxy)pyridine; APNE, N-acetyl-DL-phenylalanine-b-naphthylester; Atrovastat , HMG-CoA reductase inhibitor; A77 1726, Leflunomide metabolite; BAY117082, E3((4-methylphenyl)-sulfonyl)-2-propenenitrile; BAY117083, E3((4-t-butylphenyl)-sulfonyl)-2-propenenitrile; BTEE, N-benzoyl L-tyrosine-ethylester; CaMKK, Calcium/calmodulin-dependent kinase kinase; Compound 26, 2-amino-3-cyano-4-aryl-6-(2-hydroxy-phenyl)pryridine analog; D609, phosphatidylcholine-phospholipase C inhibitor; Decoy oligonucleotides, synthetic decoys which "compete" with transcription factors for binding to their consensus sequences; DCIC, 3,4-dichloroisocoumarin; DFP, diisopropyl fluorophosphate; DQ 65-79, aa 65-79 of the α helix of the α-chain of the class II HLA molecule DQA03011; DTD, (4,10-dichloropyrido [5,6:4,5] thieno[3,2-d':3,2- d]-1, 2; E3330, (2E)-3-[5-(2,3-Dimethoxy-6-methyl-1,4-benzoquinoyl)]-2-nonyl-2-propenoic acid; EGTA, Ethylene Glycol Tetraacetic Acid; FK506, Tarcolimuc; HBEGEF, Heparin-binding EGF-like GF; hCG, human choriogonadotropin; HSCO, Hepatoma Substrated-cDNA library clone one; KT-90, synthetic derivative of morphine; LF15-0195, analog of 15-deoxyspergualine; LLM, N-acetyl-leucinyl-leucynil-methional; LY294002, wortmannin, PI3-kinase inhibitor; Metals, examples are chromium, cadmium, gold, lead, mercury, zinc, arsenic, and titanium;MG101, also ALLnL, N-acetyl-leucinyl-leucinyl-norleucinal; MG115, also Z-LLnV, carbobenzoxyl-leucinyl-leucynil-norvalinal; MG132, also Z-LLL, carbobenzoxyl-leucinyl-leucynil-leucynal; MOL294, methyl(4R/S)-4-hydroxy-4-[(5S,8S)/(5R,8R))-8-methyl-1,2-diox-2-phenyl-2,5,5,8-tetrahydro-1H[1,2,4]triazolo[1,2-9]-pyridazin-5-yl]-2-butynoate; MnSOD, Mangenese superoxide dismutase; α-MSH, alpha-Melanocyte stimulating hormone; Nitrosylcobalamin, vitamin B12 analog; p202a, IFN- inducible protein; PDTC, Pyrrolnedithiocarbamate; PG-15-deoxy-Δ(12,14)-PGJ(2),Prostaglandin 15-deoxy-Delta(12,14)-PGJ(2); Phenolic antioxidants,example is Hydroquinone; PMC, (2,2,5,7,8-pentamethyl-6-hydroxychromane); PS341, Velcade, bortezolimb; Quinadril, ACE inhibitor; RO31-8220,PKC inhibitor; SB203580, p38 MAPK inhibitor; SC236, COX-2 inhibitor; Serotonin derivative, N-(p-coumaroyl) serotonin; Siah2, Seven in abstentia homolog2; SLPI, Secretory leukocyte protease inhibitor; SspH1 and IpaH9.8, Leucine-rich effector proteins of Salmonella & Shigella; T-614, methanesulfoanilide anti-arthritis inhibitor; THI 52, 1-naphthylethyl-6,7-dihydroxy-1,2,3,4-tetrahydroisoquinoline; TLCK, N-a-tosyl-L-lysine chloromethyl ketone; TNP-470, angiogenesis inhibitor; TPCK, N-a-tosyl-L-phenylalanine chloromethyl ketone; U0126, MEK inhibitor; VEGF, vascular endotheliail growth factor; YopJ, encoded by Yersinia pseudotuberculosis; ZAS3 protein, zinc finger protein which binds NF-κB site.

[+] Investigated in author's laboratory.

of Alzheimer's disease. Rheumatoid arthritis is a chronic inflammatory disease characterized by persistent joint swelling and progressive destruction of cartilage and bone. NF-κB plays an essential role in transcriptional activation of TNF and IL-1. Together they form a positive regulatory cycle that may amplify and maintain the rheumatoid disease process.

HOW TO INHIBIT NF-κB ACTIVATION?

Because of the role of NF-κB in a wide variety of diseases, inhibitors of NF-κB activation are extensively sought (TABLE 4). Different steps in the NF-κB activation pathway are being targeted to block NF-κB. These include inhibitors of proteosome that mediate IκBα degradation, inhibitors of kinase (IKK), which mediate IκBα phosphorylation, decoy peptides from IκBα, IKK, and p65 proteins. The double-stranded oligodeoxynucleotides (ODNs) that possess consensus NF-κB sequence as transcription factor decoys (TFDs) also have been found to inhibit NF-κB binding to native DNA sites. Examples of proteasome blockers include peptide aldehydes such as ALLnL, LLM, Z-LLnV, and Z-LLL, lactacystine, PS-341, ubiquitin ligase inhibitors, and cyclosporine A. Several cytokines that are produced by Th2 have been found to suppress NF-κB activation. These include IL-4,[26] IL-13,[27] and IL-10.[28] Additionally, endocrine hormones such as HCG,[29] LH, MSH,[30] and GH[31] have been shown to abrogate NF-κB activation. Both IFN-α and IFN-β, which exhibit antiviral, antiproliferative, and immunosuppressive activities, also abolish NF-κB activation.[32] Several phytochemicals from different plants have been identified that can suppress NF-κB activation effectively.[33–46] These include curcumin (turmeric), resveratrol (red grapes), guggulsterone (guggul), ursolic acid (from holy basil), betulinic acid (birch trees), eugenol (cloves), gingerol (ginger), oleandrin (oleander), silymarin (artichoke), emodin (aloe), capsaicin (red chili), anethol (anise), and others. All these blockers of NF-κB have potential in the treatment of a wide variety of diseases. Pharmacological safety, bioavailability, and efficacy *in vivo* will determine their therapeutic potential in particular diseases.

CONCLUSION

This minireview shows that NF-κB is an important transcription factor that is activated by a wide variety of stimuli, controls the expression of a large number of genes, mediates pathogenesis of various diseases, and can be suppressed by numerous agents. NF-κB activation, however, is required for the proper function of the immune system. Proliferation of T cells and B cells, activation of macrophages, proliferation and survival of dendritic cells, and activation of T cells are dependent on NF-κB activation. Some recent evidence, however, indicates that while NF-κB1 mediates an inflammatory response, NF-κB2 mediates an immune response.[47] This suggests that suppression of the NF-κB1 pathway that controls inflammation may have less effect on the immune system. This remains to be determined. That NF-κB activation has been linked with most diseases is not too surprising considering that as many as 98% of all diseases are proinflammatory. Thus, the thesis that NF-κB is a "smoke-detector" that is activated by cigarette smoke[48] or a "stress-signal" is quite appropriate.

REFERENCES

1. AGGARWAL, B.B. 2004. Nuclear factor-kappaB: the enemy within. Cancer Cell **6:** 203–208.
2. AGGARWAL, B.B. 2003. Related articles, links abstract signalling pathways of the TNF superfamily: a double-edged sword. Nat. Rev. Immunol. **3:** 745–756.
3. TAKADA, Y., F.R. KHURI & B.B. AGGARWAL. 2004. Protein farnesyltransferase inhibitor (SCH 66336) abolishes NF-kappaB activation induced by various carcinogens and inflammatory stimuli leading to suppression of NF-kappaB-regulated gene expression and up-regulation of apoptosis. J. Biol. Chem. **18:** 26287–26299.
4. TAKADA, Y. & B.B. AGGARWAL. 2004. TNF activates Syk protein tyrosine kinase leading to TNF-induced MAPK activation, NF-kappaB activation, and apoptosis. J. Immunol. **173:** 1066–1077.
5. TAKADA, Y., X. FANG, M.S JAMALUDDIN, *et al.* 2004. Genetic deletion of glycogen synthase kinase-3beta abrogates activation of IkappaBalpha kinase, JNK, Akt, and p44/p42 MAPK but potentiates apoptosis induced by tumor necrosis factor. J. Biol. Chem. **279:** 39541–3954.
6. OZES, O.N., L.D. MAYO, J.A. GUSTIN, *et al.* 1999. NF-kappaB activation by tumour necrosis factor requires the Akt serine-threonine kinase. Nature **401:** 82–85.
7. YANG, J., Y. LIN, Z. GUO, *et al.* 2001. The essential role of MEKK3 in TNF-induced NF-kappaB activation. Nat. Immunol. **2:** 620–624.
8. FUNAKOSHI-TAGO, M., Y. SONODA, S. TANAKA, *et al.* 2003. Related articles, links free full text tumor necrosis factor-induced nuclear factor kappaB activation is impaired in focal adhesion kinase-deficient fibroblasts. J. Biol. Chem. **278:** 29359–29365.
9. MANNA, S.K., H.J. ZHANG, T. YAN, *et al.* 1998. Overexpression of manganese superoxide dismutase suppresses tumor necrosis factor-induced apoptosis and activation of nuclear transcription factor-kappaB and activated protein-1. J. Biol. Chem. **273:** 13245–13254.
10. MANNA, S.K., M.T. KUO & B.B. AGGARWAL. 1999. Overexpression of gamma-glutamylcysteine synthetase suppresses tumor necrosis factor-induced apoptosis and activation of nuclear transcription factor-kappa B and activator protein-1. Oncogene **18:** 4371–4382.
11. MANNA, S.K., N.K. SAH & B.B. AGGARWAL. 2000. Protein tyrosine kinase p56lck is required for ceramide-induced but not tumor necrosis factor-induced activation of NF-kappa B, AP-1, JNK, and apoptosis. J. Biol. Chem. **275:** 13297–13306.
12. TAKADA, Y., A. MUKHOPADHYAY, G.C. KUNDU, *et al.* 2003. Hydrogen peroxide activates NF-kappa B through tyrosine phosphorylation of I kappa B alpha and serine phosphorylation of p65: evidence for the involvement of I kappa B alpha kinase and Syk protein-tyrosine kinase. J. Biol. Chem. **278:** 24233–24241.
13. MUKHOPADHYAY, A., S.K. MANNA & B.B. AGGARWAL. 2000. Pervanadate-induced nuclear factor-kappaB activation requires tyrosine phosphorylation and degradation of IkappaBalpha. Comparison with tumor necrosis factor-alpha. J. Biol. Chem. **275:** 8549–8555.
14. SINGH, S., B.G. DARNAY & B.B. AGGARWAL. 1996. Site-specific tyrosine phosphorylation of IkappaBalpha negatively regulates its inducible phosphorylation and degradation. J. Biol. Chem. **271:** 31049–31054.
15. FAN, C., Q. LI, D. ROSS & J.F. ENGELHARDT. 2003. Tyrosine phosphorylation of I kappa B alpha activates NF kappa B through a redox-regulated and c-Src-dependent mechanism following hypoxia/reoxygenation. J. Biol. Chem. **278:** 2072–2080.
16. DEVARY, Y., C. ROSETTE, J.A. DIDONATO & M. KARIN. 1993. NF-kappa B activation by ultraviolet light not dependent on a nuclear signal. Science **261:** 1442–1445.
17. LEE, S.J., A. DIMTCHEV, M.F. LAVIN, *et al.* 1998. A novel ionizing radiation-induced signaling pathway that activates the transcription factor NF-kappaB. Oncogene **17:** 1821–1826.
18. LI, N. & M. KARIN. 1998. Ionizing radiation and short wavelength UV activate NF-kappaB through two distinct mechanisms. Proc. Natl. Acad. Sci. USA **95:** 13012–13017.
19. GIL, J., J. RULLAS, M.A. GARCIA, *et al.* 2001. The catalytic activity of dsRNA-dependent protein kinase, PKR, is required for NF-kappaB activation. Oncogene **20:** 385–394.

20. DIGICAYLIOGLU, M. & S.A. LIPTON. 2001. Erythropoietin-mediated neuroprotection involves cross-talk between Jak2 and NF-kappaB signalling cascades. Nature **412:** 641–647.
21. WARIS, G., A. LIVOLSI, V. IMBERT, *et al.* 2003. Hepatitis C virus NS5A and subgenomic replicon activate NF-kappaB via tyrosine phosphorylation of IkappaBalpha and its degradation by calpain protease. J. Biol. Chem. **278:** 40778–40787.
22. KUMAR, A., Y. TAKADA, A.M. BORIEK & B.B. AGGARWAL. 2004. Nuclear factor-kappaB: its role in health and disease. J. Mol. Med. **82:** 434–448.
23. AGGARWAL, B.B., Y. TAKADA, S. SHISHODIA, *et al.* 2004. Nuclear transcription factor NF-kappa B: role in biology and medicine. Indian J. Exp. Biol. **42:** 341–353.
24. SHISHODIA, S. & B.B. AGGARWAL. 2004. Nuclear factor-kappaB activation mediates cellular transformation, proliferation, invasion angiogenesis and metastasis of cancer. Cancer Treat. Res. **119:** 139–173.
25. GARG, A. & B.B. AGGARWAL. 2002. Nuclear transcription factor-kappaB as a target for cancer drug development. Leukemia **16:** 1053–1068.
26. MANNA, S.K. & B.B. AGGARWAL. 1998. Interleukin-4 down-regulates both forms of tumor necrosis factor receptor and receptor-mediated apoptosis, NF-kappaB, AP-1, and c-Jun N-terminal kinase. Comparison with interleukin-13. J. Biol. Chem. **273:** 33333–33341.
27. MANNA, S.K. & B.B. AGGARWAL. 1998. IL-13 suppresses TNF-induced activation of nuclear factor-kappa B, activation protein-1, and apoptosis. J. Immunol. **161:** 2863–2872.
28. BHATTACHARYYA, S., P. SEN, M. WALLET, *et al.* 2004. Immunoregulation of dendritic cells by IL-10 is mediated through suppression of the PI3K/Akt pathway and of IkappaB kinase activity. Blood **104:** 1100–1109.
29. MANNA, S.K., A. MUKHOPADHYAY & B.B. AGGARWAL. 2000. Human chorionic gonadotropin suppresses activation of nuclear transcription factor-kappa B and activator protein-1 induced by tumor necrosis factor. J. Biol. Chem. **275:** 13307–13314.
30. MANNA, S.K. & B.B. AGGARWAL. 1998. Alpha-melanocyte-stimulating hormone inhibits the nuclear transcription factor NF-kappa B activation induced by various inflammatory agents. J. Immunol. **161:** 2873–2880.
31. HAEFFNER, A., N. THIEBLEMONT, O. DEAS, *et al.* 1997. Inhibitory effect of growth hormone on TNF-alpha secretion and nuclear factor–kappaB translocation in lipopolysaccharide-stimulated human monocytes. J. Immunol. **158:** 1310–1314.
32. MANNA, S.K., A. MUKHOPADHYAY & B.B. AGGARWAL. 2000. IFN-alpha suppresses activation of nuclear transcription factors NF–kappa B and activator protein 1 and potentiates TNF-induced apoptosis. J. Immunol. **165:** 4927–4934.
33. SHISHODIA, S. & B.B. AGGARWAL. 2004. Related articles, links free full text guggulsterone inhibits NF-kappaB and IkappaBalpha kinase activation, suppresses expression of anti-apoptotic gene products, and enhances apoptosis. J. Biol. Chem. **279:** 47148–47158.
34. BHARTI, A.C., Y. TAKADA & B.B. AGGARWAL. 2004. Curcumin (diferuloylmethane) inhibits receptor activator of NF-kappa B ligand-induced NF-kappa B activation in osteoclast precursors and suppresses osteoclastogenesis. J. Immunol. **172:** 5940–5947.
35. TAKADA, Y. & B.B. AGGARWAL. Flavopiridol inhibits NF-kappaB activation induced by various carcinogens and inflammatory agents through inhibition of IkappaBalpha kinase and p65 phosphorylation: abrogation of cyclin D1, cyclooxygenase-2, and matrix metalloprotease-9. J. Biol. Chem. **279:** 4750–4759.
36. TAKADA, Y. & B.B. AGGARWAL. 2003. Betulinic acid suppresses carcinogen-induced NF-kappa B activation through inhibition of I kappa B alpha kinase and p65 phosphorylation: abrogation of cyclooxygenase-2 and matrix metalloprotease-9. J. Immunol. **171:** 3278–3286.
37. SHISHODIA, S., S. MAJUMDAR, S. BANERJEE & B.B. AGGARWAL. 2003. Ursolic acid inhibits nuclear factor-kappaB activation induced by carcinogenic agents through suppression of IkappaBalpha kinase and p65 phosphorylation: correlation with down-regulation of cyclooxygenase 2, matrix metalloproteinase 9, and cyclin D1. Cancer Res. **63:** 4375–4383.

38. ASHIKAWA, K., S. MAJUMDAR, S. BANERJEE, *et al.* 2002. Piceatannol inhibits TNF-induced NF-kappaB activation and NF-kappaB-mediated gene expression through suppression of IkappaBalpha kinase and p65 phosphorylation. J. Immunol. **169:** 6490–6497.
39. BHARTI, A.C., N. DONATO, S. SINGH & B.B. AGGARWAL. 2003. Curcumin (diferuloylmethane) down-regulates the constitutive activation of nuclear factor-kappa B and IkappaBalpha kinase in human multiple myeloma cells, leading to suppression of proliferation and induction of apoptosis. Blood **101:** 1053–1062.
40. MAJUMDAR, S., B. LAMOTHE & B.B. AGGARWAL. 2002. Thalidomide suppresses NF-kappa B activation induced by TNF and H2O2, but not that activated by ceramide, lipopolysaccharides, or phorbol ester. J. Immunol. **168:** 2644–2651.
41. MANNA, S.K., N.K. SAH, R.A. NEWMAN, *et al.* 2000. Oleandrin suppresses activation of nuclear transcription factor-kappaB, activator protein-1, and c-Jun NH2-terminal kinase. Cancer Res. **60:** 3838–3847.
42. CHAINY, G.B., S.K. MANNA, M.M. CHATURVEDI & B.B. AGGARWAL. 2000. Related articles, links abstract anethole blocks both early and late cellular responses transduced by tumor necrosis factor: effect on NF-kappaB, AP-1, JNK, MAPKK and apoptosis. Oncogene **19:** 2943–2950.
43. MANNA, S.K., A. MUKHOPADHYAY & B.B. AGGARWAL. 2000. Resveratrol suppresses TNF-induced activation of nuclear transcription factors NF-kappa B, activator protein-1, and apoptosis: potential role of reactive oxygen intermediates and lipid peroxidation. J. Immunol. **164:** 6509–6519.
44. MANNA, S.K., A. MUKHOPADHYAY, N.T. VAN & B.B. AGGARWAL. 1999. Silymarin suppresses TNF-induced activation of NF-kappa B, c-Jun N-terminal kinase, and apoptosis. J. Immunol. **163:** 6800–6809.
45. KUMAR, A., S. DHAWAN & B.B. AGGARWAL. 1998. Emodin (3-methyl-1,6,8-trihydroxyanthraquinone) inhibits TNF-induced NF-kappaB activation, IkappaB degradation, and expression of cell surface adhesion proteins in human vascular endothelial cells. Oncogene **17:** 913–918.
46. CHATURVEDI, M.M., A. KUMAR, B.G. DARNAY, *et al.* 1997. Sanguinarine (pseudochelerythrine) is a potent inhibitor of NF-kappaB activation, IkappaBalpha phosphorylation, and degradation. J. Biol. Chem. **272:** 30129–30134.
47. SHISHODIA, S. & B.B. AGGARWAL. 2004. Nuclear factor-kappaB: a friend or a foe in cancer? Biochem. Pharmacol. **68:** 1071–1080.
48. ANTO, R.J., A. MUKHOPADHYAY, S. SHISHODIA, *et al.* 2002. Cigarette smoke condensate activates nuclear transcription factor-kappaB through phosphorylation and degradation of IkappaB(alpha): correlation with induction of cyclooxygenase-2. Carcinogenesis **23:** 1511–1518.

Thiolsulfinate Allicin from Garlic

Inspiration for a New Antimicrobial Agent

ROGER HUNTER, MINO CAIRA, AND NASHIA STELLENBOOM

Department of Chemistry, University of Cape Town, Rondebosch, 7701, Cape Town, South Africa

ABSTRACT: Consideration of the underlying features responsible for garlic-allicin's antimicrobial activity as well as its instability has prompted an investigation into substituted *S*-aryl alkylthiolsulfinates as a class of garlic mimic with enhanced stability. Synthesis of the targets has inspired the development of new methods for synthesizing unsymmetrical aralkyl disulfides, which are then oxidized to the targets. Some simple representatives have been synthesized, setting the scene for a full SAR study of this relatively unexplored class of thiolsulfinate.

KEYWORDS: allicin; unsymmetrical disulfide synthesis; *S*-aryl alkylthiolsulfinate

INTRODUCTION

The folklore and therapeutic benefits of garlic date back about 5,000 years to the Middle and Far East and probably originated in the advanced civilizations of the Indus valley, from where garlic was imported to China before spreading to Egypt, Greece, and throughout the Roman Empire into Europe.[1–3] In more recent times, however, modern science has established the fascinating chemistry that goes on when a garlic clove is crushed or attacked by a pathogen. Under such circumstances, the compartment separating an *S*-(allyl)-*L*-cysteine sulfoxide substrate, named alliin, from its enzyme alliinase, is destroyed, resulting in the two substances coming into intimate contact with one another. Rapid elimination ensues, producing 2-propenesulfenic acid (Equation 1), which self-condenses (Equation 2) to produce an *S*-(allyl)-thiolsulfinate called allicin, which accounts for about 70–80% of the organic material produced initially (FIG. 1).[4]

Allicin 1 is a fascinating substance in that it has potent antimicrobial activity, in part because it is unstable and therefore has not been worked out by the biosphere. Allicin was first isolated and studied in the laboratory by Cavallito[5,6] in 1944. He later demonstrated[7] the synthetic material obtained from selective oxidation of diallyl disulfide using perbenzoic acid to be identical to that isolated from freshly crushed garlic. Interestingly, Cavallito carried out antibacterial assays on allicin and

Address for correspondence: Roger Hunter, Department of Chemistry, University of Cape Town, Rondebosch, 7701, South Africa. Voice: +27 (021) 650 2544; fax: +27 (021) 689 7499. roger@science.uct.ac.za

Ann. N.Y. Acad. Sci. 1056: 234–241 (2005). © 2005 New York Academy of Sciences. doi: 10.1196/annals.1352.011

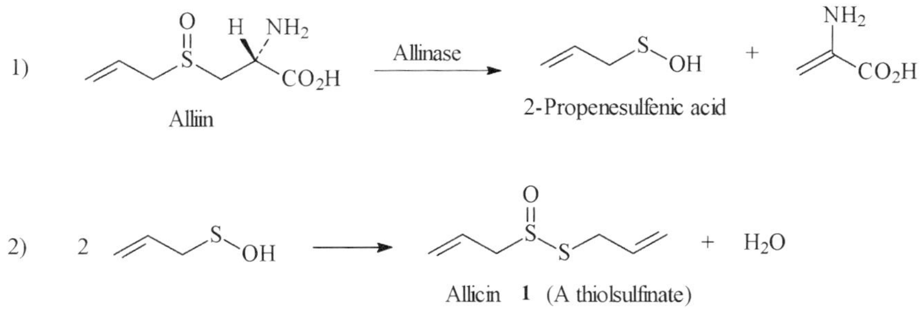

1) Alliin —Allinase→ 2-Propenesulfenic acid + (NH$_2$, CO$_2$H)

2) 2 (S–OH) → Allicin **1** (A thiolsulfinate) + H_2O

FIGURE 1. Production of allicin in the garlic clove.

a number of its congeners, demonstrating them to be potent, yet labile, antimicrobial agents.[7] In the years that followed, several *in vitro* studies demonstrated allicin to have potent antibacterial, antifungal, and antiparasitic activity against a range of microorganisms including methicillin-resistant *Staphylococcus aureus*.[4,8] Antiviral activity has also been demonstrated.[9] Despite its potent activity, synthetic allicin, obtained either biomimetically or chemically from diallyl disulfide, has not been developed for human use *in vivo,* undoubtedly because of its instability and also because pharmaceutical companies cannot patent it in view of its having been in the public domain for too long.[4] Various companies do sell it as an aqueous solution, but the purity of such allicin is questionable since allicin is unstable, the degree of instability depending on a range of parameters including solvent, pH, concentration, and the presence of additives.[10] Similarly, commercial garlic supplements can lose their potency as a result of alliinase inactivity.[11] Thus, despite allicin's potent antimicrobial activity, its benefits to humans remain predominantly at the culinary level with no available quantitative data concerning its ability to fight human disease.[12]

This landscape, particularly given the situation in South Africa concerning the need for affordable treatments for opportunistic infections in AIDS, prompted us to return to the drawing board and explore whether the essential motif in allicin could be exploited in a more stable system towards developing a new class of antimicrobial agent. The two questions that we focused on were: (1) How does allicin exert its biological effect? and (2) What structural features in allicin are responsible for its instability? The answer to the first question is generally accepted as being that the thiolsulfinate grouping acts as an electrophilic site for sulfhydryl groups of thiols, resulting in their oxidation to disulfides.[13,14] Attack by the soft thiol group would first take place at the softer sulfenyl sulfur followed by a second attack at the harder 2-propenesulfenic acid sulfur that is expelled.[15] The overall result is oxidation of 2-mol equivalents of thiol, using one of allicin (FIG. 2).

Evidence for this hypothesis dates back to the work of Cavallito, who was the first researcher to demonstrate that cysteine is oxidized to *S*-(thioallyl)-cysteine by allicin in the respective molar stoichiometry of 2:2:1 just mentioned.[6] Subsequently, other workers have corroborated the findings from this experiment. It therefore emerges that the thiolsulfinate grouping is the dominant pharmacophore and presents itself as a desirable starting template for substitution to be built around. Regarding allicin's

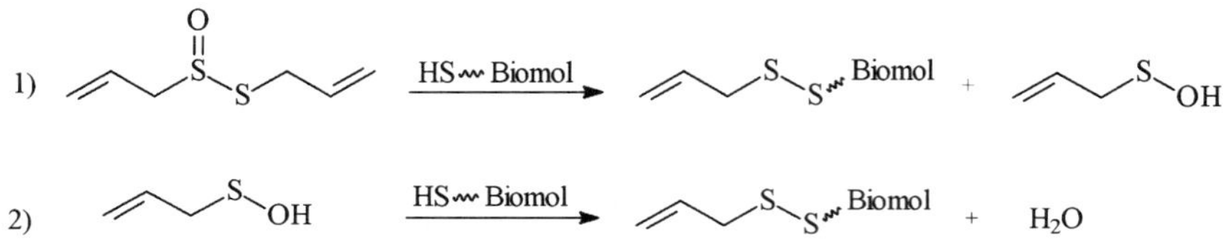

1) HS~Biomol → Biomol + OH

2) OH HS~Biomol → Biomol + H_2O

FIGURE 2. Oxidative action of allicin.

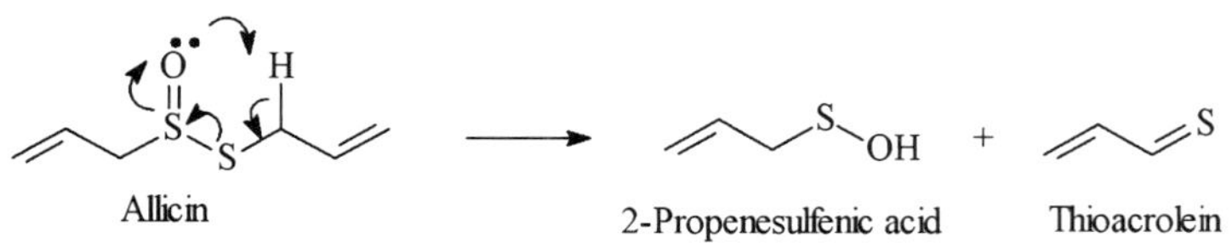

Allicin → 2-Propenesulfenic acid + Thioacrolein

FIGURE 3. Mechanism of decomposition of allicin.

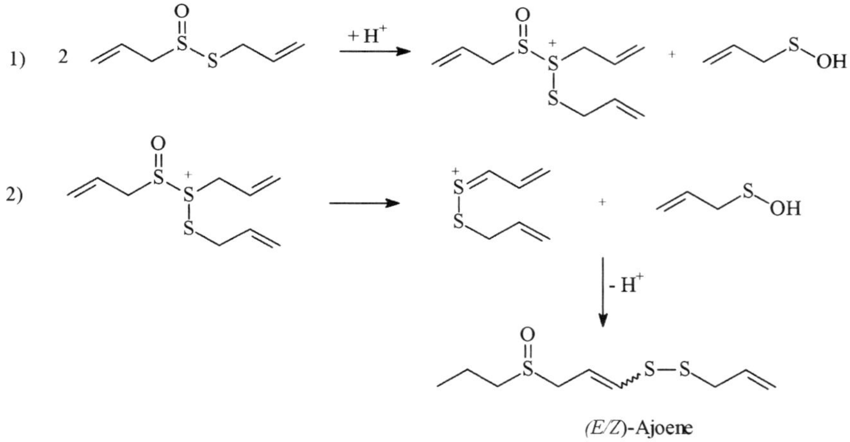

1) 2 + H+ → + OH

2) → + OH

- H+

(*E*/*Z*)-Ajoene

FIGURE 4. Rearrangement of allicin to (*E*/*Z*)-ajoene.

instability, the answer has been elegantly provided by the work of Block, who in some seminal papers[16,17] in the 1970s and 1980s reported that allicin fragments into thioacrolein and 2-propenesulfenic acid, the former subsequently dimerizing to 1,2- and 1,3-dithiins. Of crucial importance regarding this fragmentation reaction is the presence of an allylic hydrogen adjacent to the sulfenyl sulfur (FIG. 3).

Alternatively, allicin may participate in a pathway initiated by reaction with itself, with the end product being the interesting product ajoene, which has gained interest as an anticancer[18] and antithrombotic agent over the last 20 years since its discovery (FIG. 4).

Given this backdrop, our intention was to use this mechanistic information towards developing novel garlic "mimics" by retaining the thiolsulfinate grouping

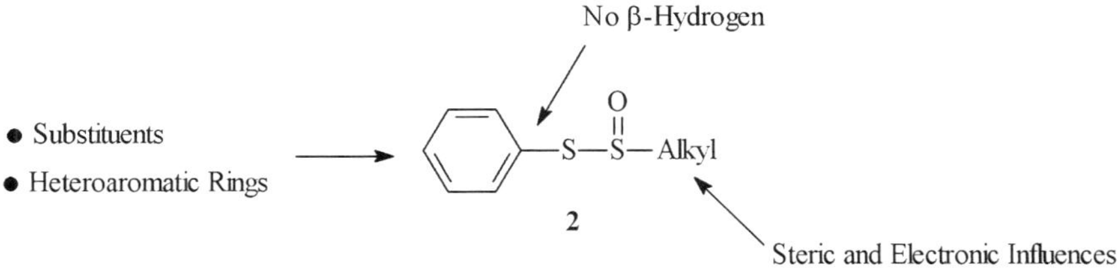

FIGURE 5. Template for new allicin "mimics."

and replacing the sulfenyl allyl group with an aromatic ring lacking a β-hydrogen in order to inhibit the initial fragmentation. The substitution would invite the possibility of aromatic ring substitutions for attenuating electrophilicity at the sulfenyl sulfur; heteroaromatic changes could also be considered in the same vein. On the sulfinyl sulfur side, it was decided to stay with an aliphatic group in view of the known tendency of diarylthiolsulfinates to disproportionate.[19] FIGURE 5 summarizes our structure-activity-relationship (SAR) motif for the target *S*-aryl alkylthiolsulfinates **2.**

A literature survey of the class of compound identified in FIGURE 5 revealed that a full structure–activity study had not been carried out[20–22] even though simple members were known.

SYNTHESIS OF TARGETS AND PRELIMINARY RESULTS

A retrosynthetic analysis of target **2** identified two straightforward disconnections as being selective mono-oxidation of the unsymmetrical disulfide[23] and nucleophilic substitution by a thiol of an alkanesulfinyl chloride.[24,25] Both these synthetic strategies are well documented in the literature. However, since the preparation of alkanesulfinyl chlorides can be nontrivial,[26] it was decided to pursue the former approach. Furthermore, the over-oxidation product of **2**, the thiolsulfonate **3**, was also a desirable target for biological testing. Unsymmetrical disulfide synthesis[27] has been a focus of many research groups over the years in view of its relevance to biological systems. The majority of methods involve nucleophilic substitution of a sulfenyl derivative by a thiol or its derivative. Preparation of the sulfenyl intermediate suffers from the drawback of often involving a number of steps or the use of a toxic chlorinating agent such as $SOCl_2$ or Cl_2. In trying to develop a new method for the transformation, which is both green-friendly and involves a one-pot procedure, it was decided to explore the possibility of using 1-chlorobenzotriazole[28] as the oxidant, which is easily obtained as an air-stable crystalline solid by the oxidation of 1,2,3-benzotriazole by sodium hypochlorite in acetic acid. After several experiments, gratifyingly it was eventually established that the addition of an aromatic thiol (1 eq) to BtCl (1.5 eq) in CH_2Cl_2 as solvent at –78°C resulted in conversion to BtSR (by tlc) via in situ formation of RSCl. BtH (1 eq) was added to the reaction to facilitate trapping of RSCl to form BtSR over its conversion to the homodimer RSSR. On addition of an aliphatic thiol (1.5 eq) at –20°C, BtSR was converted to

TABLE 1. Yields of aralkyl disulfides

R	R_1	Yield %
(a) *p*-MeOPh	*n*-C_3H_7	82
(b) *p*-CH_3Ph	*n*-C_3H_7	90
(c) *o*-CO_2MePh	*t*-C_4H_9	95
(d) *p*-MeOPh	Allyl	60

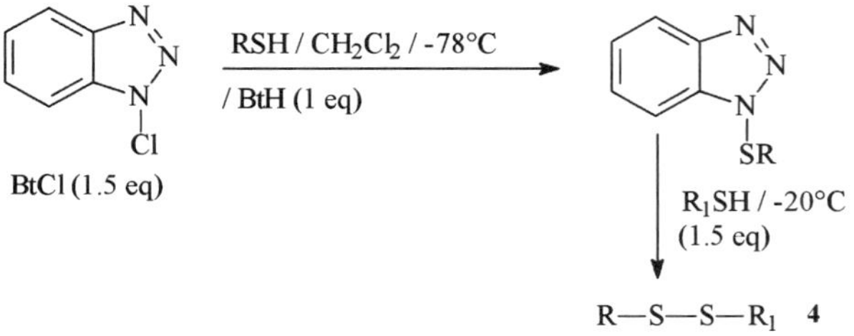

FIGURE 6. Unsymmetrical disulfide synthesis using BtCl.

the unsymmetrical disulfide by nucleophilic substitution at sulfur with expulsion of BtH. After destruction of excess reagent (BtCl) with aq. $Na_2S_2O_3$ and conventional work-up, column chromatography allowed isolation of the desired aralkyl unsymmetrical disulfide **4** in high yield. The exception appears to occur when using 2-propenethiol in the second step to generate the allicin *S*-allyl functionality, in which case the yield was only moderate. No interference from homodimer disulfides was observed with the garlic mimics (aralkyl systems). However, preliminary results indicate that this method is unsuitable for aliphatic disulfides in that an aliphatic thiol is not converted to the crucial BtSR intermediate in the first step. Furthermore, preparation of unsymmetrical aromatic disulfides, although not targets in the current study, did give rise to the formation of some homodimer disulfide, which in most cases could be separated effectively by chromatography. The full scope of this new one-pot procedure will be published elsewhere. FIGURE 6 and TABLE 1 summarize a few of the preliminary results relevant to synthesis of targets **2**.

With an easy-to-use, versatile method in hand, we next turned attention to the oxidation step to thiolsulfinate target. It was anticipated that the sulfur bound to the aliphatic group would be more nucleophilic in view of a reduced nucleophilicity of the other sulfur due to resonance. In the event, using standard reagents for this transformation like peracetic acid or *m*-chloroperbenzoic acid (1 eq), both the thiolsulfinate **2** as well as the thiolsulfonate **3** products were obtained. Results for entries (a) and (b) in TABLE 1 are shown in FIGURE 7.

The thiolsulfinates **2a,b** and thiolsulfonates **3a,b** could be distinguished by their NMR spectra.[29] In the ^1H and ^{13}C NMR spectra, the protons and carbon α to the oxygenated sulfur appeared more downfield for the thiolsulfonate **3** compared to

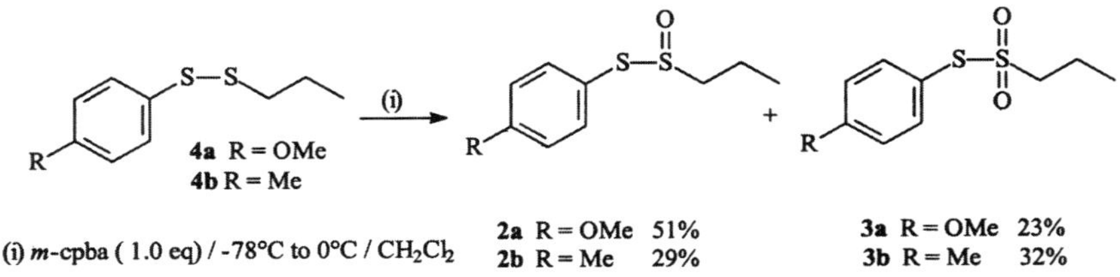

4a R = OMe
4b R = Me

(i) *m*-cpba (1.0 eq) / -78°C to 0°C / CH_2Cl_2

2a R = OMe 51%
2b R = Me 29%

3a R = OMe 23%
3b R = Me 32%

FIGURE 7. Oxidation of unsymmetrical disulfides.

TABLE 2. Relative stability of 1, 2a, and 5

Compound	Stability
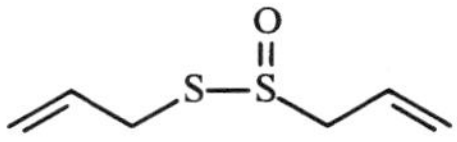 **1**	(1) Unstable to column chromatography at 23°C (rt) (2) Half-life = 16 h at rt, neat
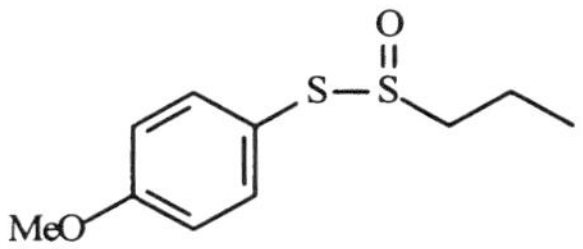 **2a**	(1) Stable to column chromatography at rt (2) <5 % change after 3 weeks at 10°C (in $CDCl_3$)
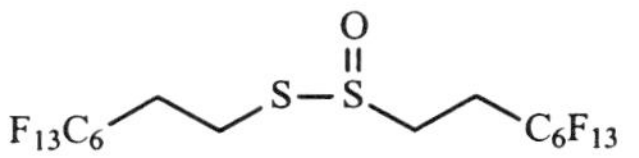 **5**	(1) Stable to recrystallization at 50°C

those of the thiolsulfinate **2** as observed by other workers. Despite being fairly simple structures, **2a** and **2b** are actually new compounds[a] at this point in time and are therefore attractive prototypes for our structure–activity study. A preliminary stability study was carried out and details are given in TABLE 2 comparing thiolsulfinate **2a** with literature data for allicin as well as a recent fluorinated thiolsulfinate **5**. The breakdown of thiolsulfinate **2a** was evaluated by tlc and ^{1}H NMR.

[a]Full spectroscopic and analytic data for **2a/3a** follows. (**2a**) IR ν_{max} ($CHCl_3$)/ cm^{-1} 1030 (S=O stretch); δ_H (400 MHz, $CDCl_3$) 1.09 (3H, t, CH_3), 1.88 (2H, m, CH_2), 3.04 (2H, m, CH_2), 3.83 (3H, s, CH_3), 6.93 (2H, d, Ar), 7.53 (2H, d, Ar); δ_C (100.57 MHz, $CDCl_3$) 13.3 (CH_3), 17.2 ($-CH_2\underline{C}H_2CH_3$), 55.5 ($-OCH_3$), 57.7 ($-\underline{C}H_2CH_2CH_3$), 115.1, 119.3 (C-S), 137.4, 161.6 (C-O); HRMS: *m/z* 230.0410 (M^+), $C_{10}H_{14}O_2S_2$ requires 230.0435. (**3a**) IR ν_{max} ($CHCl_3$)/ cm^{-1} 1320 (Asym. SO_2 stretch), 1130 (Sym. SO_2 stretch); δ_H (400 MHz, $CDCl_3$) 1.05 (3H, t, CH_3), 1.94 (2H, m, CH_2), 3.14 (2H, m, CH_2), 3.85 (3H, s, CH_3), 6.96 (2H, d, Ar), 7.59 (2H, d, Ar); δ_C (100.57 MHz, $CDCl_3$) 12.7 (CH_3), 17.3 ($-CH_2\underline{C}H_2CH_3$), 55.5 ($-OCH_3$), 60.7 ($-\underline{C}H_2CH_2CH_3$), 115.4, 118.7 (C-S), 138.0, 162.4 (C-O); HRMS: *m/z* 246.0363 (M^+), $C_{10}H_{14}O_3S_2$ requires 246.0384. The analytical data for compounds **2b/3b** will be reported elsewhere.

CONCLUSIONS AND FUTURE WORK

As revealed in TABLE 2, our *S*-(4-methoxyphenyl) propylthiolsulfinate **2a** shows a marked increase in stability compared to allicin. Of interest in this regard is the recent report by Brace[30] on the fluorinated thiolsulfinate **5** shown in TABLE 2, which shows increased stability compared to its hydrogenated analog. Another option available, which we have pursued with ajoene,[31] is to include the thiolsulfinate into a cyclodextrin[32] to further enhance stability as well as to influence bioavailability.

In summary, we have developed new methods for unsymmetrical aralkyl disulfide synthesis that allows us to now embark on a comprehensive structure–activity study of *S*-aryl alkylthiolsulfinates regarding the influence of thiolsulfinate structure on both antimicrobial activity as well as stability. The lipophilicity of the aromatic ring may present a barrier to organism cell-wall entry, and appropiate substitution will therefore have to be made to ensure an appropriate balance between lipophilicity and hydrophilicity for effective overall passage of the antimicrobial agent. However, it is hoped that patterns will emerge that uncover new motifs for further developing the thiolsulfinate grouping as a key pharmacophore in the discovery of new antimicrobial agents.

ACKNOWLEDGMENTS

The work was supported by the National Research Foundation, Pretoria, South Africa.

REFERENCES

1. THOMSON, M. & M. ALI. 2003. Garlic [*Allium sativum*]: a review of its potential use as an anticancer agent. Curr. Cancer Drug Targets **3:** 67–81.
2. KAMEL, A. & M. SALEH. 2000. Recent studies on the chemistry and biological activities of the organosulfur compounds of garlic (*Allium sativum*). Stud. Nat. Prod. Chem. **23** (Bioactive Natural Products [Part D]): 455–485.
3. LAWSON, L.D. 1998. Garlic: a review of its medicinal effects and indicated active compounds. ACS Symposium Series **691** (Phytomedicines of Europe): 176–209.
4. ANKRI, S. & D. MIRELMAN. 1999. Antimicrobial properties of allicin from garlic. Microbes Infect. **2:** 125–129.
5. CAVALLITO, C.J. & J.H. BAILEY. 1944. Allicin, the antibacterial principle of *Allium sativum*. I. Isolation, physical properties and antibacterial action. J. Am. Chem. Soc. **66:** 1950–1951.
6. CAVALLITO, C.J., J.S. BUCK & C.M. SUTER. 1944. Allicin, the antibacterial principle of *Allium sativum*. II. Determination of the chemical structure. J. Am. Chem. Soc. **66:** 1952–1954.
7. SMALL, L.V.D., J. BAILEY & C.J. CAVALLITO. 1947. Alkyl thiolsulfinates. J. Am. Chem. Soc. **69:** 1710–1713.
8. CUTLER, R.R. & P. WILSON. 2004. Antibacterial activity of a new, stable, aqueous extract of allicin against methicillin-resistant *Staphylococcus aureus*. Br. J. Biomed. Sci. **61:** 71–74.
9. WEBER, N.D., D.O. ANDERSEN, J.A. NORTH, *et al.* 1992. *In vitro* virucidal effects of *Allium sativum* (garlic) extract and compounds. Planta Med. **58:** 417–423.
10. FREEMAN, F. & Y. KODERA. 1995. Garlic chemistry: stability of *S*-(2-propenyl)-2-propene-1-sulfinothioate (allicin) in blood, solvents, and simulated physiological fluids. J. Agric. Food Chem. **43:** 2332–2338.

11. LAWSON, L. & Z.J. WANG. 2001. Low allicin release from garlic supplements: a major problem due to the sensitivities of alliinase activity. J. Agric. Food Chem. **49:** 2592–2599.
12. MIRON, T., M. MIRONCHIK, D. MIRELMAN, *et al.* 2003. Inhibition of tumor growth by a novel approach: in situ allicin generation using targeted alliinase delivery. Mol. Cancer Ther. **2:** 1295–1301.
13. MIRON, T., A. RABINKOV, D. MIRELMAN, *et al.* 2000. The mode of action of allicin: its ready permeability through phospholipid membranes may contribute to its biological activity. Biochim. Biophys. Acta **1463:** 20–30.
14. GILES, G., K.M. TASKER & C. JACOB. 2002. Oxidation of biological thiols by highly reactive disulfide-S-oxides. Gen. Physiol. Biophys. **21:** 65–72.
15. KICE, J.L. 1968. Electrophilic and nucleophilic catalysis of the scission of the sulfur-sulfur bond. Accounts Chem. Res. **1:** 58–64.
16. BLOCK, E. & J. O'CONNOR. 1974. Chemistry of alkyl thiosulfinate esters. VII. Mechanistic studies and synthetic applications. J. Am. Chem. Soc. **96:** 3929–3944.
17. BLOCK, E., S. AHMAD, J.L. CATALFAMO, *et al.* 1986. The chemistry of alkyl thiosulfinate esters. IX. Antithrombotic organosulfur compounds from garlic: structural, mechanistic, and synthetic studies. J. Am. Chem. Soc. **108:** 7045–7055.
18. HASSAN, H.T. 2004. Ajoene (natural garlic compound): a new anti-leukaemia agent for AML therapy. Leuk. Res. **28:** 667–671.
19. KICE, J.L., C.G. VENIER, G.B. LARGE & L. HEASLEY. 1969. Mechanisms of reactions of thiosulfinates (sulfenic anhydrides). III. Sulfide-catalyzed disproportionation of aryl thiosulfinates. J. Am. Chem. Soc. **91:** 2028–2035.
20. ISENBERG, N. & M. GRDINIC. 1973. Thiolsulfinates. Int. J. Sulfur Chem. **8:** 307–320.
21. LACOMBE, S.M. 1999. Oxysulfur compounds derived from disulfides: stability and reactivity. Rev. Heteroatom. Chem. **21:** 1–41
22. SHEN, C., H. XIAO & K.L. PARKIN. 2002. *In vitro* stability and chemical reactivity of thiolsulfinates. J. Agric. Food Chem. **50:** 2644–2651.
23. FREEMAN, F. & C. ANGELETAKIS. 1985. Formation of elusive vic-disulfoxides and OS-sulfenyl sulfinates during the *m*-chloroperoxybenzoic acid (*m*-cpba) oxidation of alkyl aryl disulfides and their regioisomeric sulfinothioic acid *S*-esters. J. Org. Chem. **50:** 793–798.
24. BACKER, H.J. & H. KLOOSTERZIEL. 1954. Thiolsulfinic esters. Rec. Trav. Chim. Pays-Bas Belg. **73:** 129–139.
25. HARPP, D.N., T. AIDA & T.H. CHAN. 1983. Organosulfur chemistry. Part 45. A general, high-yield preparation of thiosulfinate esters using organotin precursors. Tetrahedron Letts. **24:** 5173–5176.
26. SCHWAN, A.L., R.R. STRICKLER, R. DUNN-DUFAULT & D. BRILLON. 2001. Oxidative fragmentations of 2-(trimethylsilyl)ethyl sulfoxides: routes to alkane-, arene-, and highly substituted 1-alkenesulfinyl chlorides. Eur. J. Org. Chem. **9:** 1643–1654.
27. BAO, M. & M. SHIMIZU. 2003. *N*-Trifluoroacetyl arenesulfenamides, effective precursors for synthesis of unsymmetrical disulfides and sulfenamides. Tetrahedron **59:** 9655–9659.
28. REES, C.W & R.C. STORR. 1968. 1-Chlorobenzotriazole: a new oxidant. Chem. Commun. 1305–1306.
29. FREEMAN, F., C.N. ANGELETAKIS & T.J. MARICICH. 1981. Proton NMR and carbon-[13] NMR spectra of disulfides, thiosulfinates and thiosulfonates. Org. Magn. Reson. **7:** 53–58.
30. BRACE, N.O. 2000. Oxidation chemistry of perfluoroalkyl-segmented thiols, disulfides, thiosulfinates, and thiosulfonates. The role of the perfluoroalkyl group in searching out new chemistry. J. Fluorine Chem. **105:** 11–23.
31. CAIRA, M.R., R. HUNTER, S.A. BOURNE & V.J. SMITH. 2004. Preparation, thermal behaviour and solid-state structures of inclusion complexes of permethylated-β-cyclodextrin with the garlic-derived antithrombotics (*E*)- and (*Z*)-ajoene. Supramol. Chem. **16:** 395–403.
32. NIKOLIC, V., M. STANKOVIC, A. KAPOR, *et al.* 2004. Allylthiosulfinate: beta-cyclodexatrin inclusion complex: preparation, characterization and microbiological activity. Die Pharmazie **59:** 845–848.

Effect of Rhythmic Breathing (Sudarshan Kriya and Pranayam) on Immune Functions and Tobacco Addiction

VINOD KOCHUPILLAI,[a] PRATIK KUMAR,[b] DEVINDER SINGH,[a] DHIRAJ AGGARWAL,[a] NARENDRA BHARDWAJ,[a] MANISHA BHUTANI,[a] AND SATYA N. DAS[c]

Department of [a]Medical Oncology, [b]Medical Physics Unit, Institute Rotary Cancer Hospital, and [c]Department of Biotechnology, All India Institute of Medical Sciences, New Delhi 110029, India

ABSTRACT: Stress, a psychophysiological process, acts through the immune-neuroendocrine axis and affects cellular processes of body and immune functions, leading to disease states including cancer. Stress is also linked to the habit of tobacco consumption and substance abuse, which in turn also leads to diseases. Sudarshan Kriya (SK) and Pranayam (*P*), rhythmic breathing processes, are known to reduce stress and improve immune functions. Cancer patients who had completed their standard therapy were studied. SK and P increased natural killer (NK) cells significantly (*P* <0.001) at 12 and 24 weeks of the practice compared to baseline. Increase in NK cells at 24 weeks was significant (*P* <0.05) compared to controls. There was no effect on T-cell subsets after SK and P either in the study group or among controls. SK and P helped to control the tobacco habit in 21% of individuals who were followed up to 6 months of practice. We conclude that the inexpensive and easy to learn and practice breathing processes (SK and P) in this study demonstrated an increase in NK cells and a reduction in tobacco consumption. When confirmed in large and randomized studies, this result could mean that the regular practice of SK and P might reduce the incidence and progression of cancer.

KEYWORDS: rhythmic breathing; Sudarshan Kriya and Pranayam; immune function; tobacco addiction

INTRODUCTION

The newer field of psychoneuroimmunology (PNI) highlights the link between behavior, neuroendocrine function, immune response, and health.[1,2] Physical and/ or psychological stimuli may set a pattern of neurotransmitters, hormones, and cytokines that may act on receptors of immune cells, altering the effector population and

Address for correspondence: Prof. Vinod Kochupillai, Chief, Institute Rotary Cancer Hospital (IRCH), All India Institute of Medical Sciences (AIIMS), New Delhi, 110029, India. Voice: +91-11-2658 9821; fax: +91-11-2658 8663.
vinodkochupillai@yahoo.com

Ann. N.Y. Acad. Sci. 1056: 242–252 (2005). © 2005 New York Academy of Sciences.
doi: 10.1196/annals.1352.039

functions. In turn, cytokines can act on brain cells and endocrine cells through their receptors to modulate their function. Blalock and Smith[3] in 1980 demonstrated that while secreting gamma interferon (IFN-γ), human leukocytes also produced a peptide that was recognized by antisera to adrenocorticotropic hormone (ACTH). The molecular weight and biological activity of this leukocyte-derived ACTH was subsequently shown to be similar to its pituitary-derived counterpart. Thus, the concept of a reciprocal PNI dialogue was derived.

Woloski *et al.*[4] suggested that bidirectional communication is mediated by direct innervation of lymphoid tissues and by neuropeptides and cytokines released by the cells of both nervous and immune systems. A host of cytokines previously thought to be limited to the lymphoid tissues, including IFN-γ, interleukin(IL)-1, IL-6, tumor necrosis factor (TNF), and IL-10, were recently found to be produced in the CNS by neurons and astroglial cells.[5] In addition, peripheral nerves and endocrine hormones and peptides may directly affect the activity of lymphocytes.[6] Similarly, peripherally released immune cytokines can penetrate the blood brain barrier and modulate central neuroendocrine activities, notably at the level of the hypothalamic-pituitary axis.[5]

Pert,[7] in a book entitled "Molecules of Emotions," suggested that mind, spirit, and emotions are unified with the physical body through the neuropeptides and their receptors, named by her as molecules of emotions. These molecules, according to Pert, form the base for awareness and consciousness. Acting intelligently, these molecules communicate information over a network linking all the systems and organs, emotions, thoughts, and spirit. This body-wide network is ever changing, dynamic, and infinitely flexible. This network, however, gets blocked by suppressed emotions or during a state commonly referred to as "stress."

The phenomenon called stress is a psychophysiological process experienced as a negative emotional state. It is the product of an appraisal of a situation and the resulting coping ability available to the individual.[1] Stress is associated with a distinct set of physiological changes, most notably on the immune-neuroendocrine axis. Direct adrenergic innervations of peripheral lymphoid tissue lead to rapid changes in the activities and distribution of subsets of immune cells.[8] Rapid translocation of peripheral memory T lymphocytes and natural killer (NK) cells from the spleen and peripheral lymph nodes to the microvasculature of the lung, skin, and bone marrow appears to be an intrinsic component of the acute response to stress.[9] Posttraumatic stress is associated with a smaller number of lymphocytes, T cells, NK cell activity, and the total amount of IFN-γ and IL-4. Long-lasting immune suppression and its long-term implications on health have also been observed among such individuals.[10] Exam stress in healthy medical students led to cytokine dysregulation.[11] In addition to dysregulating the immune system, stress modulates apoptosis (genetically programmed cell death) and alters DNA repair processes.[2,12] Oxidative stress is also aggravated by psychosocial stress.[13]

These effects of stress on cellular processes of the body lead to many disease states including infection, atopic diseases, viral infections,[1,11] asthma,[1,14] cardiovascular diseases,[15,16] nonorganic visual disturbances in children,[17] irritable bowel syndrome, and other gastrointestinal disorders[18,19] as well as autoimmune diseases such as rheumatoid arthritis.[20] Psychosocial factors may also play a role in the progression of tumors such as hepatocellular carcinoma[21] and squamous intraepithelial lesions in human immunodeficiency virus-infected women.[22]

Stress is also linked to tobacco use. Individuals in distress or depression-leading stressful lives are more likely to be smokers.[22,23] Adolescent smoking is associated with stressful life events including negative peer and school-related events.[23–25] Analysis after the September 11th terrorist attack on New York City showed an increase in cigarette smoking by 10% and an increase in alcohol and marijuana use.[26] Tobacco dependence is a well-known risk factor for many diseases including cardiovascular disease,[24] bronchitis, and cancer.[27]

Studies discussed so far indicate that psychosocial stressors could alter immunological and cellular processes that induce disease conditions including cancer and may induce tobacco dependence. Hence, behavioral interventions in the form of relaxation techniques have been studied to determine their influence. Our earlier studies utilized rhythmic breathing processes—Sudarshan Kriya and Pranayam (SK & P)—as stress elimination techniques.[28,29] A fall in blood lactate levels, an increase in antioxidant defense,[28] and an electroencephalographic (EEG) pattern consistent with relaxed alertness[29] were observed. T-cell subsets and NK cells were studied in those practicing SK & P regularly for at least 2 years, and the results were compared with those in normal controls and cancer patients in remission. Total T cells and T-helper subsets were significantly higher in those practicing SK & P and in normal controls compared to cancer patients. A significant difference was found in NK cells, which were significantly higher (P <0.001) in those practicing SK and P than in normal controls and cancer patients.[30]

In the current study, we evaluated the role of SK & P on the immune function of cancer patients in remission. We also evaluated the effect of these breathing processes on tobacco addiction.

PATIENTS AND METHODS

Sudarshan Kriya and Pranayam: Breathing Processes

Yoga, meditation, and Pranayam are centuries old, time-tested processes that are known to relax the mind and energize the body. Recently, HH Sri Sri Ravi Shankar introduced SK, an innovative yogic technique. The process is introduced to participants through a 22–24-hour structured program called the Art of Living (AOL) workshop spread out over 6 days. SK, a rhythmic cyclical breathing of slow, medium, and fast cycles, is preceded by Ujjayi Pranayam, slow breathing of three cycles per minute, a forced inspiration and expiration against airway resistance; Bhastrika Pranayam, the rapid inhalation and exhalation at 20–30 cycles per minute; and brief chanting. These processes are practiced while sitting with the eyes closed and the awareness focused on breathing. This process ends with 10 minutes of rest in a tranquil supine position. It can be followed by 20 minutes of meditation.[28,29]

Effect of Sudarshan Kriya and Pranayam on T-Cell Subsets and Natural Killer Cells in Cancer Patients in Remission

T-cell subsets and NK cells were studied in cancer patients. Twenty-seven cancer patients registered at the Institute Rotary Cancer Hospital (IRCH), All India Institute of Medical Sciences (AIIMS), were included in the study. These patients had already been treated: 22 were in complete remission (CR) at the time of study; 3 had stable disease (SD): and 2 had progressive disease (PD).

All patients were offered an AOL workshop to learn SK & P. Those having learned SK & P were advised to practice it regularly at home and come for weekly followup sessions. However, 21 patients (9 with breast cancer; 5 with multiple myeloma [MM]); 3 with lymphoma; and 1 each with acute myeloid leukemia, acute lymphoblastic leukemia, gastric carcinoma, and germ cell tumor of the ovary) agreed to attend the workshop; these were taken as the study group. In the study group, 16 of 21 were in CR, the remaining patients having SD or PD. The median age of this group was 51 years (range 22–65).

The remaining six patients, 4 with breast cancer and 1 each with MM and lymphoma, were unwilling or unable to undertake the workshop; hence, they were used as the controls. All 6 patients in the control group were in CR at the time of study. The median age of this group was 45 years (range 35–50).

Immune function studies including $CD3^+$, $CD4^+$, $CD8^+$, and NK cells were carried out by multicolor flow cytometry in all patients. The fluorescence signal acquired in the flow cytometer was analyzed using CELL QUEST software (Beckton Dickinson). The minimum gap between the study and the last course of chemotherapy was 2 months. Peripheral blood samples were drawn at baseline (day 0), week 1, week 12, and week 24. Statistical analysis was done using computer software SPSS-10

Effect of Rhythmic Breathing on Tobacco Addiction

This study was conducted on 82 tobacco users who attended the AOL workshops. The subjects were advised to practice SK & P daily at home and to come for weekly followups.

A questionnaire was designed to collect information on the pattern and amount of their tobacco consumption. The response to the questionnaire was collected on day 0 (baseline), day 6, week 2, week 3, month 2, and month 6. A spreadsheet database was created and data were analyzed.

RESULTS

Immune Studies on Cancer Patients in Remission

No significant differences were noted in the population of T-cell subsets between the control group and the study group at baseline and subsequently. There were no differences in the NK cell population in the control group between day 0 and week 4. However, the NK cell population of the study group showed a statistically significant increase at week 12 ($P = 0.001$) and week 24 ($P = 0.0001$) compared to baseline. The difference was significant ($P < 0.05$) at week 24 between the study arm and the control arm (FIG. 1).

Effect of Rhythmic Breathing Processes on Tobacco Addiction

Most subjects (62%) started tobacco use when they were in the age range of 15–25 years (FIG. 2). Cigarette smoking was the most prevalent form of tobacco use

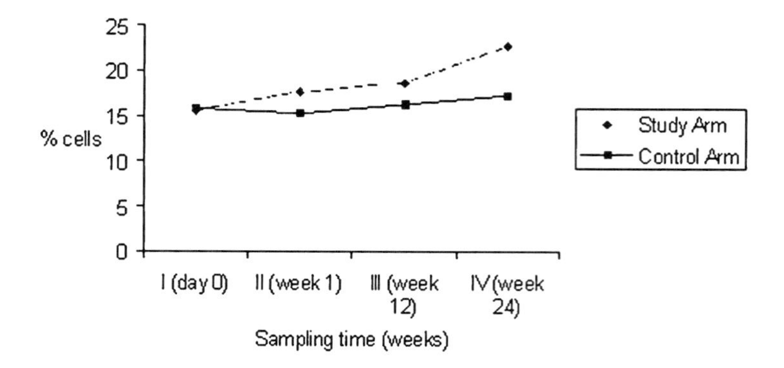

FIGURE 1. NK cell activity in the control and study arms (those practicing Sudarshan Kriya and Pranayam) at various time periods. The P values at various time periods were: Study Arm: I *vs*. II, P not significant; I *vs*. III, $P = 0.001$; II *vs*. III, $P = 0.0001$; II *vs*. IV, $P = 0.0001$; III *vs*. IV, $P = 0.0001$. Control Arm: I, II, III, and IV *vs*. each other, P not significant; I, II, and III of study arm *vs*. I, II, and III of control arm, P not significant. IV of study arm *vs*. IV of control arm, $P < 0.05$.

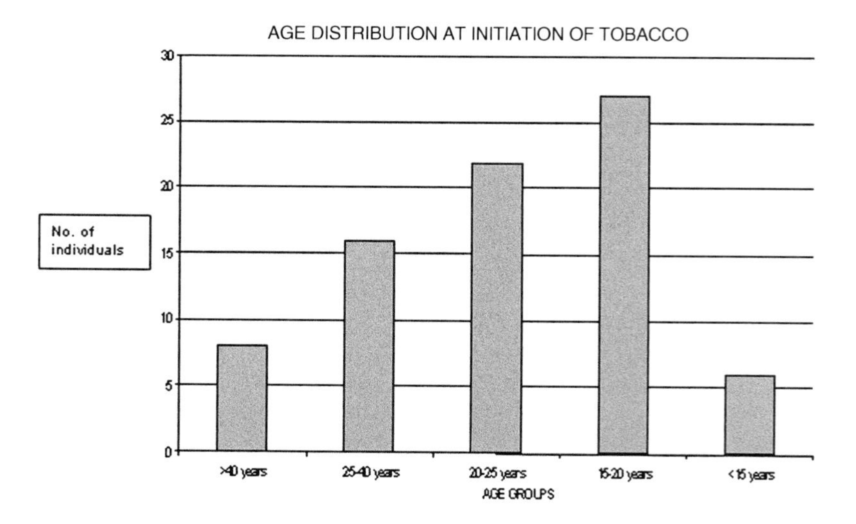

FIGURE 2. Age distribution at initiation of tobacco use among subjects.

(54%); other methods included *bidi* smoking and chewing tobacco directly or through different brands such as *Pan Masala*. In most cases, initiation came from friends (83%), mainly for fun's sake; the majority (97%) knew that tobacco was harmful. Of the tobacco users 91% wished to stop the habit; 83% made an attempt; however, only 36% could stay without tobacco for less than a week.

Of those who attended the AOL workshop to the last day(day 6), 53 of 82 (65%) remained without tobacco (complete cessation); those remaining reported reduction in tobacco use by 50–90%. Fifty-six percent observed withdrawal symptoms in the form of restlessness, difficulty in concentration, and anger being the most frequent. The majority (95%), however, felt that regular practice of SK & P would help them to quit tobacco, as they thought that the urge to smoke was less when they practiced SK & P.

Individuals were followed up at weekly to monthly intervals. At month 6, complete cessation of tobacco use was reported by 17 of 34 persons available for followup. However, among the entire group of 82, the success rate was 17 of 82 (21%).

DISCUSSION

There is considerable evidence of the efficacy of relaxation techniques in the treatment of coronary artery disease, hypertension, headache, insomnia, incontinence, chronic low backache, and arthritis.[31] Immune enhancement via behavioral intervention has also been documented.[32,33] Forty-five healthy adults were assigned to one of the three protocols (relaxation training, social contact, or no intervention). Subjects who underwent relaxation training showed significant enhancement in NK cell activity at the end of 1 month's intervention with a concomitant decrease in distress-related symptoms in comparison with nonsignificant changes in the other two groups.[33] This was the first well-documented study to demonstrate immune enhancement via behavioral intervention. Subsequent studies[34–36] further substantiated that stress-reducing interventions can improve immune functioning.

One study[37]included patients with stage I or II malignant melanoma who had not received any treatment after surgical excision of the tumor. One group received structured 6 weeks of group intervention, which included health education, relaxation techniques, and psychological support. Reduced psychological distress, a significant increase in the percentage of NK cells, as well as an increase in NK cell cytotoxic activity were demonstrated in the intervention group compared to the controls.[37] Six years' followup of these patients showed a trend towards greater recurrence as well as higher mortality rates among patients in the control group than in the intervention group.[38]

In another study,[39] a year of weekly supportive group therapy with self-hypnosis for pain was associated with extended survival time in women with metastatic breast cancer. A similar study by Goodwin *et al.*, [40] however, found no difference in survival among patients with metastatic breast cancer; however, the support group did exhibit improved mood and perception of pain compared with the controls.

In the current study, we evaluated T-cell subsets $CD3^+$, $CD4^+$, and $CD8^+$ and NK cells in cancer patients after completing the standard therapy. The malignant process by itself and the therapeutic processes, including radiotherapy and chemotherapy, are known to suppress/alter immune functions. Impaired immune function may be associated with the development and progression of cancer.[21,41,42] Hence, any method that improves the immune status of an individual is likely to reduce the incidence of development and progression of cancer. This study examines the effects of rhythmic breathing processes (SK &P) on the immune parameters of cancer patients who have already undergone the standard recommended therapy and are on routine observation during followup.

Techniques employing conscious breathing are extremely powerful. Changes in the rate and depth of breathing produce changes in the quantity and kind of peptides that are released from the brain stem. By bringing the process of breathing into consciousness and doing something to alter it; either holding the breath or breathing extra fast, one causes the peptides to diffuse rapidly throughout the cerebrospinal fluid. Since many of these peptides are endorphins, the body's natural opiates, one soon achieves diminution of pain.[7] Peptide-respiratory link is well documented: virtually any peptide found any where else can be found in the respiratory center.[7] This may provide the scientific rationale for the powerful healing effects of consciously controlled breath pattern, including that observed following the practices of SK & P.

The T-cell subsets (cytotoxic/suppressor T cells $CD3^+CD8^+$; helper T cells $CD3^+$ $CD4^+$) are responsible for the cell-mediated immune response to intracellular parasites, viruses, fungi, and bacteria. Many tumors, despite progressive growth and metastases, contain immune T cells with specific antitumor reactivity (tumor-infiltrating lymphocytes). The NK cells ($CD3^-$ $CD16.56^+$) constitute the innate immunity of the body and play an important role in a variety of immune functions including defense against viral infections and surveillance of tumor cells.[43] NK cells also form the first line of defense against metastatic spread of cancer. The decreased number and impaired function are associated with the development as well as the progression of cancer.[37] The NK cell function is adversely affected by stressful events. For instance, NK cell activity was found depressed in bereaved spouses.[44] Lower levels of NK cell cytotoxicity were documented among the spouses of cancer patients and among those who reported lower levels of social support.[45] In a study of newly wed couples, those who were more negative or hostile during a discussion of marital problems with the spouse showed a greater downward change in NK cell activity.[46] Studies using rodent models showed that stresses decrease NK cell cytotoxicity and enhance metastatic spread of transplantable tumors.[47,48]

Owing to the known antitumor activity of T cells and NK cells, we analyzed the changes in the population of these cells in response to SK & P. The changes over time were compared with the baseline as well as with the control arm of the patients who did not practice or learn SK & P.

There was a statistically significant increase in the NK cell population in the cancer patients undergoing SK & P. However, the percentage of NK cells between day 0 and week 1 of SK & P practice was not statistically different; at week 12 of practice, a significant increase in the percentage of NK cells was seen compared to day 0 and week 1 ($P = .001$). Similarly, in week 24 samples, a further increase in the percentage of NK cells ($P = .0001$) was seen. Compared with week 12 samples, the NK cell population continued to increase over time with continued practice of SK & P. The lack of increase in the percentage of of NK cells by week 1 may be due to the long half-life of lymphocytes. A comparison of the NK cell changes in the study group *vs.* the control group that did not practice SK & P revealed a significant difference in the percentage of NK cells by 24 weeks (P <0.05). This suggests that a rise in NK cell function in this study was indeed the effect of SK & P practice and not merely due to recovery from the prior immunosuppressive effects of chemotherapy. No significant difference was noted in the $CD3^+$, $CD4^+$, and $CD8^+$ cell population in the study arm or in the control arm.

With NK cells forming the backbone of the body's known defenses against cancer, documentation of their rise in the response to SK & P in cancer patients may turn out to be a significant step in the management of cancer patients. SK & P are easily available techniques, are cost effective, and have no known side effects; hence, they may have an adjunctive role to other anticancer therapies such as surgery, chemotherapy, and radiotherapy. The addition of SK & P may help to reduce the chances of recurrence among those who practice it regularly.

Although this study is encouraging and suggests that SK & P may be used as an adjunct in cancer management with the likely possibility of reducing the chance of recurrence, the study has several lacunae that need to be taken into account while proposing further studies. The study is prospective and controlled but not randomized. The patients included in the study have different types of cancer; the status of cancer is also not entirely uniform. For validation of the results of this study, randomized controlled studies using a single type of cancer with similar disease status are needed.

Tobacco consumption was responsible for 100 million deaths during the last century. If this trend continues, deaths may increase to one billion.[49,50] Reduction in tobacco addiction (smoking and/or chewing) will accrue benefits in heart disease, neonatal and maternal health, stroke, and peripheral vascular disease as well as in a variety of cancers, including that of the lung.[27] Despite intense efforts over last 2 decades to control tobacco use, 1.2 billion individuals continue to smoke. Death from smoking is projected to increase and 70% of deaths are likely to be in developing countries. In India, for instance, 45% of men smoke or chew tobacco compared to 28% in the Western world.

The United States Food and Drug Administration approved medications used to treat tobacco dependence (bupropion and nicotine replacement therapy) are effective in only a fraction of smokers.[51] Moreover, nicotine is known to produce side effects such as muscular twitching, respiratory difficulty, abdominal cramps, and confusion. Moreover, nicotine by itself is highly addictive, and individuals trying to cut back or quit nicotine use may suffer withdrawal symptoms including depression, frustration or anger, irritability, trouble sleeping, difficulty in concentrating, and restlessness. Controlled deep breathing has been useful in relieving symptoms of smoking withdrawal.[52]

We have used breathing processes (SK & P) as stress elimination techniques to study their effect on tobacco cessation. Fifty-three of 82 individuals abstained from smoking during a 6-day workshop. Fifty-six percent did observe withdrawal symptoms in the form of restlessness, difficulty in concentrating, and anger; the majority (95%), however, felt that regular practice of SK & P would help them to quit tobacco use. By 6 months, 34 individuals had been followed up; 17 had still not smoked, despite the fact that many were not practicing SK & P on a regular basis. Calculated from the original number of 82 individuals, the success rate is 21%. The continuous abstinence rates at 12 months with bupropion were reported to be 21% compared to 11% for the placebo group.[51]

A controlled study of a larger number of subjects and better followup are necessary to document the results accurately. This study, however, does suggest that easy, cost effective, and nontoxic breathing processes (SK & P) help reduce tobacco consumption.

ACKNOWLEDGMENT

HH Sri Sri Ravi Shankar, who has rediscovered and popularized the process of Sudarshan Kriya and Pranayam, inspired the study. The study was carried out in collaboration with Ved Vignan Mahavidyapeeth, Bangalore. Terry Fox Foundation, Canada, funded the study.

REFERENCES

1. AGGARWAL, S.K. & G.D. MARSHALL, JR. 2001. Stress effects on immunity and its application to clinical immunology. Clin. Exp. Allergy **31:** 25–31.
2. BOVDJERG, D.H. 1991. Psychoneuroimmunology. Implications for oncology. Cancer **67:** 828–832.
3. BLALOCK, J.E. & E.M. SMITH. 1980. Human leukocyte interferon: structural and biological relatedness to adrenocorticotrophic hormone and endorphins. Proc. Natl. Acad. Sci. USA **77:** 5972–5974.
4. WOLOSKI, B.M. *et al.* 1985. Corticotropin-releasing activity of monokines. Science **230:** 1035–1037.
5. WATKINS, L.R., S.F. MAIER & L.E. GOEHLER. 1995. Cytokine to brain communication: a review & analysis of alternative mechanisms. Life Sci. **57:** 1011–1026.
6. JOHNSON, H.M., M.O. DOWNS & C.H. PONTZER. 1992. Neuroendocrine peptide hormone regulation of immunity. Chem. Immunol. **52:** 49–83.
7. PERT, C.B. 1997. Molecules of Emotions: the Science Behind Mind-Body Medicine. Simon & Schuster, Inc. New York.
8. BELLINGER, D.L. *et al.* 1992. Innervation of lymphoid organs and implication in development, aging and autoimmunity. Int. J. Immunopharmacol. **14:** 329–344.
9. KRADIN, R. & H. BENSON. 2000. Stress, the relaxation responses and immunity. Mod. Asp. Immunobiol. **1:** 110–113.
10. KAWAMURA, N., Y. KIM & N. ASUKAI. 2001. Suppression of cellular immunity in men with a past history of posttraumatic stress disorder. Am. J. Psychiatry **158:** 484–486.
11. MARSHALL, G.D., JR. *et al.* 1998. Cytokine dysregulation associated with exam stress in healthy medical students. Brain. Behav. Immun. **12:** 297–307.
12. KIECOLT-GLASER, J.K. *et al.* 2002. Psycho-oncology and cancer: psychoneuroimmunology and cancer. Ann. Oncol. **13:** 165–169.
13. SCARPELLINI, F., M. SBRACIA & L. SCARPELLINI. 1994. Psychological stress and lipoperoxidation in miscarriage. Ann. N.Y. Acad. Sci. **709:** 210–213.
14. SANDBERG, S. *et al.* 2004. Asthma exacerbations in children immediately following stressful life events: a Cox's hierarchical regression. Thorax **59:** 1046–1051.
15. THOMAS, K.S. *et al.* 2004. Job strain, ethnicity, and sympathetic nervous system activity. Hypertension **44:** 891–896.
16. TAICH, A. *et al.* 2004. Prevalence of psychosocial disturbances in children with nonorganic visual loss. J. AAPOS **8:** 457–461.
17. LEA, R. & P.J. WHORMELL. 2004. Psychological influences on the irritable bowel syndrome. Minerva Med. **95:** 443–450.
18. LOCKE, G.R., 3RD *et al.* 2004. Psychosocial factors are linked to functional gastro intestinal disorders: a population based nested case-control study. Am. J. Gastroenterol. **99:** 350–357.
19. ZAUTRA, A.J. *et al.* 2004. Immune activation and depression in women with rheumatoid arthritis. J. Rheumatol. **31:** 457–463.

20. STEEL, J. *et al.* 2004. The role of psychosocial factors in the progression of hepatocellular carcinoma. Med. Hypotheses **62:** 86–94.
21. PEREIRA, D.B. *et al.* 2003. Life stress and cervical squamous intraepithelial lesions in women with human papillomarvirus and human immunodeficiency virus. Psychosom. Med. **65:** 427–434.
22. YOUNG, L.E., A.D. JAMES & S.L. CUNNIGHAM. 2004. Lone motherhood and risk for cardiovascular disease. The National Population Health Survey (NPHS), 1998–99. Can. J. Public Health **95:** 329–335.
23. LA ROSA, E. *et al.* 2004. Psychological distress and stressful life antecedents associated with smoking. A survey of subjects consulting a preventive health center. Presse Med. **33:** 919–926.
24. BOOKER, C.L. *et al.* 2004. Stressful life events, smoking behavior and intentions to smoke among and multiethnic sample of sixth graders. Ethn. Health **9:** 369–397.
25. VLAHOV, D. *et al.* 2004. Consumption of cigarettes, alcohol and marijuana among New York City residents six months after the September 11 terrorist attack. Am. J. Drug Alcohol Abuse **30:** 385–407.
26. STEWART, B.W. & A.S. COATES. 2005. Cancer prevention: a global perspective. J. Clin. Oncol. **23:** 392–403.
27. SHARMA, H. *et al.* 2003. Sudarshan Kriya practitioners exhibit better antioxidant status and lower blood lactate levels. Biol. Psychol. **63:** 281–291.
28. BHATIA, M. *et al.* 2003. Electrophysiologic evaluation of Sudarshan Kriya: an EEG, BAER, and P300 study. Ind. J. Physiol. Pharmacol. **47:** 157–163.
29. DAS, S.N. & V. KOCHUPILLAI. 2002. Flow cytometric study of T cell subsets and natural killer cells in peripheral blood of "Art of Living teachers", cancer patients and normal individuals. Presented at Science of breath: an international symposium on Sudarshan Kriya, Pranayam and consciousness. New Delhi, India, March 2 & 3.
30. ASTIN, J.A. *et al.* 2003. Mind–body medicine: state of the science, implications for practice. J. Am. Board Fam. Pract. **16:** 131–147.
31. KIECOLT-GLASER, J.K. & R. GLASER. 1992. Psychoneuroimmunology: can psychological interventions modulate immunity? J. Consult. Clin. Psychol. **60:** 569–575.
32. KIECOLT-GLASER, J.K. *et al.* 1985. Psychosocial enhancement of immunocompetence in a geriatric population. Health Psychol. **4:** 25–41.
33. ZACHARIAE, R. *et al.* 1990. Effect of psychological intervention in the form of relaxation and guided imagery on cellular immune function in normal healthy subjects. Psychother. Psychosom. **54:** 32–39.
34. MCGRADY, A. *et al.* 1992. The effects of biofeedback-assisted relaxation on cell-mediated immunity, cortisol, and white blood cell count in healthy adult subjects. J. Behav. Med. **15:** 343–354.
35. GREEN, M.L., R.G. GREEN & W. SANTORO. 1988. Daily relaxation modifies serum and salivary immunoglobulins and psychophysiologic symptom severity. Biofeedback Self. Regul. **13:** 187–199.
36. FAWZY, F.I. *et al.* 1990. A structured psychiatric intervention for cancer patients. II. Changes over time in immunological measures. Arch. Gen. Psychiatry **47:** 729–735.
37. FAWZY, F.I. *et al.* 1993. Malignant melanoma. Effects of an early structured psychiatric intervention, coping and affective state on recurrence and survival 6 years later. Arch. Gen. Psychiatry **50:** 681–689.
38. SPIEGEL, D. *et al.* 1989. Effect of psychosocial treatment on survival of patients with metastatic breast cancer. Lancet **2:** 888– 891.
39. GOODWIN, P.J. *et al.* 2001. The effect of group psychosocial support on survival in metastatic breast cancer. N. Engl. J. Med. **345:** 1719–1726.
40. PENN, I. & T.I. STARZL. 1972. A summary of the status of de novo cancer in transplant recipients. Transplant Proc. **4:** 719–732.
41. LUECKEN, L.J. & B.E. COMPAS. 2002. Stress, coping and immune function in breast cancer. Ann. Behav. Med. **24:** 336–344.
42. HERBERMAN, R.B. & J.R. ORTALDO. 1981. Natural killer cells: their roles in defenses against disease. Science **214:** 24–30.
43. IRWIN, M. *et al.* 1988. Plasma cortisol and natural killer cell activity during bereavement. Biol. Psychiatry **24:** 173–178.

44. BARON, R.S. *et al.* 1990. Social support and immune function among spouses of cancer patients. J. Pers. Soc. Psychol. **59:** 344–352.
45. KIECOLT-GLASER, J.K. *et al.* 1993. Negative behavior during marital conflict is associated with immunological down-regulation. Psychosom. Med. **55:** 395–409.
46. STEFANSKI, V. 2001. Social stress in laboratory rats: behavior, immune function and tumor metastasis. Physiol. Behav. **73:** 385–391.
47. WU, W. *et al.* 2000. Social isolation stress enhanced liver metastasis of murine colon 26–L5 carcinoma cells by suppressing immune responses in mice. Life Sci. **66:** 1827–1838.
48. VINEIS, P. *et al.* 2004. Tobacco and cancer: recent epidemiological evidence. J. Natl. Cancer Inst. **96:** 99–106.
49. LERMAN, C., F. PATTERSON & W. BERRETTINI. 2005. Treating tobacco dependence: state of the science and new directions. J. Clin. Oncol. **23:** 311–323.
50. AMERICAN CANCER SOCIETY. Prevention and early detection. Quitting smokeless tobacco. Available http://www.cancer.org/docroot.
51. TONNESEN, P. *et al.* 2003. A multicentre, randomized, double blind, placebo-controlled, 1-year study of bupropion SR for smoking cessation. J. Intern. Med. **254:** 184–192.
52. MCCLERNON, F.J., E.C. WESTMAN & J.E. ROSE. 2004. The effects of controlled deep breathing on smoking withdrawal symptoms in dependent smokers. Addict. Behav. **29:** 765–772.

Mediterranean Diet and Cardiovascular Health

DARIO GIUGLIANO AND KATHERINE ESPOSITO

Division of Metabolic Diseases, Department of Geriatrics and Metabolic Diseases, Cardiovascular Research Centre, Second University of Naples, Naples, Italy

ABSTRACT: Diets that are high in fruits, vegetables, legumes, and whole grains and include fish, nuts, and low-fat dairy products have protective health effects. The traditional Mediterranean diet encompasses these dietary characteristics. Other compounds of the Mediterranean diet, the antioxidants, which exist in abundance in vegetables, fruit, beverages, and also virgin olive oil, may contribute to the prevention of coronary heart disease and possibly several forms of cancer and other diseases, thus providing a plausible explanation for its apparent benefits. It may be misleading to focus on a single element of the diet; this may explain, at least in part, the disappointing and frustrating results obtained in trials with vitamin supplementation, prematurely thought to be "the magic bullet" for preventing a myriad of chronic diseases. The results of intervention studies aimed at evaluating whether Mediterranean-type diets are superior to classic diets in the secondary prevention of coronary heart disease have all been encouraging. The biologic mechanisms by which these compounds might exert their effects include, among others, antioxidant functions and induction of detoxification enzymes. However, from a public health perspective it is not essential to wait for elucidation of every mechanism underlying health promotion activities and interventions; given the simplicity of the diet quality score, increasing the intake of recommended foods represents a practical recommendation for improving health.

KEYWORDS: Mediterranean diet; coronary heart disease; reactive oxygen species; antioxidants; vegetable foods

INTRODUCTION

Nearly 50 years ago, Keys[1] recognized the enormously divergent rates of heart disease around the world, even after adjusting for differences in age. Although coronary disease was and remains the leading cause of death in the United States and many developed and developing countries, it was almost nonexistent in the traditional cultures of Crete and Japan. The concept of the Mediterranean diet originated from the Seven Countries Study initiated by Keys in the 1950s. The study showed that despite a high fat intake, the population of the island of Crete in Greece had very low rates of coronary heart disease and certain types of cancer and had a long life expectancy. The traditional dietary patterns typical of Crete, much of the rest of Greece, and

Address for correspondence: Dario Giugliano, Division of Metabolic Diseases, Department of Geriatrics and Metabolic Diseases, Cardiovascular Research Centre, Second University of Naples, Naples, Italy. Voice and fax: ++39 081 5665054.
dario.giugliano@unina2.it

Ann. N.Y. Acad. Sci. 1056: 253–260 (2005). © 2005 New York Academy of Sciences.
doi: 10.1196/annals.1352.012

southern Italy in the early 1960s were considered to be largely responsible for the good health observed in these regions. Again, results from the Seven Countries Study indicate that the two Greek and the three Italian cohorts had the lowest death rates from all causes among the European and U.S. cohorts in the study. Moreover, results of the 25-year follow-up study indicate that the occurrence of cardiovascular disease was much lower in southern European countries than in northern European countries.[2] Although ecological interpretations of these data are beset by difficulties, a potential explanation for the lower cardiovascular mortality rates among Mediterranean populations is their traditional diet.

Mortality statistics from the WHO database covering the period 1960 to 1990 provide intriguing evidence that something unusual has been affecting the health of the Mediterranean population in a beneficial way, particularly with respect to coronary heart disease. Even though health care for many of these populations was inferior to that available to people in Northern Europe and North America and the prevalence of smoking was unusually high, death rates in the Mediterranean region were generally lower and adult life expectancy generally higher than those of the economically more developed countries, particularly among men.[3] In a recent large prospective survey involving about 22,000 adults from Greece (the Greek cohort of the EPIC study), an inverse correlation was shown between a greater adherence to the Mediterranean diet and death.[4] In particular, approximately a 2/9 increment in the Mediterranean diet score was associated with a 25% reduction in total mortality and a 33% reduction in mortality from coronary heart disease. These associations were independent of sex, smoking status, level of education, body mass index, and level of physical activity. Moreover, the relation was significant among participants 55 years of age or older, but not among younger participants, indicating increasing cumulative exposure to a more healthy diet. In other studies, adherence to similar healthful lifestyle practices was found to be associated with an 83% reduction in the rate of coronary disease, a 91% reduction in diabetes in women, and a 71% reduction in colon cancer in men.[5]

WHAT IS THE MEDITERRANEAN DIET?

Defining a Mediterranean diet is challenging, given the broad geographic region, including at least 16 countries, that borders the Mediterranean Sea. There are cultural, ethnic, religious, economic, and agricultural production differences that result in different dietary practices in these areas and that preclude a single definition of a Mediterranean diet. Over the years, the Mediterranean diet has been promoted as a model for healthy eating. The diet, however, has been surrounded by as much myth as scientific evidence. There is no single "Mediterranean diet." Moreover, the differences that Keys observed in mortality from coronary disease among different populations could well be attributable to confounding by other lifestyle-related factors, such as physical activity. Nonetheless, the traditional Mediterranean diet, based on food patterns typical of many regions in Greece and Southern Italy in the early 1960s, may be thought of as having the components indicated in TABLE 1. The main characteristics of the Mediterranean diet include an abundance of plant food (fruits, vegetables, whole grain cereals, nuts, and legumes); olive oil as the principal source of fat; fish and poultry consumed in low to moderate amounts; relatively low con-

TABLE 1. Dietary pattern characteristic of the traditional Mediterranean diet

High consumption of fruits (3–4 servings/day)
High consumption of vegetables (2–3 servings/day)
High monounsaturated-to-saturated fat ratio (2)
Daily consumption of cereals (whole grains) and legumes
Moderate consumption of fish and nuts (4–5 servings/week)
Moderate consumption of wine (1–2 wineglasses/day)
Low consumption of meat and meat products (4–5 servings/month)
Daily consumption of low-fat dairy products

sumption of red meat; and moderate consumption of wine, normally with meals. So, the dietary patterns that prevail in the Mediterranean region have many common characteristics: total lipid intake may be high, as in Greece (around or in excess of 40% of total energy intake), or moderate, as in Italy (around 30% of total energy intake); the Italian variant of the Mediterranean diet is characterized by a higher consumption of pasta, whereas in Spain, fish consumption is particularly high. In all instances, however, the ratio of monounsaturated to saturated fats is much higher than that in other places of the world, including northern Europe and North America. It has become customary to represent the Mediterranean diet in the form of a triangle (pyramid), the base of which refers to foods that are to be consumed most frequently and the top to those to be consumed rarely, with the remaining foods occupying intermediate positions.

OXIDANTS AND ANTIOXIDANT DEFENSE

Until recently, evaluation of the Mediterranean diet focused on the low content of saturated fatty acids and the high content of composite carbohydrates and dietary fiber. Recent studies imply that other compounds of the Mediterranean diet, the antioxidants, which exist in abundance in vegetables, fruit, beverages, and also virgin olive oil, may contribute to the prevention of coronary heart disease and possibly several forms of cancer and other diseases, thus providing a plausible explanation for its apparent benefits.

Oxidants are products of normal aerobic metabolism and the inflammatory response. Paradoxically, aerobiosis is inevitably associated with the generation of reactive oxygen species (ROS) that damage biologic macromolecules such as DNA, carbohydrates, lipids, and proteins. Some ROS are radicals, that is, atoms or molecules that contain at least one unpaired electron in their external orbital. This particular configuration confers a very high chemical reactivity owing to the presence of an electromagnetic field, normally absent when an electron pair is present in the external orbital. Examples of reactive radical compounds are the hydroxyl radical ($HO^{\cdot}$), the peroxyl radical (ROO^{-}), and the superoxide anion radical (O_2^{-}). Reactive nonradical compounds are hydrogen peroxide (H_2O_2), peroxynitrite ($ONOO^{-}$), and singlet oxygen ($^1O_2$).

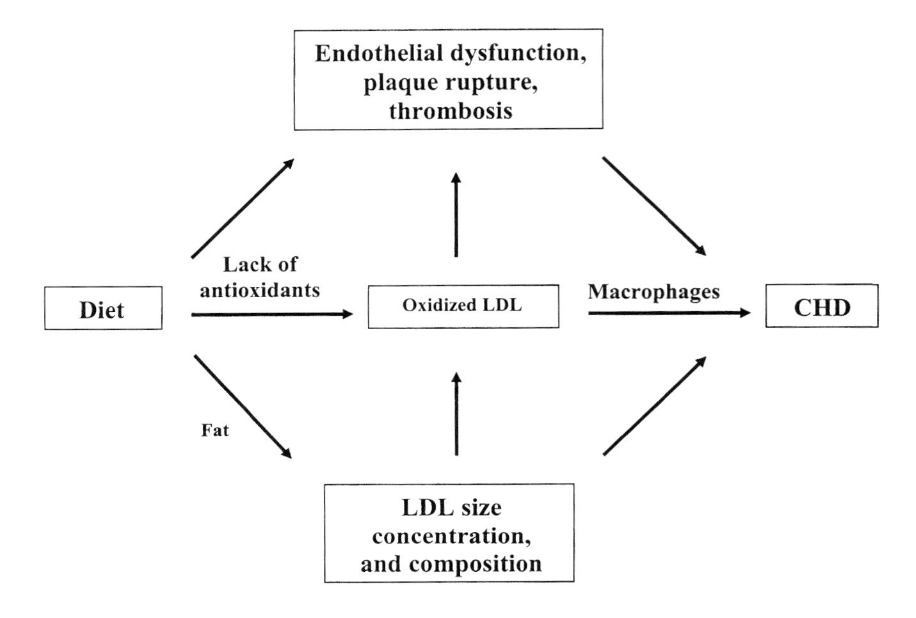

FIGURE 1. Some possible mechanisms through which a lack of antioxidants in the diet may promote coronary heart disease (CHD).

The organism defends itself against ROS species. The main enzymes directly involved in the detoxification of ROS are superoxide dismutase, catalase, and glutathione peroxidases. The antioxidant defense also includes a variety of exogenously (diet) derived compounds. The most prominent dietary antioxidants are vitamin C, vitamin E, and beta carotene. An imbalance between ROS production and antioxidant defense systems may increase the oxidative burden (oxidative stress) and lead to the damage of macromolecules.[6]

A shortage of antioxidants in the diet might promote coronary heart disease (CHD) through mechanisms other than the accumulation of oxidized low-density lipoproteins (LDLs) in macrophages. The occurrence of CHD depends not only on the rate at which atherosclerotic plaques grow, but also on endothelial function, smooth muscle cell proliferation, thrombosis, and plaque rupture (FIG. 1). For example, a single high-fat meal rich in saturated fat (Western meal) impairs endothelial functions in healthy subjects; this does not occur when the same subjects eat an isocaloric high-carbohydrate meal (pizza), as evidenced by the unchanged plasma levels of intercellular adhesion molecule-1 (ICAM-1) and vascular cell adhesion molecule-1 (VCAM-1), proinflammatory cytokines tumor necrosis factor-α (TNF-α) and interleukin-6, and the unchanged vascular response to L-arginine, the natural precursor of nitric oxide.[7,8]

FROM ALIMENTS TO ELEMENTS AND BACK

Because oxidative processes are important in the development of atherosclerosis and in the light of epidemiologic associations that provide support for the potential health benefits of foods rich in natural antioxidants, antioxidant-vitamin supplementation has been proposed for the treatment and prevention of coronary disease. Epidemiologic associations do not indicate causality, however. The protective effects of vegetables and fruit are not observed with pharmacologic doses of plant foods or their constituents. Currently, there is little direct experimental evidence from randomized trials in a primary prevention setting. The encouraging controversial results of short-term trials in participants with coronary atherosclerosis[9] were followed by two mega-trials with vitamin E in a postmyocardial infarction population[10] and in patients at high cardiovascular risk[11] who failed to show clinical benefits. Recently, Miller and colleagues[12] reported the results of a carefully conducted meta-analysis of clinical trials of vitamin E supplementation. They conclude that high doses of this agent increase the risk of death. Their meta-analysis involved data from 19 randomized trials, which recorded 12,504 deaths. Overall, being randomly assigned to receive vitamin E had no effect, either positive or negative. However, the data suggested a decreased risk of death associated with vitamin E in trials that used lower doses (<400 IU) and showed a statistically significant trend towards increased risk at doses of 400 IU and above.

It may be misleading to focus on a single element of the diet; this may explain, at least in part, the disappointing and frustrating results obtained in trials with vitamin supplementation, prematurely thought to be "the magic bullet" for preventing a myriad of chronic diseases. Guidelines from some professional or governmental panels recommend that attempts be made to obtain vitamins and minerals from food sources rather than supplements.[13]

INTERVENTION STUDIES

The first clinical-trial evidence in support of the health benefits of the Mediterranean diet came from the Lyon Diet Heart Study, in which 605 patients who had had a myocardial infarction were randomly assigned to a "Mediterranean-style" diet or a control diet resembling the American Heart Association Step I diet. The aim of the Lyon Diet Heart Study was to test whether a Mediterranean type of diet was superior to the classic dietary counseling given by cardiologists in the secondary prevention of coronary disease.[14] The Mediterranean diet model supplied 30% of energy from fats and less than 10% of energy from saturated fatty acids. Regarding essential fatty acids, the intake of 18:2 (n-6) (linoleic acid) was restricted to 4% of energy and the intake of 18:3 (n-3) (alpha-linolenic acid) provided more than 0.6% of energy. In practical terms, the dietary instructions could be summarized as follows: more grains and bread, more cereals, legumes, and beans, more fresh vegetables and fruits, more fish, less meat (beef, lamb, and pork), sausages, and luncheon meats (to be replaced by poultry), and no butter and cream, to be replaced by canola oil-based margarine with a *trans* fatty acid content of less than 5%. Finally, the oils recommended for salad and cooking were exclusively olive and canola (erucid acid–free rapeseed oil) oils. In the study, the most frequent nonfatal events were new acute myocardial infarction and episodes of unstable angina that are commonly due to rupture of an atherosclerotic plaque. The risk of these two endpoints was reduced by about 70% by the Mediterranean diet, indicating that biologic changes associated with it resulted in a significant local anti-inflammatory effect. The number of new episodes of heart failure was also reduced in the Mediterranean group. Total cholesterol, systolic blood pressure, leukocyte count, female gender, and aspirin were each significantly and independently associated with recurrence, indicating that the Mediterranean diet does not alter the usual relations between risk factors of coronary disease and recurrence, at least quantitatively.

Singh *et al.*[15] tested an "Indo-Mediterranean diet" in 1,000 patients in India with existing coronary disease or at high risk for coronary disease. As compared with the control diet, the intervention diet—characterized by increased intake of mustard or soybean oil, nuts, vegetables, fruits, and whole grains—reduced the rate of fatal myocardial infarction by one third and the rate of sudden death from cardiac causes by two thirds.

Knoops and colleagues[16] from the Netherlands, France, Spain, and Italy showed that in European men and women aged 70 through 90 years adherence to a Mediterranean diet pattern, moderate alcohol consumption, nonsmoking status, and physical activity each were associated with a lower rate of all-cause mortality. Taken together, the combination was associated with a mortality rate of about one third that of those with none or only one of these protective factors. These healthful behaviors were not extreme; for example, the physical activity criterion could be met by half an hour of walking daily.

Esposito and colleagues[17] explored the possible mechanisms underlying dietary intervention. The investigators randomized 180 patients (99 men, 81 women) with the metabolic syndrome to a Mediterranean-style diet (instructions about increasing daily consumption of whole grains, vegetables, fruits, nuts, and olive oil; $n = 90$) vs. a cardiac-prudent diet with fat intake less than 30% ($n = 90$). Physical activity increased equally in both groups. After 2 years, body weight decreased more in the intervention group than in the control group, but even after controlling for weight loss, inflammatory markers and insulin resistance declined more in the intervention than in the control group, whereas endothelial function improved. Only 40 patients in the intervention group still had the metabolic syndrome after 2 years compared with 78 patients on the control diet. These results suggest a plausible mechanism for the beneficial effects of the Mediterranean diet.

SUMMARY AND CONCLUSIONS

Despite increased public awareness of the importance of diet in decreasing the risk of chronic disease, large gaps remain in food-based recommendations and the actual dietary practice of the population. A substantial body of knowledge demonstrates that the abundant consumption of food of plant origin, including vegetables, fruit, and whole grain, and the dietary patterns rich in these foods conveys a markedly lower risk of coronary disease. From a public health perspective it is not essential to wait for elucidation of every mechanism underlying health promotion activities and interventions; given the simplicity of the diet quality score, increasing the intake of certain foods represents a practical recommendation for improving health.

A recent statement from the American Heart Association declares that the most prudent and scientifically supportable recommendation for the general population is to consume a balanced diet with emphasis on antioxidant-rich fruits and vegetables and whole grains.[13] When diet provides a sufficient supply of antioxidants, there is no need for supplements. The non-Mediterranean populations should probably take advantage of the opportunity to dramatically lower the risk of cardiovascular disease by widely consuming a diet that features a dietary pattern that includes fruits, root vegetables, leafy green vegetables, breads and cereals, fish, and food high in linolenic acid, such as vegetable oils (olive, canola) and nuts and seeds.

Higher levels of consumption of olive oil are considered to be the hallmark of the traditional Mediterranean diet. For centuries, olive oil has been treasured in Greece and other Mediterranean countries for its healing and nutritional properties. The use of olive oil now extends beyond the Mediterranean region. Cumulative evidence suggests that olive oil may have a role in the prevention of coronary disease and several types of cancer because of its high levels of monounsaturated fatty acids and polyphenolic compounds.

Other important plant-based sources of monounsaturated fatty acids include nuts and rapeseed (canola) oil. Monounsaturated fat, whether from olive oil or other sources, may have the same beneficial effects on blood lipids and oxidative stress, but this possibility has not been fully studied. Both the Lyon Diet Heart Study and the recent Indian study have emphasized rapeseed, or canola, oil as a source of linolenic acid. Thus, a Mediterranean-type diet, when translated into other cultures, can use food options beyond olive oil for increasing the intake of monounsaturated fats and polyunsaturated fats at the expense of saturated and *trans* fats and refined carbohydrates. It is worth noting that traditional diets from Mediterranean and Asian countries share most dietary characteristics, such as a relatively high intake of fruits, vegetables, nuts, legumes, and minimally processed grains, despite the use of different sources of plant oils.

Dietary patterns in Greece and other Mediterranean countries are changing rapidly, with increased consumption of saturated fat and refined carbohydrates. The healthy Mediterranean diet in Italy is being abandoned by the population. Keys and others have lamented the loss of the "good Mediterranean diet" and called for the reversal of the current dietary trend.

From a historic perspective,[18,19] the association of the Mediterranean diet with some of the greatest ancient civilizations–Greek, Etruscan, and Roman–may have been coincidental, although the pioneering British nutritionist John Waterloo has argued that: "It is difficult to conceive how the Greeks and Romans could have achieved such remarkable feats, which involved far more than a small elite, if they had not in general had an adequate and nourishing diet."

REFERENCES

1. Nestle, M. 1995. Mediterranean diets: historical and research overview. Am. J. Clin. Nutr. **61:** 1313S–1320S.
2. Menotti, A. 2000. Coronary heart disease incidence in northern and southern European populations: a reanalysis of the Seven Countries Study for a European coronary risk chart. Heart **84:** 238–244.
3. Willett, W.C. 1994. Diet and health: what we should eat? Science **264:** 532–537.

4. TRICHOPOULOU, A. *et al.* 2003. Adherence to a Mediterranean diet and survival in a Greek population. N. Engl. J. Med. **348:** 2599–2608.
5. RIM, E.B. & R.J. STAMPFER. 2004. Diet, lifestyle, and longevity: the next steps? JAMA **292:** 1490–1492.
6. HALLIWELL, B. 1996. Antioxidants in human health and disease. Ann. Rev. Nutr. **16:** 33–50.
7. GIUGLIANO, D. *et al.* 2001. Pizza and vegetables don't stick to the endothelium. Circulation **104:** e34–35.
8. NAPPO, F. *et al.* Postprandial endothelial activation in healthy subjects and type 2 diabetic patients: role of fat and carbohydrate meals. J. Am. Coll. Cardiol. **39:** 1145–1150.
9. RIMM, E.R. & M.J. STAMPFER. 1997. The role of antioxidants in preventive cardiology. Curr. Opin. Cardiol. **12:** 188–194.
10. GISSI-PREVENZIONE INVESTIGATORS (Gruppo Italiano per lo Studio della Sopravvivenza nell'Infarto miocardico). 1999. Dietary supplementation with n-3 polyunsaturated fatty acids and vitamin E after myocardial infarction: results of the GISSI-Prevenzione trial. Lancet **354:** 447–455.
11. THE HEART OUTCOMES PREVENTION EVALUATION STUDY INVESTIGATORS. 2000. Vitamin E supplementation and cardiovascular events in high-risk patients. N. Engl. J. Med. **342:** 154–160.
12. MILLER, E.R. *et al.* 2005. Meta-analysis: high-dosage vitamin E supplementation may increase all-cause mortality. Ann. Intern. Med. **142:** 37–46.
13. KRIS-ETHERTON, P.M. *et al.* for the Nutrition Committee of the American Heart Association Council on Nutrition, Physical Activity, and Metabolism. 2004. Antioxidant vitamin supplements and cardiovascular disease. Circulation **110:** 637–641.
14. DE LORGERIL, M. *et al.* 1999. Mediterranean diet, traditional risk factors and the rate of cardiovascular complications after myocardial infarction. Final report of the Lyon Diet Heart Study. Circulation **99:** 779–785.
15. SINGH, R.B. *et al.* 2002. Effect of Indo-Mediterranean diet on progression of coronary disease in high risk patients: a randomized single blind trial. Lancet **360:** 1455–1461.
16. KNOOPS, K.T.B. *et al.* 2004. Mediterranean diet, lifestyle factors, and 10-year mortality in elderly European men and women: the HALE Project. JAMA **292:** 1433–1439.
17. ESPOSITO, K. *et al.* 2004. Effect of a Mediterranean-style diet on endothelial dysfunction and markers of vascular inflammation in the metabolic syndrome: a randomized trial. JAMA **292:** 1440–1446.
18. GIUGLIANO, D. *et al.* 2000. The Mediterranean Diet: origins and myths. Idelson-Gnocchi Publishers. Reddick, FL.
19. GIUGLIANO, D. *et al.* 2001. The Way They Ate. Origins of the Mediterranean diet. Idelson-Gnocchi Publishers. Reddick, FL.

Antiulcer and Antioxidant Activity of *Asparagus racemosus* WILLD and *Withania somnifera* DUNAL in Rats

MAHEEP BHATNAGAR,[a] SIDDHRAJ S. SISODIA,[b] AND REKHA BHATNAGAR[c]

[a]*Department of Zoology, University College of Science, Mohan Lal Sukhadia University, Udaipur-313001, India*

[b]*Department of Pharmacology, B.N. College of Pharmacy, Udaipur-313001, India*

[c]*Department of P.S.M., R.N.T. Medical College, Udaipur-313001, India*

ABSTRACT: Comparative study of the antiulcer and antisecretory activity of *Asparagus racemosus* Willd (Shatawari) and *Withania somnifera* Dunal (Ashwagandha) root extract with a standard drug, ranitidine, in various models of gastric ulcer in rats is presented. Ulcer was induced by the indomethacin (NSAID) and swim (restraint) stress treatment. Results demonstrated that *A. racemosus* as well as *W. somnifera* methanolic extract (100 mg/kg BW/day p.o.) given orally for 15 days significantly reduced the ulcer index, volume of gastric secretion, free acidity, and total acidity. A significant increase in the total carbohydrate and total carbohydrate/protein ratio was also observed. Study also indicated an increase in antioxidant defense, that is, enzymes superoxide dismutase, catalase, and ascorbic acid, increased significantly, whereas a significant decrease in lipid peroxidation was observed. *A. racemosus* was more effective in reducing gastric ulcer in indomethacin-treated gastric ulcerative rats, whereas *W. somnifera* was effective in stress-induced gastric ulcer. Results obtained for both herbal drugs were comparable to those of the standard drug ranitidine.

KEYWORDS: cytoprotection; gastric ulcer; mucosal defense; ulcer protection; *Asparagus racemosus*; *Withania somnifera*; antioxidant defense

INTRODUCTION

Peptic ulcer is a very prevalent gastrointestinal disorder, characterized by disruption of the mucosal integrity attributed to various aggressive factors (acid, pepsin, stress, *Helicobacter pylori,* and NSAIDs) and defensive factors (mucus bicarbonate, blood flow, prostaglandins, etc). Clinically, regulation of gastric acid secretion is a major therapeutic target in the management of the disease.[1]

Address for correspondence: Dr. Maheep Bhatnagar, Department of Zoology, University College of Science, Mohan Lal Sukhadia University, Udaipur-313001, India. Voice: 91-0294-2413955 (O), 0294-2441250 (R).
mbhatnagar@yahoo.com

**Ann. N.Y. Acad. Sci. 1056: 261–278 (2005). © 2005 New York Academy of Sciences.
doi: 10.1196/annals.1352.027**

Most drugs currently used in ulcer treatment either counteract the effects of gastric acid secretion or exert cytoprotective effects. As they differ in speed of healing, time taken to achieve relief of symptoms, rate of ulcer relapse, tolerability, and their adverse effects on these, they are ideal therapeutically.[2] Certain medicinal plants such as plantain banana roots and the leaves of *Panax ginseng,*[5] shilajit,[6] *Cinnamon cassia,*[7] and ginger[8] have been reported to possess antiulcer[3,4,6] properties. Herbal formulation UL409 containing six medicinal plants including *Glycyrhiza glabra, Seauscerea lappa, Aegle marmelos, Foeniculum vulgare, Rosa damascena,* and *Santallum album* also demonstrated similar protective effects.[9] Antiulcer activity of *Azadirachta indica,*[10] *Melia azedarach,*[11] *Oscimum sanctum,*[12,13] *Centella asiatica,*[14] *Selageinella bryopteris,*[15] *Emblica officianalis,*[16] *Enatia clorontha,*[17] *Tephrosa purpurea,*[18] and *Mikania laevigata*[19] have also been reported. Bapna and Balraman[20] reported antiulcer and antioxidant properties of a herbomineral formulation. De *et al.*[21] for the first time showed the use of Shatawari fruit juice, restraint stress, and cystamine-induced gastric ulcer. Shatawari mandur also showed significant protection against acute gastric ulcer with little effect on acid and pepsin secretion.[22] Although the direct action of *Withania somnifera* on gastric ulcer has not been investigated, the drug is commonly prescribed as an antistress agent[23–25] and an antioxidant.[26–28] The cytoprotective actions of ashwagandha[29–31] have also been reported.

In view of these observations, the current study was carried out to compare the antiulcer and antisecretory activities of the root extracts of *Asparagus racemosus* Willd (Shatawari) and *Withania somnifera* Dunal (Ashwagandha) commonly used in the traditional and Ayurvedic medicinal systems in India, with the standard antiulcer drug ranitidine, in two experimental gastric ulcer rat models.

MATERIAL AND METHODS

Animals. Young adult (BW 2,255 gm) Wistar rats of an inbred strain and of either sex were used in the study. Animals were housed at controlled conditions of a light and dark cycle (12 h/12h) and a temperature of 22 ± 2°C. They were given a standard laboratory diet and water *ad libitum.*

Drug Preparation and Dose Schedule. Fresh roots of *A. racemosus* and *W. somnifera* were collected, cut into small pieces, and dried in a shed. After 7 days of drying, the roots were powdered and sieved. Powder was than macerated with a methanol and distilled water mixture for 24 hours, and then using the successive solvent methanol extraction procedure, extract was prepared. After vacuoevaporation, crude extract was suspended in carboxymethylcellulose. Animals were given a 100 mg/kg/BW daily dose orally using an oral feeding tube. Treatment was continued for 15 days.

Treatment. Gastric ulcer was induced by treatment with indomethacin and restraint swim stress.

a. Indomethacin-induced gastric ulcer (Mc Cofferty et al.[32]*)*: Rats were fasted for 24 hours, but had free access to water and were then treated with two doses of 10 mg/kg/BW of indomethacin at an interval of 15 hours.

b. Restraint swim (water immersion) stress: The method described by Brodie *et al.*[33] was used. Rats were fasted for 12 hours. They were immersed in water up to the xiphoid process for 6 hours daily up to 15 days. The temperature of the water was maintained at 242°C. Care was taken not to allow coprophagy.

Pyloric Ligation. Pyloric ligation was carried out to collect the gastric contents. Rats were anesthetized with pentobarbitone sodium (40 mg/kg i.p) and the abdomen was opened by a small midline incision below the xiphoid process. The pyloric portion of the stomach was slightly lifted out and ligated. After ligation, the stomach was placed inside carefully, and the abdominal wall was sutured.

Rats were sacrificed at the end of experiment. Animals were decapitated after 4 hours of pyloric ligation. The abdomen was opened, the cardiac end of the stomach was dissected out, and the contents were drained into a tube. IThe inner surface of the gastric region was also examined for gastric lesions.

Experimental Groups.

Group I. Rats were given only vehicle (0.5% carboxymethylcellulose) each day.

Group II. A: Rats were treated with indomethacin (20 mg/kg/BW); B: rats were subjected to restraint swim stress for 6 hours.

Group III. A: Rats were given *A. racemosus* extract followed by indomethacin; B: rats were given drug extract followed by swim stress.

Group IV. A: Rats were given *W. somnifera* extract followed by indomethacin; B: rats were given drug extract followed by swim stress.

Group V. Rats were given only *A. racemosus* extract.

Group VI Rats were given only *W. somnifera* extract.

Group VII. A: Rats were given only ranitidine followed by indomethacin; B: rats were given only ranitidine followed by swim stress.

Parameters. Histologic study: At autopsy, pieces of tissue, including the ulcer area, were fixed in 10% neutral chilled formalin for 12 hours. After rinsing with distilled water (3×2 changes) the next day, they were dehydrated, cleared in xylene, and embedded in paraffin wax. Section5 mm in thickness were cut and stained with hematoxylin and eosin.

Physical parameters: Ulcer index (score and mucosal area) was determined by the methods of Ganguli and Bhatnagar[35] and Goyal.[36]

Biochemical parameters: Volume of gastric secretion was determined by the method of Parmar *et al.*[37] Free acidity (Hawk[38]), total acidity,[38,39] total carbohydrate,[40] total protein,[41] TC/TP ratio,[38,39] superoxide dismutase,[42] catalase,[43] malondialdehyde,[44] and ascorbic acid[45] were determined.

Statistics. All data are expressed as mean ± SEM. Statistical analysis was carried out using the unpaired *t* test.

RESULTS

Ulcer Index. After 3 days of indomethacin treatment, gastric ulcers were developed with an incidence of 90%. Pretreatment of rats with *A. racemosus* and *W. somnifera* extract up to 15 days in both indomethacin- and stress-induced ulcer rats significantly reduced the gastric ulcer index, that is, both score and mucosal area. In the indomethacin-treated group, the effects of *A. racemosus* extract in reducing the ulcer score (FIG. 1a) and mucosal area (FIG.1b) were more significant than those of the *W. somnifera* extract (FIG. 2a, score; FIG. 2b, mucosal area). In the stress group, *W. somnifera* showed more significant effects than did *A. racemosus* and ranitidine, but in both cases ranitidine was significantly more effective than the *A. racemosus* and *W. somnifera* extracts.

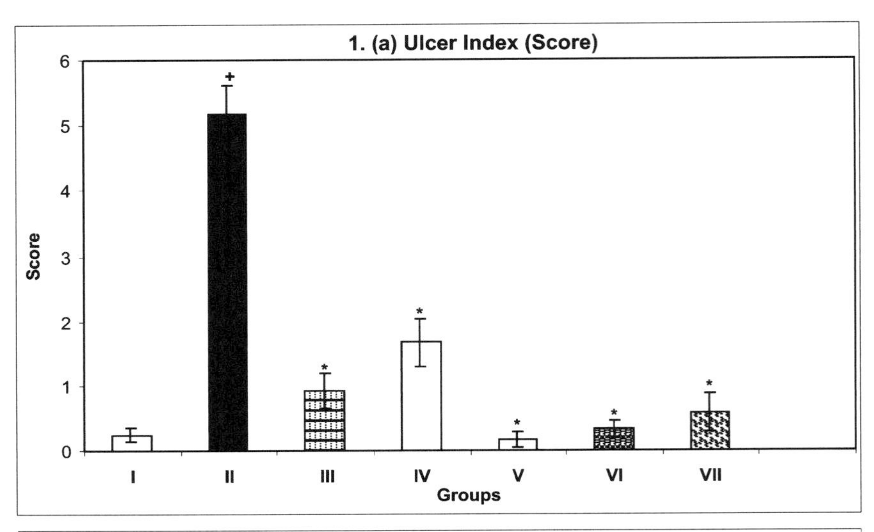

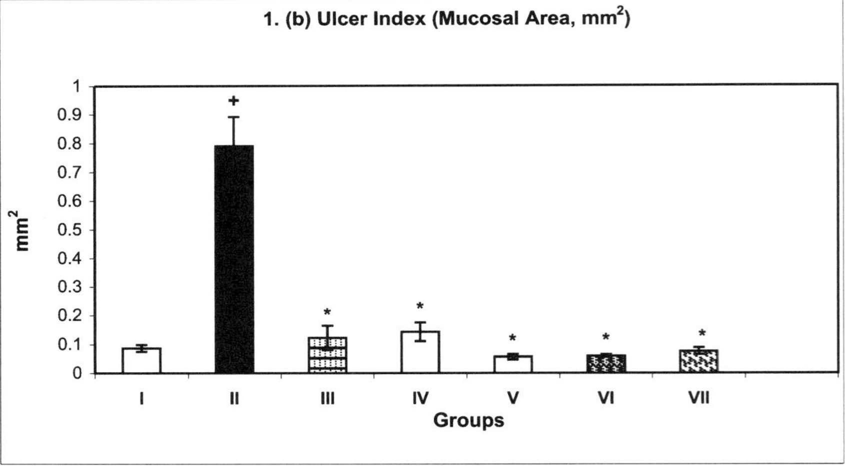

FIGURE 1. (**a**) Effects of *A. racemosus* and *W. somnifera* on ulcer index (score) in indomethacin-induced gastric ulcerative rats. All values are represented as mean ± SEM ($n = 6$). Groups: I. Control; II. Indo./Stress; III. Indo./Str. + *A.racemosus*; IV. Indo./Stress + *W. somnifera*; V. Only *A. racemosus*; VI. Only *W. somnifera*; VII. Indo./Stress + Ranitidine. (**b**) Effects of *A. racemosus* and *W. somnifera* on ulcer index (mucosal area) in indomethacin-induced gastric ulcerative rats. All values are represented as mean ± SEM ($n = 6$).

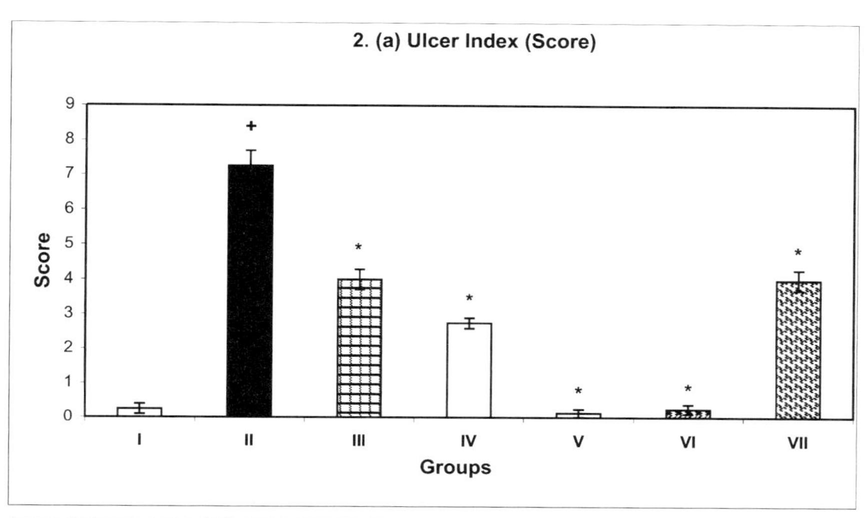

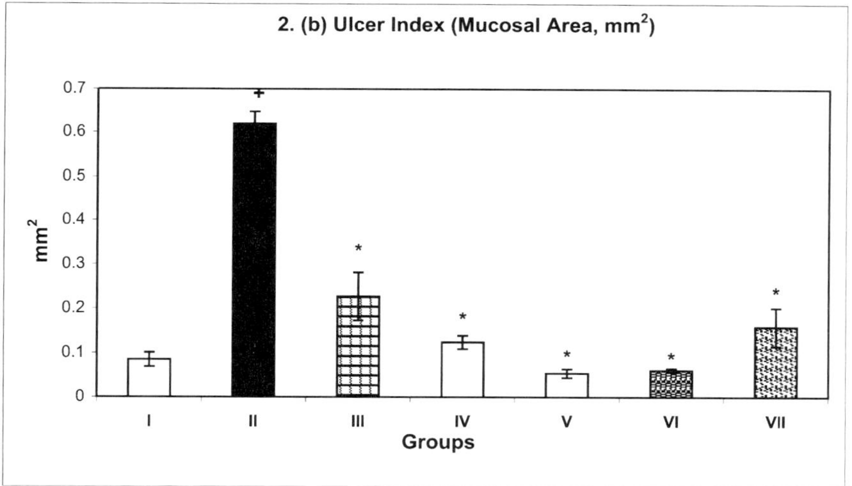

FIGURE 2. (**a**) Effects of *A. racemosus* and *W. somnifera* on ulcer index (score) in stress-induced gastric ulcerative rats. All values are represented as mean ± SEM ($n = 6$). (**b**) Effects of *A. racemosus* and *W. somnifera* on ulcer index in stress-induced gastric ulcerative rats. All values are represented as mean ± SEM ($n = 6$).

Volume of Gastric Contents and Free and Total Acidity. In the indomethacin group (FIG. 3a), although the effects of both herbal extracts and ranitidine treatment on gastric content were on par, in the stress group *W. somnifera* treatment showed a more significant decrease in gastric contents (FIG. 3b) than did *A. racemosus* and ranitidine treatment. In the indomethacin group, free and total acidity (FIG. 4a) also decreased significantly after treatment with both herbal extracts and ranitidine as well, but ranitidine treatment showed a more significant decrease than did the drug

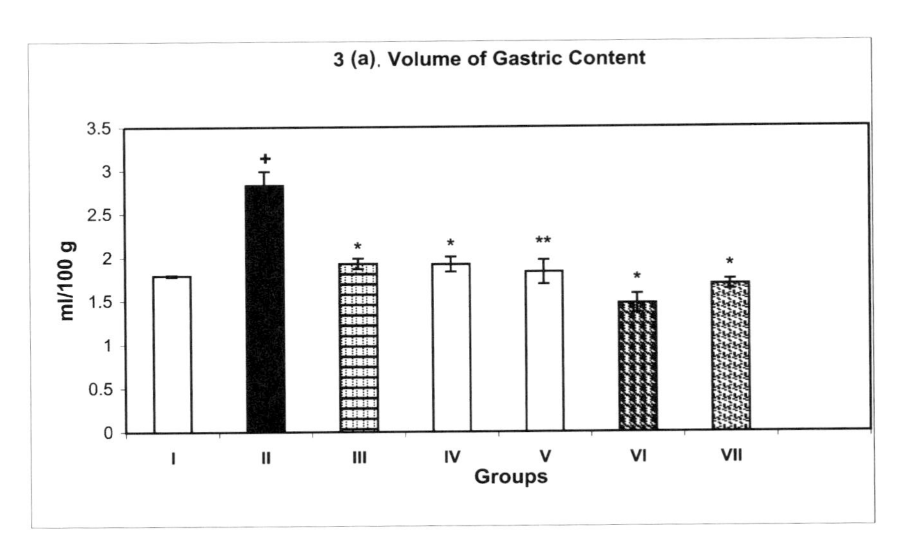

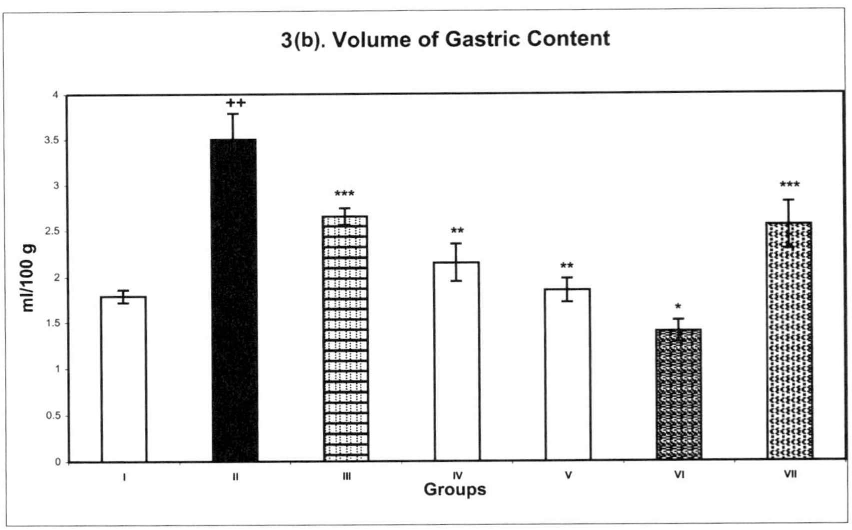

FIGURE 3. (**a**) Effects of *A. racemosus* and *W. somnifera* on gastric contents in indomethacin-induced gastric ulcerative rats. All values are represented as mean ± SEM (*n* = 6). (**b**) Effects of *A. racemosus* and *W. somnifera* on volume of gastric contents in stress-induced gastric ulcerative rats. All values are represented as mean ± SEM (n = 6).

extracts. Similarly, in the stress group, free and total acidity (FIG. 4b) showed a significant decrease after *W. somnifera* treatment compared to ranitidine and *A. racemosus* treatment.

Total Carbohydrate (TC), Total Protein (TP). and TC/TP Ratio. A significant increase in TC (FIG. 5a) and TC/TP (FIG. 6a) ratio was observed in the indomethacin-treated group after treatment with both the drug extracts and ranitidine, whereas

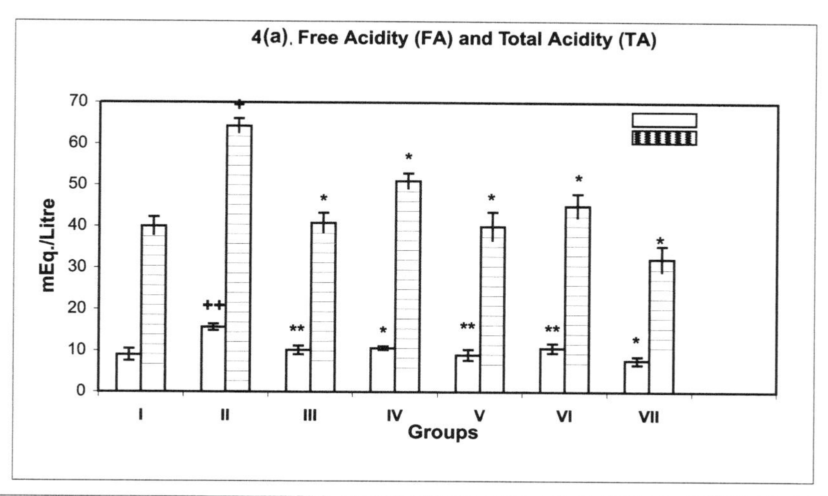

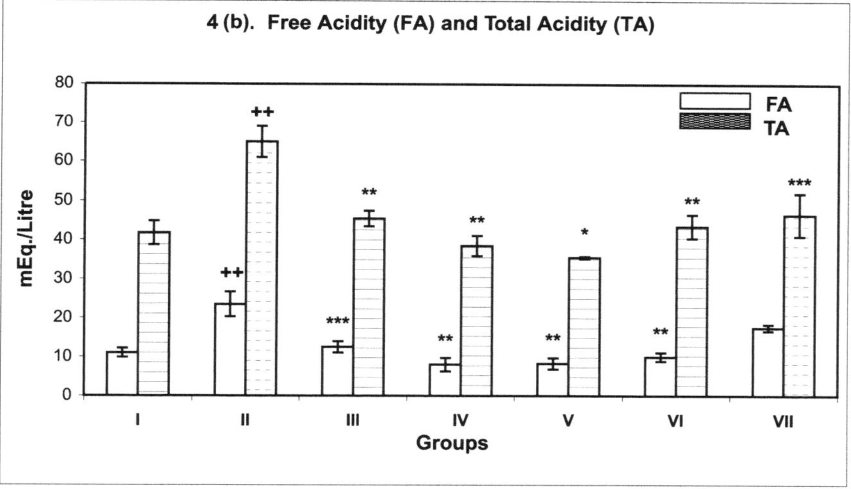

FIGURE 4. **(a)** Effect of *A. racemosus* and *W. somnifera* on volume of free acidity and total acidity in indomethacin-induced gastric ulcerative rats. All values are represented as mean ± SEM ($n = 6$). **(b)** Effect of *A. racemosus* and *W. somnifera* on volume of free acidity and total acidity in stress-induced gastric ulcerative rats.All values are represented as mean ± SEM (n = 6.)

in the stress-treated group, TC (FIG. 5b) showed a significant increase after treatment with drug extracts (III and IV) and ranitidine (VII) as compared to the stress group (II), but TP did not show any significant change in these groups compared with the stress group (II). The TC/TP ratio also showed a significant increase in all three groups (III, IV, and VII) compared with the stress group (II) in the indomethacin-treated group (FIG. 6a), but in the stress-treated group (FIG. 6b) a decrease in the ratio

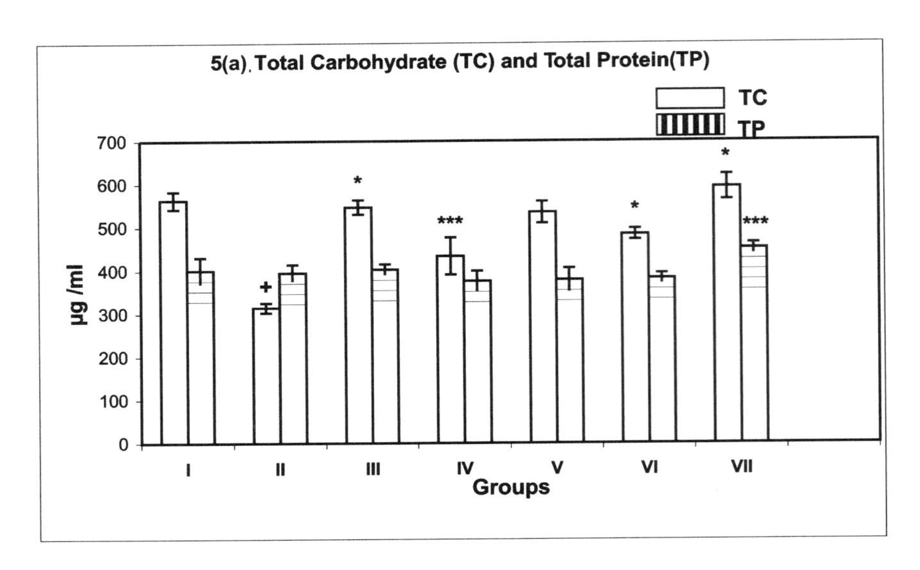

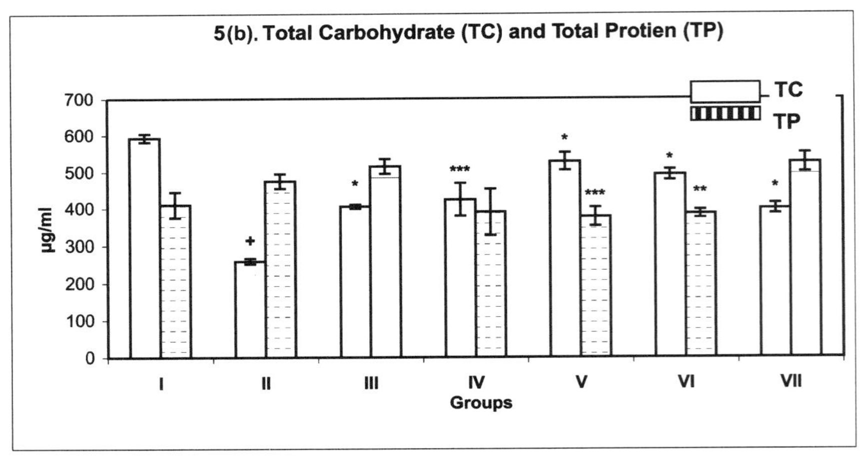

FIGURE 5. (**a**) Effects of *A. racemosus* and *W. somnifera* on TC and TP in indomethacin-induced gastric ulcerative rats All values are represented as mean ± SEM (n = 6). (**b**) Effects of *A. racemosus* and *W. somnifera* on TC and TP in stress-induced gastric ulcerative rats. All values are represented as mean ± SEM (n = 6).

was on par in *A. racemosus*- and ranitidine-treated groups compared with the stress group (II) and the *W. somnifera*-treated group (IV).

Antioxidant Activities. Superoxide dismutase (SOD): A significant increase in SOD was observed in both stomach and liver (FIG. 7a) after treatment with *A. racemosus* and *W. somnifera* extracts and ranitidine compared with the indomethacin-treated group (II). No significant difference was observed in SOD activity in both stomach and liver when these treatment groups (III, IV, and VIII) were compared. In

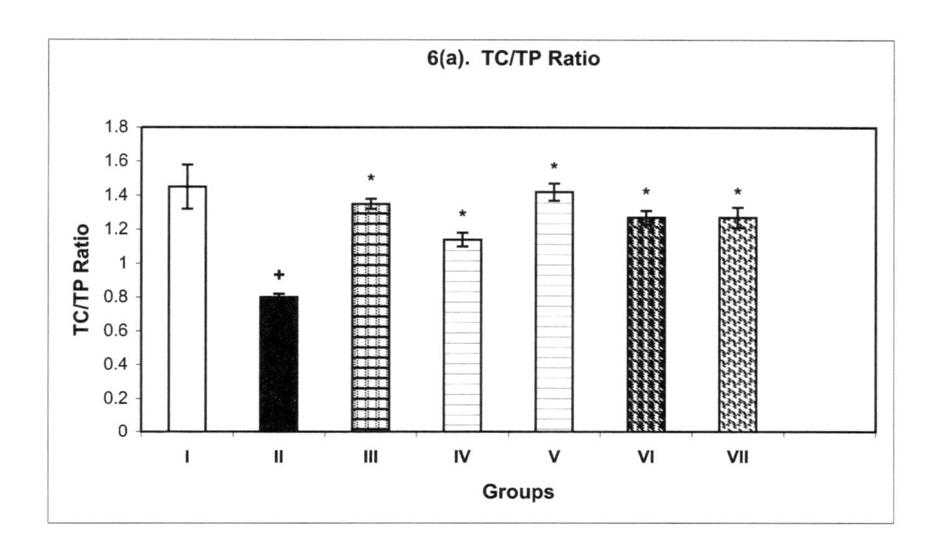

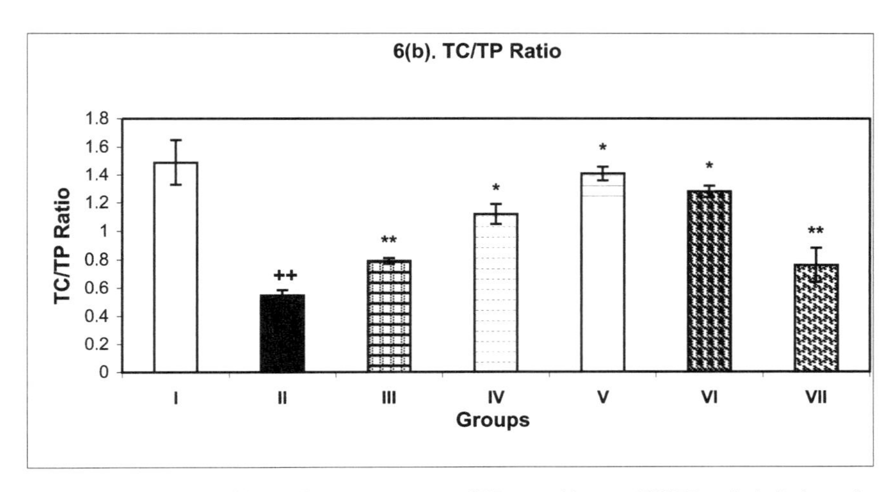

FIGURE 6. (**a**) Effects of *A. racemosus* and *W. somnifera* on TC/TP ratio in indomethacin-induced gastric ulcerative rats All values are represented as mean ± SEM (n = 6). (**b**) Effects of *A. racemosus* and *W. somnifera* on TC,TP and TC/TP ratio in stress-induced gastric ulcerative rats.All values are represented as mean ± SEM (n = 6).

the stress group (FIG. 7b), *W. somnifera* treatment showed a significant increase in SOD compared with both *A. racemosus* extract and ranitidine.

Catalase (CAT): In the indomethacin-induced ulcer group, CAT activity was significantly reduced in both the stomach and liver (FIG. 8a). After treatment with *A. racemosus* and *W. somnifera* extracts and ranitidine, a significant increase in CAT activity was observed. This increase was more significant in the *A. racemosus* group than in the *W. somnifera* and ranitidine-treated groups in the liver, but in the stomach

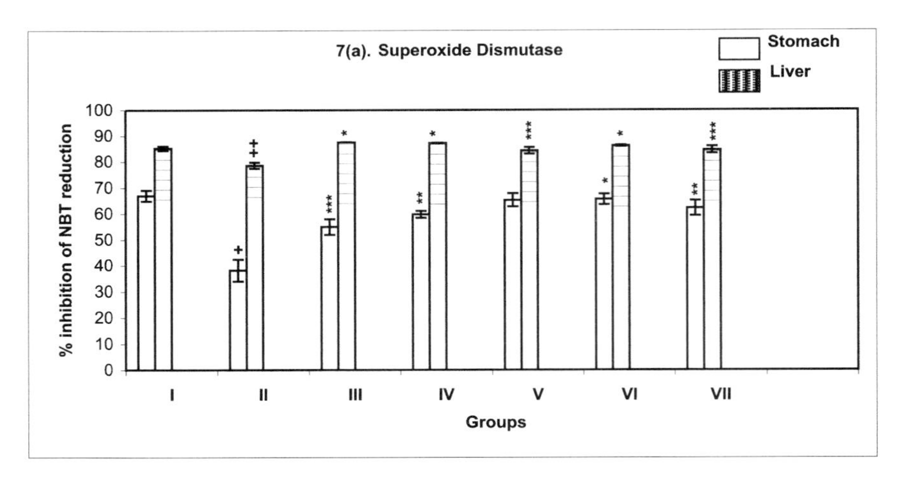

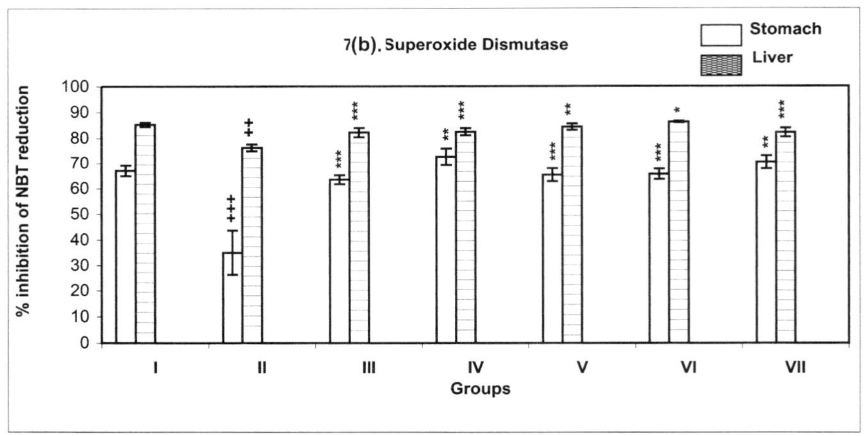

FIGURE 7. (**a**) Effects of *A. racemosus* and *W. somnifera* on SOD in indomethacin-induced gastric ulcerative rats. (**b**) Effects of *A. racemosus* and *W. somnifera* on SOD in stress-induced gastric ulcerative rats.

all three treatment groups (III, IV, and VII) showed an increase in CAT activity as compared to the ulcer group (II). In the stress group (FIG. 8b), in both liver and stomach, CAT activity was significantly reduced after stress treatment (II) compared with the controls. Although a significant increase in CAT activity was observed in all three treatment groups (III, IV, and VII), as in the indomethacin group, activity in the liver was on par in these groups. In the stomach after treatment with *A. racemosus* extract, CAT activity showed a significant increase compared with the control. *W. somnifera* and ranitidine treatment also showed a significant increase, but it was less significant than that with *A. racemosus*.

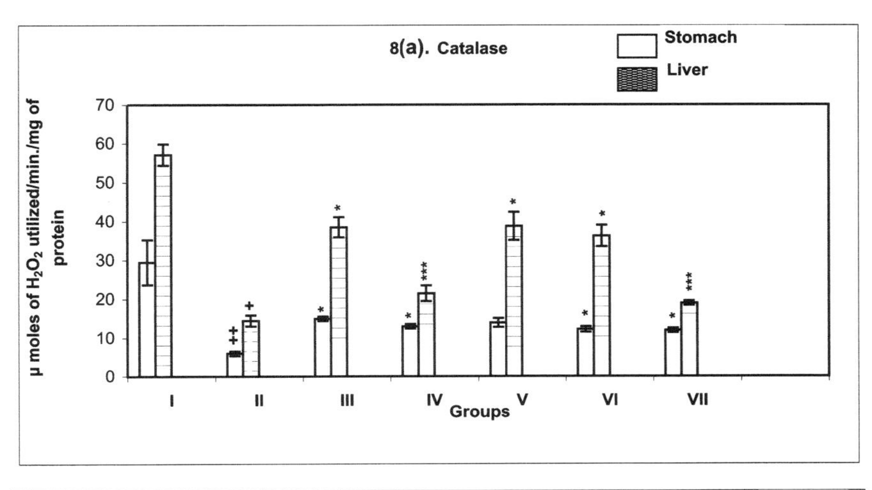

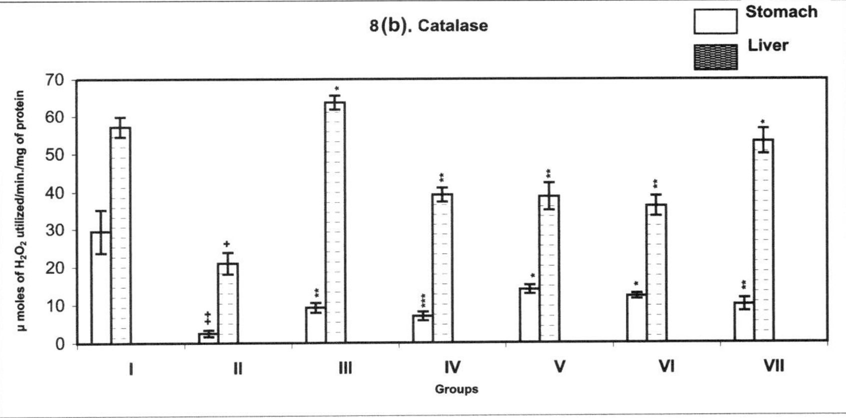

FIGURE 8. (**a**) Effects of *A. racemosus* and *W. somnifera* on CAT in indomethacin-induced gastric ulcerative rats. (**b**) Effects of *A. racemosus* and *W. somnifera* on CAT in stress-induced gastric ulcerative rats. All values are represented as mean ± SEM ($n = 6$).

Lipid peroxidation: MDA significantly increased in the stomach and liver after indomethacin (FIG. 9a) as well as stress (FIG. 9b) treatment. In the indomethacin group (FIG. 9a), ranitidine showed a more significant decrease in MDA in both the stomach and liver than did the herbal drugs. *A. racemosus, W. somnifera,* and ranitidine treatment significantly reduced the MDA level in both the stomach and liver, but changes were more pronounced in the stomach. Both *A. racemosus* and *W. somnifera* extract showed more significant changes in the stress group (FIG. 9b) than in the ranitidine group.

Ascorbic acid: Ascorbic acid also decreased significantly in both the stomach and liver after indomethacin (FIG. 10a) and stress (FIG. 10b) treatment compared to the control. In both of these ulcer groups, after treatment with A. *racemosus, W. somnifera,* and ranitidine, a significant increase in ascorbic acid levels was observed in both the stomach and liver. This increase was more significant with *A. racemosus* in the indomethacin-treated group (Fig. 10a) and with *W. somnifera* in the stress group (FIG. 10b).

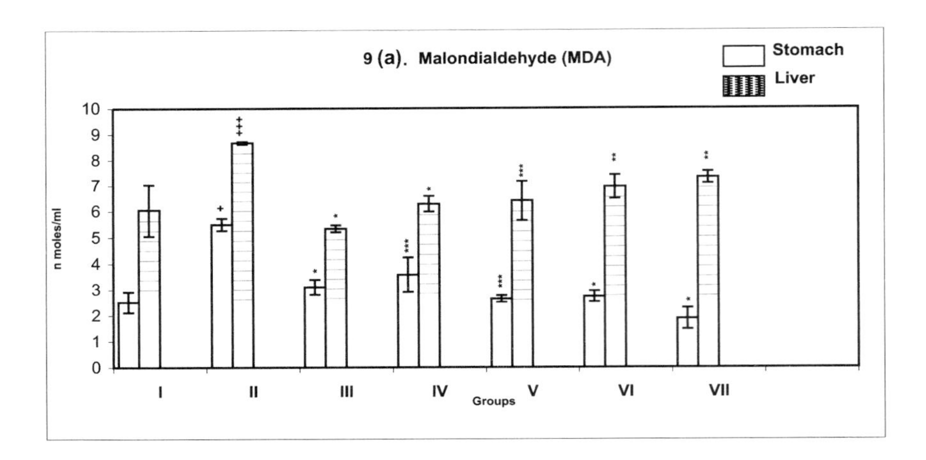

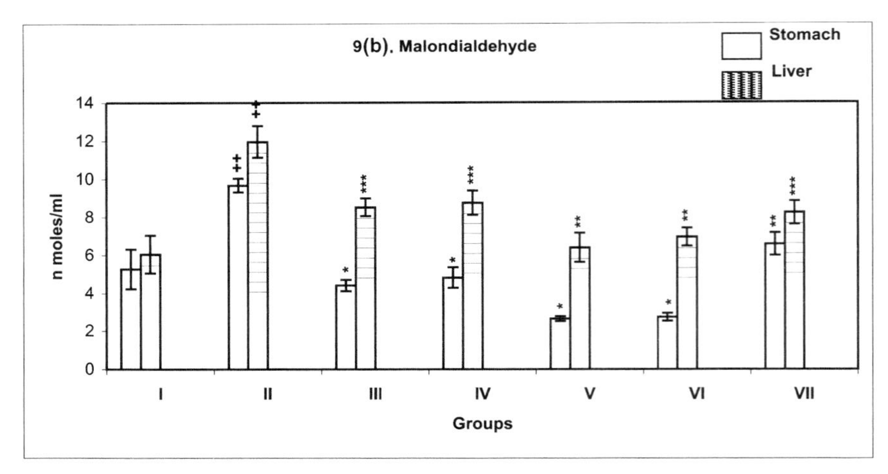

FIGURE 9. (**a**) Effects of *A. racemosus* and *W. somnifera* on MDA in indomethacin-induced gastric ulcerative rats. (**b**) Effects of *A. racemosus* and *W. somnifera* on MDA in stress-induced gastric ulcerative rats. All values are represented as mean ± SEM (n = 6).

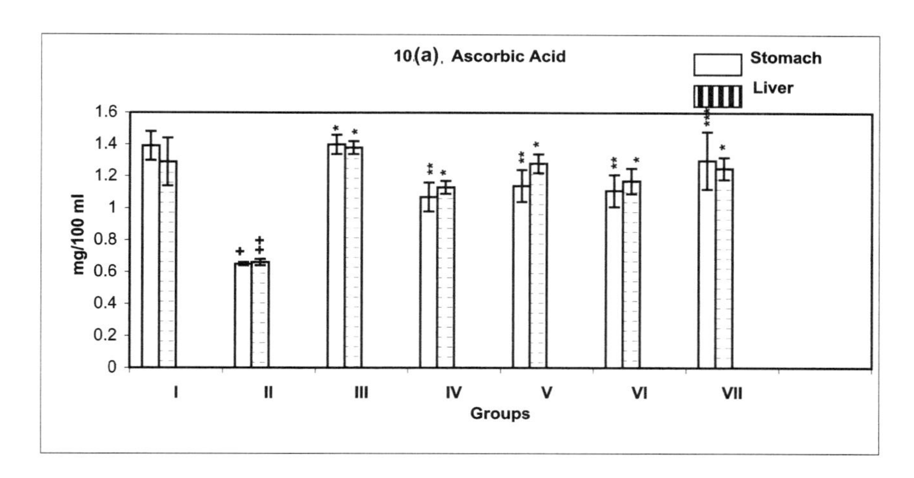

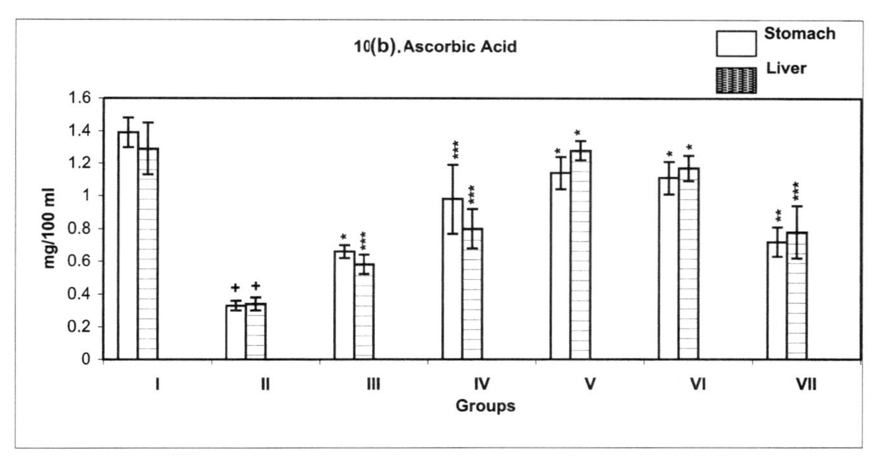

FIGURE 10. (**a**) Effects of *A. racemosus* and *W. somnifera* on ascorbic acid in indomethacin-induced gastric ulcerative rats. (**b**) Effects of *A. racemosus* and *W. somnifera* on ascorbic acid in stress-induced gastric ulcerative rats. All values are represented as mean ± SEM ($n = 6$).

DISCUSSION

This study compares both the antiulcer and the antioxidant activity of methanolic extracts of the roots of both *A. racemosus* Willd (Shatawari) and *W. somnifera* Dunal (Ashwagandha) with that of the commonly used conventional antiulcer drug ranitidine. Ulcer was induced in young adult rats by indomethacin treatment and a restraint swim stress protocol. Results demonstrate that indomethacin treatment as well as stress induces gastric ulcer with an incidence of 90%, and pretreatment of these rats with *A. racemosus* and *W. somnifera* root extracts daily for up to 15 days significantly affected the gastric ulcer index, that is, both score and mucosal area, as well as various secretory parameters such as gastric contents, free and total acidity, carbohydrates, TC/TP ratio along with antioxidant parameters such as SOD, CAT enzymes, MDA level, and ascorbic acid level. In the indomethacin group, *A. racemosus* showed a significant decrease in the score and mucosal area comparable to that of ranitidine, but in the stress group *W. somnifera* showed a more pronounced decrease than did *A. racemosus* and ranitidine. These changes are indicative of the effective antiulcer activity of *A. racemosus* and *W. somnifera* comparable to that of the standard drug ranitidine.

Pretreatment with the methanolic extract of both drugs also significantly reduced the volume of gastric secretion, free acidity, and total acidity. Decreased ulcer index and gastric contents as well as secretory parameters can be implicated with protective effects of the drugs. Histopathologic observations of the gastric surface also showed the healing activity, as the lesioned area was significantly reduced on the fifteenth day of treatment.

Studies of antioxidant properties of both the drugs and ranitidine also indicate antiulcer effects. A significant increase in SOD and CAT was observed after treatment with *A. racemosus, W. somnifera* extracts, as well as ranitidine in both the indomethacin- and stress-treated ulcer group. Lipid peroxidation significantly increased in both indomethacin- and stress-treated groups as indicated by a significant rise in the MDA level. Interestingly, *A. racemosus, W. somnifera,* and ranitidine treatment significantly reduced the MDA level in both stomach and liver tissues although changes were more pronounced in the stomach after ranitidine treatment. A significant increase in SOD and CAT activity and ascorbic acid contents after treatment with the *A. racemosus* and *W. somnifera* extract as well as ranitidine, both in stomach and liver tissue, and a significant decrease in lipid peroxidation (MDA) indicate an increase in antioxidant defense. The decrease in MDA levels in these tissues is suggestive of reduced tissue damage due to failure of the formation of excessive free radicals. Ascorbic acid regenerates vitamin E, and both are free radical scavengers.

Ascorbic acid showed the tissue vitamin status. After treatments with *A. racemosus, W. somnifera* extracts, and ranitidine, a significant increase in ascorbic acid levels was observed in both the stomach and liver of both indomethacin- and stress- treated ulcer groups. This increase was more significant after *A. racemosus* treatment in the indomethacin ulcer group, but more significant in the stress group after *W. somnifera* treatment.

Changes in gastric contents, free and total acidity, total carbohydrate, total protein, and TC/TP ratio are indicative of secretory activity. Results of the current study also demonstrate antisecretory activity of *A. racemosus* and *W. somnifera* root extracts as well as ranitidine. Both indomethacin- and stress-treated ulcer groups showed a significant decrease in the volume of gastric contents as well as free and total acidity after treatment with *A. racemosus, W. somnifera,* and ranitidine. In the indomethacin group, although the effects of both herbal extracts and ranitidine treatment were on par, in the stress group *W. somnifera* treatment showed a more significant decrease. A significant increase in TC, TP, and TC/TP ratio was also observed in the indomethacin and stress group after treatment with both drug extracts and ranitidine. In the indomethacin group, TC showed a significant increase after treatment with ranitidine and *A. racemosus* compared with TP. In the stress ulcer group, TC decreased significantly compared with TP. The TC/TP ratio showed a significant increase in both drug extract- and ranitidine-treated groups. In the indomethacin-treated ulcer group, *A. racemosus* treatment showed a significant increase in the TC/TP ratio, but in the stress-treated group the decrease in ratio was significant in the *W. somnifera*-treated group. Total carbohydrate is indicative of the mucus-secreting index, and TC/TP is taken as a reliable marker for mucin secretion,[47,48] which makes a protective layer over the stomach wall.

Earlier studies have shown that prostaglandins form a vital component in the gastric mucus defense. They are found throughout the gut in high concentration and are the major stimulus for synthesis in a gastric cell.[49] They may also act directly on parietal cells.[50] Robert[51] recognized that prostaglandins inhibit gastric acid secretion. Indomethacin inhibits prostaglandin synthesis, which results in severe gastric ulceration. Prostaglandin E_2 and I_2 are predominantly synthesized by the gastric mucosa and are known to inhibit the secretion of gastric acid but to stimulate secretion of mucus and bicarbonate. Hydrophobic surfactant-like phospholipid secretion in gastric epithelial cells is also stimulated by prostaglandins.[52,53] Protection by these drugs can therefore be implicated in the generation of prostaglandins. Observations in the current study clearly suggest that in the healing process of both *A. racemosus* and *W. somnifera* extracts, inhibition of acid enhances the healing of ulcers. In addition, antisecretory activity of these drugs may be due to coupled cytoprotective effects, which may be mediated through increases in prostaglandin secretion. In this study we also compared the healing efficacy of herbal drugs with the standard anti-ulcer drug ranitidine. Interestingly, it was observed that in the stress-treated group, *W. somnifera* was more effective than *A. racemosus*. This shows that the antistress activity of the *W. somnifera* could be useful in reducing the changes of stress-induced ulceration. Although healing efficacy was very similar with both herbal drugs and ranitidine, *A. racemosus* showed more protection in the indomethacin-induced ulcer group. Parmar and Parmar[54] and Deshpande *et al.*,[18] using plant extract of *Tephrosia purpurea,* recently demonstrated that flavonoids possess significant anti-ulcer activity. *A. racemosus* is rich in flavonoids,[55] and therefore our observation of the significant antiulcer and antisecetory activity of *A. racemosus* is well supported. Although ranitidine is commonly prescribed for ulcer treatment because H_2-receptor antagonists are accepted as being extremely safer drugs, fewer side effects, which include diarrhea, constipation, headache, drowsiness, fatigue, and muscular pain, occur as common complaints.[56,57] Valle[58] discussed reversible systemic toxicities of H_2-receptor antagonists, which include pancytopenia, neutropenia, anemia, and thrombocytopenia. In view of these, the herbals used in the current study provide a better alternative as antiulcer agents.

ACKNOWLEDGMENTS

We wish to thank the Department of Biotechnology, Government of India, for awarding a major research project to one of the authors (M.B.)

REFERENCES

1. Decktor, D.L., R.G. Pendelton, A.T. Kellner, *et al.* 1989. Acute effects of rantidine, famotidine and omneprazole on plasma gastrin in rats. J. Pharm. Exp. Thera. **249:** 1–5.

2. OKABE, S., K. INOUE, K. TAKAGI, *et al.* 1994. Comparative studies of antisecretory, antiulcer and mucosal protective effects of NL-1300-0-3, cimetidine and omeprezole in rats. Tera. Res.**15:** 349–366.
3. SANYAL, A.K., K.K. GUPTA & N.K. CHOUDHRY. 1964. Studies on peptic ulceration. Part I. Role of banana in phenylbutazone-induced ulcers. Arch. Int. Pharmacodyn. **149:** 339–400.
4. ELLIOT, R.C. & G.J.F. HOPEWOOD. 1976. The effects of banana supplemented diet on gastric ulcers in mice. Pharmacol Res. Commun. **8:** 167–171.
5. SUN, XIAO.BO., T. MATSUMOTO & H. YAMADA. 1992. Purification of an anti-ulcer polysaccharide from the leaves of Panax ginseng. Planta Med. **58:** 432–435.
6. GHOSAL, S., J.P. REDDY & V.K. LAL. 1976. Shilajit I. Chemical constituents. J. Pharm. Sci. **65:** 772.
7. AKIRA, T., S. TANAKA. & M. TABATA. 1986. Pharmacological studies on the antiulcerogenic activity of chianese cinnamon. Plant Med. **2:** 440–443.
8. YOSHIKAWA, M., S. HATAKEYAAMA, K. TANAGUCHIH. *et al.* 1992. 6-gingesulphonic acid a new antiulcer principle and ginger glycolipids. A, B and C, three new monoacyldigalactosylglycerols from Zingiberis rhizome originating in Taiwan. Chem. Pharm. Bull. **40:** 2239–2241.
9. KULKARNI, S.K. & R.K. GOYAL. 1996. Gastric anti-ulcer of UL-409 in rats. Ind. J. Exp. Biol. **34:** 683–686.
10. PILLAI, N.R. & G. SANTHAKUMARI. 1984. Effects of nimbdin on acute and chronic gastroduodenal ulcer models in experimental models. Planta Med. **46:** 143–146.
11. MOURSI, S.A.H & M.H. AL-KHATIB. 1984. Effect of Melia azedarach fruits on gipsing restraint stress induced ulcers in rats. Japan J. Pharmacol. **36:** 527–533.
12. MANDAL, S., D.N. DAS, K. DE, *et al.* 1993. Ocimum sanctum Linn A study on gastric ulceration and gastric secretion in rats. Ind. J. Physiol. Pharmacol. **37:** 91–92.
13. RAJ, K. & S. KAPOOR. 1999. Flavonoids. A review of biological activity. Ind. Drugs. **36:** 668–678.
14. CHATTERJEE, T.K, A. CHAKRABORTY, M. PATHAK & G.C. SENGUPTA. 1992. Effect of plant extract of Centella asiatica Linn on cold restraint stress ulcers in rats. Ind. J. Exp. Biol. **30:** 889–891.
15. PANDEY, S., M.M. KHAN, K. SHANKAR & N. SINGH. 1993. An experimental study on anti stress and antioxidant activity of *Selaginnnella bryopteris*. J. Biol. Chem. Res. **12:** 128–129.
16. MATHEW. S.M., S.B. RAO, G.R. NAIR & C.R.S. NAIR. 1995. Antiulcer activity of amla extract. Collected Abstracts of Int. Sem. on Recent Trends in Pharma. Sci. (Otacmund) :7, Feb. 18–20.
17. TAN, P.V., B. NYASSE., G.E. ENOW–OROCK, *et al.* 2000. Prophylactic and healing properties of new anti-ulcer compound from Enantia chlorentha in rats. Phytomedicine **7:** 291–296.
18. DESHPANDE, S.S., G.B. SHAH & N.S. PARMAR. 2003. Antiulcer activity of *Tephrosa purpurea* in rats. Ind. J.Pharmacol. **35:** 168–172.
19. BIGHETTI, A.E., M.A. ANTONIO, L.K. KOHN, *et al.* 2005. Anti-ulcerogenic activity of a criude hydroacloholic extract and coumarin isolated from *Mikania laevigata* Schultz Bip. Phytomed. **12:** 72–77.
20. BAPNA, P.A & R. BALRAM. 2005. Anti-ulcer and anti-oxadant activity of pepticare, a herbomineral formulation. Phytomedicine **12:** 264–270.
21. DE, B., R.N. MAITI, V.K. JOSHI, *et al.* 1997. Effect of some sitaviraya drugs on gastric secretion and ulceration. Ind. J. Exp. Biol. **34:** 97–101.
22. DUTTA, G.K., K. SAIRAM, S. PRIYAMBADA, *et al.* 2002. Antiulcerogenic activity of Stavri mandur, an Ayurvedic herbo mineral preparation. Ind. J. Exp. Biol. **40:** 1173–1177.
23. ARCHNA, R. & A. NAMASIVAYAUM. 1999. Antistressor effects of *Withania somnifera* . J. Ethanopharmacol. **64:** 91–93.
24. SINGH, N., S.S. ABBAS., V. SINGH & A. SINGH. 2002. Adaptogens antistress agents: a study focussing on Indian plants. Antiseptic **90:** 243–249.
25. Mc QUAID, K.R. 2004. Drugs used in the treatment of gastrointestinal disease. Basic and 9th Clinical Pharmacology. G.K. Bertram, Ed. :1038–1039. McGraw-Hill Companies. London.

26. BHATTACHARYA, S.K., K.S. SATYAN & S. GHOSAL. 1997. Antioxidant activity of glycowithnolides from *Withania somnifera*. Ind. J. Exp. Biol. **35:** 236–239.
27. BHATTACHARYA, A., S. GHOSAL & S.K. BHATTACHARYA. 2001. Antioxidant effect of *W. somnifera* glycowthanolides in chronic foot shock stress induced perturbations of oxidative free radical scavenging enzymes and lipid peroxidation in rat frontal cortex and striatum. J. Ethanopharmacol. **74:** 1–6.
28. RAJPUT, A.S., A.K. CHOUDHARY & D.A. PANDEY. 2001. Holistic approach to preserve health through the rasayana chikitsa. Nat. Symp. on Ancient Ind. Sc. New Delhi. Dec 115. pp 246.
29. DEVI, P. 1996. *Withania somnifera* Dunal (Ashwagandha): potential plant source of a promising drug for cancer chemotherapy and radiosensitization. Ind. J. Exp. Biol. **34:** 927–932.
30. SHULKA, D., S. JAIN, K. SHARMA & M. BHATNAGAR. 2000. Effects of Semecarpus anacardium Linn on neuron cell bodies in hippocampal sub regions of the stressed female rats. Ind. Drugs **37:** 379–382.
31. JAIN, S., S.D. SHUKLA, S.D. SHARMA & M. BHATNAGAR. 2000. Neuroprotective effects of *Withania somnifera* Dunal in hippocampal subregions of female albino rats. Phyto. Ther. Res. **15:** 544–548.
32. MC COFFERTY, D.M., D.N. GRANGER & J.L. WALLACE. 1995. Indomethacin induced gastric injury and leukocyte adherence in arthritic versus healthy rats. Gastroenterology **109:** 1173–1180.
33. BRODIE, D.A. 1966. The mechanism of gastric hyperacidity produced by pyloric ligation in the rat. Am. J. Dig. Dis. **11:** 231–241.
34. SHAH, C.S.& J.S. QADRY. 2004. A Text Book of Pharmacognosy, 12th Ed. B.S. Shah, Ed. :328–388. Prakashan. Ahmedabad.
35. GANGULI, A.K. & O.P. BHATNAGAR. 1973. Effect of bilateral adrenalectomy on production of restraint ulcer in the stomach of albino rats. Can. J. Physiol. Pharmacol. **51:** 748–750.
36. GOEL, R.K. & K. SAIRAM. 2002. Anti-ulcer drugs from indigenous sources with emphasis on *Musa sapientum, Tamrabhasma, Asperagus racemosus* and *Zingiber officinale*. Ind. J. Pharmacol. **34:** 100–110.
37. PARMAR, N.S., M. TARIQ & A.M. AGEEL. 1987. Studies on the gastroulcerogenic effect of pylorus ligated, hypothermic restraint stress, reserpine, indomethacine and aspirin in morphine dependent rats. Alc. Drug. Res. **7:** 734–739.
38. HAWK, A. 1965. Hawk's Physiological Chemistry, 14 Ed. :483–485. McGraw-Hill Book Co.
39. GOYAL, R.K. 2003. Practicals in Pharmacology, 3rd Ed. :128–129. B.S. Shah Prakashan. Ahmedabad.
40. NAIR, B.R. 1976. Investigation of the venoum of South Indian scorpion Heterometrus scaper. Ph.D. Thesis, University of Kerala, Trivendrum.
41. LOWRY, O.H., N.J. ROSENBOROUGH, A.L.FARR & R.J. RANDAL. 1951. Protein measurement with folin phenol reagent. J. Biol. Chem. **193:** 265–275.
42. WINTERBOURN, C.C., R.E. Hawkins, M. BRIAN & R.W. CARRELL. 1975. The estimation of red cell super oxide dismutase activity. J. Lab. Clin. Med. **85:** 337–341.
43. SINHA, A.K.1972. Colorimetric assay of catalase. Anal. Biochem. **47:** 389–390.
44. BUEGE, J.A. & S.D. AUST. 1978. The thiobarbituric acid assay. Meth. Enzymol. **52:** 306–307.
45. NATELSON, S.1971. Techniques of Clinical Chemistry, 3rd Ed. :286. Charles C Thomas. Springfield, IL.
46. PRANJOTHY, M.B. & C.B. PATIL. 2000. Vitamin E in antioxidant formulation. Ind. Drugs **37:** 236–238.
47. MOZSIK, G.Y., B. KISS, J. JAVOR, *et al.* 1969. Effect of cholinesterase inhibitor treatment on the phosphorous and nucleic acid metabolism in stomach wall. Pharmacology **2:** 45–59.
48. SANYAL, A.K., P.K. MITRA & R.K. GOYAL. 1983. A modified method to estimate dissolved mucosubstances in gastric juice. Ind. J. Exp. Biol. **21:** 78–84.
49. DOMISCHKE, W. & S. DOMISHCKE. 1978. Prostaglandin stimulated gastric mucus secretion in man. Acta Hepatol. Gastroenterol. **78:** 292–294.

50. ROBERT, A. 1981. Prostaglandins and gastrointestinal tract. *In* Physiology of the Gastrointestinal Tract. L.R. Johnson, Ed. :1407–1434. Raven Press. New York.
51. ROBERT, A. 1979. Cytoprotection by prostaglandins. Gasteroenterology **77:** 761–767.
52. SHAY, H., S.A. KOMAROV., S.E. FELS, *et al.* 1945. A simple method for the uniform production of gastric ulceration in the rats. Gastroenterology **5:** 43–61.
53. ALY, A. 1985. Prostaglandins in clinical treatment of gastroduodenal mucosal lesions: A review. Scand. J. Gastroenterol. **20:** 543–553.
54. ALY, A. 1987. Prostaglandins in clinical treatment of gastroduodenal mucosal lesions: A review. Scand. J.Gastroenterol. Suppl.137: 43–49.
55. PARMAR. N.S. & S. PARMAR. 1998. Anti-ulcer potentials of flavonoids. Ind. J. Pharmacol. **42:** 343–351.
56. SAXENA, V.K. & S. CHOURASIA. 2001. A new isoflavone from the roots of *Asperagus racemosus*. Fitotherapia. 72:307–309.
57. WILLEMIJNTJE, A.H. & J.P. PANKAJ. 2001. Agents used for control of gastric acidity and treatment of peptic ulcers and gastroesophageal reflux disease. *In* The Pharmacological Basis of Therapeutics. G.H. Joel & E.L. Lee, Eds. :1010–1011. Xth Ed. McGraw-Hill Medical Publishing Division. Britain.
58. VALLE, J.D. 2001. Peptic ulcer disease and related disorders. *In* Harrison's; Principles of Internal Medicine. E. Brounwald, A.S. Fauci, D.L. Kasper, S.L. Hauser, D.L. Longo & J.L. Jameson, Eds. :1656–1680. The McGraw-Hill Companies. Britain.

Novel Drugs and Vaccines Based on the Structure and Function of HIV Pathogenic Proteins Including Nef

AHMED A. AZAD

Faculty of Health Sciences, Medical School, University of Cape Town, Observatory,7925, Cape Town, South Africa

ABSTRACT: Evidence is presented to suggest that HIV-1 accessory protein Nef could be involved in AIDS pathogenesis. When present in extracellular medium, Nef causes the death of a wide variety of cells *in vitro* and may therefore be responsible for the depletion of bystander cells in lymphoid tissues during HIV infection. When present inside the cell, Nef could prevent the death of infected cells and thereby contribute to increased viral load. Intracellular Nef does this by preventing apoptosis of infected cells by either inhibiting proteins involved in apoptosis or preventing the infected cells from being recognized by CTLs. Neutralization of extracellular Nef could prevent the death of uninfected immune cells and thereby the destruction of the immune system. Neutralization of intracellular Nef could hasten the death of infected cells and help reduce the viral load. Nef is therefore a very important molecular target for developing therapeutics that slow progression to AIDS. The N-terminal region of Nef and the naturally occurring bee venom mellitin have very similar primary and tertiary structures, and they both act by destroying membranes. Chemical analogs of a mellitin inhibitor prevent Nef-mediated cell death and inhibit the interaction of Nef with cellular proteins involved in apoptosis. Naturally occurring bee propolis also contains substances that prevent Nef-mediated cell lysis and increases proliferation of CD4 cells in HIV-infected cultures. These chemical compounds and natural products are water soluble and nontoxic and are therefore potentially very useful candidate drugs.

KEYWORDS: novel drugs; vaccines; Nef; HIV; pathogenic proteins

INTRODUCTION

In the last quarter century, HIV/AIDS has devastated the health and the economies of much of the developing world. Having caused over 3.1 million deaths last year, it is now the leading cause of death from all infectious diseases. Over 40 million people are now living with HIV/AIDS. Over 90% of the deaths and infections occur in countries of the developing world that are least able to manage the epidemic or afford the costly combination of drugs that have succeeded in keeping

Address for correspondence: Ahmed A. Azad, Faculty of Health Sciences, Medical School, University of Cape Town, Anzio Road, Observatory, 7925, Cape Town, South Africa.
a_azad05@yahoo.com.au

**Ann. N.Y. Acad. Sci. 1056: 279–292 (2005). © 2005 New York Academy of Sciences.
doi: 10.1196/annals.1352.013**

the disease under control in the developed countries of the West. The most affected areas are sub-Saharan Africa followed by South and Southeast Asia.

The best possible option for the developing world is prevention through public education and the use of affordable and effective vaccines. No such vaccines are currently available, and it is highly unlikely that any will become available in the near future. A vast majority of infected people are not even aware of their status and will unwittingly pass on the virus to many more. Therefore, the number of infected persons will continue to rise for a long time in the future in the absence of an effective and affordable preventive vaccine to bring down the infection rate. During this time, therapeutic intervention remains the only option available to persons living with HIV/AIDS.

In recent years, highly active antiretroviral therapy (HAART), which involves the use of a combination of antiviral drugs, has caused a dramatic decline in AIDS-related morbidity and mortality in the Western world and remains the only effective therapy at the present moment. HAART has been highly beneficial in reducing mother-to-child transmission of HIV and in emergency situations, but serious problems are associated with the long-term use of HAART. On prolonged use, treatment failure occurs in over 50% of patients due to the emergence of drug-resistant strains, and the situation is made worse by noncompliance due to severe side effects, new disease conditions, and inconvenient life-style restrictions. There is also life-long persistence of active virus in HAART-resistant reservoirs in the body. The price still remains high despite concerted international efforts to bring it down, and it is simply not a practical option for most of the developing countries as a life-long universal therapy for all infected individuals.

There is, therefore, an urgent and continuing need for drugs that are more affordable and patient friendly and that also can arrest the destruction of the immune system. The acceptance that HIV/AIDS has to be managed as a chronic infection will allow the strategic development of drugs and therapeutic vaccines that can slow progression to full-blown AIDS.

AIDS

Pathogenesis

The defining feature of AIDS, the end stage of HIV infection, is the rapid decline in the number of CD^+ T lymphocytes,[1] the natural host for HIV. During the so-called clinical latency period, which lasts about 10 years, the number of CD^+ cells in the peripheral blood (2% of total) remains steady and the amount of virus in the blood remains low. Most HIV researchers believe that this steady state results from the direct cytopathic effects of the virus[2,3] and the continuous production of $CD4^+$ cells and the elimination of infected cells. In the end stage of the disease, the virus wins the protracted battle, and there is a rapid decline in the number of $CD4^+$ T lymphocytes concomitantly with acute viremia.

Contrary to what appears to be the case in the peripheral blood, however, the depletion of $CD4^+$ cells is not confined to the terminal stage. It is very prominent in lymphoid tissues, where 80% of $CD4^+$ cells reside, even during the long, so-called clinical latency period.[4,5] During the course of HIV infection, the architecture of the lymph nodes and other lymphoid organs is gradually but irreversibly destroyed[6] due to the elimination of $CD4^+$ and other cells in the lymphoid tissues by apoptosis and necrosis. The elimination of immune cells with resultant lymphoid atrophy is what causes the severe immunodeficiency seen in AIDS patients. Similar cell depletion and tissue atrophy in the brain and in the central nervous system (CNS) cause AIDS dementia.

The majority opinion in the HIV/AIDS research community certainly favors the concept of direct cytopathic effects of the virus,[2,3] which holds that HIV kills only the cells that it infects. However, this view does not adequately explain[7,8] why less than 1% of the cells are infected and yet the majority of uninfected $CD4^+$ T lymphocytes and other cells that cannot be infected by HIV, such as $CD8^+$ T lymphocytes, B lymphocytes, and nonimmune cells, are progressively eliminated during the course of the infection (reviewed in Refs. 9 and 10). This "bystander effect" is totally dependent on HIV infection, suggesting a role for HIV gene products outside the infected cells. But while the uninfected immune cells are eliminated, the virus finds clever ways to prolong the survival of its host, the infected $CD4^+$ T lymphocytes. The end result is an increased viral load and the progressive destruction of the immune system. An effective therapeutic intervention to slow progression to AIDS should not only target the virus, as the currently used HAART does, but also aim to prevent the destruction of the immune system and promote the elimination of HIV-infected cells.

Virus Proteins Involved in AIDS Pathogenesis

Research in my own laboratory[9–25] (as well as unpublished results presented here) has focused on identifying viral agents involved in the elimination of immune cells and the prolonged survival of HIV-infected cells and in developing strategies to reverse these effects. Our studies have suggested that HIV-1 accessory proteins Nef and Vpr may be responsible for the aforementioned effects. Nef and Vpr, which do not form a part of the virus structure and do not have any enzyme activity, are encoded by the HIV genome. In the past, these two viral proteins were not considered to be biologically important as they were not required for virus replication in established cell lines. Recent studies clearly show that both of these proteins are indispensable for disease progression. In this presentation, I only concentrate on Nef.

NEF

Role of Nef in AIDS Pathogenesis

Nef is a 27-kDa myristoylated virion-associated protein that is found exclusively in primate lentiviruses HIV and SIV. HIV and SIV are the only two viruses known to cause AIDS in humans and monkeys, respectively. The intact *nef* gene has been shown to be essential for persistent infection and progression to AIDS in rhesus monkeys,[26] and *nef* gene deletion has been observed in HIV-infected long-term nonprogressors.[27] Nef has been assigned more functions than most other viral proteins, but these will not be addressed here.

The following is a summary of Nef research results obtained in my own laboratory[11–25] (and unpublished results) and other closely related work.

Role of Extracellular Nef in Destruction of Immune System

Nef and N-terminal fragments of Nef are membrane-active and disrupt cell membranes. When present in the extracellular medium in the myristoylated form, Nef and N-terminal fragments of Nef destroy all types of human and mammalian cells such as $CD4^+$ T lymphocytes, $CD8^+$ T lymphocytes, B lymphocytes, PBLs, monocytes, RBCs, and progenitor cells, as would be expected of an agent involved in the elimination of immune cells in the lymphoid tissues.

The myristoylated N-terminal fragment of Nef rapidly destroys $CD4^+$ cells as measured by the release of lactic dehydrogenase into the extracellular medium (FIG. 1). This is not seen with the non-myristoylated form of the same peptide or a control myristoylated peptide from another region of Nef.

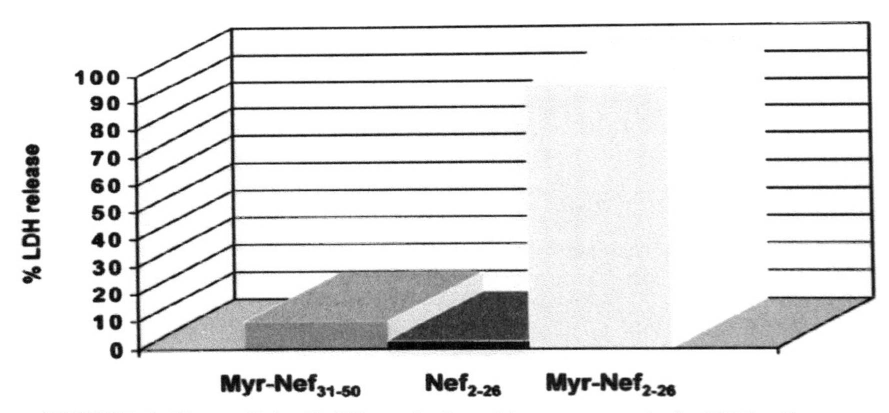

FIGURE 1. Extracellular Nef N-terminal peptides are cytotoxic for CEM cells.

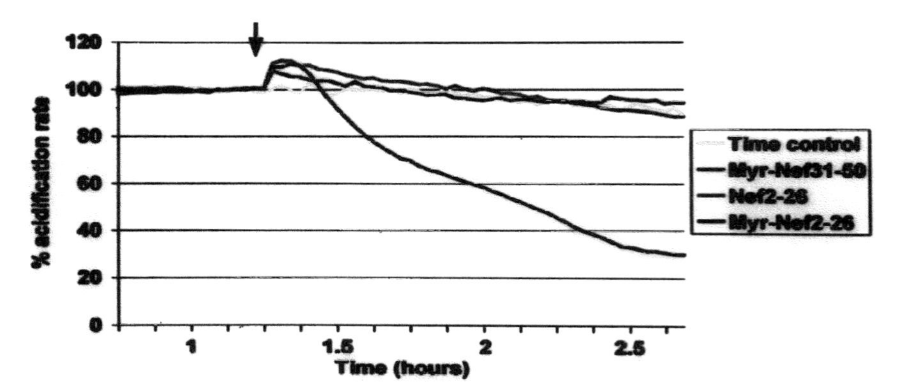

FIGURE 2. Effect of extracellular N-terminal Nef peptides on CEM cells in the cytosensor microphysiometer.

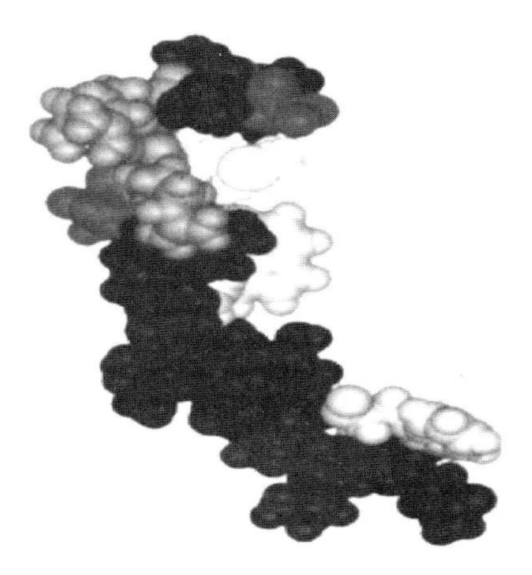

FIGURE 3. The cytolytic domain of the Nef protein is resident in the N-terminal region of the molecule whose primary and tertiary structures are shown here.

$$\text{Myr-G}_2\text{G}\underline{\text{K}}\text{WS}\underline{\text{K}}\text{SS}_9\text{VIGWPAVR}_{16}\text{ERMRR}_{22}\text{-OH}$$

$\leftarrow$Random$\rightarrow$ $\leftarrow$ α-Helix $\rightarrow$

In the Cytosensor Microphysiometer, the myristoylated N-terminal peptide of Nef causes a very rapid decline in the metabolism of $CD4^+$ cells (FIG. 2), which is not seen with the non-myristoylated peptide or the control myristoylated peptide.

The cytolytic domain of the Nef protein is resident in the N-terminal region of the molecule whose primary and tertiary structures are shown in FIGURE 3.

The cell-killing activity of Nef is critically dependent on the N-terminal myristoyl group and the two positively charged lysines in the adjacent N-terminal random structure. The membrane-active α-helix following the random structure enhances cell lysis in the presence of the other determinants, but it has no lytic activity of its own. The primary sequence is also important, as an internal region of Nef with very similar structural motifs is nonlytic.

There is strong circumstantial evidence to suggest that Nef and N-terminal fragments of Nef are involved in the elimination of immune cells, the destruction of lymphoid tissues, and the progression to AIDS.

Nef and N-terminal fragments of Nef have been shown to be present in the extracellular medium and on the surface of HIV-infected cells (reviewed in Ref. 6). In SCID mice transplanted with human liver/thymus or PBLs, Nef causes severe depletion of $CD4^+$ T lymphocytes.[38,39] In mice made transgenic with the *nef* gene, there is T-cell depletion, atrophy of lymphoid tissues, and severe AIDS-like pathologies.[40–42] In monkeys, the cell-killing activity of Nef is critically dependent on the N-terminal myristoyl group and on the two positively charged lysines in the adjacent N-terminal random structure. The membrane-active-helix following the random structure enhances cell lysis in the presence of the other determinants but has no lytic activity of its own. The primary sequence is also important, as an internal region of Nef with very similar structural motifs is nonlytic.

Strong circumstantial evidence suggests that Nef and N-terminal fragments of Nef are involved in the elimination of immune cells, the destruction of lymphoid tissues, and the progression to AIDS. Nef and N-terminal fragments of Nef have been shown to be present in the extracellular medium and on the surface of HIV-infected cells (reviewed in Ref. 9). In SCID mice transplanted with human liver/thymus or PBLs, Nef causes severe depletion of $CD4^+$ T lymphocytes.[28,29] In mice made transgenic with the *nef* gene, there is T-cell depletion, atrophy of lymphoid tissues, and severe AIDS-like pathologies.[30–32] In monkeys infected with SHIV (SIV in which the SIV *nef* gene has been replaced with the HIV nef gene), HIV Nef causes simian AIDS.[33,34] The aforementioned studies strongly suggest a role for Nef in disease progression.

Interaction of Nef with Host Cell Proteins

Earlier studies had shown that during HIV infection of lymphoid tissues, cells that were undergoing apoptosis were rarely infected.[35] On the other hand, HIV-infected cells were rarely undergoing apoptotic death,[35] suggesting that something within the infected cells was keeping them alive longer. There is now strong circumstantial evidence to suggest that intracellular Nef could be the agent responsible for the prolonged survival of HIV-infected cells.

Full-length Nef, but not N-terminal deleted Nef, downregulates the surface expression of CD4 and IL2-R, which are involved in MHC II antigen recognition and cell proliferation, respectively. This suggests that Nef might interfere with cell-signaling mechanisms inside infected cells. Full-length Nef and N-terminal fragments of Nef bind directly to CD4, $p56^{lck}$, tumor suppressor p53, and MAPK (*Erk*2) and also inactivate the enzyme activities of $p56^{lck}$ and MAPK. All these four proteins that Nef binds to are also involved in promoting apoptosis, suggesting that Nef might interfere with the apoptotic pathway of infected cells.

ELISA measurements clearly show (FIG. 4) that full-length Nef and an N-terminal 57 amino acid fragment of Nef bind to the tumor suppressor p53 in a dose-dependent manner, but this interaction does not occur with a Nef protein in which the N-terminal 19 amino acids are missing.

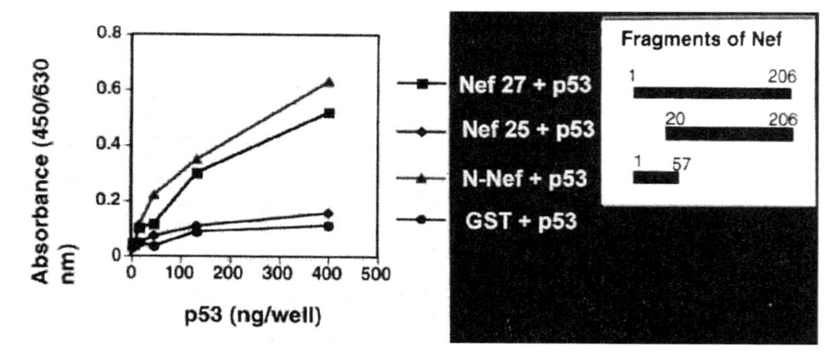

FIGURE 4. N-terminus of Nef binds to p53.

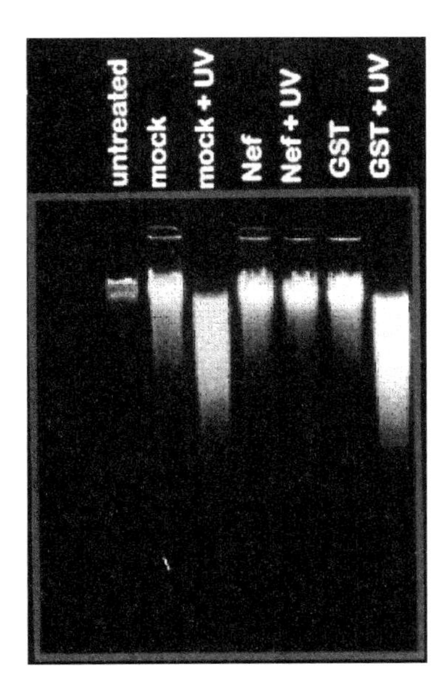

FIGURE 5. Nef protects against UV-induced apoptosis.

Because p53 promotes apoptosis and Nef binds to p53, it was of interest to see if Nef inhibits p53-mediated apoptosis. UV-induced apoptosis of cells is mediated by p53, and a diagnostic of apoptosis is DNA fragmentation. Gel analysis of DNA from UV-treated $CD4^+$ cells clearly shows (FIG. 5) that DNA fragmentation is seen in cells electroporated with buffer or the human protein GST but not in cells electroporated with full-length Nef.

The absence of DNA fragmentation in the presence of intracellular Nef in UV-treated $CD4^+$ cells clearly shows that intracellular Nef inhibits apoptosis. This agrees with results retained with HIV-infected cell cultures and lymphoid tissues in which the amount of apoptosis in infected cells was much lower than that in uninfected cells. Nef also inhibits $p56^{lck}$-mediated apoptosis of $CD4^+$ T lymphocytes.

By inhibiting apoptosis of infected cells by two independent mechanisms, Nef could help prolong their survival, allowing increased virus production. Inactivation of intracellular Nef could therefore help to hasten the apoptotic death of HIV-infected cells.

There are a number of other ways by which intracellular Nef has been suggested to prolong the survival of HIV-infected cells. One of these is the evasion of $CD8^+$ CTL-mediated apoptosis of HIV-infected $CD4^+$ T lymphocytes that is responsible for suppressing acute viremia after the initial infection. During the long clinical latency period, when there is vigorous virus production in the lymphoid tissues, there is also continuous generation and persistence of HIV-specific $CD8^+$ CTLs. So why do the existing HIV-specific CTLs fail to recognize infected cells during the later stages of the disease and thereby prevent progression to AIDS?

$CD8^+$ CTLs eliminate virus-infected cells by recognizing MHC I molecules containing viral peptides on the surface of HIV-infected $CD4^+$ T lymphocytes. This results in the apoptotic death of the virus-containing cells. Studies have shown that Nef downmodulates the surface expression of MHC I molecules on the surface of $CD4^+$ cells so that they cannot be recognized or eliminated by HIV-specific $CD8^+$ CTLs.[36–39]

THERAPEUTIC STRATEGIES AND DRUG DEVELOPMENT

Drugs or antibodies that bind to and inactivate the N-terminal domain of Nef could help to stop the killing of uninfected $CD4^+$ cells and other immune cells and thus prevent the destruction of lymphoid tissues. Such drugs could be new chemical entities (NCEs) based on the structures and functions of the cytotoxic domain of Nef, or they could be based on lead compounds identified from chemical or natural product libraries. DNA vaccines or live expression vectors that express small amounts of the cytotoxic peptides of Nef in immunogenic form over long periods could produce specific antibodies that neutralize or inactivate the cytotoxic domain of Nef in the extracellular medium.

Anti-Nef compounds that can enter infected cells could bind to and inactivate $p56^{lck}$ and the tumor suppressor p53 and thus allow the infected cells to proceed to apoptotic death. The inactivation of Nef inside infected cells could also reverse the Nef-mediated suppression of MHC I molecule display on the surface of infected cells, so that these cells can be recognized and eliminated by preexisting HIV-specific $CD8^+$ CTLs. If this approach were to work, then it would be the best possible therapeutic "vaccination" strategy. Both of the aforementioned approaches should help to decrease the viral load and the amount of extracellular Nef that destroys the uninfected immune cells and thus help to slow progression to AIDS.

Thus, Nef is a very important target for drug development as anti-Nef compounds in the extracellular medium could prevent the elimination of uninfected immune cells, while their availability inside infected cells could hasten the apoptotic death of these cells.

Anti-Nef Compounds Based on Mellitin Inhibitor

There is striking similarity between the primary and tertiary structures of the N-terminal region of Nef and the honeybee venom mellitin (FIG. 6). Nef and mellitin also act similarly by disrupting cell membranes.

^{1}GGKW SKSSV IGWPAVRERMRR^{21} **Nef**

GIGAV LKVLTTG LPALIS WI KR *Mellitin*

Because of the structural and functional similarity between the N-terminus of Nef and mellitin, it was worth investigating the ability of any known mellitin inhibitor to also inhibit Nef-mediated cell lysis. A small peptide inhibitor of mellitin, identified from a hexapeptide combinatorial library, was found to be moderately successful in inhibiting Nef-mediated cell lysis. However, this peptide was highly toxic and insoluble.

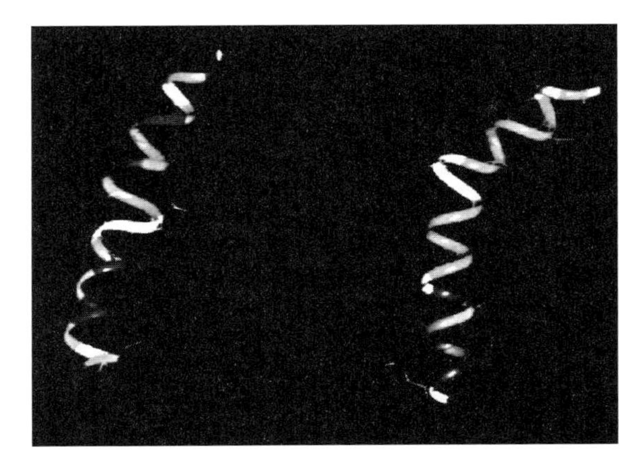

FIGURE 6. Striking similarity between the primary and tertiary structures of the N-terminal region of Nef and the honeybee venom mellitin.

The anti-mellitin hexapeptide was used as a lead compound for the synthesis of analogue molecules with the hope of producing more effective inhibitors of Nef that would also be less toxic and more soluble. Each of these NCEs was screened in two separate assays.

One assay (FIG. 7), based on the Nef-induced lysis of CD4$^+$ cells and release of lactic dehydrogenase (LDG) into the extracellular medium, measures LDH by ELISA.

The second assay (FIG. 8) was based on the uptake of a fluorescent dye by CD4$^+$ cells and the Nef-induced release of this dye into the extracellular medium and its measurement in a fluorometer.

About 500 NCEs were synthesized and subjected to both assays to avoid false-positive results and to ensure that they themselves were not cytotoxic. A small number of them were very potent inhibitors of Nef-induced cell lysis, while not exhibiting any cytotoxicity of their own (FIG. 9). The result on the extreme right is for Bee Propolis.

Some of these anti-Nef compounds were soluble, and a few of them were also capable of binding to tumor suppressor p53 (FIG. 10) and also to p56^{lck}.

It will be interesting to determine if any of these molecules can enter cells and reverse the Nef-mediated inhibition of apoptosis and if they could allow the re-display of MHC I molecules on the surface of infected cells so that they can be recognized and eliminated by CD8$^+$ CTLs.

Bee Propolis Inhibits Nef-Mediated Cell Killing

Bee propolis consists of the salivary secretion of bees and a very heterogeneous mixture of resinous plant material. Bees use this to physically strengthen their hives. Hives coated with propolis also remain microbiologically sterile. Bee propolis, which is nontoxic and readily available in health food stores, has long been used in Chinese medicine in the treatment of patients with low immunity. (Bee propolis used

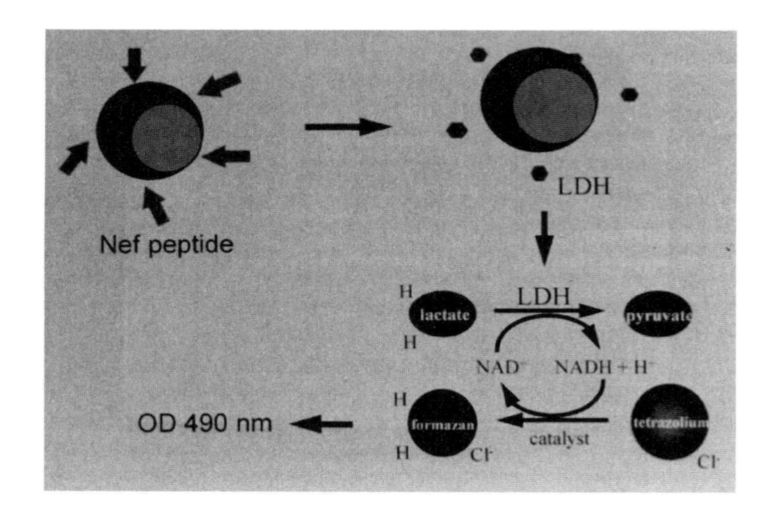

FIGURE 7. LDH assay.

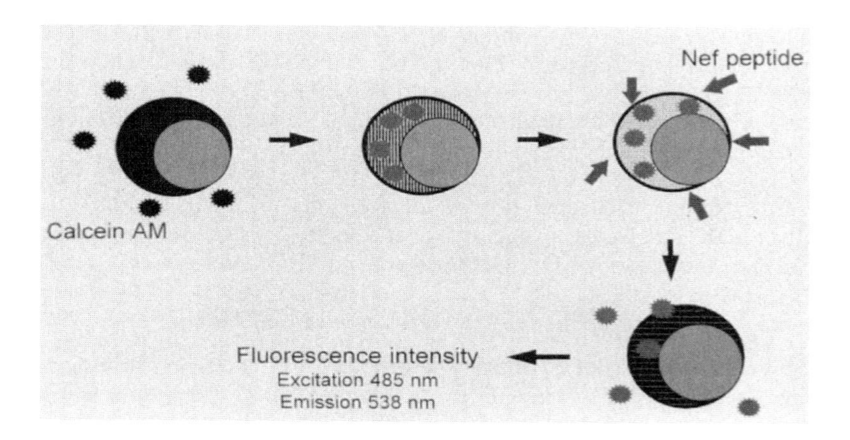

FIGURE 8. Calcein assay.

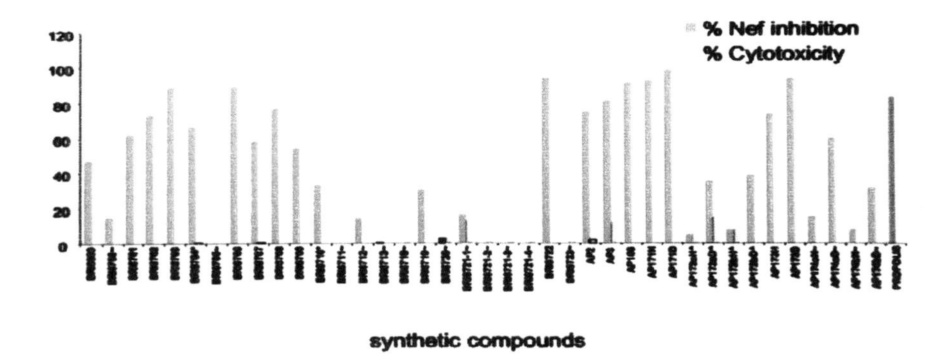

FIGURE 9. Inhibition of Nef-mediated cytolysis and cytotoxicity of synthetic compound.

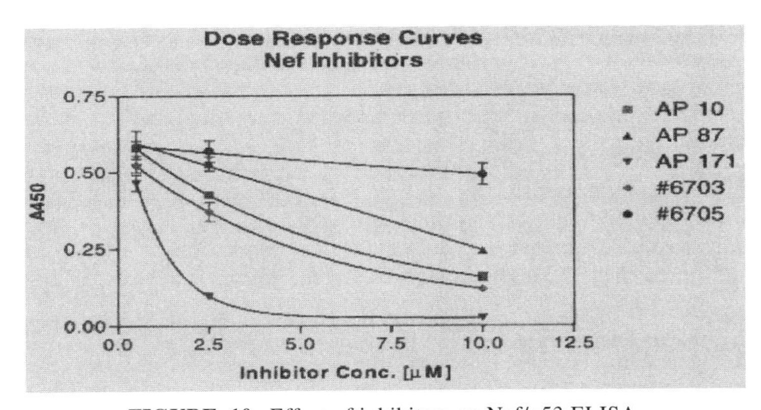

FIGURE 10. Effect of inhibitors on Nef/p53 ELISA.

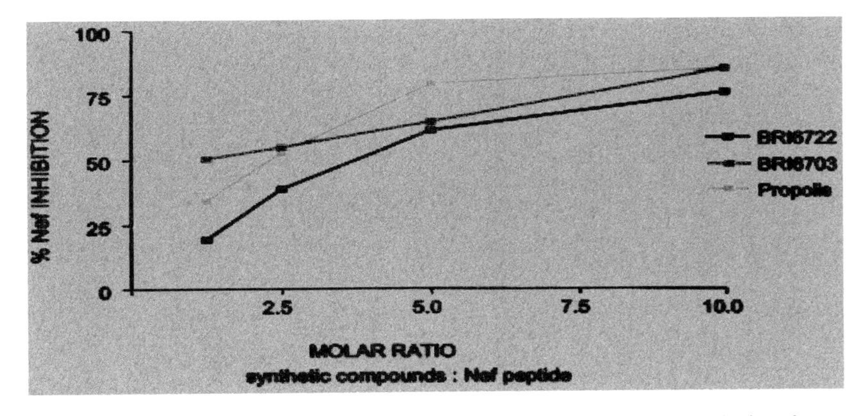

FIGURE 11. Dose-response curve of α-Nef compounds measured by calcein release.

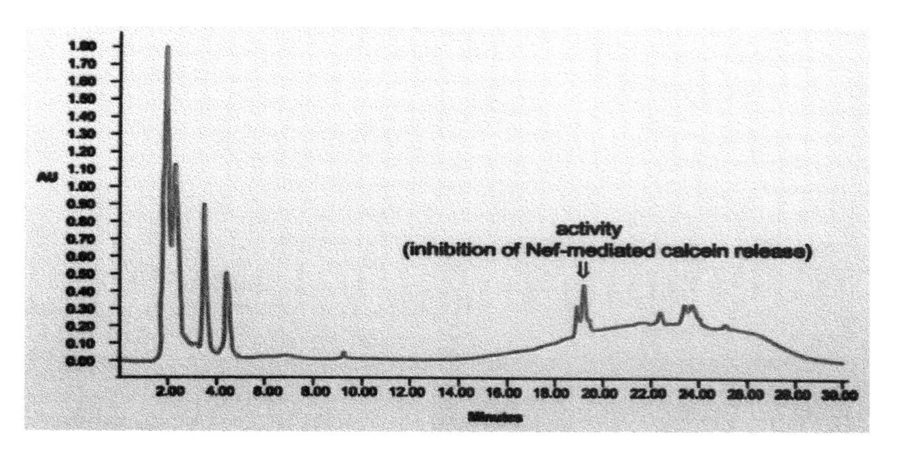

FIGURE 12. Preparative HPLC of soluble fraction of been propolis (22 × 250 mm Econosphere C18 column; 40–80% MeOH/H_2O).

in these studies was a gift from a pharmaceutical company in Hangzou, China, and had been subjected to extensive toxicologic and pharmacologic testing by the company. It was supplied to us in the form of a powder.)

The water-soluble fraction of bee propolis was highly active in inhibiting Nef-mediated lysis of $CD4^+$ T lymphocytes (FIG. 9) and is just as active as the best of the NCEs described earlier (FIG. 11).

All of the anti-Nef activity of bee propolis is contained in the water-soluble fraction that consists of a heterogeneous mixture of low molecular weight compounds. The water-soluble fraction has no deleterious effects on tonsil histocultures even after continuous exposure; in fact, it very strongly supports the growth and proliferation of lymphocytes in culture. This material could therefore be used without further purification in clinical trials to see if it can help maintain the architecture of lymphoid organs and prevent the destruction of the immune system.

The anti-Nef activity in the water-soluble fraction could be isolated as a single peak of activity (FIG. 12). This isolated material can he used as a lead compound for further development into a "modern medicine" if required.

ACKNOWLEDGMENTS

I wish to gratefully acknowledge the contributions of my colleagues, especially Ian Macreadie, whose efforts over the years have resulted in the publications from my laboratory listed below (Refs. 11–25). There are too many people to thank, and I hope they will understand that I am not able to name them all individually. I would like to thank Amod Kulkarni from the Division of Medical Virology, IIDMM, University of Cape Town, Cape Town, South Africa, for his time and effort in preparing this manuscript.

REFERENCES

1. LEVY, J.A. 1993. Pathogenesis of human immunodeficiency virus infection. Microbiol. Rev. **57:** 183–289.
2. WEI, X. *et al.* 1995. Viral dynamics in human immunodeficiency virus type 1 infection. Nature **373:** 117–122.
3. HO, D.D. *et al.* 1995. Rapid turnover of plasma virions and CD4 lymphocytes in HIV-1 infection. Nature **373:** 123–126.
4. PANTALEO, G. *et al.* 1993. HIV infection is active and progressive in lymphoid tissue during the clinically latent stage of disease. Nature **362:** 355–358.
5. EMBERTSON, J. *et al.* 1993. Massive covert infection of helper T lymphocytes and macrophages by HIV during the incubation period of AIDS. Nature **362:** 359–362.
6. PANTALEO, G. *et al.* 1994. Role of lymphoid organs in the pathogenesis of human immunodeficiency virus (HIV) infection. Immunol. Rev. **140:** 105–130.
7. ASCHER, M.S. *et al.* 1995. HIV results in the frame. Paradox remains. Nature **375:** 196.
8. ANDERSON, R.W. *et al.* 1998. Direct HIV cytopathicity cannot account for CD4 decline in AIDS in the presence of homeostasis: a worst-case dynamic analysis. Acquired Immun. Defic. Syndr. Hum. Retroviruses **17:** 245–252.
9. AZAD, A.A. 2000. Could Nef and Vpr proteins contribute to disease progression by promoting depletion of bystander cells and prolonged survival of HIV-infected cells. Biochem. Biophys. Res. Commun**. 267:** 677–685.
10. AZAD, A.A. 2000. The possible involvement of HIV accessory Nef and Vpr in bystander cell death and prolonged survival of HIV-infected cells. *In* Advances in Animal Virology. S. Jameel & L. Villareal, Eds. :297–307. Oxford & IBH Publishing Co. Pvt. Ltd. New Delhi and Calcutta.
11. MACREADIE, I.G. *et al.* 1993. Yeast-derived HIV-1 Nef protein lacks G-protein activities. Mol. Biol. (Life Sci. Adv.) **12:** 99–105.
12. MACREADIE, I.G. *et al.* 1993. Expression of HIV-1 *nef* in yeast: The 27 kDa nef protein is myristylated and fractionates with the nucleus. Yeast **9:** 565–573.
13. AZAD, A.A. *et al.* 1994. Large scale production and characterisation of recombinant human immunodeficiency virus type 1 *Nef.* J. Gen. Virol. **75:** 651–655.
14. CURTAIN, C.C. *et al.* 1994. Fusogenic activity of amino-terminal region of HIV type 1 Nef protein. AIDS Res. Hum. Retroviruses **10:** 1231–1240.
15. GREENWAY, A.L. *et al.* 1994. Nef27, but not the Nef25 isoform of human immunodeficiency virus-type pNL4.3 down-regulates surface CD4 and IL-2R expression in peripheral blood mononuclear cells and transformed cells. Virology **198:** 245–258.
16. GREENWAY, A.L., A.A. AZAD & D. MCPHEE. 1995. Human immunodeficiency virus type 1 Nef protein inhibits activation pathways in peripheral blood mononuclear cells and T-cell lines. J. Virol. **69:** 1842–1850.
17. MACREADIE, I.G. *et al.* 1995. Stress- and sequence-dependent release into the culture medium of HIV-1 Nef produced in *Saccharomyces cerevisiae.* Gene **162:** 239–243.
18. GREENWAY, A., A.A. AZAD, J. MILLS & D. MCPHEE. 1996. Human immunodeficiency virus type 1 Nef binds directly to Lck and mitogen-activated protein kinase, inhibiting kinase activity. J. Virol. **70:** 6701–6708.
19. CURTAIN, C.C. *et al.* 1997. Cytotoxic activity of the amino-terminal region of HIV type 1 Nef protein. AIDS Res. Human Retroviruses **13:** 1213–1220.
20. BARNHAM, K.J. *et al.* 1997. Solution structure of a polypeptide from the N-terminus of the HIV protein Nef. Biochemistry **36:** 5970–5980.
21. MACREADIE, I.G. *et al.* 1997. Cytotoxicity resulting from addition of HIV-1 Nef N-terminal peptides to yeast and bacterial cells. Biochem. Biophys. Res. Commun. **232:** 707–711.
22. MACREADIE, I.G. *et al.* 1998. Expression of HIV-1 *nef* in yeast causes membrane perturbation and release of the myristylated Nef protein. J. Med. Sci. **5:** 203–210.
23. MACREADIE, I.G. *et al.* 1998. Human immunodeficiency virus type 1 Nef protein causes death in stressed yeast due to determinants near the N-terminus and elsewhere in Nef. Biochem. Mol. Biol. Int. **46:** 277–286.
24. CURTAIN, C.C. *et al.* 1999. Structural requirements for the cytotoxicity of the N-terminal region of HIV Type 1 Nef. AIDS Res. Human Retroviruses **14:** 1543–1551.

25. GREENWAY, A.L. *et al.* 2002. Human immunodeficiency virus type 1 Nef binds to tumor suppressor p53 and protects cells against p53-mediated apoptosis. J. Virol. **76:** 2692–2702.
26. KESTLER, H.W. *et al.* 1991. Importance of the nef gene for maintenance of high virus loads and for development of AIDS. Cell **65:** 651–662.
27. DEACON, N.J. *et al.* 1995. Genomic structure of an attenuated quasi species of HIV-1 from a blood transfusion donor and recipients. Science **270**: 988–991.
28. JAMIESON, B.D. *et al.* 1994. Requirement of human immunodeficiency virus type 1 nef for in vivo replication and pathogenicity. J.Virol. **68**: 3478–3485.
29. GULIZA, R.J. *et al.* 1997. Deletion of nef slows but does not prevent CD4-positive T-cell depletion in human immunodeficiency virus type 1-infected human-PBL-SCID mice. J. Virol. **71:** 4161–4164.
30. SKOWRONSKI, J. *et al.* 1993. Altered T cell activation and development in transgenic mice expressing the HIV-1 nef gene. EMBO J. **12:** 703–713.
31. LINDEMAN, D. *et al.* 1994. Severe immunodeficiency associated with a human immunodeficiency virus 1 NEF/3′-long terminal repeat transgene. J. Exp. Med. **179:** 797–807.
32. HANNA, Z. *et al.* 1998. Nef harbors a major determinant of pathogenicity for an AIDS-like disease induced by HIV-1 in transgenic mice. Cell **95:** 163–175.
33. MANDELL, C.P. *et al.* 1999. SIV/HIV Nef recombinant virus (SHIVnef) produces simian AIDS in rhesus macaques. Virology **265:** 235–251.
34. SAWAI, E.T. *et al.* 2000. Pathogenic conversion of live attenuated simian immunodeficiency virus vaccines is associated with expression of truncated Nef. J. Virol. **74:** 2038–2045.
35. FINKEL, T.H. *et al.* 1995. Apoptosis occurs predominantly in bystander cells of HIV- and SIV-infected lymph nodes. Nature Med. **1:** 129–134.
36. SCHWARTZ, O. *et al.* 1996. Endocytosis of major histocompatibility complex class I molecules is induced by the HIV-1 Nef protein. Nature Med. **2:** 328–342.
37. COLLINS, K.L. *et al.* 1998. HIV-1 Nef protein protects infected primary cells against killing by cytotoxic T lymphocytes. Nature **391:** 397–401.
38. PIQUET, V. *et al.* 1999. The downregulation of CD4 and MHC-I by primate lentiviruses: a paradigm for the modulation of cell surface receptors. Immunol. Rev. **168**: 51–63.
39. COLLINS, K.L. & D. BALTIMORE. 1999. HIV's evasion of the cellular immune response. Immunol. Rev. **168**: 65–74.

Anti-HIV, Anti-Poxvirus, and Anti-SARS Activity of a Nontoxic, Acidic Plant Extract from the *Trifollium* Species Secomet-V/anti-Vac Suggests That It Contains a Novel Broad-Spectrum Antiviral

GIRISH J. KOTWAL,[a] JENNIFER N. KACZMAREK,[a]
STEVEN LEIVERS,[d] YOHANNES T. GHEBREMARIAM,[a]
AMOD P. KULKARNI,[a] GABRIELE BAUER,[b] CORENA DE BEER,[c]
WOLFGANG PREISER,[c] AND ABDU RAHMAN MOHAMED[a]

[a]*Division of Medical Virology, Institute of Infectious Diseases and Molecular Medicine, Faculty of Health Sciences, University of Cape Town, Observatory, Cape Town, South Africa*

[b]*Department of Medical Virology, NHLS, Tygerberg Business Unit, Cape Town, South Africa*

[c]*Institute of Medical Virology, J.W. Goethe-Universitat, Frankfurt am Main, Germany*

[d]*Quantum House, Technopark, Stellenbosch, 6090, Uniedal, South Africa*

ABSTRACT: Enveloped animal viruses such as human immunodeficiency virus (HIV), hepatitis B virus, hepatitis C virus, human papillomavirus, Marburg, and influenza are major public health concerns around the world. The prohibitive cost of antiretroviral (ARV) drugs for most HIV-infected patients in sub-Saharan Africa and the serious side effects in those who have access to ARV drugs make a compelling case for the study of complementary and alternative therapies. Such therapies should have scientifically proved antiviral activity and minimal toxic effects. A plant extract, Secomet-V, with an anecdotal indication in humans for promise as an anti-HIV treatment, was investigated. Using a previously described attenuated vaccinia virus vGK5, we established the antiviral activity of Secomet-V. Chemical analysis showed that it has an acidic pH, nontoxic traces of iron (<10 ppm), and almost undetectable levels of arsenic (<1.0 ppm). The color varies from colorless to pale yellow to dark brown. The active agent is heat stable at least up to sterilizing temperature of 121°C. The crude plant extract is a mixture of several small molecules separable by high-pressure liquid chromatography. The HIV viral loads were significantly reduced over several months in a few patients monitored after treatment with Secomet-V. Secomet-V was also found to have antiviral activity against the SARS virus but not against the West Nile virus. Secomet-V, therefore, is a

Address for correspondence: Girish J. Kotwal, Division of Medical Virology, IIDMM, Faculty of Health Sciences, University of Cape Town, Observatory, Cape Town 7925, South Africa. Voice: +27-21-406-6676; fax: +27-21-406-6018.
gjkotw01@yahoo.com

**Ann. N.Y. Acad. Sci. 1056: 293–302 (2005). © 2005 New York Academy of Sciences.
doi: 10.1196/annals.1352.014**

broad-spectrum antiviral, which possibly works by neutralizing viral infectivity, resulting in the prevention of viral attachment.

KEYWORDS: **anti-HIV; anti-poxvirus; Secomet-V; anti-SARS; broad-spectrum antiviral; acidic plant extract; nontoxic; enveloped viruses**

INTRODUCTION

Enveloped viruses such as the human immunodeficiency virus (HIV), hepatitis B virus, and influenza virus remain a major threat to human health worldwide, infecting over 1 billion people. Together, these three viruses affect at least 5–10 million individuals, or roughly 25% of the South African population, causing considerable morbidity and mortality. In 2004, HIV resulted in more than 3 million total deaths worldwide, with an estimated 5.3 million people infected in South Africa alone (data provided by UNAIDS). To combat diseases caused by HIV and other enveloped-viruses, a sustained multipronged approach is needed such as the one described previously.[1] One of the most powerful weapons against HIV has been the life-prolonging drug cocktails for AIDS treatment. Unfortunately, the availability of such drug cocktails is limited to a very small fraction of those infected because of the high cost of treatment. To make matters worse, there are also severe side effects seen in people taking the drugs. Therefore, there is an urgent need to continue to search for new therapies, which may be just as effective as new antiviral drugs. Other enveloped viruses such as variola virus, the causative agent of smallpox, are considered to be prime candidates for bioweapons in the aftermath of terrorist attacks. In addition, the emergence of monkeypox as a disease-causing agent in the Democratic Republic of Congo (DRC) in central Africa once again emphasizes a need to continue to search for an antiviral agent against enveloped viruses. The promise of chemotherapy for poxviruses (summarized in FIG. 1) and complications from vaccination against smallpox with vaccinia virus seem quite bleak because of the severe side effects.[2] In addition, the looming threat of an influenza pandemic and lack of reliable treatment leave the world vulnerable to the upcoming emergence of new virulent strains. The spread of the SARS-causing virus was nipped in the bud by the end of 2004, but that does not mean we can be complacent and ready to dispense a repertoire of antiserum treatments. Antiviral drugs would be the best ammunition to bolster the preparedness against this agent. For centuries, humans have looked at finding natural solutions to health problems. Recently, plants have been a focus of several investigators around the world in search of antiviral agents.[3–5] Based on interesting anecdotal evidence suggesting properties protective against viral infections, a potentially promising plant extract was developed. Reliable scientific evidence was needed to indicate that the extract was truly worth considering as a potential treatment for viral infections and doing more research was justified. With a well-established infrastructure and the methodology to cultivate and to titer viruses accurately[6] to evaluate antiviral effects, it was possible to show that indeed a small volume of the plant extract termed Secomet-V was able to inactivate approximately 1 million virus particles of the attenuated recombinant vaccinia virus vGK5[7] in 1 minute consistently and reproducibly. The recombinant attenuated virus vGK5 was chosen because it grows as well as the wild-type Western Reserve strain of vaccinia

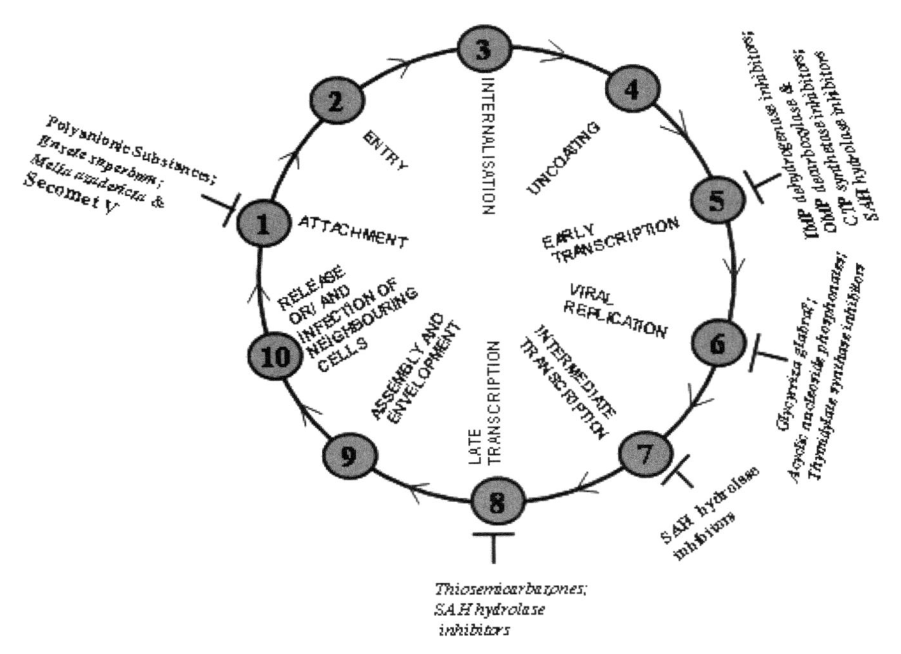

FIGURE 1. Life cycle of poxviruses and sites of inhibition by natural antivirals.

virus in cell culture. The use of the highly attenuated vGK5 strain is also relatively safe, because of the lack of a neurovirulence factor[8] attributed to its inability to replicate in the brain and cause complications.[9] Anecdotal evidence in humans had indicated that the plant extract had beneficial effects against common colds, influenza virus, and HIV. Because Secomet-V is registered as herbal medicine, it was provided by the Secomet company only to patients with no access to antiretroviral (ARV) drugs and under the care of a certified physician. The viral loads of all patients decreased significantly and to nearly minimal levels among the compliant patients, indicating that the extract does indeed have a positive effect on viremia. The levels do not decrease as rapidly as with ARV therapy or highly active antiretroviral therapy (HAART); however, there were no significant side effects with Secomet-V, and the disease symptoms were reduced considerably soon after treatment. It is clear that the amount of active ingredient in the plant extract needs to be identified and increased. Besides increasing the efficacy of the active ingredient by identifying, purifying, and characterizing it, the hope is to reduce any other component that may cause toxicity. It has become evident that changing the process of plant extraction (and avoiding the addition of charcoal), as well as using ultrafiltration with a 3-kDa cutoff filter, greatly reduces both the long-term and short-term toxicity. Secomet-V is a broad-spectrum agent that has the potential to be a safe and highly efficacious antiviral therapy. Identifying the active ingredient would characterize the extract to the fullest extent possible and have a major impact on the development of a new synthetic antiviral originating from a natural source. Another advantage of identifying the bioactive ingredient would be an understanding of the mechanism of action of Secomet-V. Currently, all the evidences indicate that the mechanism of action is through rendering the virus noninfectious and subsequently blocking viral attachment.

METHODOLOGY

Antiviral Activity Assay

The attenuated recombinant vaccinia virus vGK5 was produced in BSC-1 cells or chicken eggs in required quantities and titered in the presence or absence of the plant extract or the fractions from high-pressure liquid chromatography (HPLC) columns as described earlier.[6] Essentially, approximately 1 million virus particles were mixed with 10–20 μl of Secomet-V.

Assay of HIV

Two different cell lines, CCRF CEM and CEM 174, were used and infected with a subtype B HIV virus (R482). Plant extracts were diluted with phosphate-buffered saline. For the experiment, 12-well plates were used and in each plate the following was added: 1 ml RPMI-1640 medium (with Pen and Strep), 100 μl infected CCRF CEM (424.4 pg/ml), or 100 μl infected CEM 174 (20.3 pg/ml) and varying amounts of plant extract. For negative controls, the uninfected cell lines were set up in parallel with the infected cell lines. The positive controls were the infected cell lines on the days that the supernatant was collected. To look at the direct effects of the bioactive molecules on HIV infectivity, a pure virus stock of a million virus particles with varying amounts of the purified bioactive molecules was treated and compared with untreated samples. The p24 assay of surviving virus with and without treatment was also compared. Additional anti-HIV assays were performed directly with Secomet-V with direct contact with HIV at Virologic, Inc. (South San Francisco, CA).

Short-Term Toxicity Assay

Varying amounts of African green monkey kidney cell lines (BSC-1 cells) in a monolayer on 96-well confluent plates were treated with varying amounts of the purified fractions and then stained with crystal violet to determine cell cytoxicity as judged by the integrity of the monolayer.

RESULTS AND DISCUSSION

Chemical Properties of Secomet-V

Secomet-V varies in color from almost colorless, when filtered with a 500-Da cutoff filter, to dark brown when unfiltered. The bioactive agent therefore is smaller than or close to 500 Da in size. Autoclaving does not eliminate the activity, suggesting that the bioactive agent is heat stable up to at least 121°C. The pH of the undiluted extract is approximately 2.0. The virus comes in contact with the extract at a

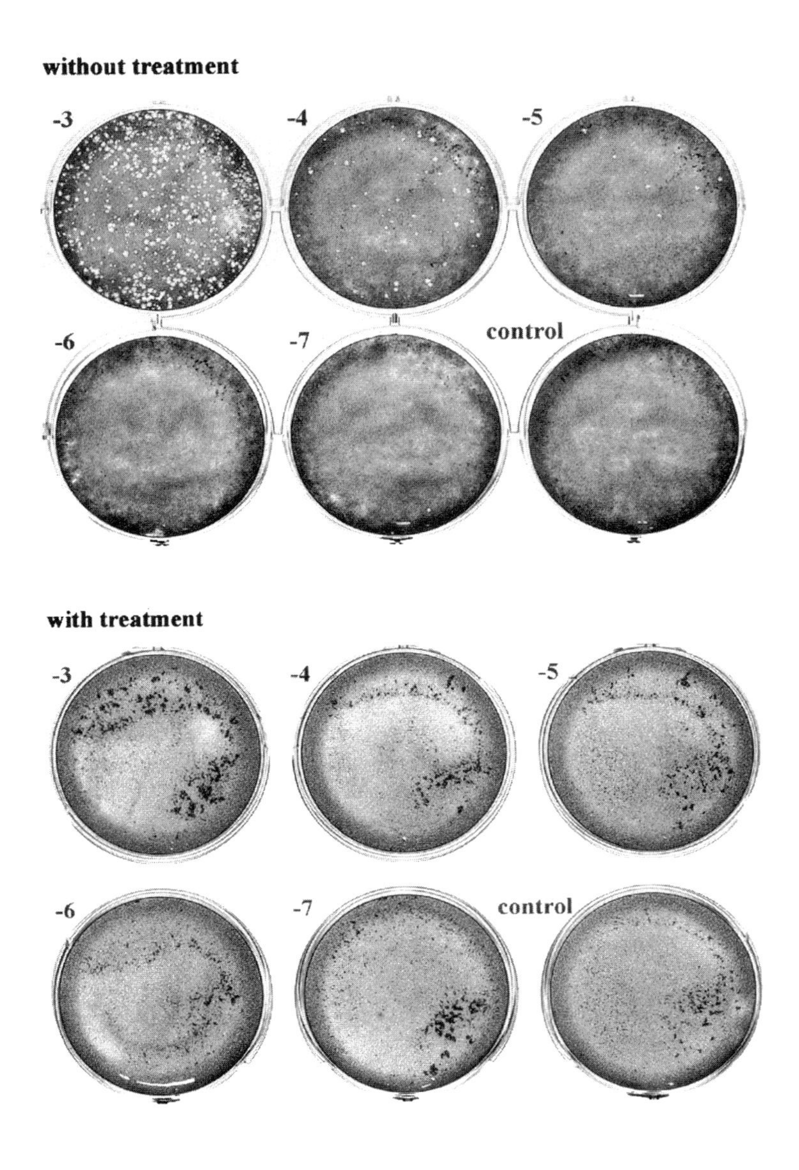

FIGURE 2. Effect of Secomet-V on viral infectivity (as seen from inhibition of plaque formation).

buffered pH of the medium (RPMI-1640) of approximately 3.5. The dry weight varies from approximately 3 mg/ml to 40 mg/ml based on the ultrafiltration cutoff. The arsenic (undetectable) and iron (<10 ppm) content was found to be well below toxic levels, within limits prescribed in the National Formulary, USA (analyzed by using Perkin-Elmer Atomic Absorption Spectrophotometer at the forensic chemistry lab, Western Cape Police Department, Cape Town, South Africa). Subjects have not complained about its taste, suggesting that it has at least a tolerable taste.

Antiviral Activity of Secomet-V

Secomet-V, an extract of an African plant also found elsewhere in Asia, has been found to have potent antiviral activity against a poxvirus (vaccinia virus), rendering about 1 million particles noninfectious in 1 min with a 50th of a milliliter in *in vitro* assays (FIG. 2). Independent reports of anti-HIV testing done by Virologic, Inc. by mixing the virus with the extract have shown that the HIV particles are neutralized by Secomet-V. The same extract failed to cause any reduction in virus titers if the cells were first infected and then directly followed by addition of the plant extract, indicating that Secomet-V neutralized viral infectivity but did not inhibit any of the post-entry steps in the life cycle of the virus. HIV-infected cells treated with plant extract showed no significant effect on the viral levels (TABLE 1). Therefore, apparently there is not any effect on the life cycle of HIV once it has entered cells. This was also observed with vGK5 infection. If Secomet-V is added after the virus is allowed to attach and internalize for 2 h, there is no significant effect on the virus replication and plaque formation, once again confirming the hypothesis that one or

TABLE 1. Anti-HIV activity

	Plant extracts					
	Dilution	Volume	CCRF CEM 26/12 pg/ml	CEM 174 26/12 pg/ml	CCRF CCEM 26/ 12 pg/ml	CEM 174 26/12 pg/ml
Neg	Uninfected culture		Neg	Neg	Neg	Neg
Pos	Infected culture		342.509	11.476	270.061	8.243
Pos	Infected culture				282.896	11.023
A1	1:1	20 µl	260.242	14.711	220.145	10.648
A2	1:10	20 µl	309.229	9.171	230.496	9.566
A3	1:100	20 µl	350.264	9.739	232.879	7.424
A4	1:1,000	20 µl	264.138	16.528	206.609	9.465
B1	1:1	20 µl	305.567	26.243	218.981	10.833
B2	1:10	20 µl	309.51	20.649	231.429	12.939
B3	1:100	20 µl	289.007	14.979	221.897	9.08
B4	1:1,000	20 µl	313.627	17.4	225.606	10.76

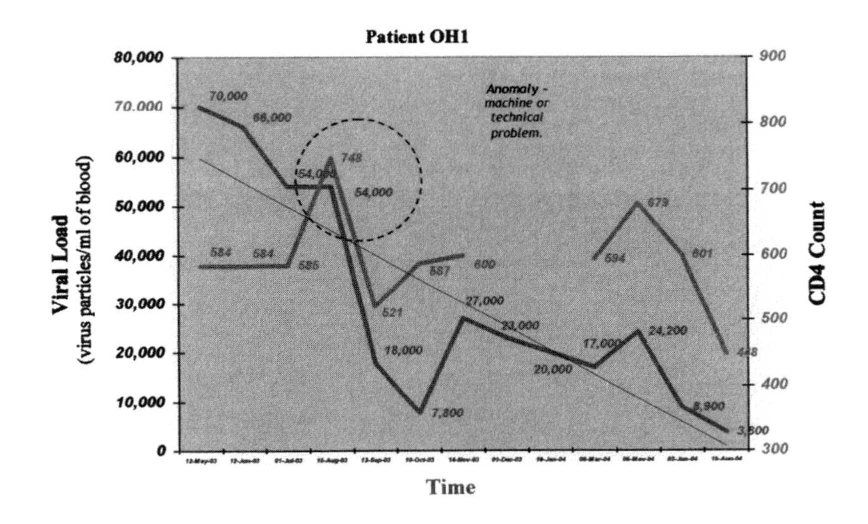

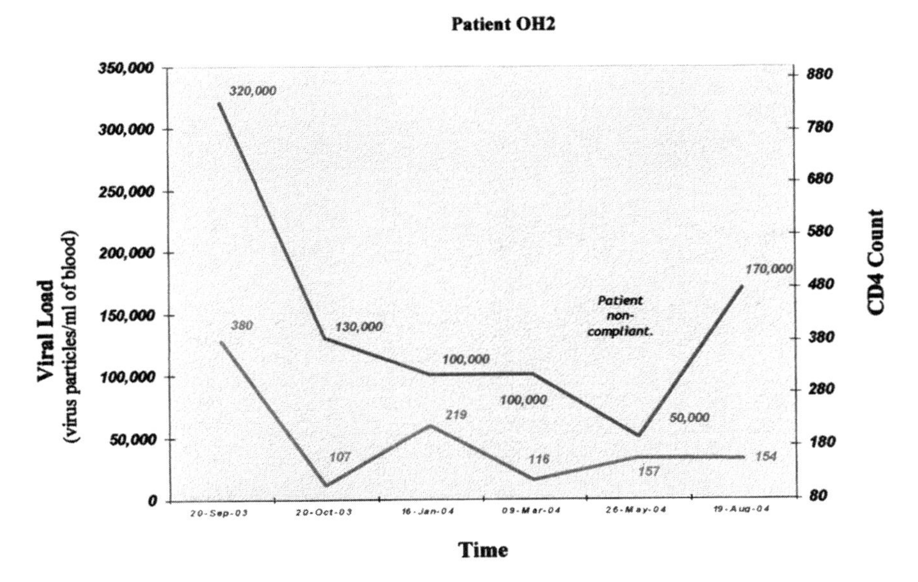

FIGURE 3. Viral load and CD4 profile of HIV-infected patients over time after treatment with Secomet-V.

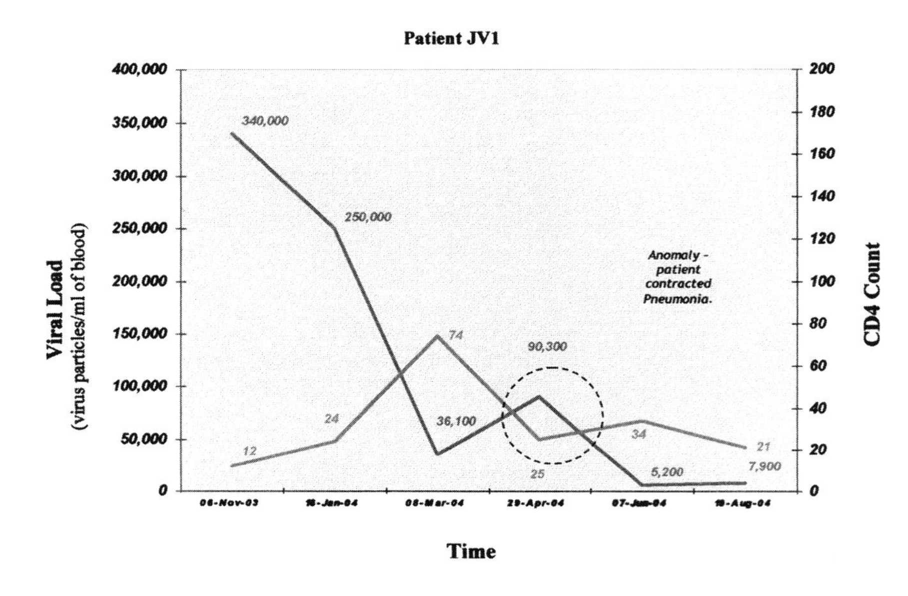
Patient JV1
400,000
350,000
300,000
250,000
200,000
150,000
100,000
50,000
0
Viral Load
(virus particles/ml of blood)
200
180
160
140
120
100
80
60
40
20
0
CD4 Count
340,000
250,000
36,100
90,300
5,200
7,900
12
24
74
25
34
21
Anomaly - patient contracted Pneumonia.
06-Nov-03
08-Mar-04
29-Apr-04
Time

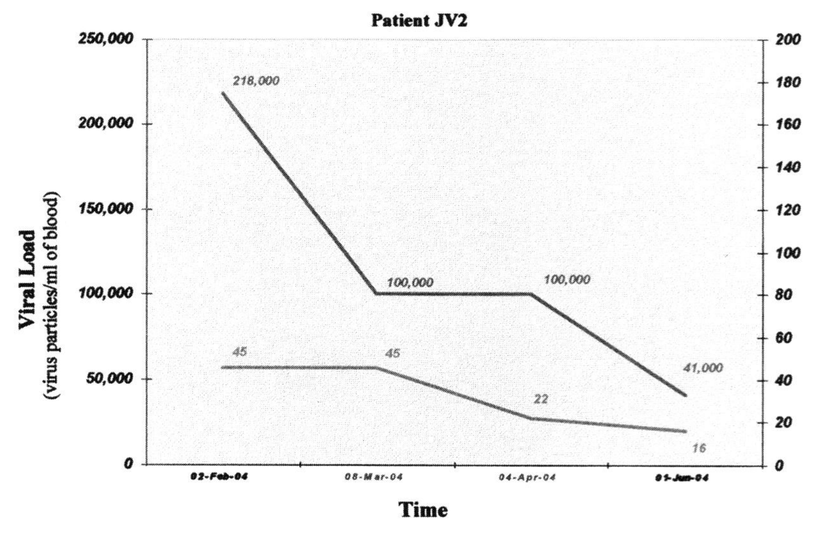
Patient JV2
250,000
200,000
150,000
100,000
50,000
0
Viral Load
(virus particles/ml of blood)
200
180
160
140
120
100
80
60
40
20
0
CD4 Count
218,000
100,000
100,000
41,000
45
45
22
16
02-Feb-04
08-Mar-04
04-Apr-04
Time

FIGURE 3B.

more of the small molecules in the plant extract is, by some as yet unknown mechanism, rendering the virus noninfectious. This suggests that once the virus enters the cells, the extract is no longer effective. This key finding indicates that it is most likely Secomet-V acting as a broad-spectrum inhibitor of viral entry. Because of this novel mechanism of inhibition, it would not have the same problems of resistance that are common with ARVs, as well as the problems of viral evasion due to changes in a surface glycoprotein amino acid sequence. To determine whether Secomet-V was able to reduce viremia, four HIV patients infected with HIV-1 subtype C and who had full-blown AIDS were given the plant extract. These patients had no access to ARVs, and their CD4 counts had decreased quite significantly prior to treatment with Secomet-V. Viral loads were found to have been significantly reduced within a few months as observed (FIG. 3A, B). The SARS corona virus was also found to have been rendered inactive by Secomet-V (data not shown) when tested along with other agents as a coded panel. The West Nile virus was similarly tested, but Secomet-V had no effect on the West Nile virus at concentrations at which other viruses were inactivated. The dry weight of 1 ml of the crude plant extract is approximately 22 mg and that of the ultrafiltered (with a 5-kDa cutoff) plant extract is 3 mg. There was no difference in the effectiveness of the plant extract in rendering vaccinia virus noninfectious whether it was autoclaved or not, suggesting that the bioactive agent is most likely but not necessarily a heat-stable compound and not a small peptide. A preliminary HPLC separation of the ultrafiltered plant extract on a C18 column showed 3 major peaks, 3 minor peaks, and approximately 12 quite tiny peaks. Previously, it has been shown that the trifollium species has flavanoids and salicylic acid.

Short-Term Toxicity

Besides testing for efficacy, preliminary studies were completed to determine the short-term and long-term toxicity of the extract. In a simple assay in which varying amounts of the extract were added to 3 million African green monkey kidney cells (BSC-1 cells), it was found that the 10 μl of the extract rendered half of the million virus particles inactive, and when 100 μl of the plant extract was added, the monolayer became detached and up to 50 μl of the extract was tolerated with no cytotoxicity present. Therefore, at 10% of the concentration of the toxic dose, 50% of the virus was rendered inactive. When the extract was filtered through a 3-kDa filter, the toxicity diminished, but the antiviral activity remained intact. Similar studies have been conducted on potential long-term effects of the crude extract (Ballardin *et al.*, this volume) that could be carcinogenic. Secomet-V was tested for mutagenicity using the Ames gene mutation test in *Salmonella* and the chromosome damage (clastogenic) micronucleus test in human lymphocytes. The crude extract presents weak clastogenic activity but powerful mutagenic activity in the Ames test with the addition of exogenous metabolic activation. The purification of the extract without the addition of charcoal results in a drastic reduction of the extract's mutagenicity, which has been virtually reduced by means of further ultrafiltration (cutoff <3,000 Da). If ultrafiltration almost eliminates the short-term and long-term toxic molecules, then pure bioactive molecules are likely to be free of all toxicity associated with the crude extract. These findings open new possibilities of HIV therapy because Secomet-V, devoid of mutagenic activity, should reduce the possibility of the induction of resistant viral strains by previous ARV drugs.

ACKNOWLEDGMENTS

We thank Abubakar Jacobs and Andre Baard, Forensic Chemistry Laboratory, Cape Town Police Department, for their help in analyzing the samples for arsenic and iron. G.J.K. is currently Senior International Wellcome Trust Fellow for biomedical sciences in South Africa. A.P.K. and Y.T.G. are recipients of the Poliomyelitis Research Foundation (PRF) of South Africa and Senior Entrance Fellowship and International Students' Fellowship at the University of Cape Town, South Africa. This work was not supported by funding from Secomet pvt. Ltd. and is an independent study that was in no way influenced by Secomet.

REFERENCES

1. Kotwal, G.J. 2004. HIV treatment and eradication in South Africa. J. R. Soc. Med. **97:** 1–2.
2. De Clercq, E. 2001. Vaccinia virus inhibitors as a paradigm for the chemotherapy of poxvirus infections. Clin. Microbiol. Rev. **2:** 382–397.
3. Jacob, J.R. *et al.* 2004. Korean medicinal plant extracts exhibit antiviral potency against viral hepatitis. J. Alt. Compl. Med. **10:** 1019–1026.
4. Notka, F. *et al.* 2004. Concerted inhibitory activities of *Phyllanthus amarus* on HIV replication *in vitro* and *ex vivo*. Antiviral Res. **64:** 93–102.
5. Neurath, A.R. *et al.* 2004. *Punica granatum* (Pomegranate) juice provides an HIV-entry inhibitor and candidate topical microbicide. BMC Infect. Dis. **4:** 41–52.
6. Kotwal, G.J. & M.-R. Abrahams. 2004. Growing poxviruses and determining virus titer. *In* Vaccinia Virus and Poxvirology, Methods and Protocols, S.N. Isaacs, Ed. :101–112. Vol. 269. Humana Press. Totowa, NJ.
7. Jonczy, E.A., J. Daly & G.J. Kotwal. 2000. A novel approach using an attenuated recombinant vaccinia virus to test the antiviral effects of handsoaps. Antiviral Res. **45:** 149–153.
8. Kotwal, G.J., A.W. Huegin & B. Moss. 1989. Mapping and insertional mutagenesis of a vaccinia virus gene encoding a 13,800 Da secretory protein. Virology **171:** 579–587.
9. Billings, B., S.A. Smith, Z. Zhang, *et al.* 2004. Lack of N1L gene expression results in significant decrease in vaccinia virus replication in mice brain. Ann. N.Y. Acad. Sci. **1030:** 297–302.

In Vitro Mutagenicity Studies of the Antiretrovirals AZT, Didanosine, and 3TC and a Plant Antiviral Extract Secomet-V Derived from the *Trifollium* Species

MICHELA BALLARDIN,[a] ROBERTO SCARPATO,[a] GIRISH J. KOTWAL,[b] AND ROBERTO BARALE[a]

[a]*Dipartimento di Scienze dell'Uomo e dell'Ambiente, University of Pisa, Pisa, Italy*

[b]*Division of Medical Virology, IIDMM, University of Cape Town, Medical School, Observatory, Cape Town 7925, South Africa*

ABSTRACT: The plant extract Secomet-V has previously been shown by Kotwal *et al.* to have potent antiviral activity. It was tested for mutagenecity with the Ames gene mutation test in *Salmonella* and the chromosome damage (clastogenic) micronucleolus (MN) test in human lymphocytes. These tests predict long-term carcinogenesis activity of the agents tested. Secomet-V (with charcoal added) demonstrated weak clastogenic activity, but powerful mutagenic activity in the Ames test with the addition of exogenous metabolic activation. The mutagenic activity of the conventional antiretroviral drugs AZT, Didanosine (DID), and 3TC alone and in dual combinations was also assessed for the first time for Salmonella mutagenicity without any mutagenic effects. AZT, DID, and 3TC have also been tested for MN induction; DID and 3TC resulted negatively, whereas AZT was positive in a dose-related manner. The dual combinations of AZT and DID, 3TC and DID plus 3TC did not result in any additive or synergistic effect. Purification in the absence of charcoal results in a drastic reduction in extract mutagenicity, which is almost reduced completely by further ultrafiltration (cutoff <3,000 Da). This fraction, which is a mixture of molecules of less than 3,000 Da, still possesses the capability to induce sister chromatid exchanges in human lymphocytes. This could be due to residual mutagenicity or, more likely, to the slowdown of the DNA replication process. These findings open new possibilities for HIV therapy, because this antiviral activity of Secomet-V purified in the absence of charcoal and further filtered through a 3,000-Da filter is devoid of mutagenic activity and therefore safe for long-term use.

KEYWORDS: mutagenicity; complex mixtures; human lymphocytes; plant extract; Salmonella Ames test; micronuclei; sister chromatid exchanges

Address for correspondence: Roberto Barale, Via S. Giuseppe, n. 22, 56100-Pisa, Italy. Voice: ++39(050)836224; fax: ++39(050) 551290.
r.barale@geog.unipi.it

Ann. N.Y. Acad. Sci. 1056: 303–310 (2005).
doi: 10.1196/annals.1352.029

INTRODUCTION

Toxicology of Complex Mixtures of Herbal Origin

Herbal extracts offer significant potential in the treatment of several disorders such as diabetes, hypertension, and AIDS. However, therapeutic utility of the extracts of herbal origin could be hampered by the presence of harmful chemical compounds, such as heavy metals, proteins, and other contaminants. The separation of these toxic compounds from active ingredients poses a great challenge. The toxic effect of these compounds may range from mild adverse effects (e.g., hyperacidity) to malignant cancer depending on the nature and duration of treatment. It is important to separate active compounds from toxic ones eventually present into a therapeutically effective plant crude extract. Also, it is essential to screen these herbal extracts for their safety before their use in actual therapy. Mutagenicity of these extracts, especially if they are intended for chronic use, is an important indicator that needs to be studied. Bioassay-directed fractionation of a complex mixture into several parts according to their chemicophysical properties and subsequent analysis for toxicity are the most convenient approaches to developing safe, yet effective therapeutic molecules.[1]

GENETIC TOXICOLOGY

Substances that have a toxic effect on genetic material are said to be genotoxic. Such substances increase the error rate in the reduplication of the genome and induce mutations by damaging the organism's DNA. Mutations in germ cells are passed onto the organism's offspring and can cause congenital or hereditary defects. Mutations in somatic cells can result in cell death, an increased risk of disease, and even cancer. Drug regulatory authorities require that new substances undergo different tests for genotoxicity and mutagenicity. Both bacteria and mammalian cells are employed to achieve prompt detection of gene mutations and chromosomal aberrations. In fact, genotoxicity represents a unique case in toxicology, because the effects can be extrapolated to all living organisms. According to the directives of the European Community for the registration of traditional herbal medicine products, a test of gene mutation (i.e., Ames test) and one of chromosome aberration (i.e., micronucleus test) are required.[2]

In addition, by excluding prokaryotes, the biologic machinery involved in the distribution of genetic information during mitosis and meiosis, the spindle apparatus is about the same in all organisms, thus allowing the detection of aneugenic activity, giving rise to genomic mutations, namely, aneuploidies and polyploidies, by using suitable eukaryotic cells.

The battery for genotoxicity testing is based on *in vitro* and *in vivo* assays to be used, stepwise, according to the results obtained and the drug to be tested. The "first line" is represented by two *in vitro* assays: a point mutation assay, such as the Ames test, in *Salmonella typhimurium*, and a chromosome aberration assay, such as the micronucleus assay or chromosome assay in mammalian cells.

In the current study, we report the mutagenic activity of a widespread antiviral agent of natural origin, Secomet-V, used against HIV infections and tested as a plant crude extract following purification procedures. Actually, many of the current antiviral drugs have mutagenic activity, which can represent a carcinogenic risk for the patient,[3] but also may compromise the results of therapy because changing the DNA of the virus can speed up the appearance of drug-resistant mutants.

MATERIAL AND METHODS

Plant extract Secomet-V was provided by Secomet Pvt. ltd. It was used both as is and after several purification steps including the addition of charcoal and after passing the extract through a 3,000-Da cutoff Centriplus filter (Amicon, Danvers, MA, USA).

Salmonella Mutagenicity Tests. The standard Ames plate test was applied by using *S. typhimurium* TA98 and TA100 strains with and without exogenous metabolic activation. S9 (10% in the mix), obtained from Sprague-Dawley male rats and pretreated with Arochlor 1254 according to Maron and Ames, was used for metabolic activation.[4] The indirect mutagen 2-amminofluorene (2AF) was used at a dose of 5.0 μg/plate for checking both strain sensitivity and S9 efficiency.

Lymphocyte Micronucleus Assay. The micronucleus (MN) assay in human lymphocytes was carried out according to the cytochalasin B-block method.[5] Plant extracts of 50 and 100 μL were added, and the mixture was incubated for about 72 h. S9 mix was added to the incubation mixture at a final concentration of 5% for 3 h. MN were scored in binucleated cells, whereas the cell proliferation index (PI) was assessed by evaluating the ratio between mono- and binucleated cells.

Sister Chromatid Exchange Assay (SCE). SCE induction in human lymphocytes was evaluated according to a standard chromatid exchange assay. Cells were exposed for about 72 h for this study.[6] The cell proliferation index (PI) was assessed by evaluating the percentage of first, second, and third metaphases of cell culture. Statistical analysis was performed using STAGRAPHICS *Plus,* Version 2 (Plus Ware, Rockville, MD, USA).

RESULTS AND DISCUSSION

Mutagenicity of Raw Extract: Salmonella Assay

The raw extract Secomet-V possesses notable toxicity towards the bacterial cells even at lower doses. With the addition of exogenous metabolic activation (rat S-9), toxicity is reduced, notably allowing the expression of powerful mutagenic activity, particularly on strain TA98 (38 his$^+$ revertants/L/plate), which is in conformity with the presence of a frameshift mutagen (FIG. 1a). It was difficult to assess whether the metabolism activated some pro-mutagens present in the mixture and/or elicited generalized detoxifying activity, which allowed the replication of bacterial cells and therefore the expression of mutants. Strain TA100 was found to be less responsive than strain TA98, although the mutagenic effect of the raw extract is evident in this strain (10, 7 his$^+$ revertants/L/plate) (FIG. 1b). The addition of S-9 showed similar detoxifying effects to those of strain TA98. The antiretrovirals AZT, DID, and 3TC tested alone, in the range of doses of 0.15–1.25, 7.8–62.5, and 187–1500 g/plate, respectively, and in dual combinations, were completely negative.

a)

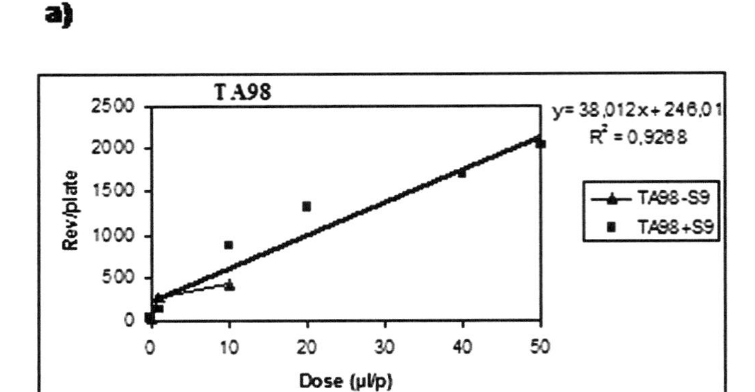

b)

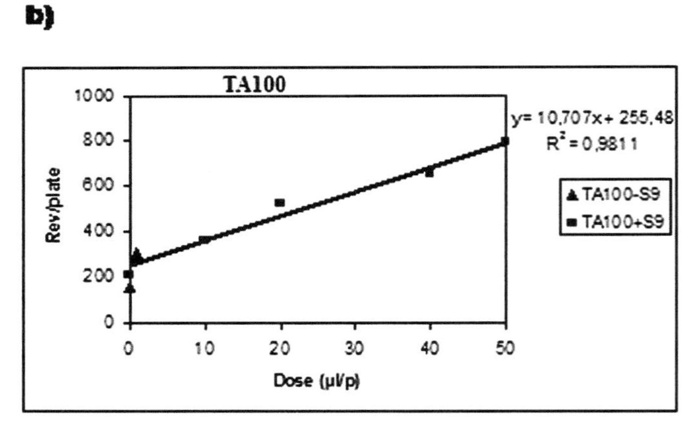

FIGURE 1. Mutagenicity of Secomet-V crude plant extract in the Ames test with and without metabolic activation (S9). (**a**) TA98 *Salmonella* strain; (**b**) TA100 *Salmonella* strain. Linear regression and correlational analysis is reported.

Mutagenicity of Raw Extract: Micronucleus Assay

Two different plant crude extracts revealed the induction of a weak, but a statistically significant, increase of MN in human lymphocytes after an exposure of 72 h (FIG. 2). The addition of S9 resulted in a significant increase in the level of MN after 3 h of treatment. These data demonstrate that the raw extract introduces weak, but significant, genotoxicity on the chromosome and/or on the mitotic spindle of human lymhocytes. The use of citric acid or molasses to produce the crude extract resulted in the abolition of the clastogenic activity found in the original extract that was pro-

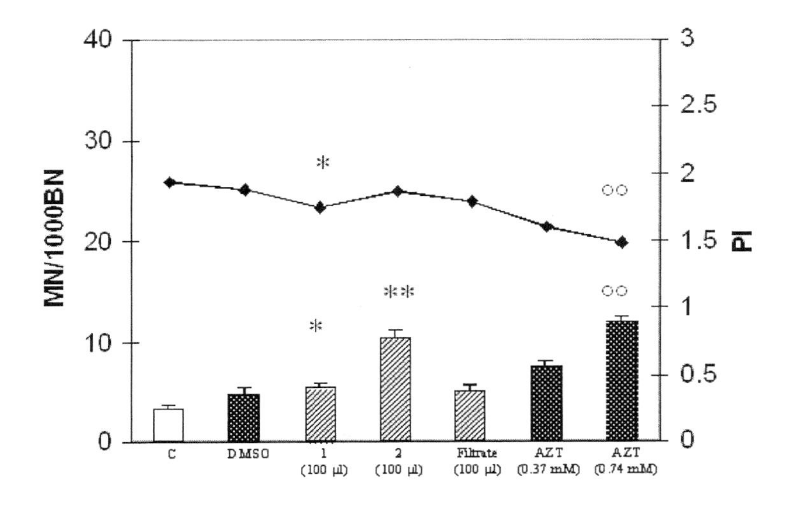

FIGURE 2. MN induction in human lymphocytes by crude and charcoal filtrate of Secomet-V (*dashed bars*) compared with control C and by AZT (*dark bars*) compared with controls with the addition of DMSO. Proliferation indexes (PIs) are also reported. *Statistically different versus control at P <.05. **Statistically different versus control at P <.01. Statistically different versus DMSO at P <.05. °°Statistically different versus DMSO at P <.01.

duced with the addition of charcoal (FIG. 2). On the other hand, AZT induced a significant increase of MN in a dose-related manner. DID and 3TC did not induce a significant increase of MN, and the combination with AZT did not result in any potentiation of AZT genotoxicity (data not reported). We performed these assays because of the synergistic effects in inducing point mutations in human cells by combinations of AZT plus DID that were previously observed.[8]

The withholding of charcoal caused a drastic reduction in the toxicity and mutagenicity of the original extract in both *Salmonella* strains used in the study. However, metabolic activity was responsible for the mutagenicity of the original extract (1, 5 his^+ revertants/L/plate). Purification without using charcoal resulted in the reduction of mutagenicity of the original extract about 30-fold in the TA98 strain. Later, the extract was passed through a 3-kDa cutoff filter and separated into two fractions. The fraction >3 kDa maintained the same mutagenicity (1.06 his^+ revertants/L/plate) as that of the purified fraction with added charcoal, whereas the fraction <3 kDa showed a statistically insignificant effect (0, 46 his^+ revertants/L/plate) on the mutagenicity (FIG. 4). However, the antiviral effect of the extract filtered using a 3-kDa cutoff filter (fraction with <3,000 Da) was the same as that of the unpurified extract. These results demonstrate that the molecular weight of the active antiviral compound(s) is less than 3,000 Da, and the compound(s) is(are) devoid of the mutagenic effect. The SCE assay of the fractions revealed a dose-dependent effect of the extract on the SCEs. The mechanism of formation of SCEs is not yet clear. It is known that SCEs can be induced by agents able to damage DNA and chromosomes, but particularly by agents that stop DNA replication, by inhibiting the DNA topoisomerases.[7] The antiviral activity of the extract can therefore be attributed to its inhibitory effect on DNA replication.

In conclusion, the current investigation revealed that the mutagenicity of the plant extract can be reduced successfully by filtration of the extract using a 3-kDa cutoff filter, without any reduction in antiviral activity. It is also evident that the molecular weight of the active compound is less than 3 kDa. However, the plant extract should be thoroughly studied for its toxicity and safety using *in vivo* and in *vitro* models. This might add to the therapeutic value of the extract. Furthermore, insight into the chemical nature of the compound is also necessary for rational drug development.

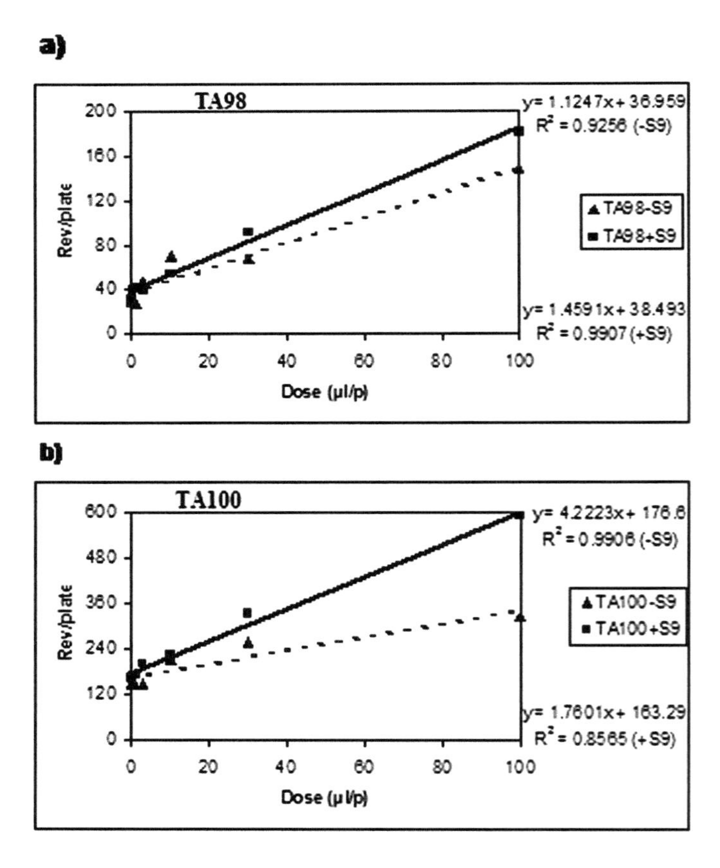

FIGURE 3. Mutagenicity of charcoal-purified Secomet-V in the Ames test with and without metabolic activation (S9). (**a**) TA98 *Salmonella* strain; (**b**) TA100 *Salmonella* strain. Linear regression and correlational analysis are reported.

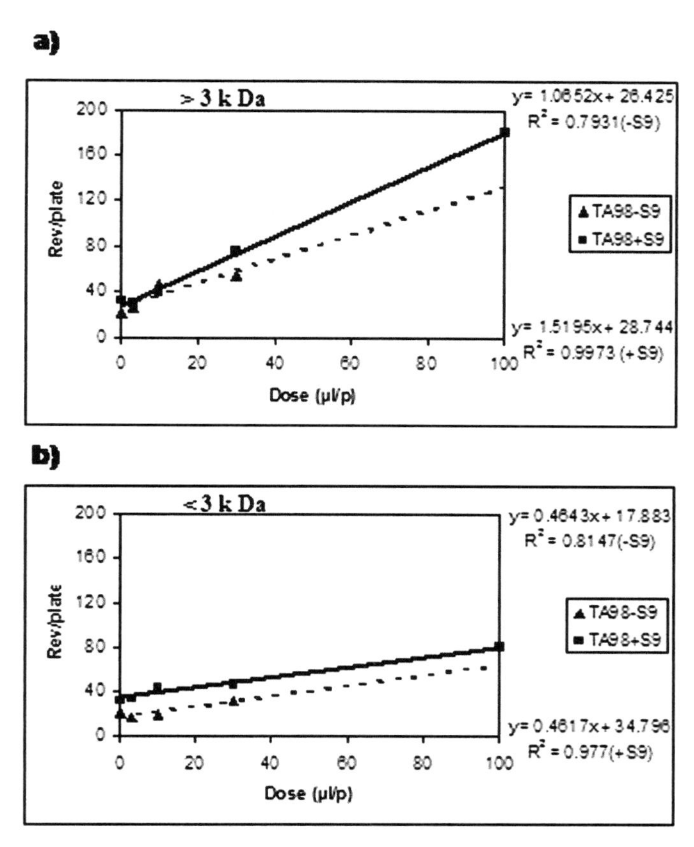

FIGURE 4. Comparison of the mutagenicity of Secomet-V extracts obtained by passing through a 3-kDa cutoff filter assessed using the TA98 *Salmonella* strain with and without metabolic activation (S9). (**a**) Fraction above 3,000 Da (>3 kDa); (**b**) fraction below 3,000 Daltons (<3 kDa). Linear regression and correlational analysis are reported.

ACKNOWLEDGMENTS

We gratefully acknowledge the careful review and revision of the manuscript by Amod Kulkarni, Division of Medical Virology, University of Cape Town, Cape Town, South Africa. G.J.K. is currently Senior International Wellcome Trust Fellow for Biomedical Sciences in South Africa.

REFERENCES

1. Krewski, D. & R.D. Thomas. 1992. Carcinogenic mixtures. Risk Anal. **12:** 105–113.
2. Directive 2001/83/EC of the European Parliament and of the Council of 6 November 2001 on the Community code relating to medicinal products for human use (Official Journal L 311, 28/11/2001 :67–128).

3. WUTZLER, P. & R. THUST. 2001. Genetic risks of antiviral nucleoside analogues: a survey. Antiviral Res. **49:** 55–74.
4. MARON, D.M. & B.N. AMES. 1983. Revised methods for the Salmonella mutagenicity test. Mutat. Res. **113:** 173–215.
5. FENECH, M. 1993. The cytokinesis-block micronuceus technique: a detailed description of themethod and its application to genotoxicity studies in human populations. Mutat. Res. **285:** 35-44.
6. LATT, S.A. *et al.* 1981. Sister chromatid exchanges: a report of the U.S. EPA's Gene-Tox Program. Mutat. Res. **87:** 17–62.
7. PINERO, J. *et al.* 1996. Sister chromatid exchange induced by DNA topoisomerases poisons in late replicating heterochromatin: influence of inhibition of replication and transcription. Mutat. Res. **354:** 195–201.

Punica granatum (Pomegranate) Juice Provides an HIV-1 Entry Inhibitor and Candidate Topical Microbicide

A. ROBERT NEURATH, NATHAN STRICK, YUN-YAO LI, AND
ASIM K. DEBNATH

New York Blood Center, New York, New York 10021, USA

ABSTRACT: For ~24 years the AIDS pandemic has claimed ~30 million lives, causing ~14,000 new HIV-1 infections daily worldwide in 2003. About 80% of infections occur by heterosexual transmission. In the absence of vaccines, topical microbicides, expected to block virus transmission, offer hope for controlling the pandemic. Antiretroviral chemotherapeutics have decreased AIDS mortality in industrialized countries, but only minimally in developing countries. To prevent an analogous dichotomy, microbicides should be acceptable, accessible, affordable, and accelerative in transition from development to marketing. Already marketed pharmaceutical excipients (inactive materials of drug dosage forms) or foods, with established safety records and adequate anti-HIV-1 activity, may provide this option. Therefore, fruit juices were screened for inhibitory activity against HIV-1 IIIB using CD4 and CXCR4 as cell receptors. The best juice was tested for inhibition of: (1) infection by HIV-1 BaL, utilizing CCR5 as the cellular coreceptor, and (2) binding of gp120 IIIB and gp120 BaL, respectively, to CXCR4 and CCR5. To remove most colored juice components, the adsorption of the effective ingredient(s) to dispersible excipients and other foods was investigated. A selected complex was assayed for inhibition of infection by primary HIV-1 isolates. The results indicate that HIV-1 entry inhibitors from pomegranate juice adsorb onto corn starch. The resulting complex blocks virus binding to CD4 and CXCR4/CCR5 and inhibits infection by primary virus clades A to G and group O. Therefore, these results suggest the possibility of producing an anti-HIV-1 microbicide from inexpensive, widely available sources, whose safety has been established throughout centuries, provided that its quality is adequately standardized and monitored.

KEYWORDS: pomegranate juice; *Punica granatum*; human immunodeficiency virus (HIV-1); virus entry inhibitors; CD4; CXCR4 CCR5; receptors for HIV-1; coreceptors for HIV-1; microbicides

BACKGROUND

The global acquired immunodeficiency syndrome (AIDS) epidemic has proceeded relentlessly for ~24 years with no promising prophylactic intervention in sight. In

Address for correspondence: A. Robert Neurath, Biochemical Virology Laboratory, Lindsley F. Kimball Research Institute, 310 East 67th Street, New York, NY 10021, USA; Voice: 212-570-2275; fax: 212-570-3299.
arneurath@att.net

Ann. N.Y. Acad. Sci. 1056: 311–327 (2005).
doi: 10.1196/annals.1352.015

2004, there were 4.9 million new HIV infections and 3.1 million AIDS deaths.[1] To date, the number of individuals living with human immunodeficiency virus type 1 (HIV-1) infection/AIDS has reached 39.4 million,[1] and ~28 million people have already died from AIDS since the beginning of the pandemic.[1,2] Most new infections have been acquired by the mucosal route, heterosexual transmission playing the major (~80%) role. Although the incidence of transmission per unprotected coital act is estimated to be low (0.0001–0.004), but strikingly increased when acutely infected individuals are involved,[3,4] the cumulative effect is overwhelming.

Anti-HIV-1 vaccines applicable to global immunization programs are not expected to become available for many years. Therefore, other prevention strategies are urgently needed. This includes educational efforts and the application of mechanical and/or chemical barrier methods. The latter correspond to microbicides, that is, topical formulations designed to block HIV-1 infection (and possibly transmission of other sexually transmitted diseases) when applied vaginally (and possibly rectally) before intercourse.[3,5–7] Conceptually, it is preferred that the active ingredient(s) of microbicide formulations (1) block virus entry into susceptible cells by preventing HIV-1 binding to the cellular receptor CD4, the coreceptors CXCR4/CCR5, and to receptors on dendritic/migratory cells (capturing and transmitting virus to cells that are directly involved in virus replication), respectively,[3,8–11] and/or (2) are virucidal. The formulations must not adversely affect the target tissues and should not cause them to become more susceptible to infection after microbicide removal.[12,13]

Treatment with antiretroviral drugs has decreased mortality from AIDS in industrialized countries but so far has had a minimal effect in developing countries.[14] To avoid a similar dichotomy with respect to microbicides, they should be designed and selected to become affordable and widely accessible, while shortening the time between research and development and their marketing and distribution as much as possible. This would be facilitated if mass manufactured products with established safety records were found to have anti-HIV-1 activity. Qualifying candidates to be considered for microbicide development may possibly be discovered by screening pharmaceutical excipients ("inactive" ingredients of pharmaceutical dosage forms) and foods, respectively, for antiviral properties.

While exploring the possibility that chemical modification of food proteins may lead to the generation of compounds with anti-HIV-1 activity, we discovered in 1994 that bovine β-lactoglobulin (the major protein of whey) modified by 3-hydroxyphthalic anhydride (3HP-β-LG) blocked infection by HIV-1 and herpesviruses, both *in vitro* and in animal model systems.[15–24] Considering its antiviral potency, ease of preparation, and practically unlimited and inexpensive source (the worldwide production of whey is approximately 86 billion kg annually), 3HP-β-LG appeared to represent an excellent candidate microbicide for prevention of the sexual transmission of HIV-1. By coincidence, an epidemic of bovine spongiform encephalopathy (BSE) was ongoing at the same time in the United Kingdom and considered to cause a new variant of Creutzfeldt-Jakob disease (vCJD) in humans. This raised questions related to the safety of bovine milk. However, scientific research results indicate that BSE cannot be transmitted by cow's milk even if the milk comes from a cow with BSE, because no detectable infectivity in milk from BSE-infected animals could be demonstrated. Evidence from other animal and human transmissible spongiform encephalopathy (TSE) studies suggests that milk does not transmit these diseases.

Milk and milk products, even from countries with a high incidence of BSE, are therefore considered safe.[25] Consequently, the United States Food and Drug Administration has exempted milk-derived products from restrictions applied to their use as pharmaceutical ingredients.[26] In accordance with this, some bovine milk-derived products are being generally recognized as safe (GRAS).[27] In addition, the current risk of acquiring vCJD from eating beef and beef products is approximately 1 case per 10 billion servings in the United Kingdom and likely to be smaller in other countries.[28] Notwithstanding these unequivocal conclusions, the World Health Organization recommends that the pharmaceutical industry should avoid the use of materials from animal species in which TSEs naturally occur.[29] Based on these negative recommendations from the WHO, further development of 3HP-β-LG as a topical microbicide had to be abandoned.

We initiated the screening of pharmaceutical excipients for compounds with anti-HIV activity. This led to the discovery that cellulose acetate 1,2-benzenedicarboxylate used for coating enteric tablets and capsules has anti-HIV activity and represents a promising candidate microbicide.[30–34] Here we report the outcome of screening fruit juices, many of which have been reported to provide health benefits.[35–53] All juices were neutralized to pH ≈7 to discount nonspecific effects caused by acidity.

The results presented here and the corresponding methods are freely available online at http://www.biomedcentral.com/content/pdf/1471-2334-4-41.pdf.[54]

RESULTS

Anti-HIV-1 Activity of Pomegranate Juice

Serial twofold dilutions of juices (apple, black cherry, blueberry, coconut milk, cranberry, elderberry, grape [red], grapefruit, honey, lemon, lime, pineapple, pomegranate, and red beet [10% reconstituted dry powder]) were assayed for inhibition of infection by HIV-1 IIIB of cells expressing the CD4 and CXCR4 receptors and coreceptors. Most juices (diluted fourfold) had no inhibitory activity, except blueberry, cranberry, grape, and lime juice, respectively (endpoints for 50% inhibition of infection [ED_{50}] between 1/16 and 1/64). Consistently, pomegranate juice (PJ) from distinct geographic areas had the highest inhibitory activity (FIG. 1; vertically shaded area). Since HIV-1 viruses utilizing CCR5 as coreceptor (= R5 viruses) are predominantly transmitted sexually,[3,55] it was important to test whether PJ can inhibit not only infection by HIV-1 IIIB, a virus utilizing CXCR4 as coreceptor (= X4 virus), but also infection by an R5 virus, HIV-1 BaL. Results in FIGURE 1 (horizontally shaded area) show that infection by the latter virus is also inhibited, albeit less effectively, than that by HIV-1 IIIB.

Blocking virus entry is a primary target for microbicide development.[3,8–11] Therefore, it was of interest to determine whether PJ inhibited the binding of the HIV-1 envelope glycoprotein gp120 to CD4, the common receptor for both X4 and R5 viruses. Pretreatment of both gp120 IIIB and BaL by PJ inhibited subsequent binding of soluble labeled CD4 (FIG. 2). This suggested that one or more PJ ingredients bound strongly or irreversibly to the CD4 binding site on gp120. These results, obtained in an enzyme-linked immunosorbent assay (ELISA) using gp120 immobilized on polystyrene plates, were confirmed in another assay in which both

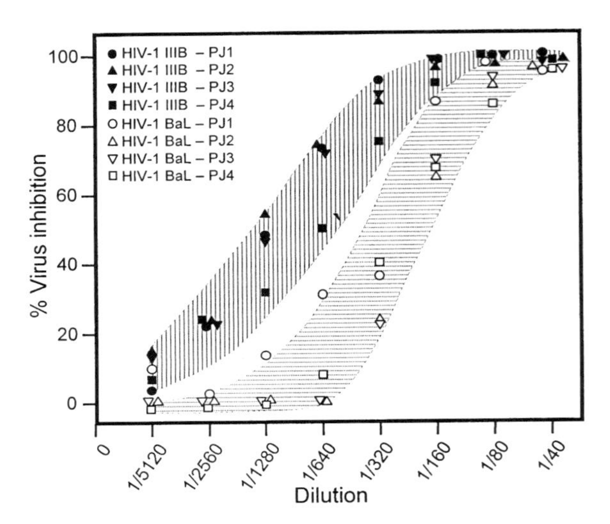

FIGURE 1. Inhibition of HIV-1 infection of HeLa-CD4-LTR-β-gal and U373-MAGI-CCR5E cells, respectively, by pomegranate juice (PJ). LTR = long terminal repeat; *vertically shaded area* = HIV-1 IIIB; *horizontally shaded area* = HIV-1 BaL. Four distinct PJs (PJ1 to PJ4) were tested. Infection was monitored by measuring β-galactosidase.

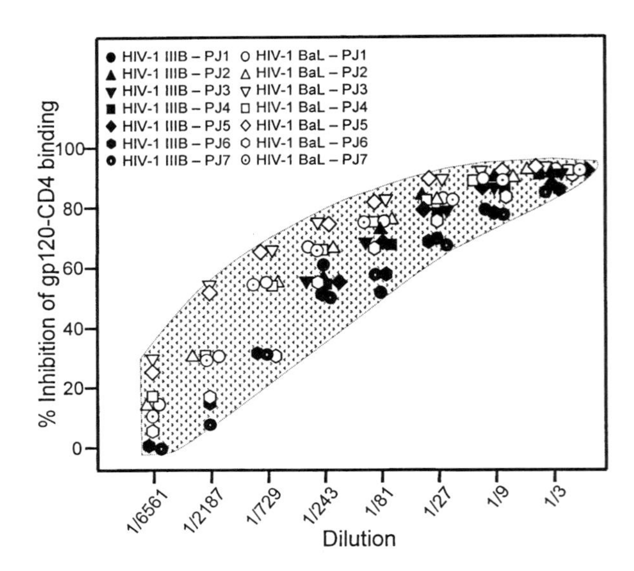

FIGURE 2. Inhibition of CD4 binding to recombinant gp120 IIIB and BaL, respectively, by pomegranate juice (PJ). Recombinant gp120 coated wells were incubated with dilutions of the PJ for 1 h at 37°C. After removal of the juice and washing the wells, biotinyl-CD4 was added, and its binding to the wells was measured by ELISA.

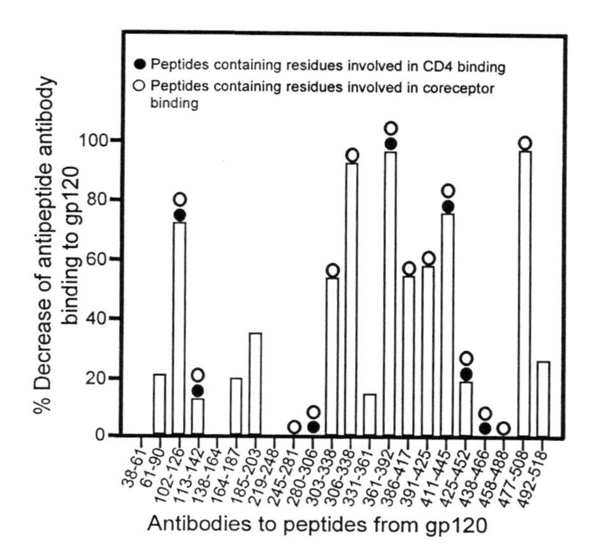

FIGURE 3. Inhibition by pomegranate juice (PJ) of binding to gp120 of antibodies to synthetic peptides from the gp120 sequence. Wells of polystyrene plates coated with gp120 IIIB were incubated with fourfold diluted PJ for 1 h at 37°C. After removal of PJ, the wells were washed, and 50-fold diluted anti-peptide antisera[106] were added. Bound IgG was quantitated by ELISA. PJ was not added to control wells. Decreases of absorbance, as compared to the respective control wells, are plotted.

gp120 and CD4 were in soluble form (data not shown). In reverse experiments, pretreatment of CD4 with PJ failed to block subsequent gp120 binding. Other juices having anti-HIV-1 activity (blueberry, cranberry, grape, and lime) failed to block gp120–CD4 binding.

To delineate sites on gp120 blocked by the PJ inhibitor(s), the inhibitory effect of PJ on binding to gp120 IIIB of antibodies to peptides derived from the amino acid sequence of gp120 was studied. The binding of antibodies to peptides (102-126), (303-338), (306-338), (361-392), (386-417), (391-425), (411-445), and (477-508) was significantly (≥50%) inhibited (FIG. 3). The binding to gp120 IIIB of monoclonal antibodies 9284 and 588D, specific for the gp120 V3 loop (residues 303–338) and the CD4 binding site, respectively,[56,57] was each inhibited by 97%. Some of the relevant peptides contain residues involved in CD4 binding,[58–60] whereas all discerned peptides include residues involved in coreceptor binding.[61–66] The locations of the peptides and of residues involved in receptor/coreceptor binding on the X-ray crystallographic structure of gp120 are shown in FIGURE 4. These results suggest that the PJ inhibitor(s) may also block gp120–coreceptor binding. This will be addressed subsequently.

Separation of Anti-HIV-1 Inhibitor(S) from Pomegranate Juice

Pomegranate juice is intensely colored; therefore, it cannot be directly formulated into a microbicide because it would stain clothing, which is unacceptable. Attempts were made to separate or isolate the active ingredient(s) from PJ. After striving

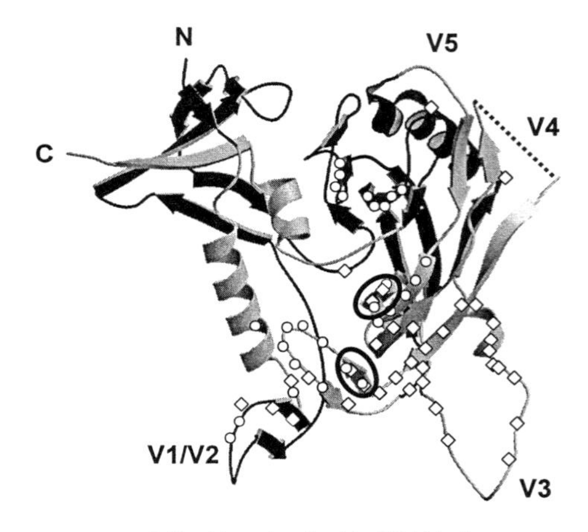

FIGURE 4. Location on the gp120 structure of segments corresponding to antipeptide antibodies whose attachment to gp120 is inhibited by ≥50 % in the presence of pomegranate juice (*gray*) and of amino acid residues involved in CD4 and CXCR4/CCR5 coreceptor binding, respectively. *Black portions* of the structure correspond to anti-peptide antibodies whose attachment to gp120 is not significantly inhibited by PJ. The CD4 domains and the antigen-binding fragment of a neutralizing antibody were excised from the structure of the gp120–CD4–antibody complex[58] (1gc1 retrieved from the Protein Data Bank (pdb) [http://www.rcsb.org/pdb/). The V3 loop, generated by homology modeling, was added to the gp120 structure as described[31] The figure was generated by Molscript[107] and Raster3D.[108,109] The locations of gp120 variable loops (V1 – V5) and of the N- and C-termini of the sequence are indicated.

intermittently for over 4 years to accomplish this, it was discovered that the inhibitor(s) of gp120–CD4 binding can be adsorbed effectively (≥99%) onto a selected brand of corn starch, PURITY® 21 corn starch NF grade (National Starch and Chemical Company, Bridgewater, NJ; S21) (FIG. 5), resulting in a nearly colorless product, designated as PJ-S21. PJ-S21 suspended in water or unbuffered 0.14 M NaCl had a pH of 3.2 (due to adsorbed PJ ingredients because starch provides a neutral pH) compatible with the acidic vaginal environment in which it would remain stabile after application (see below). Inhibitors of gp120–CD4 binding could be eluted from PJ-S21 by extraction with ethanol/acetone 6:4. Drying of the extract followed by gravimetry indicated that the extract contained 3.17 mg solids per gram of PJ-S21.

PJ-S21, to the same extent as the original PJ, inhibited the binding of gp120 IIIB–CD4 complexes to cells expressing CXCR4, as determined by flow cytometry (FIG. 6). Similarly, binding of a gp120 BaL–CD4 fusion protein to cells expressing CCR5 was blocked by PJ and PJ-S21, as detemined by a cell-based ELISA[67] (FIG. 7). Therefore, PJ-S21 is an inhibitor of both X4 and R5 virus binding to the cellular receptor CD4 and coreceptors CXCR4/CCR5. PJ-S21 also inhibited gp120 binding to peripheral blood mononuclear cells as determined by flow cytometry (FIG. 8). To confirm that

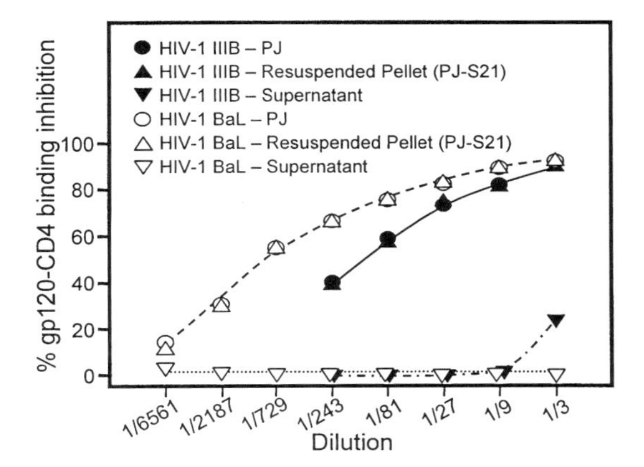

FIGURE 5. Adsorption onto corn starch of gp120–CD4 binding inhibitor(s) from pomegranate juice (PJ). Corn starch (PURITY® 21, NF grade; 200 mg/ml) was added to PJ prefiltered to remove particulates. After mixing for 1 h at ~20°C, the starch was allowed to settle and the supernatant fluid was removed by aspiration. The pellets, resuspended (200 mg/ml) in phosphate-buffered saline, and the supernatant fluids were tested at serial dilutions for inhibition of CD4 binding to gp120 IIIB as described in the legend for FIGURE 2. The inhibitory activity of the resuspended pellet against gp120 BaL–CD4 binding was then confirmed. Control starch did not inhibit gp120–CD4 binding.

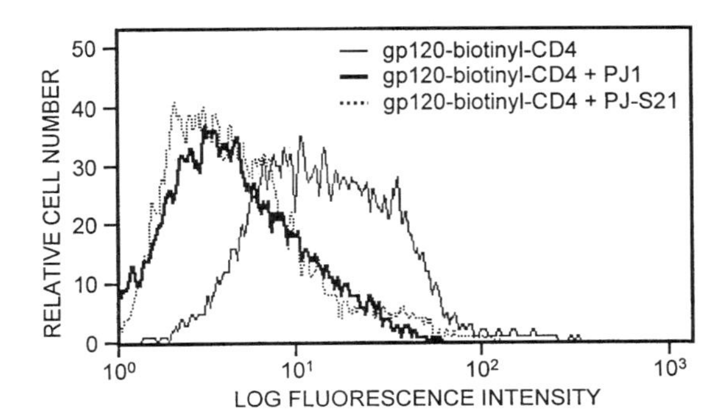

FIGURE 6. Inhibition by pomegranate juice (PJ) and PJ-S21, respectively, of gp120 IIIB–CD4 complex binding to cells expressing CXCR4 coreceptors. HIV-1 IIIB gp120 (5 μg) and biotinyl-CD4 (2.5 μg) were added to 100 μl phosphate-buffered saline (PBS) containing 100 μg bovine serum albumin (BSA) (PBS-BSA) and PJ (final threefold dilution) or PJ-S21 (67 mg corresponding to 212 μg solids from PJ adsorbed onto starch). After 1 h at 20°C, the respective mixtures were added to 10^6 MT-2 cells. After 30 min, the cells were washed 3 times with PBS-BSA and PE-streptavidin (a fluorescent label specific for biotin; 0.1 μg) was added. After 20 min, the cells were washed and fixed by 1% formaldehyde in PBS. Flow cytometry analysis was performed in a FACSCalibur flow cytometer (Becton Dickinson Immunocytometric Systems, San Jose, CA). The median relative fluorescence values for cells exposed to gp120–CD4; gp120–CD4 + PJ; gp120–CD4 + PJ-S21; and control cells were: 13.7, 4.0, 4.3, and 2.1, respectively.

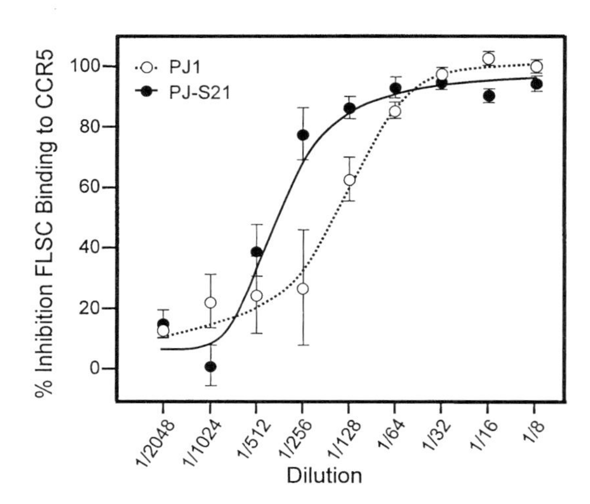

FIGURE 7. Inhibition by pomegranate juice (PJ) and PJ-S21, respectively, of FLSC binding to CCR5 expressing Cf2Th/synCCR5 cells. FLSC is a full-length single chain protein consisting of BaL gp120 linked with the D1D2 domains of CD4 by a 20 amino acid linker. The inhibitory effect was quantitated using a cell-based ELISA.[48] The starting concentration of PJ-S21 was 200 mg/ml, corresponding to 634 μg/ml solids adsorbed onto starch from PJ.

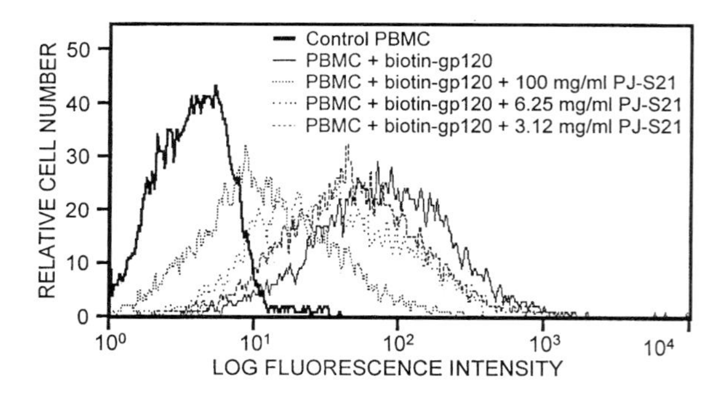

FIGURE 8. Inhibition by PJ-S21 of biotinyl-gp120 IIIB binding to peripheral blood mononuclear cells (PBMCs). HIV-1 IIIB biotinyl–gp120 (5 μg) was added to 100 μl of PBS-BSA containing graded quantities of PJ-S21. After 1 h at 20°C, the respective mixtures were added to 10^6 PBMCs. After 30 min, the cells were washed 3× with PBS-BSA and PE-streptavidin (0.1 μg was added). Subsequently, the procedures described in the legend to FIGURE 6 were used. The median relative fluorescence values for control cells and cells exposed to biotinyl-gp120 in the absence and presence of PJ-S21 (100, 6.25, and 3.12 mg/ml) were 4.1, 81.31, 12.2, 35.2, and 50.0, respectively. 100 mg of PJ-S21 corresponds to ~320 μg solids adsorbed from PJ onto starch.

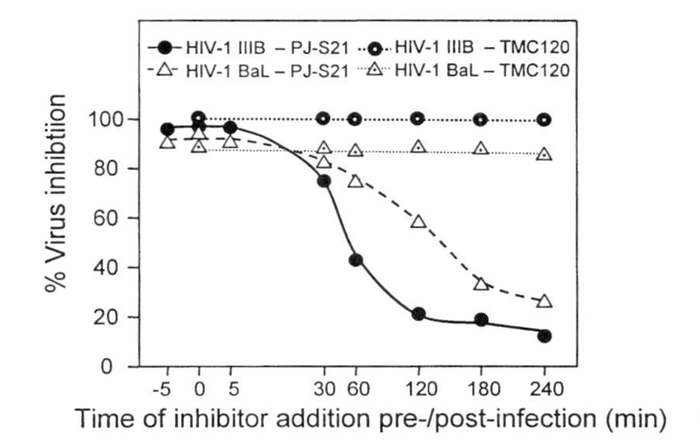

FIGURE 9. Inhibition of HIV-1 IIIB or BaL replication depends on the time of PJ-S21 addition pre- or postinfection. For comparison, the inhibition of infection by the nonnucleoside reverse transcriptase inhibitor TMC-120, added to cells at distinct intervals after HIV-1, was determined (*dotted lines*). Virus infection was measured by quantitation of β-galactosidase.

PJ-S21 functions as a virus entry inhibitor, the complex was added to cells at time intervals before and after infection of cells by HIV-1 IIIB and BaL, respectively. Results shown in FIGURE 9 demonstrate that PJ-S21 interferes with early steps of the virus replicative cycle.

To be considered as a topical microbicide, PJ-S21 must be formulated to withstand storage in a tropical environment. Accelerated thermal stability studies revealed that a water suspension of PJ-S21 maintained only 4, 11, and 33%, respectively, of its original activity (measured by inhibition of gp120–CD4 binding) when stored for 30 minutes at 60°C and 1 week at 50°C or 40°C. However, a dried PJ-S21 powder remained fully active after storage at 50°C for 12 weeks (the longest time used in the evaluation). Consequently, anhydrous formulations should be preferred for further development.

Three such formulations were prepared: two kinds of suppositories, melting at 37°C, and a tablet. The inhibitory activity of PJ-S21 was fully preserved after 12 weeks of storage at 50°C within tablets and at 30°C within the suppositories (the highest temperature considered to prevent melting). Data showing the inhibition of infection by HIV-1 IIIB and BaL, respectively, by PJ-S21 and its formulations (except the tablets that also contain anti-HIV-1 inhibitors other than PJ-S21, that is, Carbopol 974P[33]) are summarized in FIGURE 10. Their inhibitory activities against HIV-1 IIIB and BaL were similar, unlike the inhibitory activities of the original PJs (FIG. 1). These formulations were also virucidal, albeit at concentrations higher than those sufficient for inhibition of infection. These experiments also revealed that PJ-S21 was not cytotoxic under the experimental conditions used. The inhibitory/ virucidal activities were maintained in the presence of seminal fluid at a 1:1 (w/w) ratio of seminal fluid to PJ-S21 (data not shown).

A microbicide can be considered potentially successful only if it displays antiviral activity against primary virus isolates belonging to distinct virus clades and

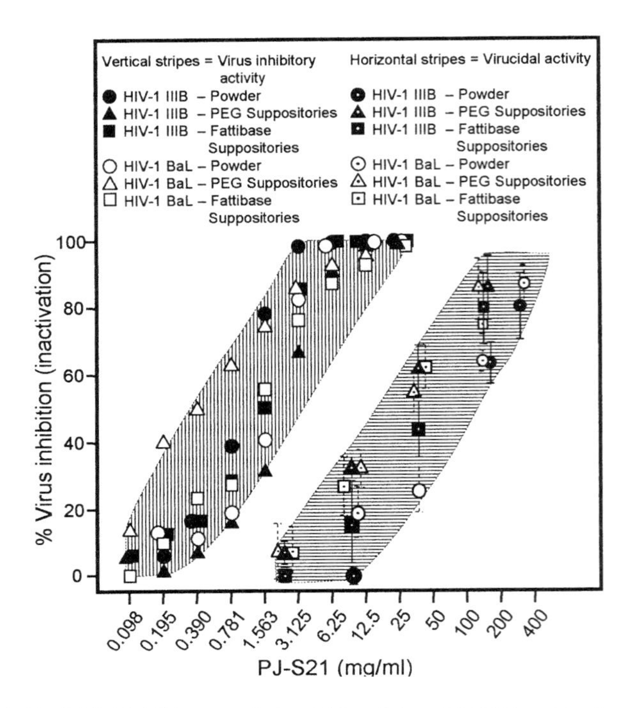

FIGURE 10. HIV-1 inhibitory and virucidal activity of PJ-S21 and its formulations. Inhibition of infection by HIV-1 IIIB and BaL, respectively, was determined as described in the legend for FIGURE 1. To measure virucidal activity, the respective viruses were mixed with graded quantities of PJ-S21 for 5 min at 37°C. After low speed centrifugation, the viruses were separated by precipitation with polyethylene glycol (PEG 8000) and centrifugation. The resuspended pellets and control untreated viruses were serially diluted, and the dilutions assayed for infectivity. The concentration range given on the abscissa corresponds to 0.31–1,268 μg solids adsorbed from PJ to starch.

phenotypes. PJ-S21 meets this requirement, because it inhibited infection by primary HIV-1 strains of all clades tested having R5 and X4R5 (dual-tropic) phenotypes (TABLE 1).

DISCUSSION

Pomegranates have been venerated for millennia for their medicinal properties[68–73] and considered sacred by many of the world's major religions. In deference to pomegranates, the British Medical Association and several British Royal Colleges feature the pomegranate in their coat of arms. The Royal College of Physicians of London had adopted the pomegranate in their coat of arms by the middle of the sixteenth century.[68] The best known literary reference to the contraceptive power of pomegranate seeds is classical Greek mythology. Ironically, this report shows that

TABLE 1. Inhibitory activity of PJ-S21 on infection by primary HIV-1 strains

Primary strain	Subtype, Coreceptor use	ED_{50} (mg/ml)[a]	ED_{90} (mg/ml)[a]
92RW008	A, R5	0.50 ± 0.05	2.76 ± 0.28
94UG103	A, X4R5	1.42 ± 0.54	3.42 ± 0.98
92US657	B, R5	0.62 ± 0.11	2.86 ± 0.33
93IN101	C, R5	3.56 ± 1.10	8.87 ± 2.55
93MW959	C, R5	1.02 ± 0.19	3.54 ± 0.90
92UG001	D, X4R5	0.62 ± 0.17	2.94 ± 0.85
93THA051	E, X4R5	0.86 ± 0.01	4.09 ± 0.08
93BR020	F, X4R5	4.25 ± 0.78	8.31 ± 1.04
RU570	G, R5	0.42 ± 0.09	1.54 ± 0.16
BCF02	Group O, R5	0.59 ± 0.29	3.92 ± 0.27

[a] $ED_{50(90)}$ = effective dose(s) of PJ-S21 for 50% (90%) inhibition of infection. One gram of PJ-S21 contains approximately 3.2 mg of the inhibitors adsorbed to starch from pomegranate juice.

pomegranate juice contains HIV-1 entry inhibitors targeted to the virus envelope corresponding to a class of antiretroviral drugs still scarce in development.[74]

Pomegranate juice contains several ingredients[75,76] that, isolated from natural products other than PJ, were reported to have anti-HIV activity, such as caffeic acid,[77] ursolic acid,[78] catechin, and quercetin[79,80] and also anti-herpes simplex virus (HSV) activity.[81,82] However, these compounds, in purified form, obtained commercially, did not block (at 200 μg/ml) gp120–CD4 binding as measured by the ELISA as just described and did not adsorb to corn starch, unlike the entry inhibitor(s) from PJ. In fact, the supernatant after treatment of PJ with starch and removal of the entry inhibitors retained anti-HIV-1 activity and also inhibited HSV-1, whereas the HIV-1 entry inhibitors that adsorbed onto starch did not inhibit HSV. Thus, the antiviral activities in the supernatant appeared to be nonspecific and probably similar to those of extracts from pomegranate rind[83,84] and were not characterized further. Additional information[85–88] has revealed that the findings apply to crude extracts from pomegranate rind prepared at elevated temperatures under conditions that destroy the HIV-1 entry inhibitor described here.

The inhibitor(s) interfering with gp120 binding to CD4 (FIGS. 2 and 5) blocked additional sites on gp120 (FIG. 3) involved in interaction with the CXCR4/CCR5 coreceptors (FIGS. 4, 6, and 7). This was not completely expected and can be explained either by the presence of multiple inhibitors with distinct or overlapping specificities in PJ-S21 or by induction of gp120 conformational changes[89] resulting in blockade of both CD4 and CXCR4/CCR5 binding sites on gp120. Similar effects have been noted for other small molecule inhibitors.[90] Simultaneous blocking of more than a single site on HIV-1 involved in virus entry is expected to increase the effectiveness of candidate microbicides.[11] The target sites for the inhibitor(s) are likely to be located within the protein moiety of gp120, because binding of labeled *Galanthus nivalis* lectin (specific for terminal mannose residues[91]) and other lectins to gp120 oligosaccharides was not diminished in the presence of PJ or PJ-S21 (data not shown).

Blocking of CD4 binding sites on HIV-1 gp120 by monoclonal antibodies or a CD4-IgG2 recombinant protein has been shown to be sufficient to inhibit HIV-1 infection of human cervical tissue *ex vivo*[11] and in preventing virus transmission to macaque monkeys when applied vaginally.[92] Therefore, it seems likely that PJ-S21 will be similarly effective, an expectation that remains to be confirmed in an *in vivo* macaque model system and in human clinical trials, as a candidate topical microbicide. This anticipation would be strengthened if drinking of PJ decreases the HIV-1 viral load in already infected individuals, an issue to be explored.

The application of PJ-S21 as a topical anti-HIV-1 microbicide requires reasonable uniformity among batches produced at distinct times and locations. Similarities in gp120–CD4 binding inhibitory activity among distinct freshly prepared and commercial juices stored for unknown periods (FIG. 2) suggest that this should be feasible. Pasteurization of juice for 30 seconds at 85°C resulted in complete loss of inhibitory activity. A commercial PJ concentrate exposed to 61°C and two other concentrates, presumably prepared by evaporation at elevated temperatures, had no or drastically diminished activity. The gp120–CD4 inhibitory activity from PJ3 (juice with fructose and citric acid added) failed to bind to starch. Separate experiments revealed that these compounds interfere with inhibitor binding to corn starch. Therefore, PJs intended for production of the PJ-S21 complex must be sterilized by filtration and be free of additives.

Particular attention must be devoted to the selection of starch, a pharmaceutical excipient generally used in vaginal formulations,[93] for effective binding of the virus entry inhibitors from PJ. Among a dozen starches tested, the best results were obtained with S21. With other brands, the adsorption of the inhibitors was either incomplete or their binding did not result in a complex having activity in the ELISA measuring gp120–CD4 binding inhibition (ARGO® corn starch), presumably, because of irreversible binding of the PJ inhibitors. Interestingly, only a few references are available regarding the use of starch as an adsorbent for different compounds: flavors,[94,95] dyes,[96-98] low-molecular mass saccharides,[99] lipids,[100,101] proteins,[102] and iodine.[103]

The intended dose of PJ-S21 for vaginal application is 1.0 to 1.5 g (= 3.17 – 4.76 mg solids from PJ adsorbed onto starch), that is, ≥100-fold higher than the dose needed for blocking HIV-1 infection *in vitro* (FIG. 10, TABLE 1) and therefore expected to meet requirements for likely *in vivo* protection against vaginal challenge.[104] This quantity of PJ-S21 is produced from 5 to 7.5 ml of PJ, that is, ≤ 5% of a single (150 ml) serving of juice, attesting to the safety, feasibility, and economy of this proposed candidate topical microbicide.

In an alternative approach to formulation development, PJ-S21 can be incorporated into a water-dispersible film (similar to the widely available "breath control" strips) or into water-dispersible sponges,[105] which are converted into a gel following topical application.[34] Each of the aforementioned formulations would meet the following requirements: (1) minimization of waste disposal problems associated with the use of applicators needed for delivery of microbicidal gels/creams; (2) simplicity; (3) small packaging and discretion related to purchase, portability, and storage; (4) low production costs; (5) amenability to industrial mass production at multiple sites globally; and (6) potential application as rectal microbicides. Furthermore, it would remain possible to produce for local use PJ-S21–based gel formulations with a limited shelf life, avoiding the costs of producing dry PJ-S21 powders via appropriate low temperature drying processes. Whichever of these formulations is selected, adequate quality control will be needed to assure uniform anti-HIV-1 activity of the final product(s) as well as to establish reproducible conditions for manufacture.

CONCLUSIONS

PJ-S21 can be classified as an AAAA candidate microbicide: acceptable; accessible; affordable; and accelerative in transition from development to marketing. Therefore, PJ-S21 would be expected to circumvent some problems associated with antiretroviral drugs and possibly some of the other candidate microbicides, that is, uncertainty related to potential side effects, investment and time needed to establish inexpensive large scale production, and monopoly of supply.

REFERENCES

1. UNAIDS/WHO. 2004. Global summary of the HIV and AIDS epidemic. December 2004. [http://www.unaids.org/EN/resources/epidemiology/epicore.asp].
2. WHO/SEARO CDS HIV/AIDS. 2004. HIV/AIDS Facts and Figures: Facts about HIV/AIDS-Global. [http://w3.whosea.org/en/Section10/Section18/Section348.htm].
3. Shattock, R.J. & J.P. Moore. 2003. Inhibiting sexual transmission of HIV-1 infection. Nat. Rev. Microbiol. **1:** 25–34.
4. Pilcher, C.D. *et al.* 2004. Brief but efficient: acute HIV infection and the sexual transmission of HIV. J. Infect. Dis. **189:** 1785–1792.
5. Stone, A. 2002. Microbicides: a new approach to preventing HIV and other sexually transmitted infections. Nat. Rev. Drug. Discovery **1:** 977–985.
6. Shattock, R. & S. Solomon. 2004. Microbicides–aids to safer sex. Lancet **363:** 1002–1003.
7. Brown, H. 2004. Marvellous microbicides. Intravaginal gels could save millions of lives, but first someone has to prove that they work. Lancet **363:** 1042–1043.
8. Moore, J.P. & R.W. Doms. 2003. The entry of entry inhibitors: a fusion of science and medicine. Proc. Natl. Acad. Sci. USA **100:** 10598–10602.
9. Pierson, T.C. & R.W. Doms. 2003. HIV-1 entry inhibitors: new targets, novel therapies. Immunol. Lett. **85:** 113–118.
10. Davis, C.W. & R.W. Doms. 2004. HIV Transmission: closing all the doors. J. Exp. Med. **199:** 1037–1040.
11. Hu, Q. *et al.* 2004. Blockade of attachment and fusion receptors inhibits HIV-1 infection of human cervical tissue. J. Exp. Med. **199:** 1065–1075.
12. Fichorova, R.N., L.D. Tucker & D.J. Anderson. 2001. The molecular basis of nonoxynol-9-induced vaginal inflammation and its possible relevance to human immunodeficiency virus type 1 transmission. J. Infect. Dis. **184:** 418–428.
13. Fichorova, R.N. *et al.* 2004. Interleukin (IL)-1, IL-6 and IL-8 predict mucosal toxicity of vaginal microbicidal contraceptives. Biol. Reprod. **71:** 761–769.
14. Weiss, R. 2001. AIDS: unbeatable 20 years on. Lancet **357:** 2073–2074.
15. Neurath, A.R. *et al.* 1995. Blocking of CD4 cell receptors for the human immunodeficiency virus type 1 (HIV-1) by chemically modified bovine milk proteins: potential for AIDS prophylaxis. J. Mol. Recognition **8:** 304–316.
16. Jiang, S. *et al.* 1996. Chemically modified bovine-lactoglobulin blocks uptake of HIV-1 by colon- and cervix-derived epithelial cell lines. J. Acquir. Immune Defic. Syndr. Hum. Retrovirol. **13:** 461–462.
17. Neurath, A.R. *et al.* 1996. A herpesvirus inhibitor from bovine whey. Lancet **347:** 1703.

18. NEURATH, A.R. *et al.* 1997. 3-Hydroxyphtaloyl–lactoglobulin: I. Optimization of production and comparison with other compounds considered for chemoprophylaxis of mucosally transmitted human immunodeficiency virus type 1. Antivir. Chem. Chemother. **8:** 131–139.
19. NEURATH, A.R. *et al.* 1996. Bovine-lactoglobulin modified by 3-hydroxyphthalic anhydride blocks the CD4 cell receptor for HIV. Nat. Med. **2:** 230–234.
20. JIANG, S. *et al.* 1997. Virucidal and antibacterial activities of 3-HP--LG. *In* Vaccines 97: Molecular Approaches to the Control of Infectious Diseases. F. Brown *et al.* Eds. :327–330. Cold Spring Harbor Laboratory Press. New York.
21. NEURATH, A.R. *et al.* 1997. 3-Hydroxyphtaloyl–lactoglobulin: II. Anti-human immunodeficiency virus type 1 activity in *in vitro* environments relevant to prevention of sexual transmission of the virus. Antivir. Chem. Chemother. **8:** 141–148.
22. NEURATH, A.R., N. STRICK & Y.-Y. LI. 1998. 3-Hydroxyphthaloyl-lactoglobulin: III. Antiviral activity against herpesviruses. Antivir. Chem. Chemother. **9:** 177–184.
23. KOKUBA, H., L. AURELIAN & A.R. NEURATH. 1998. 3-Hydroxyphthaloyl -lactoglobulin: IV. Antiviral activity in the mouse model of genital herpesvirus infection. Antivir. Chem. Chemother. **9:** 353–357.
24. WYAND, M.S. *et al.* 1999. Effect of 3-hydroxyphthaloyl–lactoglobulin on vaginal transmission of simian immunodeficiency virus in rhesus monkeys. Antimicrob. Agents Chemother. **43:** 978–980.
25. FDA'S CENTER FOR FOOD SAFETY AND APPLIED NUTRITION (CFSCAN). 2004. Commonly asked questions about BSE in products regulated by FDA's Center for Food Safety and Applied Nutrition (CFSAN) [http://vm.cfsan.fda.gov/~comm/bse-faq.html%20].
26. CHIU, Y.-Y. 2002. An Update CDER Biotechnology BSE Activities: BSE in manufacturing FDA regulatory actions. [http://www.temple.edu/pharmacy_QARA/fda_conf_Chiu.ppt].
27. U.S. FOOD AND DRUG ADMINISTRATION CfFSaAN. 2001. Agency response letter GRAS Notice No. GRN 000077. [http://vm.cfsan.fda.gov/~rdb/opa-g077.html].
28. CDC NCfID. 2004. Update 2002: Bovine spongiform encephalopathy and variant Creutzfeldt-Jakob disease. [http://www.cdc.gov/ncidod/diseases/cjd/bse_cjd.htm].
29. WORLD HEALTH ORGANIZATION. 2002. Bovine spongiform encephalopathy (Fact Sheet No. 113). [http://www.who.int/mediacentre/factsheets/fs113/en/].
30. NEURATH, A.R. *et al.* 1999. Design of a "microbicide" for prevention of sexually transmitted diseases using"inactive" pharmaceutical excipients. Biologicals **27:** 11–21.
31. NEURATH, A.R. *et al.* 2001. Cellulose acetate phthalate, a common pharmaceutical excipient, inactivates HIV-1 and blocks the coreceptor binding site on the virus envelope glycoprotein gp120. BMC Infect. Dis. **1:** 17. [http://www.biomedcentral.com/content/pdf/1471-2334-1-17.pdf].
32. NEURATH, A.R. *et al.* 2002. Anti-HIV-1 activity of cellulose acetate phthalate: Synergy with soluble CD4 and induction of "dead-end" gp41 six-helix bundles. BMC Infect. Dis. **2:** 6. [http://www.biomedcentral.com/content/pdf/1471-2334-2-6.pdf].
33. NEURATH, A.R., N. STRICK & Y.-Y. LI. 2002. Anti-HIV-1 activity of anionic polymers: a comparative study of candidate microbicides. BMC Infect. Dis. **2:** 27. [http://www.biomedcentral.com/content/pdf/1471-2334-2-27.pdf].
34. NEURATH, A.R., N. STRICK & Y.-Y. LI. 2003. Water dispersible microbicidal cellulose acetate phthalate film. BMC Infect. Dis. **3:** 27. [http://www.biomedcentral.com/content/pdf/1471-2334-3-27.pdf].
35. AVIRAM, M. *et al.* 2000. Pomegranate juice consumption reduces oxidative stress, atherogenic modifications to LDL, and platelet aggregation: studies in humans and in atherosclerotic apolipoprotein E-deficient mice. Am. J. Clin. Nutr. **71:** 1062–1076.
36. PEHOWICH, D.J., A.V. GOMES & J.A. BARNES. 2000. Fatty acid composition and possible health effects of coconut constituents. West Indian Med. J. **49:** 128–133.
37. HAMMERSTONE, J.F., S.A. LAZARUS & H.H. SCHMITZ. 2000. Procyanidin content and variation in some commonly consumed foods. J. Nutr. **130:** 2086S–2092S.
38. AVIRAM, M. & L. DORNFELD. 2001. Pomegranate juice consumption inhibits serum angiotensin converting enzyme activity and reduces systolic blood pressure. Atherosclerosis **158:** 195–198.

39. JOSHI, S.S., C.A. KUSZYNSKI & D. BAGCHI. 2001. The cellular and molecular basis of health benefits of grape seed proanthocyanidin extract. Curr. Pharm. Biotechnol. **2:** 187–200.
40. LIU, Y. *et al.* 2001. Citrus pectin: characterization and inhibitory effect on fibroblast growth factor-receptor interaction. J. Agric. Food Chem. **49:** 3051–3057.
41. KAPLAN, M. *et al.* 2001. Pomegranate juice supplementation to atherosclerotic mice reduces macrophage lipid peroxidation, cellular cholesterol accumulation and development of atherosclerosis. J. Nutr. **131:** 2082–2089.
42. HOWELL, A.B. 2002. Cranberry proanthocyanidins and the maintenance of urinary tract health. Crit. Rev. Food Sci. Nutr. **42:** 273–278.
43. MILBURY, P.E. *et al.* 2002. Bioavailablility of elderberry anthocyanins. Mech. Ageing Dev. **123:** 997–1006.
44. WANG, X.H., L. ANDRAE & N. J. ENGESETH. 2002. Antimutagenic effect of various honeys and sugars against Trp-p-1. J. Agric. Food Chem. **50:** 6923-6928.
45. SUN, J. *et al.* 2002. Antioxidant and antiproliferative activities of common fruits. J. Agric. Food Chem. **50:** 7449–7454.
46. CAVANAGH, H.M., M. HIPWELL & J.M. WILKINSON. 2003. Antibacterial activity of berry fruits used for culinary purposes. J. Med. Food **6:** 57–61.
47. SANCHEZ-MORENO, C. *et al.* 2003. Anthocyanin and proanthocyanidin content in selected white and red wines. Oxygen radical absorbance capacity comparison with nontraditional wines obtained from highbush blueberry. J. Agric. Food Chem. **51:** 4889–4896.
48. POLAGRUT, J.A. *et al.* 2003. Effects of flavonoid-rich beverages on prostacyclin synthesis in humans and human aortic endothelial cells: association with ex vivo platelet function. J. Med. Food **6:** 301–308.
49. GHELDOF, N., X.H. WANG & N.J. ENGESETH. 2003. Buckwheat honey increases serum antioxidant capacity in humans. J. Agric. Food Chem. **51:** 1500–1505.
50. AVIRAM, M. *et al.* 2004. Pomegranate juice consumption for 3 years by patients with carotid artery stenosis reduces common carotid intima-media thickness, blood pressure and LDL oxidation. Clin. Nutr. **23:** 423–433.
51. JIRATANAN, T. & R.H. LIU. 2004. Antioxidant activity of processed table beets (*Beta vulgaris* var, conditiva) and green beans (*Phaseolus vulgaris* L.). J. Agric. Food Chem. **52:** 2659–2670.
52. HUANG, H.Y. *et al.* 2004. Antioxidant activities of various fruits and vegetables produced in Taiwan. Int. J. Food Sci. Nutr. **55:** 423–429.
53. NINFALI, P. *et al.* 2005. Antioxidant capacity of vegetables, spices and dressings relevant to nutrition. Br. J. Nutr. **93:** 257–266.
54. NEURATH, A.R. *et al.* 2004. *Punica granatum* (pomegranate) juice provides an HIV-1 entry inhibitor and candidate topical microbicide. BMC Infect. Dis. **4:** 41 [http://www.biomedcentral.com/content/pdf/1471-2334-4-41.pdf].
55. SHATTOCK, R.J. & R.W. DOMS. 2002. AIDS models: microbicides could learn from vaccines. Nat. Med. **8:** 425.
56. SKINNER, M. A. *et al.* 1988. Characteristics of a neutralizing monoclonal antibody to the HIV envelope glycoprotein. AIDS Res. Hum. Retrovir. **4:** 187–197.
57. LAAL, S. & S. ZOLLA-PAZNER. 1993. Epitopes of HIV-1 glycoproteins recognized by the human immune system. *In* Immunochemistry of AIDS, Chemical Immunology, Vol. 56. E. Norrby, Ed. :91-111. Karger. Basel.
58. KWONG, P.D. *et al.* 1998. Structure of an HIV gp120 envelope glycoprotein in complex with the CD4 receptor and a neutralizing human antibody. Nature **393:** 648–659.
59. XIANG, S.H. *et al.* 2002. Mutagenic stabilization and/or disruption of a CD4-bound state reveals distinct conformations of the human immunodeficiency virus type 1 gp120 envelope glycoprotein. J. Virol. **76:** 9888–9899.
60. PANTOPHLET, R. *et al.* 2003. Fine mapping of the interaction of neutralizing and non-neutralizing monoclonal antibodies with the CD4 binding site of human immunodeficiency virus type 1 gp120. J. Virol. **77:** 642–658.
61. WESTERVELT, P., H.E. GENDELMAN & L. RATNER. 1991. Identification of a determinant within the human immunodeficiency virus 1 surface envelope glycoprotein critical for productive infection of primary monocytes. Proc. Natl. Acad. Sci. USA **88:** 3097–3101.

62. WESTERVELT, P. *et al.* 1992. Macrophage tropism determinants of human immunodeficiency virus type 1 *in vivo*. J. Virol. **66:** 2577–2582.
63. RIZZUTO, C.D. *et al.* 1998. A conserved HIV gp120 glycoprotein structure involved in chemokine receptor binding. Science **280:** 1949–1953.
64. CORMIER, E. G. & T. DRAGIC. 2002. The crown and stem of the V3 loop play distinct roles in human immunodeficiency virus type 1 envelope glycoprotein interactions with the CCR5 coreceptor. J. Virol. **76:** 8953–8957.
65. SUPHAPHIPHAT, P. *et al.* 2003. Effect of amino acid substitution of the V3 and bridging sheet residues in human immunodeficiency virus type 1 subtype C gp120 on CCR5 utilization. J. Virol. **77:** 3832–3837.
66. LIU, S., S. FAN & Z. SUN. 2003. Structural and functional characterization of the human CCR5 receptor in complex with HIV gp120 envelope glycoprotein and CD4 receptor by molecular modeling studies. J. Mol. Model **9:** 329–336.
67. ZHAO, Q., G. ALESPEITI & A.K. DEBNATH. 2004. A novel assay to identify entry inhibitors that block binding of HIV-1 gp120 to CCR5. Virology **326:** 299–309.
68. LANGLEY, P. 2000. Why a pomegranate? BMJ **321:** 1153–1154.
69. NAVARRO, V. *et al.* 1996. Antimicrobial evaluation of some plants used in Mexican traditional medicine for the treatment of infectious diseases. J. Ethnopharmacol. **53:** 143–147.
70. LEE, J. & R.R. WATSON. 1988. Pomegranate: a role in health promotion and AIDS? *In* Nutrients and Foods in AIDS. R.R. Watson RR. Ed. :213–216. CRC. Boca Raton.
71. PRASHANTH, D., M.K. ASHA & A. AMIT. 2001. Antibacterial activity of *Punica granatum*. Fitoterapia **72:** 171–173.
72. MOUHAJIR, F. *et al.* 2001. Multiple antiviral activities of endemic medicinal plants used by Berber peoples of Morocco. Pharm. Biol. **39:** 364–374.
73. NEGI, P.S., G.K. JAYAPRAKASHA & B.S. JENA. 2003. Antioxidant and antimutagenic activities of pomegranate peel extracts. Food Chem. **80:** 393–397.
74. GREENE, W.C. 2004. The brightening future of HIV therapeutics. Nat. Immunol. **5:** 867–871.
75. POYRAZOGLU, E., V. GOEKMEN & N. ARTIK. 2002. Organic acids and phenolic compounds in pomegranates (*Punica granatum* L.) grown in Turkey. J. Food Composition and Analysis **15:** 567–575.
76. MODULE **2:** Phytochemicals (minerals, phytamins, and vitamins). 2003. [http://www.ars-grin.gov/duke/syllabus/module2.htm].
77. MAHMOOD, N. *et al.* 1993. Inhibition of HIV infection by caffeoylquinic acid derivatives. Antivir. Chem. Chemother. **4:** 235–240.
78. MA, C. *et al.* 1998. Inhibitory effects of ursolic acid derivatives from cynomorium songaricum, and related triterpenes on human immunodeficiency viral protease. Phytother. Res. **12:** S138–S142.
79. MAHMOOD, N. *et al.* 1996. The anti-HIV activity and mechanisms of action of pure compounds isolated from *Rosa damascena*. Biochem. Biophys. Res. Commun. **229:** 73–79.
80. DETOMMASI, N. *et al.* 1998. Anti-HIV activity directed fractionation of the extracts of Margyricarpus setosus. Pharma. Biol. **36:** 29–32.
81. ZHANG, J. *et al.* 1995. Antiviral activity of tannin from the pericarp of *Punica granatum* L. against genital Herpes virus in vitro. Zhongguo Zhong Yao Za Zhi **20:** 556–8, 576.
82. LI, Y. *et al.* 2004. Antiviral activities of medicinal herbs traditionally used in Southern Mainland China. Phytother. Res. **18:** 718–722.
83. Pomegranates could help in battle against AIDS. 1996. Reuters NewMedia, Inc. [http://www.aegis.com/news/re/1996/RE960310.html].
84. Medical breakthrough. 1996. British Muslims Monthly Survey IV (**3**),6. [http://artsweb.bham.ac.uk/bmms/1996/03March96.html#Medical%20breakthrough].
85. JASSIM, S.A.A., S.P. DENYER & G.S.A.B. STEWART. 1998. Antiviral or antifungal composition comprising an extract of pomegranate rind or other plants and method of use. US Patent 5,840,308. November 24, 1998.
86. SHEHADEH, A.A. 2000. Herbal extract composition and method with immune-boosting capability. US Patent 6,030,622. February 29, 2000.

87. JASSIM, S.A.A., S.P. DENYER & G.S.A.B. STEWART. 2001. Antiviral or antifungal composition and method. US Patent 6,187,316. February 2, 2001.
88. JASSIM, S.A.A. & S.P. DENYER. 2002. Antiviral or antifungal compositon and method. US Patent Application 20020064567. May 30, 2002.
89. HSU, S.-T. & A.M.J.J. BONVIN. 2004. Atomic insight into the CD4 binding-induced conformational changes in HIV-1 gp120. Proteins **55:** 582–593.
90. NEURATH, A.R. *et al.* 1994. Tin protoporphyrin IX used in control of heme metabolism in humans effectively inhibits HIV-1 infection. Antivir. Chem. Chemother. **5:** 322–330.
91. HAMMAR, L. *et al.* 1995. Lectin-mediated effects of HIV type 1 infection in vitro. AIDS Res. Hum. Retrovir. **11:** 87–95.
92. VEAZEY, R.S. *et al.* 2003. Prevention of virus transmission to macaque monkeys by a vaginally applied monoclonal antibody to HIV-1 gp120. Nat. Med. **9:** 343–346.
93. GARG, S. *et al.* 2001. Compendium of pharmaceutical excipients for vaginal formulations. Pharma. Techn. Drug Deliv. Sept. 14–24.
94. YAO, W.H. 2002. Adsorbent characteristics of porous starch. Starch/Starke **54:** 260–263.
95. WHISTLER, R.L. 1991. Microporous granular starch matrix compositions. US Patent 4,985,082. January 15, 1991.
96. BERSET, C., H. CLERMONT & S. CHEVAL. 1995. Natural red colorant effectiveness as influenced by absorptive supports. J. Food Sci. **60:** 858–861, 879.
97. STUTE, R. & H.U. WOELK. 1974. Interaction between starch and reactive dyes. New technique for the investigation of starch. II. Influence on fixation reaction of starch. Starch/Starke **26:** 1–9.
98. SEGUCHI, M. 1986. Dye binding to the surface of wheat starch granules. Cereal Chem. **63:** 518–520.
99. TOMASIK, P., Y.-J. WANG & J.L. JANE. 1995. Complexes of starch with low-molecular saccharides. Starch/Starke **47:** 185–191.
100. ZHANG, G., M.D. MALADEN & B.R. HAMAKER. 2003. Detection of a novel three component complex consisting of starch, protein, and free fatty acids. J. Agric. Food Chem. **51:** 2801–2805.
101. JOHNSON, J.M., E.A. DAVIS & J. GORDON. 1990. Lipid binding of modified corn starches studies by electron spin resonance. Cereal Chem. **67:** 236–240.
102. TOMAZIC-JEZIC, V.J., A.D. LUCAS & B.A. SANCHEZ. 2004. Binding and measuring natural rubber latex proteins on glove powder. J. Immunoassay Immunochem. **25:** 109–123.
103. CONDE-PETIT, B. *et al.* 1998. Comparative charaterization of aqueous starch dispersions by light microscopy, rheometry, and iodine binding behavior. Starch/Starke **50:** 184–192.
104. MOORE, J. *et al.* 2004. Development of fusion/entry inhibitors as topical microbicides. Presented at Microbicides 2004. London. March 28–31 2004. [http://www.microbicides2004.org.uk/progtue.html].
105. NEURATH, A.R. & N. STRICK. 2003. Biodegradable microbicidal vaginal barrier device. US Patent 6,572,875. June 3, 2003.
106. NEURATH, A.R., N. STRICK & S. JIANG. 1992. Synthetic peptides and anti-peptide antibodies as probes to study interdomain interactions involved in virus assembly: The envelope of the human immunodeficiency virus (HIV-1). Virology **188:** 1–13.
107. KRAULIS, P. J. 1991. MOLSCRIPT: a program to produce both detailed and schematic plots of protein structures. J. Appl. Cryst. **24:** 946–950.
108. BACON, D.J. & W.F. ANDERSON. 1988. A fast algorithm for rendering space-filling molecule pictures. J. Mol. Graphics **6:** 219–220.
109. MERRITT, E.A. & D.J. BACON. 1997. Raster3D: photorealistic molecular graphics. Meth. Enzymol. **277:** 505–524.

Strategies for Human Papillomavirus Therapeutic Vaccines and Other Therapies Based on the E6 and E7 Oncogenes

V.A. GOVAN

Division of Medical Virology, Department of Clinical Laboratory Sciences and Institute of Infectious Diseases and Molecular Medicine, Faculty of Health Sciences, University of Cape Town, Observatory, Cape Town, South Africa

Abstract: High-risk human papillomaviruses (HR-HPVs) are one of the most devastating oncogenic viruses worldwide and have been causally linked with the development of human cervical cancer. Several prophylactic and therapeutic clinical HPV vaccine trials are in progress. Although prophylactic vaccines are useful in preventing the incidence of cervical cancer, the elimination of existing HPV infections needs to be addressed, because cervical cancer is the leading female cancer in developing countries. Several different and encouraging strategies have been investigated in a preclinical and clinical setting for the treatment and elimination of existing HPV-induced infection. This review summarizes the therapeutic clinical trials and the different preclinical research strategies that are under investigation whereby HR-HPV E6 and E7 oncogenes are delivered in a nucleic acid form, in viral and bacterial vectors, or as peptide- and protein-based vaccines.

Keywords: therapeutic strategies; HPV E6 and E7 vaccines; cervical cancer

INTRODUCTION

Epidemiologic and experimental data have identified an unequivocal association of high-risk human papillomavirus (HR-HPV) infection and the development of cervical cancer.[1] More than 100 different types of HPVs have been identified,[2] which can be divided into low-risk, nononcogenic, or high-risk oncogenic types. The predominant low-risk types are HPV6 and 11, and they cause 90% of genital warts (condyloma acuminata), whereas HPV16 and 18 are the most prevalent high-risk types, and they cause 70% of cervical cancer and cervical intraepithelial neoplasia (CIN).[3–5] Notably, although the prevalence of HPV DNA among the general population is high, most infected individuals are able to eliminate the virus without intervention and do not develop cancer.[6] Thus, an effective host immune response and the genetic makeup of the individual may be an important determinant in the suscepti-

Address for correspondence: V.A. Govan, Room S3.26.1, Wernher and Beit, South Wing, Institute of Infectious Diseases and Molecular Medicine, Faculty of Health Sciences, University of Cape Town, Medical School, Observatory, Cape Town, South Africa. Voice: +27-21-4066366; fax: +27-21-4066018.

vgovan@curie.uct.ac.za

Ann. N.Y. Acad. Sci. 1056: 328–343 (2005). © 2005 New York Academy of Sciences.
doi: 10.1196/annals.1352.016

bility to cervical cancer.[7] Consequently, several studies have investigated genetic host factors and immune responses, to elucidate the association between genital HPV infection and cervical cancer.[8–10] The studies yielded conflicting results and highlight the possible variability of genetic host factors in different population groups and should not be viewed in isolation.[9] Nevertheless, the life cycles of these papillomaviruses (PVs) are linked to the differentiation stages of the host epithelial cells, and replication is restricted exclusively to the stratified squamous epithelium.[10]

Because PVs are host and tissue specific, animal PV models have been developed and have provided important information for vaccine development. Furthermore, studies in animal models have afforded much insight into the molecular mechanisms that regulate normal cell growth, together with a better understanding of the processes involved in cancerous changes in cells.[11]

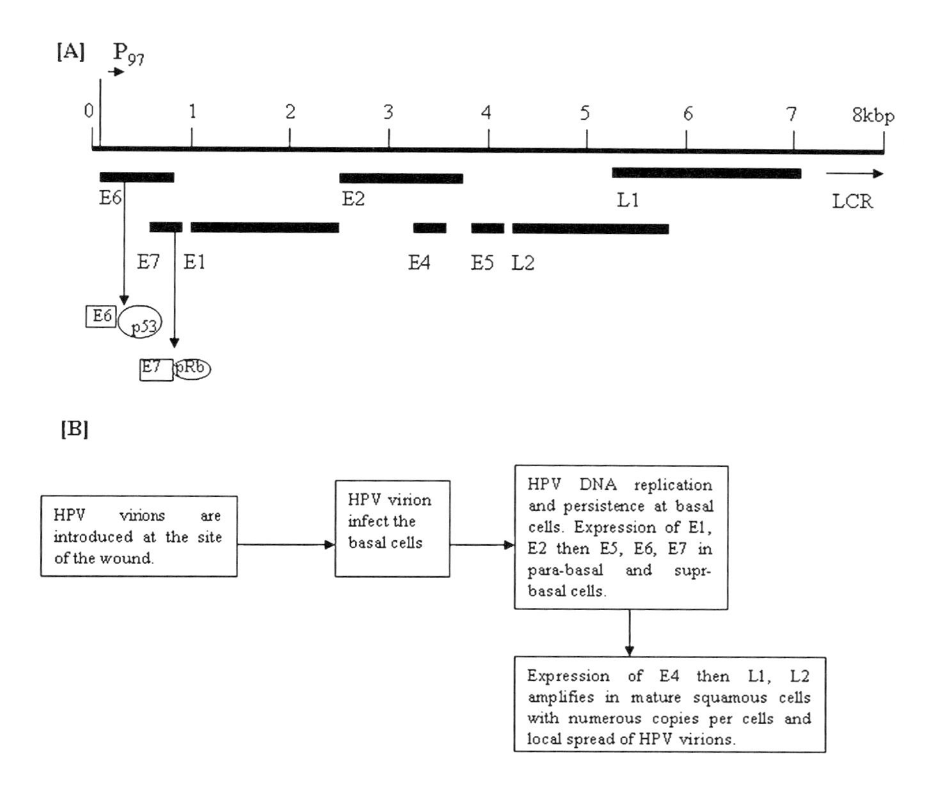

FIGURE 1. (**A**) Linearized HPV 16 genome organization showing the early and late genes. HPV16 E6 and E7 mRNA are expressed from a single promoter, P97, which distinguishes it from low-risk types. HPV16 E6 and E7 bind to the tumor suppressor proteins, p53 and pRb, facilitating cell immortalization. (**B**) Schematic representation of the main stages of HPV life. (Modified from MAN, S. 1998. Human cellular immune responses against human papillomaviruses in cervical neoplasia. Exp. Rev. Mol. Med. July 3.)

An understanding of the molecular pathogenesis of PV-associated cancer is important for selecting vaccine targets, because PVs are able to evade the host immune system in the early stages of carcinogenesis. PVs are small, nonenveloped viruses containing an 8-kb double-stranded closed circular DNA genome, encoding six early proteins (E1, E2, E5, E6, E7, and E8), two late proteins (L1 and L2), and a noncoding regulatory region, the long-control region (LCR) (FIG. 1).[1] The early genes contribute to transformation and viral replication, the late genes provide capsid proteins, and the LCR contains the origin of replication.[1]

Expression of the late L1 gene *in vitro* results in assembly of virus-like particles (VLPs) and induces high titers of virus-neutralizing antibodies when administered as an immunogen.[12,13] Because of the encouraging Phase II clinical efficacy trials being well tolerated and resulting in high antibody titers, three HPV-VLP vaccines have entered Phase III trials as potential prophylactic vaccines.[14–16] Although these results are encouraging for the prevention of infection with HPV, there is an urgent need for the development of therapeutic strategies for existing HPV infection, particularly in developing countries where cervical cancer is the most common cancer in women. In addition, the L1 and L2 capsid antigens are not expressed in precancerous and cancerous tissue and therefore may not be significantly beneficial for therapeutic purposes.[17] Furthermore, the current therapies for HPV-induced lesions rely on surgical removal of the infected tissues, cryotherapy, laser evaporation therapy, or excision procedures such as loop electrosurgical excision procedure (LEEP) or cold-knife conization. However, the current treatments are invasive, costly, limited, uncomfortable, and inefficient and recurrence is a common outcome,[18] contributing to a public health burden worldwide. Therefore, the development of a successful, noninvasive therapeutic strategy is essential to reduce the incidence and recurrence of existing disease. This review summarizes a few clinical trials that have targeted HR-HPV E6 and E7 genes and also discusses the advantages and disadvantages of different E6- and E7-directed therapeutic vaccination strategies that may assist in the development of future cancer vaccines for the regression and elimination of existing HPV infections and their precursors.

HR-HPV E6 AND E7 AS TARGET PROTEINS FOR THERAPY

The E6 and E7 oncoproteins of HR-HPV are approximately 150 and 100 amino acids, respectively, and their functions are pleiotropic.[1] There is a profusion of experimental data that provide evidence that E6 and E7 are primarily expressed in human keratinocytes, and only the intact expression of these oncogenes efficiently maintains the immortalization of the cells transfected with HR-HPV.[19,20] In addition, E6 and E7 are the only HPV proteins that are expressed consistently in cervical cancer cells[21] and often are referred to as the hallmark of cervical cancer.[22] Of interest, their zinc finger structures are unique and are involved in binding zinc.[23] Furthermore, it has been shown that the E6 and E7 oncoproteins are able to complex with the tumor suppressor gene products, p53 and retinoblastoma (pRb), respectively. This interaction facilitates the deregulation of the cell cycle control mechanisms in which p53 and pRb play a fundamental role as they have tumor-suppressive and cell cycle growth inhibitory properties.[24] Specifically, similar to the simian virus 40 large antigen (SV40 TAg) and the adenovirus 5 E1A proteins, HR-HPV E7 can bind to pRb, resulting in the hyperphosphorylation of pRb, and mediates the release of the transcription factor E2F, which activates the genes for cell proliferation.[25] In comparison, the HR-HPV E6 protein, SV40 Tag, and the adenovirus 5 E1B 55-kDa protein are able to complex with p53, which is mediated by E6-associated protein ligase (E6-AP) and targets p53 for ubiquitinylation and proteosomal degradation.[26]

One of the main differences that distinguish high-risk and low-risk HPV types is the structure of the mRNA of E6 and E7 and the process by which it is expressed.[23] The E6 and E7 mRNA of high-risk types such as HPV16 and HPV18 is expressed from a single promoter P97 and P105, respectively, and the mRNA of E6 and E7 low-risk types such as HPV6 and 11 is expressed from two independent promoters.[23] The main promoters of the HR-HPV types are located in front of the E6 open reading frame (ORF), and all produce differentially spliced polycistronic transcripts, which encode the early genes.[27] Conversely, the main promoter of the low-risk HPV types is located in the E6 ORF, in front of the E7 ORF, and the two independent promoters that express the E6 and E7 genes are found in front of the E6 ORF.[21] Consequently, the disparity between the high-risk and low-risk types of HPV in their oncogenic potential is caused by the production of transcripts that encode the E7 protein as the first ORF.[27] Furthermore, the binding properties of the low-risk and high-risk HPV E6 and E7 proteins differ in several biochemical and functional properties. The low-risk HPV E7 binds pRB at a lower efficiency, and the E6 protein of the low-risk types have minimal or no transformation activity.[23]

Evidently, the aforementioned oncogenes are ideal targets for therapeutic vaccines and antiviral therapies.

THERAPEUTIC STRATEGIES

Advances in the knowledge of choosing targets for vaccine design have largely been determined by preclinical *in vitro* and *in vivo* studies. Although the main target for prophylactic vaccines has focused on the L1 gene, which has been shown to be effective in generating high titers of serum antibodies because of its restricted expression, it will not be able to induce significant therapeutic effects for HPV-induced infections. There is, however, evidence that the two HR-HPV oncoproteins E6 and E7 are immunogenic, with the production of both humoral and cell-mediated responses, which would be ideal for therapeutics.[28] In addition, it has been reported that immunization with E7 can hinder the progress of E7-expressing tumors in animal models and humans.[29,30] Therefore, currently various preclinical strategies are being investigated and Phase I/II clinical trials are ongoing for inducing the regression and elimination of established HPV infection in which the two oncogenes E6 and E7 are administered in different delivery systems.

Therapeutic HPV Vaccines in Clinical Trials

The main focus of therapeutic vaccination is to elicit a strong, long-lasting, and specific cell-mediated immune response for the reduction and elimination of established HPV-induced infections. Recently, several therapeutic vaccines have entered Phase I and II clinical trials and have produced encouraging as well as discouraging clinical responses. Thus, the promising results generated from certain human studies will probably enter the next stage of clinical trials, whereas the disappointing outcomes achieved in other trials need to be investigated to determine why the preclinical models performed suboptimally in the human situation. Only the clinical trials that generated encouraging results justifying further clinical investigations will be discussed.

A Phase I dose-escalation safety study using the therapeutic agent ZYC101 was conducted in patients with HPV16-associated anal dysplasia and women with CIN 2/3 who were HPV16 and HLAA2-positive.[31,32] ZYC101 is a bacterial expression plasmid DNA that expresses a segment (codon 83-95) of the HPV16 E7 protein which is fused to secretory leader sequence derived from the human HLA-DRA*0101 locus.[31,32] The constructs then were encapsulated in biodegradable poly (D,L-lactide-co-glycolide) microparticles for delivery.[33] After receiving four intramuscular injections at 3-week intervals, 10 of 12 patients showed augmented peptide-specific immune responses, as demonstrated by direct ELISPOT. Moreover, the 10 patients demonstrated an increased immune response 6 months after initial therapy with no adverse effects.[31] Because the ZYC101 performed well in this trial, a multicenter, double-blind randomized, placebo-controlled phase trial was conducted to assess the safety and efficacy of a similar agent, ZYC101a, in 127 women with histologically proved CIN 2/3.[34] Subjects were injected intramuscularly once every 3 weeks with either placebo or ZYCO101a (100 μg DNA or 200 μg DNA). The results of this phase trial demonstrated that the ZYC101a drug was well tolerated at all dose levels tested, and no systemic side effects were described. Interestingly, the CIN 2/3 patients who were younger than 25 years of age ($n = 43$) presented with a convincing increase in the resolution of CIN2/3 compared with their placebo control counterparts (70% vs. 23%; $P = 0.007$), and ZYCO101a activity was not limited to HPV16- and 18-positive lesions.[34] These results show promise and warrant further investigation as a possible therapeutic vaccination strategy for CIN 2/3 patients. In addition, as the ZYC101 incorporates DNA and is encapsulated within small particles, this approach is suitable for intracellular delivery of nucleic acids to the phagocytic antigen-presenting cells (APCs). This process would elicit an HPV-specific cellular immune response,[31,33] and cellular immunity is the primary requirement for therapeutic vaccines, making this strategy favorable for further clinical testing.

A Phase I trial was conducted to assess the safety and immunogenicity of a mixture of HPV16 E6E7 fusion protein and ISCOMATRIX™ adjuvant (HPV16 Immunotherapeutic) in 31 CIN patients.[35] HPV Immunotherapeutic is a combination of bacterially derived recombinant HPV E6E7 fusion protein[36] and ISCOMATRIX™ adjuvant.[37] The vaccine and placebo formulations were administered three times at three weekly intervals. The vaccine dosages included mixtures of 120 μg ISCOMATRIX™ adjuvant plus 20 μg, 60 μg, or 200 μg HPV16 E6E7 and were found to be safe. Eighty percent of patients elicited a specific cell-mediated (CMI) response. A total of 7 of 14 subjects who were HPV16-positive at recruitment tested negative for HPV DNA at the end of the study, with significant reductions observed in each of the other subjects. The trial participant who received the immunotherapy developed specific T-cell responses to HPV16 E6E7 protein and was also higher in the delayed type hypersensitivity responses, *in vitro* cytokine release, and CD8 T-cell responses compared with the placebo recipients. The use of ISCOMATRIX™ adjuvant in combination with the HPV16 Immunotherapeutic shows promise, because it is safe and immunogenic and was found to influence the reduction of HPV viral load in the cervical epithelium. Indeed, ISCOMATRIX™ adjuvant would be attractive in vaccine regimens because they have been reported to be safe, highly immunogenic, easy to produce, and robust and can be manufactured on a large scale,[38] but further studies and efficacy trials are warranted. Furthermore, the authors of the article[10] recommend that the reduction in viral load in the cervix should be studied as a surrogate marker for vaccine efficacy in early investigations.

Previously, a preclinical study demonstrated that the elimination of established TC1 and C3 tumors in C57BL6 mice, vaccinated with GlaxoSmithKline (GSK) adjuvant and HPV16 E7, was successfully achieved.[39] Thus, a multicenter Phase I/II trial was studied to test the immunogenicity of an HPV16-E7 protein-based vaccine in 11 patients with HPV-positive CIN.[40] The vaccine (PD-E7/GSKAS02B) consisted of a fusion protein (PD-E7) with a mutated HPV16 E7 protein linked to the first 108 amino acids of *Haemophilus influenza* protein D, formulated in an adjuvant system (GlaxcoSmithKline Biologicals AS02B) containing MPL, QS-21, and oil-in-water emulsion. A total of seven CIN 1/3 patients were vaccinated intramuscularly three times at 2 weekly intervals with the PD-E7/GSKAS02B vaccine, and three CIN1 patients were injected with a 0.9% NaCl solution as placebo. Within 8 weeks of vaccination, all the vaccinated patients (seven of seven) developed significant levels of E7-specific plasmatic IgG, and six of seven also produced anti-PD IgG.[40] Furthermore, it was shown that five of seven vaccinated patients (71%) demonstrated elevated IFN-γ $CD8^+$ T-cell responses upon PD-E7 stimulation, and two patients generated long-term T-cell immunity toward the vaccine antigen and E7. Although the sample size was small, the humoral and cellular immune responses elicited were promising, and thus larger trials should be evaluated in CIN patients to determine the efficacy of the PD-E7/AS02B vaccine.

Preclinical studies have been conducted in a tumor mouse model in which vaccination with HPV E6 and E7 peptides and incomplete Freund's adjuvant induced protection against tumors induced by HPV16-transformed cells.[41] Based on these encouraging results, a Phase I trial was conducted in women with CIN or vulvar intraepithelial neoplasia (VIN), and all subjects were HLA-A2– and HPV16-positive.[42] A total of 18 patients were vaccinated four times at 3-week intervals with escalating doses of the E7 peptide (amino acids 12–20) and emulsified in incomplete Freund's adjuvant. In addition, beginning with the eleventh patient, an eight–amino acid E7 peptide linked to PADRE-965.10 helper peptide with a covalently linked lipid tail was added to the vaccine regimen. Only 3 of 18 patients demonstrated complete regression of dysplastic lesions and cytokine release, and cytolysis assays revealed that 10 of 16 patients had increases in E7-specific reactivity. In addition, albeit 12 of 18 patients were able to clear the virus from cervical scrapings by the fourth vaccination, all biopsy samples were still positive for HPV viral RNA after vaccination as determined by *in situ* RNA hybridization. This trial performed reasonably as three patients demonstrated complete regression and nine patients had partial regression; a more stringent and better delivery system that evokes a strong cell-mediated response and would induce complete viral clearance in all patients, except the control placebo group, would be desirable.

Therapeutic Preclinical Studies

Although several Phase I/II clinical therapeutic trials are currently in progress, it is unknown whether the general outcome in the final human Phase III efficacy trials will be successful. In addition, several challenges and potential shortcomings are evident in the current strategies in clinical trials, necessitating better approaches. Evidently, focus should be on issues related to delivery systems, the induction of specific cellular immune responses, the selection of proper control placebo groups, and vaccine efficacy before studies can be tested further in human clinical trials. In addition, the stage of the virus life cycle and disease should be considered, which would include precancerous lesions and preferably be before viral integration, so that the proper antiviral response is stimulated. Consequently, several therapeutic strategies are currently being investigated for their use as potential vaccines and therapy targeted for the regression and eradication of HPV-induced infections and show promise for clinical research. Only the preclinical research that shows promise is discussed in this section.

DNA Vaccines

Several successful studies using naked DNA vaccines have been reported in animal models.[43] Because DNA-based vaccines are stable, are easy to deliver, can either stably integrate into the genome or be maintained in an episomal form, are generally safe, and have been shown to elicit both cytotoxic T lymphocytes (CTLs) and antibody responses to HPV antigens, they are attractive candidates for therapeutic vaccines.[17,44] However, the drawback of naked DNA vaccine is its inability to amplify and spread *in vivo*, limiting its potency.[45]

Various studies have evaluated different approaches to improve the potency of DNA vaccines, which include fusing antigens to chemokines,[46] coadministration with CpG oligonucleotides,[47] targeting of antigens for rapid degradation,[48] or enhancement of intracellular spreading to increase the number of cells expressing antigen.[49] However, the most promising delivery of DNA has been shown in preclinical studies using the gene gun approach. Essentially, DNA-coated gold beads are projected into the epidermis, targeting DNA directly to professional APCs for processing and presentation to antigen-specific $CD8^+$ T cells.[43,45,50] Recently, Lin *et al.* demonstrated that vaccination with DNA encoding endoplasmic reticulum (ER) chaperone molecules linked to HPV16 E7 led to a significant increase in the frequency of E7-specific $CD8^+$ T cells.[45] In addition, strong protective and therapeutic antitumor effects were achieved in vaccinated mice.[45] It would be interesting to determine whether the increased antigen-specific cellular responses and the strong protection achieved in preclinical trials would be translated in human trials, and therefore further investigations are justifiable.

Nucleic Acids

RNAi

RNA interference (RNAi) is a phenomenon that has been shown in insects, plants, worms, fungi, and mammalian cells and has been linked to post-transcriptional gene silencing.[51] RNAi is a process by which long dsRNA are cleaved into short-interfering RNA (siRNA), by a RNase III–like class enzyme, termed Dicer.[52] The siRNA then associates with a protein–RNA effector nuclease complex, termed RNAi-induced silencing complex (RISC). This complex then is guided to catalyze the sequence-specific degradation of the target mRNA. There are several features that make the RNAi phenomenon attractive. It is sequence specific and a single molecule can be used to inhibit the expression of specific targeted genes without disrupting the expression of normal host genes. In addition, a functional immune system is not required for this approach and therefore can be effective in immunocompromised individuals.

Several groups have applied RNAi as a therapeutic strategy to target HPV16 or 18 E6 and E7 oncogenes and were able to demonstrate partial suppression of the expression of these genes *in vitro*[11,53,54,55] and *in vivo*.[56] Jiang and colleagues, however, followed a different approach of inducing apoptosis of human cancer cells by using a gel-based delivery of siRNA.[57] Essentially, SiHa (cervical carcinoma cells) or HCT116 (colorectal carcinoma cells) were established, grown, and overlaid with molten, low melting point agarose, liposome, and siRNA formulation. This study aimed at elucidating the role of siRNA in a gel-based formulation as a possible topical delivery therapeutic strategy for HPV-induced cervical cancer and other external and internal human disorders and diseases.[57]

Although some studies have reported that the selectivity of the siRNA was inadequate[58] and the activation of the interferon system had been induced by siRNA,[59,60] these investigations were not supported by *in vivo* studies. Nevertheless, taking together the successful *in vivo* studies that have been reported, the RNAi approach shows potential as immunotherapy for HPV-induced infections. RNAi would be an ideal therapeutic strategy for HPV immunotherapy because only the oncogenes are targeted for destruction and hence only the tumor cells are targeted, without affecting the healthy cells. However, this approach requires a more robust delivery system for *in vivo* studies, and further investigations should focus on safety and biologic and pharmacologic activity of the therapy before this approach can be investigated in the clinical setting.

Antisense Strategies

The antisense approach involves the construction of a molecule that is complementary to the mRNA of the DNA sequence of the disease gene that needs to be targeted for destruction, thus preventing the translation of that specific gene into protein. Several studies have demonstrated the inhibition of HPV E6 and E7 function *in vitro* using plasmid-based antisense-RNA delivery, delivering the antisense-RNA by adenoviral and retroviral vectors[61,62] and by antisense oligodeoxyribonucleotide.[63] Others have determined an antisense exposed region within HPV16 E6 mRNA and developed an anitsense phosphorothioated oligodeoxyribonuleotide (PO-ODN) analogue to the anti-E6–exposed region, which when tested in an *in vivo* system produced inhibition of tumor growth in nude mice.[64] In addition, PO-ODN has been tested in several other human clinical trials.[65]

The antisense approach has the advantage as a possible therapeutic strategy for the elimination of established HPV infection, because it is able to induce the specific decrease of the viral target oncoprotein so that disease progression is prevented. However, this approach justifies further *in vitro* and *in vivo* investigations for safety and robust delivery systems so that it can be achieved in clinical studies.

Ribozymes

Ribozymes (RZs) are catalytic RNA molecules that have been shown to inhibit viral replication, modulate tumor progression, and analyze cellular gene function.[66,67] RZs have been shown to exist in various categories of naturally occurring catalystic RNAs.[68] The hammerhead and the hairpin RZ motifs[22] have been used as antiviral agents to cleave and eliminate E6 and E7 transcripts of different papillomaviruses.[69] In addition, others have directed the hammerhead RZ to the translational start sites of HPV16 E6 and E7 and demonstrated the successful cleavage and disruption of gene expression in an *in vitro* assay.[70] Similarly, the disruption of HPV18 transcripts was achieved by hammerhead RZs targeted against E6 and E7.[71] RZs potentially could be a promising tool for gene therapy because of their intrinsic simplicity and small size for allowing it to be included in different flanking sequence motifs, without altering the specificity of cleavage.[22] However, extensive studies are required to further evaluate the efficacy of this agent before it can be tested in clinical trials.

Bacterial Vector Delivery Systems

Salmonella

Recombinant attenuated *Salmonella* (facultative intracellular bacteria) strain has been used widely as mucosal vaccine vectors to deliver heterologous antigens. The advantages of using this bacterial delivery system include the ease of expressing different viral proteins; it can deliver engineered plasmids and is highly immunogenic. It has been demonstrated that attenuated *Salmonella* expressing the E7 oncogene of HPV16[72] or the HPV16 E7 epitopes inserted into hepatitis B virus core antigen particles[73] is able to elicit a strong immune response. In addition, Salmonella-based vaccines are able to induce humoral, secretory, and cell-mediated immune responses, including cytotoxic T cells, against the heterologous antigens they express.[72] Furthermore, it has been shown that the $PhoP^c$ phenotype of *Salmonella* is essential for the efficient induction of antiviral immune responses.[74] However, others have shown that the use of recombinant *Salmonella* as a vaccine delivery vehicle against HPV infection was able to revert back to the single attenuating mutation.[75] Although this delivery system shows promise in terms of its ability to induce a strong immune response and the ease of delivery of foreign genes, several challenges face this approach, which need to be addressed. The most critical factor is safety and this should be examined stringently in preclinical settings. Indeed, understanding the mechanism of proper presentation of relevant antigens, mounting an efficient long-lasting immune response, and the ability to potently reduce and eliminate established tumors are equally important.

Listeria

Listeria monocytogenes is a facultative intracellular bacterium and has been used extensively as a possible vaccine delivery vehicle against HPV-associated diseases. The life cycle of *L. monocytogenes* makes it an ideal delivery vector for vaccine development, because it is taken up mainly by APCs, where in the phagosome most of the bacteria are digested and the antigens are targeted to the MHC class II pathway for antigen processing and presentation.[76] In parallel, a few bacteria escape into the cytosol through the actions of listeriolysin O (LLO) protein and presented in an MHC class I–restricted manner to $CD8^+$ T cells for the direct destruction of tumor cells and the removal of virally infected host cells.[77,78] In addition, when *Listeria* is unable to lyse the phagosome membrane, the antigens are targeted to the MHC class II pathway for presentation to $CD4^+$ T cells.[78] Both the $CD4^+$ and $CD8^+$ T cells are important for a strong cell-mediated immune response. It has been demonstrated that the route of vaccination plays an important role in antigen delivery as seen in various *in vivo* studies that show recombinant *L. monocytogenes* expressing the HPV16 E7 can lead to regression of established tumors when administered intraperitoneally[79] or orally.[80] Although these encouraging *in vivo* studies could potentially be translated into clinical trials, caution should be taken, because it is unclear whether the cellular immune responses to the vaccines will be efficacious. Recently, these cautions have been reiterated by studies focusing on immune responses induced by *L. monocytogenes* expressing HPV16 E7.[76,81] It has been demonstrated in a tumor mouse model that the ability of recombinant *L. monocytogenes* expressing HPV16 E7 to induce regression of established TC1 tumors is dependent on tumor sensitivity to IFN-γ[81] and induced $CD4^+$ T cells, which are suppressive in nature and correlate with impaired tumor regression by this vaccine.[76] These results highlight the importance of analyzing cellular immune responses and safety of these delivery systems before human clinical trials are established.

Viral Vector Delivery Systems

Alphaviruses

Several viral vectors expressing the HPV E6 and E7 have been extensively studied as possible HPV vaccine strategies. The Semliki Forest virus (SFV) belongs to the alphavirus genus and has been shown to induce a robust cellular immune response. The recombinant SFV expressing the HPV16 E6 and E7 oncoproteins elicited strong cytotoxic T-cell responses, eliminated pre-existing HPV16-induced tumors in mice, and is dependent of the route of immunization.[82] Others have vaccinated mice with replication-defective Sindbis virus replicon particles encoding herpes simplex virus type 1 VP22 fused to HPV16 E7,[83] which generated E7-specific $CD8^+$ T-cell immune response and eliminated pre-existing tumors. These results show promise and justify further clinical studies because SFV is safe and elicits a strong and long-lasting CTL response that can eliminate established tumors.

Adenoviruses

Recombinant adeno-associated virus expressing the HPV16 E7 oncoprotein fused to heat shock protein 70 (HSP 70) demonstrated an enhanced CD4- and CD8-dependent CTL response plus significant antitumor activity *in vitro*.[84] Others have utilized the adenovirus vector to deliver IL-12 (AdIL-12) as an adjuvant for an HPV16 E7 subunit vaccine *in vivo*.[85] The combined injection of AdIL-12 plus E7 as a therapeutic strategy was shown to eliminate established tumor in a tumor mouse model and induced enhanced E7-specific antibody responses and a significant E7-specific $CD8^+$ T-cell response. The use of viral vectors as vaccine vehicles has several advantages, as they are highly immunogenic and can be easily constructed and manipulated to meet specific expression requirements for delivery and safety. This strategy, however, requires further investigations to determine whether viral integration will be inhibited and most importantly to determine the efficacy of this system to induce a potent and long-lasting immune response for the eradication of established HPV-induced infection.

Delivery Systems and Combination Strategies

One of the formidable problems facing therapeutic vaccines is that of enhanced vaccine potency. Although several studies on HPV-associated anti-tumor responses have been successfully investigated in preclinical studies, the translation in clinical trials was only modestly successful. Therefore, efforts to enhance vaccine potency have focused on exploring better delivery systems to elicit effective cell-mediated anti-tumor immune responses. A good antigen delivery system should direct the targeted antigen to organized lymph tissue such as the draining lymph nodes or the spleen.[86] In addition, as immune responses are initiated by APCs in draining lymph nodes[50] and the plasmid DNA encoding the tumor antigen should be directed into the lymph nodes and spleen, resulting in a 100- to 1,000-fold increase in vaccine potency.[87] Moreover, loading of antigen into the cytoplasm of DC extends the duration of its presentation on MHC class I molecules and enhances CTL responses.[88] The liposome-based DNA delivery system named liposome-polycation-DNA (LPD) indeed addresses the aforementioned issues. LPD is a clinically proven safe particle and when injected systemically it is distributed to all major organs (including the spleen) and is able to induce the rapid production of Th1 cytokines.[89] Recently, it was demonstrated in preclinical studies that LPD particles incorporating the HPV16 E7 protein as antigen elicited both strong cellular and antibody immune responses and vaccination induced complete regressions of established tumors in a tumor mouse model.[90] It is suggested that LPD could potentially be a potent vaccine carrier and/or adjuvant for many antigens and definitely warrants further investigations.

Other approaches that have showed enhanced immunity in preclinical studies are prime-boost strategies. It has been shown that prime-boost strategies using combinations of priming with DNA plasmid or viral vector or protein and boosting with a heterologous viral vector encoding the immunogen have been effective in several viral, parasite, or tumor antigen vaccinations.[91] In addition, several combination therapy strategies have been successfully investigated in HPV-associated preclinical therapeutic research studies and include CpG-stimulated DCs, recombinant viruses, and peptide-pulsed DC.[43] This strategy justifies further investigations as indicated by an ongoing and potentially promising human trial, which involves combining a vaccinia-based vaccine containing HPV16/18 E6/E7 (TA-HPV) and fusion protein of HPV16 L2E6E7 (TA-CIN) in a heterlogous prime-boost strategy.[92,93] Although this trial is ongoing and immune responses and efficacy will be available only at the end of this study, it is noteworthy that similar approaches should be tested in different stages (CIN2/3) of HPV-induced infections so that an effective, strong, and robust immune response can be activated.

FUTURE AND CONCLUDING REMARKS

The last two decades have certainly produced excellent advancement in our understanding of HPV-associated infections. In terms of prophylactic vaccines the future is indeed very bright as the first-generation vaccines are in the last stages of clinical trials and have been shown to be well tolerated, eliciting high titers of neutralizing antibodies. This bodes well for the prevention of HPV-associated infections.

However, therapeutic vaccines are urgently required to reduce or eliminate existing HPV-induced infections. This is particularly important in developing countries where cervical disease continues to be a public health burden. The ideal therapeutic vaccine would elicit the complete elimination of established HPV-induced cervical lesions without affecting normal cells and elicit a sustained and robust CTL response, while being cost effective and safe. Despite the difficult challenges, there is growing confidence in several therapeutic strategies that have been investigated in preclinical studies and have generated impressive results using HR-HPV E6 and E7 oncogenes in different delivery systems. However, most therapeutic clinical trials of these vaccines have been disappointing. A possible reason for this is that patients recruited in the studies were at an advanced stage of cervical disease, which is known to be associated with genetic instability, viral immune escape, viral antigen tolerance, and downregulation of MHC class I alleles. Future therapeutic clinical trials should focus on including women who are precancerous (with CIN 3) before viral integration has taken place. Since it is well known that not all CIN 3 progresses to carcinoma, HPV viral load could be used as a marker for persistent infection prior to vaccination.

In summary, the development of a therapeutic vaccine is indeed a major challenge but will be achievable in the not too distant future. After all, prophylactic vaccines were but a dream two decades ago.

ACKNOWLEDGMENTS

V.A.G. is supported by the Faculty of Health Sciences, University of Cape Town. This review is not intended to discuss all therapeutic vaccine strategies; therefore, apologies to authors that have not been cited.

REFERENCES

1. zur Hausen, H. 2000. Papillomaviruses causing cancer: evasion from host-cell host in early events in carcinogenesis. J. Natl. Cancer Inst. **92:** 690–698.
2. Bernard, H.U. 2005. The clinical importance of the nomenclature, evolution and taxonomy of human papillomaviruses. J. Clin. Virol. **32S:** S1–S6.
3. de Villiers, E.M. *et al.* 2004. Classification of papillomaviruses. Virology **324:** 17–27.
4. Bosch, F.X. *et al.* 2003. Chapter 1: human papillomavirus and cervical cancer-burden and assessment of causality. J. Natl. Cancer Inst. Monogr. **31:** 3–13.
5. von Krogh, G. 2001. Management of anogenital warts (*Condylomata acuminata*). Eur. J. Dermatol. **11:** 598–604.
6. Koutsky, L.A. 1997. Epidemiology of genital human papillomavirus infection. Am. J. Med. **102:** 3–8.
7. Magnusson, P.K.E. *et al.* 2000. Cervical cancer risk: is there a genetic component? Mol. Med. Today **6:** 145–148.
8. Stanckzuk, G.A. *et al.* 2001. Cancer of the uterine cervix may be significantly associated with a gene polymorphism coding for increased IL-10 production. Int. J. Cancer **94:** 792–794.

9. GOVAN, V.A. *et al.* 2003. Ethnic differences in allelic distribution of IFN-γ in South African women but no link with cervical cancer. J. Carcinog. **2:** 3.
10. DESHPANDE, A. *et al.* 2005. TNF-a promoter polymorphisms and susceptibility to human papillomavirus 16-associated cervical cancer. J. Infect. Dis. **191:** 969–976.
11. FRAZER, I. 2004. Prevention of cervical cancer through papillomavirus vaccination. Nat. Rev. Immunol. **4:** 46–54.
12. MILNER, J. 2003. RNA interference for treating cancers caused by viral infection. Expert Opin. Biol. Ther. **3:** 459–467.
13. HINES, J.F. *et al.* 1994. The expression of L1 proteins of HPV-1, HPV-6, and HPV-11 display type-specific epitopes with native conformation and reactivity with neutralizing and non-neutralizing antibodies. Pathobiology **62:** 165–171.
14. CHRISTENSEN, N.D. *et al.* 1994. Assembled baculovirus-expressed human papillomavirus type 11 L1 capsid protein virus-like particles are recognized by neutralizing monoclonal antibodies and induce high titres of neutralizing antibodies. J. Gen. Virol. **75:** 2271–2276.
15. KOUTSKY, L.A. 2004. Prophylactic HPVL1 virus-like particle (VLP) vaccines. Abstract number 047. Presented at the 21st International Human Papillomavirus Conference, Mexico City, Mexico.
16. VILLA, L. *et al.* 2005. Prophylactic quadrivalent human papillomavirus (types 6, 11, 16, and 18) L1 virus-like particle vaccine in young women: a randomised double-blind placebo-controlled multicentre phase II efficacy trial. Lancet Oncol. **6:** 271–278.
17. BRINKMAN, J.A. *et al.* 2005. The impact of anti HPV vaccination on cervical cancer incidence and HPV induced cervical cancer lesions: consequences for clinical management. Eur. J. Oncol. **26:** 129–142.
18. ROBINSON, W. III. 2001. Management of cervical cancer neoplasia. Cancer Treat. Res. **104:** 287–302.
19. HAWLEY-NELSON, P. *et al.* 1989. HPV 16 E6 and E7 proteins cooperate to immortalize human foreskin keratinocytes. EMBO J. **8:** 3905–3910.
20. MUNGER, K. *et al.* 1989. The E6 and E7 gene of human papillomavirus type 16 together are necessary and sufficient for transformation of primary human keratinocytes. J. Virol. **63:** 4417–4421.
21. SMOTKIN, D. & F.O. WETTSTEIN. 1986. Transcription of HPV16 early genes in a cervical cancer and a cervical cancer derived cell line, and the identification of the E7 protein. Proc. Natl. Acad. Sci. USA **83:** 4680–4686.
22. DIPAOLO, J.A. & L.M. ALVAREZ-SALAS. 2004. Advances in the development of therapeutic nucleic acids against cervical cancer. Expert Opin. Biol. Ther. **4:** 1–14
23. HOWLEY, P.M. 1996. Papillomavirinae: the viruses and their replication. *In* Fields Virology, 3rd ed. B.N. Fields *et al.*, Eds.: 2045–2073. Lippincott-Raven Publishers. Philadelphia.
24. WEINBERG, R.A. 1991. Tumor suppressor genes. Science **254:** 1138–1146.
25. DYSON, N. 1998. The regulation of E2F by pRb-family proteins. Genes Dev. **12:** 2245–2262.
26. SCHEFFNER, M. *et al.* 1993. The HPV-16 and E6-AP complex functions as a ubiquitin-protein ligase in the ubiquitination of p53. Cell **75:** 495–505.
27. ROSENSTIERNE, M.W. *et al.* 2003. Identification and characterization of a cluster of transcription start sites located in the E6 ORF of human papillomavirus type 16. J. Gen. Virol. **84:** 2909–2920.
28. TINDLE, R.W. & I.H. FRAZER. 1994. Immune responses to human papillomaviruses and the prospects for human papillomavirus-specific immunisation. Curr. Top. Microbiol. Immunol. **186:** 217–253.
29. TINDLE, R.W. *et al.* 1997. Vaccines against papillomaviruses. *In* New Generation Vaccines. Vol. 50. 2nd Ed. R.W. Tindle, I.H. Frazer, A. Levine, G. Woodrow, X. Kaper & X. Cobon, Eds. :769–783. Marcel Decker. New York.
30. GISSMANN, L. *et al.* 2001. Therapeutic vaccine for human papillomavuiruses. Intervirology **44:** 167–175.
31. KLENCKE, B. *et al.* 2002. Encapsulated plasmid DNA treatment for human papillomavirus 16-associated anal dysplasia: a Phase I study of ZYC101. Clin. Cancer Res. **8:** 1028–1037.

32. SHEETS, E.E. *et al.* 2003. Immunotherapy of human cervical high-grade cervical intraepithelial neoplasia with microparticle-delivered human papillomavirus 16 E7 plasmid DNA. Am. J. Obstet. Gynecol. **188:** 916–926.
33. HEDLEY, M.L., J. CURLEY & R. URBAN. 1998. Microspheres containing plasmid-encoded antigens elicit cytotoxic T-cell responses. Nat. Med. **4:** 365–368.
34. GARCIA, F. *et al.* 2004. ZYC101a for treatment of high-grade cervical intraepithelial neoplasia: a randomized controlled trial. Obstet. Gynecol. **103:** 317–326.
35. FRAZER, I.H. *et al.* 2004. Phase 1 study of HPV16-specific immunotherapy with E6E7 fusion protein and ISCOMATRIX adjuvant in women with cervical intraepithelial neoplasia. Vaccine **23:** 172–181.
36. EDWARDS, S.J. *et al.* 1998. Design of a candidate recombinant therapeutic vaccine for cervical cancer. Recent Res. Dev. Biotech. Bioeng. **1:** 343–356.
37. WINDON, R.G. *et al.* 2001. Local immune responses to influenza antigen are synergistically enhanced by the adjuvant ISCOMATRIX. Vaccine **20:** 490–497.
38. PEARSE, M.J. & D. DRANE. 2005. ISCOMATRIX adjuvant for antigen delivery. Adv. Drug Deliv. Rev. **57:** 465–474.
39. GERHARD, C.M. *et al.* 2001. Therapeutic potential of protein and adjuvant vaccinations on tumour growth. Vaccine **19:** 2583–2589.
40. HALLEZ, S. *et al.* 2004. Phase I/II trial of immunogenicity of a human papillomavirus (HPV) type 16 E7 protein-based vaccine in women with oncogenic HPV-positive cervical intraepithelial neoplasia. Cancer Immunol. Immunother. **53:** 642–650.
41. FELTKAMP, M.C. *et al.* 1993. Vaccination with cytolytic T lymphocyte epitope containing peptide protects against a tumor induced by human papillomavirus type 16 transformed cells. Eur. J. Immunol. **23:** 2242–2249.
42. MUDERSPACH, L.I. *et al.* 2000. A phase I trial of a human papillomavirus (HPV) peptide vaccine for women with high-grade cervical and vulvar intraepithelial neoplasia who are HPV 16 positive. Clin. Cancer Res. **6:** 3406–3416.
43. CHRISTENSEN, N.D. 2005. Emerging human papillomavirus vaccines. Expert Opin. Emerg. Drugs **10:** 5–19.
44. EIBEN, G.L. *et al.* 2003. Cervical cancer vaccines: recent advances in HPV research. Viral Immunol. **16:** 111–121.
45. LIN, C.-T. *et al.* 2005. Enhancement of DNA vaccine potency through linkage of antigen gene to ER chaperone molecules, ER-60, tapasin, and calnexin. J. Biomed. Sci. **12:** 279–287.
46. BIRAYAN, A. *et al.* 1999. Genetic fusion of chemokines to a self tumor antigen induces protective, T-cell dependent antitumor immunity. Nat. Biotechnol. **17:** 253–258.
47. KLINMAN, D.M. *et al.* 1997. Contribution of CpG motifs to the immunogenicity of DNA vaccines. J. Immunol. **158:** 3635–3639.
48. RODRIGUEZ, F. *et al.* 1998. DNA immunisation with minigenes: low frequency of memory cytotoxic T lymphocytes and inefficient antiviral protection are rectified by ubiquitination. J. Virol. **72:** 5174–5181.
49. HUNG, C.F. *et al.* 2002. Improving DNA vaccine potency by linking Marek's disease virus type 1 VP22 to an antigen. J. Virol. **76:** 2676–2682.
50. CONDON, C. *et al.* 1996. DNA-based immunization by in vivo transfection of dendritic cells. Nat. Med. **2:** 1122–1128.
51. CULLEN, B.R. 2004. Derivation and function of small interfering RNAs and microRNAs. Virus Res. **102:** 3–9.
52. BERSTEIN, E. *et al.* 2001. Role of ribonuclease in the initiation step of RNA interference. Nature **409:** 363–366
53. STEVENSON, M. 2004. Therapeutic potential of RNA interference. N. Engl. J. Med. **351:** 1772–1777.
54. HALL, A.H.S. & K.A. ALEXANDER. 2003. RNA interference of human papillomavirus type 18 E6 and E7 induces senescence in HeLa cells. J. Virol. **77:** 6066–6069.
55. JIANG, M. & J. MILNER. 2002. Selective silencing of viral gene expression in HPV-positive human cervical carcinoma cells treated with siRNA, a primer of RNA interference. Oncogene **21:** 6041–6048.
56. YOSHINOUCHI, M. *et al.* 2003. In vitro and in vivo growth suppression of human papillomavirus 16-positive cervical cancer cells by E6 siRNA. Mol. Ther. **8:** 762–768.

57. JIANG, M. *et al.* 2004. Gel-based application of siRNA to human epithelial cancer cells induces RNAi-dependent apoptosis. Oligonucleotides **14:** 239–248.
58. JACKSON, A. *et al.* 2003. Expression profiling reveals off-target gene regulation by RNAi. Nat. Biotechnol. **21:** 635–637.
59. SLEDZ, C.A. *et al.* 2003. Activation of the interferon system by short-interfering RNAs. Nat. Cell Biol. **5:** 834–839.
60. KARIKO, K. *et al.* 2004. Small interfering RNAs mediate sequence-independent gene suppression and induce immune activation by signaling through toll-like receptor 3. J. Immunol. **172:** 6545–6549.
61. CHOO, C.K. *et al.* 2000. Retrovirus-mediated delivery of HPV16E7 antisense RNA inhibited tumorigenecity of CaSki cells. Gynecol. Oncol. **78:** 293–301.
62. CHO, C.W. *et al.* 2002. HPV E6 antisense induces apoptosis in CaSki cells via suppression of E6 splicing. Exp. Mol. Med. **34:** 159–166.
63. TAN, T.M. & R.C. TING. 1995. In vitro and in vivo inhibition of human papillomavirus type 16 E6 and E7 genes. Cancer Res. **55:** 4599–4605.
64. ALVAREZ-SALAS, L.M. *et al.* 1998. Inhibition of HPV16 E6/E7 immortalisation of normal keratinocytes by hairpin ribozymes. Proc. Nat. Acad. Sci. USA **95:** 1189–1194.
65. BENNETT, C.F. 1998. Antisense oligonucleotides: is the glass half full or half empty? Biochem. Pharmacol. **55:** 9–19.
66. PAVCO, P.A. *et al.* 2000. Antitumor and antimetastatic activity of ribozymes targeting the messenger RNA of vascular endothelial growth factor receptors. Clin. Cancer Res. **6:** 2094–2103.
67. MORRISSEY, D.V. *et al.* 2002. Characterization of nuclease-resistant ribozymes directed against hepatitis B virus RNA. J. Viral Hepat. **9:** 411–418.
68. ECKSTEIN, F., A.R. KORE & K.L. Nakamaye. 2001. In vitro selection of hammerhead ribozyme sequence variants. Chembiochem **2:** 629–635.
69. HE, Y.K. *et al.* 1993. In vitro cleavage of HPV16 E6 and E7 RNA fragments by synthetic ribozymes and transcribed ribozymes from RNA-trimming plasmids. FEBS Lett. **322:** 21–24.
70. LU, D. *et al.* 1994. Ribozyme-mediated in vitro cleavage of transcripts arising from the major transforming genes of human papillomavirus type 16. Cancer Gene Ther. **1:** 267–277.
71. CHEN, Z. *et al.* 1995. Effectiveness of three ribozymes for cleavage of an RNA transcript from human papillomavirus type 18. Cancer Gene Ther. **2:** 263–271.
72. KRUL, M.R. *et al.* 1996. Induction of an antibody response in mice against human papillomavirus (HPV) type 16 after immunisation with recombinant Salmonella strains. Cancer Immunol. Immunother. **43:** 44–48.
73. LONDONO, L.P. *et al.* 1996. Immunization of mice using *Salmonella typhimurium* expressing human papillomavirus type 16 E7 epitopes inserted into hepitis B virus core antigen. Vaccine **14:** 545–552.
74. BALMELLI, C. *et al.* 1998. Nasal immunisation of mice with human papillomavirus type 16 virus-like particles elicits neutralising antibodies in mucosal secretions. J. Virol. **72:** 8220–8229.
75. MILLER, S.I. & J.J. MEKALANOS. 1990. Constitutive expression of the *phoP* regulon attenuates *Salmonella* virulence and survival within macrophages. J. Bacteriol. **172:** 2485–2489.
76. HUSSAIN, S.F. *et al.* 2005. What is needed for effective antitumor immunotherapy? Lessons learned using *Listeria monocytogenes* as a live vector for HPV-associated tumors. Cancer Immunol. Immunother. **54:** 577–586.
77. PORTNOY, D.A. *et al.* 1992. Molecular determinants of *Listeria monocytogenes* pathogenesis. Infect. Immun. **60:** 1263–1267.
78. PAMER, E.G. *et al.* 1997. MHC class I antigen processing of *Listeria monocytogenes* proteins: implications for dominant subdominant CTL responses. Immunol. Rev. **158:** 129–136.
79. GUNN, G.R. *et al.* 2001. Two *Listeria monocytogenes* vaccine vectors that express different molecular forms of human papilloma virus-16 (HPV-16) E7 induce qualitatively different T cell immunity that correlates with their ability to induce regression of established tumors immortalized by HPV-16. J. Immunol. **167:** 6471–6479.

80. LIN, C.W. *et al.* 2002. Oral vaccination with recombinant *Listeria monocytogenes* expressing human papillomavirus type 16 E7 can cause tumor growth in mice to regress. Int. J. Cancer **102:** 629–637.
81. DOMINIECKI, M.E. *et al.* 2005. Tumor sensitivity to IFN-γ is required for successful antigen-specific immunotherapy of a transplantable mouse tumor model for HPV-transformed tumors. Cancer Immunol. Immunother. **54:** 477–488.
82. DAEMEN, T. *et al.* 2004. Superior therapeutic efficacy of alphavirus-mediated immunization against human papilloma virus type 16 antigens in a murine tumour model: effects of the route of immunization. Antivir. Ther. **9:** 733–742.
83. CHENG, W.F. *et al.* 2002. Cancer immunotherapy using Sindbis virus replicon particles encoding a VP22-antigen fusion. Hum. Gene Ther. **13:** 553–568.
84. LIU, D.W. *et al.* 2000. Recombinant adeno-associated virus expressing human papillomavirus type 16 E7 peptide DNA fused with heat shock protein DNA as a potential vaccine for cervical cancer. J. Virol. **74:** 2888–2894.
85. AHN, W.S. *et al.* 2003. A therapy modality using recombinant IL-12 adenovirus plus E7 protein in a human papillomavirus 16 E6/E7-associated cervical cancer animal model. Hum. Gene Ther. **14:** 1389–1399.
86. KARRER, U. *et al.* 1997. On the key role of secondary lymphoid organs in antiviral immune responses studied in alymphoplastic (aly/aly) and spleenless (Hox11–/–) mutant mice. J. Exp. Med. **185:** 2157–2170.
87. MALOY, K.J. *et al.* 2001. Intralymphatic immunization enhances DNA vaccination. Proc. Natl. Acad. Sci. USA **98:** 3299–3203.
88. WANG, R.F. *et al.* 2002. Enhancement of antitumor immunity by prolonging antigen presentation on dendritic cells. Nat. Biotechnol. **20:** 149–154.
89. DILEO, J. *et al.* 2003. Lipid-protamine-DNA-mediated antigen delivery to antigen-presenting cells results in enhanced anti-tumor immune responses. Mol. Ther. **7:** 640–648.
90. CUI, Z. *et al.* 2005. Liposome-polycation-DNA (LPD) particle as a carrier and adjuvant for protein-based vaccines: therapeutic effect against cervical cancer. Cancer Immunol. Immunother. Epub.
91. WOODLAND, D.L. 2004. Jump-starting the immune system: prime-boosting comes of age. Trends Immunol. **25:** 98–104.
92. VAN DER BURG, S.H. *et al.* 2001. Pre-clinical safety and efficacy of TA-CIN, a recombinant HPV 16L2E6E7 fusion protein vaccine, in homologous and heterologous prime-boost regimens. Vaccine **19:** 3652–3660.
93. SMYTH, L.J. *et al.* 2004. Immunological responses in women with HPV16 associated lower genital intraepithelial neoplasia (LGIN) following induced heterologous prime-boost HPV 16 oncogene vaccination. Clin. Cancer Res. **10:** 2954–2961.

ProtEx™: A Novel Technology to Display Exogenous Proteins on the Cell Surface for Immunomodulation

NARENDRA P. SINGH,[a,b] ESMA S. YOLCU,[a] NADIR ASKENASY,[c] AND HAVAL SHIRWAN[a]

[a]*Institute for Cellular Therapeutics and Department of Microbiology and Immunology, University of Louisville, Louisville, Kentucky 40202, USA*

[b]*Department of Pharmacology & Toxicology, Virginia Commonwealth University, Richmond, Virginia 23298, USA*

[c]*The Leah and Edward M. Frankel Laboratory of Bone Marrow Transplantation, Schneider Children's Medical Center of Israel, Petach Tikva, Israel*

ABSTRACT: Gene therapy as an immunomodulatory approach has the potential to treat various inherited and acquired immune-based human diseases. However, its clinical application has several challenges, varying from the efficiency of gene transfer, control of gene expression, cell and tissue targeting, and safety concerns associated with the introduction of exogenous DNA into cells/tissues. Gene therapy is also a time- and labor-intensive procedure. As an alternative, we recently developed a novel technology, ProtEx™, that allows for rapid, efficient, and durable display of exogenous proteins on the surface of cells, tissues, and organs without detectable toxicity. This technology exploits the strong binding affinity ($K_d = 10^{-15}$ *M*) of streptavidin with biotin and involves generation of chimeric molecules composed of the extracellular portions of immunological proteins of interest and a modified form of streptavidin, biotinylation of biological surfaces, and decoration of the modified surface with chimeric proteins. Biotin persists on the cell surface for weeks both *in vitro* and *in vivo,* thereby providing a platform to display exogenous proteins with extended cell surface kinetics. Two chimeric proteins, rat FasL (SA-FasL) and human CD80 (CD80-SA), were generated and tested for cell surface display and immunomodulatory functions. SA-FasL and CD80-SA molecules persisted on the surface of various cell types for extended periods, varying from days to weeks *in vitro* and *in vivo*. The cell surface kinetics, however, were protein and cell type dependent. SA-FasL showed potent apoptotic activity against Fas^+ cells as a soluble protein or displayed on the cell surface and effectively blocked alloreactive responses. The display of CD80-SA on the surface of tumor cells, however, converted them into antigen-presenting cells for effective stimulation of autologous and allogeneic T-cell responses. ProtEx technology, therefore, represents a practical and effective alternative to DNA-based gene therapy for immunomodulation.

Address for correspondence: Narendra P. Singh, Department of Pharmacology & Toxicology, Virginia Commonwealth University, Richmond, VA 23298. Voice: 804-828-2072; fax: 804-828-2117.

npsingh@vcu.edu

Ann. N.Y. Acad. Sci. 1056: 344–358 (2005).
doi: 10.1196/annals.1352.036

KEYWORDS: gene therapy; ProtEx^{TM} technology; cell surface display; decoration; immunomodulation

INTRODUCTION

Gene therapy has the potential to treat various inherited and acquired immune disorders.[1–3] However, many barriers remain to the routine application of gene therapy in clinical settings. These include inefficient delivery of the gene of interest, low expression efficiency of the introduced gene, and safety concerns associated with the introduction of exogenous foreign DNA into cells/tissues.[4] In addition, gene therapy is a time- and effort-intensive procedure that requires significant financial resources. In this context, the search for alternative methods to achieve the goals of gene therapy remains elusive. The display of exogenous proteins of interest on the cell surface represents a potential alternative to gene therapy for immunomodulation. A variety of immunological disorders may possibly be treated by the display of the required proteins on the cell surface instead of their expression from exogenously introduced DNA. In this context, we developed the ProtEx technology that exploits the high-affinity interaction ($K_d = 10^{-15}$ *M*) of streptavidin with biotin to display exogenous proteins chimeric with streptavidin on the surface of cells, tissues, and organs for immunomodulation.[5–7] Streptavidin also exists as stable tetramers and oligomers under physiological conditions, [5,8] and as such it may not only allow the durable display of chimeric proteins on the cell surface, but also enhance signal transduction by facilitating aggregation of the proteins on the cell surface.[9]

To test the efficacy of ProtEx technology, we generated two chimeric molecules with streptavidin, rat apoptotic FasL (SA-FasL) and human costimulatory CD80 molecule (CD80-SA). The choice of FasL was due to its role in immune homeostasis and establishment of peripheral tolerance by downregulating the immune responses.[10,11] CD80, however, is critical to the initiation of immune responses by interacting with CD28 and generating signal 2.[12] FasL, in recent years, has extensively been exploited for the induction of tolerance to transplant and autoimmune antigens for the purpose of preventing allograft rejection and treating autoimmune diseases.[13–17] CD80, however, has been used to stimulate immune responses to infections and tumors.[18–21] Tumor cells genetically modified to express CD80 were shown to induce effective antitumor immune responses with therapeutic efficacy in various animal cancer models and limited clinical settings.[19,22–28]

We demonstrated that both SA-FasL and CD80-SA chimeric molecules can be durably displayed on the surface of various primary and tumor cell lines *in vitro* and *in vivo*. SA-FasL showed potent apoptotic activity against cells expressing the Fas receptor as a soluble protein or displayed on the cell surface. Importantly, splenocytes displaying SA-FasL blocked alloreactive T-cell responses when used as antigen-presenting cells (APCs). Similarly, we observed that tumor cells decorated with CD80-SA served as APCs for the generation of effective antitumor immune responses *ex vivo*. Taken together, our data demonstrate that ProtEx technology provides an effective alternative method to gene therapy for immunomodulation with broad research and therapeutic implications.

METHODS

Animals and Cells

BALB/c (H-2^d) and C57BL/6 (H-2^b) mice were purchased from Jackson Laboratory (Bar Harbor, ME). PVG.1U ($RT1^u$) and PVG.R8 ($RT1.A^aB^uD^uC^u$) rats were bred in our animal colony at the University of Louisville. ACI ($RT1^a$) rats were obtained from Harlan Sprague Dawley (Indianapolis, IN). The animals were cared for and maintained in accordance with Institutional and National Institutes of Health guidelines.

A20, a B-cell lymphoma line derived from BALB/c mice, and HEC-1-A, derived from human endometrial cancer, were used in this study. Both cell lines were purchased from American Type Culture Collection (Rockville, MD) and cultured as suggested by ATCC guidelines.

Cloning and Expression of SA-FasL and CD80-SA Chimeric Molecules

Generation of protein chimeric with streptavidin depends on the type of protein, and accordingly the target gene is positioned 5′ or 3′ of the streptavidin gene in an expression vector. This strategy allows the gene to remain functional and keep the streptavidin region of the protein free to bind with biotin. As such, FasL, type II protein, was cloned 3′ and CD80, type I protein, 5″ to the streptavidin gene (FIG. 1). The details of cloning of streptavidin, SA-FasL, and CD80-SA and their expression using an inducible vector in insect cells have previously been described.[29,30]

Biotinylation and Decoration of the Cell Membrane with Chimeric Proteins

For biotinylation, lymphocytes or tumor cells were first incubated in 15 μM freshly prepared EZ-Link™ Sulfo-NHS-LC-Biotin (Pierce) in phosphate-buffered saline solution (PBS) for 30 minutes at room temperature. After washing twice with PBS, the biotinylated cells were suspended in PBS supplemented with 50–100 ng of SA-FasL or CD80-SA per 10^6 cells and incubated at 4°C for 20–30 min with intermittent mixing. Cells were then washed twice with cold PBS and analyzed in flow cytometry using APC-labeled streptavidin, MFL4 labeled with PE, and PE-labeled CD80 mAb (clone L307.4) to assess the levels of biotin, SA-FasL, and CD80-SA on the cell surface, respectively.

Display Kinetics of Biotin, SA-FasL, and CD80-SA on the Cell Surface

Various cell types were first biotinylated with 15 μM freshly prepared EZ-Link™ Sulfo-NHS-LC-Biotin and then either not decorated or decorated with SA-FasL or CD80-SA (50–100 ng/10^6 cells). The cells were cultured *in vitro* for various time points. The presence of biotin, SA-FasL, and CD80-SA on the cell surface was determined by flow cytometry using streptavidin-APC for biotin, MFL4-APC mAb for SA-FasL, and PE-labeled anti-human CD80 mAb (clone L307.4) for CD80-SA. Unmodified cells served as negative controls. For *in vivo* experiments, biotinylated splenocytes were labeled with CFSE (2.5 μM; Molecular Probes, Eugene, OR), and either not decorated or decorated with SA-FasL or CD80-SA. Forty to fifty million cells displaying biotin or SA-FasL or CD80-SA were intravenously injected into syngeneic animals. Spleens and lymph nodes were harvested at various times post-injection and analyzed for the presence of biotin, SA-FasL, and CD80-SA on the surface of CFSE-positive cells using flow cytometry with appropriate controls.

Apoptosis Assays

To perform apoptotic assays, mouse A20 cells (1×10^6/ml) were incubated in the presence of 100 ng of S2 supernatant or SA-FasL for 8–24 h at 37°C in a 5% CO_2 incubator.

Similarly, FasL-decorated rat (ACI) splenocytes were coincubated with ConA-activated PVG.1U lymphocytes at 1–5 effector:target ratios for 8–24 h. Cells were harvested, washed, and stained with propidium iodide or 7-amino actinomycin D (7-AAD) as markers for dead cells and annexin V labeled with appropriate fluorochromes as markers for apoptotic cells. To determine apoptosis in target cells, these cells were stained with fluorochrome-conjugated antibodies against a cell-specific marker (antibody against $RT1.A^u$ for PVG.1U rat).

Immunomodulation and Mixed Lymphocyte Reactions

Lymphocytes harvested from naïve animals were used in *in vitro* proliferation assays.[31] Responder and stimulator cells (1×10^5 cells/well) were cocultured intriplicate in complete mixed lymphocyte reaction (MLR) medium for 3–5 days and pulsed with 1 μCi of [^{3}H]thymidine (NEN Life Sciences Products, Boston, MA) for the last 16–18 h of culture. Similarly, mixed lymphocyte tumor reactions (MLTRs) were performed using peripheral blood lymphocytes (PBLs) collected from healthy volunteers or cancer patients and suspended in complete human MLTR medium (RPMI-1640 supplemented with 10 mM HEPES, 100 U/ml penicillin and 100 μg/ml streptomycin, 274 μM L-arginine, and 5% pooled human serum) and used as responders against irradiated (10,000 cGy) tumor cell lines of various antigenic disparities. Blood and tumor samples were collected from subjects with informed consent under guidelines and procedures approved by the institutional review board of the University of Louisville. All MLR and MLTR assays were performed in three replicate wells per data point, and results are presented as mean ± SD of triplicate wells of representative experiments. Responses to syngeneic and third-party antigens served as controls in MLRs, whereas responses to autologous cells and phytohemagglutinin served as controls in MLTR assays.

RESULTS

Display Kinetics of Biotin on the Cell Surface

We first determined biotinylation of the cell membrane as a function of various biotin concentrations and tested the effect of biotinylation on the proliferation of lymphocytes in response to a T-cell mitogen to establish optimal conditions for the display of exogenous proteins without major effect on the viability and antigen-presenting capacity of target cells. Rat or mouse splenocytes were first biotinylated with different concentrations (5–150 μM) of Sulfo-NHS-LC-Biotin and analyzed using APC-labeled SA with flow cytometry. There was a direct relationship between the biotin concentration and the level of biotin displayed on the cell surface, 5–15 μM biotin giving the desired range.

We next biotinylated various primary and established cell lines and determined the kinetics of biotin on the cell surface. Splenocytes decorated with 15 μM biotin were cultured *in vitro* for various time points in the presence or absence of ConA. Cells were harvested and analyzed using flow cytometry. Biotin remained present on the cell surface for an extended period with $t_{1/2}$ of >20 and 18 days for resting (FIG. 1A) and actively dividing splenocytes (FIG. 1B). However, we observed faster turnover of biotin on the surface of dividing cells than nondividing cells (FIG. 1A and B). On day 20, 53% of the nondividing cells were positive for biotin (FIG. 1A), whereas only 36% of dividing cells scored positive for biotin (FIG. 1B). *In vivo* studies demonstrated similar kinetics for cell

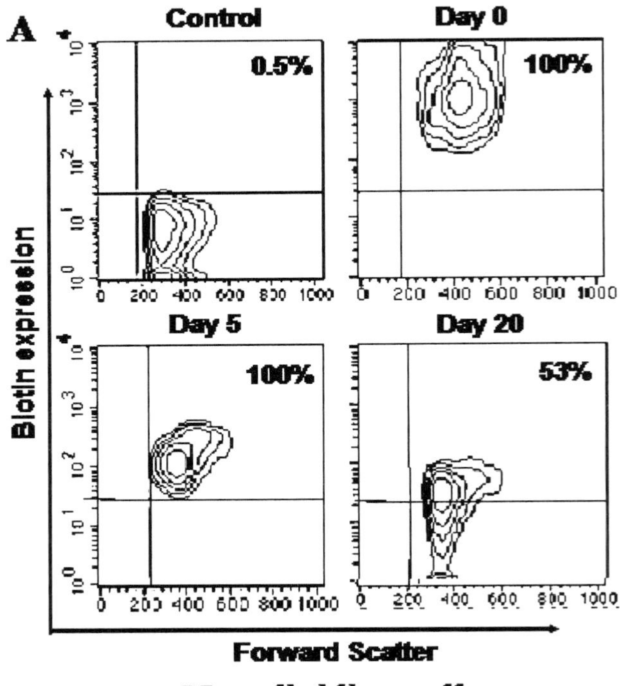

FIGURE 1. Display of biotin on the cell surface. (**A** and **B**) Presence of biotin on the surface of dividing and nondividing cells. Rat splenocytes were biotinylated with 15 μM EZ-Link™ biotin and then cultured in the absence (no ConA [**A**]) or presence of 2.5 μg/ml ConA (ConA [**B**]). The cells were harvested at various time points, and the level of biotin on the cell surface was determined using flow cytometry after staining with streptavidin-APCs. The nondividing cells possessed about 53% biotin on day 20 (FIG. A), whereas dividing cells possessed about 36% biotin (FIG. B) on their cell surface. (**C**) Presence of biotin on the surface of endothelial cells. Rat endothelial cells were biotinylated (15 μM EZ-Biotin) and then cultured for various time periods. Cells were harvested at different time points by trypsinization and analyzed by flow cytometry for the presence of biotin on the cell surface. Data in **A**, **B**, and **C** are representative of at least three to five independent experiments.

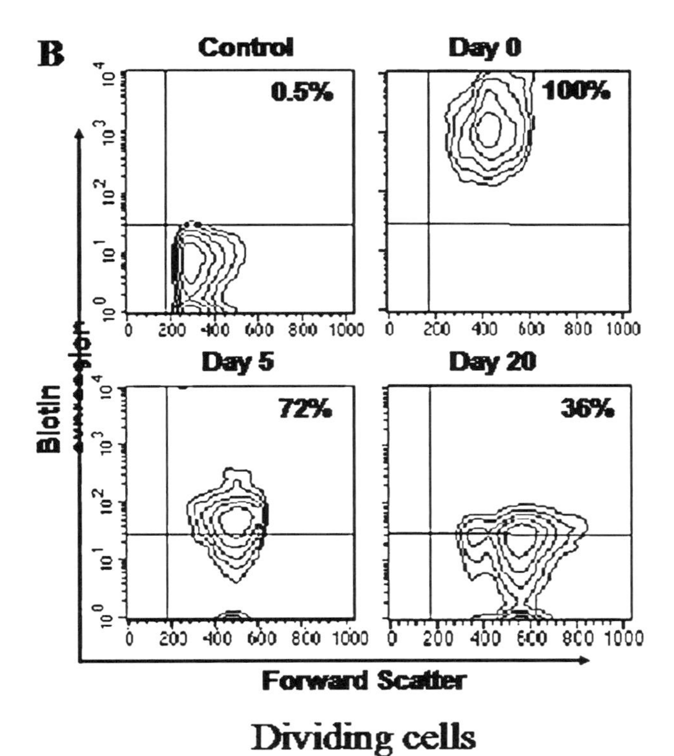

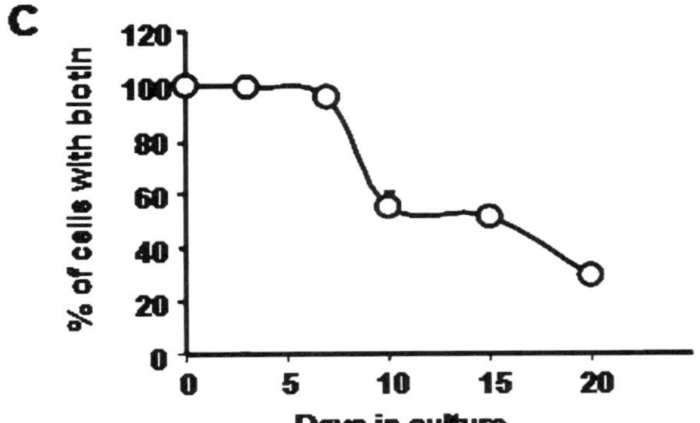

C

% of cells with biotin

Days in culture

FIGURE 1. *Continued.*

surface persistence of biotin (data not shown). The long-term persistence of biotin on the cell surface was general and not cell type-specific, since rat aortic endothelial cells showed retention of biotin on the surface for extended periods of time (FIG. 1C). However, the kinetics of biotin persistence on endothelial cells were biphasic with an initial estimated $t_{1/2}$ of 7–7.5 days followed by a longer $t_{1/2}$ of more than 20 days (FIG. 1C). The initial faster degradation of biotin on endothelial cells might be due to active cell division in the initial culture, but it slowed down when the culture became confluent.

Effect of Biotinylation on the Proliferative Response of Lymphocytes

To determine whether biotinylation affects the function of lymphocytes, we performed *in vitro* proliferation assays. Rat splenocytes were first biotinylated with different concentrations (1.5–150 μM) of EZ-Biotin and then cultured in the presence of 2.5 μg ConA/ml. Cells were pulsed with [^{3}H] thymidine for the last 18 h of the culture and harvested at various time points to determine proliferation. Minimal inhibition of proliferation by biotin was noted as compared with the unbiotinylated cells at all biotin concentrations tested during the first 2 days of culture. This inhibitory effect was not apparent on day 3 of the culture because all cultures showed similar levels of proliferation (FIG. 2). Taken together, these data suggest that biotinylation of the cell surface molecules does not have a significant effect on the function of the cell vis-à-vis proliferation in response to a mitogen. It, however, remains to be determined if biotinylation is totally innocuous or affects some other functions of the cells, besides proliferation, that can be detected using more sensitive tests.

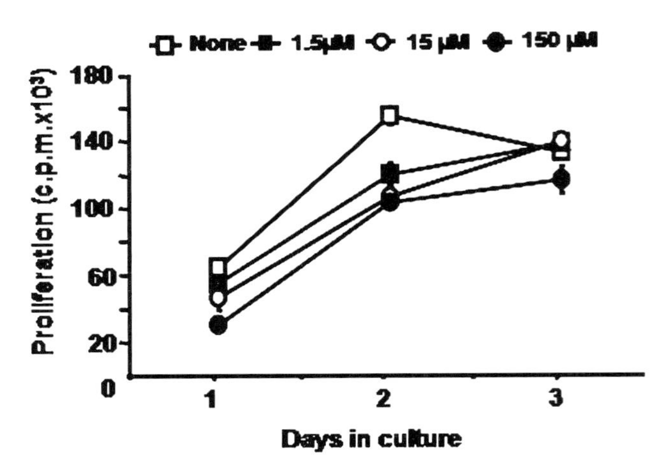

FIGURE 2. Effect of biotin binding on proliferation of rat splenocytes. Splenocytes harvested from PVG.1U rats were biotinylated with different concentrations of EZ-biotin (1.5–150 μM) and then the biotinylated splenocytes were activated with ConA for 1–3 days at 37°C, 5% CO_2. Proliferation of cells was determined by adding 1 μCi of [^{3}H]thymidine 16-18 h before harvesting the cells. Biotin showed no adverse effects on the proliferation of splenocytes. Data shown are the mean ± SD of three independent experiments.

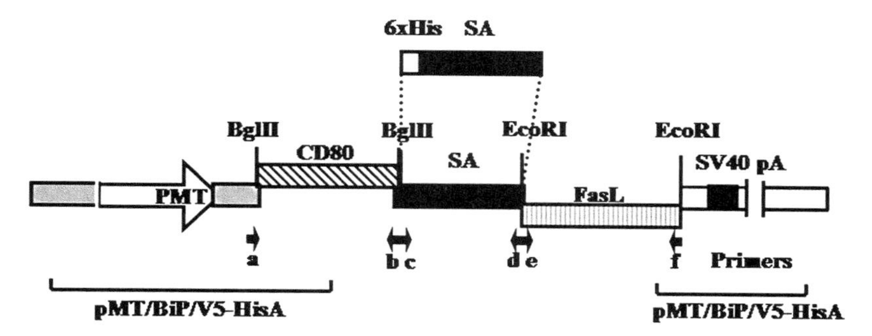

FIGURE 3. Schematic diagram showing construction of SA-FasL and CD80-SA in PMT/BiP/V5-HisA vector. Gene for type I proteins are inserted 5′ to streptavidin, whereas genes for type II proteins are positioned 3′ to streptavidin to facilitate the correct structure.

Display Kinetics of SA-FasL and CD80-SA on the Cell Surface

To test whether ProtEx technology can be used to display exogenous proteins chimeric with SA, we performed *in vitro* assays using SA-FasL or CD80-SA molecules that were expressed using an insect expression system (FIG. 3).[29,30] Rat or mouse splenocytes were biotinylated with 15 μM Sulfo-NHS-LC-Biotin, decorated with either SA-FasL or CD80-SA, and then cultured *in vitro*. The cells were collected at various time points and analyzed using flow cytometry. Both SA-FasL and CD80-SA were successfully displayed on the surface of biotinylated cells and persisted on the cell surface for weeks (FIG. 3). *In vivo* kinetics for SA-FasL displayed monophasic loss with a $t_{1/2}$ of >3.5 days (data not shown), whereas CD80-SA showed patterns similar to the kinetics observed *in vitro*, with a $t_{1/2}$ of >10 days (data not shown). The display kinetics for SA-FasL was cell type dependent, as the $t_{1/2}$ on the surface of the heart vasculature was 9.5 days.[32] Taken together, these data demonstrate that primary cells, established cells, and organs such as the heart can be effectively modified to display exogenous proteins chimeric with streptavidin, and the turnover kinetics of proteins are protein and cell type dependent with $t_{1/2}$ ranging from days to weeks.

SA-FasL Causes Apoptosis in Primary and Activated T Cells

To test whether SA-FasL is functional and retains its apoptotic function, apoptotic assays were performed using Fas+ mouse A20 lymphoma cell line and mitogen-activated lymphocytes. A20 cells cultured in the presence of SA-FasL (100 ng) showed more than 69% cells undergoing apoptosis (FIG. 4A). Cell culture containing the same concentration of S2 supernatant caused minimal (7%) apoptosis (FIG. 4A). We next tested whether SA-FasL displayed on the cell surface can induce apoptosis in Fas+ lymphocytes. PVG.1U rat splenocytes were activated with ConA for 3 days, purified using Ficoll gradients, labeled with CFSE, and used as target for SA-FasL displaying ACI splenocytes at various effector-to-target ratios. SA-FasL on the surface of splenocytes resulted in significant apoptosis in activated T cells in a dose-dependent manner (>80–90% apoptosis at 5:1 effector:target ratio (FIG. 4B). Taken together, these data demonstrate that SA-FasL possesses potent apoptotic activity against Fas^+ established or primary cells either as a soluble molecule or displayed

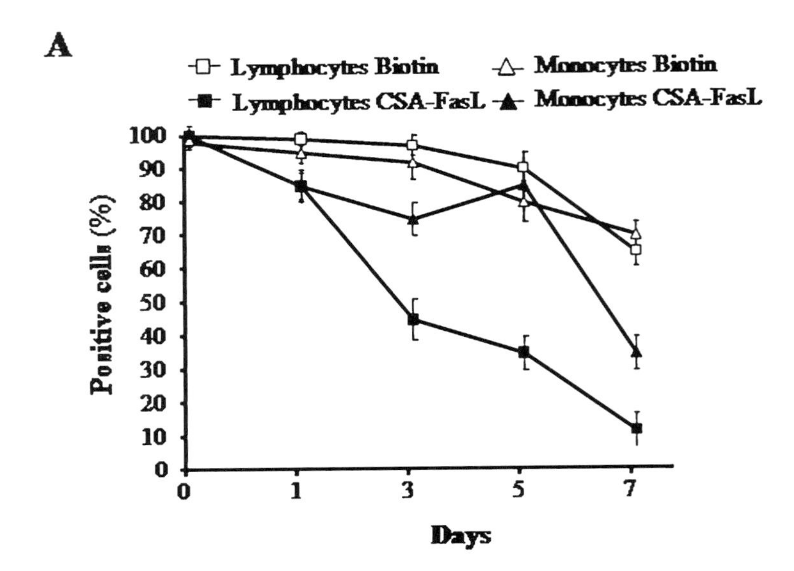

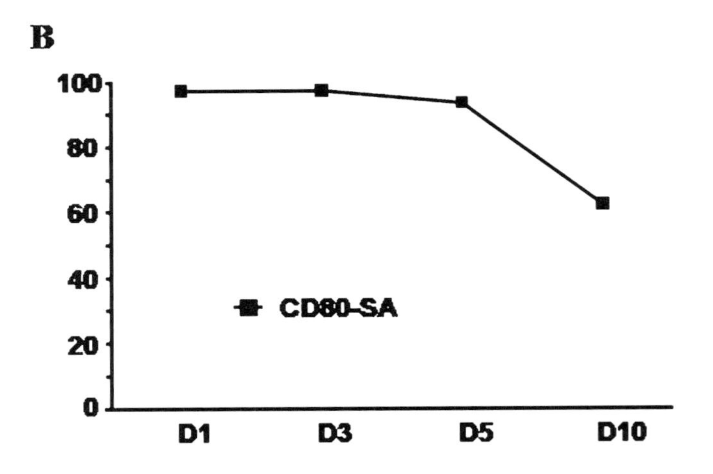

FIGURE 4. Kinetics of SA-FasL and CD80-SA on the cell surface. Biotinylated (15 μM) rat splenocytes were decorated with either SA-FasL or CD80-SA (50–100 ng/10^6 cells) and cultured at 37°C, 5% CO_2. Cells were harvested at different time points, and the display of chimeric proteins on the cell surface was determined using an APC-labeled anti-FasL antibody (MFL4) and PE-labeled human CD80 antibody (clone L307.4) in flow cytometry.

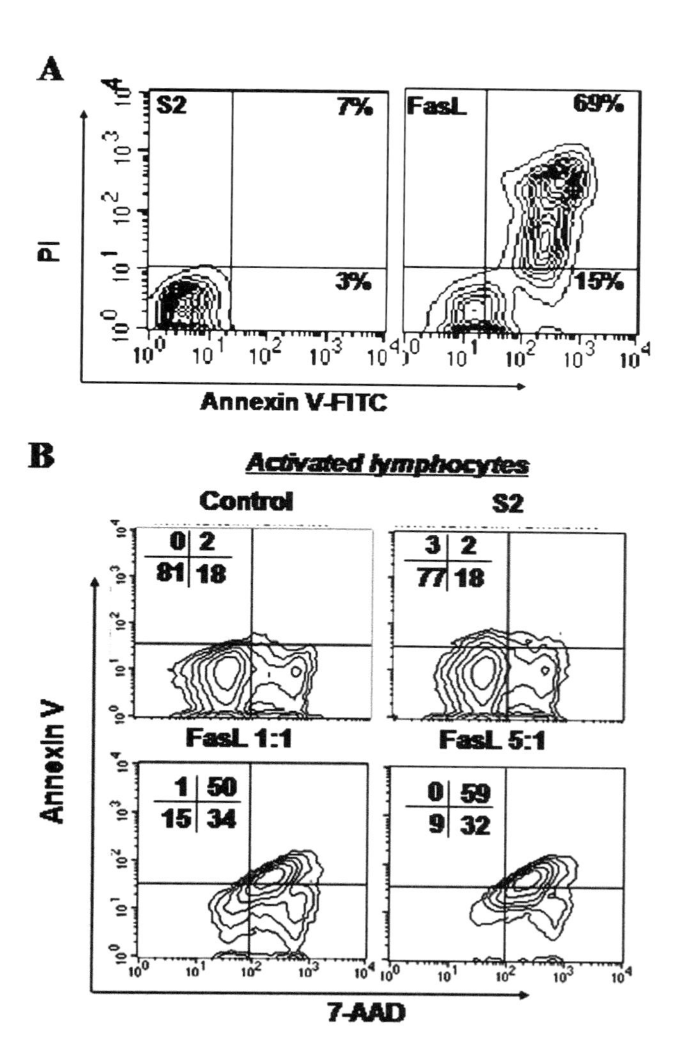

FIGURE 5. SA-FasL induces apoptosis in Fas$^+$ cells and activated lymphocytes. (**A**) Apoptosis in Fas$^+$ mouse A20 lymphoma cell line. A20 cells were incubated overnight with ~100 ng SA-FasL in culture supernatant. Apoptosis was determined by staining with propidium iodide (PI) as a marker for total cell death and annexin V-FITC for apoptotic cell death in flow cytometry. Cells incubated in S2 control supernatant served as negative control. SA-FasL apoptosis of A20 cells. (**B**) Splenocytes displaying SA-FasL induce apoptosis in activated T cells. PVG.1U splenocytes were activated with ConA for 3 days, and T cells were then purified by Ficoll gradient and maintained ininterleukin (IL)-2 supplemented culture for 3 days. T cells were labeled with CFSE and used as target for SA-FasL-displaying irradiated ACI splenocytes at 1:1 and 5:1 effector:target ratios. Apoptosis was determined by flow cytometry using PE-labeled annexin V and 7-AAD after 16 h of incubation. Analyses were performed by gating on CFSE-positive cells. Unmodified ACI splenocytes (Control) and biotinylated ACI splenocytes treated with S2 culture supernatant (S2) served as controls.

on the cell surface. This potent apoptotic activity may be due to the existence of this molecule as tetramers and oligomers, because the naturally occurring trimeric soluble form of FasL has been shown to have minimal apoptotic activity.[33]

Immunomodulation with SA-FasL Inhibits Alloreactive Responses

To test whether SA-FasL-decorated cells can be used as APCs to inhibit alloreactive responses, ACI splenocytes were biotinylated (5 mM), decorated SA-FasL (50–100 ng/10^6 cells), and used as stimulators in a standard mixed lymphocyte reaction (MLR) assay.[31] FasL-decorated splenocytes significantly blocked ($P <.005$) the proliferative response induced by alloantigens (FIG. 5). The inhibition was SA-FasL-specific, because biotinylated splenocytes displaying none or similar levels of core streptavidin (SA) had no inhibitory effect.

Immunomodulation with CD80-SA Generates Auto/alloreactive Responses

To test whether chimeric CD80-SA is functional and can generate T-cell immune responses, we performed MLTR assays using primary tumor cells or established tumor cell line displaying CD80-SA as an APC. Human tumor cell line (HEC-1-A, endometrial cancer) decorated with CD80-SA generated potent proliferative responses in peripheral blood lymphocytes from healthy volunteers (data not shown). The proliferative response was specific to CD80-SA, because unmanipulated tumor cells or biotinylated cells treated with a culture supernatant of S2 cells transfected with a nonfunctional gene, denoted control S2, had no or a minimal proliferative response. Furthermore, the observed proliferative response was more associated with $CD8^+$ T cells than $CD4^+$ T cells (FIG. 6). Similarly, primary tumor cells obtained from

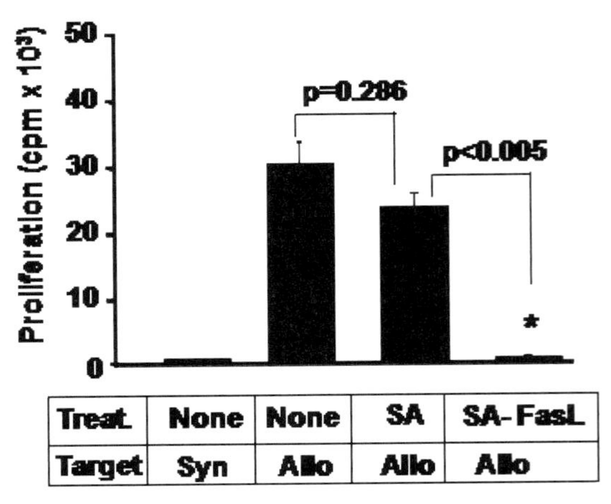

FIGURE 6. Splenocytes expressing SA-FasL block alloreactive responses *in vitro*. (**A**) PVG.1U (RT1u) rat lymph node cells (1× 10^5) were cocultured with irradiated (2,000 cGy) ACI rat splenocytes (None) and biotinylated splenocytes treated with S2 supernatant (S2) or decorated with FasL (SA-FasL). Cultures were maintained for 5 days at 37°C in a 5% CO_2 incubator and pulsed with 1 μCi of [^{3}H] thymidine/well 16–18 h before harvesting. Data shown are from one of three representative experiments.

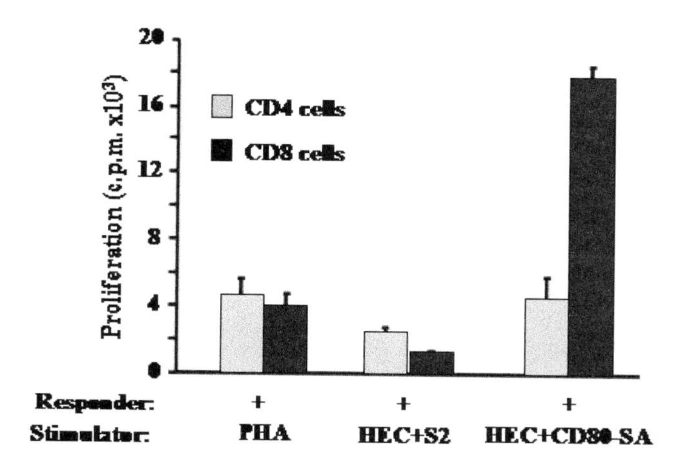

FIGURE 7. CD80-SA induces preferably better proliferation of CD8 cells than CD4 cells. $CD4^+$ and $CD8^+$ T cells were sorted from PBLs harvested from healthy person using flow cytometry. Purified $CD4^+$ and $CD8^+$ (1×10^5) T cells were cocultured for 5–6 days with an equal number of irradiated (10,000 cGy) HEC-1A (a human cell line) biotinylated and treated with S2 supernatant (S2) or decorated with CD80-SA at 37°C in a 5% CO_2 incubator and pulsed with 1 μCi of [^{3}H] thymidine/well 16–18 h before harvesting the cells.

patients and decorated with CD80-SA also generated autologous immune responses (data not shown). These results indicate that CD80-SA generates effective costimulatory signals that lead to T-cell proliferation.

DISCUSSION

Gene therapy is being used extensively for immunomodulation to upregulate the immune responses against tumors and various infectious agents[34,35] or to downregulate the immune responses to treat autoimmune diseases or rejection of foreign grafts.[15–17,20,36] With all the successes of gene therapy in experimental settings, many hurdles exist for its routine application to the clinic for therapeutic purposes. Therefore, there is an imminent need to discover an alternative approach that can allow rapid and long-term display of exogenous proteins on the cell surface for the purpose of immunomodulation. The ProtEx technology reported herein has the potential to be an alternative approach to gene therapy for immunomodulation. This technology involves generation of protein chimeric with streptavidin, biotinylation of biological surfaces, and decoration with chimeric proteins under physiological conditions.

We demonstrated that biotin persists on the cell surface *in vitro* and *in vivo* for weeks. The long-term persistence of biotin on the cell surface is critical to the use of ProtEx as a broad platform technology to display various exogenous proteins of interest on the surface of cells, tissues, and organs for therapeutics. The success of this technology also depends on the decoration of any biological surfaces with exogenous proteins and the display of these proteins for a period of time critical to the manipulation of the immune responses. Using two exogenous proteins, FasL and CD80, with opposite functions in immune responses, we demonstrated that these proteins can be displayed on the surface of various cell types in durable and functional forms. The kinetics of protein turnover on the cell surface were dependent on the protein, cell type, and metabolic stage of the cell and varied from days to weeks *in vivo*. For example, SA-FasL has a $t_{1/2}$ of 3.5 days on the surface of splenocytes *in vivo* and 9.5 days when displayed on the surface of heart vascular endothelium.[32] In contrast to SA-FasL, CD80-SA had significantly slower turnover kinetics with $t_{1/2}$ of >10 days on the surface of splenocytes. Biotinylation and decoration with SA-FasL and CD80-SA showed no toxic effects on the cells and did not interfere with the functions of the cells.

In recent years, there have been reports demonstrating the use of various approaches to display exogenous proteins on the cell surface, but all these approaches were efficient only in the short term (1–3 days) *in vitro*.[17,37–40] Some of these approaches were very cumbersome[38] and others involved random integration of proteins into the cell membrane in a way that may interfere with the anticipated functions of the proteins [37] and therefore may not be suitable for therapeutic applications in clinics.

The ProtEx technology has several unique advantages over other reported protein display approaches. First, due to the inherent physicochemical properties of streptavidin, proteins chimeric with streptavidin possess the ability to exist as tetramers and even higher-order structures.[6] We assume that proteins as tetramers or higher order structures may enhance the ability of the proteins to perform their natural functions. This is especially more important for the molecules that are involved in immunological mechanisms, as coaggregation of receptor/ligand interactions is a critical factor to provide effective signal transduction. [9] In this respect, ProtEx technology not only allows rapid display of exogenous proteins on the cell surface but may also augment their functions. Our observations that the SA-FasL molecule possesses robust apoptotic activity and blocks alloantigen immune responses and that CD80-SA generates T-cell immune responses against tumor auto-/alloantigens support this notion (FIGS. 4A and B; 5A and B; and 6). Second, ProtEx technology is simple and performed in a short time (< 2 h) without detectable cellular toxicity. Third, the amount of chimeric proteins on the cell surface can easily be adjusted by varying the concentration of biotin. This technology also allows the display of more than one chimeric protein simultaneously without any adverse effects. This characteristic of ProtEx technology is very important, as depending on the nature of immunological functions, there is the requirement of more than one protein (tumor immune responses). This technology also provides immediate availability of a functional protein on the cell surface and thus reduces the long period required for gene therapy. In conclusion, ProtEx technology not only allows rapid and durable display of proteins on the cell surface but also possesses simplicity, safety, and efficacy, and therefore this technology can be used as an alternative to gene therapy in the treatment of various immunodeficient diseases in clinical settings.

ACKNOWLEDGMENTS

This work was funded by grants from American Heart Association Postdoctoral Fellowship (0120396B to E.S.Y.), National Institutes of Health (R21 DK61333, R01

AI47864), Juvenile Diabetes Research Foundation (1-2001-328), Kentucky State Lung Cancer Research Grant, the Commonwealth of Kentucky Research Challenge Trust Fund (H.S.), the Frankel Trust for Experimental Bone Marrow Transplantation, and the Daniel M. Soref Charitable Trust (N.A.).

REFERENCES

1. BLAESE, R.M. 1995. Steps toward gene therapy. 2. Cancer and AIDS. Hosp. Pract. **30:** 37.
2. BORDIGNON, C., L.D. NOTARANGELO, N. NOBILI, *et al.* 1995. Gene therapy in peripheral blood lymphocytes and bone marrow for ADA- immunodeficient patients. Science **270:** 470.
3. CAVAZZANA-CALVO, M., S. HACEIN-BEY, G. DE SAINT-BASILE, *et al.* 2000. Gene therapy of severe combined immunodeficiencies. Transfus. Clin. Biol. **7:** 259.
4. ANDERSON, W.F. 1998. Human gene therapy. Nature **392:** 25.
5. GREEN, N.M. 1990. Avidin and streptavidin. Methods Enzymol. **184:** 51.
6. PAHLER, A., W.A. HENDRICKSON, M.A. KOLKS, *et al.* 1987. Characterization and crystallization of core streptavidin. J. Biol. Chem. **262:** 13933.
7. SANO, T., S. VAJDA, C.L. SMITH & C.R. CANTOR. 1997. Engineering subunit association of multisubunit proteins: a dimeric streptavidin. Proc. Natl. Acad. Sci. USA **94:** 6153.
8. REZNIK, G.O., S. VAJDA, T. SANO & C.R. CANTOR. 1998. A streptavidin mutant with altered ligand-binding specificity. Proc. Natl. Acad. Sci. USA **95:** 13525.
9. GRAKOUI, A., S.K. BROMLEY, C. SUMEN, *et al.* 1999. The immunological synapse: a molecular machine controlling T cell activation. Science **285:** 221.
10. GRIFFITH, T.S., T. BRUNNER, S.M. FLETCHER, *et al.* 1995. Fas ligand-induced apoptosis as a mechanism of immune privilege. Science **270:** 1189.
11. WU, J., T. ZHOU, J. ZHANG, *et al.* 1994. Correction of accelerated autoimmune disease by early replacement of the mutated lpr gene with the normal Fas apoptosis gene in the T cells of transgenic MRL-lpr/lpr mice. Proc. Natl. Acad. Sci. USA **91:** 2344.
12. SCHWARTZ, R.H. 1992. Costimulation of T lymphocytes: the role of CD28, CTLA-4, and B7/BB1 in interleukin-2 production and immunotherapy. Cell **71:** 1065.
13. ALLISON, J., H.M. GEORGIOU, A. STRASSER & D.L. VAUX. 1997. Transgenic expression of CD95 ligand on islet beta cells induces a granulocytic infiltration but does not confer immune privilege upon islet allografts. Proc. Natl. Acad. Sci. USA **94:** 3943.
14. GEORGE, J.F., S.D. SWEENEY, J.K. KIRKLIN, *et al.* 1998. An essential role for Fas ligand in transplantation tolerance induced by donor bone marrow. Nat. Med. **4:** 333.
15. LAU, H.T., M. YU, A. FONTANA & C.J. STOECKERT, JR. 1996. Prevention of islet allograft rejection with engineered myoblasts expressing FasL in mice. Science **273:** 109.
16. SWENSON, K.M., B. KE, T. WANG, *et al.* 1998. Fas ligand gene transfer to renal allografts in rats: effects on allograft survival. Transplantation **65:** 155.
17. ZHANG, F., W. G. SCHMIDT, Y. HOU, *et al.* 1992. Spontaneous incorporation of the glycosyl-phosphatidylinositol-linked protein Thy-1 into cell membranes. Proc. Natl. Acad. Sci. USA **89:** 5231.
18. LANG, S., Y. ATARASHI, Y. NISHIOKA, *et al.* 2000. B7.1 on human carcinomas: costimulation of T cells and enhanced tumor-induced T-cell death. Cell Immunol. **201:** 132.
19. MARTIN-FONTECHA, A., M. MORO, M.C. CROSTI, *et al.* 2000. Vaccination with mouse mammary adenocarcinoma cells coexpressing B7-1 (CD80) and B7-2 (CD86) discloses the dominant effect of B7-1 in the induction of antitumor immunity. J. Immunol. **164:** 698.
20. MIN, W.P., R. GORCZYNSKI, X.Y. HUANG, *et al.* 2000. Dendritic cells genetically engineered to express Fas ligand induce donor-specific hyporesponsiveness and prolong allograft survival. J. Immunol. **164:** 161.
21. SANTRA, S., D.H. BAROUCH, A.H. SHARPE & N.L. LETVIN. 2000. B7 co-stimulatory requirements differ for induction of immune responses by DNA, protein and recombinant pox virus vaccination. Eur. J. Immunol. **30:** 2650.

22. ANTONIA, S.J., J. SEIGNE, J. DIAZ, *et al.* 2002. Phase I trial of a B7-1 (CD80) gene modified autologous tumor cell vaccine in combination with systemic interleukin-2 in patients with metastatic renal cell carcinoma. J. Urol. **167:** 1995.
23. BRIONES, J., J.M. TIMMERMAN, D.L. PANICALLI & R. LEVY. 2003. Antitumor immunity after vaccination with B lymphoma cells overexpressing a triad of costimulatory molecules. J.. Natl Cancer Inst. **95:** 548.
24. GILLIGAN, M.G., P. KNOX, S. WEEDON, *et al.* 1998. Adenoviral delivery of B7-1 (CD80) increases the immunogenicity of human ovarian and cervical carcinoma cells. Gene Ther. **5:** 965.
25. GUINN, B.A., M.A. DEBENEDETTE, T.H. WATTS & N.L. BERINSTEIN. 1999. 4-1BBL cooperates with B7-1 and B7-2 in converting a B cell lymphoma cell line into a long-lasting antitumor vaccine. J. Immunol. **162:** 5003.
26. HORIG, H., D.S. LEE, W. CONKRIGHT, *et al.* 2000. Phase I clinical trial of a recombinant canarypoxvirus (ALVAC) vaccine expressing human carcinoembryonic antigen and the B7.1 co-stimulatory molecule. Cancer Immunol. Immunother. **49:** 504.
27. ROSENBERG, S.A. 2001. Progress in human tumour immunology and immunotherapy. Nature **411:** 380.
28. SUMIMOTO, H., K. TANI, Y. NAKAZAKI, *et al.* 1997. GM-CSF and B7-1 (CD80) co-stimulatory signals co-operate in the induction of effective anti-tumor immunity in syngeneic mice. Int. J. Cancer **73:** 556.
29. SINGH, N.P., E.S. YOLCU, D.D. TAYLOR, *et al.* 2003. A novel approach to cancer immunotherapy: tumor cells decorated with CD80 generate effective antitumor immunity. Cancer Res. **63:** 4067.
30. YOLCU, E.S., N. ASKENASY, N.P. SINGH, *et al.* 2002. Cell membrane modification for rapid display of proteins as a novel means of immunomodulation: FasL-decorated cells prevent islet graft rejection. Immunity **17:** 795.
31. SHIRWAN, H., L. BARWARI & D.V. CRAMER. 1997. Rejection of cardiac allografts by T cells expressing a restricted repertoire of T-cell receptor V beta genes. Immunology **90:** 572.
32. ASKENASY, N., E.S. YOLCU, Z. WANG & H. SHIRWAN. 2003. Display of Fas ligand protein on cardiac vasculature as a novel means of regulating allograft rejection. Circulation **107:** 1525.
33. NAGATA, S. & P. GOLSTEIN. 1995. The Fas death factor. Science **267:** 1449.
34. BAROUCH, D.H., S. SANTRA, J. E. SCHMITZ, *et al.* 2000. Control of viremia and prevention of clinical AIDS in rhesus monkeys by cytokine-augmented DNA vaccination. Science **290:** 486.
35. WIERDA, W.G. & T.J. KIPPS. 2000. Gene therapy of hematologic malignancies. Semin. Oncol. **27:** 502.
36. MATSUE, H., K. MATSUE, M. WALTERS, *et al.* 1999. Induction of antigen-specific immunosuppression by CD95L cDNA-transfected 'killer' dendritic cells. Nat. Med. **5:** 930.
37. CHEN, A., G. ZHENG & M.L. TYKOCINSKI. 2000. Hierarchical costimulator thresholds for distinct immune responses: application of a novel two-step Fc fusion protein transfer method. J. Immunol. **164:** 705.
38. MCHUGH, R.S., S.N. AHMED, Y.C. WANG, *et al.* 1995. Construction, purification, and functional incorporation on tumor cells of glycolipid-anchored human B7-1 (CD80). Proc. Natl. Acad. Sci. USA **92:** 8059.
39. VAN BROEKHOVEN, C.L., C.R. PARISH, G. VASSILIOU & J.G. ALTIN. 2000. Engrafting costimulator molecules onto tumor cell surfaces with chelator lipids: a potentially convenient approach in cancer vaccine development. J. Immunol. **164:** 2433.
40. WAHLSTEN, J.L., C.D. MILLS & S. RAMAKRISHNAN. 1998. Antitumor response elicited by a superantigen-transmembrane sequence fusion protein anchored onto tumor cells. J. Immunol. **161:** 6761.

Development and Evaluation of a Suppository Formulation Containing *Lactobacillus* and Its Application in Vaginal Diseases

VINITA V. KALE, RASHMI V. TRIVEDI, SANJAY P. WATE, AND KISHOR P. BHUSARI

Nagpur College of Pharmacy, Wanadongri, Hingna Road, Nagpur-441110, Maharashtra, India

ABSTRACT: *Lactobacillus* has long been considered the protective flora in the vagina that displaces and kills vaginal pathogens. Lactic acid, H_2O_2, and antibacterial agents such as lactocin and bacitracin produced by *Lactobacillus* act against the vaginal pathogens. The first objective of this research was to develop a local application pharmaceutical formulation of a vaginal suppository containing lyophilized culture of *Lactobacillus*. The second objective was to establish its *in vivo* performance by developing *in vitro* methods of evaluation. Lyophilized culture of *Lactobacillus sporogenes* was selected for this study. Three formulations of the suppositories were prepared by the molding method. Formulations I, II, and III contained cocoa butter, glycerinated gelatin, and PEG 1000 base, respectively. The prepared suppositories were characterized for physical properties. Assembly to simulate the application site was designed. Methods to evaluate the viability, production of lactic acid, and H_2O_2 produced by the released *Lactobacillus* at the application site were developed and the antagonistic activity was demonstrated. From the physical characteristics of the suppository formulations, the glycerinated gelatin suppository (formulation II) containing lyophilized *Lactobacillus* was found to be satisfactory. The developed assembly was satisfactory in simulating the application site. The *Lactobacillus* released was viable and exhibited the production of lactic acid, hydrogen peroxide, and antagonistic activity against the uropathogen. The suppository formulation containing *Lactobacillus* and the methods of its evaluation were successfully developed in this research work and have several applications in the vaginal diseases of women.

KEYWORDS: lactobacillus; probiotic; bacterial vaginosis; lactic acid; hydrogen peroxide; antagonistic activity

INTRODUCTION

The vagina represents a dynamic ecosystem mainly dominated by certain species of *Lactobacillus* that exert a significant influence on the microbiology of the

Address for correspondence: Vinita V. Kale, Nagpur College of Pharmacy, Wanadongri, Hingna Road, Nagpur-441110, Maharashtra, India. Voice: 07104-236352 or 235084; fax: 07104-235084.
kvinita@rediffmail.com

**Ann. N.Y. Acad. Sci. 1056: 359–365 (2005). © 2005 New York Academy of Sciences.
doi: 10.1196/annals.1352.017**

vagina.[1,2] However, this flora is often compromised by a series of endogenous or exogenous factors such as diabetes, use of contraceptive devices, unprotected sex, and oral antibiotics, leading to serious complications in the form of bacterial vaginosis (BV) or vaginal cancer.[3,4] The dominant species of *Lactobacillus* suppresses the growth of other pathogenic bacteria in the vagina through the production of lactic acid, hydrogen peroxide (H_2O_2), and antimicrobial substances such as lactocin and bacitracin.[5]

Lactic acid production, which maintains the vaginal pH at about 4.5 or less, is considered the major mode of action of *Lactobacillus* against vaginal infections.[6] Besides lactic acid, H_2O_2 inhibits the growth of vaginal pathogens, particularly those that lack or have low levels of catalase or peroxidase enzymes.[7,8] Bacitracin and lactocin are proteins that have bactericidal activity against the BV pathogen *G. vaginalis* and uropathogen *Escherichia coli.*[9,10] To prevent vaginal infections, it is important to keep the vagina colonized by *Lactobacillus,* and the local application of products that contain *Lactobacillus* is likely to reduce vaginal infections as confirmed by clinical studies.[11,12]

Currently the products for vaginal delivery of *Lactobacillus* include dairy products (yogurt, acidophilus milk, etc) and commercially available powders and tablets containing *Lactobacillus*. These products often have poor patient compliance for several reasons such as irritation, discomfort, and leakage at the application site.[13,14] In that case, a suppository dosage form containing *Lactobacillus* would be a better option. However, only a few reports are available on the formulation of *Lactobacillus* in the form of a vaginal suppository. [15] Limited information is available in the literature on the effects of a suppository base, the breaking strength, or other processing parameters of a suppository on *in vitro* release of *Lactobacillus* from the formulation and its *in vivo* viability after release.

A vaginal suppository has certain advantages: dose uniformity can be maintained, insertion into the vagina without irritation is possible, and a large volume of dissolution fluid is not required for the release of active substance. Consequently, the current investigation was undertaken with two objectives: to develop local application pharmaceutical formulation in the form of a vaginal suppository containing *Lactobacillus* in lyophilized form; to establish its *in vivo* performance by developing *in vitro* methods of evaluation.

MATERIAL AND METHODS

The lactic acid bacillus (*Lactobacillus sporogenes*) culture was selected from commercially available products containing different *Lactobacillus* species. The criteria in the selection were that the *Lactobacillus* should produce lactic acid and H_2O_2 and should show antagonistic activity against the uropathogen *E. coli 1 2004*. The lyophilized culture of lactic acid bacillus was obtained from Uni-Sankyo India Ltd., Chiplun, India. The uropathogen *E. coli 1 2004* was obtained from Kakde Pathology Lab, Nagpur, India. The deMan Rogosa Sharpe (MRS) medium, used for cultivation of *Lactobacillus* and uropathogen, was obtained from Himedia Laboratories Pvt. Ltd., Mumbai, India. Suppository base materials such as glycerin, gelatin PEG 1000, and cocoa butter were purchased from S.D. Fine Chemical, Vadodra, India. For

TABLE 1. Formulation for 10 suppositories of lactic acid bacillus

Formulation	Base	Lactic acid bacillus (lyophilized)
I	Cocoa butter 13.5 g	9 g
II	Glycerin 7.5 ml Gelatin 17 g Purified water 2.5 ml	9 g
III	Polyethylene glycol 14 g (PEG) 1000	9 g

determination of H_2O_2, tetramethylbenzidine and peroxidase were obtained from SIGMA. All other chemicals and reagents used were of an analytic grade.

Preparation of Suppository

Three different compositions of suppositories, each containing lyophilized *Lactobacillus* culture, were prepared by taking the appropriate quantity of the base as given in TABLE 1. The three bases used were cocoa butter, glycerinated gelatin, and polyethylene glycol (PEG) in formulations I, II, and III, respectively. The suppositories were prepared by the fusion method. [16] The displacement value of the lyophilized culture of *Lactobacillus* was determined in the respective bases. Formulations I and III were prepared by melting the appropriate bases in a water bath at 70°C. The lyophilized culture of *Lactobacillus* was added to the melted base when cooled to 45°C and thoroughly mixed. The mixed mass was immediately poured into previously lubricated metal molds (of six suppositories) and cooled. While preparing formulation II, glycerin and gelatin were melted together and to the molten mass was added the suspension of lyophilized *Lactobacillus* in water when cooled to 45°C and mixed. The suppositories were prepared as just described. All the suppository formulations were refrigerated for storage.

Physical Characterization of the Suppository

The following parameters were considered for the physical characterization of suppository formulations I, II, and III.

Uniformity of Weight. From each batch of formulation, 20 suppositories were weighed individually. The average weight and percentage deviation values were calculated.

Surface Texture. The surface texture of the suppository was observed visually. Formation of smooth or rough and uniform surface was recorded.

Breaking Strength (Hardness) Test. The test was performed on the cylindrical portion obtained by cutting the thick slices of the middle hard portion of the suppository using the Monsanto hardness tester (Campbell Electronics Ltd., India). Breaking strength values were calculated by an average of six suppositories.

In Vitro *Release Time Test.* The United States Pharmacopeia (USP) rotating basket dissolution apparatus was used for *in vitro* release time testing. The dissolution medium was 250 ml of phosphate buffer of pH 6.2, whose temperature was maintained at 37°C. The suppository was placed in a metal basket, which was spun at

50 rpm. The time taken for the sample to melt, that is, disperse completely in the surrounding fluid, was determined.

Development of Simulated Application Site and Methods to Evaluate Production of Lactic Acid, H_2O_2, and Antagonistic Activity

A simple assembly was designed for the release studies based on a method reported in the literature.[17] The assembly (condenser) was fabricated. The diameter of the inner tube of the condenser was 1 cm. All the components of the assembly were sterilized before use, and aseptic conditions were maintained throughout the working period. Vaginal physiology was simulated by means of a cellophane tube 5 cm in length and 2 mm in thickness. The cellophane tube was placed inside the condenser at the lower end. The condenser was positioned at a 140° angle from the surface. Each end of the tube was kept open. Water maintained at 37°C was circulated through the condenser. The test suppository was inserted into the cellophane tube and 5 ml of MRS broth medium, corrected to pH 4.5 with 85% w/v lactic acid, was dropped into the cellophane tube over the suppository. A container was placed under the condenser to collect the discharged (melted) liquid. The experiment was continued for 2 hours to ensure complete melting of the suppository. The collected mass was mixed thoroughly, and the mixed mass that was maintained in a liquified condition at 40°C in a water bath was used to determine whether the released *Lactobacillus* produced lactic acid and H_2O_2 and showed antagonistic activity.

Lactic Acid Production. A change in pH was considered an indicator in determining lactic acid production by *Lactobacillus*. About 1.5 ml of mixed mass was inoculated into MRS broth at its formulated pH of 6.2 and incubated for 24 hours at mild CO_2 tension at 37°C. The pH was measured with a pH meter (Elico digital pH meter, India) using a calibrated glass electrode.

H_2O_2 Production. The qualitative determination of H_2O_2 produced by *Lactobacillus* was demonstrated using a method reported in the literature.[2] A loopful of mixed mass was spotted on the MRS agar plate containing 25 mg/100 ml of tetramethylbenzidine as an indicator and 1 mg/100 ml peroxidase as an enzyme. Petri dishes were incubated at 37°C for 24 hours under mild CO_2 tension. Plates were then exposed to ambient air at room temperature, and blue-colored colonies were observed.

Antagonistic Activity Test. The antagonistic activity of the *Lactobacillus* against the uropathogen *E. coli 1 2004* was studied using a method described in the literature with slight modification.[18] The mixed mass (1 ml) was transferred in MRS broth medium and incubated at 37°C in a mild CO_2 condition for 24 hours.

The seeded plates were prepared by transferring 0.2 ml of culture of pathogenic bacteria *E. coli 1 2004* (OD 620 nm as 0.45– 0.55) in a Petri dish. To it was added 30 ml of molten MRS agar medium cooled to 45°C. The contents were mixed gently by rotating the Petri dish. After the solidification of medium the Petri dishes were incubated at 37°C for 2 hours and punched aseptically using a sterile well cutter. An aliquot (0.3 ml) of the *Lactobacillus* growth of mixed mass was poured aseptically into the well. The Petri dish was incubated at 37°C for 24 hours. Formation of the zone of inhibition around the well was noted.

In Vitro *Adherence Test*. The cellophane tube was removed from the assembly after 2 hours and cut into pieces. One piece was placed in a Petri dish followed by

TABLE 2. Physical properties of suppository

Parameter	Formulation		
	I	II	III
Content uniformity	2 ± 0.52	2 ± 0.21	1.98 ± 0.087
Breaking strength	–	1.4–1.5 kg/cm^2	1.9–2.1 kg/cm^2
Surface texture	Rough with surface fracture	Uniform with smooth surface	Uniform with dry surface
In vitro release time test	8 min	22 min	15 min

15 ml of molten MRS agar cooled to 45°C. After solidification the Petri dish was incubated at 37°C for 24 hours. The formation of typical colonies of *Lactobacillus* over the cellophane piece was observed.

RESULTS AND DISCUSSION

Physical Characteristics of the Suppositories

The physical characteristics of suppository formulations I, II, and III are given in TABLE 2. The weight variation of formulation II was within the acceptable limit of 100% ± 5%. The cocoa butter suppositories (formulation I) showed a deviation of more than 20%. The PEG 1000 suppository was smaller than the other formulation suppositories, perhaps because of the contractility of the PEG 1000 base after solidification.

The surface of cocoa butter suppositories was rough with several fractures, whereas the surface of PEG 1000 suppositories was smooth but appeared waxy. The glycerinated gelatin suppository had a uniform and smooth surface.

The breaking strength of the suppository was between 1.3 and 1.4 kg/cm^2 for formulation II and between 1.9 and 2.1 kg/cm^2 for formulation III. The cocoa butter suppositories (formulation I) were soft, and hence their breaking strength could not be determined. Cocoa butter has a melting point around 32°C and hence might have become soft at room temperature. The PEG 1000 suppositories were harder than those of gelatin perhaps because the PEG base is devoid of water and remains solid even at temperatures of 37–40°C.

The *in vitro* release time test measures the liquefaction time of the suppository. The *in vitro* release time test revealed that the time required to melt the whole suppository was minimal for formulation I and maximal for formulation II. This difference was mainly due to the melting characteristics of the three bases. Cocoa butter melts at body temperature (melting range 33.5–35°C).[19] The melting temperature of PEG 1000 is slightly higher than body temperature, requiring more time for release of active substance compared to cocoa butter.[16] On the contrary, glycerinated gelatin does not melt at body temperature, but slowly dissolves in the secretions of the body cavity.[20]

From the foregoing findings, the physical characteristics of formulation II containing lyophilized *Lactobacillus* in a glycerogelatin base were most satisfactory and were selected for further studies.

Evaluation of Production of Lactic Acid, H_2O_2, and Antagonistic Activity

Sufficient growth of the *Lactobacillus* (10^7 colony-forming units/ml) was observed when grown on a standard MRS medium plate. Colony characteristics and gram staining confirmed the presence of *Lactobacillus*. This indicates that the viability of the *Lactobacillus* was not affected during preparation of the formulation.

The pH of the MRS broth that was initially at 6.2 changed to 3.9 ± 0.52, perhaps due to the formation of lactic acid in the medium. Production of lactic acid by the probiotic is the fundamental requirement for treating vaginal infections. Few investigators have demonstrated that a pH value of less than 4.5 was indicative of the presence of *Lactobacillus* in a range of 10^8 bacilli/ml. Also, it is known that a higher vaginal pH (4.7–5.9) characterizes the vagina of women with BV, suggesting that the *Lactobacillus* released from the formulation would acidify the environment that can inhibit the growth of other organisms.

A qualitative determination of H_2O_2 produced by the *Lactobacillus* was demonstrated as blue-colored colonies that appeared on the plate. The result demonstrated that acid pH was required for H_2O_2 production, which coincided with the findings of other workers.[21]

In the antagonistic test, *Lactobacillus* showed a moderate antibacterial effect (zone size between 3 and 5 mm) against the uropathogen *E. coli 1 2004*. The inhibition could be the combined effect of acidification of the medium and production of the antimicrobial substance by the *Lactobacillus*.

Adherence Test

As suggested by a few investigators,[5] the biocell mass in the vaginal environment was responsible for the pH changes, that is, the quantity of acid produced. This means that to maintain the healthy vaginal ecosystem (of low pH), organic acids must be produced, and this was possible only if active growth of the *Lactobacillus* was present in the vagina. Active growth of bacillus was possible only when it adhered to the wall of the vagina. The *in vitro* adherence test developed in this research indicates that *Lactobacillus* released from the dosage form adhered to the cellophane membrane. Further gram staining confirmed its presence.

CONCLUSION

The dosage formulation containing lyophilized *L. sporogenes* appears to be a good candidate for probiotic prophylaxis and treatment of vaginal infections. The developed assembly was satisfactory in simulating the application site. The suppository completely dissolved at the application site. The viability of *L. sporogenes* was not affected during preparation of the suppository. The *Lactobacillus* released exhibited the production of lactic acid, hydrogen peroxide, and antagonistic activity against the uropathogen. Thus, the suppository formulation containing *Lactobacillus* and the methods of its evaluation developed in this research work may be beneficial in preventing bacterial vaginosis. It may also exert an antitumor effect and prevent the acquisition of HIV and other sexually transmitted diseases. The *Lactobacillus* may modulate the immune response by interfering with the inflammatory cascade that leads to preterm birth. Similarly, it can degrade lipids and enhance cytokine levels that might promote embryo development.

REFERENCES

1. MARDH, P.A. 1991. The vaginal ecosystem. Am. J. Obstet. Gynecol. **165:** 1163–1168.
2. ANGELES-LOPEZ, M., E.G.C RAMOS & C.A. SANTIAGO. 2001. Hydrogen peroxide production and resistance to nonoxinol-9 in *Lactobacillus* spp. isolated from the reproductive age women. Microbiolgia **43:** 171–176.
3. ESCHENBACH, D.A. 1993. History and review of bacterial vaginosis. Am. J. Obstet. Gynecol. **169:** 441–445.
4. BAUER, G. 2001. *Lactobacillus* mediated control of vaginal cancer through specific reactive oxygen species interactions. Med. Hypotheses **57:** 252–257.
5. ARONTCHEVA, A. *et al.* 2001. Defence factor of vaginal lactobacilli. Am. J. Obstet. Gyncol. **185:** 375–379.
6. BOSKEY, E.R. *et al.* 1999. Acid production by vaginal flora *in vitro* is consistent with the rate and extent of vaginal acidification. Infect. Immun. **67:** 5170–5175.
7. KLEBANOFF, S.J. *et al.* 1991. Control of the microbial flora of the vagina by H_2O_2 generating lactobacilli. J. Infect. Dis. **165:** 19–25.
8. ESCHANBACH, D.A. *et al.* 1989. Prevalence of hydrogen peroxide producing lactobacillus specie in normal women and women with bacterial vaginosis. J. Clin. Microbiol. **27:** 251–256.
9. KAEHAMMER, T.R. 1988. Bacitracin of lactic acid bacteria. Biochmie **70:** 337–349.
10. MCGROARTY, J.A & G. REID. 1988. Detection of lactobacillus substance that inhibits *Escherichia coli*. Can. J. Microbiol. **34:** 974–978.
11. HALLEN, A., C. JARSTRAN & C. PAHLSON. 1992. Treatment of bacterial vaginosis with lactobacilli. Sex Transm. Dis. **19:** 146–148.
12. SHALEV, E. *et al.* 1996. Ingestion of yogurt containing *Lactobacillus acidophilus* compared with pasteurised yogurt as prophylaxis for recurrent Candidial vaginitis and bacterial vaginosis. Arch. Fam. Med. **5:** 593–596.
13. DESHPANDE, A.A., C.T. RHODES & M. DAHISH. 1992. Intravginal drug delivery. Drug Dev. Ind. Pharm. **18:** 1225–1279.
14. RICHARDSON, J.L. & L. ILLUM. 1992. Routes of delivery: case studies. The vaginal route of peptide and protein drug delivery. Adv. Drug Delv. Rev. **8:** 341–366.
15. LITSCHGI, M.S. *et al.* 1980. Effectiveness of *Lactobacillus* vaccine on *Trichomonos* infections in women. Fortschr. Med. **98:** 1624–1627.
16. LAHMAN, L., H. LIBERMAN & J.L. KANIZ. 1975. Theory and Practice of Industrial Pharmacy. 3rd Ed. Vergese Publishing House. Mumbai.
17. SETNIKAR, I. & S. FANTELI. 1962. Liquefaction time of rectal suppository. J. Pharm. Sci. **51:** 566–578.
18. GUPTA, P.K & B.K. MITAL. 1996. Antagonistic activity of *Lactobacillus acidophilus* fermented milk against different pathogenic bacteria. Exp. Biol. **34:** 1245–1247.
19. AHMED, M.G. *et al.* 2000. Formulation, release characteristics and evaluation of nimusilide suppositories. Indian J. Pharm. Sci. **62:** 196–199.
20. CARTER, S.J. 1975. Dispensing for Pharmaceutical Students, 3rd Ed. Pitman Medical. London.
21. FONTAINE, E.A & D. TAYLOR-ROBINSON. 1990. Comparison of quantitative and qualitative methods of detecting hydrogen peroxide produced by human vaginal strins of *Lactobacillus*. J. Appl. Bacteriol. **69:** 326–331.

Novel Bacterial Delivery System with Attenuated *Salmonella typhimurium* Carrying Plasmid Encoding *Mtb* Antigen 85A for Mucosal Immunization

Establishment of Proof of Principle in TB Mouse Model

SHREEMANTA K. PARIDA,[a,b,c] KRIS HUYGEN,[b] BERNHARD RYFFEL,[d] AND TRINAD CHAKRABORTY[a]

[a]*Institute of Medical Microbiology, Justus-Liebig University, Giessen, Germany*

[b]*Pasteur Institute of Brussels, Brussels, Belgium*

[c]*Armauer Hansen Research Institute, Addis Ababa, Ethiopia*

[d]*Institute for Infectious Diseases and Molecular Medicine, Cape Town, South Africa*

ABSTRACT: Tuberculosis (TB), the leading killer of young adults worldwide, newly affects one person every second and kills one in every 15 seconds. The recent increase of TB in developing countries has been exacerbated by many causes including pandemic HIV, war and political instability, drug resistance, and increasing poverty. Genetic immunization has emerged with tremendous potential in vaccination against TB with success in animal models with naked DNA encoding different genes such as Ag85A, Pst3, and hsp65. However, there are shortcomings in translating this success into reality in human clinical trials due to limitations at the level of delivery, quality, and quantity of DNA to be administered, which can be circumvented by using an attenuated bacteria delivery system for transgene vaccination for mucosal immunization targeting the inductive sites of the immune system. We compare this novel delivery system using an attenuated Salmonella Δ*aroA* strain through a mucosal route with classic intramuscular DNA delivery using a potential protective antigen, Ag85A, of *Mycobacterium tuberculosis* in a mouse infection virulent challenge model. We show an immune response and superior protection in the mice at the level of the lungs as well as the spleen against a virulent challenge after intranasal immunization by recombinant *Salmonella typhimurium* carrying a eukaryotic expression plasmid encoding Ag85A rather than the classic DNA immunization and at par with the protection conferred by BCG. This study establishes the proof of principle of this system for further exploitation of this platform for vaccine development, which is being pursued for postexposure vaccine development for TB.

Address for correspondence: Dr. Shreemanta K. Parida, Armauer Hansen Research Institute, PO Box 1005, Addis Ababa, Ethiopia. Voice: +251 1 21 37 62; fax: + 251 1 21 15 63. skparida@onetel.com

Ann. N.Y. Acad. Sci. 1056: 366–378 (2005). © 2005 New York Academy of Sciences. doi: 10.1196/annals.1352.030

KEYWORDS: DNA vaccine; *Mycobacterium tuberculosis;* Ag85A; bacterial delivery; transgenic vaccination; attenuated *Salmonella typhimurium*; mucosal immunization

INTRODUCTION

Tuberculosis (TB), the leading killer of young adults worldwide due to a single infectious agent, newly affects one person every second and kills one in every 15 seconds. According to the latest WHO report quoting epidemiologic data for 2003, TB causes about 8.8 million new cases annually with about 2 million deaths worldwide. In Africa, the number of cases per 100,000 population is the highest at 350 cases with 83 deaths.[1] The recent increase in TB in developing countries has been exacerbated by many causes including pandemic HIV, war and political instability, drug resistance, and increasing poverty. HIV causes a 6% annual increase in the number of TB cases across sub-Saharan Africa that bears the brunt of the HIV-fueled TB epidemic. Despite available effective chemotherapy and BCG vaccine, TB poses a global health emergency. Directly observed chemotherapy (DOTs) with multiple drugs is effective, but only 77% of the global population had access to DOTs by the end of 2003, and it has so far reached about 70% coverage despite all efforts from WHO and other international agencies.[1] BCG vaccine has been safe for human use since 1921. It has been effective in controlling childhood TB including TB meningitis, but it does not confer protection against adult TB of endogenous or exogenous infection. Several factors have been attributed to the failure of BCG to be protective in adults including differences in BCG vaccine strains, differences in genetic predisposition within and between human populations, differences in virulence between *Mycobacterium tuberculosis (Mtb)* strains, interference with or masking of protection by environmental mycobacterial infection, as well as the result of cumulative deletion of genes (16 deletions comprising 129 open reading frames [ORFs] in BCG vaccine strains).

Development of an effective vaccine still remains the most promising interventions means of controlling infectious diseases. In recent years, considerable progress in understanding the immune mechanisms involved in pathology and protection has led to numerous innovations in vaccine design. An ideal vaccine should be inexpensive, safe and easy to administer through natural routes, and effective. Genetic immunization has emerged as a technique with tremendous potential in vaccination against infectious diseases. Indeed, for *Mtb*, vaccination with naked DNA encoding different genes such as Ag85A, Pst3, and hsp-65 has been reported to be highly efficacious in a mouse infection model.[2–5] Nevertheless, limitations exist at the level of delivery, quality, and quantity of DNA to be administered. This can be overcome by the use of a bacterial delivery system using live attenuated bacteria as carriers of the plasmid DNA carrying the gene(s) of interest. Live attenuated bacterial carriers expressing heterologous antigens are attractive vehicles for the mucosal delivery of vaccines. An attenuated strain of *Salmonella typhimurium (Stym)* encoding truncated forms of ActA or listeriolysin (two virulence factors of *Listeria monocytogenes*) has been a highly effective vehicle for oral delivery of DNA vaccines, eliciting responses in all three effector compartments of the immune system, namely, cytotoxic $CD8^+$ T cells, $CD4^+$ T cells, and antibodies.[6] In many studies this approach in different disease models has been demonstrated to be effective in several rodent models of infectious diseases and tumors.[7]

The current study was aimed at examining this delivery strategy with DNA encoding Ag85A for evaluating the protective immune response in an established animal model of a disease of great public health importance, that is, tuberculosis, and comparing it with the previously well-characterized antigen with the naked DNA approach alone to establish proof of principle of this approach. Targeted bacterial delivery systems are versatile, simple, and potentially economical to produce. The same technology can be applied to vaccine development of other major infectious diseases such as malaria, HIV, and toxoplasmosis as well as cancer.

MATERIAL AND METHODS

Strains and Plasmids. Attenuated *S. typhimurium* Δ*aro*A strain SL7207 (*Stym)* 2337-65 derivative *hisG*46, Δ407 [*aroA*::Tn10{Tc-s}] was kindly provided by Prof. Stocker, Stanford University, California, USA. The bacterial strains were routinely grown at 37°C in Luria-Bertini broth or agar (Difco, Germany) supplemented with 100 μg/ml of ampicillin, when required. Plasmid pCMVβ was obtained from Clontech, and Vical plasmid VR1020 was obtained from Vical Inc., San Diego, CA, USA. Plasmids pCMVβ and VR1020 had ampicillin and kanamycin resistance genes, respectively, as selection markers. These plasmids were maintained in *Escherichia coli* DH5α strain.

Amplification of the Ag85A Gene from Mtb and Construction of the Recombinant Plasmid. A clinical specimen of *Mtb* was obtained from Lowenstein-Jensen slant culture, and genomic DNA was then extracted as previously described.[2] According to the complete DNA sequence of Ag85A from the *M. tuberculosis* complete genome, a set of custom-made primers containing restriction site *Not*I at the 5′ end were designed (sense primer, ATAAGAATGCGGCCGCCATGCTGGTCGGCGCC-GTCGGTGGC; antisense primer, ATAAGAATGCGGCCGCTTAGGCGCCCTG-GGGCGCGGGCCC). The designed primers were used to amplify Ag85A in a total volume of 100 μl under the following conditions: at 94°C for 5 min, then 25 cycles at 94°C for 1 min, 55°C for 30 s, and 72°C for 1 min, 30 s, followed by 5 min at 72°C. The polymerase chain reaction (PCR) product was then analyzed on 0.8% agarose gel stained with ethidium bromide.

PCR products were separated using a QIAquick gel extraction kit (QIAGEN, Germany). Purified Ag85A DNA fragments were subcloned into TOPO cloning vector (PCR vector 2.1; Invitrogen, Germany). Recombinant clones were characterized for positive clones by PCR for amplification of the insert, and two of the positive clones were sequenced for the Ag85A by ALF Express. The clones were also digested with *Eco*RI for further characterization. Fragments of *Not*I-digested PCR2.1–Ag85A were cloned into the *Not*I site of the eukaryotic expression vector pCMVβ through a series of enzyme digestions and ligation reactions (after digesting with *Not*I followed by *Eco*RV to separate β-gal from the plasmid before purification of fragments and ligation). FIGURE 1 depicts the genetic map of pCMVCβ::Ag85A. The recombinant pCMVβ-Ag85A was then confirmed by PCR with a set of designed primers for the vector pCMVβ as well as insert Ag85A and restriction enzyme digestion. The insert was sequenced using a new set of primers for Ag85A labeled with

Cy5 at the 5′ end for sequencing pCMVβ with the flanking region of the insert to determine the correct orientation of the insert.

Construction of Recombinant Attenuated S. typhimurium *Carrying the* M. tuberculosis *Ag85A Gene.* Competent *S. typhimurium ΔaroA* SL7207 was prepared, and the extracted recombinant plasmid pCMVβ::Ag85A was used to transform the final host strain SL7207 by electroporation. The attenuated *Stym* carrying pCMVβ plasmid encoding the Ag85A gene was cultured in LB medium with ampicillin and characterized by PCR and restriction digestion analysis. The recombinant vector was maintained in *E. coli,* and the recombinant was checked by PCR and nucleotide sequencing for the proper orientation of the gene and the correct reading frame.

In Vitro *Transfection.* The recombinant pCMVβ::Ag85A was transfected into Hela cells to detect protein expression. Hela cells were cultured in 24-well plates with sterile glass coverslips at 37°C in Dulbecco's modified Eagle's medium (DMEM) supplemented with 10% FBS (Sigma, Taufkirchen, Germany), ampicillin, and gentamycin. Before transfection, cells were washed with a new plain DMEM medium, and a mixture of pCMVβ::Ag85A and FuGene transfection reagent (Boehringer Mannheim, Germany) was added to the cells. Thirty-six hours after transfection, coverslips were removed from the wells and washed in phosphate-buffered saline (PBS) solution followed by fixation with 3.7% formaldehyde in PBS. The coverslips with the cells were first treated with 2% Triton X-100 in PBS for 1 min at room temperature to permeabilize the cell membrane before staining by specific monoclonal antibodies to Ag85A (TD-17; *source*: Dr. Kris Huygen, Pasteur Institute of Brussels, Brussels, Belgium) for 30 min at room temperature followed by washing. The coverslips were stained with goat–anti-mouse IgG coupled to FITC for 30 min at room temperature before washing. The coverslips were them mounted using mounting medium (BioRad, Germany) and sealed on the slides before being examined under a confocal microscope (Leica, Germany).

Studying the Immune Response in a Mouse Model. The recombinant *Salmonella* has been examined for immunogenicity and efficacy in a stringent mouse model of C57BL/6 (haplotype H-2^b) at the Pasteur Institute of Brussels by the oral as well as the intranasal route. The immune response in mice was evaluated and compared to the naked DNA approach alone with appropriate controls.

Immunization. C57BL/6 mice 6–8 weeks old were immunized orally or intranasally with 10^8 and 10^7 recombinant *Stym,* respectively, at an interval of 3 weeks three times. *Salmonella* with an empty cassette was used as an appropriate control. V1JN-ns.tPA plasmid encoding Ag85A (from Dr. Kris Huygen) was given intramuscularly into the quadriceps in both hind legs (2×50 μg) at a dose of 100 μg every 3 weeks three times for comparison with the bacterial delivery system.

Humoral Immune Response. Serum samples of the marked mice were obtained at day −1, 20, 41, and 62 days and tested for specific antibodies against the native Ag85 in ELISA.

Cell-Mediated Immune Response. Immune analysis was performed in three mice from each group at a time point of weeks 3, 6, and 9, at 3 weeks after each immunization. Spleen cells from the immunized mice were restimulated with antigens (purified protein derivative of BCG, PPD; native Ag85 antigen complex, Ag85; native culture filtrate antigens, CF; and pokeweed mitogen, PWM *in vitro*). Supernatants were collected after 24 hours for interleukin (IL)-2 and after 72 hours for gamma interferon (IFN-γ). IFN-γ in the supernatants was measured using ELISA.

Pulmonary Challenge with Mtb. Recombinant luciferase reporter *M. tuberculosis H37RV* wild-type (made available by Prof. Douglas Young and colleagues) was used to challenge immunized mice 8 weeks after the third immunization by the intravenous route through the lateral tail vein with 10^6 colony-forming units (CFUs) of *Mtb.* Mice were sacrificed 4 and 12 weeks after challenge to determine the bacterial load in the tissue. Lungs and spleen from individual animals (4–6 in each group) were dissected and homogenized in PBS supplemented with penicillin (1 µl/ml) and amphotericin B (fungizone 2 µl/ml), and serial threefold dilutions were plated on Middlebrook 7H11 agar supplemented with OADC. Plates were incubated at 37°C in sealed plastic bags, and the number of CFUs was determined after 4–5 weeks. Results are presented as the mean $\log_{10}$ CFUs per lung or spleen (SEM).

The number of bioluminescent organisms was determined by assessing the bioluminescence of the homogenized organs after lysing the erythrocytes in relative light units (RLUs) by a Turner Design 20/20 luminometer and 1% *n*-decyl-aldehyde in ethanol as substrate. A previous report had established RLU measurement as a convenient and fast alternative method to time-consuming CFU enumeration, which had the potential problems of fungal contamination during the long incubation period.[8] We performed several control experiments to assess the correlation of CFUs with RLUs and had established that 1 RLU was equivalent to 10^4 CFUs.

Statistical Analysis. The bacterial load, determined by RLUs, was individually converted to a CFU load per organ, taking our standard of 1 RLU = 10^4 CFUs and converting that to a $\log_{10}$ value. For statistical analysis, the individual $\log_{10}$ values in each group were compared by analysis of variance. *P* values for the overall vaccine effect on the lungs at weeks 4 and 12 are 0.025 and <0.0001 and 0.019 and <0.0001 for the spleen, respectively.

RESULTS

The feasibility and efficacy of the bacterial delivery strategy with DNA encoding a well-characterized protective antigen (Ag85A) from *Mtb* were evaluated in this study using an attenuated *aro*A mutant of *S. typhimurium.*

Expression of Ag85A by the Recombinant Vector. The recombinant plasmid pCMVβ::Ag85A was constructed as detailed in the Material and Methods section. Recombinant *Stym* was also checked by PCR for Ag85A, and expression of the plasmid was assessed by immunofluorescence following transfection of Hela cells with the recombinant plasmid. Strong expression was observed intracellularly. Expression was monitored *in vitro* by detection of Ag85A using monoclonal antibodies specific for the antigen using immunofluorescence and confocal microscopy. Staining for Ag85A was observed to be localized and appeared to cluster in structures within the cytoplasm and nucleus (FIG. 2).

Immunogenicity. Intranasal immunization resulted in better immunogenicity than did oral immunization with recombinant Salmonella carrying plasmid encoding Ag85A, as assessed by IL-2 and IFN-γ levels in the supernatants of restimulated splenocytes from immunized animals (data not shown). We measured the total antibodies to Ag85A in the sera of individually immunized mice 3 weeks after immunization by ELISA with native antigen. We also cloned Ag85A into Vical plasmid VR1020 and made recombinant *Stym* carrying Vical plasmid encoding Ag85A. We

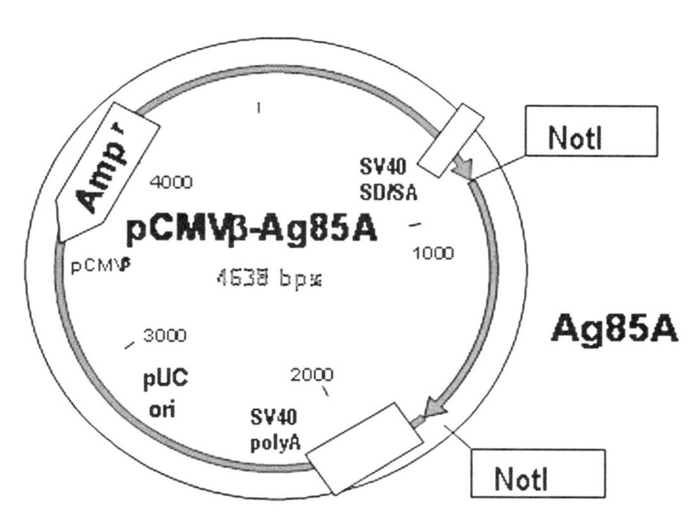

FIGURE 1. Genetic map of the recombinant pCMVβ-Ag85A with Ag85A clones at the *Not*I site in place of β-galactosidase. The vector backbone has a tPA single sequence, pUC.

explored the possibility of targeting the eukaryotically expressed antigen to the endoplasmic reticulum using the tPA signal peptide in the Vical plasmid VR1020 vector using Ag85A, which gave a better immune response in the animal studies than did the pCMV::Ag85A vector (data not shown). Both transcription and translation efficiencies are enhanced with the tPA signal. Intranasal immunization with this construct induced an enhanced CTL response as well as a local lung response. Based on these findings, we did challenge studies using intranasal immunization with recombinant *Stym* ::VR1020 plasmid encoding Ag85A.

Protection Studies against Challenge of Wild-Type M. tuberculosis. Protection experiments were performed by challenging the immunized C57BL/6 mice with various vaccine candidates including BCG and naked DNA 8 weeks after the last booster immunization at the end of week 14 intravenously with recombinant live *M. tuberculosis* H37RV with luciferase reporter enzymes. Five mice from each group were sacrificed 4 and 12 weeks after challenge, and the lungs and spleen homogenates were quantitatively monitored for the presence of bacteria using the RLU-based bacterial cell count assay.

The mycobacterial loads in the lungs and spleen were quantified by measuring relative luminescence in the tissue homogenate of the organs from the individual mouse. The results of the protection experiment indicated that the recombinant Salmonella-based bacterial delivery system provides protection that is superior to naked DNA. In our experience, *S. typhimurium* harboring expression plasmid of Ag85A with the tPA signal has given stronger protection (more than fivefold) to that of naked DNA immunization. The protection was more than eightfold higher than that of the control mice immunized with recombinant Salmonella encoding the empty pVR1020 Vical plasmid (FIG. 3).

At 4 weeks after challenge, mice immunized with *S. typhimurium* harboring the expression plasmid of Ag85A with tPA signal immunization were significantly pro-

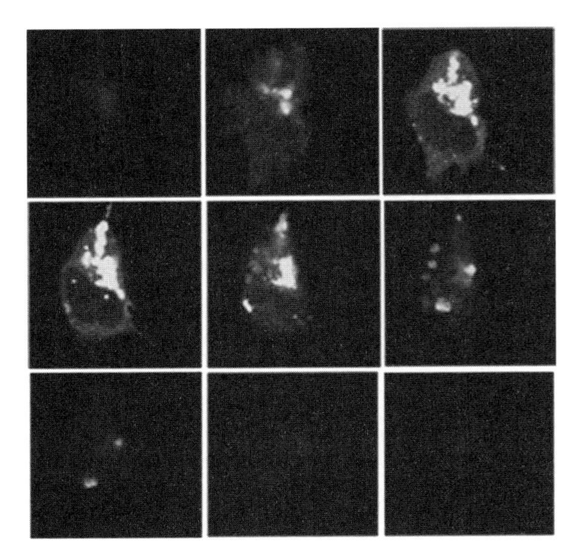

FIGURE 2. Ag85A expression level assessed by confocal microscopy after staining with monoclonal antibodies to Ag85A in Hela cells after transfection with the recombinant vector using FuGene 6 transfection reagent (Boehringer Mannheim).

tected in terms of the bacterial load in the lungs in comparison with intramuscular DNA immunization over naive unimmunized control and controls immunized with empty vector in *Stym*. The significance level from the statistical analysis between the groups is depicted in TABLE 1. The level of significance was further enhanced at 12 weeks post challenge as seen by differences in the protection efficacy between these groups. The protective efficacy in the spleen was only significant between the groups immunized with *Stym* harboring the expression plasmid of Ag85A vs. the intramuscular DNA immunized group and vs. the naive unimmunized control and the group immunized with *Stym* harboring empty expression plasmid. Some degree of protection has been observed only with *Stym* with empty expression plasmid, which can be explained by nonspecific immunity from the bacterial vector. In parallel experiments, BCG provided 0.8- to 1.0-log protection at the level of the CFU burden in the organs. A similar level of protection was afforded with *Stym* carrying plasmid encoding Ag85A in the range of 0.8- to 1.3-log protection in three independent sets of experiments (FIG. 4).

DISCUSSION

This study was aimed at using attenuated *S. typhimurium* to target recombinant eukaryotic expression plasmids with the *Mtb* antigen Ag85A to mucosal surfaces relevant to tuberculosis in an animal model and to assess its effectiveness in comparison to vaccine delivery employing naked DNA. Among the many new approaches currently being investigated for tuberculosis, such as subunit polypeptide vaccines,

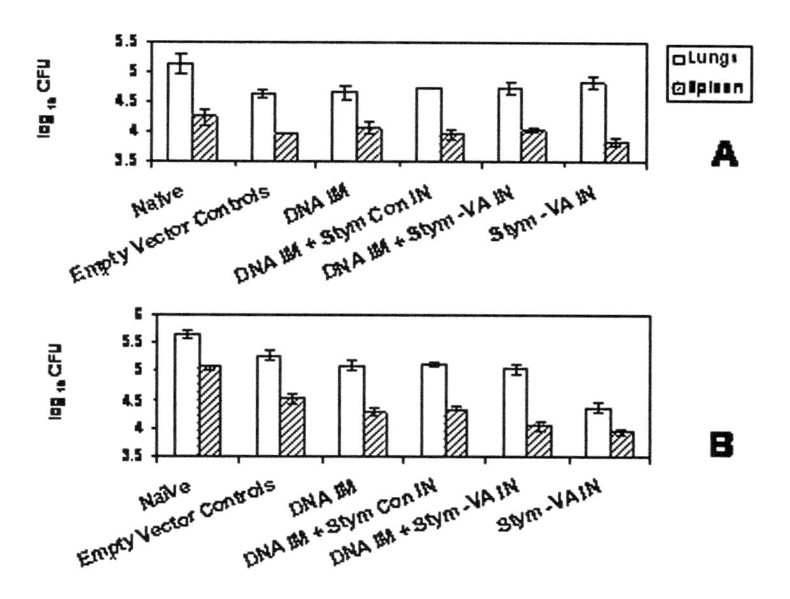

FIGURE 3. (**A** and **B**) Bacterial load from the lungs and spleen of immunized C57BL/6 mice at 4 and 12 weeks, respectively, after challenge with virulent *M. tuberculosis* H37RV with the luciferase construct showing levels of protection. Data show mean ± SEM of the $\log_{10}$ CFU value from 5 mice in each group. Immunization with *S. typhimurium (Stym) SL7207* carrying VR 1020-*Ag85A* (VA) was done intranasally, and DNA encoding Ag85A was given i.m. (3 doses at 3-wk intervals). Mice rested for 6 weeks before challenged with 1×10^6 of *M. tb lux* i.v. Empty vector controls = empty DNA plasmid i.m. + *Stym* with empty plasmid intranasally. DNA represents naked DNA given i.m. Con = control.

TABLE 1. ***P* values from statistical analysis of the individual bacterial load between each group**

	Lungs		Spleen	
	Week 4	Week 12	Week 4	Week 12
DNA vs. StymAg85	0.0109	<0.0001	0.0352	<0.0001
DNA vs. naive control	0.5078	0.5331	0.6458	0.253
StymAg85 vs. naive control	0.0044	0.0003	0.1059	<0.0001
StymAg85 vs. naive control and empty Stym	0.0024	<0.0001	0.1059	<0.0001
DNA vs. naive control and empty Stym	0.6359	0.4034	0.9371	0.0043

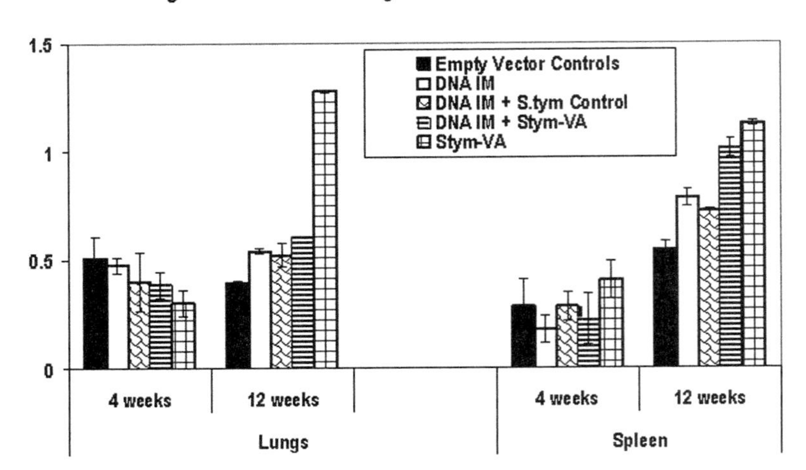

FIGURE 4. Level of protection in log scale with immunization from a representative experiment. Y-axis denotes level of protection in $\log_{10}$ CFU levels (difference between mean values of CFUs in naive challenged mice and mean values from named groups of immunized mice).

attenuated *M. tuberculosis* vaccines, and improved BCG in terms of live-vectored vaccines and nucleic acid vaccines, DNA-based vaccines have shown substantial promise. Prior immunization with nucleic acid vaccines with plasmid DNA encoding secreted Ag85A, PstS3, and Hsp-65 has shown protection *in vivo* in mice against subsequent challenge with *M. tuberculosis*[2,4,5] and in guinea pigs for Ag85A.[3] A major hindrance in translating the success of DNA vaccination from pre-clinical animal experiments to human clinical trials has been the low uptake of DNA from traditional administration by intramuscular or subcutaneous routes due to low numbers of antigen-presenting cells (APCs) present in muscle and degradation of the majority of delivered DNA resulting in poor immune responses. Plasmids with tPA or ubiquitin signal sequence have been shown to generate a 2- to 10-fold higher level of immune response than have the plasmids without such sequence.[9] The bacterial delivery system offers targeted delivery of DNA vaccines specific to APCs.[10]

Live bacterial vector vaccines consisting of attenuated strains of Salmonella or Shigella expressing heterologous antigen(s) have been shown to be a promising platform for vaccine development to elicit protective immune responses in infectious and neoplastic diseases.[11–16] Attenuated *Salmonella* strains are attractive as potential live vectors, as these can be administered mucosally and stimulate effective humoral and cell-mediated immune responses. Strains attenuated by deletion of genes involved in the production of metabolites essential for cell wall synthesis or replication within host cells have been produced and tested previously. The resulting mutants retain their ability to infect their hosts, but they cannot undergo subsequent replication and die within the target cells due to the lack of essential metabolites. These strains target APCs such as dendritic cells and macrophages. After phagocytosis, the bacteria survive within the target cells to a limited extent by either modifying the phagosomal compartment or breaching the cytoplasm. Once these bacteria undergo lysis, the plasmid content is released, leading to expression of the encoded protein by the host translational system, leading to production in the cytoplasm and hence effective presentation and generation of immune response in all three effector compartments of the immune system, namely, cytotoxic $CD8^+$ cells, $CD4^+$ T cells, and antibodies.[6]

Studies comparing orogastric and intranasal administration of *Stym* elicited stronger antibody responses as well as proliferative responses against both the bacterial and the heterologous antigen after 28 days of intranasal immunization.[17] The presence of live vaccine vectors was observed in nasal associated lymphoid tissue (NALT), lungs, and Peyer's patches after delivery. It was suggested that the induction of immune responses in a murine model depends not only on the host specificity of the vaccine strain but also on a complex cascade of events stemming from the ability of the vaccine's organisms to reach the appropriate inductive sites where they can be internalized by M cells, be carried to other organs, replicate, persist, and express the foreign antigen. M cells are specialized epithelial cells that transcytose particles across epithelial barriers to an intraepithelial lymphoid pocket created by the modification of the basolateral surface of the M cell. M cells in the lungs facilitate the uptake of *Stym* and deliver the plasmid rapidly to the draining lymph nodes helping in the development of effective protective immune responses to infection.[18]

The mucosal vaccine for transgene vaccination developed in this study used a live attenuated *aro*A mutant strain of *S. typhimurium*, which cannot grow *in vivo,* and carried the gene of interest under the control of a eukaryotic promoter. The mutant strain D*aro*A has an auxotrophic mutation in a key biosynthetic pathway and as a result fails to grow in tissue as it is deficient in folate biosynthesis and uptake.[19] The live attenuated bacteria carry the DNA through the airway tract, and *Salmonellae* enter macrophages and dendritic cells where they die because of their attenuating mutation, liberating multiple copies of the DNA vaccine directly into the phagocytes to generate effective immune responses.

Results with the bacterial delivery system in *Salmonella* in tuberculosis have been very promising in terms of an enhanced cellular immune response as evidenced by elevated levels of IFN-γ and the presence of specific antibodies to Ag85 following intranasal immunization. This is at least as good as that after immunization with naked DNA. The immune response generated by the recombinant Salmonella-DNA vaccine given through the intranasal route has been better at both the systemic and the local levels, that is, the lungs, than that generated by DNA vaccine given through the intramuscular route. These responses were obtained with a recombinant expressing Ag85A under the control of the CMV early promoter. The intrinsic non-specific adjuvant effect of the vector itself contributed to the overall specific protective responses observed with recombinant *Stym::V*Ag85A immunization, which is superior to intramuscular DNA encoding the same antigen in the same plasmid backbone. The protective efficacy of the approach is almost on par with the level of protection given by BCG in the animal model. Further studies with larger sets of vaccinated animals are required to examine and document the apparent superiority of the recombinant *Stym* encoding Ag85A over BCG. Nevertheless, this has clearly established the proof of principle of this approach for TB vaccine development.

In the last several years, different groups have used bacteria-based DNA delivery with some success and promise, but progress has not been great enough to reach the stage of human clinical trials. Several newly developed mutants in Salmonella have been found to improve the safety of the live attenuated vector, but the magnitude of the immune response has been limited because of the severe instability of transformants carrying these expression plasmids as well as the anti-vector response generated by the primary immunization, which affects the boosting effect of the response to the heterologous antigen. Newly designed expression vectors employing stable low copy number replicons have been able to significantly improve the potency and versatility of Salmonella-mediated DNA vaccination.[20] Several publications have reported the generation of potential vaccine vectors with attenuations in unique genes using the specific pathogenicity island (SPI-2) and type III secretion system in *S. typhi* and *S. typhimurium*. Some of these mutant strains with D*aroC* and D*ssaV* have been successfully tried and well tolerated in human volunteers.[21] Attenuated *Listeria monocytogenes* has also been used as a potential bacterial vaccine vector for heterologous antigens in several viral diseases as well as cancers. Safety and shedding of such strains have been tested in human volunteers in a dose escalation study with success.[22] Existing antivector immunity has been shown not to diminish the therapeutic efficacy of the Listeria vaccine vector.[23] The anti-vector response issues could also be addressed by using a prime-boost approach using the heterologous vector system. Thus, there is ample scope to further improve this delivery system with optimization of the recombinant strains by the addition of factors such as cytokines or co-stimulatory molecules or molecules enhancing antigen presentation and the memory T-cell response.

One third of the world's population and two thirds of the population in endemic countries harbor non-replicating persistent (NRP)-latent *M. tuberculosis* having the potential to reactivate, causing active clinical disease after any immune suppression. Post-exposure vaccine strategy to boost T-cell immune responses targeting the NRP *Mtb* to eliminate or reduce those from asymptomatic infected individuals would significantly prevent the risk of reactivation, leading to shrinkage of the biggest reservoir of infection and the eventual impact on breaking the chain of transmission of this threatening disease. We have embarked on using this bacterial-based delivery of DNA encoding putative protective antigens for the development of a post-exposure vaccine for NRP-TB using antigen(s) expressed during the NRP (latent) state of the bacilli. These studies are in progress in pre-clinical animal models of latent TB.[24]

A generic vaccine approach using attenuated bacterial vectors such as Salmonella and Listeria can be exploited as a potential vaccine platform to deliver heterologous protective antigens. The resulting multivalent vaccines would have the potential to alleviate the suffering from infectious diseases in the developing world. Thanks to the concerted efforts of many global forums, we have a number of candidate TB vaccines in development pipelines with three already in Phase I clinical trial with the hopes of at least one reaching the population in the next decade.

ACKNOWLEDGMENTS

We express our thanks to Dr. Eugen Domann, Torsten Hain, Soudabeh Djafari, and Alexandra Amend for their constructive comments and technical support during the cloning, Kamiel Palfliet for assistance during the animal experiments, and Dr. William Reece for his help in statistical analysis.

We wish to acknowledge the financial support of an EMBO short-term fellowship to S.K.P. for his work at IPB, Brussels, and a DIF grant from WHO/TDR to T.C. and S.K.P.

REFERENCES

1. Global Tuberculosis Control WHO Report 2005.
2. HUYGEN, K. *et al.* 1996. Immunogenicity and protective efficacy of a tuberculosis DNA vaccine. Nat. Med. **2:** 893–898.
3. BALDWIN, S.L. *et al.* 1998. Evaluation of new vaccines in the mouse and guinea pig model of tuberculosis. Infect. Immun. **66:** 2951–2959.
4. TANGHE, A. *et al.* 1999. Immunogenicity and protective efficacy of tuberculosis DNA vaccines encoding putative phosphate transport receptors. J. Immunol. **162:** 1113–1119.
5. LOWRIE, D.B. *et al.* 1999. Therapy of tuberculosis in mice by DNA vaccination. Nature **400:** 269–271.
6. DARJI, A. *et al.* 1997. Oral somatic transgene vaccination using attenuated *S. typhimurium*. Cell **91:** 765–775.
7. DIETRICH, G. *et al.* 2003. Live attenuated bacteria as vectors to deliver plasmid DNA vaccines. Curr. Opin. Mol. Ther. **5:** 10–19.
8. SNEWIN, V.A. *et al.* 1999. Assessment of immunity to mycobacterial infection with luciferase reporter constructs. Infect. Immun. **67:** 4586–4593.
9. LI, Z. *et al.* 1999. Immunogenicity of DNA vaccines expressing tuberculosis proteins fused to tissue plasminogen activator signal sequences. Infect. Immun. **67:** 4780–4786.
10. BERZOFSKY, J.A., J.D. AHLERS & I.M. BELYAKOV. 2001. Strategies for designing and optimizing new generation vaccines. Nat. Rev. Immunol. **1:** 209–219.
11. GRILLOT-COURVALIN, C. *et al.* 1998. Functional gene transfer from intracellular bacteria to mammalian cells [see comments]. Nat. Biotechnol. **16:** 862–866.
12. SIZEMORE, D.R., A.A. BRANSTROM & J.C. SADOFF. 1995. Attenuated Shigella as a DNA delivery vehicle for DNA-mediated immunization. Science **270:** 299–302.
13. MEDINA, E. *et al.* 1999. Salmonella vaccine carrier strains: effective delivery system to trigger anti-tumor immunity by oral route. Eur. J. Immunol. **29:** 693–699.
14. PAGLIA, P. *et al.* 1998. Gene transfer in dendritic cells, induced by oral DNA vaccination with Salmonella typhimurium, results in protective immunity against a murine fibrosarcoma. Blood **92:** 3172–3176.
15. DIETRICH, G. *et al.* 1999. Delivery of DNA vaccines by attenuated intracellular bacteria. Immunol. Today **20:** 251–253.
16. PAGLIA, P. *et al.* 2000. In vivo correction of genetic defects of monocyte/macrophages using attenuated Salmonella as oral vectors for targeted gene delivery. Gene Ther. **7:** 1725–1730.
17. PASETTI, M.F. *et al.* 2000. A comparison of immunogenicity and in vivo distribution of Salmonella enterica serovar Typhi and Typhimurium live vector vaccines delivered by mucosal routes in the murine model. Vaccine **18:** 3208–3213.
18. TEITELBAUM, R. *et al.* 1999. The M cell as a portal of entry to the lung for the bacterial pathogen Mycobacterium tuberculosis. Immunity **10:** 641–650.
19. HOISETH, S.K. & B.A. STOCKER. 1981. Aromatic-dependent Salmonella typhimurium are non-virulent and effective as live vaccines. Nature **291:** 238–239.
20. BAUER, H. *et al.* 2005. Salmonella-mediated oral DNA vaccination using stabilized eukaryotic expression plasmids. Gene Ther. **12:** 364–372.
21. BOTHA, T. & B. RYFFEL. 2002. Reactivation of latent tuberculosis by an inhibitor of inducible nitric oxide synthase in an aerosol murine model. Immunology **107:** 350–357.
22. HINDLE, Z. *et al.* 2002. Characterization of Salmonella enterica derivatives harboring defined aroC and Salmonella pathogenicity island 2 type III secretion system (ssaV) mutations by immunization of healthy volunteers. Infect. Immun. **70:** 3457–3467.

23. ANGELAKOPOULOS, H. *et al.* 2002. Safety and shedding of an attenuated strain of Listeria monocytogenes with a deletion of actA/plcB in adult volunteers: a dose escalation study of oral inoculation. Infect. Immun. **70:** 3592–3601.
24. STARKS, H. *et al.* 2004. Listeria monocytogenes as a vaccine vector: virulence attenuation or existing antivector immunity does not diminish therapeutic efficacy. J. Immunol. **173:** 420–427.

TRAIL (TNF-Related Apoptosis-Inducing Ligand) Induces Necrosis-Like Cell Death in Tumor Cells at Acidic Extracellular pH

OLIVIER MEURETTE, LAURENCE HUC, AMELIE REBILLARD, GWENAELLE LE MOIGNE, DOMINIQUE LAGADIC-GOSSMANN, AND MARIE-THERESE DIMANCHE-BOITREL

INSERM U620, Faculté de Pharmacie, Université de Rennes 1, 35043 Rennes cedex, France

ABSTRACT: How tumor microenvironment, more specifically low extracellular pH (6.5), alters cell response to TNF-related apoptosis-inducing ligand (TRAIL)-based cancer therapy has yet to be determined. The aim of the current work was to test the effect of acidic extracellular pH on TRAIL-induced cell death in human HT29 colon carcinoma and HepG2 hepatocarcinoma cell lines as well as in human primary hepatocytes. We found an increase in TRAIL sensitivity at low extracellular pH, which is partially inhibited by Bcl-2 expression in HT29 cells. At low extracellular pH, TRAIL induced a new form of cell death, sharing necrotic and apoptotic features in tumor cells. By contrast, human primary hepatocytes were resistant to TRAIL-induced cell death even at acidic extracellular pH.

KEYWORDS: TRAIL; pH; apoptosis; necrosis; hepatocarcinoma cells; colon cancer cells; human primary hepatocytes

INTRODUCTION

TRAIL (TNF-α related apoptosis-inducing ligand) is a cytokine belonging to the tumor necrosis factor (TNF)-α superfamily and a promising anticancer agent. In fact, TRAIL induces apoptosis in many cancer cells but not in most normal cells.[1] Moreover, TRAIL administration in nude or SCID mice reduces growth of colon carcinoma or mammary adenocarcinoma xenografts without any toxic effects.[1,2]

Apoptosis is a genetically regulated cell death that plays an essential role in development, tissue homeostasis, and elimination of damaged cells. Disruption of molecular mechanisms controlling apoptosis can lead to oncogenesis and chemoresistance of transformed cells. Two main apoptotic pathways leading to caspase activation have been characterized: the intrinsic mitochondria-mediated pathway[3] and the extrinsic death receptor-mediated pathway.[4] TRAIL-mediated cell death

Address for correspondence: Marie-Therese Dimanche-Boitrel, INSERM U620, Faculté de Pharmacie, Université Rennes 1, 2 av Prof Léon Bernard, 35043 Rennes cedex, France. Voice: (33)2.23.23.48.37; fax: (33)2.23.23.47.94.
marie-therese.boitrel@rennes.inserm.fr

Ann. N.Y. Acad. Sci. 1056: 379–387 (2005).
doi: 10.1196/annals.1352.018

occurs essentially through activation of the death receptor apoptotic pathway and involves interaction of the intracytoplasmic death domain of TRAIL receptors (TRAIL-R1 [DR4] and TRAIL-R2 [DR5]) with the adaptor molecule FADD (Fas-associated death domain). In turn, FADD recruits procaspase-8 to form the DISC (death-inducing signaling complex), in which caspase-8 is activated by autoproteolysis, and initiates the proteolytic caspase cascade that triggers activation of downstream effector caspases. Two complementary pathways were shown to lead to apoptosis following caspase-8 activation.[5] The first one, in which caspase-8 directly activates caspase-3, does not involve the mitochondria and escapes the resistance mediated by overexpression of Bcl-2 and related antiapoptotic proteins. The second pathway involves the cleavage of Bid, a BH3-domain-only proapoptotic protein that belongs to the Bcl-2 family.[6] The truncated Bid (tBid) induces mitochondrial changes, including the translocation of proapoptotic proteins such as cytochrome *c*, Smac/DIABLO, and apoptosis-inducing factor (AIF), from the mitochondria to either cytosol, where they activate the caspase cascade, or to nucleus, where they trigger DNA fragmentation and chromatin condensation.[6,7] This second pathway is inhibited by Bcl-2 and related antiapoptotic proteins.

All cancer cells, however, are not sensitive to TRAIL, and several molecular mechanisms have been described to explain this resistance.[8] Moreover, how TRAIL efficacy is modulated in the context of tumor microenvironment remains poorly understood. A high rate of glycolytic metabolism frequently induces high proton production, which contributes to acidification of the extracellular environment.[9] Furthermore, intracellular pH might be an important regulator of cell death.[10] In the current work, we lowered the extracellular pH to a value of 6.5 in HT29 and HepG2 cells as well as in normal human primary hepatocytes to study the effect of such a maneuver on cell sensitivity to TRAIL-induced cell death. The effect of Bcl-2 expression in HT29 cells was also investigated in response to TRAIL-induced cell death at low extracellular pH.

MATERIAL AND METHODS

Chemicals and Antibodies

The recombinant human soluble Flag-tagged TRAIL was from Alexis Biochemicals (Coger, Paris, France; http://www.alexis-corp.com). The anti-Flag M2 was from Sigma-Aldrich (Saint-Quentin Fallavier, France). The anti-Flag M2 was used to crosslink the ligand TRAIL, leading to the formation of oligomers that are more effective in inducing cell death. Briefly, increased concentrations of the soluble Flag-tagged TRAIL were incubated with 2 μg/ml anti-Flag M2 for 5 minutes at room temperature before cell treatment. The pan-caspase inhibitor (z-VAD-fmk), the caspase-8 inhibitor (z-IETD-fmk), and the caspase inhibitor negative control (z-FA-fmk) were from Calbiochem (France Biochem, Meudon, France). Antagonistic mouse monoclonal antibodies directed against the extracellular domain of human TRAIL-R1 (DR4) and TRAIL-R2 (DR5) were from Alexis Biochemicals. The antagonistic mouse monoclonal antibody directed against the Fas receptor (ZB4) was from Immunotech (Coulter, Marseille, France).

Cell Lines and Cytotoxic Assay

The HT29 human colon carcinoma cell line and the HepG2 human hepatocarcinoma cell line were obtained from ATCC (Rockville, MD, USA) and cultured in Eagle's minimum essential medium (Eurobio, Les Ulis, France) or Williams E medium (GibcoBRL, Life Technologies, Cergy Pontoise, France), respectively, supplemented with 10% (v/v) fetal calf serum (GibcoBRL), glutamine (2 mM; HT29 and HepG2), and bovine serum albumin (0.1 mg/ml), bovine insulin (1 µg/ml), and hydrocortisone (0.25 µg/ml; HepG2) at 37°C under a 5% CO_2 atmosphere. The HT29 neo/Bcl-2 cells were previously described.[11]

Microscopic detection of apoptosis or necrosis was carried out in both floating and adherent cells recovered after TRAIL treatment using nuclear chromatin staining with 1 µg/ml Hoechst 33342 and 1 µg/ml propidium iodide (PI) for 15 minutes at 37°C. Apoptotic cells (i.e., those with condensed blue chromatin or fragmented blue nuclei) or necrotic cells (i.e., those with red nuclei) were counted and compared with those of the total population ($n = 300$ cells).

Primary Human Hepatocyte Cultures and Cytotoxic Assay

Human hepatocytes from adult donors undergoing resection for primary and secondary tumors were obtained by perfusion using a collagenase solution as described previously.[12] Cells were seeded at a density of 3×10^4 cells/cm^2 in Williams E medium supplemented with glutamine (2 mM), bovine serum albumin (0.2 mg/ml), bovine insulin (10 µg/ml), and 10% (v/v) fetal calf serum. The next day, hydrocortisone (10^{-7} M) was added. Hepatocytes from six adult donors were isolated, and treatments were performed in triplicate under the same conditions as just described for human cancer cell lines.

Modification of Extracellular pH (pH_e) and Measurement of Intracellular pH (pH_i)

To modify the value of extracellular pH, a growth medium without sodium hydrogenocarbonate was used. To set the extracellular pH of this medium to either 7.4 or 6.5, we supplemented it with 23 mM or 3 mM sodium hydrogenocarbonate, respectively, in an atmosphere of 5% CO_2. The pH_i of cells cultured on glass coverslips was monitored with the pH-sensitive fluorescent probe carboxy-SNARF-1 (carboxy-seminaphtorhodafluor; Molecular Probes, Invitrogen, Cergy-Pontoise, France).[13] Cells were loaded with SNARF by incubating them in a 5-µM solution of the acetoxy-methyl ester for 20 minutes at 37°C just prior to recording pH_i. SNARF-loaded cells were placed in a continuously perfused recording chamber (at a temperature of 37 ± 1°C) mounted on the stage of an epifluorescent microscope (Nikon Diaphot). The necessary setup to produce and detect fluorescence has previously been described.[14] The emission ratio 640/590 (corrected for background fluorescence) detected from intracellular SNARF was calculated and converted to a linear pH scale using *in situ* calibration obtained by the nigericin technique described elsewhere.[13]

RESULTS

Extracellular pH is lower in tumor than in normal tissue due to the intense glycolytic activity of cancer cells, whereas intracellular pH is little, if not affected. This gives rise to a cellular transmembrane pH gradient difference between these tissues with a tumor-specific pH gradient that might enhance the cytotoxicity of chemotherapeutics.[15] To verify that such a gradient exists under our experimental conditions, we first measured the intracellular pH when cells were cultured in an acidic extracellular medium. We observed a decrease of about 0.2 pH unit of the intracellular pH in HT29 and HepG2 cell lines cultured at pH_e 6.5 (FIG. 1A), which clearly indicated an increase in the transmembrane pH gradient. We then studied the effect of such a modification on cell sensitivity to TRAIL-induced cell death. HT29 and HepG2 cells that were initially resistant became very sensitive following a 24-hour treatment with TRAIL at low extracellular pH (FIG. 1B). Surprisingly, the dead cells did not exhibit the typical apoptotic morphology with condensed blue chromatin and/or fragmented blue nuclei observed when cells were treated with TRAIL at normal (7.4) extracel-

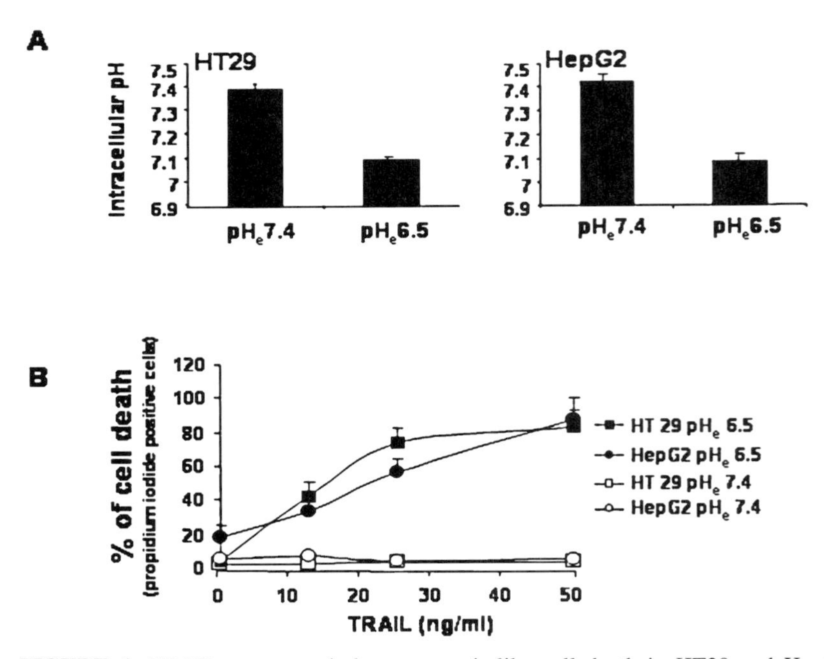

FIGURE 1. TRAIL treatment induces necrosis-like cell death in HT29 and HepG2 cells at acidic extracellular pH (6.5). (**A**) Intracellular pH values were determined by microspectrofluorimetry using a carboxy-SNARF-1-AM at normal (7.4) and acidic (6.5) extracellular pH in HT29 and HepG2 cells as described in **Material and Methods**. (**B**) Cells were exposed to increased concentrations of TRAIL (12.5, 25, and 50 ng/ml crosslinked with 2 μg/ml anti-Flag M2) for 24 h at normal (7.4) or acidic (6.5) extracellular pH before identifying dead cells by staining nuclear chromatin with Hoechst 33342 and propidium iodide. Three hundred cells per point were counted. The percentage of necrotic cell deaths was estimated as the percentage of propidium iodide–positive cells. Results are the mean ± SD of three independent experiments.

lular pH (approximately 15% of apoptotic cells in HT29 cells treated with TRAIL [50 ng/ml] for 24 hours at pH_e 7.4; data not shown). Indeed, they underwent an early plasma membrane permeabilization to propidium iodide, which is a common feature of necrosis (approximately 80% of necrotic cells in HT29 cells treated with TRAIL [50 ng/ml] for 24 hours at pH_e 6.5; FIG. 1B). Nevertheless, this cell death was dependent on caspase activation since a pan-caspase inhibitor (z-VAD-fmk) or a caspase-8 inhibitor (z-IETD-fmk) blocked TRAIL-induced cell death at low extracellular pH in HT29 cells (FIG. 2A). The caspase inhibitor negative control (z-FA-fmk) had no effect on TRAIL-induced cell death at pH_e 6.5 (FIG. 2A). To determine

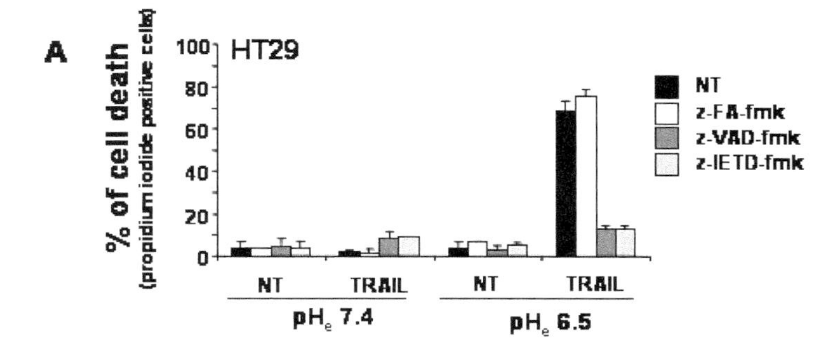

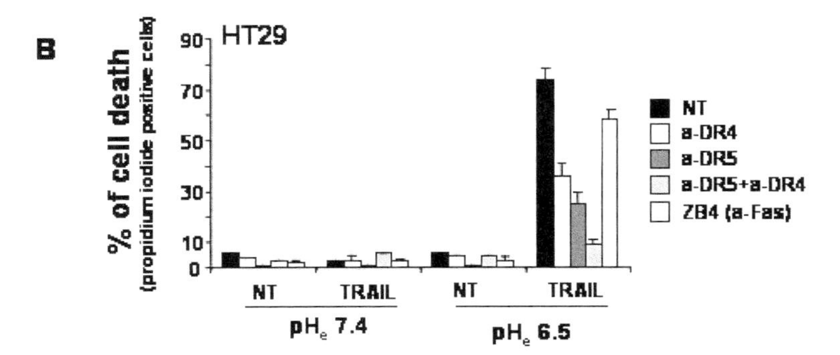

FIGURE 2. TRAIL-induced cell death at acidic extracellular pH depends on caspase activation and on TRAIL death receptor (DR4 and DR5) pathway. (**A**) HT29 cells were treated for 24 h with TRAIL (50 ng/ml crosslinked with 2 μg/ml anti-Flag M2) at normal (7.4) or acidic (6.5) extracellular pH in the presence or absence of caspase inhibitors z-VAD-fmk and z-IETD-fmk, at a concentration of 10 μM, respectively. Z-FA-fmk (10 μM) was used as a caspase inhibitor negative control. (**B**) HT29 cells were treated for 24 h with TRAIL (50 ng/ml crosslinked with 2 μg/ml anti-Flag M2) at normal (7.4) or acidic (6.5) extracellular pH in the presence or absence of antagonistic antibodies anti-DR4 (10 μg/ml) and/or anti-DR5 (10 μg/ml). An antagonistic anti-Fas receptor antibody (ZB4, 10 μg/ml) was used as a negative control. In **A** and **B**, the percentage of cell deaths was estimated as the percentage of propidium iodide positive cells as described in FIGURE 1. Results are the mean ± SD of three independent experiments.

if this cell death pathway was dependent on TRAIL death receptors, we used antagonistic antibodies directed against TRAIL-R1 (DR4) or TRAIL-R2 (DR5). Both DR4 and DR5 were required for TRAIL-induced cell death at an acidic extracellular pH because a combination of antagonistic antibodies directed against the extracellular domain of each receptor was necessary to completely inhibit TRAIL-mediated cell death in HT29 cells, whereas only partial inhibition was observed when either antibody was used alone (FIG. 2B). When an antagonistic antibody directed against a Fas receptor (ZB4) was used as a negative control, no effect on TRAIL-induced cell death at low extracellular pH was detected (FIG. 2B). As an acidic extracellular pH is usually a specific characteristic of tumor microenvironment, we tested the sensitivity of normal human primary hepatocytes to TRAIL-induced cell death under acidic extracellular conditions. Under these experimental conditions, normal human primary hepatocytes remained resistant to TRAIL-induced cell death (FIG. 3), sug-

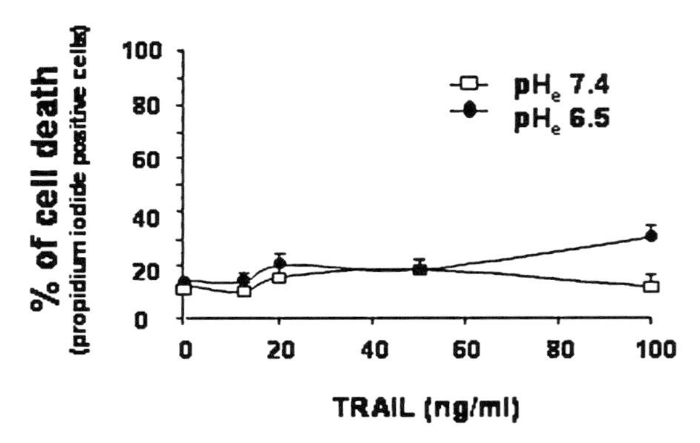

FIGURE 3. TRAIL is not cytotoxic towards human primary hepatocytes even at acidic extracellular pH. Human primary hepatocytes were treated as in FIGURE 1B, and the percentage of cell deaths was estimated as the percentage of propidium iodide–positive cells. Results are the mean ± SD of six independent experiments.

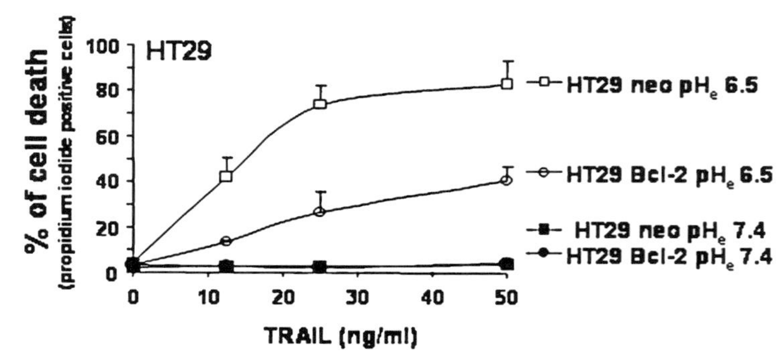

FIGURE 4. Bcl-2 expression in HT29 cells delays TRAIL-induced cell death at acidic extracellular pH. HT29 neo/Bcl-2 cells are treated as in FIGURE 1B, and the percentage of cell deaths was estimated as the percentage of propidium iodide–positive cells. Results are the mean ± SD of three independent experiments.

gesting that normal cells even in the close tumor microenvironment should not be sensitive to TRAIL-induced cell death. Finally, the effect of Bcl-2 expression on TRAIL-induced cell death at a low extracellular pH was investigated in HT29 cells. We demonstrated that Bcl-2 expression delayed this cell death (FIG. 4), thus indicating that Bcl-2 might be a resistant factor towards this cell death pathway.

DISCUSSION

An acidic extracellular pH is a characteristic of numerous solid tumors.[9] Furthermore, the tumor-specific pH gradient could play an important role in chemosensitivity.[16,17] We show here that an acidic extracellular environment led to a high sensitization to TRAIL-induced cell death in human colon carcinoma and hepatocarcinoma cells. Our data agree with the observations of Lee *et al.*[18] who showed that human prostate carcinoma cells could be sensitized to TRAIL-induced apoptosis under an acidic extracellular pH. Surprisingly, under our experimental conditions, this cell death appeared not to be apoptotic. In fact, the plasma membrane was permeabilized early, and the chromatin was not fragmented, pointing to a necrosis-like cell death. Interestingly, we observed that TRAIL-induced cell death at low extracellular pH was blocked by caspase inhibitors (z-IETD-fmk or z-VAD-fmk), thus suggesting that caspases, more specifically caspase-8, are necessary to this particular cell death pathway. Such caspase-8 involvement has also been described in necrotic cell death of human lung adenocarcinoma A549 cells exposed to hyperoxia.[19] Furthermore, Bcl-2 expression in HT29 cells appeared to delay TRAIL-induced cell death at low extracellular pH, suggesting that mitochondria might be involved in this process. An increased interaction between t-Bid and Bax has been correlated with the increase in TRAIL sensitivity under acidic extracellular conditions in human colorectal carcinoma CX-1 cells, also indicating a role of the mitochondrial death pathway.[18] In this regard, it is worth noting that the antiapoptotic members of the Bcl-2 superfamily have been shown to inhibit both necrosis and apoptosis.[20] Recently, in prostate adenocarcinoma cells, TRAIL treatment was reported to induce a necrosis process insensitive to a pan-caspase inhibitor (z-VAD-fmk).[21] In the same way, a necrotic death pathway independent of caspase activation has been described after triggering other death receptors, such as TNFR1 or Fas.[22,23] Necrosis has long been considered an accidental cell death as opposed to programmed cell death, apoptosis. However, it appears now that multiple programmed cell death pathways exist,[24,25] and necrosis appears to be part of those playing a role in embryogenesis, normal tissue renewal, and immune response.[20] In this regard, it is also noteworthy that the inflammatory response caused by necrosis may activate an innate immune response under some pathologic conditions such as cancer.[20] This point is important to emphasize because, depending on the extent of such a response, necrosis could improve the efficacy of chemotherapy by activating an anticancer cell immune response. Currently, the molecular mechanisms involved in TRAIL-induced cell death at acidic extracellular pH are under investigation. In fact, it is of clinical relevance to understand how this cell death pathway occurs after TRAIL treatment *in vivo*. We also showed that human primary hepatocytes were not sensitized to TRAIL-induced cell death under acidic conditions, suggesting that targeting cellular pH could be a potential approach to selectively sensitizing tumor cells to TRAIL-based cancer therapy. In this regard, cellular pH has been considered a promising molecular target for cancer chemotherapy.[26] However, Bcl-2 might be an important resistant factor.

ACKNOWLEDGMENTS

We thank the Biological Resource Centre (BRC) of Rennes for supplying of isolated human primary hepatocytes. We also thank the Ligue Nationale Contre le Cancer (the Morbihan, Côte d'Armor, and Ille et Vilaine Committees), Rennes Métropole, and the Région Bretagne for their financial support.

REFERENCES

1. Ashkenazi, A., R.C. Pai, S. Fong, *et al.* 1999. Safety and antitumor activity of recombinant soluble Apo2 ligand. J. Clin. Invest. **104:** 155–162.
2. Walczak, H., R.E. Miller, K. Ariail, *et al.* 1999. Tumoricidal activity of tumor necrosis factor-related apoptosis- inducing ligand *in vivo*. Nat. Med. **5:** 157–163.
3. Zimmermann, K.C. & D.R. Green. 2001. How cells die: apoptosis pathways. J. Allergy Clin. Immunol. **108:** S99–103.
4. Ashkenazi, A. 2002. Targeting death and decoy receptors of the tumour-necrosis factor superfamily. Nat. Rev. Cancer **2:** 420–430.
5. Scaffidi, C., S. Fulda, A. Srinivasan, *et al.* 1998. Two CD95 (APO–1/Fas) signaling pathways. Embo J. **17:** 1675–1687.
6. Li, H., H. Zhu, C.J. Xu, *et al.* 1998. Cleavage of BID by caspase 8 mediates the mitochondrial damage in the Fas pathway of apoptosis. Cell **94:** 491–501.
7. Kroemer, G. & J.C. Reed. 2000. Mitochondrial control of cell death. Nat. Med. **6:** 513–519.
8. Zhang, L. & B. Fang. 2005. Mechanisms of resistance to TRAIL-induced apoptosis in cancer. Cancer Gene Ther. **2:** 228–237.
9. Wike-Hooley, J.L., A.P. van den Berg, J. van der Zee, *et al.* 1985. Human tumour pH and its variation. Eur. J. Cancer Clin. Oncol. **21:** 785–791.
10. Lagadic-Gossmann, D., L. Huc & V. Lecureur. 2004. Alterations of intracellular pH homeostasis in apoptosis: origins and roles. Cell Death Differ. **11:** 953–961.
11. Lacour, S., O. Micheau, A. Hammann, *et al.* 2003. Chemotherapy enhances TNF-related apoptosis-inducing ligand DISC assembly in HT29 human colon cancer cells. Oncogene **22:** 1807–1816.
12. Guguen-Guillouzo, C., J.P. Campion, P. Brissot, *et al.* 1982. High yield preparation of isolated human adult hepatocytes by enzymatic perfusion of the liver. Cell Biol. Int. Rep. **6:** 625–628.
13. Buckler, K.J. & R.D. Vaughan-Jones. 1990. Application of a new pH-sensitive fluoroprobe (carboxy-SNARF-1) for intracellular pH measurement in small, isolated cells. Pflugers Arch. **417:** 234–239.
14. Lagadic-Gossmann, D., M. Rissel, M. Galisteo, *et al.* 1999. Intracellular pH alterations induced by tacrine in a rat liver biliary epithelial cell line. Br. J. Pharmacol. **128:** 1673–1682.
15. Kozin, S.V., P. Shkarin & L.E. Gerweck. 2001. The cell transmembrane pH gradient in tumors enhances cytotoxicity of specific weak acid chemotherapeutics. Cancer Res. **61:** 4740–4743.
16. Tannock, I.F. & D. Rotin. 1989. Acid pH in tumors and its potential for therapeutic exploitation. Cancer Res. **49:** 4373–4384.
17. Raghunand, N. & R.J. Gillies. 2001. pH and chemotherapy. Novartis Found. Symp. **240:** 199–211; discussion 265–198.

18. Lee, Y.J., J.J. Song, J.H. Kim, *et al.* 2004. Low extracellular pH augments TRAIL-induced apoptotic death through the mitochondria-mediated caspase signal transduction pathway. Exp. Cell Res. **293:** 129–143.
19. Wang, X., S.W. Ryter, C. Dai, *et al.* 2003. Necrotic cell death in response to oxidant stress involves the activation of the apoptogenic caspase-8/bid pathway. J. Biol. Chem. **278:** 29184–29191.
20. Proskuryakov, S.Y., A.G. Konoplyannikov & V.L. Gabai. 2003. Necrosis: a specific form of programmed cell death? Exp. Cell Res. **283:** 1–16.
21. Kemp, T.J., J.S. Kim, S.A. Crist, *et al.* 2003. Induction of necrotic tumor cell death by TRAIL/Apo-2L. Apoptosis **8:** 587–599.
22. Denecker, G., D. Vercammen, W. Declercq, *et al* 2001. Apoptotic and necrotic cell death induced by death domain receptors. Cell Mol. Life Sci. **58:** 356–370.
23. Holler, N., R. Zaru, O. Micheau, *et al.* 2000. Fas triggers an alternative, caspase-8-independent cell death pathway using the kinase RIP as effector molecule. Nat. Immunol. **1:** 489–495.
24. Jaattela, M. 2004. Multiple cell death pathways as regulators of tumour initiation and progression. Oncogene **23:** 2746–2756.
25. Leist, M. & M. Jaattela. 2001. Four deaths and a funeral: from caspases to alternative mechanisms. Nat. Rev. Mol. Cell Biol. **2:** 589–598.
26. Izumi, H., T. Torigoe, H. Ishiguchi, *et al.* 2003. Cellular pH regulators: potentially promising molecular targets for cancer chemotherapy. Cancer Treat. Rev. **29:** 541–549.

Cloning and Expression of a Functionally Active Truncated *N*-Glycosylated KSHV ORF4/KCP/Kaposica in the Methylotrophic Yeast *Pichia pastoris*

NEUZA A. GOMES PEREIRA,[a] MARIA A. JULIANO,[b] ADRIANA K. CARMONA,[b] EDWARD D. STURROCK,[a] AND GIRISH J. KOTWAL[a]

[a]*Division of Medical Virology, Institute of Infectious Diseases and Molecular Medicine, Faculty of Health Sciences, University of Cape Town, Observatory -7925, Cape Town, South Africa*

[b]*Department of Biophysics, Division of Nephrology, Escola Paulista de Medicina, Universidade Federal de São Paulo, Rua 3 de Maio 100, São Paulo, 04044-020, Brazil*

ABSTRACT: Kaposi's sarcoma herpesvirus (KSHV) is a typical DNA virus that is associated with a number of proliferative diseases including Kaposi's sarcoma. The KSHV open reading frame (ORF) 4 encodes a complement regulatory protein (Kaposi complement control protein, KCP) that binds complement components and inhibits the complement-mediated lysis of cells infected by the virus, thus providing a strategy for evasion of the host complement system. Based on primary sequence analysis and comparison with other functionally and structurally similar proteins, oligonucleotide primers were designed to amplify by polymerase chain reaction (PCR) three regions of the predicted ORF 4 from human herpes virus-8 (HHV-8) DNA isolated from a primary effusion lymphoma cell line. The PCR products were inserted by ligation into the expression vector pPIC9 to generate three recombinant plasmids for heterologous expression in the yeast, *Pichia pastoris*, to produce separately the four N-terminal sushi domains (KCP-S, small), KCP protein lacking the putative transmembrane-binding domain (KCP-M, medium), and the full-length protein (KCP-F, full). Expression of the viral proteins was confirmed by SDS-PAGE, immunologic detection, and Western blot analyses using a rabbit polyclonal antibody directed against a selected peptide region that is common to all three recombinant KCPs. KCP-S directly from expression media could inhibit complement-mediated lysis of sensitized sheep erythrocytes by approximately 60% in a hemolysis assay. This result confirms previous reports that recombinant KCP is twice as efficient in inhibiting the classic pathway-mediated lysis of erythrocytes than is the vaccinia virus complement control protein, which also contains four sushi domains.

KEYWORDS: cloning; Kaposi sarcoma herpesvirus (KSHV); open reading frame (ORF) 4; yeast; *Pichia pastoris*; vaccinia virus complement control protein (VCP); Kaposi's sarcoma complement control protein (KCP)

Address for correspondence: Girish J. Kotwal, Division of Medical Virology, IIDMM, Faculty of Health Sciences, University of Cape Town, Observatory 7925, Cape Town, South Africa. Voice: +27 21 406 6676; fax: +27 21 406 6018.

gjkotw01@yahoo.com

Ann. N.Y. Acad. Sci. 1056: 388–404 (2005). © 2005 New York Academy of Sciences. doi: 10.1196/annals.1352.019

INTRODUCTION

Tumor viruses are currently regarded as essential tools in cancer research, as many have led to the discovery of critical cell regulatory proteins such as p53, p300, and E2F, just to name a few. Kaposi's sarcoma-associated herpesvirus (KSHV) is rapidly becoming one of the prime models of viral transformation because its genome has been fully sequenced and annotated. KSHV is also known as human herpesvirus 8 (HHV-8) because it is the eighth human herpesvirus identified to date.[1] This viral strain was initially discovered from an acquired immunodeficiency syndrome-Kaposi's sarcoma (AIDS-KS) skin lesion in 1993 by a molecular biologic technique, representational difference analysis (RDA).[2]

KSHV, the most recently identified human herpesvirus closely related to Epstein-Barr virus, is a large, enveloped, double-stranded DNA virus and a member of the subfamily *Gammaherpesvirinae.*[3] Currently, the proliferative diseases unmistakably related to KSHV infection include a tumor of endothelial cell origin known as Kaposi's sarcoma (KS), some plasma cell forms of multicentric Castleman's disease, which is a B-cell lymphoproliferative disorder, and a body cavity-based or primary effusion lymphoma (BCBL or PEL), which also infects B cells.[4]

The complement system, which is part of the innate immune defense system, is composed of a group of approximately 30 serum proteins that interact with each other in a cascade-type reaction, causing the splitting of subsequent proteins into fragments.[5] The powerful nature of the complement system in either the presence or the absence of antibodies can lead to virus neutralization and opsonization, lysis of virus-infected cells, and amplification of inflammatory and specific immune responses.[6] Considering the potentially destructive nature of the complement proteins, which do not have the ability to differentiate between self and non-self and therefore destroy any cell including host cells,[7] the host tissues therefore have specific inhibitors/regulators of the complement system on their cell surface. Viruses are obligate intracellular parasites and ultimately depend on the cell machinery of the host to survive and propagate. Because the complement proteins implicated in the cascade activation process are unable to discriminate between self and non-self, viruses are highly vulnerable to the complement system. Therefore, during co-evolution, viruses such as KSHV have developed and acquired clever strategies to overcome the complement system of the host in order to succeed as pathogens.[8] These strategies can be divided into two main categories, namely, active strategies, in which the virus has obtained immunomodulatory genes, and passive strategies, such as error-prone replication, which allows for rapid antigenic evolution[9] and, therefore, consequent evasion of the host immune system.

KSHV, like other large DNA viruses, encodes genes whose protein products explicitly undermine immune system responses, including the complement cascade, thereby leading to persistent infection and pathogenesis within the host. One such protein product was predicted by sequence analysis to be encoded by the ORF 4, also shown to share homology with cellular complement control proteins (CCPs) known as regulators of complement activation (RCA).[1] The RCA protein family is characterized by the presence of short consensus repeat (SCR) modules, also referred to as CCP or sushi domains. Each CCP module is comprised of approximately 60 amino acids and consists of a conserved motif made up of four disulfide-linked cysteines, prolines, tryptophan, and many other residues that together form a compact core in a beadlike structure.[10] KSHV ORF 4 gene is expressed on the surface of KSHV-infected cells[8] during the lytic life cycle of the virus.[11] Analysis of the primary sequence as deduced from the ORF 4 revealed that KCP is predicted to be 551 amino acid (aa) residues in size, including a 15-aa signal peptide. KCP is a type 1 membrane-bound protein, and structural analysis based on bioinformatics revealed that KCP is comprised of a number of distinct regions, four N-terminal tandemly repeating CCP domains projecting into the extracellular space, a proline-rich region of about 70 aa, ending in a dicysteine motif, a 202-aa serine/threonine-rich region speculated to be heavily *O*-glycosylated, and a carboxy terminal hydrophobic region of 26 aa, which was identified as a putative transmembrane domain that anchors the protein to the cell membrane.[8]

Transfected mammalian cell lines have been shown to express the KSHV ORF 4 gene as a full-length transcript and as two alternatively spliced transcripts, which only differ in length between the CCP domains and the potential transmembrane region.[8] Functional assays have confirmed the KCP-dependent mechanism by which KSHV can subvert complement attack by the host. The purified native protein was shown to inhibit human complement-mediated lysis of erythrocytes, prevent the deposition of C3b on the cell surface, and act as a cofactor for factor I-mediated cleavage of C3b and C4b. The latter reveals that the KSHV inhibitor of the complement cascade regulates the formation of C3 convertase by turning its subunits, C3b and C4b, into their inactive forms.[12] Furthermore, other studies have revealed that recombinant KCP is twice as efficient in inhibiting the classic pathway-mediated lysis of erythrocytes and approximately five times stronger than **V**accinia virus **c**omplement control **p**rotein (VCP) in inhibiting the alternative pathway-mediated lysis of erythrocytes.[13] The work presented here confirms these results in a sense that when the same expression system but a different vector is used, the *P. pastoris* expression system can be a reliable means for the production of glycosylated bioactive KCP at biologically active significant concentrations. Furthermore, the presence of glycans on other proteins belonging to the RCA family has been shown to extensively influence their biologic activities. For example, the **m**embrane **c**ofactor protein isoforms with a larger *O*-glycosylation domain was found to bind C4b more efficiently than the smaller and less glycosylated isoforms.[14] Perhaps the high potency of recombinant KCP-S in inhibiting complement-mediated lysis of antibody-coated sheep erythrocytes may be attributed to posttranslational modification of the protein such as glycosylation. Furthermore, *N*-glycosylation analysis with peptide-*N*-glycosidase F showed a shift in molecular weight of the protein between the untreated and the treated protein sample.

MATERIAL AND METHODS

Recombinant DNA Technology and Construction of the Expression Plasmids

The three ORF 4 proteins encompassing SCRs 1 to 4 (FIG. 1) were amplified by polymerase chain reaction (PCR) with the Expand™ High Fidelity PCR System (Roche, Mannheim, Germany). KSHV DNA isolated from a KSHV-infected primary effusion lymphoma (PEL) cell line (kindly provided by Prof. Chris Boshoff, University College, London, UK) was used as a template with specific primers 5′-

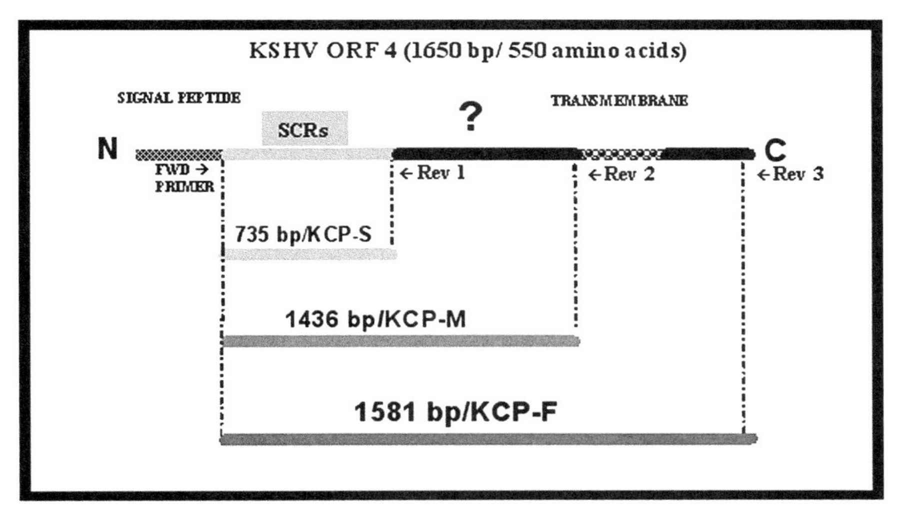

FIGURE 1. Schematic diagram of the recombinant KSHV ORF 4 transcript map. The figure represents the KSHV ORF 4 gene showing the N- and C-terminus as well as the signal peptide (*small dots*), the SCRs-containing region (*light grey*), and the putative transmembrane region (*large spheres*). Three sets of primers were carefully designed consisting of a sequence-specific forward and three consequent reverse primers in order to flank three different-sized DNA transcripts as represented by the three different shades of grey, respectively.

GAATTCAAGTGTTCCCAAAAAACC-3′ (Fwd), 5′-GCGGCCGC**TTA**-CAAAA-CAC-ACTTAGG-3′ (Rev1), 5′-GCGGCCGCCGATTTTAGACGC**TTA**CGGTG-GCTG-3′ (Rev2), and 5′-GCGGCCGC**CTA**ACGAAAGAACAG-3’ (Rev3). The sense primer introduced a 5′ *Eco*RI site and the antisense primers introduced a 3′ *Not*I site as well as stop codons, as represented in boldface. The amplified DNA fragments (FIG. 2) were digested with *Eco*RI and *Not*I restriction enzymes and ligated into the secretory expression plasmid vector pPIC9 (Invitrogen, Carlsbad, CA) among *Eco*RI (5′ end) and *Not*I (3′ end) to generate the pPIC9 constructs, namely, pPIC9/KSHV ORF 4 (735), pPIC/KSHV ORF 4 (1436), and pPIC9/KSHV ORF 4 (1581) (FIG. 3) (TABLE 1). Before ligation into *Eco*RI/*Not*I digested pPIC9 vector, the PCR fragments were cloned into pGEMT-Easy (Promega, Madison, WI) vector (TABLE 1) and digested with the same restriction endonucleases to produce compatible overhangs.

Transformation of P. pastoris *and Selection for Positive Clones*

The pPIC9 plasmid constructs were linearized with *Sac*I prior to transformation of *P. pastoris* GS115 *his*4 competent cells with the *Pichia* Comp^{RT} transformation kit. After transformation, cells were plated on regeneration glucose medium (RDB) (1 M sorbitol, 1% glucose, 1.34% yeast nitrogen base, 0.00004% biotin, and 0.005% amino acid mix containing glutamic acid, methionine, lysine, leucine, and isoleucine). Recombinant colonies auxotrophic for histidine (His$^+$) were selected by

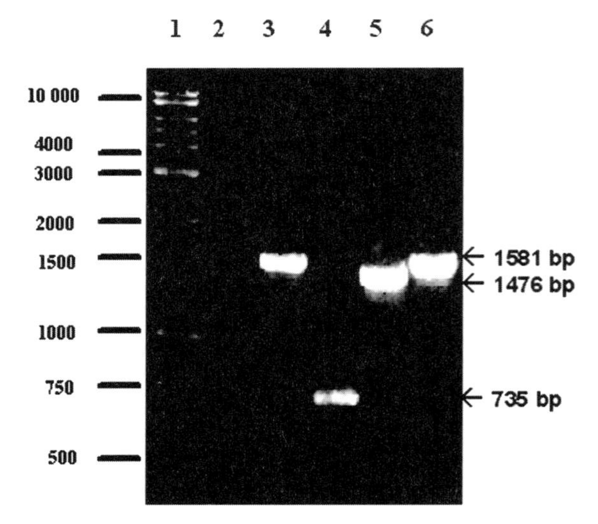

FIGURE 2. Agarose gel of the KSHV ORF 4 PCR products. Ethidium bromide staining of electrophoretically resolved (1% [w/v] agarose gel) PCR products. *Lane 1*: 1-kbp DNA ladder molecular weight marker (10,000, 8,000, 6,000, 5,000, 4,000, 3,000, 2,500, 2,000, 1,500, 1,000, 750, 500, and 250/253 bp [*top to bottom]*). *Lane 2*: VCP amplified from vaccinia virus DNA with the VCP-specific forward and reverse primers used as a PCR-positive control and a faint band of expected fragment size; 700 bp can be visualized. *Lane 3*: KSHV ORF 4-specific 1597-bp PCR product amplified from KSHV DNA template with the forward primer 03-1420 and the reverse primer 03-1421. *Lane 4*: KSHV ORF 4-specific 735 bp PCR product amplified from KSHV DNA template with the forward primer 03-0668 and the reverse primer 03-0669. *Lane 5*: KSHV ORF 4-specific 1436-bp PCR product amplified from KSHV DNA template with the forward primer 03-0668 and the reverse primer 03-0670. *Lane 6*: KSHV ORF 4-specific 1581-bp PCR product amplified from KSHV DNA template with the forward primer 03-0668 and the reverse primer 03-0.671.

their ability in the absence of this amino acid to grow on RDB plates incubated at 30°C for 2–4 days. Screening for the two different possible phenotypes, $His^+ Mut^+$ and $His^+ Mut^s$ was performed by plating the His^+ transformants on minimal glucose (MD) (1.34% yeast nitrogen base, 0.00004% biotin, and 0.5% glucose) *vs* minimal methanol (MM) (1.34% yeast nitrogen base, 0.00004% biotin, and 0.5% methanol) plates. Additionally, genomic DNA was isolated from *P. pastoris* integrants as described by the manufacturer (QIAGEN, Valencia, CA), and the presence of inserts in the yeast genome was analyzed by PCR with 5′-GACTGGTTCCAATTGACAAGC- 3′ and 5′- GCAAATGGCATTCTGACATCC-3′ specific for the *AOX1* promoter and terminator sequences, respectively.

Expression of the Recombinant KCP Proteins in P. pastoris

Small-scale experiments were carried out to test whether the transformed *P. pastoris* recombinant colonies were suitable for KCP expression. Transformed *P. pastoris* colonies were cultured for 16–18 h at 30°C in 5 ml buffered glycerol-

TABLE 1. Plasmids and *Pichia* strains used throughout the study

Plasmid	Relevant characteristics	Source of reference
pGEMT-Easy	3.0 kb, Amp®, T7 promoter, SP6 promoter	Promega
pGEMT/KSHV ORF4 (735)	3.8 kb, Amp®, T7 promoter, SP6 promoter, 735-bp KCP gene	This study
pGEMT/KSHV ORF4 (1436)	4.5 kb, Amp®, T7 promoter, SP6 promoter, 1436-bp KCP gene	This study
pGEMT/KSHV ORF4 (1581)	4.6 kb,Amp®, T7 promoter, SP6 promoter, 1581-bp KCP gene	This study
pPIC9	8.0 kb, Amp®^{R}, HIS4, P_{AOX1}, α-factor signal seq	Invitrogen
pPIC9/KSHV ORF4 (735)	8.8 kb, Amp®, HIS4, P_{AOX1}, α-factor signal seq, 735-bp KCP gene	This study
pPIC9/KSHV ORF4 (1436)	9.5 kb, Amp®, HIS4, P_{AOX1}, α-factor signal seq, 1436-KCP gene	This study
pPIC9/KSHV ORF4 (1581)	9.6 kb, Amp®, HIS4, P_{AOX1}, α-factor signal seq, 1581-bp KCP gene	This study

Strains	Relevant characteristics	Source of reference
GS115	Wild-type, genotype *his*4, phenotype Mut$^+$	Invitrogen
GS115/pPIC9	GS115 transformed with pPIC9	This study
GS115/pPIC9/KSHV ORF4 (735)	GS115 transformed with pPIC9/KSHV ORF4 (735)	This study
GS115/pPIC9 KSHV ORF4 (1436)	GS115 transformed with pPIC9/KSHV ORF4 (1436)	This study
GS115/ pPIC9/KSHV ORF4 (1581)	GS115 transformed with pPIC9/KSHV ORF4 (1581)	This study

complex medium ([BMGY] 0.1% yeast extract, 0.2% peptone, 100 mM potassium phosphate pH 6.0, 1.34% yeast nitrogen base, 0.00004% biotin, and 1% glycerol). Cells were harvested at 3,000 × g for 5 minutes at room temperature and resuspended to an OD_{600} of 1.0 in buffered methanol-complex medium ([BMMY] 0.1% yeast extract, 0.2% peptone, 100 mM potassium phosphate buffer, pH 6.0, 1.34% YNB, 4 $\times 10^{-5}$ biotin, and 1% methanol) for induction of the *AOX*1 promoter. These cultures were maintained for 4 days (30°C and 250 rpm) and supplemented daily with methanol to reach a final concentration of 0.5%. Selected colonies from the small-scale expressions were chosen for medium-scale (100 ml) production of the recombinant KCP proteins. At various time-points, 0, 24, 48, 72, and 96 h after induction, 1-ml aliquots of the cultures were harvested, and both the cell pellets and the supernatant were stored at –80°C prior to SDS-PAGE analysis. After 4 days, the cells were harvested for 5 min at 3,000 × *g*. The supernatant was further centrifuged at 10,000 rpm for 1 h and clarified by passing through an 0.22-m filter to remove contaminating particles. The clarified solution was then concentrated by ultrafiltration through a Pellicon®XL filtration unit (Millipore, Billerica, MA) with a molecular weight ex-

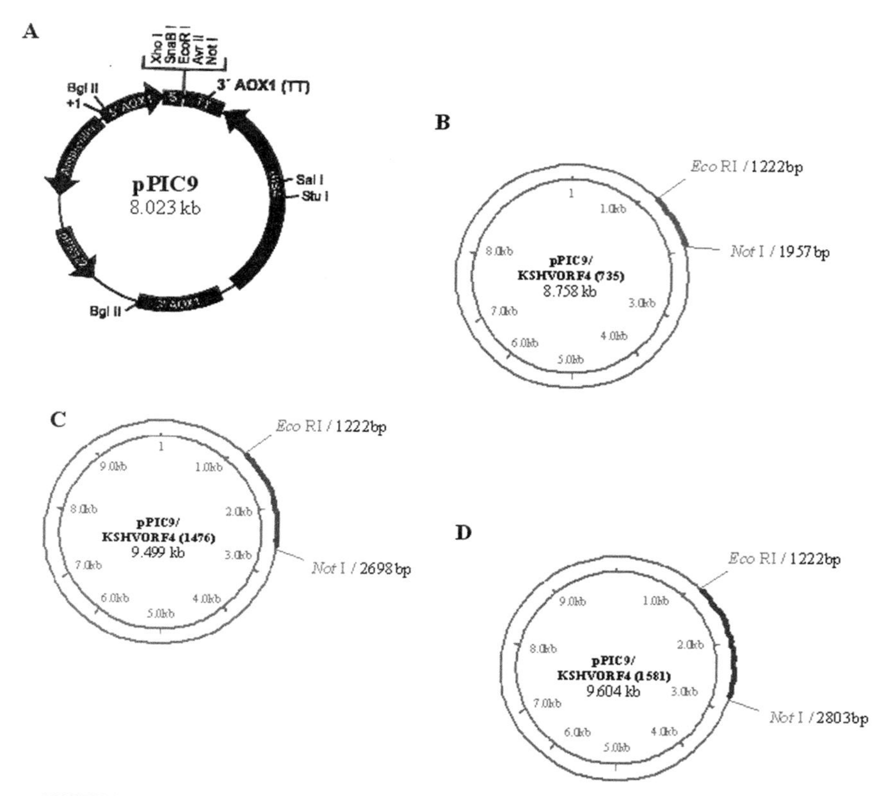

FIGURE 3. Restriction plasmid maps of the pPIC9 vector and constructs. (**A**) Vector circle map showing the location and size of each feature of pPIC9, including a multiple cloning site for easy ligation of the genes of interest to the vector. (**B**) Double-restriction endonuclease digestion of the 8758-bp pPIC9/KSHVORF 4 (735) construct with the REs *Eco*RI and *Not*I produces two fragments of sizes 8023 and 735 bp. (**C**) Double-restriction endonuclease digestion of the 9499-bp pPIC9/KSHVORF 4 (1436) construct with the REs *Eco*RI and *Not*I produces two fragments of sizes 8023 bp and 1436 bp. (**D**) Double-restriction endonuclease digestion of the 9604-bp pPIC9/KSHVORF 4 (1581) construct with the REs *Eco*RI and *Not*I produces two fragments of sizes 8023 bp and 1581 bp.

clusion limit of 10 kDa, which reduced the volume of the supernatant to approximately 15 ml. Supernatants from cultures of colonies transformed with pPIC9 (empty vector) were used as negative controls for experiments after the same induction and concentration steps.

Peptide Synthesis and Purification

Peptides were synthesized by the 9-fluoroenylmethoxycarbonyl technique[15] based on the method described by Atherton and Sheppard[16] with an automated benchtop simultaneous multiple solid-phase peptide synthesizer (PSSM 8 system;

Shimadzu, Tokyo, Japan). Peptides were purified by high-performance liquid chromatography with a C-R7A Shimadzu UV-vis detector and a Shimadzu RF-535 fluorescence detector coupled to a Vydac C_{18} analytical column. Amino acid analysis was carried out in a Beckman 6,300 amino acid analyzer following hydrolysis in 6 N HCl with 5% phenol at 110°C for 48 h. Matrix-assisted laser deabsorption ionization-mass spectrometry was performed on a TofSpec E instrument from Micromass, with a matrix of -cyano-4-hydroxycinnamic acid.

Preparation of Rabbit Anti-KCP Peptide Antisera

The KCP peptide was coupled to Keyhole Limpet Hemocyanin (KLH) using methodology described previously. Once coupled, the peptide-carrier protein conjugate was used in a schedule of injections to elicit an immune response in rabbits against the peptide sequence.

Immunologic Detection of the Recombinant KCP Proteins, SDS-PAGE, and Western Blot

The production of the proteins of interest in the supernatant of the yeast culture medium was analyzed by slot-blot using the manifold II slot-blot system (Schleicher & Schuell Bioscience Inc., Keene, NH) and a vacuum pump. Aliquots (100 µl) of supernatants were vacuum-blotted onto nitrocellulose Hybond-ECL membranes (Amersham, Buckinghamshire, UK), which were, in turn, incubated overnight at 4°C in blocking solution (5% [w/v] non-fat milk powder in tris-buffered saline [TBS] [50 mM Tris-HCl pH 7.5, 150 mM NaCl]). After blocking, the membranes were incubated with rabbit anti-KCP peptide polyclonal antisera at 1:1000 in tris-buffered saline Tween-20 (TBST) (0.1% [v/v] Tween-20 in TBS). After two washes in TBST followed by two washes with blocking solution (10 min each), the membranes were incubated for 30 minutes with horseradish peroxidase-conjugated anti-rabbit IgG (Roche, Mannheim, Germany) at 1:12 000 in blocking solution. The membranes were rinsed thoroughly four times with TBST for 15 minutes each, and bound antibodies were detected using the chemiluminescence Western blotting kit (mouse/rabbit) system (Roche, Mannheim, Germany).

Prepared protein samples were resolved by discontinuous (0.1% [w/v]) SDS and [12% (w/v)] PAGE according to the method of Laemmli[17] and O'Farrel,[18] using a Bio-Rad electrophoresis set.

For Western blot analysis, samples resolved by discontinuous SDS-PAGE were transferred to a nitrocellulose membrane (Hybond-ECL) using the mini trans-blot protean II cell apparatus set (Bio-Rad, Hercules, CA) according to the manufacturer's instructions. Detection of the KCP proteins was carried out as described above.

Biologic Activity Test: Hemolysis Assay

The activities of the heterologously expressed recombinant proteins were determined by testing the ability of the proteins to inhibit complement-mediated lysis of sensitized sheep red blood cells (ssRBCs) using a well-described hemolysis assay.[19] Pure protein was not available for this analysis and therefore different volumes (µl) of the media containing the proteins under investigation were used instead, as a more qualitative approach, in order to determine whether or not the *P. pastoris*-expressed recombinant KCP proteins retained their functional activity. Supernatants from *Pichia* cultures transformed with pPIC9 (no insert) were used as a negative control for the hemolysis assay.

N-Glycosylation Analysis

The supernatant (60–90 μl) from cultures of recombinant strains containing approximately 30 μg of protein of interest was made up to 100 μl with 50 mM sodium phosphate solution, pH 7.5, and the protein sample was concentrated using a SpeedVac apparatus. The concentrated protein was then resuspended in 25 μl of 50 mM sodium phosphate solution, pH 7.5, after which 2.5 μl of denaturation buffer (0.2% SDS with 100 mM β-mercaptoethanol) was added. The reaction mixture was denatured by heating at 100°C for 10 minutes followed by incubation with 3 μl of peptide-*N*-glycosidase F (PNGase F) enzyme solution (500 units/ml) (an enzyme that hydrolyzes all types of *N*-glycan chains from glycoproteins) for 3–18 h at 37°C to allow deglycosylation to occur and then stopped by heating to 100°C for 5 minutes. The samples were then assessed by SDS-PAGE. Purified VCP (from the WR strain), which does not contain any *N*-linked carbohydrate sites, was used as a negative control to show that any shift in molecular weight was not due to contaminating proteases but to PNGase F activity.

RESULTS AND DISCUSSION

Recombinant DNA Technology and Construction of the Expression Plasmids

The three DNA portions of the KSHV ORF 4 encompassing SCRs 1 to 4 were successfully amplified from a KSHV DNA template by PCR, using a KSHV ORF 4-specific forward primer and three consecutive KSHV ORF 4-specific reverse primers. The PCR products (10 μl) were resolved by gel electrophoresis on 1% (w/v) agarose gel (FIG. 2).

For the heterologous production of the KCPs in *P. pastoris*, the PCR products were ligated to the expression vector pPIC9 to generate three recombinant plasmids, pPIC9/KSHV ORF 4 (735), pPIC9/KSHV ORF 4 (1436), and pPIC9/KSHV ORF 4 (1581) coding for a soluble protein comprising the four sushi domains (S-), a larger protein lacking the putative transmembrane-binding domain (M-), and the full-length ORF 4 (F-), respectively (FIG. 3).

Double-stranded plasmid DNA template, pPIC9/KSHVORF 4 (735), pPIC9/KSHVORF 4 (1436), and pPIC9/KSHVORF 4 (1581), isolated from *Escherchia coli* transformants was subjected to restriction endonuclease digestion, and the DNA fragments were analyzed by agarose gel electrophoresis (FIG. 4). The *Eco*RI- and *Not*I-cut pPIC9/KSHVORF 4 (735) (FIG. 4, lanes 2 and 4) resolved into two visible bands corresponding to the expected 8.0 kbp vector and the 735-bp DNA fragment, while that of the *Eco*RI- and *Not*I-cut pPIC9/KSHVORF 4 (735) (FIG. 4, lane 3) resolved into only one visible band corresponding to the 8,000-bp vector DNA, showing that two of the three constructs contained the 735-bp gene of interest. Furthermore, the *Eco*RI- and *Not*I-cut pPIC9/KSHVORF 4 (1436) (FIG. 4, lanes 5–7) resolved into two visible bands corresponding to the expected 8,000-bp vector and

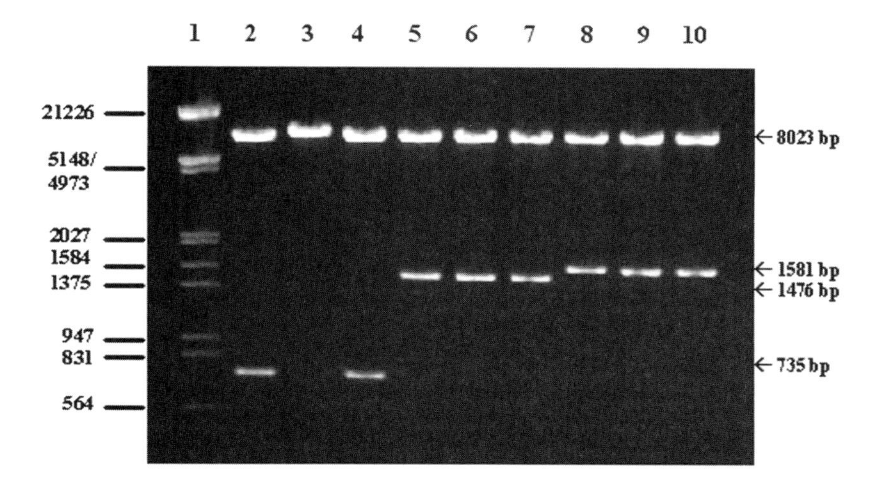

FIGURE 4. Agarose gel of the pPIC9/KSHV ORF 4 constructs RE analysis. Ethidium bromide staining of electrophoretically resolved (1% [w/v] agarose gel) *Eco*RI and *Not*I digested pPIC9/KSHV ORF 4 DNA constructs. *Lane 1*: λ DNA/*Eco*RI and *Hind*III molecular weight marker (21226, 5148/4973, 4268, 3530, 2027, 1904, 1584, 1375, 947, 831, 564, 125 bp [from *top to bottom*]). *Lane 2*: *Eco*RI and *Not*I digested pPIC9/KSHVORF 4 (735) into which the KSHV ORF 4 *Eco*RI/*Not*I 735-bp DNA fragment was successfully ligated. *Lane 3*: *Eco*RI and *Not*I digested pPIC9/KSHVORF 4 (735) into which the KSHV ORF 4 *Eco*RI/*Not*I 735-bp DNA fragment was not ligated. *Lane 4*: Same as lane 2. *Lanes 5-7*: *Eco*RI and *Not*I digested pPIC9/KSHVORF 4 (1436) into which the KSHV ORF 4 *Eco*RI/*Not*I 1436-bp DNA fragment was successfully ligated. *Lanes 8-10*: *Eco*RI and *Not*I digested pPIC9/KSHVORF 4 (1581) into which the KSHV ORF 4 *Eco*RI/*Not*I 1581-bp DNA fragment was successfully ligated.

the 1436-bp DNA fragment, showing that all three recombinants under investigation were transformed with pPIC9 containing the 1436-bp insert of interest. Similarly, the *Eco*RI- and *Not*I-cut pPIC9/KSHVORF 4 (1581) (FIG. 4, lanes 8–10) resolved into two visible bands corresponding to the expected size of the 8.0 kbp vector and the 1581-bp DNA fragment, showing again that the three picked *E. coli* recombinants were successfully transformed with pPIC9 containing the 1581-bp gene of interest.

Transformation of P. pastoris *and Selection for Positive Clones*

Plasmid vector pPIC9 and pPIC9-constructs were digested with *Sac*I, and after visualizing the expected products by agarose gel electrophoresis, they were used to transform *P. pastoris* GS115 competent cells. More than 50 colonies of transformed *P. pastoris* selected for their ability to grow in the absence of histidine (His^+) were obtained in all transformations carried out.

To confirm that the three different-sized KSHV ORF 4 gene had integrated into the *P. pastoris* genome, genomic DNA isolated from colonies transformed with empty pPIC9 or KCP-containing vectors was analyzed by PCR with *AOX* 1 external primers. PCR products corresponding to fragments of the expected sizes of the empty vector (492 bp) and the pPIC9/KSHV ORF 4 (735), (1436), and (1581)-containing transformants (1227, 1968, and 1581 bp, respectively) were observed after agarose gel analysis (data not shown).

Expression and Detection of the Recombinant KCP Proteins in P. pastoris

Recombinant colonies with the KSHV ORF 4 gene integrated in their genomes were used for small expression experiments (5 ml) in BMMY medium. Expression of the KCP-M and KCP-F viral proteins was confirmed by immunoblot analysis after blotting 100 μl of culture media supernatant onto nitrocellulose Hybond-ECL membranes (FIGS. 6A and 7A, respectively).

Small-scale expression was followed by medium-scale (100 ml) production of KSHV ORF 4 proteins from recombinant *P. pastoris* colonies transformed with the three *Sac*I-digested pPIC9/KSHV ORF 4 plasmid constructs. Culture media supernatants from colonies transformed with the pPIC9 empty vector digested with *Sac*I

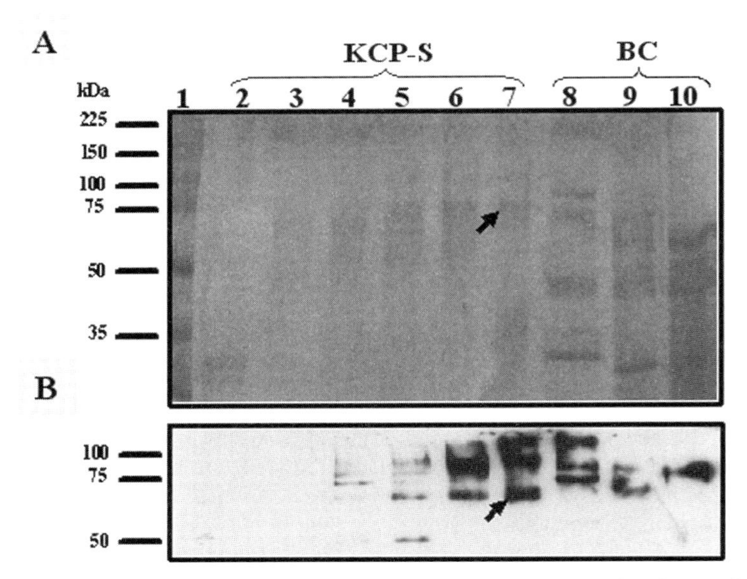

FIGURE 5. Expression and detection of the recombinant KCP–S protein. (**A**) Coomassie-stained SDS-PAGE gel of the induction profile of the KSHV ORF 4-S recombinant protein. *Lane 1*: Broad range protein molecular weight marker: 225, 150, 100, 75, 50, 35, 25, 15, and 10 kDa (*top to bottom*). *Lane 2*: Supernatant before induction. *Lane 3*: 24 h post-induction. *Lane 4*: 48 h postinduction. *Lane 5*: 72 h postinduction. *Lane 6*: 96 h postinduction. *Lane 7*: 96-h postinduction concentrated supernatant using the Pellicon XL device. *Lane 8*: BC pre-induction. *Lane 9*: BC 72 h postinduction. *Lane 10*: BC 96 h postinduction. (**B**) Western blot detection of the expression profile using rabbit anti-KCP peptide polyclonal antibody. In addition to the protein of interest, the antibody recognized several larger and smaller bands that may represent multimer aggregates and degradation products, respectively. The time-course expression of the protein was monitored every 24 h from 0 to 96 h (*lanes 2–6*) after induction, loading 30 μl of clarified supernatant per well. *Arrows* indicate maximal level of protein.

pre- and 96 h postinduction were included as a negative control in all of the protein expression and detection experiments.

The time-course of expression of the KCP-S, KCP-M, and KCP-F proteins was monitored by SDS-PAGE (FIGS. 5A, 6B, and 7B, respectively), and in the first two recombinant proteins, the induction profile was also followed by Western blot analysis, which showed that the protein could first be detected at 48 h postinduction and that the amount of protein continued to build until 72 and 96 h (FIGS. 5B and 6C, respectively). In addition, the antibody recognized several larger and smaller bands that may represent multimer aggregates and degradation products, respectively.

Intensities of SDS-PAGE bands were scanned with a densitometer (ChemiImager™ 5500), and the concentration of the *P. pastoris*-expressed KCP protein was quantitively estimated by comparison with a standard curve obtained with known

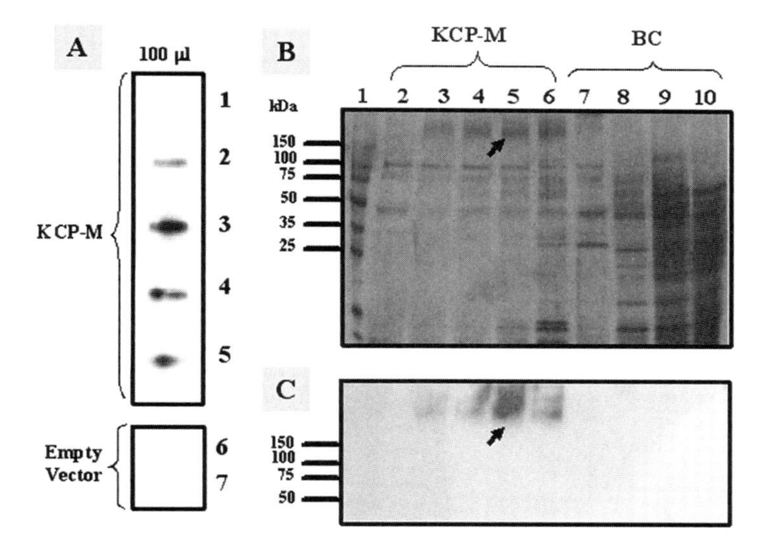

FIGURE 6. Expression and detection of the recombinant KCP–M protein. (**A**) Slot-blot incubated with rabbit polyclonal anti-KCP peptide antibody after medium-scale KCP-M protein expression with a *P. pastoris* recombinant colony obtained upon transformation with *Sac*I-digested pPIC9/KSHV ORF 4 (1436 bp) DNA construct. 100 µl of supernatant from the pre-induction sample (*row 1*) and from the different time-points postinduction (*rows 2-5*) were blotted as indicated. Equal volumes of supernatant of a recombinant colony transformed with *Sac*I-digested pPIC9 empty vector (Background control [BC]) pre-induction (*row 6*) and 96 h postinduction (*row 7*) samples were also included. (**B**) Coomassie-stained SDS-PAGE gel of the induction profile of the KCP-M recombinant protein. *Lane 1:* Broad range protein molecular weight marker: 225, 150, 100, 75, 50, 35, 25, 15, and 10 kDa (from *top to bottom*). *Lane 2*: Supernatant before induction. *Lane 3*: 24 h postinduction. *Lane 4:* 48 h post-induction; *Lane 5*: 72 h post-induction; *Lane 6:* 96 h post-induction; *Lane 7*: BC before induction. *Lane 8*: BC 48 h postinduction. *Lane 9*: BC 72 h postinduction. *Lane 10:* BC 96 h postinduction. (**C**) Western blot detection of the expression profile using rabbit anti-KCP peptide polyclonal antibody. The time-course expression of the protein was monitored every 24 h from 0 to 96 h (*lanes 2–6*) after induction, loading 30 µl of clarified supernatant per well. *Arrows* indicate maximal level of protein.

amounts of protein. Under our conditions, the yield of *P. pastoris*-expressed KCP-S, KCP-M, and KCP-F was about 5, 6.4, and 1.76 μg/ml of yeast culture medium, respectively. This protein concentration estimation correlates with the first two recombinant proteins, KCP-S and KCP-M, obtained from medium-scale expression (100 ml), whereas the full-length KCP-F protein was obtained from a small-scale experiment (5 ml). In the latter, the amount of detectable protein was very low, and as a result, only upon concentration of 100 μl of protein sample was a successful Western blot of the protein obtained (FIG. 7C).

Transformants expressing detectable protein varied dramatically among different clones, even from the same transformation procedure. Several recombinant colonies

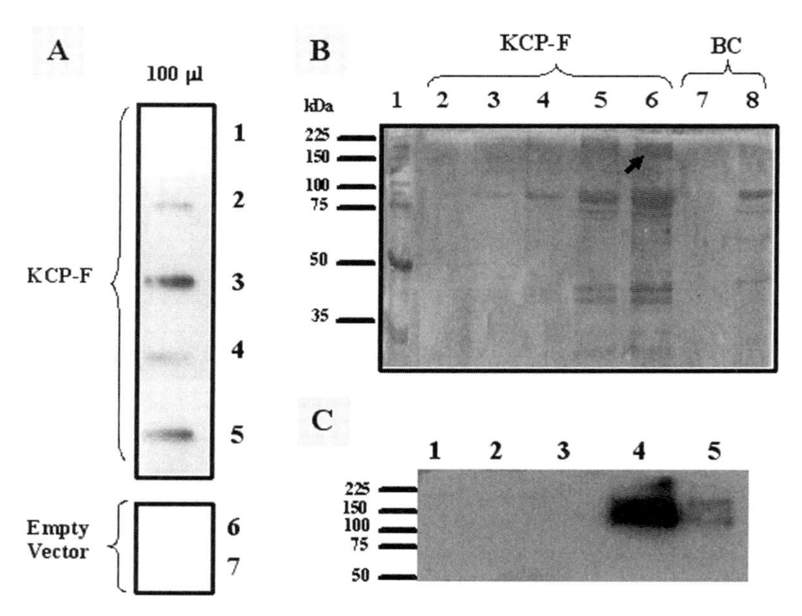

FIGURE 7. Expression and detection of the recombinant KCP–F protein. (**A**) Slot-blot incubated with polyclonal anti-KCP peptide antibody after medium-scale KSHV ORF 4-F protein expression with a *P. pastoris* recombinant colony obtained upon transformation with *Sac*I-digested pPIC9/KSHV ORF 4 (1581 bp) DNA construct. 100 ml of supernatant from the pre-induction sample (*row 1*) and from the different time-points postinduction (*rows 2-5*) were blotted as indicated. Equal volumes of supernatant of a recombinant colony transformed with *Sac*I-digested pPIC9 empty vector (background control [BC]); pre-induction (*row 6*) and 96 h postinduction (*row 7*) samples were also included. (**B**) Coomassie-stained SDS-PAGE gel of the KSHV ORF 4-F recombinant protein induction profile. *Lane 1:* Broad range protein molecular weight marker: 225, 150, 100, 75, 50, 35, 25, 15, and 10 kDa (*top to bottom*). *Lane 2*: Supernatant before induction. *Lane 3:* 24 h postinduction. *Lane 4*: 48 h postinduction. *Lane 5*: 72 h postinduction. *Lane 6:* 96 h postinduction. *Lane 7*: BC before induction. *Lane 8*: BC 96 h postinduction. (**C**) Western blot detection of the expression profile using anti-KCP peptide polyclonal antibody. *Lane 1*: Same as *lane 1* above. *Lane 2*: Supernatant before induction (20 ml). *Lane 3*: 72 h postinduction (20 ml). *Lane 4*: 72 h postinduction (100 ml concentrated with the speedvac). *Lane 5:* KCP-M 96 h postinduction supernatant used as a positive control for Western blot analysis.

were analyzed for protein expression, and although the presence of insert was confirmed by PCR, not all clones expressed detectable amounts of protein. However, the small-scale expressions showed that both Mut^+ and Mut^S recombinant *P. pastoris* transformants were able to express the KSHV ORF 4 protein provided that a good clone was selected and used for the expression. According to the Invitrogen manual, Mut^+ colonies are capable of producing larger quantities of protein in shorter periods of time; therefore, Mut^+ transformants were preferably used throughout this study.

Biologic Activity Test: Hemolysis Assay

To determine whether or not the *P. pastoris*-expressed recombinant KCP proteins retained their functional activity, the ability of these proteins in media to inhibit the complement classic pathway-mediated lysis of antibody-coated sheep erythrocytes was investigated. Our results noticeably demonstrate that unlike the ammonium-sulfate–precipitated supernatant from a *Pichia* culture transformed with pPIC9/KCP-M and F, the ammonium-sulfate–precipitated supernatant from a *Pichia* culture transformed with pPIC9/KCP-S contained a functional protein capable of binding complement proteins in a concentration-dependent fashion (FIG. 8). If taking into account the low percentage complement inhibition caused by the same volume of background control, in other words, supernatants from *Pichia* culture transformed with pPIC9 (no insert) that underwent the same induction and concentration steps as the test samples, it is clear that a KSHV functional regulator of human complement was successfully expressed in *P. pastoris* using the pPIC9/KCP-S construct. The

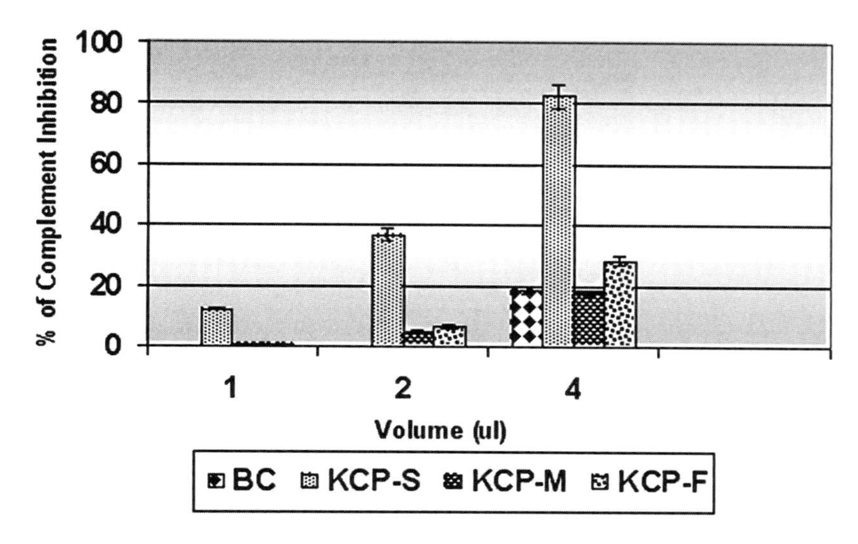

FIGURE 8. Percentage inhibition of complement-mediated lysis of ssRBC by KCP. The relative effects of the different recombinant KCP proteins, KCP-S, KCP-M, and KCP-F, on the classic pathway-mediated lysis of antibody-coated sheep erythrocytes were examined as previously described.[17] BC, supernatant from *Pichia* culture transformed with pPIC9 (no insert), was included as a negative control.

lack of biologic activity obtained in the case of the other two recombinant KCPs may be attributed to the fact that unfractionated media were used containing many native contaminating bands (FIGS. 6B and 7B), and therefore it is possible that something in the yeast media may have inhibited the activity of KCP-M and KCP-F. Another possibility is that these two recombinant KCPs may have been folded differently before being secreted into the media because of their considerable molecular sizes, and therefore perhaps some of the complement binding sites may have been covered or modified in some way, causing insufficient biologic activity. Furthermore, the rKCP-F may have undergone shedding, in other words, cleavage of the ectodomain may have taken place, resulting in the release of a soluble but inactive protein. This well recognized but not fully understood biochemical process represents an important and efficient strategy of activity regulation in a number of transmembrane proteins.[20]

N-Glycosylation Analysis

To examine rKCP for *N*-linked glycosylation, the supernatants from cultures of the recombinant strains were treated with PNGase F and subject to SDS-PAGE

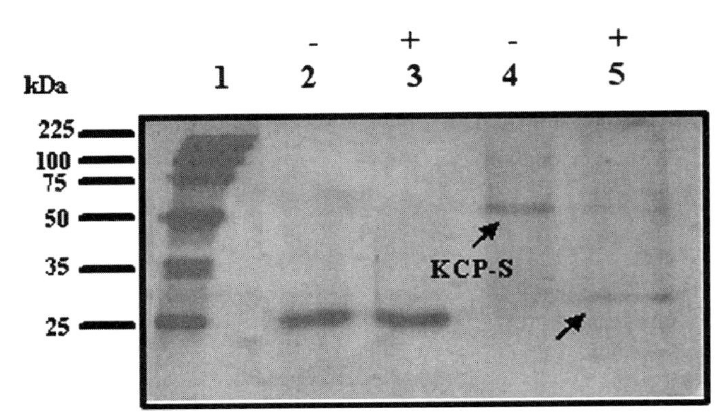

FIGURE 9. SDS-PAGE analysis of enzymatic deglycosylation with PNGase F shows that the rKCPs are *N*-glycosylated. (**A**) Coomassie-stained SDS-PAGE gel of the PNGase F enzyme digestion of the KCP-S. *Lane 1*: Broad range protein molecular weight marker: 225, 150, 100, 75, 50, 35, 25, 15, and 10 kDa (*top to bottom*). *Lanes 2 and 3*: VCP from the WR strain, which does not contain any *N*-linked carbohydrate sites was used as a PNGase F deglycosylation negative control. *Lane 4*: Supernatant of culture from pPIC9/KSHV ORF 4 (735 bp) recombinant strain. *Lane 5*: Supernatant of culture from pPIC9/KSHV ORF 4 (735bp) recombinant strain after incubation with PNGase F. (**B**) Coomassie-stained SDS-PAGE gel of the PNGase F enzyme digestion of the KCP-M. *Lane 1*: Same as *lane 1* above. *Lanes 2 and 3*: Same as *lanes 4 and 5* above, respectively (positive control). *Lane 4*: Supernatant of culture from pPIC9/KSHV ORF 4 (1436 bp) recombinant strain. *Lane 5*: Supernatant of culture from pPIC9/ KSHV ORF 4 (735 bp) recombinant strain after incubation with PNGase F. (**C**) Coomassie-stained SDS-PAGE gel of the PNGase F enzyme digestion of the KCP-F. *Lanes 1 and 6*: Same as lane 1 above. *Lanes 2 and 3*: Same as lanes 4 and 5 (FIG. 9A), respectively (positive control). *Lane 4*: Supernatant of culture from pPIC9/KSHV ORF 4 (1581 bp) recombinant strain. *Lane 5*: Supernatant of culture from pPIC9/KSHV ORF 4 (1581 bp) recombinant strain after incubation with PNGase F. Plus (+) and negative (–) symbols indicate with and without PNGase F treatment, respectively. *Arrows* indicate the *N*-deglycosylated protein of interest.

analysis. The results showed considerable size difference between treated and untreated rKCP-S proteins (FIG. 9A) and a size difference, but to a lesser extent, between treated and untreated rKCP-M and rKCP-F (FIG. 9B and C, respectively), indicating the presence of *N*-glycosylation in the rKCPs.

Furthermore, Western blot analysis with the rabbit polyclonal antibody directed against a selected KSHV ORF 4 peptide common to all three KCP recombinant proteins showed that the *P. pastoris* heterologously expressed soluble-, medium-, and full-length KCPs migrated electrophoretically as higher bands, which most likely is due to glycosylation of the proteins as, in addition to the presence of potential *N*- and *O*-linked carbohydrate sites in the protein's primary sequence, diffused bands were obtained in all the Western blots carried out, which is characteristic of glycoproteins.

In broad terms, knowledge acquired from this study may be implemented in health and disease in the sense that inadequate activation and regulation of the complement system are regarded as the main causes of tissue destruction. This, in turn, has been shown to lead to the progress of many medical conditions such as Alzheimer's disease, traumatic brain injury, ischemia, arthritis, and rejection of xenogeneic transplants,[21–24] just to name a few. Therefore, the use of CCP and its derivatives as inhibitors of undesirable complement activation is currently a very attractive area of research. More specifically, knowledge acquired from this expression study may provide some new insight into the utility and limitations of the *P. pastoris* system for the production of recombinant KCP, and perhaps with the use of different vectors, expression cassettes, or even different transformation techniques, the heterologous production of KCP may be greatly enhanced, which can then be used in various structure-function studies. It is of major significance to understand how CCPs, such as the KCP protein, perform their biologic functions at the structural level in order to gain insight into the development of new and more efficient therapeutic inhibitors of complement.

ACKNOWLEDGMENTS

Prof. Chris Boshoff of the University College London, UK, is gratefully acknowledged for KSHV DNA from PEL. Dr. Odutayo Odunuga is acknowledged for helpful discussions. N.A.G.P. is a recipient of the Poliomyelitis Research Foundation (PRF) Bursary Grant and the National Research Foundation (NRF)/Department of Labour (Dol) scarce skills scholarship. G.J.K. is currently a Senior International Welcome Trust Fellow for Biomedical Sciences in South Africa.

REFERENCES

1. RUSSO, J.J., R.A. BOHENZKY, M.-C. CHIEN, *et al.* 1996. Nucleotide sequence of the Kaposi sarcoma-associated herpesvirus (HHV8). Proc. Natl. Acad. Sci USA **93:** 1482–1486.
2. CHANG, Y., E. CESARMAN, M.S. PESSIN, *et al.* 1994. Identification of herpesvirus-like DNA sequences in AIDS-associated Kaposi's sarcoma. Science **266:** 1865–1869.
3. JENNER, R.G. & C. BOSHOFF. 2002. The molecular pathology of Kaposi's sarcoma-associated herpesvirus. Biochim. Biophys. Acta **1602:** 1–22.
4. MOORE, P.S. & Y. CHANG. 2001. Kaposi's sarcoma-associated herpesvirus. *In* Fields Virology, 4th ed., Vol. 2. D.M. Knipe & P.M. Howley, Eds. :2803–2833. Lippincott-Williams & Wilkins. Philadelphia, PA.

5. Walport, M.J. 2001. Complement. N. Engl. J. Med. **344:** 1140–1144.
6. Blom, A.M. 2004. Strategies developed by bacteria and virus for protection from the human complement system. Scand. J. Clin. Lab. Invest. **64:** 479–496.
7. Sahu, A. & J.D. Lambris. 2001. Structure and biology of complement protein C3, a connecting link between innate and acquired immunity. Immunol. Rev. **180:** 35–48.
8. Spiller, O.B., M. Robinson, E. O'Donnell, *et al.* 2003. Complement regulation by Kaposi's sarcoma-associated herpesvirus ORF4 protein. J. Virol. **77:** 592–599.
9. Chaston, T.B. & B.A. Lidbury. 2001. Genetic "budget" of viruses and the cost to the infected host: a theory on the relationship between the genetic capacity of viruses, immune evasion, persistence and disease. Immunol. Cell Biol. **79:** 62–66.
10. Barlow, P.N., A. Steinkasserer, D.G. Norman, *et al.* 1993. Solution structure of a pair of complement modules by nuclear magnetic resonance. J. Mol. Biol. **232:** 268–284.
11. Jenner, R.G., M.M. Alba, C. Boshoff & P. Kellam. 2001. Kaposi's sarcoma-associated herpesvirus latent and lytic gene expression as revealed by DNA arrays. J. Virol. **75:** 891–902.
12. Spiller, O.B., D.J. Blackbpourn, L. Mark, *et al.* 2003. Functional activity of the complement regulator encoded by Kaposi's sarcoma associated herpesvirus. J. Biol. Chem. **278:** 9283–9289.
13. Mullick, J., J. Bernet, A.K. Singh, *et al.* 2003. Kaposi's sarcoma-associated herpesvirus (human herpesvirus 8) open reading frame 4 protein (Kaposica) is a functional homolog of complement control proteins. J. Virol. **77:** 3878–3881.
14. Liszewski, M.K., M.K. Leung & J.P. Atkinson. 1998. Membrane cofactor protein: importance of *N*- and *O*-glycosylation for complement regulatory function. J. Immunol. **161:** 3711–3718.
15. Hirata, I.Y., M.H.S. Cezari, C.R. Nakaie, *et al.* 1994. Internally quenched fluorogenic protease substrates: solid phase synthesis and fluorescence spectroscopy of peptides containing ortho-amino benzoyl/dinitrophenyl groups as donor-acceptor pairs. Lett. Pept. Sci. **1**: 299–308.
16. Atherton, B. & R.C. Sheppard, Eds. 1989. Solid Phase Peptide Synthesis: A Practical Approach. Oxford University Press. Oxford, England.
17. Laemmli, U.K. 1970. Cleavage of structural proteins during the assembly of the head of bacteriophage T4. Nature **227:** 680–685.
18. O'Farrel, P.H. 1975. High resolution two-dimensional electrophoresis of proteins. J. Biol. Chem. **250:** 4007–4021.
19. Kotwal, G.J., S.N. Isaacs, R. McKenzie, *et al.* 1990. Inhibition of the complement cascade by the major secretory protein of vaccinia virus. Science **250:** 827–830.
20. Werb, Z. & Y. Yan. 1998. A cellular striptease act. Science **282:** 1279–1280.
21. Daly, J. & G.J. Kotwal. 1998. Pro-inflammatory complement activation by the A peptide of Alzheimer's disease is biologically significant and can be blocked by vaccinia virus complement control protein. Neurobiol. Aging **19:** 619–627.
22. Hicks, R.R., K.L. Keeling, M.-Y. Yang, *et al.* 2002. Vaccinia virus complement control protein enhances functional recovery after traumatic brain injury. J. Neurotrauma **19:** 705–714.
23. Low, J.M. & T.L. Moore. 2005. A role for the complement system in rheumatoid arthritis. Curr. Pharm. Des. **11:** 655–670.
24. Sacks, S.H., P. Chowdhury & W. Zhou. 2003. Role of the complement system in rejection. Curr. Opin. Immunol. **15:** 487–492.

Antiesophageal Cancer Activity from Southern African Marine Organisms

CATHERINE E. WHIBLEY,[a] ROBERT A. KEYZERS,[b] ANDREW G. SOPER,[b] MICHAEL T. DAVIES-COLEMAN,[b] TOUFIEK SAMAAI,[c] AND DENVER T. HENDRICKS[a]

[a]*Division of Medical Biochemistry, University of Cape Town, Medical School, Observatory, Cape Town, 7925, South Africa*

[b]*Department of Chemistry, Rhodes University, Grahamstown, 6140, South Africa*

[c]*Council for Scientific and Industrial Research, Durban, 4013, South Africa*

ABSTRACT: Squamous cell esophageal cancer presents a significant health burden in many developing countries around the world. In South Africa, this disease is one of the most common causes of cancer-related deaths in black males. Because this cancer is only modestly responsive to available chemotherapeutic agents, there is a need to develop more effective therapeutic agents for this cancer. Marine organisms are currently regarded as a promising source of unique bioactive molecules because they display a rich diversity of secondary metabolites. Some of these compounds have significant anticancer activity, with a few of these currently in phase I and II clinical trials. We report here an ongoing program to screen marine organisms collected from subtidal benthic communities off the coast of southern Africa for activity against cultured esophageal cancer cells. Of the 137 extracts tested, 2.2% displayed high activity (score = 3) and 11.7% displayed moderate activity (score = 2) against cultured esophageal cancer cells. Our results suggest that sponges had a higher hit rate (21.9%) than ascidians (7.1%). Using activity-directed purification, seven previously described compounds and four novel compounds, with varying activity against esophageal cancer cell lines, were isolated from the sponges *Axinella weltneri*, *Aplysilla sulphurea*, and *Strongylodesma aliwaliensis*. The results of this study suggest that subtidal benthic marine organisms collected off the coast of southern Africa hold potential for identifying possible drug leads for the development of agents with activity against esophageal cancer.

KEYWORDS: esophageal squamous cell carcinoma (ESCC); southern Africa; marine organisms; anticancer agents

Address for correspondence: Denver Hendricks, Division of Medical Biochemistry, University of Cape Town, Medical School, Anzio Road, Observatory, Cape Town, 7925, South Africa. Voice: +27 (0)21 406-6266; fax: +27 (0)21 406 6061.
hendricd@curie.uct.ac.za

Ann. N.Y. Acad. Sci. 1056: 405–412 (2005). © 2005 New York Academy of Sciences.
doi: 10.1196/annals.1352.031

INTRODUCTION

Localized areas of significantly elevated levels of esophageal squamous cell carcinoma (ESCC) occur in many parts of the world. These regions largely occur in developing countries including areas in China, Africa, Iraq, and South America.[1] In South Africa, this disease is the second most common cancer in black males, with an age standardized rate of 16.22.[2] ESCC is nearly always detected at a late stage because of the asymptomatic nature of the early stages of this disease, resulting in a poor prognosis.

Currently, chemotherapy remains an important component of the therapeutic options used to treat this devastating disease. Cisplatin and 5-fluorouracil are widely used, and if treatment is initiated at an early stage (which rarely occurs), complete remission occurs in 20–30% of treated patients.[3] For the most part, ESCC is only moderately responsive to the most commonly used chemotherapeutic drug (cisplatin), and treatment regimens are frequently interrupted because of severe emetic, nephrotoxic, and neurotoxic side effects.[3] The modest success rate of chemotherapeutic agents in the treatment of ESCC holds considerable promise, because it is unlikely that other treatment regimens (radiotherapy and surgery) will yield improved response rates in the near to slightly distant future, unless significant advances are made in diagnosing the disease at an early stage. Clearly a need exists for a new generation of chemotherapeutic agents that are substantially more effective in destroying esophageal cancer cells than the agents currently available.

Many marine organisms produce biologically active natural products (secondary metabolites) for reasons including chemical defense against predation or response to interspecies competition for limited resources. Marine organisms frequently contain novel chemical structures rarely expressed in terrestrial organisms, underscoring the importance of this resource as a repository of potential drug leads with wide ranging applications. Biologically active marine natural products have provided many important lead compounds for both the pharmaceutical and agrochemical industries.[4] Blunt *et al.*[5] recently showed that sponges closely followed by microorganisms, coelenterates, and tunicates are the primary source of bioactive metabolites in the marine environment. Interestingly, anticancer activity dominates the bioactivity reported for the secondary metabolites isolated from all four of these phyla, and there are currently 21 sponge and tunicate metabolites in preclinical or clinical trials as potential anticancer treatments.

In the southern African context, the approximately 3,000-km long coastline, stretching from Namibia in the west to southern Mozambique in the east, is characterized by high levels of endemism.[6] Associated with high levels of marine invertebrate endemism are concomitant high levels of novel secondary metabolite structural diversity. The discovery of marine natural products from southern African marine invertebrates is not unprecedented. Two important groups of anticancer compounds, the cephalostatins and the spongistatins, were isolated from two southern African marine invertebrates, a marine worm, *Cephalodiscus gilchristi*, and the sponge *Spirastrella spinispirulifer*, respectively.[7,8] We have thus embarked on a multi-institutional and multidisciplinary program to identify and characterize novel drug leads extracted from marine invertebrates collected off the shore of southern Africa that show activity against esophageal cancer.

EXPERIMENTAL STUDY

Marine Invertebrate Collection and Extraction

Bulk samples (ca. 1 kg each) of the marine invertebrates were collected by hand using SCUBA at depths ranging from 3–40 m. Samples were frozen immediately after collection and freeze-dried prior to extraction. Each lyophilized marine invertebrate was steeped in MeOH for several days and the MeOH extract partitioned between H_2O and EtOAc. Both aqueous and organic partition fractions were evaporated to dryness *in vaccuo* and selected fractions bioassayed for antiesophageal cancer activity.

Cell Culture

Human esophageal cancer cell lines WHCO1, WHCO6, and KYSE30 were routinely maintained at 37°C, 5% CO_2. Cells were cultured in DMEM supplemented with 10% (WHCO1 and WHCO6) or 5% (KYSE30) fetal calf serum, 100 U/ml penicillin, and 100 μg/ml streptomycin.

Crystal Violet Assay

Initial screening for cytotoxicity was carried out as described previously[9] by plating cells at 1,500 cells/well in 90 μl DMEM in CellStar 96-well plates. After 24 hours of incubation, test samples were added in 10 μl DMEM to a final concentration of 1, 10, and 50 μg/ml, with solvent (DMSO) at 0.2%. After 48 hours of incubation, plates were processed for staining[10] and read at 595 nm on an Anthos microplate reader 2001.

MTT Assay

IC_{50} determinations were carried out using the MTT kit from Roche (Cat #1465007). Briefly, 1,500 cells were seeded per well in 90 μl DMEM in Cellstar 96-well plates. Cells were incubated for 24 hours, and test samples were plated at a range of concentrations in 10 μl DMEM, with solvent (DMSO) at 0.2%. After 48 hours of incubation, plates were processed according to the manufacturer's instructions and read at 595 nm on an Anthos microplate reader 2001.

General Isolation Procedure for the Purification of Bioactive Marine Natural Products

A portion of the biologically active fractions (ca. 2 g) was dissolved in either Me_2CO or MeOH (250 ml) and loaded onto a column of HP-20® resin pre-equilibrated with Me_2CO (250 ml) and MeOH (250 ml). The HP-20 column was sequentially eluted with 250-ml aliquots of aqueous Me_2CO (0, 25, 50, 75, and 100% Me_2CO in H_2O). Each fraction was assayed for biologic activity. If the activity was concentrated in the 25 or 50% $Me_2CO_{(aq)}$ fractions, then these fractions were concentrated to dryness, and final purification was achieved using C18 Sep-Pak® and/or C18 RP-HPLC (MeOH/H_2O). If the bulk activity from the HP-20 column was partitioned in the 75 or 100% Me_2CO fractions, these fractions were concentrated to dryness and chromatogrammed on a silica column and/or silica HPLC using mixtures of EtOAc/Hexane to afford the final pure compounds.

Identification of Pure Compounds

Identification of pure compounds was achieved using a variety of standard spectroscopic techniques including nuclear magnetic resonance (NMR), Fourier transform infrared (FT-IR), and UV spectroscopy, low resolution mass spectrometry (LR-MS) and high resolution mass spectrometry (HR-MS), optical rotation, gas chromatography (GC), and chemical derivatization.

RESULTS AND DISCUSSION

In this report, extracts from 55 sponges, 70 ascidians (tunicates), and 12 soft corals (coelenterates) were screened for activity against esophageal cancer cell lines. Sponge, ascidian, and soft coral extracts were chosen for screening because of the high probability that extracts from these organisms would contain anticancer activity and because of the availability in South Africa of taxonomic expertise for these groups of organisms. The organisms were collected from subtidal benthic communities off southern Mozambique, the Aliwal Shoal (Kwazulu-Natal, South Africa), Algoa Bay, and the Tsitsikamma National Park (Eastern Cape, South Africa), as shown in FIGURE 1. The predominance of ascidians in the Algoa Bay samples and sponges in the Mozambique samples reflects a targeted collection strategy for these areas rather than differences in distribution or frequency of these taxa at the indicated locations.

Extracts were screened at three different concentrations against the esophageal cancer cell line WHCO1, and the activity of the extracts was rated on a scale of 0 to

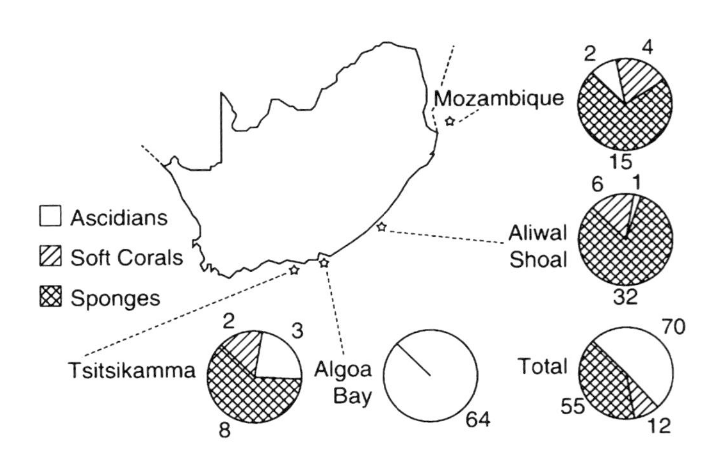

FIGURE 1. Sample collection sites. Sponge, ascidian, and soft coral samples were collected from four localities (Tsitsikamma, Algoa Bay, Aliwal Shoal, and Mozambique) off the coast of southern Africa. The number of samples collected for each taxon is indicated.

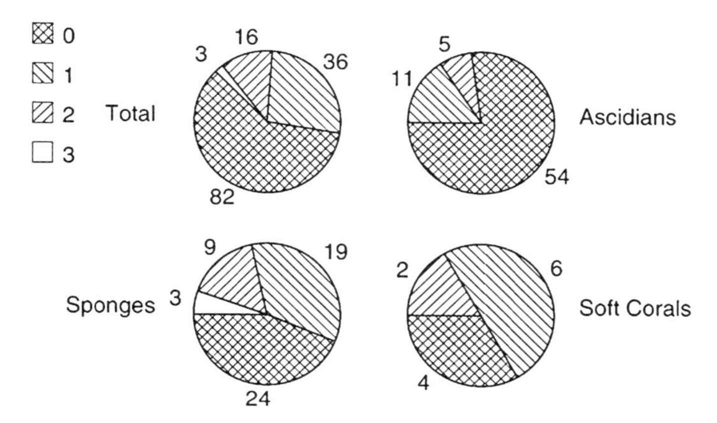

FIGURE 2. Activity of the extracts from collected samples. Extracts from the collected samples were tested for antiproliferative activity against the cultured esophageal cancer cell line WHCO1. The activity of the extracts was scored on a scale of 0–3 (0 = no activity at the concentrations tested, 1 = activity at 50 μg/ml, 2 = activity at 10 μg/ml, and 3 = activity at 1 μg/ml). All samples are grouped into their respective taxa, and sample numbers are indicated.

3 (0 = no activity at the concentrations tested, 1 = activity at only 50 μg/ml, 2 = activity at 50 and 10 μg/ml, and 3 = activity at 50, 10, and 1 μg/ml). Activity was defined as a reduction in cell number to within 10% of baseline. The highest activity level (cell death induced at 1 μg/ml) was observed only in sponge extracts (FIG. 2). Furthermore, sponges exhibited higher hit rates than did ascidians, with 22% of the sponges displaying moderate to high activity (scores ≥ 2) compared to only 7% of the ascidian samples. The results in this study suggest that sponges represent an important source of bioactive metabolites, supporting observations made elsewhere.[5] Our sampling size for soft corals is too small to draw any conclusions from the hit rate for these extracts.

Of the samples tested, several were selected to enter activity-directed purification, with the results from three extracts presented here. The EtOAc partitioned fraction from the sponge *Axinella weltneri* had moderate activity (score = 2) in our assay and was fractionated on a silica gel column to yield 7 fractions. The most active of these fractions was subjected to semipreparative HPLC purification, yielding 5 fractions. One of these, exhibiting significant activity against WHCO1, was determined by NMR to be the triterpene sodwanone A (**1**) (FIG. 3). This compound had previously been identified[11] and its anticancer activity reported.[12] **1** displayed substantial activity (IC_{50} of 1.2–1.7 μM) against three esophageal cancer cell lines tested in our laboratory (TABLE 1), results comparable to those reported previously for a lung carcinoma cell line (IC_{50} = 12 μM).[12]

The EtOAc extract of the sponge *Aplysilla sulphurea* collected in Mozambique also exhibited moderate activity (score = 2) in our assay. A portion of this extract was fractionated using HP-20 resin. The 50% Me_2CO fraction was found to be pure (+)-curcudiol (**2**) (FIG. 3), previously reported concurrently from two sponges.[13,14] The 75% Me_2CO fraction from the HP-20 column was crystalline. Recrystallization

TABLE 1. IC_{50} values (μM) for pure compounds against esophageal cancer cell lines WHCO1, WHCO6, and KYSE30

	IC_{50} values (in μM)		
	WHC01	WHC06	KYSE30
Sodwanone A (**1**)	1.2	1.7	1.4
(+)-Curcudiol (**2**)	96	>200	>200
Aplysulphurin-1 (**3**)	3.5	9.6	5.7
9, 11-Dihydrogracilin A (**4**)	1.1	1.1	1.5
Damirone C (**5**)	55.7	78.3	77.8
Makaluvamine M (**7**)	0.7	6.8	3.2
N-1-β-D-ribofuranosyldamirone C (**8**)	37.9	85.5	66
N-1-β-D-ribofuranosyl makaluvamine I (**9**)	1.6	5.7	3.2
Makaluvic acid C (**10**)	>150	>150	>150
N-1-β-D-ribofuranosyl makaluvic acid C (**11**)	60.6	>200	>200

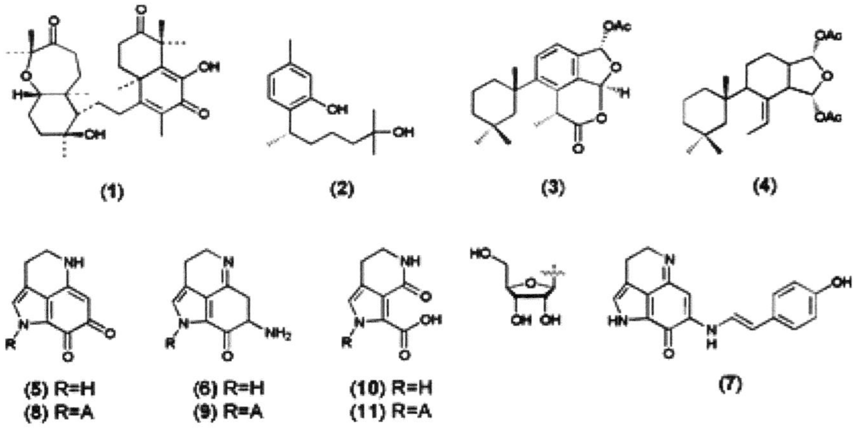

FIGURE 3. Structures of compounds isolated in this study. Sodwanone A (**1**) isolated from the sponge *Axinella weltneri*. (+)-Curcudiol (**2**), aplysulphurin-1 (**3**), and 9,11-dihydrogracilin A (**4**) isolated from the sponge *Aplysilla sulphurea*. Damirone C (**5**), makaluvamines I (**6**) and M (**7**), *N*-1-β-D-ribofuranoysldamirone C (**8**), *N*-1-β-D-ribofuranosylmakaluvamine I (**9**), makaluvic acid C (**10**), and *N*-1-β-ribofuranosylmakaluvic acid C (**11**) isolated from the sponge *Strongylodesma aliwaliensis*.

of this fraction yielded the common spongian diterpene aplysulphurin-1 (**3**) (FIG. 3), originally isolated from an Australian collection of *A. sulphurea.*[15] Finally, the supernatant mother liquor from the recrystallization was chromatogrammed using silica HPLC to yield the spongian diterpene 9,11-dihydrogracilin A (**4**) (FIG. 3), which was originally isolated from the Antarctic sponge *Dendrilla membranosa.*[16] Both **3** and **4** exhibited potent antiesophageal cancer activity against all esophageal cancer cell lines tested, whereas **2** was less active (TABLE 1).

The aqueous MeOH partition of the sponge *Strongylodesma aliwaliensis* was found to possess high activity (score = 3) in our assay. A portion of this extract was fractionated on HP-20. Two polar fractions were then further purified using a C18 Sep-Pak® followed by separation using C18 RP-HPLC to give the three known compounds, damirone C (**5**) and makaluvamines I (**6**) and M (**7**),[17] and two new compounds, *N*-1-β-D-ribofuranosyldamirone C (**8**) and *N*-1-β-D-ribofuranosylmakaluvamine I (**9**)[18] (FIG. 3). The eluent from the HP-20 column was then evaporated to dryness *in vaccuo* and separated using both C18 Sep-Pak and C18 RP-HPLC to yield the two new compounds makaluvic acid C (**10**) and *N*-1-β-D-ribofuranosylmakaluvic acid C (**11**)[19] (FIG. 3). The low yield of **6** isolated precluded bioactivity determination. All other compounds isolated from *S. aliwaliensis* were evaluated for antiesophageal cancer activity, yielding results that support the centrality of the pyrroloiminoquinone structural moiety in determining the anticancer activity of these compounds (TABLE 1).[20]

Other extracts with activity in the first round screening assays (score ≥ 2) are currently being fractionated further, while additional samples await initial screening. In conclusion, it would appear that the marine subtidal benthic communities of southern Africa hold considerable promise for identifying potential leads for effective agents against esophageal cancer.

ACKNOWLEDGMENTS

We thank M. Iqbal Parker (Medical Biochemistry, IIDMM, UCT, South Africa) for his support. This work was funded by grants from CANSA, NRF, RU, and UCT. PhD scholarships (C.W.) from NRF, DAAD, and UCT and postdoctoral research fellowships (R.K.) from NRF and RU are gratefully acknowledged.

REFERENCES

1. GABBERT, H.E. *et al.* 2000. Tumours of the oesophagus. *In* World Health Organisation Classification of Tumours. Pathology and Genetics of Tumours of the Digestive System. S.R. Hamilton & L.A. Aaltonen, Eds. :9–30. IARC Press. Lyon.
2. MQOQI, N. *et al.* 2003. National Cancer Registry of South Africa. Incidence and geographical distribution of histologically diagnosed cancer in South Africa, 1996–1997. South African Institute for Medical Research. Johannesberg.
3. LORDICK, F. *et al.* 2004. Neoadjuvant therapy for oesophagogastric cancer. Br. J. Surg. **91:** 540–551.
4. EL SAYED, K.A. *et al.* 2000. Marine natural products as leads to develop new drugs and insecticides. *In* Biologically Active Natural Products: Pharmaceuticals. S.J. Cutler & H.G. Cutler, Eds. :233–252. CRC Press. London.
5. BLUNT, J.W. *et al.* 2005. Marine Natural Products. Nat. Prod. Rep. **22:** 15–55.
6. BRANCH, G.M. *et al.* 1994. Two Oceans. 1–3. David Philip Publishers. Cape Town.
7. PETTIT, G.R. *et al.* 1988. Isolation and structure of the powerful cell growth inhibitor cephalostatin 1. J. Am. Chem. Soc. **110:** 2006–2007.
8. PETTIT, G.R. *et al.* 1993. Isolation and structure of the powerful human cancer cell growth inhibitors spongistatins 4 and 5 from an African *Spirastrella spinispirulifera* (Porifera). J. Chem. Soc. Chem. Commn. **24:** 1805–1807.
9. SAOTOME, K., H. MORITA & M. UMEDA. 1989. Cytotoxicity test with simplified crystal violet staining method using microtitre plates and its application to injection drugs. Toxicol. in Vitro **3:** 317–321.

10. RAJPUT, J. *et al.* 2004. Synthesis, characterization and cytotoxicity of some palladium(II), platinum(II), rhodium(I) and iridium(I) complexes of ferrocenylpyridine and related ligands. Crystal and molecular structure of trans-dichlorobis(3-ferrocenylpyridine)palladium (II). J. Organomet. Chem. **689:** 1553–1568.
11. RUDI, A. *et al.* 1993. Sodwanones A-C, three new triterpenoids from a marine sponge. Tetrahedron Lett. **34:** 3943–3944.
12. CARLETTI, I. *et al.* 2003. Yardenone A and B: new cytotoxic triterpenes from the Indian Ocean sponge *Axinella cf. bidderi*. J. Nat. Prod. **66:** 25–29.
13. WRIGHT, A.E. *et al.* 1987. (+)-Curcuphenol and (+)-curcudiol, sesquiterpenes phenols from shallow and deep water collections of the marine sponge *Didiscus flavus*. J.Nat. Prod. **50:** 976–978.
14. FUSETANI, N. *et al.* 1987. (+)-Curcuphenol and dehydrocurcuphenol, novel sesquiterpenes which inhibit H,K-ATPase, from a marine sponge *Epipolasis* sp. Experientia **43:** 1234–1235.
15. KARUSO, P. *et al.* 1984. The constituents of marine sponges. I. The isolation from *Aplysilla sulphurea* (Dendroceratida) of (1*R**, 1'*S**, 1"R*, 3*R**)-1-acetoxy-4-ethyl-5-(1,3,3-trimethylcyclohexyl)-1,3-dihydroisobenzofuran-1'(4),3-carbolactone and the determination of its crystal structure. Aust. J. Chem. **37:** 1081–1093.
16. MOLINSKI, T.F. & D.J. FAULKNER. 1987. Metabolites of the Antarctic sponge *Dendrilla membranosa*. J. Org. Chem. **52:** 296–298.
17. SCHMIDT, E.W. *et al.* 1995. Makaluvamines H-M and damirone C from the Pohnpeian sponge *Zyzzya fuliginosa*. J. Nat. Prod. **58:** 1861–1867.
18. KEYZERS, R.A. *et al.* 2004. Novel pyrroloquinoline ribosides from the South African Latrunculid sponge *Strongylodesma aliwaliensis*. Tetrahedron Lett. **45:** 9415-9418.
19. KEYZERS, R.A. *et al.* 2005. Makaluvic acids from the South African Latrunculid sponge *Strongylodesma aliwaliensis*. J. Nat. Prod. **68:** 506–510.
20. ANTUNES, E.M. *et al.* 2005. Pyrroloiminioquinone and related metabolites from marine sponges. Nat. Prod. Rep. **22:** 62–72.

Herbal Complement Inhibitors in the Treatment of Neuroinflammation

Future Strategy for Neuroprotection

AMOD P. KULKARNI,[a] LAURIE A. KELLAWAY,[b] AND GIRISH J. KOTWAL[a]

[a]*Division of Medical Virology and* [b]*Division of Neuroscience, University of Cape Town, Medical School, Observatory, Cape Town, 7925, South Africa*

ABSTRACT: The upregulated complement system plays a damaging role in disorders of the central nervous system (CNS). The classical and alternate pathways are two major pathways activated in neuroinflammatory disorders such as Alzheimer's disease, multiple sclerosis, traumatic brain injury, spinal cord injury, HIV-associated dementia, Parkinson's disease, and mad cow disease. Failure of currently available anti-inflammatory agents, especially cyclooxygenase inhibitors, in offering significant neuroprotection in large epidemiologic clinical trials of CNS disorders suggests an urgent need for the development of new neuroprotective agents. The positive preclinical outcomes in treating CNS disorders by complement regulatory molecules, such as vaccinia virus complement control protein, suggest the possibility of using complement-inhibitory molecules as neuroprotective agents. Several active ingredients of herbal origin are found to have complement-inhibitory activity. These herbal ingredients along with other anti-inflammatory roles might be useful in treating neuroinflammation associated with CNS disorders. Active ingredients of herbal origin with complement inhibitory ingredients are summarized and classified according to their chemical nature and specificity towards the major pathways activating the complement system. The structure activity relationship of some specific examples is also discussed in this report. This information might be helpful in formulating a natural panacea against complement-mediated neuroinflammation.

KEYWORDS: neuroprotection; neurodegeneration; complement inhibitors; herbal ingredients; alternate pathway; classical pathway; cyclooxygenases; vaccinia virus complement control protein (VCP); Alzheimer's disease (AD); HIV-associated dementia (HAD); Parkinson's disease (PD)

THE COMPLEMENT SYSTEM IN NEURODEGENERATION

The complement system is known to perform a wide range of functions in the human body. It forms an essential component of the host immune system and is

Address for correspondence: Girish J. Kotwal, Division of Medical Virology, Medical School, University of Cape Town, Anzio Road, Observatory, Cape Town-7925, South Africa. Voice: (27)-(21) 406-6676; fax (27)-(21) 406 6018.
gjkotw01@yahoo.com

Ann. N.Y. Acad. Sci. 1056: 413–429 (2005).
doi: 10.1196/annals.1352.020

associated with the clearance of molecules of foreign origin as well as the elimination of invading pathogens from the body. It plays an important role in adaptive immunity.[1] However, it is nonspecific in action and unable to distinguish between self and non-self. Under normal conditions, it is strictly regulated by complement regulatory molecules. However, in neuroinflammatory disorders, the complement regulatory molecules fail to control the activated complement components. These activated complement components then act as a double-edged sword and are responsible for the degeneration of neurons.[2] The devastating roles of the complement components in neurodegenerative disorders are well documented. Alzheimer's disease (AD),[3] multiple sclerosis (MS),[4,5] myasthenia gravis,[6] Parkinson's disease (PD),[7] traumatic brain injury (TBI),[8] and spinal cord injury (SCI)[9] are examples of a few disorders in which activated complement components play an important role. Not only do these disorders arise out of dysfunction in the normal metabolic machinery, but also neuroinflammatory disorders associated with microbial infections show involvement of complement components. In HIV-associated dementia (HAD), complement components are responsible for neuroinflammation. HIV-1 and its pathogenic proteins such as GP-120, GP-41, and Nef-1 are also associated with the activation of complement components. HIV-1 along with the aforementioned pathogenic proteins is also known to modulate the C3 promoter activity in neurons and/or astrocytes.[10] HIV has developed an effective evasion strategy involving synthesis of HIV-associated molecules from complement-mediated damage. It also utilizes the complement opsonins in the process of cell entry and replication. This whole phenomenon is outlined in a recent review by Stoiber *et al.*[11] Spongiform encepalopathies such as scrapie, are also associated with the activated complement components.[12,13]

The complement system is activated by the three major pathways, the classical pathway (CP), alternative pathway (AP), and lectin-mediated pathway. Activation of one of the pathways may lead to activation of the other pathway. The C3b component generated by spontaneous activation of the CP leads to activation of the AP by formation of the alternate pathway C3 convertase.[14] These pathways lead to the generation of opsonins and anaphylatoxins. The final common step after activation of the complement system is the generation of the membrane attack complex (MAC). The complement anaphylatoxins are responsible for the various proinflammatory events in the CNS and chemotaxis of the immunocompetent cells. The brain environment is further complicated by release of proinflammatory mediators such as nitric oxide and glutamate.[15] Mast cells are found in the brain and in the proximity of neurons. Anaphylatoxins and other mediators may activate secretion of proinflammatory molecules from the mast cell.[16] The C3a is found to be involved in the activation of the mast cells through FcγRI receptors.[17] Anaphylatoxins are also known to activate brain astrocytes through C3a and C5a receptors located on them, resulting in release of cytokines and chemokines involved in the pathogenesis of CNS disorders.[18,19] One of the complement components, C5b-9, is also known to activate prostaglandin production in glomerular epithelial cell (GEC) injury. C5b-9 upregulates the cyclooxygenase-2 enzyme (COX-2) in GEC.[20,21] COX-2 is found to be associated with the complement components in the brain. It selectively activates the C1q component of complement in the brain.[22] MAC induces concentration-dependent neuronal cell death and changes in membrane permeability to Na^+, K^+ and Ca^{2+}, release of cytokines, eicosanoids, and reactive-free radicals. These changes occur at sublytic concentrations of MAC.[23] MAC is also responsible for the demyelination of neurons in demyelinated forms of certain disorders.[24]

It is therefore well established that inflammation plays a major role in the etiology of neurodegenerative disorders and that complement components are one of the major groups of proinflammatory molecules that are involved.

CURRENT ANTI-INFLAMMATORY AGENTS

Although, inflammation plays an important role in CNS disorders, no currently available anti-inflammatory agent offers significant neuroprotection in such disorders. Anti-inflammatory agents used in therapy can be broadly classified as steroidal and nonsteroidal agents (SAIDs and NSAIDs, respectively). These drugs pose potential health hazards in long-term treatment. The role of steroidal agents such as estrogen in the treatment of dementia is controversial and is not recommended for age-associated dementia.[25] Treatment with a combination of estrogen and progestin causes a cognitive decline in postmenopausal women.[26] Prednisolone treatment also failed in clinical trials of AD.[27]

NSAIDs are either nonspecific inhibitors (mixed; inhibit both cyclooxygenases, COX-1, and COX-2) or specific inhibitors (generally target COX-2). Both are associated with adverse effects. Nonspecific NSAIDs such as indomethacin are associated with gastric mucosal damage and aggravate the problems associated with *Helicobacter pylori*.[28] Recently, a selective COX-2 inhibitor vioxx (rofecoxib) was removed from the market because of the cardiotoxicity associated with its use. Mixed NSAIDS and rofecoxib are associated with congestive heart failure (CHF) and hypertension.[29] Whether the risk of heart disease is associated with COX-2 inhibition or to the unique structure of rofecoxib still remains controversial. Further research is needed in the use of these drugs, and it is best to avoid using these agents in such disorders. COX inhibitors are associated with the risk of myocardial infarction,[30] edema, and hypertension.[31] Selective COX-2 inhibitors show similar adverse drug-related events as demonstrated by nonselective NSAIDs treatment.[32] Previously, COX-2 expression was considered inducible, but recent evidence suggests that it is constitutive in the brain. It is expressed by neurons and plays a critical role in coupling synaptic activity to neocortical blood flow.[33] COX-2 in the brain is the primary isozyme involved in memory consolidation, and COX-1 is associated with memory formation.[34]

Controversies regarding the role of COX-2 in CNS disorders, its beneficial roles in normal cognitive function as well as the adverse effects of NSAIDs in the brain are discussed in a recent review by Minghetti.[35] COX inhibitors are not only responsible for the generation of harmful prostaglandins, but also involved in the generation of PGE_2, which is known for its involvement in potential beneficial effects, such as membrane excitability and synaptic transmission in the hippocampus,[36] and neuroprotection against TNF-α.[37] Thus, inhibition of COX attenuates the potential beneficial roles of PGE_2. Recent large-scale randomized controlled clinical trials with NSAIDs in the treatment of CNS disorders yielded poor results.[38–40] However, indomethacin was found to be beneficial in mild cognitive impairment and nimesulide was also found to be effective.[41,42] Their beneficial effects can be attributed to actions other than COX inhibition.

The reason for the failure of these agents in the treatment of neuroinflammatory disorders can be attributed to their inability to target the key component involved in neuroinflammation. Also, most of them target cyclooxygenases, whose roles in the CNS are controversial. Complement plays an important role in the etiology of almost all CNS disorders, as outlined previously in this review. The role of the complement components in neuroinflammation, the interaction with other proinflammatory molecules, and the need for complement inhibition are outlined in a recent review by Kulkarni *et al.*[43]

COMPLEMENT INHIBITORS IN NEUROINFLAMMATION

Preclinical outcomes from our laboratory suggest that complement regulatory molecules might be of great help in the treatment of the aforementioned chronic neuroinflammatory disorders. Vaccinia virus complement control protein (VCP) was found to be effective in treating SCI and TBI.[44–46] It might be useful in the treatment of CNS injury associated with AD.[47] sCrry is another complement regulatory molecule found to be effective in the treatment of allergic encephylomyelitis.[48] However, no complement inhibitory molecule is currently available on the market for the treatment of neuroinflammatory disorders. Complement inhibitors are relatively new in drug therapy. Certain pharmaceutical companies are targeting the complement inhibitors for the treatment of disorders such as rheumatoid arthritis and cardiovascular disease.[49,50] The complement inhibitory molecules under development are pexelizumab and eculizumab. These monoclonal antibodies specifically target C5a, a potent anaphylatoxin, and are currently undergoing clinical trial.[49,50] Recently, pexelizumab reduced the myocardial infarction and death rate in patients who had undergone coronary artery bypass graft surgery.[51,52] However, their potential in neurodegenerative disorders and their bioavailability in the brain have yet to be investigated.

HERBAL COMPLEMENT INHIBITORS

During the last two decades, several ingredients of herbal origin have been tested for their complement inhibitory potential. To gain some perspective with a view to developing suitable complement inhibitory molecules from naturally occurring compounds, some of the active constituents of medicinal plants with complement inhibitory activity (*in vitro* in most cases unless mentioned) are discussed below. However, none of these agents has been tested clinically for its ability to offer neuroprotection.

MEDICINAL PLANTS WITH COMPLEMENT INHIBITORY INGREDIENTS

Juglans mandshurica. This plant consists of four flavonoids and two galloyl residues. The flavonoids with the most potent complement inhibitory activity found in this plant are afzelin and quercitrin. However, the galloyl residues, tetragalloyl glucose and trigalloyl glucose, were more potent than the corresponding flavonoids. The tetragalloyl residue (1, 2, 3, 4 tetragalloyl glucose) was the most potent, suggesting the importance of the galloyl moieties in complement inhibition.[53]

Glycyrrhiza glabra. This is an ancient Ayurvedic medicine known for its anti-inflammatory potential. The constituents, β-glycyrrhetinic acid and glycyrrhizin, were found to have complement inhibitory potential. Both of them induced conformational changes in C3.[54] β-glycyrrhetinic acid was more potent among them. It inhibits the CP at the level of C2 rather than C4 and C1q.[55] Glycirrhizin inhibited the C3 component of the complement anaphylatoxin C3a and C3b.[54] Apart from direct actions, these compounds are known to have some indirect anti-inflammatory activities, which make them suitable for the treatment of several inflammatory and autoimmune disorders.

Crataeva nurvala. The triterpene lupeol is an active constituent of this medicinal plant. Both lupeol and a compound synthesized from it, that is, lupeol linoleate, were found to have anti-inflammatory potential in adjuvant-induced arthritis in rats, the latter being more potent than the former. The anti-inflammatory activity can be attributed to the complement inhibitory activity of these compounds. The compounds were considered to reduce C3 convertase activity. The anti-inflammatory activity of these compounds was more than that of indomethacin; however, further elaboration of the complement inhibitory activity is essential.[56]

Ligustrum vulgare and *Phillyrea latifolia* Leaves. These belong to the *Oleaceae* family. The ethanolic extracts of both plants were found to have more complement inhibitory activity than the methanolic extract. The flavonoids apigenin, luteolin, and their glucosides, are the active constituents of these plants. The flavones showed dose-dependent inhibition of the CP. However, no such correlation was found with the AP. Further SAR studies revealed that in case of the glucosides, that is, apigenin-7-O-derivatives, complement inhibitory activity was optimum with disaccharide derivative.[57]

Morinda morindoides. This is the most popular medicinal plant in the Democratic Republic of Congo and is used traditionally to alleviate rheumatic pain. Iridoids are active ingredients of this plant. Gaertneroside, acetylgaertneroside, and gaertneric acid inhibited the CP (*in vitro* action). Gaertneroside was the most potent among these compounds. Iridoids failed to inhibit the AP.[58] Apart from iridoids, Morinda also shows quercetin and other complement inhibitory molecules. Quercetin and M_{O15} inhibited both the AP and the CP. Others showed a more pronounced and dose-dependent effect on the CP.[59]

Osbeckia aspera. The mature leaves of this Ayurvedic medicinal plant have been used traditionally to treat liver disease in Srilanka. The herb has immunosuppressive capabilities through other mechanisms apart from complement inhibition. The whole plant extract was found to have a dose-dependent effect on both the CP and the AP. The effect was more pronounced on the CP.[60]

Cedrela lilloi and *Trichilia elegans*. These medicinal plants, belonging to the Meliaceae family, grow in Argentina. The fresh leaf extracts of these two plants were found to have anticomplement activity. The extracts inhibited both the CP and the AP of complement activation. However, the chemical moieties responsible for this and the precise mechanism of complement inhibitory action of these plants are yet to be identified. These plants are also known to inhibit the phagocytosis of the peritoneal macrophages and possess strong antiproliferative activity against T cells.[61]

TABLE 1. Medicinal plants with complement inhibitory active ingredients: systematically outlined medicinal plants with complement inhibitory constituents along with their pharmacological actions on the complement system

Medicinal plants	Active ingredients	Actions on complement system	Refs.
Wedelia chinensis	Wedelosin and others	Inhibits CP and AP	66
Persicaria lapathifolia	Kaempferol glycoside and acylated quercetin glycosides	Inhibits CP at 4.3, 9.7, 3.9, and 7.6 × 10(−5) M, respectively	67
Rosemarinus officinalis	Rosmarinic acid	Inhibits CP and AP; binds C3, inhibits C5 convertase at a very high concentration	68,69,70
Petasites hybridus	Ze 339 and sesquiterpin ester petasin	Abrogates C5a-induced increase in calcium concentration	71
Soyabean oil emulsion (Intralipid)	–	Inhibits *in vitro* synthesis and secretion of C4 and C2 by guinea pig peritoneal macrophages	72
Jatropha multifida latex	Polymer proanthocyanidin	Only CPI (*in vitro*); action is mediated by Ca^{2+} depletion	73
Aloe vera (leaf parenchyma gel)	Polysaccharide with dominant mannose	Inhibits AP	74,75
W. fruticosa (Nimba arishta)	–	Complement inhibitory activity	76
Jatropha curcas L.	Curacycline A	Moderately inhibits CP	77
Centaurium spicatum	Quercetin derivatives	Complement inhibition: maximum in triacetylated compounds	78
Piper kadsura	Piperlactam S	At 1-30 μm conce, inhibits C5a-induced migration, C5a-stimulated TNF-α and TNF-β	79
Lithospermum euchromum	LR-polysaccharide IIa	Inhibits CP and AP	80
Olive (Olea europaea L.) leaves.	Apigenin and other flavonoids	Inhibits CP	81
Crataegus sinaica	Whole extract	Inhibits CP and AP	83
Osbeckia octandra, Melothria maderaspatana, Phyllanthus debilis leaves	AR-arabinogalactan	Inhibits CP and AP	84
Angelica acutiloba	Fucans	Inhibits CP and AP	85

TABLE 1. (*continued*) Medicinal plants with complement inhibitory active ingredients: systematically outlined medicinal plants with complement inhibitory constituents along with their pharmacological actions on the complement system

Medicinal plants	Active ingredients	Actions on complement system	Refs.
Ascophyllum nodosum (brown seaweed)	Aqueous leaf extract	*In vitro* complement inhibitio; tested in respiratory burst / *in vivo* zymosan-induced inflammation model	86
Trichilia glabra	Water-soluble extract: Sesquiterpenes quinines, coumarins, and flavonoids	More pronounced inhibition of CP than AP	87
Propolis V	Whole extract	Inhibits CP and AP	83

Azadirachta indica. This is commonly known as Neem in India and is well known for its medicinal value. The crude aqueous extract of the plant consists of complement inhibitory polymers NB-I and NB-II, the former being less active than the latter. Activity could be correlated with molecular weight, as NB-I, a high molecular weight compound, was less active than the low molecular weight compound NB-II. Glucose was found to be the main carbohydrate constituent.[62]

Tinospora cordifolia. This climbing shrub, commonly called gurcha, is well documented in the Ayurvedic literature and is known for its anti-inflammatory and immunomodulatory potential. These constituents were also found to mediate phagocytosis by peritoneal macrophages. However, the two compounds, syringin and cordiol, inhibited the activation of the complement by inhibiting the C-3 convertase of the CP. Like the other active constituents, these were also found to potentiate the immune response.[63]

Isopyrum thalictroides. This plant is used in Chinese medicine in the treatment of inflammatory disorders such as rheumatism, neuralgia, and silicosis. The plant with several photoberbezines and bisbezylisoquinoline (BBI) alkaloids having complement inhibitory activity may provide novel complement inhibitory molecules. However, isopyruthaline (It1), fangchinoline (It2), and isotalictrine (It3) are the three major active ingredients of this plant, whose complement inhibitory activity has been well studied. These constituents were found to inhibit the CP. It3 was the most potent inhibitor of the CP and It1 the least potent. It showed Ca^{2+}- and Mg^{2+}-dependent complement inhibition. It inhibited the formation of the first component of the complement system. The complement inhibitory actions of It2 were independent of Ca^{2+} ions, and in the case of It3, its effectiveness decreased at a very high concentration of calcium ions.[64] The BBIs affected the formation of convertase and did not inhibit the decay of the convertase. These iridoids also influenced the AP.[65] It3 was found to suppress both pathways, but It1 and It2 augmented AP hemolysis at a higher concentration.

Apart from these specific plants listed with potent complement inhibitory molecules, the active ingredients of several other medicinal plants with known anti-inflammatory activity have shown inhibition of the complement system, as outlined in TABLE 1. These herbal ingredients are systematically classified on the basis of their complement inhibitory activity (*in vitro* and/or *in vivo* in a few cases) and chemical nature. With an aim to develop either specific or nonspecific inhibitors of the complement system, these agents are also classified according to their mode of action. The compounds with novel action on the complement components are listed separately. Classification according to chemical nature might be of great help in developing structurally related potent complement inhibitory agents with neuroprotective roles.

CLASSIFICATION AS PER MODE OF ACTION

As discussed previously, the AP and CP are the two major pathways of complement activation involved in neuroinflammatory disorders. Thus, herbal complement inhibitors (HCIs) can be broadly classified as selective inhibitors (SIs) and nonselective inhibitors (NSIs) of the complement system depending on their ability to inhibit one or both pathways involved in complement activation. SIs can further be classified as classical pathway inhibitors (CPIs) or alternate pathway inhibitors (APIs). Nonselective inhibitors can be classified as strong CP–moderate AP inhibitors (SCP-MAPs), strong AP–moderate CP inhibitors (SAPMCPs), and general complement inhibitors. The ingredients included in the general complement inhibitors section may inhibit one of the pathways to a greater extent than the other, but they can be considered as potent inhibitors of both pathways. This type of classification may help to select suitable complement inhibitory molecules based on the etiology of the disease. As discussed previously, activation of one pathway leads to activation of the other. Thus, depending on disease status, SIs and NSIs can be used in therapy. These agents are outlined in FIGURE 1. Apart from these compounds, novel complement inhibitors are outlined separately in FIGURE 2.

Understanding the mode of action of the individual agent might be of great help in developing an effective combination therapy using different complement inhibitory molecules targeting different complement components. This might also prove useful in avoiding irrational combinations. Hence, the actions of herbal ingredients on the complement components are represented diagrammatically in FIGURE 3.

CLASSIFICATION ACCORDING TO CHEMICAL NATURE

The ingredients of herbal origin can be grouped into several classes, based on their chemical origin. Most compounds inhibiting the complement components are either flavonoids and their glucosides,[53,66,67,78,81] polysaccharides,[74,75,80,84] or terpenes.[56,71,87] Apart from these major chemical classes inhibiting complement, other classes of complement inhibitors include iridoids,[58] polymers,[62,73] peptides,[77] alkaloids,[79] and oils.[72] These are outlined in FIGURE 4.

Synthesis of new structurally related analogs, systematic comparative study of these compounds and study of the bioavailability to the brain will reveal the novel complement inhibitory molecules with potent antineuroinflammatory activity. These neuroprotective agents will be of great help in the treatment of complex brain disorders such as AD, HAD, and PD.

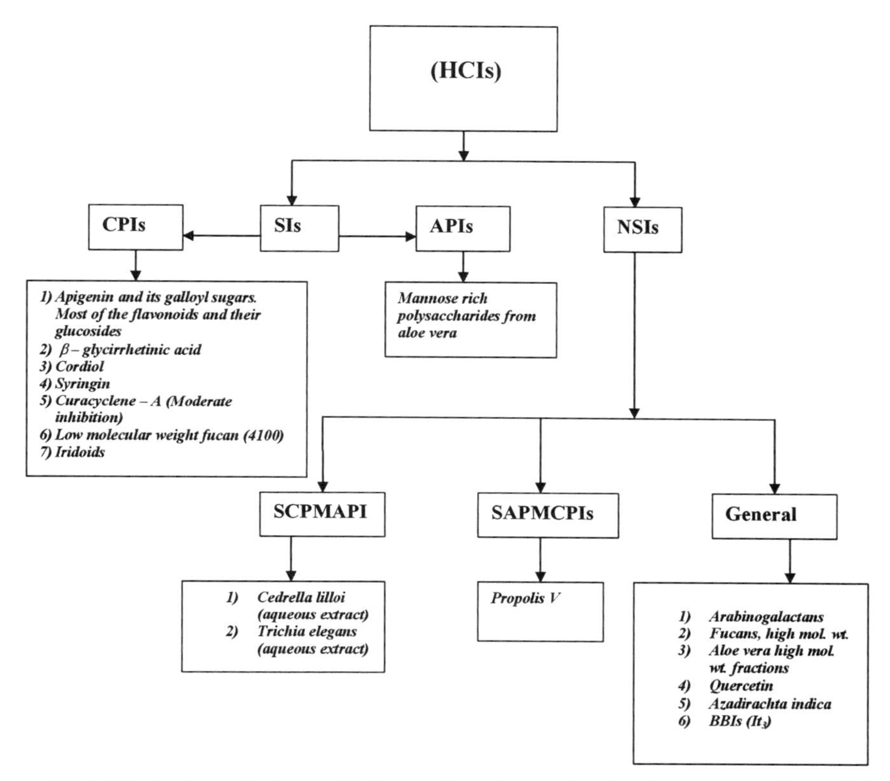

FIGURE 1. Classification of HCIs based on their mode of action and specificity. HCIs can be classified as selective and nonselective inhibitors (SIs and NSIs, respectively) based on their selectivity towards the complement pathways. The NSIs are further subclassified into different subgroups according to their ability to inhibit one pathway to a greater extent than the other.

STRUCTURE ACTIVITY RELATIONSHIP OF SOME INGREDIENTS OF HERBAL ORIGIN

Flavonoids and Derivatives

(a) The presence of the galloyl group increased the complement inhibitory activity of the flavonoids.

(b) Tetragalloyl glucose showed better activity than did trigalloyl glucose.

(c) There was an increase in anticomplement activity in inverse proportion to the number of free hydroxyl groups on the B-ring and 3,5,7-trihydroxyflavone.[53]

(d) In *L. vulgare*, triglucosides of the flavonoids were more active than those of the corresponding flavonoids.

Petasin and Ze 339: Inhibits C5a mediated calcium release from neutrophils and eosinophils

Piperlactam S: Inhibits C5a induced chemotaxis of macrophages and C5a mediated release of cytokines.

Flucans: Does not inhibit the terminal complement components

BBIs: It2 and It3 affect the formation of convertases rather than accelerating their decay

Rosmarinic acid: Binds covalently to the thioester-containing alpha'-chain of nascent C3b.

FIGURE 2. Typical actions of some HCIs. Few active ingredients of herbal origin show some typical actions on the complement system, which are different from the normal complement inhibitors. Their actions are summarized in this chart.

(e) In apigenin-7-O derivatives, optimum activity was observed in a disaccharide derivative, and activity was reduced in the trisaccharide derivative.[57]

(f) In *M. morindoides*, mixtures of compounds were more active on the CP than were individual components.[59]

Steroid-Like Compounds

(a) The only representative of this class, β-glycirrhetinic acid, showed confirmation-dependent complement inhibition. α-Glycirrhetinic acid was found to be inactive.

(b) Substitution of the hydroxyl group at the C3 position of the first ring of the compound showed a decrease in activity. The derivatives glycyrrhizin and carbenoxolone sodium were found to be less potent than β-glycirrhetinic acid.[55]

Fucans

(a) Fucan fragments (molecular weight range 4,100–21,4000) inhibited complement activation.

(b) There was an increase in complement inhibitory activity with increasing molar ratios of xylose, galactose, and glucuronic acid in flucans.

(c) Sulfate groups were necessary, but not sufficient for the complement inhibitory activity of fucans.

(d) The low molecular weight fucan (4,100) did not inhibit the AP.

(e) The glycosidic regions involved in the inhibition of the CP and AP might not be the same.[85]

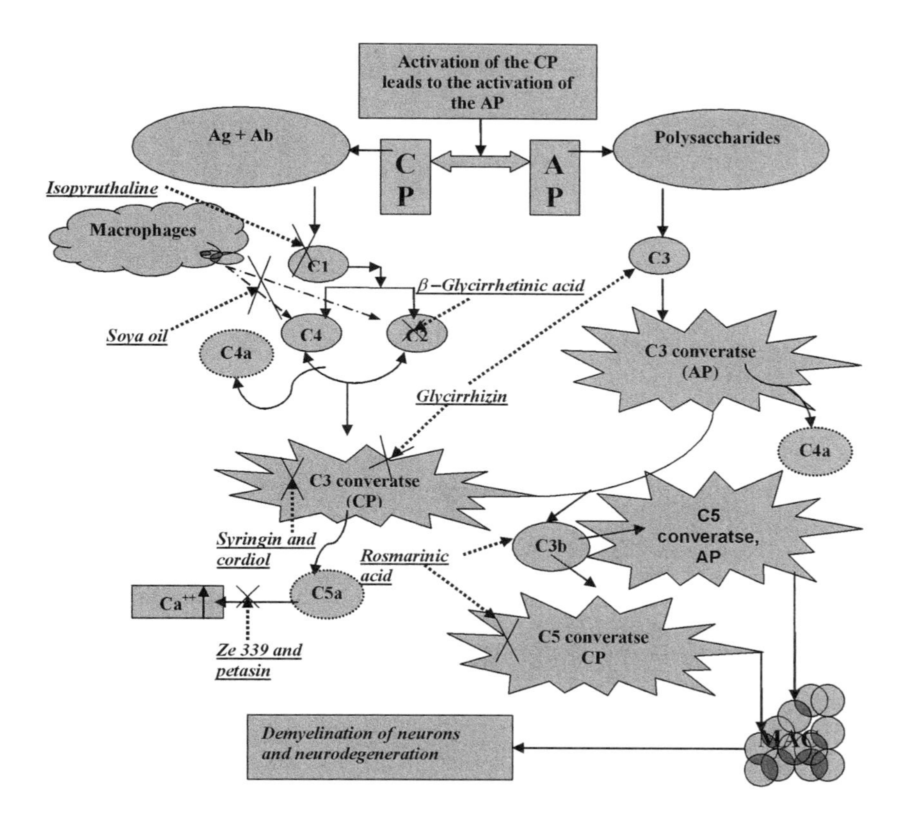

FIGURE 3. Action of HCIs on different complement components. The CP gets activated by the antigen antibody complex and the AP by the polysaccharides or other activators. Also, activation of one of the pathways may lead to the activation of the other pathway. After being activated, these pathways lead to the generation of complement opsonins (C3b, iC3b, C4b, etc.), anaphylatoxins (C3a, C4a, and C5a), and MAC. More than 30 different complement proteins are involved in the process. The constituents of herbal origin inhibit one or more components of the complement system and thereby prevent the activation of complement pathways. The complement cascade, and actions of herbal ingredients on complement components are outlined in this figure. (———→), complement pathway; (----→), synthesis and release of the complement components; (··········→), drug action; (✕), inhibition of the pathway, complement component (or binding of a drug to a particular component), and/or synthesis and release of complement component by drug molecules; (⬮) anaphylatoxins.

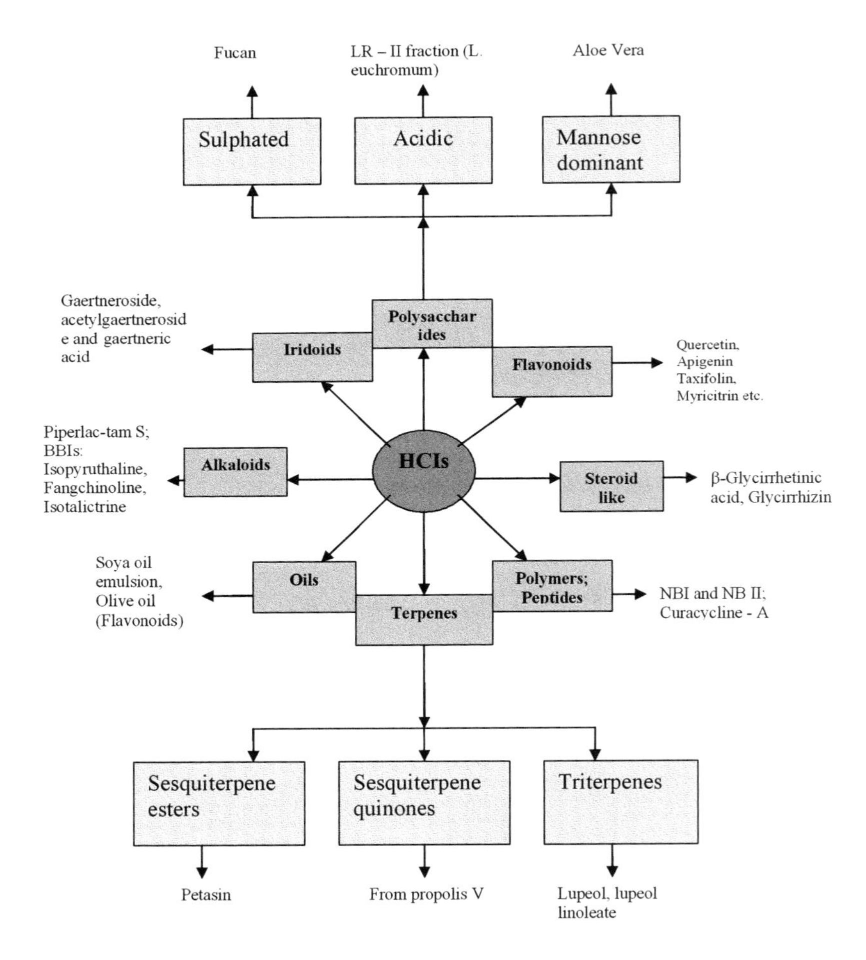

FIGURE 4. Classification of HCIs based on their chemical structure. The ingredients of herbal origin can be grouped according to their chemical structures as outlined in this figure. Majority of the compounds, which inhibit the complement system are flavonoids, terpenes and polysaccharides.

CONCLUDING REMARKS

This review focuses on the detrimental roles of the complement system in neuroinflammatory disorders and the inability of the currently prescribed NSAIDs to prevent and treat such disorders. The unavailability of a suitable complement inhibitory molecule in the market leaves an urgent need for the development of suitable complement inhibitory molecules with neuroprotective values. Complement inhibitory molecules of herbal origin, classification based on their mode of action on one or more complement components, and classification based on chemical structure and specificity and selectivity towards the particular complement component are discussed. Also, SAR studies of some of the complement inhibitory molecules are outlined. This collective information, that is, the devastating roles of the complement system along with information on herbal complement inhibitors, may shift the focus of scientists from the currently prescribed NSAIDs towards the complement inhibitory molecules. Complement inhibitory molecules may emerge as a separate class of antineuroinflammatory molecules. The SAR study outlined here may be helpful in the development of suitable complement inhibitors with neuroprotective abilities. Information on the mode of action of complement inhibitors might prove of significant help in designing a combination therapy with two or more agents with different modes of action on the complement system. However, rigorous research in this field with special emphasis on bioavailability to the brain, safety studies, separation of impurities, and clinical trials of the complement inhibitory molecules is essential in advocating herbal complement inhibitors as the next generation of neuroprotective agents.

SUMMARY

The currently existing NSAIDs are associated with side effects, and epidemiologic trials suggest their inability to offer significant neuroprotection. The positive preclinical outcome of the complement regulatory molecules, such as VCP, in treating neuroinflammatory disorders suggests their usefulness in CNS disorders, where complement plays a major role. In the last two decades, active ingredients and extracts of traditional medicinal plants with anti-inflammatory activity have been tested for their *in vitro* complement inhibitory activity. The classification, based on their mode of action (selectivity), chemical nature, action on individual complement components, and SAR studies discussed in the review may provide insight into novel complement inhibitors in the future. Systematic pharamacokinetic study of these herbal ingredients may also reveal complement regulatory compounds, which possess the ability to cross the blood-brain barrier. Thus, medicinal plants and their active ingredients with complement inhibitory activity (as discussed in this review) might prove beneficial in formulating a natural panacea against complex brain disorders.

ACKNOWLEDGMENTS

G.J.K. is currently a Senior International Wellcome Trust Fellow for Biomedical Sciences in South Africa. A.P.K. is the recipient of the Poliomyelitis Research

Foundation and Senior Entrance Merit and International Students Scholarships at the University of Cape Town.

REFERENCES

1. BARRINGTON, R. *et al.* 2001. The role of complement in innate and adaptive immune response. Immunol. Rev. **180:** 1–15.
2. GASQUE, P. *et al.* 2002. Roles of the complement system in human neurodegenerative disorders: pro-inflammatory and tissue remodeling activities. Mol. Neurobiol. **25:** 1–17.
3. TACNET-DELORME, P. *et al.* 2001. β-amyloid fibrils activate the C1 complex of complement under physiological conditions: evidence for a binding site for Aβ on the C1q globular regions. J. Immunol. **167:** 6374–6381.
4. SANDERS, M.E. *et al.* 1986. Activated terminal complement in cerebrospinal fluid in Guillain-Barre syndrome and multiple sclerosis. J. Immunol. **136:** 4456–4459.
5. SELLEBJERG, F. *et al.* 1998. Intrathecal activation of the complement system and disability in multiple sclerosis. J. Neurol Sci. **157:** 168–174.
6. ROMI, F. *et al.* 2005. The role of complement in myasthenia gravis: serological evidence of complement consumption *in vivo*. J. Neuroimmunol. **158:** 191–194.
7. MCGEER, P.L. & E.G. MCGEER. 2004. Inflammation and neurodegeneration in Parkinson's disease. Parkinsonism Relat. Disord. **10:** S3–S7.
8. SEWELL, D.L. *et al.* 2004. Complement C3 and C5 play critical roles in traumatic brain cryoinjury: blocking effects on neutrophil extravasation by C5a receptor antagonist. J. Neuroimmunol. **155:** 55–63.
9. ANDERSON, A.J. *et al.* 2004. Activation of complement pathways after contusion-induced spinal cord injury. J. Neurotrauma **21:** 1831–1846.
10. BRUDER, C. *et al.* 2004. HIV-1 induces complement factor C3 synthesis in astrocytes and neurons by modulation of promoter activity. Mol. Immunol. **40:** 949–961.
11. STOIBER, H. *et al.* 2005. Complement-opsonized HIV: the free rider on its way to infection. Mol. Immunol. **42:** 153–160.
12. MABBOTT, N.A. *et al.* 2001. Temporary depletion of complement component C3 or genetic deficiency of C1q significantly delays onset of scrapie. Nat. Med. **7:** 485–487.
13. KLEIN, M.A. *et al.* 2001. Complement facilitates early prion pathogenesis. Nat. Med. **7:** 488–492.
14. MANDERSON, A.P. *et al.* 2001. Continual low-level activation of the classical complement pathway. J. Exp. Med. **194:** 747–756.
15. BAL-PRICE, A. & G.C. BROWN. 2001. Inflammatory neurodegeneration mediated by nitric oxide from activated glia-inhibiting neuronal respiration, causing glutamate release and excitotoxicity. J. Neurosci. **21:** 6480–6491.
16. THEOHARIDES, T.C. 1990. Mast cells: the immune gate to the brain. Life Sci. **46:** 607–617.
17. WOOLHISER, M.R. *et al.* 2004. Activation of human mast cells by aggregated IgG through FcRI: additive effects of C3a. Clin. Immunol. **110:** 172–180.
18. SAYAH, S. *et al.* 2003. Two different transduction pathways are activated by C3a and C5a anaphylatoxins on astrocytes. Brain Res. Mol. Brain Res. **112:** 53–60.
19. JAUNEAU, A.C. *et al.* 2003. Complement component anaphylatoxins upregulate chemokine expression by human astrocytes. FEBS Lett. **537:** 17–22.
20. TAKANO, T. *et al.* 2001. Complement C5b-9 induces cyclooxygenase-2 gene transcription in glomerular epithelial cells. Am. J. Physiol. Renal Physiol. **281:** F841–F850.
21. TAKANO, T. & A.V. CYBULSKY. 2000. Complement C5b-9-mediated arachidonic acid metabolism in glomerular epithelial cells: role of cyclooxygenase-1 and -2. Am. J. Pathol. **156:** 2091–2101.
22. SPIELMAN, L. *et al.* 2002. Induction of the complement component C1qB in brain of transgenic mice with neuronal overexpression of human cyclooxygenase-2. Acta Neuropathol. **103:** 157–162.
23. SHEN, Y. *et al.* 1997. Characterization of neuronal cell death induced by complement activation. Brain Res. Protocols **1:** 186–194.

24. MEAD, R.J. *et al.* 2002. The membrane attack complex of complement causes severe demyelination associated with acute axonal injury. J. Immunol. **168:** 458–465.
25. ESPELAND, M.A. *et al.* 2004. Conjugated equine estrogens and global cognitive function in postmenopausal women: Women's Health Initiative Memory Study. JAMA **291:** 2959–2968.
26. HENDERSON, V.W. *et al.* 2005. Postmenopausal hormone therapy and Alzheimer's disease risk: interaction with age. J. Neurol. Neurosurg. Psychiatry **76:** 103–105.
27. AISEN, P.S. *et al.* 2000. A randomized controlled trial of prednisone in Alzheimer's disease. Alzheimer's disease cooperative study. Neurology **54:** 588–593.
28. KANATANI, K. *et al.* 2004. Effects of indomethacin and rofecoxib on gastric mucosal damage in normal and *Helicobacter pylori*-infected mongolian gerbils. J. Physiol. Pharmacol. **55:** 207–222.
29. MAMDANI, M. *et al.* 2004. Cyclo-oxygenase-2 inhibitors versus non-selective non-steroidal anti-inflammatory drugs and congestive heart failure outcomes in elderly patients: a population-based cohort study. Lancet **363:** 1751–1756.
30. GARCIA RODRIGUEZ, L.A. *et al.* 2004. Nonsteroidal antiinflammatory drugs and the risk of myocardial infarction in the general population. Circulation **109:** 3000–3006.
31. WOLFE, F., S. ZHAO & D. PETTITT. 2004. Blood pressure destabilization and edema among 8538 users of celecoxib, rofecoxib, and nonselective nonsteroidal antiinflammatory drugs (NSAID) and nonusers of NSAID receiving ordinary clinical care. J. Rheumatol. **31:** 1143–1151.
32. VERRICO, M.M. *et al.* 2003. Adverse drug events involving COX-2 inhibitors. Ann. Pharmacother. **37:** 1203–1213.
33. KIYOSHI, N. *et al.* 2000. Cyclooxygenase-2 contributes to functional hyperemia in whisker-barrel cortex. J. Neurosci. **20:** 763–770.
34. TEATHER, L.A., M.G. PACKARD & N.G. BAZAN. 2002. Post-training cyclooxygenase-2 (COX-2) inhibition impairs memory consolidation. Learn. Mem. **9:** 41–47.
35. MINGHETTI, L. 2004. Cyclooxygenase-2 (COX-2) in inflammatory and degenerative brain diseases. J. Neuropathol. Exp. Neurol. **63:** 901–910.
36. CHEN, C. & N.G. BAZAN. 2005. Endogenous PGE2 regulates membrane excitability and synaptic transmission in hippocampal CA1 pyramidal neurons. J. Neurophysiol. **93:** 929–941.
37. LEE, E.O., Y.J. SHIN & Y.H. CHONG. 2004. Mechanisms involved in prostaglandin E2-mediated neuroprotection against TNF-alpha: possible involvement of multiple signal transduction and beta-catenin/T-cell factor. J. Neuroimmunol. **155:** 21–31.
38. AISEN, P. S. *et al.* 2003. Effects of rofecoxib or naproxen vs placebo on Alzheimer disease progression: a randomized controlled trial. JAMA **289:** 2819–2826.
39. REINES, S. A. *et al.* 2004. Rofecoxib: no effect on Alzheimer's disease in a 1-year, randomized, blinded, controlled study. Neurology **62:** 66–71.
40. SCHARF, S. *et al.* 1999. A double-blind, placebo-controlled trial of diclofenac/misoprostol in Alzheimer's disease. Neurology **53:** 197–201.
41. ROGERS, J. *et al.* 1993. Clinical trial of indomethacin in Alzheimer's disease. Neurology **43:** 1609–1611.
42. AISEN, P.S., J. SCHMEIDLER & G.M. PASINETTI. 2002. Randomized pilot study of nimesulide treatment in Alzheimer's disease. Neurology **58:** 1050–1054.
43. KULKARNI, A.P. *et al.* 2004. Neuroprotection from complement-mediated inflammatory damage. Ann. N.Y. Acad. Sci. **1035:** 147–164.
44. REYNOLDS, D.N. *et al.* 2003. Vaccinia virus complement control protein modulates inflammation following spinal cord injury. Ann. N.Y. Acad. Sci. **1010:** 534–539.
45. REYNOLDS, D.N. *et al.* 2004. Vaccinia virus complement control protein reduces inflammation and improves spinal cord integrity following spinal cord injury. Ann. N.Y. Acad. Sci. **1035:** 165–178.
46. HICKS, R.H. *et al.* 2002. Vaccinia virus complement control protein enhances functional recovery after traumatic brain injury. J. Neurotrauma **19:** 705–714.
47. KOTWAL, G.J., D.K. LAHIRI & R. HICKS. 2002. Potential intervention by vaccinia virus complement control protein of the signals contributing to the progression of central nervous system injury to Alzheimer's disease. Ann. N.Y. Acad. Sci. **973:** 317–322.

48. DAVOUST, N. *et al.* 1999. Central nervous system-targeted expression of the complement inhibitor sCrry prevents experimental allergic encephalomyelitis. J. Immunol. **163:** 6551–6556.
49. WHISS, P.A. 2002. Pexelizumab Alexion (abstr). Curr. Opin. Invest. Drugs **3:** 870–877.
50. KAPLAN, M. 2002. Eculizumab (Alexion) (abstr). Curr. Opin. Invest. Drugs **3:** 1017–1023.
51. SHERNAN, S.K. *et al.* 2004. Impact of pexelizumab, an anti-C5 complement antibody, on total mortality and adverse cardiovascular outcomes in cardiac surgical patients undergoing cardiopulmonary bypass. Ann. Thorac. Surg. **77:** 942–949.
52. VERRIER, E.D. *et al.* 2004. Terminal complement blockade with pexelizumab during coronary artery bypass graft surgery requiring cardiopulmonary bypass: a randomized trial (cardiovascular surgery abstr). JAMA **291:** 2319–2327.
53. MIN, B.S. *et al.* 2003. Anti-complement activity of constituents from the stem-bark of *Juglans mandshurica*. Biol. Pharm. Bull. **26:** 1042–1044.
54. KAWAKAMI, F., Y. SHIMOYAMA & K. OHTSUKI. 2003. Characterization of complement C3 as a glycyrrhizin (GL)-binding protein and the phosphorylation of C3alpha by CK-2, which is potently inhibited by GL and glycyrrhetinic acid in vitro. J. Biochem. **133:** 231–237.
55. KROES, B.H. *et al.* 1999. Inhibition of human complement by beta-glycyrrhetinic acid. Immunology **90:** 115–120.
56. GEETHA, T. & P. VARALAKSHMI. 1999. Anticomplement activity of triterpenes from *Crataeva nurvala* stem bark in adjuvant arthritis in rats. Gen. Pharmacol. **32:** 495–497.
57. PIERONI, A. *et al.* 2000. Studies on anti-complementary activity of extracts and isolated flavones from *Ligustrum vulgare* and *Phillyrea latifolia* leaves (Oleaceae). J. Ethnopharmacol. **70:** 213–217.
58. CIMANGA, K. *et al.* 2003. Complement-Inhibiting Iridoids from *Morinda morindoides.* J. Nat. Prod. **66:** 97–102.
59. CIMANGA, K. *et al.* 1995. In vitro anticomplementary activity of constituents from *Morinda morindoides.* J. Nat. Prod. **58:** 372–378.
60. NICHOLL, D.S. *et al.* 2001. *In vitro* studies on the immunomodulatory effects of extracts of *Osbeckia aspera.* J. Ethnopharmacol. **78:** 39–44.
61. NORES, M.M. *et al.* 1997. Immunomodulatory activities of *Cedrela lilloi* and *Trichilia elegans* aqueous leaf extracts. J. Ethnopharmacol. **55:** 99–106.
62. VAN DER NAT, J.M. *et al.* 1989. Characterization of anti-complement compounds from *Azadirachta indica* (abstr). J. Ethnopharmacol. **27:** 15–24.
63. KAPIL, A. & S. SHARMA. 1997. Immunopotentiating compounds from *Tinospora cordifolia.* J. Ethnopharmacol. **58:** 89–95.
64. IVANOVSKA, N. *et al.* 1999. Complement modulatory activity of bisbenzylisoquinoline alkaloids isolated from *Isopyrum thalictroides.* I. Influence on classical pathway in human serum. Int. J. Immunopharmacol. **21:** 325–336.
65. IVANOVSKA, N., M. HRISTOVA & S. PHILIPOV. 1999. Complement modulatory activity of bisbenzylisoquinoline alkaloids isolated from *Isopyrum thalictroides.* II. Influence on C3-9 reactions *in vitro* and anti-inflammatory effect *in vivo.* Int. J. Immunopharmacol. **21:** 337–347.
66. APERS, S. *et al.* 2002. Characterisation of new oligoglycosidic compounds in two Chinese medicinal herbs. Phytochem. Anal. **13:** 202–206.
67. PARK, S.H. *et al.* 1999. Acylated flavonol glycosides with anti-complement activity from *Persicaria lapathifolia*. Chem. Pharm. Bull. **47:** 1484–1486.
68. SAHU, A., N. RAWAL & M.K. PANGBURN. 1999. Inhibition of complement by covalent attachment of rosmarinic acid to activated C3b. Biochem. Pharmacol. **57:** 1439–1446.
69. AL-SEREITI, M.R. *et al.* 1999. Pharmacology of rosemary (*Rosmarinus officinalis* Linn.) and its therapeutic potentials. Ind. J. Exp. Biol. **37:** 124–130.
70. PEAKE, P.W. *et al.* 1991. The inhibitory effect of rosmarinic acid on complement involves the C5 convertase (abstr). Int. J. Immunopharmacol. **13:** 853–857.
71. THOMET, O.A. *et al.* 2001. Role of petasin in the potential anti-inflammatory activity of a plant extract of petasites hybridus. Biochem. Pharmacol. **61:** 1041–1047.

72. STRUNK, R.C. *et al.* 1979. Inhibition of *in vitro* synthesis of the second (C2) and fourth (C4) components of complement in guinea pig peritoneal macrophages by a soybean oil emulsion. Pediatr. Res. **13:** 188–193.
73. KOSASI, S. *et al.* 1989. Inhibitory activity of *Jatropha multifida* latex on classical complement pathway activity in human serum mediated by a calcium-binding proanthocyanidin (abstr). J. Ethnopharmacol. **27:** 81–89.
74. T'HART, L.A. *et al.* 1989. An anti-complementary polysaccharide with immunological adjuvant activity from the leaf parenchyma gel of *Aloe vera* (abstr). Planta Med. **55:** 509–512.
75. HART, L.A. *et al.* 1988. Two functionally and chemically distinct immunomodulatory compounds in the gel of *Aloe vera* [abstr]. J. Ethnopharmacol. **23:** 61–71.
76. KROES, B.H. *et al.* 1993. Fermentation in traditional medicine: the impact of *Woodfordia fruticosa* flowers on the immunomodulatory activity, and the alcohol and sugar contents of *Nimba arishta* [abstr]. J. Ethnopharmacol. **40:** 117–125.
77. VAN DEN BERG, A.J. *et al.* 1995. Curcacycline A: a novel cyclic octapeptide isolated from the latex of *Jatropha curcas* L. FEBS Lett. **358:** 215–218.
78. SHAHAT, A.A. *et al.* 2003. Anticomplement and antioxidant activities of new acetylated flavonoid glycosides from *Centaurium spicatum* (abstr). Planta Med. **69:** 1153–1156.
79. CHIOU, W.F. *et al.* 2003. Anti-inflammatory properties of piperlactam S: modulation of complement 5a-induced chemotaxis and inflammatory cytokines production in macrophages. Planta Med. **69:** 9–14.
80. ZHAO, J.F. *et al.* 1993. Anti-complementary acidic polysaccharides from roots of *Lithospermum euchromum* (abstr). Phytochemistry **34:** 719–724.
81. PIERONI, A. *et al.* 1996. *In vitro* anti-complementary activity of flavonoids from olive (*Olea europaea* L.) leaves. Pharmazie **51:** 765–768.
82. SHAHAT, A.A. *et al.* 1996. Anti-complementary activity of *Crataegus sinaica* (abstr). Planta Med. **62:** 10–13.
83. THABREW, M.I. *et al.* 1991. Immunomodulatory activity of three Sri-Lankan medicinal plants used in hepatic disorders. J. Ethnopharmacol. **33:** 63–66.
84. YAMADA, H. *et al.* 1985. Studies on polysaccharides from *Angelica acutiloba*. IV. Characterization of an anti-complementary arabinogalactan from the roots of *Angelica acutiloba* Kitagawa (abstr). Mol. Immunol. **22:** 295–304.
85. BLONDIN, C. *et al.* 1996. Relationships between chemical characteristics and anticomplementary activity of fucans. Biomaterials **17:** 597–603.
86. BENENCIA, F., M.C. COURREGES & F.C. COULOMBIE. 2000. Anti-inflammatory activities of *Trichilia glabra* aqueous leaf extract. J. Ethnopharmacol. **71:** 293–300.
87. IVANOVSKA, N.D. *et al.* 1995. Immunomodulatory action of propolis. V. Anticomplementary activity of a water-soluble derivative. J. Ethnopharmacol. **47:** 135–143.

Amyloid, Cholinesterase, Melatonin, and Metals and Their Roles in Aging and Neurodegenerative Diseases

DEBOMOY K. LAHIRI,[a] DE-MAO CHEN,[a] PREETI LAHIRI,[b,c] STEVE BONDY,[a] AND NIGEL H. GREIG[d]

[a]*Department of Psychiatry, Institute of Psychiatric Research, Indiana University School of Medicine, Indianapolis, Indiana 46202, USA*

[b]*Department of Chemistry, Women's College, Banaras Hindu University, Varanasi, UP, 221005, India*

[c]*Department of Community and Environmental Medicine, University of California Irvine, Irvine, California 92697, USA*

[d]*Laboratory of Neurosciences, National Institute on Aging, Baltimore, Maryland 21224, USA*

ABSTRACT: The aging brain shows selective neurochemical changes involving several neural cell populations. Increased brain metal levels have been associated with normal aging and a variety of diseases, including Alzheimer's disease (AD). Melatonin levels are decreased in aging, particularly in AD subjects. The loss of melatonin, which is synthesized by the pineal gland, together with the degeneration of cholinergic neurons of the basal forebrain and the deposition of aggregated proteins, such as the amyloid β peptides (Aβ), are believed to contribute to the development of cognitive symptoms of dementia. Aging and its variants, such as AD, should be viewed as the result of multiple "hits," including alterations in the levels of Aβ, metals, cholinesterase enzymes, and neuronal gene expression. Herein, we present evidence in support of this theory, based on several studies. We discuss melatonin's neuroprotective function, which plays an important role in aging, prolongation of life span, and health in the aged individual. It interacts with metals and, in some cases, neutralizes their toxic effects. Dietary supplementation of melatonin restores its age-related loss. In mice, an elevated brain melatonin significantly reduced levels of potentially toxic Aβ peptides. Thus, compensation of melatonin loss in aging by dietary supplementation could well be beneficial in terms of reducing metal-induced toxicity, lipid peroxidation, and losses in cholinergic signaling. We propose that certain cholinesterase inhibitors and the NMDA partial antagonist memantine, which are FDA-approved drugs for AD and useful to boost central nervous system functioning, can be made more effective by their combination with melatonin or other neuroprotectants. Herein, we highlight studies elucidating the role of the amyloid pathway, metals, melatonin, and the cholinergic system in the context of aging and AD. Finally, melatonin is present in edible plants and walnuts, and consuming foodstuffs containing melatonin would be beneficial by enhancing the antioxidative capacity of the organisms.

Address for correspondence: Debomoy K. Lahiri, 791 Union Drive, Indiana University School of Medicine, Indianapolis, IN 46202. Voice: 317-274-2706; fax: 317-274-1365. dlahiri@iupui.edu

Ann. N.Y. Acad. Sci. 1056: 430–449 (2005). © 2005 New York Academy of Sciences. doi: 10.1196/annals.1352.008

KEYWORDS: aging; acetylcholinesterase; amyloid; amyloid precursor protein; brain; cholinergic; cobalt; copper; diet; iron; melatonin; mercury; metalloprotease; metals; mouse; neurodegeneration; neuroprotection; synaptic proteins; zinc

INTRODUCTION

There is currently a significant interest in studying age-associated biochemical changes and how these relate to brain function. Recent advances in neuroscience research and biomedical techniques have yielded several approaches to address age-related brain disorders.[1,2] Aging in animals most likely results in a reduction of several cellular and metabolic functions, such as lysosomal enzyme activities,[3] mitochondrial respiration,[4] and rates of mitochondrial and nuclear DNA repair.[5] Such declines in cellular functions during the aging process correlate, to some degree, with the cognitive impairment observed in normal aging and other age-related conditions, such as senile dementia, Alzheimer's disease (AD), and Parkinson's disease (PD). Interestingly, functional deterioration is accompanied by age-related changes in gene expression profiles.[6,7] Currently, four cholinesterase inhibitors and a NMDA partial antagonist, memantine, have been approved for the treatment of AD; however, they are mostly useful in symptomatic relief.[8] Because current approved drugs are limited in their ability to treat neurodegenerative diseases, such as AD and PD, more research is needed to understand the pathobiochemical mechanisms underlying these disorders. In this context, the pineal gland hormone melatonin (*N*-acetyl-5-methoxytryptamine,[9] FIG. 1) has been suggested to play an important role in aging and prolongation of life span.[10] Of particular relevance, melatonin may exert a beneficial action on neurodegenerative conditions in humans with debilitating diseases.[11] Serum melatonin levels decrease during aging and in AD patients.[12] On the other hand, certain metals accumulate and free radical generation increases in the aging brain. This is accompanied by changes in the cholinergic system, which is important in neuronal development, signaling, and neurotransmission. Thus, declines in melatonin and cholinergic system–mediated protective function are exacerbated by increased metal deposition in the brain. Currently, research is ongoing to understand the cellular, biochemical, and pathophysiological roles of metals, melatonin, and cholinesterase (ChE) in healthy adult and aging brains and in AD. In addition to acetylcholinesterase (AChE), its sister enzyme, butyrylcholinesterase (BChE), is also gaining impact as an important new target in AD therapy.[13] Herein, we discuss how metals, melatonin, and the cholinergic system interact with each other, and we

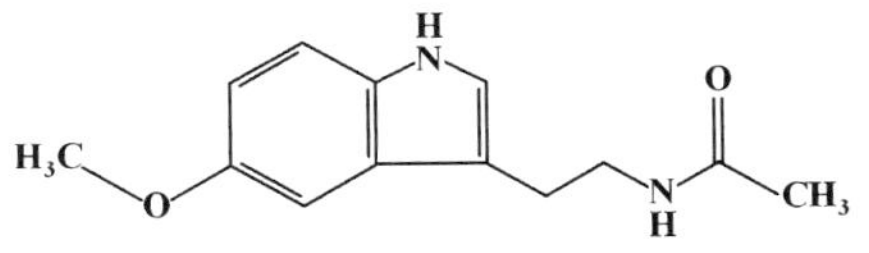

FIGURE 1. Two-dimensional structure of a melatonin molecule. Melatonin is a disubstituted tryptamine. A methoxy group is bound to the benzene ring at R_5 and an acetyl group is bound to the amine group.[9]

propose an integrated pathway/model to demonstrate how they can be optimally utilized to maximize their neuroprotective functions.

AGE-DEPENDENT CHANGES IN BRAIN CELLULAR FUNCTIONS

Age-dependent decline occurs in numerous brain cellular functions and there are concomitant alterations in cerebral enzymatic activities that are linked to mitochondria and the glutathione system.[14,15] Furthermore, age-related changes have been reported in gene expression profiles and also in relation to calorie restriction.[6,16–18] Certain important genes involved in neuronal structure and signaling, such as synaptotagmin I and apolipoprotein E (ApoE), are differentially expressed in the hypothalamus and cortex of aged animals.[18] Different proteases, which play important roles in regulating neuropeptide metabolism, APP processing, and neuronal apoptosis are upregulated in the aged brain; all these are believed to contribute significantly to brain aging.[18]

AGE-RELATED GENE EXPRESSION CHANGES AND THE EFFECT OF MELATONIN

The role of melatonin in reversing a wide variety of biochemical aspects of aging has been studied recently. Sharman *et al.*[17] tested an "aging paradigm" in normal mice in which dietary supplementation of melatonin was used. Using the gene array chip analysis technique, they analyzed changes in levels of 12,423 genes separately with respect to aging and with melatonin addition. Interestingly, melatonin could reverse age-related changes in levels of certain important genes.[17] The role of melatonin in regulating circadian and seasonal rhythms is presumably caused by its binding to MT1 and MT2 receptors, which results in the sensitization of the ubiquitous signaling enzyme, adenylate cyclase.[19] It is therefore apparent that an age-related decline in melatonin levels may be expected to induce concomitant changes in gene expression.

BIOLOGICAL ROLE OF MELATONIN

Research from several laboratories suggests that melatonin can act as (1) a regulator of aging and senescence[20] and (2) an intracellular scavenger of hydroxyl and peroxyl free radicals both *in vitro* and *in vivo*.[21] Cell culture experiments have established that melatonin prevents the death of neuroblastoma cells exposed to the Alzheimer amyloid β (Aβ) peptide.[22] Thus, melatonin has a protective role for neurons and other cells. Moreover, age-related changes in mRNA gene expression in the central nervous system (CNS) of mice can be modulated by dietary melatonin.[20,23] Notably, a recent report suggests age reversion of gene expression levels in aged mice that were given diets supplemented with melatonin.[17]

DIFFERENT COMPONENTS OF AMYLOID PLAQUES IN ALZHEIMER'S DISEASE

Aging, AD, and Down syndrome (DS) subjects are characterized by depositions of diffuse to mature senile plaques in their brain and cerebral vasculature.[24,25] Neuritic plaques are mostly composed of the toxic Aβ peptide of 4 kDa (39–43 amino acids). Besides the presence of Aβ peptides, several other protein molecules, such as AChE, BChE, and ApoE, have also been detected in plaque.[26,27] Aβ is generated from a family of large Aβ-containing precursor proteins (APPs) of 110–130 kDa (695–770 amino acids).[25] Aβ cleavage from APP depends on the activity of a C-terminal cleaving enzyme (γ-secretase) and an N-terminal cleaving enzyme (β-secretase).[8,28] Aggregates of the Aβ peptide in the oligomeric and/or in fibril form are strongly believed to be toxic to neuronal cells. For example, natural oligomers of the Aβ peptide specifically disrupt cognitive function.[29] Several factors can either promote this step of oligomerization and/or fibrillization (such as certain metals) or inhibit this process (such as melatonin). Thus, it is important to understand the interaction and role of metal and melatonin in aging and age-related disorders.

Aβ PEPTIDES, CHOLINESTERASE, AND NEURODEGENERATION

AChE Increases Aβ-Induced Toxicity

Recent findings suggest interactions between cholinergic functions and amyloid β peptides in normal animals and AD models, and these results have potential implications in the pathophysiology and treatment of AD.[30] Moreover, it has been shown that AChE which, as described, is a senile plaque component, increases amyloid fibril assembly with the formation of highly toxic Aβ–AChE complexes. Notably, the neurotoxic effect induced by Aβ–AChE complexes was higher than that induced by Aβ peptide alone, as shown via both *in vitro* and *in vivo* experiments.[31]

Effect of Aβ Peptides on Mitochondrial Permeability and Neurodegeneration

Mitochondrial permeability transition pores belong to a multiprotein complex made up of the components of both inner and outer membrane.[32] The mitochondrial pores control transport of ions and peptides in and out of mitochondria, thereby maintaining calcium homeostasis in the cell and regulating apoptosis. It has been reported that Aβ fragment 25–35 may induce pathologic activation of the permeability transition by irreversibly opening the mitochondrial pores.[33] This event is believed to be a major step in the development of neurotoxicity and neurodegeneration.

Effect of Certain ChE Inhibitors (ChEIs) on Aβ-Mediated Opening of the Mitochondrial Pore

From the perspective of cellular and neuronal development, the effect of agents on mitochondrial pores is an important aspect of mechanisms that underpin neurotoxicity. Substances that may prevent the opening of mitochondrial pores that are induced by neurotoxins may preserve mitochondrial function and, thus, may have potential as neuroprotective agents. Aβ-induced opening of mitochondrial pores can be prevented by cyclosporin A, a specific inhibitor of the permeability transition, and, notably, by an endogenous precursor of melatonin, *N*-acetylserotonin. Interestingly, the pore opening was shown to be blocked by the FDA-approved AD drug tacrine as well as by the antihistamine, dimebon, which is in development as an agent for the therapy of AD and other types of dementia.[33]

Effect of Melatonin and a ChEI on Rat Spinal Nociceptive Transmission

Mondaca and colleagues[34] recently studied the involvement of melatonin metabolites in the long-term inhibitory effect of the hormone on rat spinal nociceptive transmission. This is based on the premise that melatonin and its metabolites could bind to nuclear sites in neurons, suggesting that this hormone is able to exert long-term functional effects in the CNS via genomic mechanisms. Their results suggest that rats receiving melatonin exhibited a reduction in spinal windup activity, which was not observed in the animals receiving melatonin and eserine (physostigmine) or saline. Taken together, these data indicate a role for melatonin metabolites in long-term changes of nociceptive transmission in the rat spinal cord.

Effect of the ChEI Eserine on Melatonin Deacetylation in the Culture Medium

In an invertebrate system, it was shown that *Xenopus laevis* melanophores express a high density of high-affinity cell membrane melatonin receptors.[35] Treatment of melanophores with melatonin resulted in a loss of membrane melatonin receptors. In addition to receptor loss, a decline in the potency of melatonin to produce pigment aggregation was observed on prolonged treatment.[36] This loss of potency was much slower than the loss of receptors and was completely prevented by inclusion of eserine, which is an inhibitor of melatonin deacetylation in the culture medium. These results indicate that although receptor density declines on prolonged treatment, this is not responsible for diminishing melatonin potency, which is caused by degradation of melatonin by deacetylation and subsequent deamination in melanophores.[36]

CHOLINERGIC SYSTEM, MEMORY, AND DEVELOPMENT

Role of Acetylcholinesterase during Apoptosis and Development

Acetylcholinesterase exhibits functions unrelated to catalysis of the neurotransmitter acetylcholine (ACh), particularly during development. Although the underlying mechanism is presently unknown, a candidate peptide fragment (AChE peptide) recently was identified and shown to induce a continuum of apoptotic and necrotic neuronal cell death in rat hippocampal organotypic cultures.[37] Data from Day and Greenfield[37] suggest that nanomolar concentrations of AChE peptide exhibit pathophysiologic activity via an apoptotic pathway that could play an important role in neuronal development and neurodegeneration.

Cholinergic Signaling in the Rat Pineal Gland

Studies have suggested that cholinergic signaling also operates in the rat pineal gland.[38–40] This important observation of cholinergic signaling in the rat pineal

TABLE 1. Biologic functions of acetylcholinesterase (AChE), melatonin, and metals

AChE/melatonin/metals	Effects (biochemical, physiological)	Ref.
AChE	Increases amyloid fibril assembly	30, 31
AChE	Induces Aβ-mediated toxicity	31
Aβ(25–35)	Involved in irreversibly opening the mitochondrial pore	33
AChE inhibitor tacrine	Can block Aβ-induced mitochondrial pore opening	33
AChE inhibitor	Eserine inhibits melatonin deacetylation	36
AChE peptide	Exhibits pathophysiologic effects via an apoptotic pathway	37
Cholinergic pathway	Signals in the rat pineal gland	40, 41
Metals such as Cu	Influence APP expression and Aβ levels	47
Lead (Pb) exposure	Affects AChE activity and APP expression in developing rats	43, 44
Trace elements	Affect cognitive condition in the elderly	48
Melatonin	As intracellular scavenger of hydroxyls and peroxy radicals	10, 11
Melatonin	Prevents Aβ-mediated cell death (neuroprotection)	22
Melatonin	Acts on the nucleus accumbens to increase acetylcholine release	50
Melatonin	Prevents cobalt-mediated oxidative stress/ cytotoxicity	54
Melatonin	Binds aluminum and attenuates its toxicity	55
Melatonin	Inhibits iron-induced lipid peroxidation	68
Melatonin	Forms stable complexes with potassium, lithium, and iron	59
Melatonin	Protects copper-mediated free radical damage	64
Melatonin	Promotes neuronal differentiation	66
Melatonin	Regulates cholesterol metabolism	70
Melatonin levels	Decrease with aging in animals and the human	71, 73
Dietary melatonin addition	Restores melatonin levels and reduces Aβ peptides in mice	72

gland is based on the localization of choline acetyltransferase (ChAT) and AChE as well as muscarinic and nicotinic ACh-binding sites within the gland[41] and on observations such as the modulation of brain nicotinic receptors by nocturnal melatonin surges.[40]

Age-Related Changes in Cholinergic Neuronal Function, Learning, and Memory in Senescence-Accelerated Mice

The senescence-accelerated mouse (SAM) has been established as a murine model of accelerated aging.[42] Learning ability and memory in various tasks were studied in a SAM strain, SAMP1TA, and in a control strain of SAMR1TA at different ages.[42] The activity of ChAT and ChE in the brains of these mice was also measured and, interestingly, was lower in the striatum of SAMP1TA compared with that of SAMR1TA at ages 20 and 30 weeks. These results suggest that the strain SAMP1TA has a deficit that involves cholinergic neuronal dysfunction and has an impact on learning ability and memory, as shown by impairment of performance in latent learning and long-term memory, but not in short-term memory.[42]

METALS AND ChEI

Lead-Induced Effects on Acetylcholinesterase Activity in the Developing Rat

Exposure to low levels of lead (Pb) during early development has been implicated in behavioral abnormalities and cognitive deficits in children. A recent study focused on developmental changes in the hippocampus and cerebellum of rats after prenatal exposure to Pb.[43] In both of these brain regions, Pb exposure decreased AChE activity with an increase in age. *In vitro* studies conducted in 35-day-old rat brain showed a considerable decrease in the specific activity of AChE at high concentrations (50–100 μM) of Pb but found no changes at low Pb concentrations (5–20 μM). In the presence of eserine (physostigmine), a specific inhibitor of AChE, the inhibitory effect of Pb was potentiated, and this was more pronounced at low concentrations of Pb. The behavioral responses in open-field tests also showed a significant decrease in both Pb-exposed as well as eserine-administered rats. These data suggest that low-level postnatal Pb exposure induces alterations in the cholinergic system of the cerebellum and hippocampus of the developing brain that may contribute to behavioral and learning deficits, even after withdrawal of Pb exposure.[43] Furthermore, Basha *et al.*[44] recently showed that postnatal exposure to Pb in rats resulted in latent overexpression of APP and Aβ in the aging brain, and this supports the fetal basis of amyloidogenesis.

ROLE OF METALS IN APP PROCESSING AND AD

Metals, APP, and Inflammatory Targets

AD has been linked to inflammation and metal biology.[45] Metals, such as copper, zinc, and iron, and certain inflammatory cytokines significantly contribute to the onset of sporadic late-onset forms of the dementia. Biochemically, copper, zinc, and iron have been shown to accelerate the aggregation of the Aβ peptide and increase metal-catalyzed oxidative stress associated with amyloid plaque formation. These amyloid-associated events remain the main pathologic hallmark of AD in the cerebral cortex of AD patients. The participation of metals in the neuritic plaque has resulted in the development of specific metal chelators as a potential new therapeutic strategy for the treatment of AD. Another pathologic feature of AD is inflammation. The presence of neuroinflammatory events during AD is supported by epidemiology studies wherein use of nonsteroidal anti-inflammatory drugs was associated with a reduced risk of developing AD. Drug targets that address inflammation include the use of small molecules that prevent Aβ peptide from activating microglia, the use of cytokine suppressive anti-inflammatory drugs, and the continued search for a vaccine directed to Aβ subfragments. In addition, an iron-responsive element is present in the 5′ untranslated region (UTR) of the APP transcript (APP 5′-UTR).[45] This provides a suitable molecular tool to study the role of metals in APP metabolism, leading to the potential for development of novel drug targets for AD.

APP Processing Enzymes Are Metalloproteases

Recent studies have shown that α-secretase is a zinc metalloproteinase, and several members of the adamalysin family of proteins, tumor necrosis factor-α convertase (TACE, ADAM17), ADAM10, and ADAM9, meet some of the criteria required of α-secretase. Several studies, including those using mice in which each of the ADAMs has been knocked out, suggest that there is a group of zinc metalloproteinases able to cleave APP at the α-secretase site.[46]

Copper-Mediated APP Expression

Recent studies have illustrated the importance of copper in AD neuropathogenesis and suggested a role for APP and Aβ in copper homeostasis. To understand the role of copper in APP metabolism, Bellingham and colleagues[47] utilized human fibroblasts overexpressing the Menkes protein (MNK), a major mammalian copper efflux protein. MNK deletion fibroblasts have high levels of intracellular copper, whereas MNK overexpressing fibroblasts have severely depleted intracellular copper. They demonstrate that copper depletion significantly reduced APP protein levels and downregulated APP gene expression. Moreover, APP promoter deletion constructs identified the copper-regulatory region between –490 and +104 of the APP gene promoter both in basal MNK overexpressing cells and in copper-chelated MNK deletion cells.[47] These results favor the hypothesis that copper can regulate APP expression and support a role for APP to function in copper homeostasis.[47] Taken together, these results suggest that copper-regulated APP expression can be utilized as a potential therapeutic target for AD.

HUMAN STUDIES: EFFECT OF TRACE ELEMENTS ON COGNITIVE IMPAIRMENT IN AN ELDERLY COHORT STUDY

Different studies have shown a possible effect of micronutrients and macronutrients on cognitive function.[48] Trace elements, being routinely involved in metabolic processes and redox reactions in the CNS, could influence cognitive functions. A recent study evaluated the presence of an eventual correlation between serum trace element concentrations and cognitive function.[48] This study provides evidence of a positive correlation between cognitive function and Se, Cr, Co, and Fe serum levels, whereas a negative correlation is found between cognitive score and Cu and Al serum levels. The authors suggest that Se, Cr, and Co protect cognitive function, Cu influences the evolution of cognitive impairment, and Al contributes to the pathogenesis of AD.[48]

OVEREXPRESSION OF THE Aβ PEPTIDE OPPOSES THE AGE-DEPENDENT ELEVATIONS OF COPPER AND IRON IN THE BRAIN

Both copper and iron levels show marked increases with age and may adversely interact with the Aβ peptide, causing its aggregation and the production of neurotoxic hydrogen peroxide (H_2O_2), which may lead to AD pathogenesis.[49] APP possesses copper/zinc binding sites in its amino-terminal domain and in the Aβ domain. Overexpression of the carboxyl-terminal fragment of APP, containing Aβ, results in significantly reduced copper and iron levels in the transgenic mouse brain, whereas overexpression of the APP in Tg2576 transgenic mice results in significantly reduced copper, but not iron, levels before the appearance of amyloid neuropathology and throughout the life span of the mouse. Concomitant increases in brain manganese levels were observed with both transgenic strains. These findings, complemented by other findings of elevated copper levels in APP knockout mice, support roles for APP and Aβ in physiologic metal regulation.[49]

CHOLINERGIC SYSTEM AND MELATONIN

Melatonin Acts on the Nucleus Accumbens to Increase Acetylcholine Release

Paredes and colleagues[50] have used brain microdialysis coupled with high-performance liquid chromatography with electrochemical detection to evaluate the influence of melatonin on extracellular concentration of ACh in the nucleus accumbens (NAc) of rats. During the dialysis sessions with an activity meter, motor activity was simultaneously monitored. Melatonin and prazosin were administered locally through the dialysis probe. Melatonin dose-dependently increased ACh release in NAc. Melatonin (3 μM) decreased horizontal activity and increased vertical activity, whereas another dose (100 μM) enhanced both horizontal and vertical activity. Prazosin, a putative melatonin antagonist, blocked the effects of melatonin on both motor activity and ACh release when given 20 min before melatonin. Overall, these results suggest that melatonin modulates the release of ACh in the NAc and the pattern of motor activity in the rat.[50]

Effects of Melatonin on Cholinergic Transmission and Nicotinic Acetylcholine Receptors

Although melatonin is a hormone produced and released by the pineal gland, it is also synthesized by cells of the gastrointestinal wall, where it might be a local regulator of gut function. Barajas-Lopez *et al.*[51] investigated the possible role of melatonin as a modulator of the enteric nervous system. They showed that melatonin modulates cholinergic transmission by blocking nicotinic channels in the guinea pig submucous plexus. In another study, Markus and colleagues[52,53] showed that melatonin modulates neuronal ACh receptors located presynaptically on the sympathetic nerve terminals of the rat vas deferens and nicotinic receptors within the rat cerebellum.[40]

NEUROPROTECTIVE EFFECTS OF MELATONIN ON METAL-INDUCED ILL EFFECTS

Neutralization of the Toxic Effect of Cobalt, Aluminum, and Alkali Metals by Melatonin

Heavy metals, such as cobalt, have been shown to be elevated in brains from AD subjects as compared with controls. Cobalt-mediated oxidative stress/cytotoxicity could be prevented by pretreatment of cells with melatonin. Furthermore, a cobalt-mediated increase in the release of Aβ peptides and its associated deleterious effects could be reversed by melatonin treatment.[54] Note that although cobalt is an essential nutritional requirement for all animals and is usually bound tightly to cobalamin (vitamin B12), its unbound form could lead to neurotoxicity. In addition, melatonin has been shown to bind aluminum.[55] Aluminum toxicity is exacerbated by hydrogen peroxide and attenuated by a fragment of Aβ and melatonin. This binding suggests a potential role for this element in the etiology of AD; however, the detailed mechanism of this metal–melatonin interaction is presently not understood. Regarding other metals, it has been shown that serotonin and melatonin can form stable complexes with lithium and potassium, with serotonin preferring lithium to potassium and melatonin favoring potassium over lithium. This is important because the binding of serotonin to lithium may explain, in part, the therapeutic effects of lithium in the treatment of mood disorders.

HUMAN STUDIES: DECLINING LEVELS OF ZINC WITH AGE AND EFFECT OF MELATONIN ON ZINC TURNOVER

Zinc plasma levels decrease with advancing age, and zinc supplementation in old mice is able to restore reduced immunologic functions.[56] Melatonin treatment or pineal grafts also restores a normalized zinc turnover in aged mice: in particular, a reduced zinc plasma level is restored to normal values. These results suggest that the effect of melatonin on thymic endocrine activity and peripheral immune functions may be mediated by the zinc pool.[56] There is a link reported between zinc and melatonin with a reconstitution of thymic functions in melatonin-treated old mice.[57] After melatonin treatment, there are concomitant increments of the nocturnal peaks of zinc and melatonin, with synchronization of their circadian patterns in old mice. The beneficial effect of zinc and melatonin on thymic functions during the circadian cycle may result in prolonged survival in aging.

METAL-INDUCED LIPID PEROXIDATION AND INTERACTION OF MELATONIN

Melatonin has been found to inhibit iron-induced lipid peroxidation and neurodegeneration in rat brain.[58] Serotonin and melatonin are shown to form a complex with Fe^{3+} in the brain.[59] Melatonin reduces epileptic discharges in the brain of rats injected with discharge-inducing levels of $FeCl_3$.[60] These neuroprotective effects against iron-induced damage may be mediated by the 6-hydroxymelatonin metabolite.[61] In a situation in which melatonin levels modified by light cycle were inadequate to reduce the iron-induced neuronal death in rat cerebral cortex, a systemic infusion of melatonin was shown to be effective in reducing this damage.[62] In addition to iron–melatonin complexes, melatonin has been shown to form complexes with copper in the brain.[59] Melatonin inhibits copper-mediated LDL oxidation in the serum of postmenopausal women.[63] In addition, melatonin treatment protects against copper-mediated free radical damage in rat liver homogenates.[64] Interestingly, there is a loss of circadian rhythm in fish that have undergone chronic subtoxic copper exposure (comparable to levels found in drinking water sources that have been treated by copper-based algicides).[65] This supports a complex interaction between copper and melatonin.

ROLE OF MELATONIN ON OXIDATIVE STRESS PROTEINS AND ITS ROLE AS A SCAVENGER

Melatonin prevents the death of neuroblastoma cells exposed to Aβ,[22] and melatonin promotes neuronal growth factor (NGF)–mediated neuronal differentiation in cell cultures.[66] Melatonin likely acts as an intracellular scavenger of hydroxyl and peroxyl free radicals when administered at both physiologic and pharmacologic doses *in vitro* as well as *in vivo*.[21,67] Melatonin has also been reported to reverse the effect of ApoE ε4 on Aβ peptide.[68] Regarding an antioxidant property, melatonin, xanthurenic acid, resveratrol, epigallocatechin gallate (EGCG), vitamin C, and α–lipoic acid differentially reduce oxidative DNA damage induced by Fenton reagents.[69] Using the nucleic acid derivative 8-hydroxydeoxyguanosine (8-OH-dG), as a biomarker of oxidative DNA damage, melatonin was capable of reversing the pro-oxidant effect of resveratrol and vitamin C. It had an antagonistic effect when used in combination with EGCG, and it exhibited synergism in combination with vitamin C and with lipoic acid.[69]

EFFECT OF MELATONIN ON CHOLESTEROL METABOLISM

Recent results strongly suggest that melatonin participates in the regulation of cholesterol metabolism and in the prevention of oxidative damage to membranes. In particular, serum cholesterol and lipid peroxidation are decreased by melatonin in diet-induced hypercholesterolemic rats.[70]

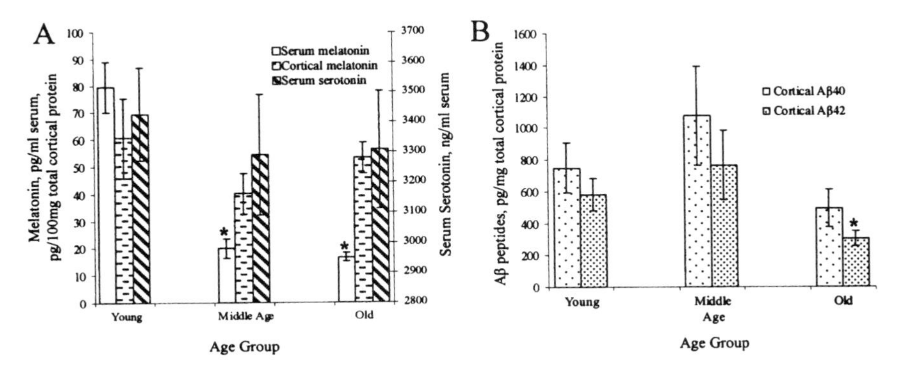

FIGURE 2. Age-related differences in levels of melatonin, serotonin, and Aβ peptides in mice. (**A**) Age-related changes in melatonin and serotonin levels in mice. Serum levels of melatonin *(white bar)* and serotonin *(diagonal hatch)* and of cortical melatonin (*horizontal stipple*) were measured and expressed as pg/ml for the serum and pg/100 mg for cortex samples. Specific samples that are significantly different at $P<0.05$ from the corresponding young mouse sample are indicated with an *asterisk.* (**B**) Age-related changes in Aβ peptide levels in mice. Mouse brain cortex Aβ40 *(diffuse stipple)* and Aβ42 *(dense stipple)* were measured by a sensitive sandwich-based ELISA. Specific samples that are significantly different at $P<0.05$ from the corresponding young mouse sample are indicated with an *asterisk.*

LEVELS OF MELATONIN DECREASE IN ANIMALS DURING AGING

We recently measured melatonin levels in three age groups of mice and reported that melatonin levels decrease with age.[71,72] Serum melatonin levels were highest in young mice, decreasing in 12-month-old and 27-month-old animals. Serum melatonin levels were reduced by ~70% more in older animals than in the young (FIG. 2). Notably, melatonin was detected in cerebral cortex tissue, which has significantly lower levels than does serum.

EFFECT OF DIETARY MELATONIN SUPPLEMENTATION

To understand its potential therapeutic benefit, we examined the effects of dietary melatonin supplementation on levels of endogenous melatonin and its sulfate derivative. We compared levels of melatonin in young (6-month-old), middle-aged (12-month-old), and senescent (27-month-old) mice before and after dietary melatonin supplementation. On measuring the effects of oral administration of melatonin in mice on plasma and cerebral cortex melatonin levels, we observed that dietary melatonin (40 ppm) resulted in a significant increase in melatonin levels in all tissues of all ages of animals tested (FIG. 3).[71,72] Thus, dietary melatonin supplementation can raise the level of endogenous melatonin present in the cerebral cortex and other tissues. Notably in mice, an elevated brain melatonin significantly reduced levels of potentially toxic Aβ peptides, which are involved in amyloid deposition and plaque formation in AD.[71] It thus is possible that administered melatonin not only can com-

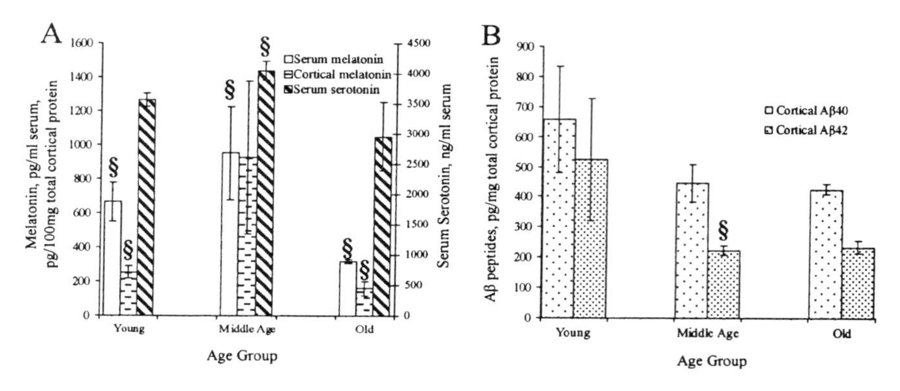

FIGURE 3. Effects of dietary melatonin supplementation on melatonin, serotonin, and Aβ peptide levels in mice. (**A**) Effects of dietary melatonin supplementation on serum melatonin and serotonin and cortical melatonin in mice. Levels of serum melatonin *(white bar)* and serotonin *(diagonal hatch)* and of cortical melatonin *(horizontal stipple)* were measured. Specific samples that are significantly different at $P < 0.05$ from the corresponding unsupplemented sample (FIG. 3A, middle-aged serum melatonin, for example) are indicated with a "§" symbol. (**B**) Effects of dietary melatonin supplementation on Aβ peptide levels in mice. Aβ40 *(diffuse stipple)* and Aβ42 *(dense stipple)* were measured. Specific samples that are significantly different at $P < 0.05$ from the corresponding unsupplemented sample (FIG. 3B, middle-aged Aβ42, for example) are indicated with a "§" symbol.

pensate for the loss of melatonin due to aging, but also appears capable of entering the brain by crossing the blood–brain and blood–cerebrospinal fluid (CSF) barriers, when it is added to the diet.

HUMAN STUDIES: LEVELS OF MELATONIN DECREASE DURING AGING

In humans, melatonin levels were decreased in postmortem CSF with aging, AD, and *APOE* ε4/4 genotype (FIG. 4).[73] This trend is consistent with our previous report in animals.[71,72] These findings together with others suggest that reductions in brain and CSF melatonin levels, which occur during aging, may contribute to a pro-amyloidogenic microenvironment in the aging brain.

PROTECTIVE EFFECTS OF MELATONIN

Current research suggests that oral administration of melatonin reduces oxidative stress and proinflammatory cytokines that are induced by Aβ peptide in rat brain.[74] Interestingly, melatonin increased survival and inhibited oxidative and amyloid pathology in a transgenic model of AD.[75] In an intriguing case report of monozygotic twins with AD who were treated with melatonin,[76] one of the treated twins showed a milder impairment of memory function, with substantial improvement of

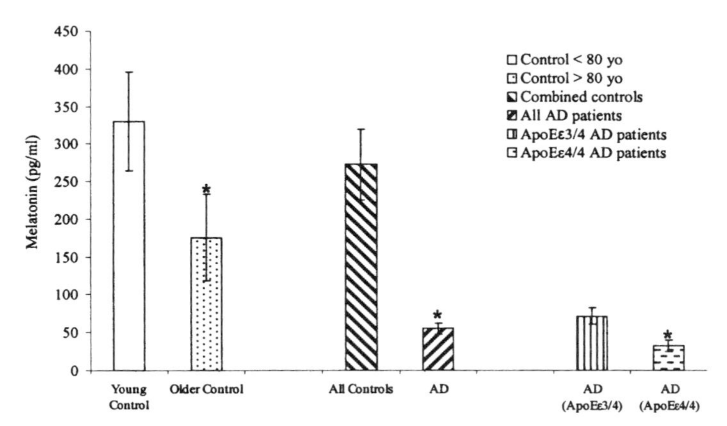

FIGURE 4. Melatonin levels in human CSF by age, AD, and ApoE genotype. Human CSF was assayed for melatonin in young control (<80 years of age), old control (>80 years of age), and AD patients.[72] Data were analyzed comparing young control *(white bar)* vs. old control *(stipple)*, combined controls *(falling diagonal)* vs. all AD patients *(rising diagonal)*, and AD patients of *ApoE* ε3/4 genotype *(vertical bar)* vs. AD patients of *ApoE* ε4/4 genotype *(horizontal stipple)*. In all cases, differences are significant to at least $P < 0.05$ (indicated by an *asterisk*). This figure was adapted from the published data of Liu *et al.*[72]

sleep quality and a reduction of sundowning. In addition to these results, there is a report of disturbed function of the pineal gland in familial amyloid polyneuropathy.[77]

PROCESSES OF NEUROPROTECTION AND NEURODEGENERATION

On the basis of the aforementioned results, we propose that Aβ, cholinesterases, melatonin, and metal homeostasis interact in ways that can lead to either neuroproliferation and neuroprotection or neurodegeneration (FIG. 5). Acetylcholinesterase can function to enhance neuronal development through its noncholinergic function or can produce "hypertoxic" fibrils with Aβ. In addition, AChE contributes to melatonin formation by inhibiting a cholinergic decrease of *N*-acetyltransferase enzyme activity in the pineal gland.[38] Melatonin normally functions to diminish reactive oxidizing species, reduces Aβ levels, reduces the proportion of Aβ in fibrillar form, and blocks metal toxicity. Metal homeostasis contributes to normal functioning of metalloproteases that include the secretases. When it is upset, metal homeostasis leads to metal toxicity, which then contributes to Aβ fibril formation. Metalloprotease activity in direct relationship to AD includes both the neuroprotective α-secretase pathway, which reduces Aβ levels and leads to neuroprotective soluble APP and P3 peptides, and the β-secretase pathway, which produces Aβ peptide. Aβ is generally nontoxic as a monomer and can polymerize into potentially toxic oligomers and fibrils, which may be facilitated by excessive levels of metals such as copper, zinc, or mercury. These fibrils can, as described, include AChE. Aβ fibrils can form amyloid plaques, which lead to neurodegeneration consequent to mechanical disruption of tissue.

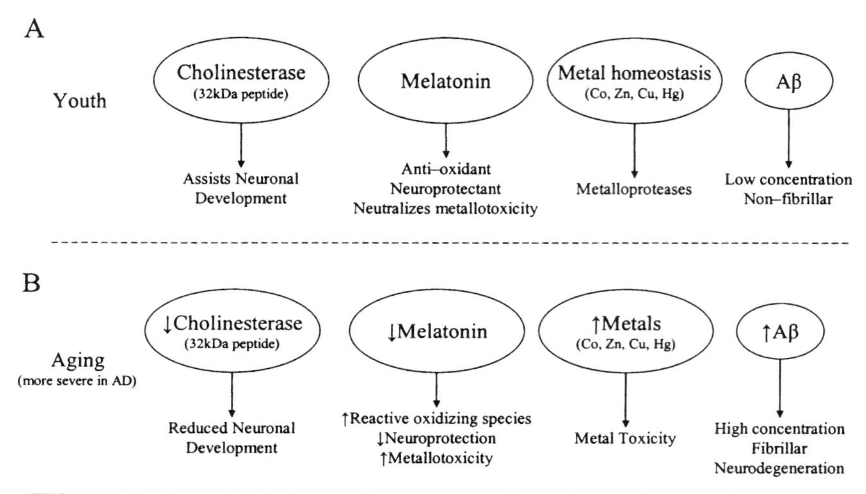

FIGURE 5. Age-related changes in neuroprotective and neurodegenerative agents and conditions. (**A**) Conditions in youthful nervous system. In a healthy youthful nervous system, AChE and/or its peptide product operates to assist neuronal development. Melatonin acts as an antioxidant and a neuroprotectant and neutralizes metal toxicity within neuronal cells. Metal homeostasis keeps levels of various inorganic form metals, such as copper, zinc, and cobalt, sufficient for metalloprotease activity, while rejecting or sequestering metals such as mercury. Aβ peptide is in low concentration and mostly nonfibrillar in form. (**B**) Effects of aging and Alzheimer's disease. In aging, selective neurochemical changes occur in the cholinergic system that reduces neuronal development. Melatonin levels are decreased, which permits an increase in reactive oxidizing species, permits greater metallotoxicity, and reduces neuroprotection. Metal homeostasis becomes unbalanced and moves toward metallotoxicity, permitting the buildup of toxic levels of metals, such as copper, cobalt, and zinc. Aβ levels increase and more of the Aβ peptide is found in its fibrillar form and plaques, leading to neurodegeneration.

PRESENCE OF MELATONIN IN EDIBLE PLANT PRODUCTS SUCH AS WALNUTS

Melatonin has been identified in different plant species, including mono- and dicotyledonous angiosperms.[78] Melatonin levels vary widely in different plants, and melatonin is unequally distributed within a given plant. Notably, high levels of melatonin are reported in traditionally medicinal plants,[79] such as feverfew (*Chrysanthemum parthenium*) and St. John's wort (*Hypericum perforatum*). For example, melatonin concentrations in leaves and flowers from St. John's wort are 1.75 and 4.39 μg/g, respectively. Low but significant levels of melatonin are present (in the ng/g tissue range) in the seeds of a number of plants, including black and white mustard (*Brassica nigra* and *B. hirta*), fenugreek (*Trigonella foenum-graecum*), and almond (*Prunus amygdalus*).[80] Recently, Reiter and colleagues have shown the presence of melatonin in walnuts. Interestingly, feeding walnuts to rats increases blood melatonin concentrations, and this increase correlates with the increased antioxidative capacity of this fluid. In this case, the authors estimated "total antioxidant power" of the serum using the trolox equivalent antioxidant capacity and ferric-reducing ability of serum methods.[81] Moreover, the indole has been found in roots, stems, leaves, fruits, and seeds.[78] Because melatonin taken in the diet gets absorbed by the gut and crosses the blood–brain barrier, the consumption of melatonin-rich foodstuffs would enhance the antioxidative capacity of the organisms.

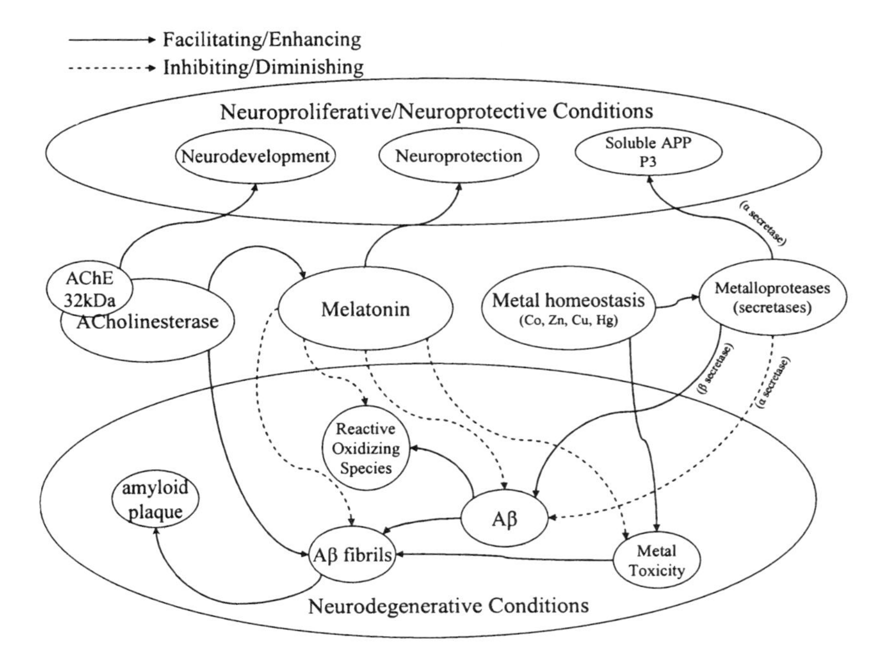

FIGURE 6. Processes of neuroprotection and neurodegeneration. Cholinesterases, melatonin, and metal homeostasis interact in ways that can lead to either neuroproliferation and neuroprotection or neurodegeneration. *Solid arrows* indicate a pathway or effect that facilitates a product or result, whereas *dashed arrows* indicate the diminution or inhibitory effect. Acetylcholinesterase can function to enhance neuronal development through its proteolytic product(s) or can produce "hypertoxic" fibrils with Aβ. In addition, AChE contributes to melatonin formation by inhibiting a cholinergic decrease of *N*-acetyltransferase enzyme activity in the pineal gland.[37] Melatonin normally functions to diminish reactive oxidizing species. It reduces Aβ levels, reduces the proportion of Aβ in fibrillar form, and blocks metal toxicity. Metal homeostasis contributes to normal functioning of metalloproteases that include the secretases. When it is upset, metal homeostasis leads to metal toxicity, which contributes to Aβ fibril formation. Metalloprotease activity in direct relationship to AD includes both the neuroprotective β-secretase pathway, which reduces Aβ levels and leads to neuroprotective soluble APP and P3 peptides, and the β-secretase pathway, which produces Aβ peptide. Aβ is neurotoxic as a monomer and can polymerize into fibrils, which may be facilitated by toxic levels of metals, such as copper, zinc, or mercury. These fibrils can include AChE to produce the described "hypertoxic" fibril. Aβ fibrils can form amyloid plaques, which lead to neurodegeneration because of their mechanical disruption of tissue.

SUMMARY

In short, age-related changes, neuroprotective and neurodegenerative agents, and conditions are discussed as depicted in FIGURE 6. In a healthy youthful nervous system, AChE and/or its peptide product operate to assist neuronal development. Melatonin acts as an antioxidant and a neuroprotectant and neutralizes metal toxicity within neuronal cells. Metal homeostasis keeps levels of various inorganic form metals, such as copper, zinc, and cobalt, sufficient for metalloprotease activity while rejecting or sequestering metals such as mercury. Aβ peptide exists in low concentration and is mostly nonfibrillar in form. In aging, the cholinergic system is changed, thus reducing neuronal development. Melatonin levels are reduced, which permits an increase in reactive oxidizing species, permits greater metallotoxicity, and reduces neuroprotection. Metal homeostasis becomes unbalanced and moves toward metallotoxicity, permitting the buildup of toxic levels of metals, such as copper, cobalt, and zinc. Aβ levels increase and more of the Aβ is found in fibrillar form and plaques, leading to neurodegeneration.

ACKNOWLEDGMENTS

This work was supported in part by grants from the National Institutes of Health (AG 18379, AG 18884, AG 16794, and ES 7992). We are sincerely thankful for the support from Yuan-Wen Ge and Bryan Maloney.

REFERENCES

1. GREIG, N.H. *et al.* 2004. New therapeutic strategies and drug candidates for neurodegenerative diseases: p53 and TNF-α inhibitors, and GLP-1 receptor agonists. Ann. N.Y. Acad. Sci. **1035:** 290–315.
2. LAHIRI, D.K. *et al.* 2004. Melatonin, metals, and gene expression: implications in aging and neurodegenerative disorders. Ann. N.Y. Acad. Sci. **1035:** 216–230.
3. NAKAMURA, Y. *et al.* 1989. Age-dependent change in activities of lysosomal enzymes in rat brain. Mech. Ageing Dev. **50:** 215–225.
4. WEI, Y.H. *et al.* 1998. Oxidative damage and mutation to mitochondrial DNA and age-dependent decline of mitochondrial respiratory function. Ann. N.Y. Acad. Sci. **854:** 155–170.
5. STEVNSNER, T. *et al.* 2002. Mitochondrial repair of 8-oxoguanine and changes with aging. Exp. Gerontol. **37:** 1189–1196.
6. LEE, C.K., R. Weindruch & T.A. Prolla. 2000. Gene-expression profile of the ageing brain in mice. Nat. Genet. **25:** 294–297.
7. CAO, S.X. *et al.* 2001. Genomic profiling of short- and long-term caloric restriction effects in the liver of aging mice. Proc. Natl. Acad. Sci. USA **98:** 10630–10635.
8. LAHIRI, D.K. *et al.* 2003. A critical analysis of new molecular targets and strategies for drug developments in Alzheimer's disease. Curr. Drug Targets. **4:** 97–112.
9. O'NEIL, M.J., A. SMITH & P. HECKELMAN. 2001. The Merck index: an encyclopedia of chemicals, drugs, and biologicals. Merck. Whitehouse Station, NJ.
10. BONDY, S.C. *et al.* 2004. Retardation of brain aging by chronic treatment with melatonin. Ann. N.Y. Acad. Sci. **1035:** 197–215.
11. REITER, R.J., D.X. TAN & M.A. PAPPOLLA. 2004. Melatonin relieves the neural oxidative burden that contributes to dementias. Ann. N.Y. Acad. Sci. **1035:** 179–196.

12. OZCANKAYA, R. & N. DELIBAS. 2002. Malondialdehyde, superoxide dismutase, melatonin, iron, copper, and zinc blood concentrations in patients with Alzheimer disease: cross-sectional study. Croat. Med. J. **43:** 28–32.
13. GREIG, N.H., D.K. LAHIRI & K. SAMBAMURTI. 2002. Butyrylcholinesterase: an important new target in Alzheimer's disease therapy. Int. Psychogeriatr. **14** (suppl. 1): 77–91.
14. BENZI, G. & A. MORETTI. 1995. Age- and peroxidative stress-related modifications of the cerebral enzymatic activities linked to mitochondria and the glutathione system. Free. Radic. Biol. Med. **19:** 77–101.
15. LOVELL, M.A., C. XIE & W.R. MARKESBERY. 2000. Decreased base excision repair and increased helicase activity in Alzheimer's disease brain. Brain Res. **855:** 116–123.
16. LEE, C.K. *et al.* 1999. Gene expression profile of aging and its retardation by caloric restriction. Science **285:** 1390–1393.
17. SHARMAN, E.H. *et al.* 2004. Age-related changes in murine CNS mRNA gene expression are modulated by dietary melatonin. J. Pineal Res. **36:** 165–170.
18. JIANG, C.H. *et al.* 2001. The effects of aging on gene expression in the hypothalamus and cortex of mice. Proc. Natl. Acad. Sci. USA **98:** 1930–1934.
19. PERSENGIEV, S.P. 1999. Multiple domains of melatonin receptor are involved in the regulation of glucocorticoid receptor-induced gene expression. J. Steroid Biochem. Mol. Biol. **68:** 181–187.
20. BONDY, S.C. *et al.* 2002. Dietary modulation of age-related changes in cerebral pro-oxidant status. Neurochem. Int. **40:** 123–130.
21. REITER, R.J. *et al.* 1999. Melatonin as a pharmacological agent against neuronal loss in experimental models of Huntington's disease, Alzheimer's disease and Parkinsonism. Ann. N.Y. Acad. Sci. **890:** 471–485.
22. PAPPOLLA, M.A. *et al.* 1997. Melatonin prevents death of neuroblastoma cells exposed to the Alzheimer amyloid peptide. J. Neurosci. **17:** 1683–1690.
23. SHARMAN, K.G. *et al.* 2002. Dietary melatonin selectively reverses age-related changes in cortical cytokine mRNA levels, and their responses to an inflammatory stimulus. Neurobiol. Aging. **23:** 633–638.
24. SELKOE, D.J. 1994. Normal and abnormal biology of the beta-amyloid precursor protein. Ann. Rev. Neurosci. **17:** 489–517.
25. SAMBAMURTI, K., N.H. GREIG & D.K. LAHIRI. 2002. Advances in the cellular and molecular biology of the beta-amyloid protein in Alzheimer's disease. Neuromol. Med. **1:** 1–31.
26. SHENG, J.G., R.E. MRAK & W.S. GRIFFIN. 1996. Apolipoprotein E distribution among different plaque types in Alzheimer's disease: implications for its role in plaque progression. Neuropathol. Appl. Neurobiol. **22:** 334–341.
27. GIACOBINI, E. 2003. Cholinergic function and Alzheimer's disease. Int. J. Geriatr. Psychiatry. **18:** S1–S5.
28. KIMBERLY, W.T. & M.S. WOLFE. 2003. Identity and function of gamma-secretase. J. Neurosci. Res. **74:** 353–360.
29. CLEARY, J.P. *et al.* 2005. Natural oligomers of the amyloid-beta protein specifically disrupt cognitive function. Nat. Neurosci. **8:** 79–84.
30. YAN, Z. & J. FENG. 2004. Alzheimer's disease: interactions between cholinergic functions and beta-amyloid. Curr. Alz. Res. **1:** 241–248.
31. INESTROSA, N.C., S. URRA & M. COLOMBRES. 2004. Amyloid-beta-peptide complexes in Alzheimer's disease. The Wnt signaling pathway. Curr. Alz. Res. **1:** 249–254.
32. NILSEN, J. & R.D. BRINTON. 2004. Mitochondria as therapeutic targets of estrogen action in the central nervous system. Curr. Drug Targets CNS Neurol. Disord. **3:** 297–313.
33. BACHURIN, S.O. *et al.* 2003. Mitochondria as a target for neurotoxins and neuroprotective agents. Ann. N.Y. Acad. Sci. **993:** 334–344; discussion, 345–349.
34. MONDACA, M. *et al.* 2004. Involvement of melatonin metabolites in the long-term inhibitory effect of the hormone on rat spinal nociceptive transmission. Pharmacol. Biochem. Behav. **77:** 275–279.
35. TEH, M.T. & D. SUGDEN. 2001. An endogenous 5-HT(7) receptor mediates pigment granule dispersion in *Xenopus laevis* melanophores. Br. J. Pharmacol. **132:** 1799–1808.

36. Teh, M.T. & D. Sugden. 2002. Desensitization of pigment granule aggregation in *Xenopus leavis* melanophores: melatonin degradation rather than receptor down-regulation is responsible. J. Neurochem. **81:** 719–727.
37. Day, T. & S.A. Greenfield. 2003. A peptide derived from acetylcholinesterase induces neuronal cell death: characterisation of possible mechanisms. Exp. Brain Res. **153:** 334–342.
38. Phansuwan-Pujito, P., M. Moller & P. Govitrapong. 1999. Cholinergic innervation and function in the mammalian pineal gland. Microsc. Res. Tech. **46:** 281–295.
39. Wagner, G., R. Brandstatter & A. Hermann. 2000. Adrenergic and cholinergic regulation of in vitro melatonin release during ontogeny in the pineal gland of Long Evans rats. Neuroendocrinology **72:** 154–161.
40. Markus, R.P. *et al.* 2003. Melatonin nocturnal surge modulates nicotinic receptors and nicotine-induced [3H]glutamate release in rat cerebellum slices. J. Pharmacol. Exp. Ther. **305:** 525–530.
41. Laitinen, J.T., K.S. Laitinen & T. Kokkola. 1995. Cholinergic signaling in the rat pineal gland. Cell. Mol. Neurobiol. **15:** 177–192.
42. Nitta, A. *et al.* 1995. Age-related changes in learning and memory and cholinergic neuronal function in senescence accelerated mice (SAM). Behav. Brain Res. **72:** 49–55.
43. Reddy, G.R. *et al.* 2003. Lead induced effects on acetylcholinesterase activity in cerebellum and hippocampus of developing rat. Int. J. Dev. Neurosci. **21:** 347–352.
44. Basha, M.R. *et al.* 2005. The fetal basis of amyloidogenesis: exposure to lead and latent overexpression of amyloid precursor protein and beta-amyloid in the aging brain. J. Neurosci. **25:** 823–829.
45. Rogers, J.T. & D.K. Lahiri. 2004. Metal and inflammatory targets for Alzheimer's disease. Curr. Drug Targets **5:** 535–551.
46. Allinson, T.M. *et al.* 2003. ADAMs family members as amyloid precursor protein alpha-secretases. J. Neurosci. Res. **74:** 342–352.
47. Bellingham, S.A. *et al.* 2004. Copper depletion down-regulates expression of the Alzheimer's disease amyloid-beta precursor protein gene. J. Biol. Chem. **279:** 20378–20386.
48. Smorgon, C. *et al.* 2004. Trace elements and cognitive impairment: an elderly cohort study. Arch. Gerontol. Geriatr. Suppl. 393–402.
49. Maynard, C.J. *et al.* 2002. Overexpression of Alzheimer's disease amyloid-beta opposes the age-dependent elevations of brain copper and iron. J. Biol. Chem. **277:** 44670–44676.
50. Paredes, D. *et al.* 1999. Melatonin acts on the nucleus accumbens to increase acetylcholine release and modify the motor activity pattern of rats. Brain Res. **850:** 14–20.
51. Barajas-Lopez, C. *et al.* 1996. Melatonin modulates cholinergic transmission by blocking nicotinic channels in the guinea-pig submucous plexus. Eur. J. Pharmacol. **312:** 319–325.
52. Markus, R.P., W.M. Zago & R.C. Carneiro. 1996. Melatonin modulation of presynaptic nicotinic acetylcholine receptors in the rat vas deferens. J. Pharmacol. Exp. Ther. **279:** 18–22.
53. Zago, W.M. & R.P. Markus. 1999. Melatonin modulation of presynaptic nicotinic acetylcholine receptors located on short noradrenergic neurons of the rat vas deferens: a pharmacological characterization. Braz. J. Med. Biol. Res. **32:** 999–1006.
54. Olivieri, G. *et al.* 2001. Melatonin protects SHSY5Y neuroblastoma cells from cobalt-induced oxidative stress, neurotoxicity and increased beta-amyloid secretion. J. Pineal Res. **31:** 320–325.
55. Lack, B., S. Daya & T. Nyokong. 2001. Interaction of serotonin and melatonin with sodium, potassium, calcium, lithium and aluminium. J. Pineal Res. **31:** 102–108.
56. Mocchegiani, E. *et al.* 1994. The immuno-reconstituting effect of melatonin or pineal grafting and its relation to zinc pool in aging mice. J. Neuroimmunol. **53:** 189–201.
57. Mocchegiani, E. *et al.* 1998. Presence of links between zinc and melatonin during the circadian cycle in old mice: effects on thymic endocrine activity and on the survival. J. Neuroimmunol. **86:** 111–122.
58. Lin, A.M. & L.T. Ho. 2000. Melatonin suppresses iron-induced neurodegeneration in rat brain. Free. Radic. Biol. Med. **28:** 904–911.

59. LIMSON, J., T. NYOKONG & S. DAYA. 1998. The interaction of melatonin and its precursors with aluminium, cadmium, copper, iron, lead, and zinc: an adsorptive voltametric study. J. Pineal Res. **24:** 15–21.
60. KABUTO, H., I. YOKOI & N. OGAWA. 1998. Melatonin inhibits iron-induced epileptic discharges in rats by suppressing peroxidation. Epilepsia **39:** 237–243.
61. MAHARAJ, D.S., J.L. LIMSON & S. DAYA. 2003. 6-Hydroxymelatonin converts Fe (III) to Fe (II) and reduces iron-induced lipid peroxidation. Life Sci. **72:** 1367–1375.
62. HAYTER, C.L., G.M. BISHOP & S.R. ROBINSON. 2004. Pharmacological but not physiological concentrations of melatonin reduce iron-induced neuronal death in rat cerebral cortex. Neurosci. Lett. **362:** 182–184.
63. WAKATSUKI, A. *et al.* 2000. Melatonin inhibits oxidative modification of low-density lipoprotein particles in normolipidemic post-menopausal women. J. Pineal Res. **28:** 136–142.
64. PARMAR, P. *et al.* 2002. Melatonin protects against copper-mediated free radical damage. J. Pineal Res. **32:** 237–242.
65. HANDY, R.D. 2003. Chronic effects of copper exposure versus endocrine toxicity: two sides of the same toxicological process? Comp. Biochem. Physiol. A: Mol. Integr. Physiol. **135:** 25–38.
66. SONG, W. & D.K. LAHIRI. 1997. Melatonin alters the metabolism of the beta-amyloid precursor protein in the neuroendocrine cell line PC12. J. Mol. Neurosci. **9:** 75–92.
67. PARLAKPINAR, H. *et al.* 2002. Physiological and pharmacological concentrations of melatonin protect against cisplatin-induced acute renal injury. J. Pineal Res. **33:** 161–166.
68. POEGGELER, B. *et al.* 2001. Melatonin reverses the profibrillogenic activity of apolipoprotein E4 on the Alzheimer amyloid Abeta peptide. Biochemistry **40:** 14995–15001.
69. LOPEZ-BURILLO, S. *et al.* 2003. Melatonin, xanthurenic acid, resveratrol, EGCG, vitamin C and alpha-lipoic acid differentially reduce oxidative DNA damage induced by Fenton reagents: a study of their individual and synergistic actions. J. Pineal Res. **34:** 269–277.
70. HOYOS, M. *et al.* 2000. Serum cholesterol and lipid peroxidation are decreased by melatonin in diet-induced hypercholesterolemic rats. J. Pineal Res. **28:** 150–155.
71. LAHIRI, D.K. *et al.* 2004. Age-related changes in serum melatonin in mice: higher levels of combined melatonin and 6-hydroxymelatonin sulfate in the cerebral cortex than serum, heart, liver and kidney tissues. J. Pineal Res. **36:** 217–223.
72. LAHIRI, D.K. *et al.* 2004. Dietary supplementation with melatonin reduces levels of amyloid beta-peptides in the murine cerebral cortex. J. Pineal Res. **36:** 224–231.
73. LIU, R.Y. *et al.* 1999. Decreased melatonin levels in postmortem cerebrospinal fluid in relation to aging, Alzheimer's disease, and apolipoprotein E-epsilon4/4 genotype. J. Clin. Endocrinol. Metab. **84:** 323–327.
74. ROSALES-CORRAL, S. *et al.* 2003. Orally administered melatonin reduces oxidative stress and proinflammatory cytokines induced by amyloid-beta peptide in rat brain: a comparative, in vivo study versus vitamin C and E. J. Pineal Res. **35:** 80–84.
75. MATSUBARA, E. *et al.* 2003. Melatonin increases survival and inhibits oxidative and amyloid pathology in a transgenic model of Alzheimer's disease. J. Neurochem. **85:** 1101–1108.
76. BRUSCO, L.I., M. MARQUEZ & D.P. CARDINALI. 1998. Monozygotic twins with Alzheimer's disease treated with melatonin: case report. J. Pineal Res. **25:** 260–263.
77. HIGA, S. *et al.* 1987. Disturbed function of the pineal gland in familial amyloid polyneuropathy. J. Neural Transm. **69:** 97–103.
78. REITER, N.J. & D.X. TAN. 2002. Melatonin: an antioxidant in edible plants. Ann. N.Y. Acad. Sci. **957**: 341–344.
79. MURCH, S.J., C.B. SIMMONS & P.K. SAXENA. 1997. Melatonin in feverfew and other medicinal plants. Lancet **350:** 1598–1599.
80. MANCHESTER, L.C. *et al.* 2000. High levels of melatonin in the seeds of edible plants: possible function in germ tissue protection. Life Sci. **67:** 3023–3029.
81. REITER, R.J., L.C. MANCHESTER & D.X. TAN. 2005. Melatonin in walnuts: influence on leveles of melatonin and total antioxidant capacity of blood. Nutrition (epub ahead of print).

Administration of Vaccinia Virus Complement Control Protein Shows Significant Cognitive Improvement in a Mild Injury Model

NIRVANA S. PILLAY,[a] LAURIE A. KELLAWAY,[b] AND GIRISH J. KOTWAL[a]

[a]*Division of Medical Virology and* [b]*Division of Neuroscience, University of Cape Town, Medical School, Observatory, Cape Town, 7925, South Africa*

ABSTRACT: Previous studies have shown that traumatic mild brain injury in a rat model is accompanied by breakdown of the blood brain barrier and the accumulation of inflammatory cells. A therapeutic agent, vaccinia virus complement control protein (VCP), inhibits both the classic and the alternative pathways of the complement system and, in so doing, prevents cell death and inflammation. With the use of a rat mild injury model, the effects of VCP on spatial learning and memory were tested. Training in a Morris water maze consisted of a total of 16 trials over a 2-day period before rats were anesthetized and subjected to mild (1.0-1.1 atm) lateral fluid percussion injury (FPI) 3.0 mm lateral to the sagittal suture and 4.5 mm posterior to bregma. Ten μl of VCP (1.7 mg/ml) was injected into the injury site immediately after FPI. Two weeks post-FPI the rats were assessed in the Morris water maze for spatial learning and memory. Neurologic motor function tests were carried out after FPI for 14 consecutive days and again after 28 days. The Morris water maze data show that FPI plus saline-injected rats spent a significantly ($P < 0.05$) larger amount of time in one of the incorrect quadrants than did the FPI plus VCP-injected group. Neurologic evaluations 24 hours postinjury revealed differences in sensorimotor function between groups. The results suggest that in a mild injury model, VCP influences neurologic outcome and offers some enhancement in spatial memory and learning.

KEYWORDS: complement control; cognition; mild injury; vaccinia virus; VCP; Morris water maze (MWM); fluid percussion injury (FPI); traumatic brain injury (TBI)

INTRODUCTION

Mild traumatic brain injury (TBI) is one of the most common neurologic disorders.[1] The vast majority of head injuries (75–90%) are clinically classified as mild.[2,3] Ten to fifteen percent of patients remain symptomatic with postconcussive symptoms for 1 year or longer.[4] These symptoms include cognitive deficits, emotional distress, behavioral disturbances, and dizziness.[4-6] Due to the subtle symp-

Address for correspondence: Girish J. Kotwal, Division of Medical Virology, Medical School, University of Cape Town, Anzio Road, Observatory, Cape Town 7925, South Africa. Voice: (27)-(21) 406-6676; fax: (27)-(21) 406 6018.
gjkotw01@yahoo.com

**Ann. N.Y. Acad. Sci. 1056: 450–461 (2005). © 2005 New York Academy of Sciences.
doi: 10.1196/annals.1352.021**

TABLE 1. Summary of the various definitions of severity using a fluid percussion device

Author/Year	Intensity/Severity	Intracranial pressure (atm)
McIntosh *et al.*[7]	Low grade	0–1.0
	Moderate	1.5–2.0
	High grade	2.5–3.6
Prins *et al.*[9]	Mild	1.35–1.45
	Moderate	2.65–2.75
	Severe	3.65–3.75
Pierce *et al.*[10]	Severe	2.7–3.1
Knoblach *et al.*[11]	Severe	2.6–2.7
Raghupathi *et al.*[12]	Mild	1.1–1.3
Hicks *et al.*[13]	Moderate	4.5
Sanders *et al.*[14]	Mild	1.4–1.5

toms after mild TBI, rates of hospital admission are low, and little is known about the short-term behavioral and cognitive deficits that may arise and persist. Consequently, animal models are used to mimic cognitive and behavioral events produced by traumatic injuries and to evaluate potential therapeutic agents. The fluid percussion brain injury (FPI) model is currently the most widely used method for producing TBI.[7] Injuries induced by lateral (parasagittal) fluid percussion in the rat closely resemble the features of head injury in humans and are therefore best suited for studying the pathophysiology of TBI and agents involved in arresting its progression. The FPI device produces a pulse (21–23 ms) of increased intracranial pressure with displacement and deformation of neural tissue.[8] This model has been shown to produce mild to severe injury that is quantifiable and reproducible.[7] The intensity of injury caused by the FPI device has been calibrated and defined according to various severity levels (TABLE 1). In this study, the severity of mild fluid percussion-induced injury compares to that first described by McIntosh *et al.*,[7] who classified low-grade injury as being 1.0 atmospheres. It is evident that the definitions of severity vary among different research laboratories (TABLE 1). This may have implications with regard to the interpretation of data, which may lead to discrepancies in findings.

Rat models of mild TBI after FPI exhibit a breakdown of the blood brain barrier,[15] microglia and macrophage activation,[16] apoptotic cell death,[12] and alterations in neutrophin expression in the hippocampus.[6] Complement components contribute to inflammatory injury in a variety of tissues, including the central nervous system (CNS). Evidence of elevated levels of complement components in the cerebrospinal fluid (CSF) and accumulation of C3 at the site of TBI in patients may implicate the contribution of complement to secondary damage caused by inflammation.[17] Recent evidence demonstrates the critical role that C3 and C5 play in TBI when secondary damage after acute TBI in C3- and C5-deficient mice is significantly reduced.[18] This supports the role of a C3 inhibitor as a candidate for modulation of inflammation. A poxvirus-derived protein that evades the consequences of complement activation was identified in 1988.[19] This immunomodulatory molecule, vaccinia virus complement control protein (VCP), inhibits both the classic and the alternative pathways of the complement system and in so doing prevents cell death and inflammation. VCP may be a candidate for preventing mild brain injury-related inflammation.

VCP is structurally similar to human complement 4b binding protein and is functionally similar to CR1, decay acceleration factor, factor H, and membrane cofactor protein.[20] VCP inhibits classic complement activation by binding C3 and C4 and acting as a cofactor for factor 1 cleavage of C3b and C4b. The alternative pathway is inhibited through the same mechanism, resulting in the cleavage of C3b into iC3b, therefore preventing the formation of the alternative pathway C3 convertase.[19] Additionally, VCP has the ability to bind heparin and heparan sulfate proteoglycans, further contributing to its anti-inflammatory function.

A previous study employing lateral fluid percussion-induced brain injury of moderate severity (4.5 atm) suggested that VCP may enhance the functional recovery of Sprague-Dawley rats.[13] In this study, we evaluate the effect of VCP administration after mild (1.0–1.1 atm) traumatic brain injury in Wistar rats in the Morris water maze (MWM) and by neurologic testing in the acute phase of recovery.

MATERIAL AND METHODS

VCP Preparation and Purification

A single colony of *Pichia pastoris* containing the VCP gene insert was first incubated in 100 ml of buffered minimal glycerol media (BMGY) for 48 hours at 30°C 220 rpm. After vigorous shaking, this starter culture was inoculated in 1 liter of BMGY and was allowed to grow for a further 48 hours. The cells were harvested by centrifugation in a Sorvall RC58 Plus centrifuge (Kendro, Connecticut, USA) for 30 minutes at 10,000 rpm. The cells were then resuspended in 250 ml of a buffered minimal methanol media (BMMY) containing 4% methanol and were then incubated under similar conditions for 48 hours. The VCP supernatant was harvested by centrifugation for 20 minutes at 15,000 rpm. VCP was first visualized by Coomassie blue staining on an SDS-PAGE (sodium dodecyl sulfate polyacrylamide) gel before further downstream processing. All VCP supernatants were pooled and then passed through a 10-K Pellicon 2 cassette that uses tangential flow for ultrafiltration (Millipore, Ireland). Five milliliters of ultrafiltrated VCP was purified by injection through a 0.22-µm filter unit into a 5-ml HiTrap heparin column (Amersham Pharmacia, Uppsala, Sweden). The protein was eluted with NaCl ranging from 100 mM NaCl to 1 M NaCl using an AKTA prime purification system (Amersham Pharmacia, Uppsala, Sweden) The protein was then concentrated using a Centriprep centrifugal filter device (YM10, Millipore, Ireland). The electrophoresis SDS-PAGE gel (FIG.1) confirms the production of VCP.

Hemolysis Assay and Determination of Protein Concentration

This assay was performed to evaluate the functional activity of VCP in hemolysis inhibition as previously described by Kotwal *et al* 1990 with a few modifications [21]. The supernatant buffer was removed from 3ml of sensitised sheep red blood cells (RBC) and was used as reaction buffer. All preparations were performed on ice. Serum was diluted in a 1:30 proportion with reaction buffer. Each reaction consisted of 75 µl of RBC, 5 µl of reaction buffer, 15 µl of serum, and 5 µl of sample to a final volume of 100 μl. The positive control excluded VCP only, and the negative control excluded the VCP and serum. The samples were incubated for 1 hour at 37°C. After centrifugation at 13,000 rpm for 1 minute, 150 μl of supernatant was transferred to a 96-well plate and was read at 405 nm on an Anthos 2010 microplate reader (Salzburg, Austria).

The protein concentration was determined using a Bio-Rad assay (Bio-Rad, California, USA), and absorbances were read at 595 nm on an Anthos 2010 microplate reader (Salzburg, Austria).

Surgical Preparation and Fluid Percussion Injury

Following training in the MWM, Wistar rats (250 g) were each randomly assigned to one of four groups: FPI + VCP-treated ($n = 6$), sham + VCP-treated ($n = 6$), FPI + saline-treated ($n = 6$), and sham + saline-treated ($n = 6$). The rats were anesthetized with Equithesin (100 mg/kg intraperitoneally). All surgical procedures were performed under semisterile conditions. When fully surgically anesthetized, the rats were immobilized in a stereotaxic frame.

A burr hole (5 mm in diameter) was constructed in the cranium, centered 3 mm lateral to the sagittal suture and 4.5 mm posterior to bregma[22] in the anteroposterior plane, leaving the underlying dura intact. A Luer-loc hub was centralized over the exposed dura and then cemented to the skull with dental cement. After the hardening of the dental cement, the Luer attachment was then coupled to a calibrated fluid percussion device via high-pressure hydraulic tubing pre-filled with isotonic saline. Mild percussion (1.0–1.1 atm) was then applied via the fluid percussion device. At the time of injury, animals were fully anesthetized as indicated by a lack of corneal reflex and the absence of a footpad-pinch withdrawal response. After the percussion shock, the dental cement was carefully removed. VCP (10 μl, 1.7 μg/ml) was injected into the rat cortex at the site of injury. Rats were then removed from the stereotaxic device, and wounds were carefully sutured and allowed to recover in separate cages from the effects of surgery. Animals were frequently checked. Control animals were treated identically, except that they received a sham injury, defined as being attached to the fluid percussion injury device (as described) but without the percussion.

Fourteen days postsurgery these animals were tested for spatial learning and memory in an MWM. These animals remained in separate cages for the entire duration of recovery. Food and water were freely available, and cages were kept in a 12h-light and 12h-dark cycle.

Cognitive Testing

In this study, spatial learning was evaluated using the MWM hidden platform version.[23] The platform was submerged under opaque water in one quadrant of the pool. The animals were released randomly and were required to find the hidden platform with the aid of visual and spatial cues (markings on posters) positioned on the walls outside the maze. The animals were placed randomly at each of the four starting points (north, south, east, and west) into the water to avoid habituating any animal to a particular area of the pool.

Preinjury Training

Preinjury training trials started 2 days before surgery. Training began the afternoon of day 1 and continued on the morning of day 2. Each rat was given eight trials in total with a 4-minute interval per trial each day. One hundred twenty seconds were allotted to finding the hidden platform in trial 1 and a maximum of 60 seconds in subsequent trials. After finding the platform, the animal was allowed to remain on the platform for 10 seconds and then lifted out of the MWM. Rats that failed to find the platform within 120 seconds in trial 1 or 60 seconds in subsequent trials were classified as having the maximum goal latency for that trial. These rats were then placed on the platform and allowed to remain there for 15 seconds so that they could visualize the extra-maze posters that serve as cues for orientation. After each trial, the rat was returned to a holding cage for 4 minutes and reintroduced to the MWM.

A probe trial that lasted 60 seconds was done at the end of the last acquisition trial on day 2 to assess the rat's spatial learning. At the end of the training session, the rat was dried and then returned to the home cage for 2 hours before undergoing surgery.

Postinjury Testing

Sham and postinjured animals were tested for memory retention on day 14 postsurgery. Eight acquisition trials were undertaken as described in the pre-training trial, using the same platform location as before. The time taken for the rat to reach the platform (goal latency) was measured. After memory retention testing, the platform was removed and a probe trial was performed. This probe trial was used to assess whether the animal exhibits a strong preference for the platform location or was using random search patterns to locate the hidden platform. Spatial learning and memory are assessed by the amount of time (in seconds) from the moment of release that the animal spends in the quadrant where the platform was previously located compared to the time spent in other quadrants. The probe test is scored as the percentage of time spent in the correct quadrant. This tests the difference in spatial and retention memory among the various groups of rats. The quadrant time indicates the amount of time (in seconds) that the rat spent in the section of the pool where the platform was previously located during training sessions.

Neurologic Evaluations

A battery of sensorimotor tests were performed by a trained observer who was blind to the injury status of the animals. Motor function was assessed by measuring the following tasks: (1) resistance to lateral pulsion; (2) forelimb flexion response during suspension by the tail; and (3) hindlimb flexion, a response of the hindlimb and toes when raised by the tail. Animals were evaluated daily for 14 consecutive days after injury for lateral pulsion (left and right), hindpaw grip (left and right), forelimb and hindlimb extension, visual placing, and tactile placing. These animals were also evaluated for their ability to stand for 5 seconds on an inclined board at 7 angles ranging from 20–50 degrees. The injury status of the animal was scored using a modified scaling system.[23] Animals were scored as follows: 20 (normal function), 15 (slightly impaired), 10 (moderately impaired), and 5 (severely impaired). To test for impaired ambulation the animals were released from the center of the inner circle on a flat surface and were required to move a full body length to the outer circle within a minute. Animals were graded according to the time taken to leave the inner circle as follows: 0–10 s (normal), 11–25 s (slightly impaired), 26–40 s (moderately impaired), and 41–60 s (severely impaired).

Data Analysis

The goal latencies to find the platform were averaged for trials 1–8 on day 1 and day 2. The probe scores 14 days postinjury were analyzed with a one-way analysis of variance (ANOVA) to determine if there were differences between the groups. When significant differences were obtained, a post-hoc multiple comparison using the Bonferroni test was done to determine which groups were significantly different from each other. Nonparametric analysis of the data was also undertaken to confirm the results revealed by parametric data analysis (those that assume normality of distribution). Nonparametric alternatives to the one-way ANOVA were performed. The Kruskal-Wallis ANOVA and Mann-Whitney U test are similar to the one-way ANOVA except that they are based on ranks rather than means.

Daily neurologic scores for each animal in both experimental and control groups were averaged and compared.

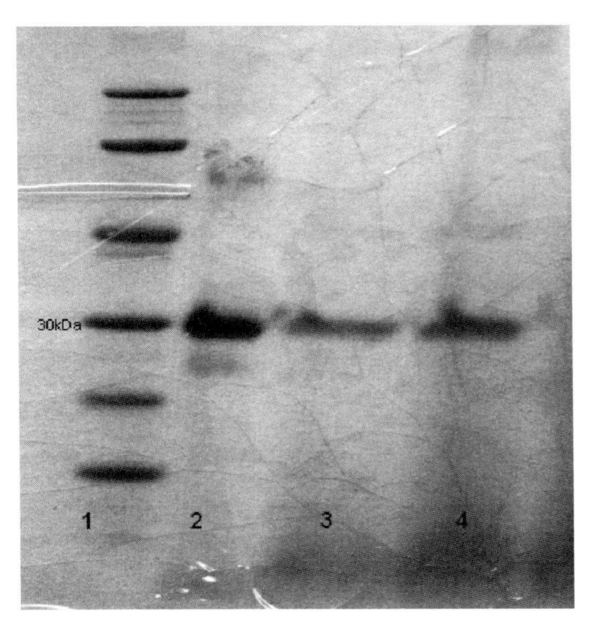

FIGURE 1. SDS-PAGE gel indicating control VCP and VCP expression in *Pichia pastoris*. *Lane 1*: Protein marker ranging from 14.4–97 kDa. *Lane 2*: VCP control. *Lane 3*: VCP expression. *Lane 4*: VCP expression after purification.

RESULTS

Vaccinia virus complement control protein production was confirmed on a sodium-dodecyl sulfate (SDS) polyacrylamide gel after purification (FIG. 1), and a hemolysis activity assay revealed 79% inhibition of lysis in RBCs. VCP was estimated to have a concentration of (1.77 μg/μl) and aliquots of 10 μl of VCP were subsequently injected in VCP-treated rats.

In the first trial of pre-training sessions, rats located the platform in an average time of 92 seconds on day 1. The goal latency of the first trial on day 2 was 54 seconds. By the last trial on day 2, animals were able to find the platform in an average of 11 seconds. The overall shorter goal latencies on day 2 than on day 1 were taken to indicate that rats were trained (FIG. 2).

Rats were then tested in the MWM 14 days after FPI and treatment to evaluate differences between groups. A one-way ANOVA was performed to evaluate differ-

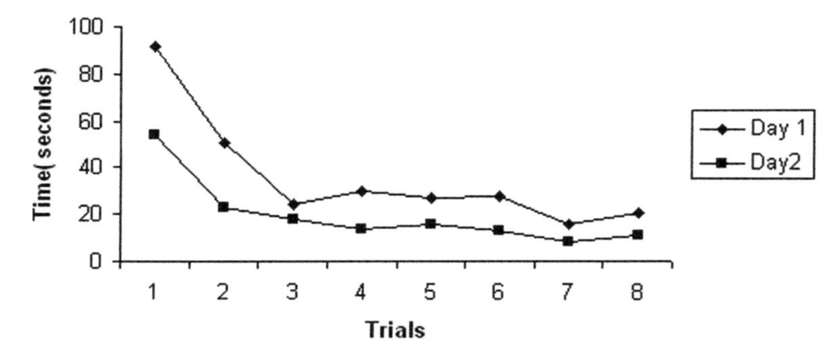

FIGURE 2. Goal latencies. Animals were randomly assigned to groups after 2 days of training. Pre-training goal latencies on day 1 ranged from 92 to 21 seconds and on day 2 from 54 to 11 seconds.

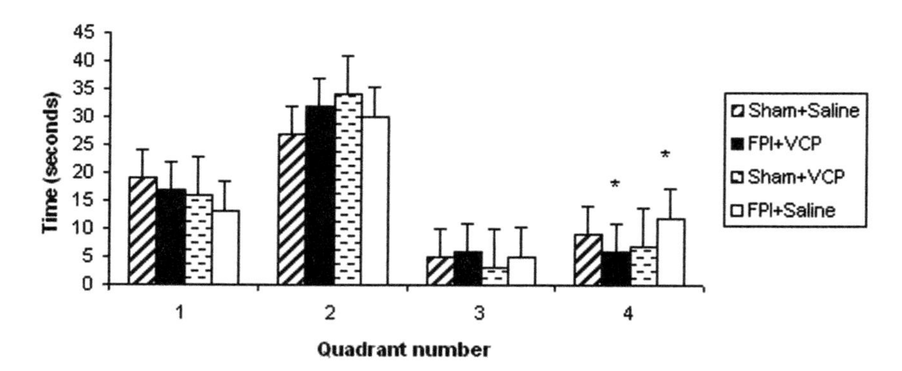

FIGURE 3. A one-way ANOVA demonstrated a significant difference (*P <0.05) in the amount of time between FPI + saline compared to FPI + VCP group, spent in the incorrect quadrant number 4 in a probe trial test 14 days postinjury. There were no significant differences in other quadrants.

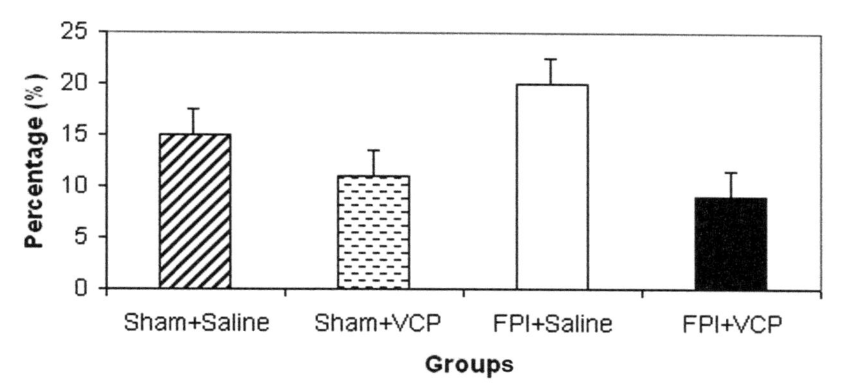

FIGURE 4. Fourteen days postinjury testing revealed that the FPI + saline groups spent 20% of time in the incorrect quadrant (number 4). FPI + VCP groups spent 9% of the time in this quadrant (*P <0.05).

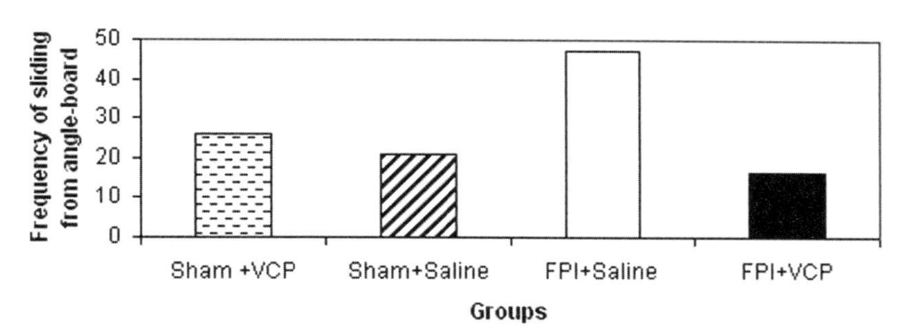

FIGURE 5. Angle board: grip test 24 hours postinjury. Sensorimotor impairment in grip was revealed by the rats' ability to stand on an inclined plane. FPI + saline group showed greater inability to grip than did other FPI and sham groups.

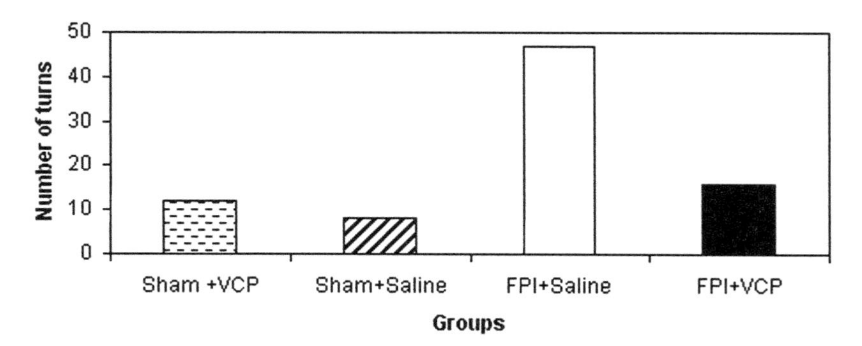

FIGURE 6. Angle board: turns 24 hours postinjury. After 24 hours of postinjury testing, sham groups and the FPI + VCP group showed no significant differences in the frequency of turns. However, the FPI + saline group turned more frequently on the inclined plane.

ences between groups in each of the four quadrants. A significant difference [$F(3,23) = 3.91$ ($P < 0.05$)] between groups was found in the amount of time spent in one of the incorrect quadrants (FIG. 3). Post-hoc analysis revealed the FPI + VCP and FPI + saline groups contributed to the significant effect in MWM testing [$F(1,10) = 11.37$, $P < 0.05$]. The FPI + VCP group spent 9% of the time in one of the incorrect quadrants compared to the FPI + saline group that spent 20% of the time in the incorrect quadrant (FIG. 4). On the basis of the neurologic scores obtained, the majority of tests on motor function were statistically not significant (data not shown), with the exception of the ambulatory and inclined plane gripping test (FIG. 5). The ambulation test was performed daily for 14 consecutive days. The FPI + VCP group scored significantly higher on the ambulation test. After 24 hours FPI, the saline group showed a higher frequency of sliding from the inclined plane than did the VCP-injected group. Frequency tests of injured rats showed that the saline group turned more frequently than did the other groups (FIG. 6). Injured rats showed more

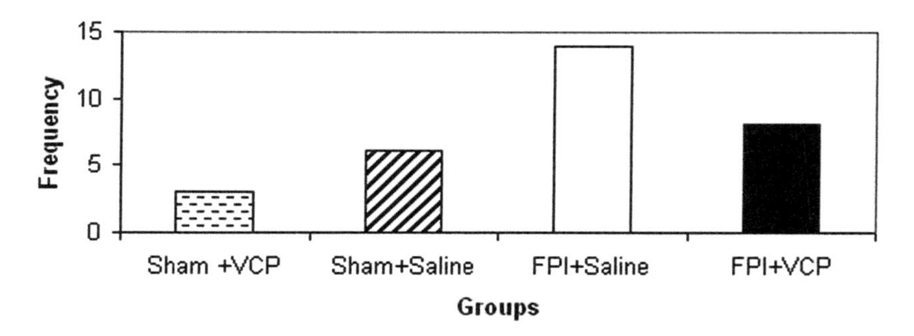

FIGURE 7. Effect of VCP on freezing behavior at the point of placement 24 hours postinjury demonstrated differences between sham and fluid percussion (FP)-injured rats. Freezing behavior was more frequent in FP-injured animals than in other groups.

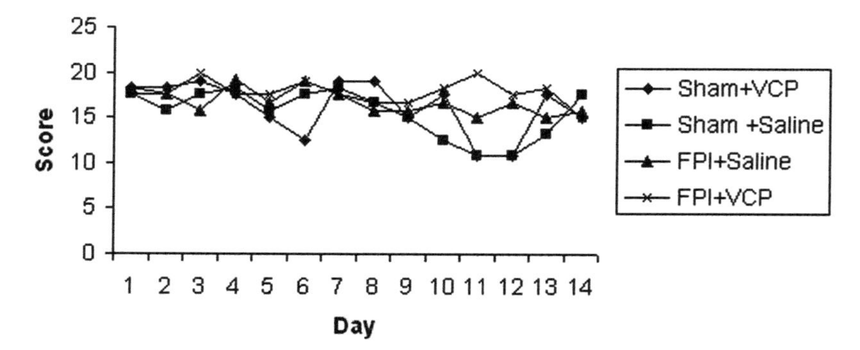

FIGURE 8. To test for impaired ambulation the animals were released from the center of the inner circle on a flat surface and were required to move a full body length to the outer circle within a minute. Daily scores were recorded for 14 consecutive days. A statistically significant difference ($P = 0.05$) in ambulation was demonstrated when sham + saline groups showed significantly more impairment than did FPI + VCP groups.

freezing behavior than did sham rats. The FPI + saline group demonstrated more freezing behavior than did the FPI + VCP group (FIG. 7).

One-way ANOVA and post-hoc comparisons revealed a significant difference [$F(3,52) = 2.69$, $P = 0.05$] in ambulation between sham + saline and FPI + VCP groups (FIG. 8).

DISCUSSION

The Morris water maze is a sensitive paradigm used to evaluate spatial learning and memory.[21] This study shows that even in mildly injured rats, subtle changes in cognition occur 2 weeks postinjury, when saline-treated injured rats spent a statistically significant amount of time in the incorrect quadrant compared to rats treated with VCP. In contrast, in a previous study using Sprague-Dawley rats, the administration of VCP in a moderate injury model revealed that spatial learning in the MWM after VCP injection was not enhanced over that of saline-treated injured controls.[13] These differences may lie in the severity of injury, strain differences, and the administered dose of VCP

The administration of VCP significantly influenced sensorimotor function. As shown in FIGURE 5, the saline-treated injured rats showed greater inability to grip than did the VCP-treated injured group. Even though there were no significant differences in motor function, a better indication of sensorimotor impairment in grip was revealed by the rats' ability to stand on an inclined plane. Injured saline-treated animals turned more frequently on the angle board than did the sham and VCP-treated injured groups. Such similar behavior was observed in a rat model characterizing the sensorimotor impairment in sham-operated and occluded middle cerebral artery rats.[24]

The basolateral amygdala has been implicated in the expression of freezing, a fear-motivated behavior in rats.[25] In this study, fluid percussion injury may affect this area because rats that were given VCP demonstrated less freezing behavior than did saline groups, as shown in FIGURE 7. VCP may offer protection of brain circuits involved in the expression of freezing behavior. To test for impaired ambulation the animals were released from the center of the inner circle on a flat surface and were required to move a full body length to the outer circle within 1 minute. The FPI + VCP group scored higher than did other groups on the ambulation test, which suggests that VCP may have a beneficial effect on motor control after brain injury.

The data presented in this study, together with limited information on the efficacy of VCP in severity-dependent brain injury, further contribute to conceptualizing the therapeutic and neuroprotective effects of VCP.

ACKNOWLEDGMENTS

N.S.P. is a recipient of the Medical Research Council (South Africa) scholarship and the Poliomyelitis Research Foundation scholarship. G.J.K. is a Senior Wellcome Trust International Fellow for biomedical sciences in South Africa. This study is partially funded by an MERC grant from the Faculty of Health Sciences, University of Cape Town.

We would like to thank Harry Hall, from the UCT workshop, for his great effort in constructing the FPI device according to the specifications of a commercially available FPI device. We would like to thank Rochelle van Wijk for being the trained observer in the study. We would also like to thank the University of Cape Town Statistics Division for verifying our results. We would also like to thank Amod P. Kulkarni, Division of Medical Virology, UCT, South Africa, for initial formatting of the manuscript.

REFERENCES

1. ALEXANDER, M.P. 2005. Mild traumatic brain injury: pathophysiology, natural history, and clinical management. Neurology **45:** 1253–1260.
2. LANGFITT, T.W. & T.A. GENNARELLI. 1982. Can the outcome from head injury be improved? J. Neurosurg. **56:** 19–25.
3. KRAUS, J.F. & P. NOURJAH. 1988. The epidemiology of mild, uncomplicated brain injury. J. Trauma **28:** 1637–1643.
4. CHAMELIAN, L. & A. FEINSTEIN. 2004. Outcome after mild to moderate traumantic brain injury: the role of dizziness. Arch. Phys. Med. Rehab. **85:** 1662–1666.
5. ZOHAR, O., S. SCHREIBER, V. GETSLEV, *et al.* 2003. Closed-head minimal traumatic brain injury produces long-term cognitive deficits in mice. Neuroscience **118:** 949–955.
6. HICKS, R.R., V.B. MARTIN, L. ZHANG & K.B. SEROOGY. 1999. Mild experimental brain injury differentially alters the expression of neurotrophin and neurotrophin receptor mRNAs in the hippocampus. Exp. Neurol. **160:** 469–478.
7. MCINTOSH, T.K., R. VINK, I. NOBLE, *et al.* 1989. Traumatic brain injury in the rat: characterization of a lateral fluid-percussion model. Neuroscience **28:** 233–244.
8. PILLAY, N.S., L.A. KELLAWAY & G.J. KOTWAL. 2004. Molecular mechanisms, emerging etiological insights and models to test potential therapeutic interventions in Alzheimer's disease. Curr. Alzheimer Res. **1:** 295–306.
9. PRINS, M.L., S.M. LEE, C.L.Y. CHENG, *et al.* 1996. Fluid percussion brain injury in the developing and adult rat: a comparative study of mortality, morphology, intracranial pressure and mean arterial blood pressure. Dev. Brain Res. **95:** 115–125.
10. PIERCE, J.E.S., D.H. SMITH, J.Q. TROJANOWSKI & T.K. MCINTOSH. 1998. Enduring cognitive, neurobehavioural and histopathological changes persist for up to one year following severe experimental brain injury in rats. Neuroscience **87:** 359–369.
11. KNOBLACH, S.M., L. FAN & A.I. FADEN. 1999. Early neuronal expression of tumor necrosis factor-[alpha] after experimental brain injury contributes to neurological impairment. J. Neuroimmunol. **95:** 115–125.
12. RAGHUPATHI, R., A.C. CONTI, D.I. GRAHAM, *et al.* 2002. Mild traumatic brain injury induces apoptotic cell death in the cortex that is preceded by decreases in cellular Bcl-2 immunoreactivity. Neuroscience **110:** 605–616.
13. Hicks, R.R., K.L. Keeling, M.Y. Yang, *et al.* 2002. Vaccinia virus complement control protein enhances functional recovery after traumatic brain injury. J. Neurotrauma **19:** 705–714.
14. SANDERS, M.J., W.D. DIETRICH & E.J. GREEN. 2001. Behavioral, electrophysiological, and histopathological consequences of mild fluid-percussion injury in the rat. Brain Res. **904:** 141–144.
15. TANNO, H., R.P. NOCKELS, L.H. PITTS & L.J. NOBLE. 1992. Breakdown of the blood-brain barrier after fluid percussion brain injury in the rat. Part 1: distribution and time course of protein extravasion. J. Neurotrauma **9:** 21–32.
16. AIHARA, N., J.J. HALL, L.H. PITTS, *et al.* 1992. Altered immunoexpression in microglia and macrophages after mild head injury. J. Neurotrauma **12:** 53–63.
17. KOSSMANN, T., P.F. STAHEL, M.C. MORGANTI-KOSSMANN, *et al.* 1997. Elevated levels of the complement components C3 and factor B in ventricular cerebrospinal fluid of patients with traumatic brain injury. J. Neuroimmunol. **73:** 63–69.

18. SEWELL, D.L., B. NACEWICZ, F. LIU, *et al.* 2004. Complement C3 and C5 play critical roles in traumatic brain cryoinjury: blocking effects on neutrophil extravasation by C5a receptor antagonist. J. Neuroimmunol **155:** 55–63.
19. KOTWAL, G.J. & B. MOSS. 1988. Vaccinia virus encodes a secretory polypeptide structurally related to complement control proteins. Nature **335:** 176–178.
20. MCKENZIE, R., G.J. KOTWAL, C.H. HAMMER & M. FRANK. 1992. Regulation of complement activity by vaccinia virus complement control protein. J. Infect. Dis. **166:** 1245–1250.
21. KOTWAL, G.J., S.T. ISSACS, R. MCKENZIE, *et al.* 1990. Inhibition of the complement cascade by the major secretory protein of vaccinia virus. Science 827–830.
22. PAXINOS, G. & C. WATSON. 1998. The Rat Brain in Stereotaxic Coordinates. 2nd Ed. Academic Press. New York.
23. MORRIS, R.G. 1984. Developments of a water-maze procedure for studying spatial learning in the rat. J. Neurosci. Methods **11:** 47–60.
24. JOSEF VAN DER STAAY, F., K.H. AUGSTEIN & E. HORVATH. 1996. Sensorimotor impairments in rats with cerebral infarction, induced by unilateral occlusion of the left middle cerebral artery: strain differences and effects of the occlusion site. Brain Res. **735:** 271–284.
25. POWER, A.E. & J.L. MCGAUGH. 2002. Cholinergic activation of the basolateral amygdala regulates unlearned freezing behavior in rats. Behav. Brain Res. **134:** 307–315.

Novel Peptides of Therapeutic Promise from Indian *Conidae*

K. HANUMAE GOWD,[a] V. SABAREESH,[b] S. SUDARSLAL,[b]
PRATHIMA IENGAR,[b] BENJAMIN FRANKLIN,[c] ANTONY FERNANDO,[c]
KALYAN DEWAN,[d] MANI RAMASWAMI,[e] SIDDHARTHA P. SARMA,[f]
SUJIT SIKDAR,[b] P. BALARAM,[b] AND K.S. KRISHNAN[a,f]

[a]*Tata Institute of Fundamental Research, Mumbai, India*

[b]*Molecular Biophysics Unit, Indian Institute of Science, Bangalore, India*

[c]*C A S in Marine Biology, Annamalai University, Porto Novo, India*

[d]*Unichem R&D Center, IISc Campus, Bangalore, India*

[e]*Institute of Neuroscience, Trinity College, Dublin, Ireland*

[f]*National Center for Biological Sciences, TIFR, Bangalore, India*

ABSTRACT: **Highly structured small peptides are the major toxic constituents of the venom of cone snails, a family of widely distributed predatory marine molluscs. These animals use the venom for rapid prey immobilization. The peptide components in the venom target a wide variety of membrane-bound ion channels and receptors. Many have been found to be highly selective for a diverse range of mammalian ion channels and receptors associated with pain-signaling pathways. Their small size, structural stability, and target specificity make them attractive pharmacologic agents. A select number of laboratories mainly from the United States, Europe, Australia, Israel, and China have been engaged in intense drug discovery programs based on peptides from a few snail species. Coastal India has an estimated 20-30% of the known cone species; however, few serious studies have been reported so far. We have begun a comprehensive program for the identification and characterization of peptides from cone snails found in Indian Coastal waters. This presentation reviews our progress over the last 2 years. As expected from the evolutionary history of these venom components, our search has yielded novel peptides of therapeutic promise from the new species that we have studied.**

KEYWORDS: **novel peptides; Conidae; molluscs; contryphans**

INTRODUCTION

Legend has it that Jeevaka defended his thesis in the seventh century BC with the following remarks. "Human diseases manifest an imbalance of bodily functions and since poisons of various kind kill by affecting one or the other bodily functions, they

Address for correspondence: K.S. Krishnan, Department of Molecular Biology, Tata Institute of Fundamental Research, Colaba, Mumbai, 40005, India
ksk@mailhost.tifr.res.in

Ann. N.Y. Acad. Sci. 1056: 462–473 (2005).
doi: 10.1196/annals.1352.022

could be drugs for treatment of disorders if used judiciously." Jeevaka was known in ancient times for some of the most astounding cures and later became Buddha's personal physician. This story in some manner points to great interest in the analysis of plant and animal toxins as starting points for drug discovery. Ayurveda and other traditional medicine systems around the world generally use plant-derived material as cures. Rare, however, are examples of animal venom and toxins as prospective drugs.

Animals use venom for both prey capture and defense. In particular, a weak and slow moving animal is more or less likely to target the enemy or prey species' nervous system because of the need for rapid immobilization. Marine cone snails represent one such class, are the largest single genus of venomous predators, estimated at about 500 species, and are widely distributed in the Indo-Pacific.[1] They prey on fish, marine worms, and molluscs.[2] Cone snails have evolved mechanisms to generate a library of toxic peptides, which act in concert to rapidly immobilize prey by simultaneously targeting several receptors and channels. Specific cabals of toxins have been identified in some cases.[2,3] In nearly 20 years, primarily due to the work of Baldomero Olivera, conotoxins have become a promising source of drug discovery in addition to becoming a major new paradigm for evolutionary studies.[4] Conotoxins target voltage-gated and ligand-gated ion channels as well as G-protein–coupled receptors with high specificity and selectivity.[5] Several conus peptides are widely used as research tools in neuroscience, such as w-conotoxin, and some are potential therapeutic agents, such as Ziconotide.[6,7] The approximately 150 conotoxins reported to date represent less than 0.3% of peptides present in the venom of conidae. Clearly, many therapeutically interesting molecules remain to be isolated and characterized from the genus conus.

The conidae of India are of particular interest because of the availability of about 77 species, 21 of which are unique.[8] We have initiated a comprehensive program of investigating the chemistry and pharmacology of conus peptides derived from the cone snails found off the coast of India. This report describes some of our research over the last 3 years.

SPECIES ABUNDANCE

From our initial surveys in India, we found that many cone snail species are available near Mumbai. During routine collection trips, we obtained live specimens of about 40 different species. Of these, 7 abundant species are the subject of our studies (FIG. 1). Cone snails are classified into three major groups according to their prey preference: the piscivorous preying upon fish (*C. achatinus*), the mollusci-vorous eating molluscs (*C. amadis, C. araneosus*), and the vermivorous feeding on polychaete worms (*C. inscriptus, C. loroisii,* and *C. hyaena hyaena*). We have also done a systematic study of the Radula-tooth of 25 of the available species. A comparative analysis showed clear-cut differences in the radula structure of piscivorous, molluscivorous, and vermivorous snails.[9] It is hoped that these studies will help to classify newly identified species (B. Franklin, A. Fernando, and K.S. Krishnan, unpublished data). An example of the comparison of radula-teeth structures of vermi-vorous and piscivorous snails is shown in FIGURE 1.

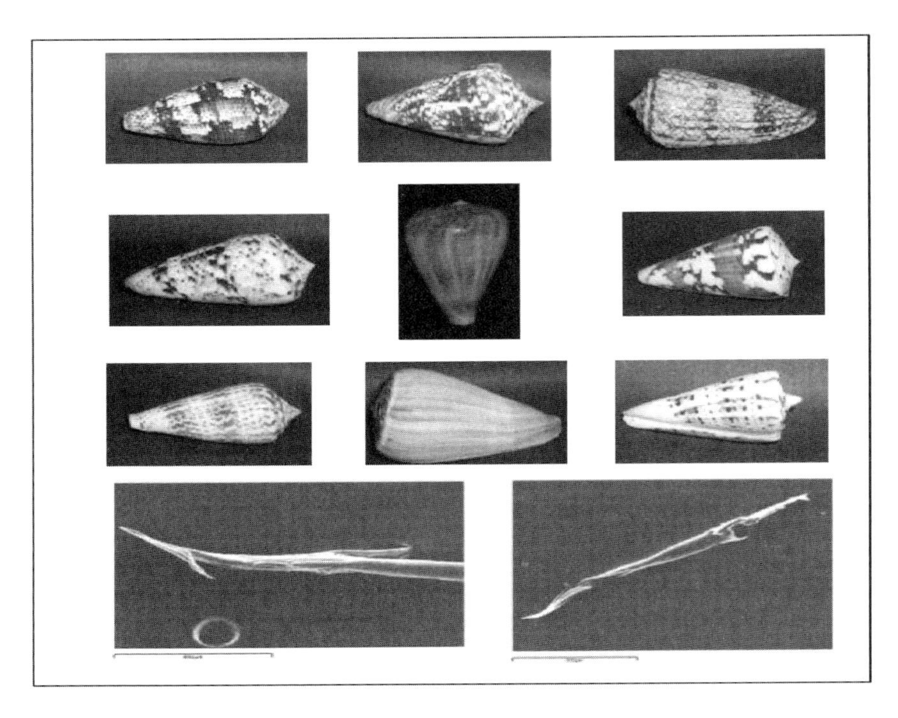

FIGURE 1. A collection of Indian cone shells. *From left to right:* (*top row*) *C. achatinus, C. amadis,* and *C. araneosus.* (*Second row*) *C. zeylanicus, C. hyaena hyaena,* and *C. malacanus.* (*Third row*) *C. inscriptus, C. loroisii,* and *C. monile.* (*Bottom row*) SEM picture of radula tooth from *C. achatinus* (piscivorous) and *C. hyaena hyaena* (vermivorous).

ISOLATION AND CHARACTERIZATION OF PEPTIDES

In the first phase of the investigation we focused primarily on peptides that can be extracted into organic solvents such as methanol and acetonitrile. The toxic components of conus venoms are mostly small peptides ranging in size from 10–50 amino acids.[5–10] A large fraction of these peptides have multiple disulfide bonds. In addition, they contain several posttranslational modifications including C-terminal amidation, proline hydroxylation, and epimerization of specific residues, in the case of contryphans.[11] MALDI-MS provides a convenient tool for fingerprinting the peptide components of conus venom and for monitoring HPLC purification of complex peptide mixtures. FIGURE 2 shows the MALDI-MS obtained by spotting the venom sample directly mixed with the matrix.

Comparison of the spectra generated in the range of 1,000–4,000 Da clearly indicates pronounced differences in the components. In this mass range, the peaks are invariably derived from peptide components. FIGURE 3 shows a typical RP-HPLC

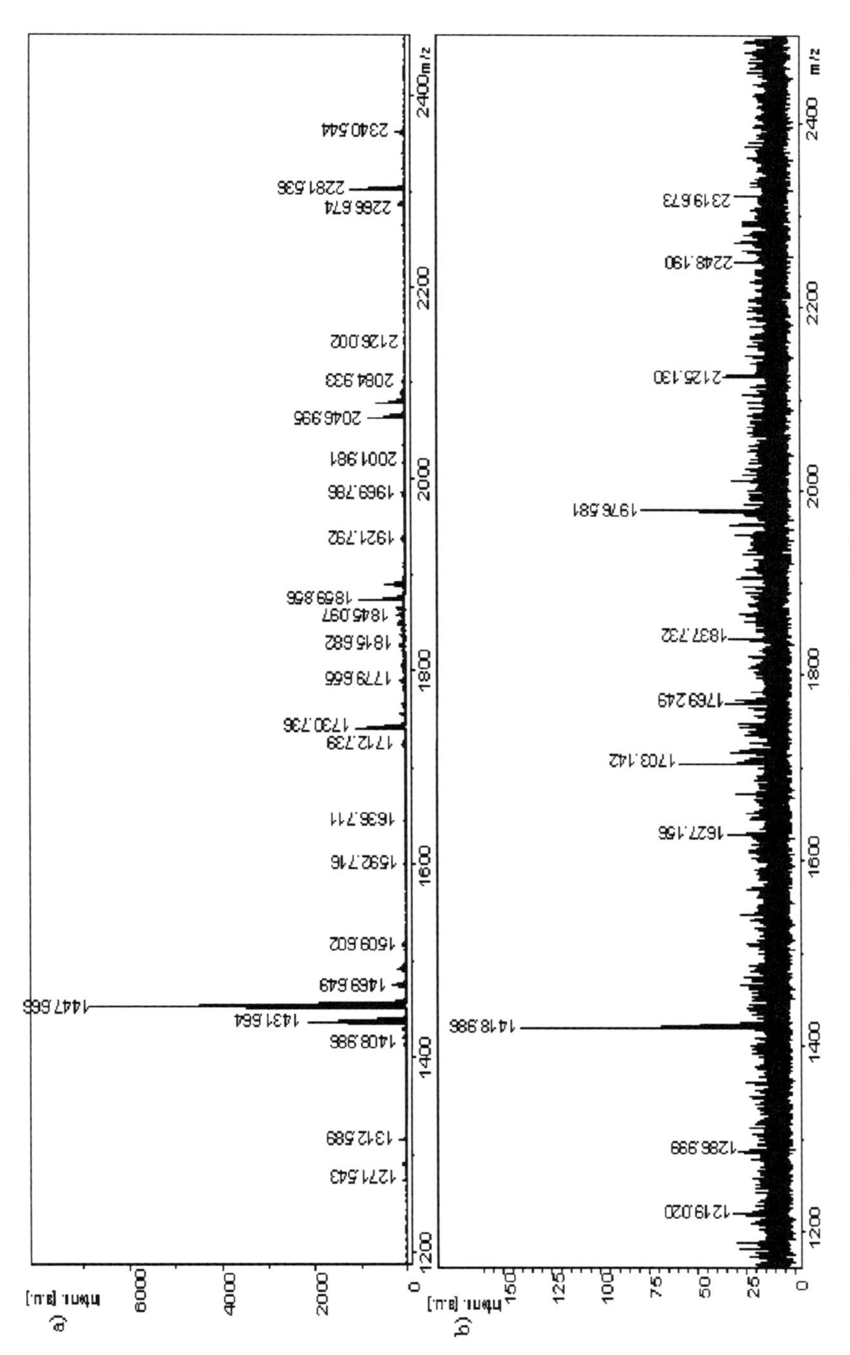

FIGURE 2. *See following page for legend.*

separation of the venom components from *C. araneosus*. Peptides widely differing in mass can coelute in a single symmetrical peak, often necessitating multiple rounds of purification.

Further characterization of venom peptides requires determination of the amino acid sequence. Tandem mass spectrometry is especially powerful for peptides of this size. In our studies, we employed both MALDI-TOF/TOF and ESI-Ion trap methods. FIGURE 4a shows a typical fragmentation spectrum obtained in a MALDI-MS/MS experiment. Before fragmentation, the peptide must be reduced (dithithreitol) and alkylated (iodoacetamide) in order to break all the covalent disulfide cross-links. Sequence determination can be achieved if suitable series of b_n and y_n ions are observed, as shown in FIGURE 4a.[12] Mass spectrometric sequencing of peptides is facilitated by using chemical modification experiments, enzymatic digestion, and multiple reduction/alkylation combinations in breaking disulfide bonds. Confirmation of the sequence usually requires examination of the fragmentation in an ion trap spectrometer. FIGURE 4b illustrates the ESI-MS/MS spectrum obtained for the peptide Ar1312 from *C. araneosus*. Inspection of FIGURE 4 reveals the presence of hydroxyproline, which results in the intense y_2 ion detected at 292 Da due to preferential cleavage of the Xxx-Hyp tertiary amide bond.

Mass spectral sequencing can lead to ambiguities in assignment of amino acid residues in some cases. The final confirmation of a determined sequence may be achieved by comparing a chemically synthesized peptide with the natural peptide. FIGURE 5a and b illustrates a comparison of the MALDI-MS/MS of reduced/alkylated natural Ar1312 and alkylated synthetic Ar1312, respectively. Both peptides have identical masses of 1,544 Da. The observed spectra revealed the identity of all major fragments, thus confirming the sequence. An analog of Ar1312, which contains asparagine in the place of hydroxyproline, with the C-terminus amidated instead of the free carboxylic acid, was synthesized. The choice of this analog was based on the frequent observation of C-terminal amidation in the conotoxins. This analog (alkylated) also has a mass of 1,544 Da. The mass spectral fragmentation pattern (FIG. 5c) reveals important differences from the natural peptide. The spectrum of the natural peptide contains a y_2 ion at 292 Da, which is absent in the Asn analog. Thus, mass spectrometry in conjugation with chemical synthesis can provide a means of establishing the amino acid sequences of peptides isolated from Conus venom. TABLE 1 provides representative examples of new sequences determined from Indian cone snails. Identification of the novel 13-residue peptide from *C. monile* (Mo 1659), which lacks disulfide bonds, provides an example of a new peptide family. This clearly indicates that exhaustive investigations of Conus venoms from diverse species may yet yield many novel sequences.

FIGURE 2. MALDI-MS of (**a**) crude venom of *C. araneosus* and (**b**) acetonitrile extract of *C. loroisii*. MALDI spectra were collected using a Bruker Daltonics, Ultraflex TOF/TOF system, in the reflectron-positive ion mode, equipped with a nitrogen laser of 337 nm. Samples were prepared by mixing equal volumes of venom or a crude acetonitril extract with the MALDI matrix (a-cyano-4-hydroxy cinnamic acid).

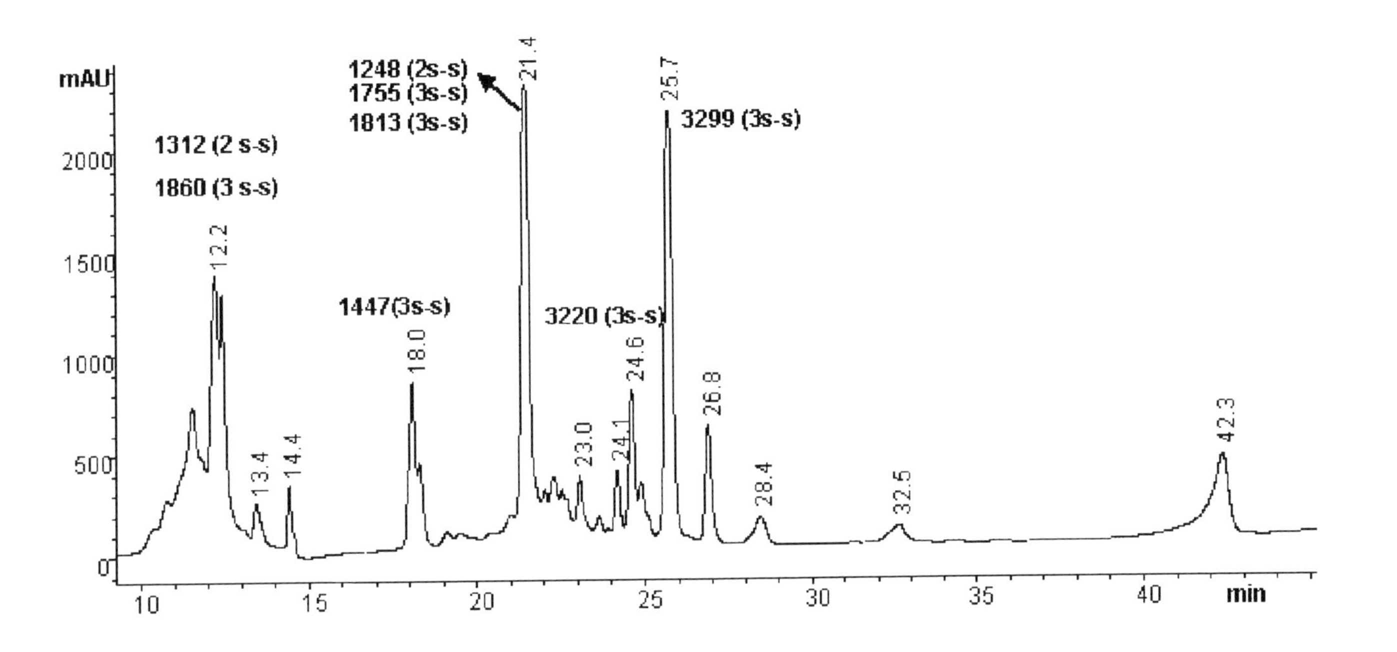

FIGURE 3. RP-HPLC purification profile of a *Conus araneosus* venom extract. Sample was injected on to a phenomenex C_{18} reverse phase HPLC column (250 × 10 mm, 4-m particle size, and 90 Å pore size) and eluted at 1 ml/min with UV absorbance at 226 nm. A linear gradient of 20–95% of acetonitrile containing 0.1% TFA over a period of 50 min was employed. The molecular masses detected by MALDI-MS are indicated above each fraction.

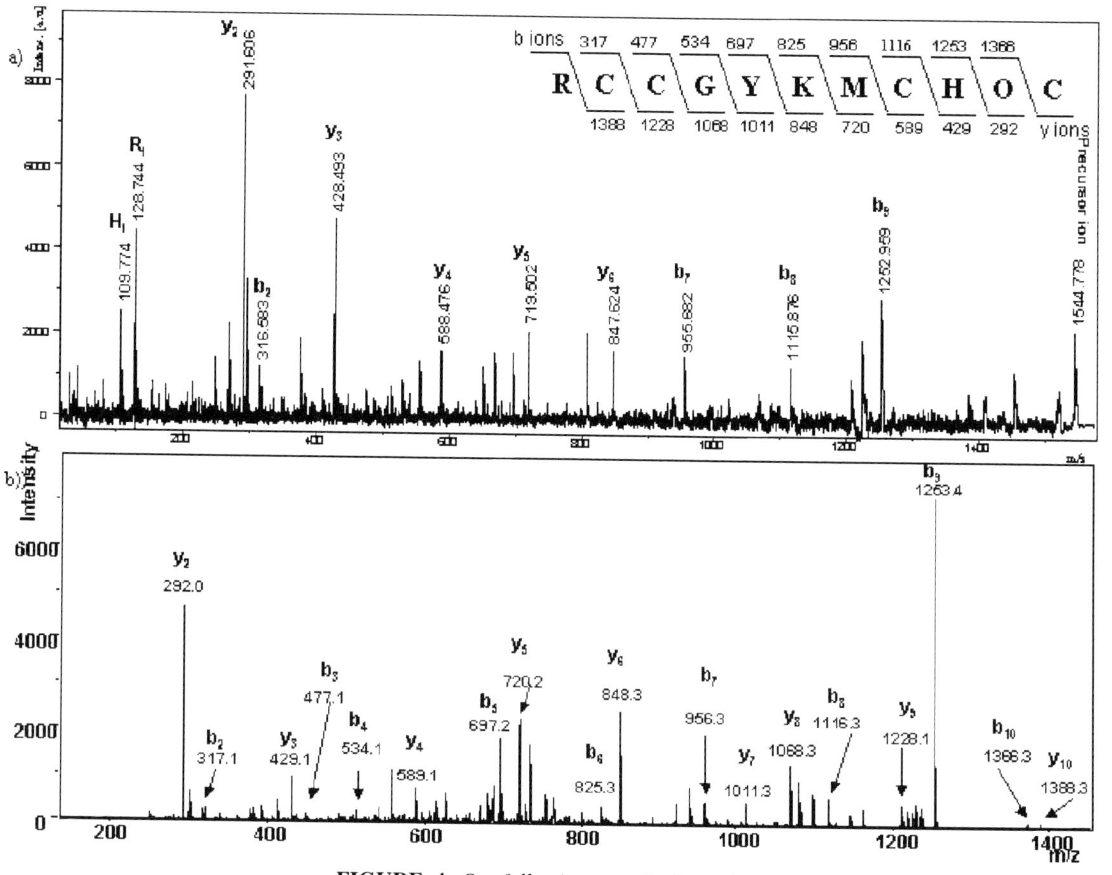

FIGURE 4. *See following page for legend.*

TABLE 1. Novel conotoxins isolated from Indian marine cone snails

Conus species	Sequence	Name of peptide	Target	Ref.
C. amadis (mollusc hunter)	CKQAGESCDIFSQNC-CVGTCAFICIE-NH_2	Am2766	Mammalian sodium channel	13
C. monile (worm eater)	FHGGSWYRFPWGY-NH_2	Mo1659	Potassium channel	14
C. loroisii (worm eater)	GCPDWDPWC-NH_2	Lo959	Under investigation	Unpublished

CONTRYPHAN

Contryphans are an interesting class of conus peptides containing a single disulfide bond. They are generally 7-9 residues in length and contain a D-amino acid residue, which is separated from the N-terminal cysteine by a single amino acid.[15] Epimerization occurs posttranslationally. Novel conformational equilibria have also been reported for contryphans.[16] Contryphans are sometimes easily detected in crude venom by the observation of masses in the range of 900–1,000 Da. In the course of our studies we determined the sequences of the contryphans from *Conus amadis*, *C. loroisii*, and *C. inscriptus*. TABLE 2 provides a comparison of mature contryphan sequences determined thus far.

The contryphan Am975 from *C. amadis* is identical to that of contryphan-P characterized from *C. purpurascens*.[16] The presence of a D-residue in the peptide from *C. loroisii* Lo959 was established by comparing the chromatographic profile of the natural peptide with two chemically synthesized analogs, which contain L-Trp or D-Trp at position 4. FIGURE 6 shows that the HPLC profile of natural Lo959 is identical to that obtained from the synthetic analog containing the D-residue at position 4. By contrast, the L-analog reveals a distinctly different HPLC profile. A curious feature of the Lo959 HPLC profile is the presence of two peaks, separated by a broad trough. Isolation of each of these peaks followed by re-injection yields the same chromatogram. Similar behavior was also obtained for the synthetic peptide containing D-Trp at residue 4. A previous study by Olivera and co-workers[17] showed that contryphan exists in two distinct conformational states, which interconvert slowly on the chromatographic time scale. This conformational process is characteristic of peptides, which contain both D-Trp and the disulfide bond. A preliminary study in our labo-

FIGURE 4. Fragment ion mass spectrum of Ar1312. (**a**) MALDI-MS/MS and (**b**) ESI-MS/MS. The masses of b and y ions according to the proposed sequence are shown above. H_i and Ri indicates the immonium ions of histidine and arginine, respectively. MALDI-MS/MS data were obtained using the LIFT option of Bruker Daltonics. Precursor ion at m/z 1554 Da was trapped and its fragmentation induced. ESI-MS/MS data were obtained on Esquire 3000 plus LC ion trap mass spectrometer (Bruker Daltonics, Germany). The singly charged precursor ion at m/z 1544 Da was selected in an ion-trap system and subjected to high-energy collision-induced dissociation (CID) with helium as a collision gas.

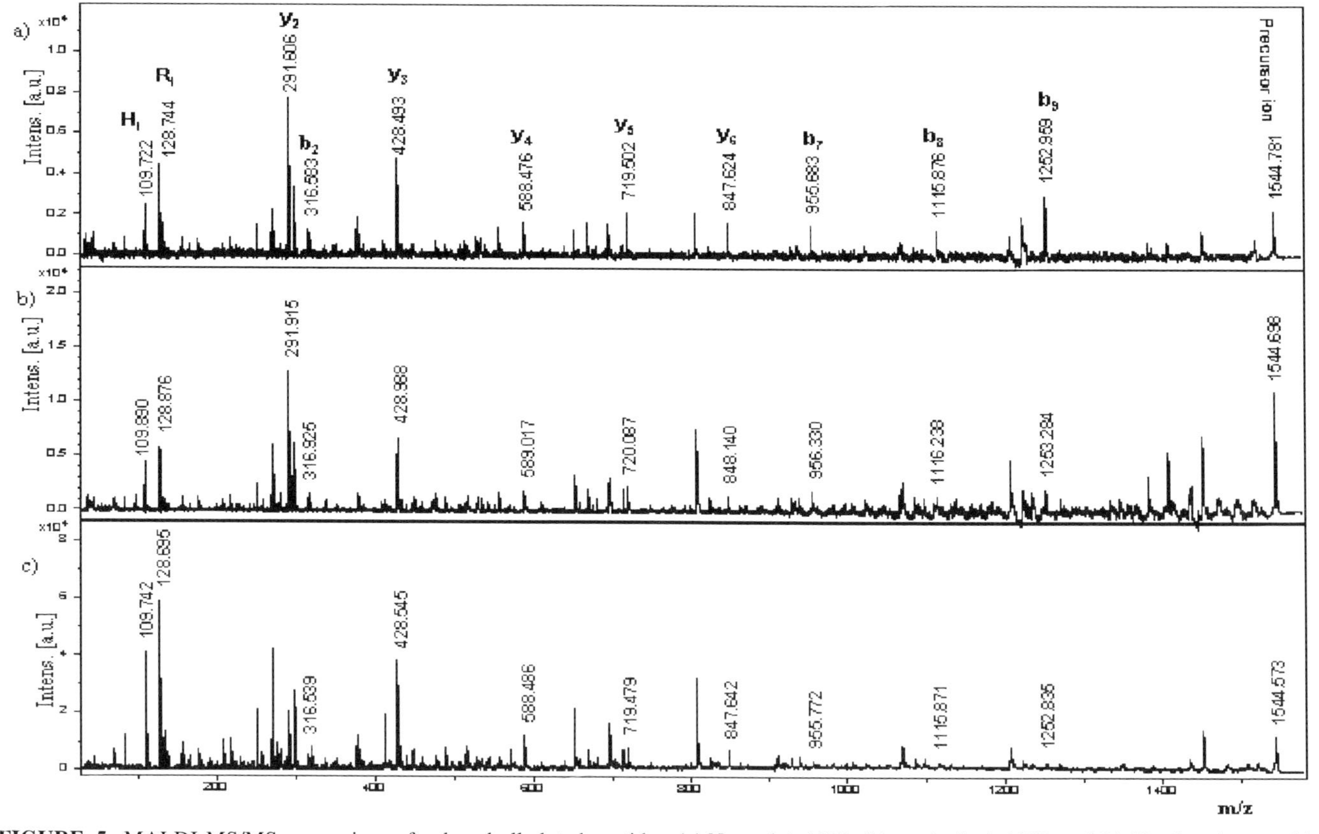

FIGURE 5. MALDI-MS/MS comparison of reduced-alkylated peptides. (**a**) Natural Ar1312, (**b**) synthetic Ar1312, and (**c**) [Asn] analog peptide. The reduction and alkylation were effected using DTT (dithiothreitol) and iodoacetamide

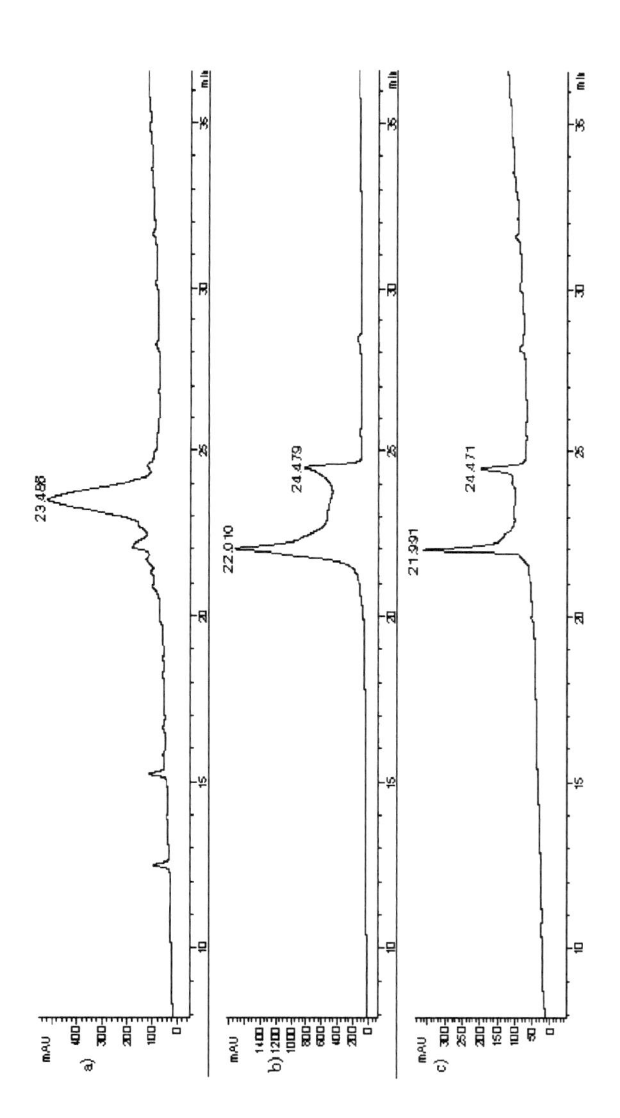

FIGURE 6. HPLC elution profiles of (**a**) synthetic Lo 959 analog containing L-Trp at position 4, (**b**) synthetic Lo 959 containing D-Trp at position 4, and (**c**) natural contryphan Lo 959. The RP-HPLC chromatograms are obtained under identical conditions with a linear gradient of acetonitrile-water using a Zorbax–C_{18} (4.6 mm × 25 cm) column.

TABLE 2. Comparison of mature contryphan sequences[a]

Peptide	Sequence	Species	Ref.
Des(Glyl)contryphan-R	CO$\underline{W}$EPWC-NH$_2$	*C. radiatus*	15
Contryphan-R	GCO$\underline{W}$EPWC-NH$_2$	*C. radiatus*	15
Bromocontryphan-R	GCO$\underline{W}$EPXC-NH$_2$	*C. radiatus*	18
Contryphan-Sm	GCO$\underline{W}$QPWC-NH$_2$	*C. stercusmuscarum*	16
Contryphan-P	GCO$\underline{W}$DPWC-NH$_2$	*C. purpurascens*	16
Contryphan-R/Tx	GCO$\underline{W}$EPWC-NH$_2$	*C. textile*	20
Contryphan-Tx	GCO$\underline{W}$QPYC-NH$_2$	*C. textile*	20
Contryphan-Vn	GDCP$\underline{W}$KPWC-NH$_2$	*C. ventricosus*	21
Leu-Contryphan-P	GCV$\underline{L}$LPWC-OH	*C. purpurascens*	17
Leu-Contryphan-Tx	CV$\underline{L}$YPWC-NH$_2$	*C. textile*	20
Glacontryphan-M	NγSγCP$\underline{W}$HPWC-NH2	*C. marmoreus*	22
Contryphan-P/Am	GCO$\underline{W}$DPWC-NH$_2$	*C. amadis*	This work
Contryphan-Lo	GCP$\underline{W}$DPWC-NH$_2$	*C. loroisii*	This work
Contryphan-In	GCV$\underline{L}$YPWC-NH$_2$	*C. inscriptus*	This work

NOTE: O, 4-trans-hydroxyproline; $\underline{W}$, D-tryptophan; $\underline{L}$, D-leucine; $\underline{X}$, bromotryptophan; γ, gamma-carboxyglutamic acid.

ratory suggests that the contryphan Am 975 acts predominantly on the high-voltage activated calcium channel.

Ongoing studies in our laboratory are focused on mass spectral determination of a specific disulfide-bonding pattern, chemical synthesis of a multiple disulfide-bonded peptide, and three-dimensional structure determination using two-dimensional NMR.[23] The oxidative folding of chemically synthesized sequences containing multiple cysteine residues results in the formation of peptides with non-native disulfide bonding.[24] We are also pursuing the determination of cDNA sequences corresponding to a specific conotoxin superfamily obtained from freshly dissected venom glands.

ACKNOWLEDGMENTS

This research was supported by a grant from the Department of Biotechnology, Government of India, at the National Center for Biological Sciences, TIFR, Molecular Biophysics Unit, IISc and Center for Advanced Studies, Annamalai University.

REFERENCES

1. ROCKELD, K.W. & A.J. KOHN. 1995. Manual of the Living Conidae. Verlag Christa Hemmen. Wiessbanden, Germany.

2. OLIVERA, B.M. 1997. Conus venom peptides, receptor and ion channel targets and drug design: 50 million years of neuropharmacology (EE Just Lecture, 1996). Mol. Biol. Cell **8:** 2101–2109.
3. OLIVERA, B.M., J.S. IMPERIAL & G. BULAJ. 2002. Cone snails and conotoxins: evolving sophisticated neuropharmacology. *In* Perspectives in Molecular Toxinology. A. Menez., Ed. Wiley. West Sussex, UK.
4. HEINRICH, T. & B.M. OLIVERA. 2004. Conus venoms: a rich source of novel ion channels-targeted peptides. Physiol. Rev. **84:** 41–68.
5. MYERS, R.A. *et al.* 1993. Conus peptides as chemical probe for receptor and ion channels. Chem. Rev. **93:** 1923–1936.
6. ADAMS, M.E. & B.M. OLIVERA. 1994. Neurotoxins: overview of an emerging research technology. TINS **17:** 151–155.
7. JONES, R.M. *et al.* 2001. Composition and therapeutic utility of conotoxins from genus Conus. Patent status 1996–2000. Exp. Opin. Ther. Patents **11:** 603–623.
8. KOHN, A.J. 2001. The Conidae of India revisited. Phuket Marine Biological Center Special Publication **25:** 357–362.
9. KOHN, A.J., M. NISHI & B. PERNET. 1999. Snail spears and scimitars: a character analysis of Conus radula teeth. J. Moll. Stud. **65:** 461–481.
10. KOHN A.J., P.R. SAUNDERS & S. WIENER. 1960. Preliminary studies on the venom of the marine snail conus. Ann. N.Y. Acad. Sci. **90:** 706–725.
11. CRAIG, A.G., P. BANDYOPADHYAY & B.M. OLIVERA. 1999. Post-translationally modified neuropeptides from conus venoms. Eur. J. Biochem. **264:** 271–275.
12. BIEMANN, K. 1990. Sequencing of peptides by tandem mass spectrometry and high-energy collision-induced dissociation. Methods Enzymol. **193:** 455–479.
13. SUDARSLAL, S. *et al.* 2003. Sodium channel modulating activity in a delta-conotoxin from an Indian marine snail. FEBS Lett. **553:** 209–212.
14. SUDARSLAL, S. *et al.* 2004. A novel 13 residue acyclic peptide from the marine snail, Conus monile, targets potassium channels. Biochem. Biophys. Res. Commun. **317:** 682–688.
15. JIMENEZ, E.C. *et al.* 1996. Contryphan is a D-tryptophan-containing Conus peptide. J. Biol. Chem. **271:** 28002–28005.
16. JACOBSEN, R. *et al.* 1998. The contryphans, a D-tryptophan-containing family of Conus peptides: interconversion between conformers. J. Pept. Res. **51:** 173–179.
17. JACOBSEN, R.B. *et al.* 1999. A novel D-leucine-containing Conus peptide: diverse conformational dynamics in the contryphan family. J. Pept. Res. **54:** 93–99.
18. JIMENEZ, E.C. *et al.* 1997. Bromocontryphan: post-translational bromination of tryptophan. Biochemistry **36:** 989–994.
19. JACOBSEN, R. *et al.* 1998. The contryphans, a D-tryptophan-containing family of Conus peptides: interconversion between conformers. J. Pept. Res. **51:** 173–179.
20. JIMENEZ, E.C. *et al.* 2001. Contryphans from Conus textile venom ducts. Toxicon **39:** 803–808.
21. MASSILIA, G.R. *et al.* 2001. Contryphan-Vn: a novel peptide from the venom of the Mediterranean snail *Conus ventricosus*. Biochem. Biophys. Res. Commun. **288:** 908–913.
22. HANSSON, K. *et al.* 2004. The first gamma-carboxyglutamic acid containing contryphan A selective L-type calcium ion channel blocker isolated from the venom of *Conus marmoreus*. J. Biol. Chem. **279:** 32453–32463.
23. SARMA, S.P. *et al.* 2005. Solution structure of d-Am2766: a hydrophobic d-conotoxin from *Conus amadis* that inhibits inactivation of neuronal voltage-gated sodium channels. Chem. Biodiversity **2:** in press.
24. CRUZ, D.R. *et al.* 2003. Detergent-assisted oxidative folding of delta-conotoxins. J. Pept. Res. **61:** 202–212.

Stabilization of Mitochondrial Membrane Potential and Improvement of Neuronal Energy Metabolism by Ginkgo Biloba Extract EGb 761

ANNE ECKERT,[a,b] UTA KEIL,[a] ISABEL SCHERPING,[a] SUSANNE HAUPTMANN,[a] AND WALTER E. MÜLLER[a]

[a]*Department of Pharmacology, Biocenter, University of Frankfurt, 60439, Frankfurt, Germany*

[b]*Neurobiology Research Laboratory, Psychiatric University Clinic, 4025 Basel, Switzerland*

ABSTRACT: Ginkgo biloba extract EGb 761 has been used for many years to treat age-related cognitive disorders including Alzheimer's disease. EGb 761 given shortly after initiating mitochondrial damage by sodium nitroprusside (nitric oxide donor) improved the mitochondrial membrane potential of PC12 cells significantly and dose dependently. Under these conditions, EGb 761 also reversed the decrease in ATP production. In addition, similar protection against oxidative damage was found in dissociated brain cells and isolated brain mitochondria after *in vitro* or *in vivo* treatment with EGb 761. Moreover, PC12 cells bearing an Alzheimer's disease-related mutation in the amyloid precursor protein, which leads to enhanced beta amyloid production, showed greater benefit from treatment with EGb 761 than did control cells. Taken together, our findings clearly show stabilization and protection of mitochondrial function as a specific and very sensitive property of EGb 761 at therapeutically relevant doses.

KEYWORDS: Ginkgo biloba; mitochondria; energy metabolism; Alzheimer's disease; aging; oxidative stress

MITOCHONDRIA, AGING, AND ALZHEIMER'S DISEASE

Mitochondria are not only essential for energy production via the respiratory chain, but also responsible for several other important cellular functions, including the tricarboxylic acid cycle and the initiation and regulation of apoptosis. Moreover, they are the main source of reactive oxygen species (ROS). A mitochondrial or free radical theory of aging is derived from data suggesting that activity of the mitochondrial respiratory chain declines with age,[1,2] mitochondria-based oxidative stress increases with age,[3–7] and furthermore mitochondrial DNA mutations accumulate

Address for correspondence: Anne Eckert, Neurobiology Research Laboratory, Psychiatric University Clinic Basel, CH-4025 Basel, Switzerland. Voice: +41 61 325 5487; fax: +41 61 325 5577.
anne.eckert@upkbs.ch

Ann. N.Y. Acad. Sci. 1056: 474–485 (2005).
doi: 10.1196/annals.1352.023

with age,[8–11] perhaps due to oxidative stress. Age-related decreases in the membrane potential of brain and liver mitochondria have also been reported.[12]

Increasing evidence suggests that mitochondrial dysfunction is a very early event in the pathogenesis of Alzheimer's disease (AD).[13–16] The activity of key mitochondrial enzymes such as α-ketoglutarate dehydrogenase,[17] pyruvate dehydrogenase, and cytochome *c* oxidase[18–20] is decreased in the AD brain. Reductions in cytochrome *c* oxidase mRNA[21] and protein expression[22] and decreased activity even in peripheral tissues[23,24] could also be observed in AD. AD can be classified into two forms: sporadic AD, which accounts for the vast majority of AD cases and in which aging represents the main risk factor, and a familial form of AD (FAD), in which rare gene mutations have been identified. The latter patients have an autosomal dominant inheritable variant of AD. Notably, patients with either sporadic AD or FAD share common clinical and neuropathologic features including amyloid beta plaques, accumulation of intracellular neurofibrillary tangles, and pronounced neuronal cell loss. The amyloid plaque is composed of amyloid beta peptide (Aβ), which is derived from amyloid precursor protein (APP) through an initial β-secretase cleavage followed by an intramembranous cut of γ-secretase. Three genes are known to be causatively linked with the pathogenesis of these early-onset FAD forms. Besides the genes encoding for presenilin 1 (PS1) on chromosome 14 and presenilin 2 (PS2) on chromosome 1, mutations in the APP gene on chromosome 21 account for these FAD cases.

In a previous study, we demonstrated that PC12 cells bearing the Swedish double mutation in the APP gene (APPsw) and exhibiting substantial Aβ levels showed decreased cytochrome *c* oxidase activity and reduced ATP levels.[25] When APPsw PC12 cells were exposed with an AD-relevant secondary insult such as oxidative stress, they showed a more pronounced decrease in mitochondrial membrane potential and ATP levels compared to that of control cells.[25] Additionally, caspase 3 activity and apoptotic cell death were increased in APPsw PC12 cells after oxidative stress.[26] Moreover, we and others have found that APP may also be mitochondrially targeted in APPsw PC12 cells[25] and in neurons from a transgenic animal model[27] and that this might lead to mitochondrial dysfunction. Exposure of isolated rat brain mitochondria to Aβ caused a decrease in mitochondrial enzyme activity, respiration, and mitochondrial membrane potential.[28–30] Aβ-binding alcohol dehydrogenase (ABAD) is a newly discovered mitochondrial protein. Lustbader *et al.*[31] could show that ABAD directly interacts with Aβ, leading to increased oxidative stress and mitochondrial dysfunction. These data suggest a central role for mitochondria in the pathology of Alzheimer's disease.

NEUROPROTECTIVE PROPERTIES OF GINKGO BILOBA EXTRACT EGB 761

Standardized Ginkgo biloba extract (EGb 761) is a valuable therapy for the treatment of memory impairment, dementia, tinnitus, and intermittent claudication. Double-blind, placebo-controlled studies showed that EGb 761 improved cognitive symptoms in the elderly and in patients with dementia.[32–36]

EGb 761 is known to exhibit antioxidant properties. It can scavenge ROS, such as hydroxy, peroxy radicals, superoxide anions, and nitric oxide.[37–39] Moreover, an increase in catalase and superoxide dismutase activities in the hippocampus, striatum, and substantia nigra of rats and a decrease in lipid peroxidation in the hippocampus were observed after treatment with EGb 761.[40] Additionally, glutathione S-transferase activity in the mouse liver was enhanced after EGb 761 treatment.[41] In human keratinocytes, EGb 761 stimulated glutathione synthesis.[42] Furthermore, EGb 761 modulated the transcription of selected genes responsible for increasing the antioxidant status of the cells.[43]

Mitochondria-protecting properties have also been described for EGb 761. Treatment with EGb 761 prevented age-associated changes in mitochondrial morphology, mitochondrial glutathione levels, and respiratory function of brain mitochondria.[12] Additionally, EGb 761 prevented glutathione oxidation in liver mitochondria and oxidative damage to brain and liver mitochondrial DNA.[44] Bilobalide, one component of EGb 761, inhibited hypoxia-induced decreases in ATP content in vascular endothelial cells *in vitro*.[45] *Ex vivo*, the respiratory control ratio of liver mitochondria was increased in rats that had been treated with bilobalide.[45] Additionally, bilobalide protected against ischemia-induced decreases in state 3 respiration of liver mitochondria by preserving the function of complexes I and III.[46] Furthermore, bilobalide caused a significant increase in mRNA levels of NADH dehydrogenase subunit 1 and a decrease in state 4 respiration.[47] In addition, an increase in cytochrome *c* oxidase subunit 3 gene expression could be observed by bilobalide.[48]

Mitochondria play a very important role in the intrinsic apoptotic pathway. Interestingly, EGb 761 protected the integrity of the mitochondrial membrane and attenuated the release of cytochrome *c* from the mitochondria, which is a critical step in the apoptotic cascade.[49] Moreover, EGb 761 led to upregulation of the antiapoptotic protein Bcl-2, and bilobalide alone led vice versa to downregulation of the proapoptotic protein Bax.[49,50] In addition, EGb 761 inhibited the activation of caspase 9 and caspase 3, and it diminished apoptotic cell death in PC12 cells after oxidative stress and in cortical cultered neurons after staurosporine treatment.[50–53] Using a neuroblastoma cell line stably expressing an AD-associated mutation, EGb 761 was able to inhibit caspase 3 activation and the formation of Aβ aggregates.[54]

GINKGO BILOBA EXTRACT EGB 761 PROTECTS MITOCHONDRIAL FUNCTION IN PC12 CELLS

In a previous study, we showed that EGb 761 can protect mitochondria of PC12 cells against hydrogen peroxide attack.[55] A significant reduction in mitochondrial membrane potential changes was found at concentrations as low as 10 μg/ml EGb 761.[55]

Mitochondrial membrane potential and ATP levels are very important markers for the function of mitochondria. A decrease in mitochondrial membrane potential is a very early event in the apoptotic cell death cascade.[56]

Mitochondrial membrane potential was measured using the fluorescence dyes rhodamine 123 (R123) and tetramethylrhodamineethylester (TMRE).[25,57,58] Transmembrane distribution of the dyes depends on the mitochondrial membrane potential. The dyes were added to the cell culture medium at a final concentration of 0.4 μM for 15 minutes. The cells were washed with HBSS, and fluorescence intensity

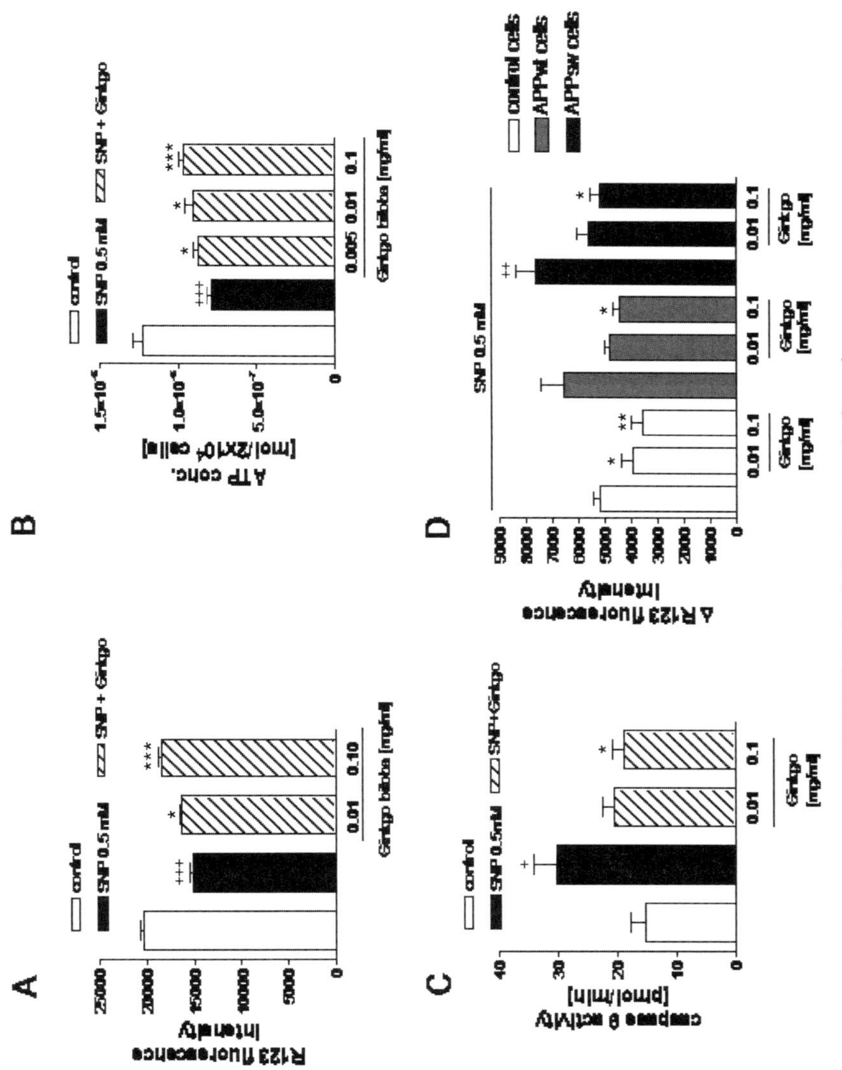

FIGURE 1. *See following page for legend.*

was determined with a Victor® multilabel counter at 490 nm (excitation) and 535 nm (emission).

The ATP levels were determined with a bioluminescence assay (ViaLight® HT).[25] The kit is based on the bioluminescent measurement of ATP. The bioluminescent method uses an enzyme, luciferase, that catalyzes the formation of light from ATP and luciferin. The emitted light is linearly related to the ATP concentration and is measured using a luminometer.[59]

First, the efficacy of EGb 761 in protecting mitochondrial function after a sodium nitroprusside (SNP) insult was tested. Sodium nitroprusside is a well known nitric oxide (NO) donor. Progressive elevation of NO production has been found in AD models. PC12 cells bearing the Swedish double mutation in the APP gene have significantly enhanced NO levels compared to control cells.[25] APPsw transgenic mice also show increased NO production in the cerebral cortex.[60] Nitric oxide is known to inhibit mitochondrial respiration and can lead to apoptotic cell death.[61] Thus, PC12 cells were treated for 24 hours with SNP (0.5 mM). EGb 761 was added 30 minutes after the onset of SNP exposure, and mitochondrial membrane potential and ATP levels were determined. EGb 761 exhibited protective efficacy against SNP-induced mitochondrial dysfunction (FIG. 1). It was able to reduce SNP-induced mitochondrial membrane potential changes in concentrations as low as 10 μg/ml (FIG. 1A), and ATP levels were already stabilized at a concentration of 5 μg/ml (FIG. 1B). In accordance with these findings, Song *et al.*[62] showed that NO-induced neurotoxicity can be reduced by bilobalide. Concerning the stabilization of ATP levels, Janssens *et al.*[45] also demonstrated protection of a hypoxia-induced ATP decrease in endothelial cells by EGb 761.

Additionally, the protective features of EGb 761 on mitochondria of APPsw-bearing PC12 cells were analyzed. Interestingly, the decrease of mitochondrial membrane potential in APPsw-bearing PC12 cells after SNP exposure is more pronounced than in control cells and wild-type APP-bearing cells (FIG. 1D). This is in accordance with previous findings showing higher vulnerability of APPsw PC12

FIGURE 1. (**A**) Treatment with EGb 761 after nitric oxide (SNP) insult improves the reduction of mitochondrial membrane potential. PC12 cells were incubated for 24 hours with 0.5 mM SNP; EGb 761 was added 30 minutes after insult. Data are expressed as means ± SEM (n = 6). ^{+++}P <0.001 vs. untreated control cells; $^{*}P$ <0.05; $^{***}P$ <0.001 vs. SNP damage, Student's unpaired t test. (**B**) Treatment with EGb 761 enhances ATP levels after nitric oxide insult. PC12 cells were incubated for 24 hours with 0.5 mM SNP; EGb 761 was added 30 minutes after insult. Data are expressed as means ± SEM (n = 6). ^{+++}P <0.001 vs. untreated control cells; $^{*}P$ <0.05; $^{***}P$ <0.001 vs. SNP damage, Student's unpaired t test. (**C**) Treatment with EGb 761 reduces the increase in caspase 9 activity after exposure to SNP. PC12 cells were pretreated for 22 hours with EGb 761, and then SNP was added for 2 hours. Data are expressed as means ± SEM (n = 5). ^{+}P <0.05 vs. untreated control cells; $^{*}P$ <0.05 vs. SNP damage, Student's unpaired t test. (**D**) Treatment with EGb 761 after nitric oxide insult strongly improves the reduction of mitochondrial membrane potential in APPsw PC12 cells. APPsw cells exhibited a significantly increased drop in SNP-induced mitochondrial membrane potential compared to control cells (^{++}P <0.01). Control cells and APP transfected PC12 cells were incubated for 24 hours with 0.5 mM SNP; EGb 761 was added 30 minutes after insult. Data are expressed as means ± SEM (n = 7). ^{++}P <0.01 vs. SNP-treated control cells; $^{*}P$ <0.05; $^{**}P$ <0.01 vs. SNP damage of the corresponding cell type, Student's unpaired t test.

TABLE 1. Protective efficacy of Ginkgo biloba extract EGb 761 on the individual complexes of the mitochondrial respiratory chain

		Protection by EGb 761 (mg/ml)	
Complex	Damage	0.01	0.10
Complex I	Rotenone	+	
Complex II	Thenoyltrifluoroacetone	+	++
Complex III	Antimycin		+
Complex IV	Natriumazide	+	++
Complex V	Oligomycin	+	++

cells against hydrogen peroxide attack.[25,26] Notably, EGb 761 also significantly reduced an SNP-induced decrease of mitochondrial membrane potential in APPsw-transfected cells (FIG. 1D). APPsw-bearing PC12 cells showed a greater benefit from treatment with EGb 761 than did control cells.

EGB 761 PROTECTS MITOCHONDRIAL RESPIRATORY CHAIN COMPLEXES IN PC12 CELLS

Defects in the mitochondrial respiratory chain seem to play a very important role in the pathogenesis of AD. Therefore, we tested the efficiency of EGb 761 to protect the individual complexes of the respiratory chain after stimulation with different specific complex inhibitors. PC12 cells were pretreated for 6 hours with EGb 761. Mitochondrial membrane potential was recorded with the fluorescence dye TMRE. TMRE exhibits a characteristic increase in fluorescence intensity at 490 nm (excitation) and 590 nm (emission) after challenging mitochondria with membrane potential decreasing drugs.[63] The basal mitochondrial membrane potential was recorded and then complex inhibitors (rotenone 2 μM, thenoyltrifluoroacetone (TTFA) 10 μM, antimycin 2 μM, oligomycin 10 μM, and sodium azide 1 mM) were added. EGb 761 was able to protect mitochondrial respiratory chain complexes II, IV, and V at a concentration of 10 μg/ml (TABLE 1). Significant protection of complexes I and III was seen with 100 μg/ml EGb 761 and greater (TABLE 1). Therefore, EGb 761 can improve the function of the mitochondrial respiratory chain. These results are in accordance with findings from Janssens *et al.*,[46] Tendi *et al.*,[47] and Chandrasekaran *et al.*,[48] also showing an improvement in mitochondrial respiratory chain function by EGb 761.

EGB 761 REDUCES CASPASE 9 ACTIVITY AFTER SNP EXPOSURE

Caspase 9 plays a very important role in the mitochondria-mediated apoptotic pathway.[64] Activation of caspase 9 leads to activation of the executor caspase 3 and finally to apoptotic cell death. Caspase 9 is activated in PC12 cells after SNP exposure with a maximum after 2 hours of SNP treatment (data not shown). Pretreatment of PC12 cells with EGb 761 for 22 hours followed by exposure to SNP for 2 hours

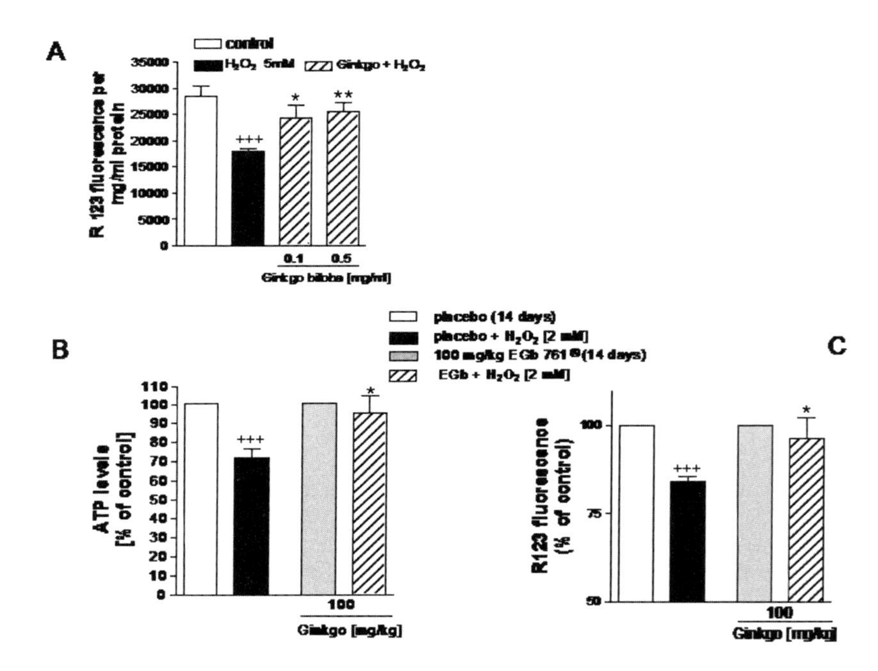

FIGURE 2. (**A**) *In vitro* treatment with EGb 761 after H_2O_2 insult improves the reduction of mitochondrial membrane potential in dissociated mouse brain cells. Brain cells were damaged with H_2O_2 (5 mM) for 1 hour; then EGb 761 was added for 6 hours. Data are expressed as means ± SEM from 8 experiments, each representing an individual animal (2–3 months old). ^{+++}P <0.01 vs. untreated control; $^{*}P$ <0.05; $^{**}P$ <0.01 vs. H_2O_2 damage, Student's unpaired *t* test. (**B**) *In vivo* treatment with EGb 761 improves the decrease of ATP levels in dissociated brain cells. Treated animals received 100 mg/kg EGb 761 p.o. once daily for 2 weeks. Control animals were treated with placebo (0.9% NaCl solution). ATP levels were measured after 2 hours' incubation of dissociated mouse brain cells with 2 mM H_2O_2. Data are expressed as means ± SEM from 7–8 experiments, each representing an individual animal (22 months old). ^{+++}P <0.01 vs. placebo control; $^{*}P$ <0.05 vs. H_2O_2 damage of brain cells from placebo-treated animals, Student's unpaired *t* test. (**C**) *In vivo* treatment with EGb 761 improves mitochondrial membrane potential of isolated mitochondria after H_2O_2 insult. Treated animals received 100 mg/kg EGb 761 p.o. once daily for 2 weeks. Control animals were treated with placebo (0.9% NaCl solution). Mitochondrial membrane potential was measured after 2 hours' incubation with 2 mM H_2O_2. Data are expressed as means ± SEM from 7–8 experiments, each representing an individual animal (22 months old). ^{+++}P <0.01 vs. placebo control; $^{*}P$ <0.05 vs. H_2O_2 damage of brain cells from placebo-treated animals, Student's unpaired *t* test.

reduced significantly the increase in caspase 9 activity (FIG. 1C). In agreement with these results, EGb 761 also reduced caspase 3 activity in PC12 cells after exposure to staurosporine.[49]

EGB 761 PROTECTS DISSOCIATED BRAIN CELLS AND ISOLATED BRAIN MITOCHONDRIA OF MICE AGAINST OXIDATIVE STRESS

The efficacy of EGb 761 in protecting dissociated brain cells and isolated brain mitochondria from mice against oxidative stress was tested. Therefore, dissociated brain cells were preincubated and damaged for 1 hour with hydrogen peroxide (H_2O_2); EGb 761 was then added *in vitro* for 6 hours and the mitochondrial membrane potential was determined. EGb 761 significantly improved mitochondrial membrane potential changes after an H_2O_2 insult at a concentration of 0.1 mg/ml (FIG. 2A). To investigate the protective effects under *in vivo* conditions, mice were

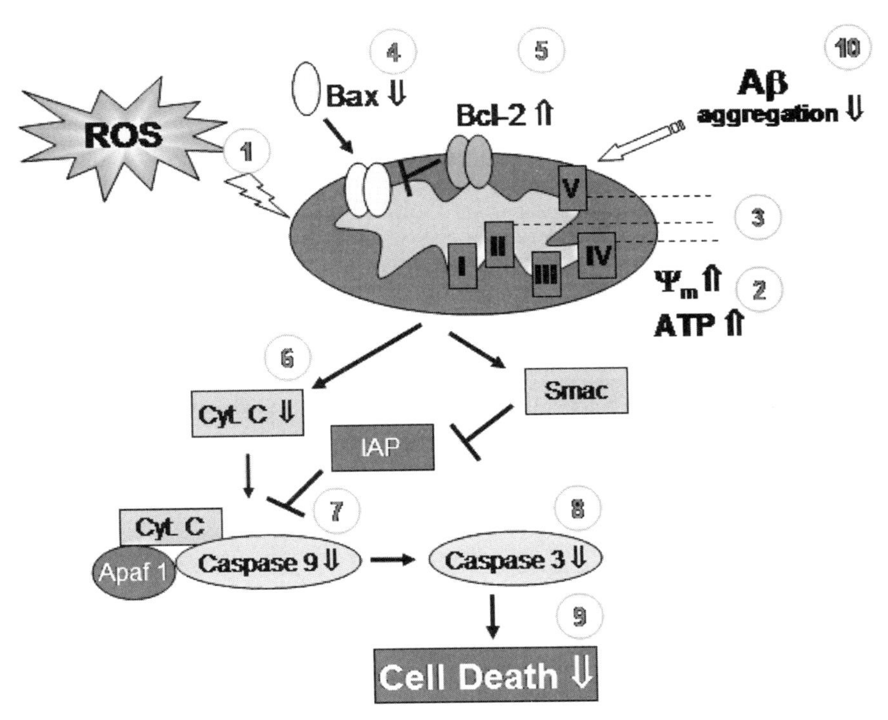

FIGURE 3. Proposed mitochondria-protecting and antiapoptotic properties of EGb 761 within mitochondrion-dependent cell death pathway: (**1**) free radical scavenging activity, (**2**) stabilization of mitochondrial membrane potential (ψ_m) and improvement of energy metabolism, (**3**) protection of mitochondrial respiratory chain complexes, particularly complexes II, IV, and V, (**4**) downregulation of proapoptotic Bax protein, (**5**) upregulation of antiapoptotic Bcl-2 protein, (**6**) inhibition of cytochrome *c* (Cyt. C) release, (**7**) reduction of caspase 9 and (**8**) caspase 3 activity after oxidative stress, (**9**) reduction of apoptotic cell death, and (**10**) inhibition of Aβ aggregation and reduction of Aβ toxicity.

treated daily for 2 weeks with 100 mg/kg EGb 761 or placebo. Dissociated brain cells and isolated brain mitochondria were injured for 2 hours with H_2O_2, and mitochondrial membrane potential and ATP levels were determined. Of note, under *in vivo* conditions, EGb 761 also improved the ATP reduction in dissociated brain cells (FIG. 2B) and the mitochondrial membrane potential decrease in isolated brain mitochondria (FIG. 2C). Therefore, we can conclude that mitochondria from mice chronically treated with EGb 761 *in vivo* appear to be protected against oxidative stress.

CONCLUSIONS

Taken together, our findings and those of others clearly show stabilization and protection of mitochondrial function as a specific and very sensitive property of EGb 761 at therapeutically relevant concentrations. EGb 761 has multiple effects on mitochondrial function and on the apoptotic pathway: stabilization of mitochondrial membrane potential and mitochondrial respiratory chain function, improvement of energy metabolism, upregulation of antiapoptotic Bcl-2 protein and downregulation of proapoptotic Bax protein, inhibition of cytochrome *c* release, reduction of caspase 9 and caspase 3 activity after oxidative stress, and reduction of apoptotic cell death (FIG. 3).

Convincing evidence indicates that mitochondrial dysfunction plays a crucial role in aging and in the pathogenesis of Alzheimer's disease. Therefore, the mitochondria-protecting mechanisms of EGb 761 can explain many of the until now rather unrelated observations of EGb 761 in brain aging and neurodegeneration.

ACKNOWLEDGMENTS

We thank Dr. W. Schwabe Arzneimittel (Karlsruhe, Germany) for support.

REFERENCES

1. COOPER, J.M., V.M. MANN & A.H. SCHAPIRA. 1992. Analyses of mitochondrial respiratory chain function and mitochondrial DNA deletion in human skeletal muscle: effect of ageing. J. Neurol. Sci. **113:** 91–98.
2. OJAIMI, J., C.L. MASTERS, K. OPESKIN, *et al.* 1999. Mitochondrial respiratory chain activity in the human brain as a function of age. Mech. Ageing Dev. **111:** 39–47.
3. HARMAN, D. 1972. The biologic clock: the mitochondria? J. Am. Geriatr. Soc. **20:** 145–147.
4. MELOV, S. 2002. Therapeutics against mitochondrial oxidative stress in animal models of aging. Ann. N.Y. Acad. Sci. **959:** 330–340.
5. LENAZ, G. 1998. Role of mitochondria in oxidative stress and ageing. Biochim. Biophys. Acta **1366:** 53–67.
6. CADENAS, E. & K.J. DAVIES. 2000. Mitochondrial free radical generation, oxidative stress, and aging. Free Radic. Biol. Med. **29:** 222–230.
7. NICHOLLS, D.G. 2002. Mitochondrial function and dysfunction in the cell: its relevance to aging and aging-related disease. Int. J. Biochem. Cell Biol. **34:** 1372–1381.
8. LIN, M.T., D.K. SIMON, C.H. AHN, *et al.* 2002. High aggregate burden of somatic mtDNA point mutations in aging and Alzheimer's disease brain. Hum. Mol. Genet. **11:** 133–145.

9. LINNANE, A.W., S. MARZUKI, T. OZAWA & M. TANAKA. 1989. Mitochondrial DNA mutations as an important contributor to ageing and degenerative diseases. Lancet **1:** 642–645.
10. WALLACE, D.C. 1992. Mitochondrial genetics: a paradigm for aging and degenerative diseases? Science **256:** 628–632.
11. GADALETA, M.N., A. CORMIO, V. PESCE, *et al.* 1998. Aging and mitochondria. Biochimie **80:** 863–870.
12. SASTRE, J., A. MILLAN, D.L.A. GARCIA, *et al.* 1998. A Ginkgo biloba extract (EGb 761) prevents mitochondrial aging by protecting against oxidative stress. Free Radic. Biol. Med. **24:** 298–304.
13. ECKERT, A., U. KEIL, C.A. MARQUES, *et al.* 2003. Mitochondrial dysfunction, apoptotic cell death, and Alzheimer's disease. Biochem. Pharmacol. **66:** 1627–1634.
14. HIRAI, K., G. ALIEV, A. NUNOMURA, *et al.* 2001. Mitochondrial abnormalities in Alzheimer's disease. J. Neurosci. **21:** 3017–3023.
15. SWERDLOW, R.H. & S.J. KISH. 2002. Mitochondria in Alzheimer's disease. Int. Rev. Neurobiol. **53:** 341–385.
16. BEAL, M.F. 1996. Mitochondria, free radicals, and neurodegeneration. Curr. Opin. Neurobiol. **6:** 661–666.
17. GIBSON, G.E., K.F. SHEU & J.P. BLASS. 1998. Abnormalities of mitochondrial enzymes in Alzheimer disease. J. Neural Transm. **105:** 855–870.
18. KISH, S.J., C. BERGERON, A. RAJPUT, *et al.* 1992. Brain cytochrome oxidase in Alzheimer's disease. J. Neurochem. **59:** 776–779.
19. MAURER, I., S. ZIERZ & H.J. MOLLER. 2000. A selective defect of cytochrome *c* oxidase is present in brain of Alzheimer disease patients. Neurobiol. Aging **21:** 455–462.
20. PARKER, W.D., JR., J. PARKS, C.M. FILLEY & B.K. KLEINSCHMIDT-DEMASTERS. 1994. Electron transport chain defects in Alzheimer's disease brain. Neurology **44:** 1090–1096.
21. CHANDRASEKARAN, K., T. GIORDANO, D.R. BRADY, *et al.* 1994. Impairment in mitochondrial cytochrome oxidase gene expression in Alzheimer disease. Brain Res. Mol. Brain Res. **24:** 336–340.
22. BEDETTI, C.D. 1985. Immunocytochemical demonstration of cytochrome *c* oxidase with an immunoperoxidase method: a specific stain for mitochondria in formalin-fixed and paraffin-embedded human tissues. J. Histochem. Cytochem. **33:** 446–452.
23. BOSETTI, F., F. BRIZZI, S. BAROGI, *et al.* 2002. Cytochrome *c* oxidase and mitochondrial F(1)F(0)-ATPase (ATP synthase) activities in platelets and brain from patients with Alzheimer's disease. Neurobiol. Aging **23:** 371–376.
24. CARDOSO, S.M., M.T. PROENCA, S. SANTOS, *et al.* 2004. Cytochrome *c* oxidase is decreased in Alzheimer's disease platelets. Neurobiol. Aging **25:** 105–110.
25. KEIL, U., A. BONERT, C.A. MARQUES, *et al.* 2004. Amyloid beta-induced changes in nitric oxide production and mitochondrial activity lead to apoptosis. J. Biol. Chem. **279:** 50310–50320.
26. MARQUES, C.A., U. KEIL, A. BONERT, *et al.* 2003. Neurotoxic mechanisms caused by the Alzheimer's disease-linked Swedish amyloid precursor protein mutation: oxidative stress, caspases, and the JNK pathway. J. Biol. Chem. **278:** 28294–28302.
27. ANANDATHEERTHAVARADA, H.K., G. BISWAS, M.A. ROBIN & N.G. AVADHANI. 2003. Mitochondrial targeting and a novel transmembrane arrest of Alzheimer's amyloid precursor protein impairs mitochondrial function in neuronal cells. J. Cell Biol. **161:** 41–54.
28. PEREIRA, C., M.S. SANTOS & C. OLIVEIRA. 1998. Mitochondrial function impairment induced by amyloid beta-peptide on PC12 cells. Neuroreport **9:** 1749–1755.
29. CANEVARI, L., J.B. CLARK & T.E. BATES. 1999. Beta-amyloid fragment 25-35 selectively decreases complex IV activity in isolated mitochondria. FEBS Lett. **457:** 131–134.
30. CASLEY, C.S., L. CANEVARI, J.M. LAND, *et al.* 2002. Beta-amyloid inhibits integrated mitochondrial respiration and key enzyme activities. J. Neurochem. **80:** 91–100.
31. LUSTBADER, J.W., M. CIRILLI, C. LIN, *et al.* 2004. ABAD directly links Abeta to mitochondrial toxicity in Alzheimer's disease. Science **304:** 448–452.
32. MIX, J.A. & W.D. CREWS, JR. 2000. An examination of the efficacy of Ginkgo biloba extract EGb761 on the neuropsychologic functioning of cognitively intact older adults. J. Altern. Complement Med. **6:** 219–229.

33. RAI, G.S., C. SHOVLIN & K.A. WESNES. 1991. A double-blind, placebo controlled study of Ginkgo biloba extract ('tanakan') in elderly outpatients with mild to moderate memory impairment. Curr. Med. Res. Opin. **12:** 350–355.
34. LE BARS, P.L., M. KIESER & K.Z. ITIL. 2000. A 26-week analysis of a double-blind, placebo-controlled trial of the ginkgo biloba extract EGb 761 in dementia. Dement. Geriatr. Cogn. Disord. **11:** 230–237.
35. LE BARS, P.L., M.M. KATZ, N. BERMAN, *et al.* 1997. A placebo-controlled, double-blind, randomized trial of an extract of Ginkgo biloba for dementia. North American EGb Study Group. JAMA **278:** 1327–1332.
36. KANOWSKI, S., W.M. HERRMANN, K. STEPHAN, *et al.* 1996. Proof of efficacy of the ginkgo biloba special extract EGb 761 in outpatients suffering from mild to moderate primary degenerative dementia of the Alzheimer type or multi-infarct dementia. Pharmacopsychiatry **29:** 47–56.
37. DEFEUDIS, F.V. & K. DRIEU. 2000. Ginkgo biloba extract (EGb 761) and CNS functions: basic studies and clinical applications. Curr. Drug Targets **1:** 25–58.
38. SMITH, J.V. & Y. LUO. 2003. Elevation of oxidative free radicals in Alzheimer's disease models can be attenuated by Ginkgo biloba extract EGb 761. J. Alzheimers Dis. **5:** 287–300.
39. SCHINDOWSKI, K., S. LEUTNER, S. KRESSMANN, *et al.* 2001. Age-related increase of oxidative stress-induced apoptosis in mice prevention by Ginkgo biloba extract (EGb761). J. Neural Transm. **108:** 969–978.
40. BRIDI, R., F.P. CROSSETTI, V.M. STEFFEN & A.T. HENRIQUES. 2001. The antioxidant activity of standardized extract of Ginkgo biloba (EGb 761) in rats. Phytother. Res. **15:** 449–451.
41. SASAKI, K., S. HATTA, K. WADA, *et al.* 2002. Effects of extract of Ginkgo biloba leaves and its constituents on carcinogen-metabolizing enzyme activities and glutathione levels in mouse liver. Life Sci. **70:** 1657–1667.
42. RIMBACH, G., K. GOHIL, S. MATSUGO, *et al.* 2001. Induction of glutathione synthesis in human keratinocytes by Ginkgo biloba extract (EGb761). Biofactors **15:** 39–52.
43. GOHIL, K., R.K. MOY, S. FARZIN, *et al.* 2001. mRNA expression profile of a human cancer cell line in response to ginkgo biloba extract: induction of antioxidant response and the Golgi system. Free Radic. Res. **33:** 831–849.
44. SASTRE, J., A. LLORET, C. BORRAS, *et al.* 2002. Ginkgo biloba extract EGb 761 protects against mitochondrial aging in the brain and in the liver. Cell Mol. Biol. **48:** 685–692.
45. JANSSENS, D., C. MICHIELS, E. DELAIVE, *et al.* 1995. Protection of hypoxia-induced ATP decrease in endothelial cells by ginkgo biloba extract and bilobalide. Biochem. Pharmacol. **50:** 991–999.
46. JANSSENS, D., E. DELAIVE, J. REMACLE & C. MICHIELS. 2000. Protection by bilobalide of the ischaemia-induced alterations of the mitochondrial respiratory activity. Fundam. Clin. Pharmacol. **14:** 193–201.
47. TENDI, E.A., F. BOSETTI, S.F. DASGUPTA, *et al.* 2002. Ginkgo biloba extracts EGb 761 and bilobalide increase NADH dehydrogenase mRNA level and mitochondrial respiratory control ratio in PC12 cells. Neurochem. Res. **27:** 319–323.
48. CHANDRASEKARAN, K., L.I. LIU, K. HATANPAA, *et al.* 1998. Stimulation of mitochondrial gene expression by bilobalide, a component of *Ginkgo biloba* extract (EGb 761). Adv. Ginkgo Biloba Extract Res. **7:** 121–128.
49. SMITH, J.V., A.J. BURDICK, P. GOLIK, *et al.* 2002. Anti-apoptotic properties of Ginkgo biloba extract EGb 761 in differentiated PC12 cells. Cell Mol. Biol. **48:** 699–707.
50. ZHOU, L.J. & X.Z. ZHU. 2000. Reactive oxygen species-induced apoptosis in PC12 cells and protective effect of bilobalide. J. Pharmacol. Exp. Ther. **293:** 982–988.
51. MULLER, W.E., U. KEIL, I. SCHERPING & A. ECKERT. 2003. Stabilization of mitochondrial membrane potential by Ginkgo biloba extract EGb 761. Soc. Neurosci. **29:** 240.
52. MULLER, W.E., U. KEIL, I. SCHERPING, *et al.* 2004. Stabilisation of mitochondrial membrane potential by Ginkgo biloba EGb 761. Focus on Alternative and Complementary Therapies **9** (Suppl. 1)**:** 34–35.
53. MASSIEU, L., J. MORAN & Y. CHRISTEN. 2004. Effect of Ginkgo biloba (EGb 761) on staurosporine-induced neuronal death and caspase activity in cortical cultured neurons. Brain Res. **1002:** 76–85.

54. Luo, Y., J.V. Smith, V. Paramasivam, *et al.* 2002. Inhibition of amyloid-beta aggregation and caspase-3 activation by the Ginkgo biloba extract EGb761. Proc. Natl. Acad. Sci. USA **99:** 12197–12202.
55. Eckert, A., U. Keil, S. Kressmann, *et al.* 2003. Effects of EGb 761 Ginkgo biloba extract on mitochondrial function and oxidative stress. Pharmacopsychiatry **36** (Suppl. 1)**:** S15–S23.
56. Green, D.R. & J.C. Reed. 1998. Mitochondria and apoptosis. Science **281:** 1309–1312.
57. Baracca, A., G. Sgarbi, G. Solaini & G. Lenaz. 2003. Rhodamine 123 as a probe of mitochondrial membrane potential: evaluation of proton flux through F(0) during ATP synthesis. Biochim. Biophys. Acta **1606:** 137–146.
58. Collins, T.J., M.J. Berridge, P. Lipp & M.D. Bootman. 2002. Mitochondria are morphologically and functionally heterogeneous within cells. EMBO J. **21:** 1616–1627.
59. Crouch, S.P., R. Kozlowski, K.J. Slater & J. Fletcher. 1993. The use of ATP bioluminescence as a measure of cell proliferation and cytotoxicity. J. Immunol. Methods **160:** 81–88.
60. Rodrigo, J., P. Fernandez-Vizarra, S. Castro-Blanco, *et al.* 2004. Nitric oxide in the cerebral cortex of amyloid-precursor protein (SW) Tg2576 transgenic mice. Neuroscience **128:** 73–89.
61. Brown, G.C. & V. Borutaite. 2002. Nitric oxide inhibition of mitochondrial respiration and its role in cell death. Free Radic. Biol. Med. **33:** 1440–1450.
62. Song, W., H.J. Guan, X.Z. Zhu, *et al.* 2000. Protective effect of bilobalide against nitric oxide-induced neurotoxicity in PC12 cells. Acta Pharmacol. Sin. **21:** 415–420.
63. Krohn, A.J., T. Wahlbrink, & J.H. Prehn. 1999. Mitochondrial depolarization is not required for neuronal apoptosis. J. Neurosci. **19:** 7394–7404.
64. Kroemer, G. & J.C. Reed. 2000. Mitochondrial control of cell death. Nat. Med. **6:** 513–519.

Low Level Laser Therapy (LLLT) as an Effective Therapeutic Modality for Delayed Wound Healing

D. HAWKINS, N. HOURELD, AND H. ABRAHAMSE

Faculty of Health, University of Johannesburg, Johannesburg, 2028, South Africa

ABSTRACT: Low level laser therapy (LLLT) is a form of phototherapy that involves the application of low power monochromatic and coherent light to injuries and lesions. It has been used successfully to induce wound healing in nonhealing defects.[1] Other wounds treated with lasers include burns, amputation injuries, skin grafts, infected wounds, and trapping injuries.[2] The unique properties of lasers create an enormous potential for specific therapy of skin diseases. As with any new device, the most efficacious and appropriate use requires an understanding of the mechanisms of light interaction with tissue as well as the properties of the laser itself.[3]

KEYWORDS: low level laser therapy (LLLT); wound healing; diabetes; stimulation

INTRODUCTION

Low level laser therapy (LLLT) or more simply known as soft laser therapy, is a dramatic therapy that has become progressively more popular in the management of a wide variety of medical conditions, such as soft tissue injuries (including sports injuries), low back pain, arthritis, and skin traumas.[2] Unlike the higher powered lasers employed in medicine, these low level lasers do not deliver enough power to damage tissue, but they do deliver enough energy to stimulate a response from the body tissues to initiate healing. Laser radiation has a wavelength-dependent capability to alter cellular behavior in the absence of significant heating. Light radiation must be absorbed to provide a biological response. The visible red and infrared portions of the spectrum have been shown to have highly absorbent and unique therapeutic effects in living tissues.

BASIC PRINCIPLES OF LLLT

Low level laser therapy applications include: acceleration of wound healing, enhanced remodeling and repair of bone, restoration of normal neural function

Address for correspondence: Heidi Abrahamse (PhD), Senior Research Fellow, University of Johannesburg, P.O. Box 17011, Doornfontein, 2028, Gauteng, South Africa. Voice: +27 11 406-8145; fax: +27 11 406-8202.
heidi@twr.ac.za

Ann. N.Y. Acad. Sci. 1056: 486–493 (2005).
doi: 10.1196/annals.1352.040

following injury, pain attenuation, and modulation of the immune system.[4] Laser therapy increases both the rate and the quality of healing, and studies show that as the healing rate increases, bacterial cultures decrease, suggesting a bioinhibitory effect upon wound infection.[2] Nussbaum *et al.*[5] analyzed the interactions between wavelength and bacterial growth of *Pseudomonas aeruginosa, Escherichia coli,* and *Staphylococcus aureus* and reported that irradiation with 1–20 J/cm^2 at a wavelength of 630 nm appeared to be commonly associated with bacterial growth inhibition, which is of considerable importance for wound healing.[5] Reports of LLLT applied to soft tissues *in vitro* and *in vivo* suggest stimulation of specific metabolic processes in healing wounds. Whereas low doses of LLLT are stimulatory, high doses of laser radiation are suppressive.[5]

Low level laser therapy irradiation includes wavelengths of between 500 and 1100 nm and typically involves the delivery of 1–4 J/cm^2 to treatment sites with lasers having output powers between 10 and 90 mW. Low-intensity radiation can inhibit as well as stimulate cellular activity as light irradiation appears to upregulate cellular metabolism and proliferation. Visible red light has long been known to promote healing in the body's cells and tissues.[6] Irradiating mitochondria with red light causes them to produce cytochromes, which increases their efficiency, and research has shown that fibroblasts and muscle cells grow five times faster when treated with red light. By decreasing the healing time of open wounds, laser therapy significantly reduces the risk of infection and other complications. It also produces a stronger repair with less unattractive scar tissue.

The effects of LLLT are photochemical, not thermal. To produce an effect, the photons must be absorbed, and different substances absorb light of different wavelengths like the cells of injured skin are more sensitive than those of intact tissue. Once the target cells have absorbed the photons, a cascade of biochemical events occurs whose ultimate result is accelerated wound healing.[8] Laser therapy is thought to work through a variety of mechanisms:

(1) Photons from a laser probe are absorbed into the mitochondria and cell membranes of the target cells.
(2) After a cell absorbs photons the energy is incorporated into the molecule to increase chemical energy, activate or deactivate enzymes, or alter physical or chemical properties of main macromolecules.[9] Photonic energy is converted to chemical energy within the cell, in the form of denosine triphosphate (ATP), which leads to normalization of cell function, pain relief, and healing.
(3) Single oxygen molecules build up, which influences the formation of ATP, which in turn leads to replication of DNA.
(4) Increased DNA leads to increased neurotransmission.
(5) A cascade of metabolic effects results in various physiological changes, which results in improved tissue repair, faster resolution of the inflammatory response, and reduction in pain.

Infra red laser light at approximately 632 nm appears to be the most effective and stimulatory frequency of laser at a cellular level with a skin penetration depth of 0.5–1 cm. A typical example of a laser used in LLLT is the helium-neon (He-Ne) laser, which can penetrate as deep as 0.5 mm into freshly excised human skin, which is regarded as sufficient for the induction of wound healing because most of the relevant target cells of low level laser irradiation are located within the epidermis and upper dermis.[5] Studies have found that laser irradiation stimulates fibroblast growth *in vitro* and also facilitates ulcer healing in the clinical situation.[7]

Karu[8] found that infrared laser (620 nm) stimulated the bacterial cell growth rate, DNA and RNA synthesis rates, enzyme activity, and cAMP levels. It is postulated that the respiratory chain is stimulated, activating ATP turnover, increasing H^+, and ultimately triggering an increase in cell proliferation. The stimulating effects of light appear to occur in "sluggish" cell cultures or during decreased activity such as trophic ulcers and indolent wounds, when low tissue oxygen concentration and pH inhibit cell growth. Conversely, where maximum regeneration is occurring naturally, laser did not appear to enhance the process.[8]

LOW LEVEL LASER THERAPY FOR WOUND HEALING

LLLT, when used appropriately, can stimulate the healing of injured tissues such as those of the dermis.[9] Investigations into the mechanisms involved have shown that many of the types of cells whose interaction results in dermal repair can be affected in a therapeutically advantageous manner by treatment with LLLT both *in vitro* and *in vivo*. Mast cells and macrophages can be stimulated to release growth factors and other substances, whereas the proliferation of fibroblasts, endothelial cells, and keratinocytes maintained in adverse conditions can also be stimulated. The development of granulation tissue is mainly controlled by growth factors released from macrophages.[9]

Wound healing involves the following phases:

- Hemostasis: platelets, endothelial cells, fibrin, and fibronectin act through growth factors and cytokines.
- Inflammation: blood clots form, bacteria are attacked, and there is an orderly recruitment of key cells into the wound site.
- Proliferation: cells necessary for wound closure multiply at the wound site to make new tissue and blood vessels.
- Remodeling: the wound is healed and the initial scar tissue is restructured.

Any device that can accelerate any of these processes (transition from hematoma to fibroplasias, development of new blood vessels, production of collagen, or even the remodelling process) could accelerate the healing process of wounds.[9] Early laser studies were confined to *in vitro* studies because little was known about the side effects of laser irradiation.[10]

Wound healing studies have focused on several types of cells including fibroblasts, lymphocytes, monocytes, macrophages, epithelial cells, and endothelial cells. The wide diversity of experimental protocols and parameters such as cell line, dose, waveform, treatment time, penetration distance, treatment area, and treatment frequency make comparison of these studies difficult. Literature indicates that laser photobioactivation accelerates inflammation, modulates the level of prostaglandin, enhances the action of macrophages, promotes fibroblast proliferation, facilitates collagen synthesis, fosters immunity, and even accelerates the healing process.[9] Using the He-Ne laser, Van Breugel and Bar[11] concluded that laser exposure time and power density determine the effect of the laser. Dependent on exposure time and power density, the laser can either stimulate or inhibit human fibroblasts *in vitro.*[11]

In the clinical situation, LLLT is an accepted, efficient, noninvasive, and painless method of treating edema, inflammation, and pain and it is used to increase circulation and promote wound healing.[12] Wound healing experiments show acceleration of healing, but these findings are often concentrated in the early phases of the healing process.[12] The effects of LLLT on wound healing are often attributed to increased cell proliferation. However, the true effect of LLLT on cell proliferation is still controversial, because of conflicting reports on the effects of visible laser light on cells in culture.

The magnitude of the laser biostimulation effect depends on the physiological state of the cell at the moment of irradiation. This explains why the effect is not always detectable as well as the variability of the results reported in the literature. In medicine, laser treatment appears to work in cases of severe damage or stress (wounding), whereas the effect of light on normally regenerating wounds may also be insignificant. Karu[8] stated that light stimulates cell proliferation if the cells are growing poorly at the time of irradiation. Thus, if a cell is fully functional, there is nothing for laser irradiation to stimulate, and therefore no therapeutic benefit will be observed.[8]

LLLT may induce positive side effects that are common also after other stimulation therapies (acupuncture). In patients with difficult or longstanding problems, LLLT can be combined very usefully with other forms of therapy (physiotherapy, acupuncture, and manipulation), relaxation therapy (self-hypnosis and meditation), medication (pharmacological, herbal, or homeopathic), and psychiatric or psychological counseling.[16] There are no absolute contraindications for LLLT; however, it is always better to be cautious when treating patients in high-risk categories. LLLT should be avoided or given with special caution in the following cases: patients with pacemakers, patients who are pregnant, patients with cancer if there is any doubt of a recurrence of metastases, and patients with labile epilepsy. It is better to avoid LLLT over the thyroid gland, ovaries, and testicles. Although LLLT has not induced cancer in any of the reported studies, the precise reactions of existing tumors to LLLT are unknown.[16] Pessoa *et al.*[17] conducted a study to investigate the effect of LLLT on the wound healing process treated with steroid and concluded that LLLT accelerated healing, caused by the steroid, acting as a biostimulative coadjuvant agent, balancing the undesirable effects of cortisone on the tissue healing process.[17] Manuskiatti and Fitzpatrick[18] conducted a study to compare the clinical response of keloidal and hypertrophic scars after treatment with interlesional corticosteroid alone or combined with 5-fluorouracil (5-FU), 5-FU alone, and the 585-nm pulsed-dye laser (PDL). There was significant improvement in keloidal and hypertrophic scars after treatment in which scar texture and erythema responded better to PDL and the long-term adverse sequelae (hypopigmentation, telangiectasia, and skin atrophy) were demonstrated in corticosteroid therapy but not in PDL.[18]

LOW LEVEL LASER THERAPY FOR DIABETES

LLLT effectively promotes wound healing without causing burn to adjacent tissue. The operative principle, known as photobiomodulation, is particularly useful in treating decubitus ulcers, typical of persons with diabetes and frail elderly patients who spend long hours in bed. Some wounds, such as decubitus ulcers, heal slowly or not at all in persons with diabetes or the frail elderly.[13] Diabetes is a chronic metabolic disorder in which utilization of carbohydrate is impaired and that of lipid and protein enhanced. It is caused by an absolute or relative deficiency of insulin. Long-term complications include neuropathy, retinopathy, generalized degenerative changes in large and small blood vessels, and increased susceptibility to infection. The consequences of leaving diabetes untreated are dialysis, heart failure, paralysis, loss of limbs, and early death. Use of a low level laser can start the healing process. Even if a wound such as a leg ulcer will not heal in all cases, pain relief is usually immediate and is the most important benefit. Stimulation of the circulation may be the primary reason that pain relief occurs after the application of LLLT to chronic wounds.[13] Laser treatment increases blood flow and raises local temperature, and no evidence has been found that laser therapy could aggravate diabetic symptoms.

In general terms, in the treatment of a chronic ulcer, a higher dose such as 3–4 J/cm^2 will be used on points along the periphery of the wound followed by a lower dose of 0.5 J/cm^2 over the open wound. The open wound needs a lower dosage than the skin-covered periphery as the laser light is not reflected or scattered but rather absorbed by the skin in the unprotected wound because it hits the uncovered cells directly.[13] Laser therapy should be recommended as an additional treatment modality for diabetic foot problems according to Kleinman *et al.*[14] However, not all results have been positive, and some studies do not support or refute the use of laser therapy as an effective therapeutic modality for diabetic ulcers (TABLE 1).

Diabetic patients have a 22-fold higher risk of nontraumatic foot amputation compared with the nondiabetic population,[15] and according to the World Health Organization, the number of patients with diabetes mellitus will double to 250 million by the year 2050. Attempts have been made to use helium neon, CO_2, and KTP lasers to encourage wound healing in diabetics. Results were inconclusive, so that further research is needed to assess the effectiveness of biostimulation for diabetic wound healing. Stadler *et al.*[16] reported that low-power laser irradiation at 830 nm significantly enhanced cutaneous wound tensile strength in a murine diabetic model, whereas Schindl *et al.*[6] reported a beneficial effect on a recalcitrant diabetic neuropathic foot ulcer. Yu *et al.*[17] used diabetic mice to compare the effect of basic fibroblast growth factor (bFGF), laser irradiation at 660 nm, and a combination of growth factor and laser therapy. Wound closure was significantly enhanced with light therapy alone or most effectively in combination with topical application of bFGF.[17] LLLT is effective in enhancing wound contraction of partial-thickness abrasions. It also facilitates wound contraction of untreated wounds, suggesting an indirect effect on surrounding tissues; however, the exact mechanism by which LLLT facilitates wound healing is largely unknown, and further investigation of the mechanism of LLLT in primary wound healing is warranted.[16]

CONCLUSION

Early laser studies were confined to *in vitro* studies because little was known about the side effects of laser irradiation.[18] More studies have therefore been performed in the area of wound healing than in any other. The majority of studies have

TABLE 1. Studies of low level laser therapy (LLLT) on open diabetic wounds

Study	Patient characteristics	Therapy	Outcomes	Comments
Shuttleworth *et al.*[20] Prospective comparative study of laser and conventional wound therapy (nonrandomized)	n = 14; age, 76.3 yr **Control group:** n = 8; 1 diabetic patient **Laser group:** n = 6; 2 diabetic patients (3 patients received both laser and conventional treatment) Leg ulcers caused by a variety of conditions Study period: 15 wk/patient	**Control group:** conventional wound care and dressings in accordance with local wound management policy **Laser group:** each laser therapy session was a maximum of 4 min using HeNe at 632.8 nm and infrared laser at 904 nm. 4 J/cm^2; patients received treatment and dressings twice a week	**Control group:** all patients showed improvement **Laser group:** 3 patients improved or healed and 3 deteriorated	Results of this study neither support nor refute the use of LLLT in wound management; further studies should incorporate a larger sample size and actively control or eliminate variables such as size of wounds
Landau *et al.*[21] Noncontrolled clinical series	n = 50 (patients had chronic diabetic foot ulcers and had not responded to conventional therapy) Age: 59 yr (±11 yr) Diabetes: Type 1: n = 14 Type 2: n = 35 Ulcer duration: 9 ± 6.6 mo Range: 2–70 mo	Hyperbaric oxygen (HBO) HBO group: 15 patients HBO and laser group: 35 patients Unilaser Scan 60, two sources of laser: HeNe at 632.8 nm and infrared laser at 904 nm 4 J/cm^2 Treatment HBO: 2–5 h, laser 20 min, 2–3×/ wk No of treatments: 25 ± 13 Range: 7– 70 Duration: 3 ± 1.8 mo Range: 1–8 mo	All patients continued medication, and antibiotic treatment was administered according to the sensitivity of the micro-organism No significant difference between groups	Topical hyperbaric oxygen alone or combined with a low level energy laser for treatment of patients with chronic diabetic foot ulcers were valuable adjuvants to conventional therapy
Gupta *et al.*[22] Double-blind, placebo-controlled study	n = 9 (12 venous ulcers) **Control group**: Age, 64.7 (±9.4 yr) Ulcer duration: 36.0 ± 21.6 wk **Intervention group**: Age: 61.0 (±7.8 yr) Ulcer duration: 105.8 ± 36.0 wk	**Control group:** placebo treatment received sham therapy from identical appearing light sources, from same delivery system **Intervention group:** 2 monochromatic optical sources: 1 (red-light) source 660 nm used over ulcer for 180 s; 1 (infra-red) source 990 nm used on periphery for ulcer for 30 s; treatments were 3/wk for 10 wk	Unhealed ulcers in control group: 87.6% Decrease in ulcer area compared to baseline: 14.7 mm^2 Unhealed ulcers in intervention group: 24.4% Decrease in ulcer area compared to baseline: 193.0 mm^2	Low level laser therapy was effective modality for treatment of venous leg ulcers No adverse effects

shown beneficial effects, and most of the work has been performed using the helium neon (He-Ne) 632.8 nm laser.[10] Research studies on the effects of low energy laser irradiation on biologic function are growing in number and scope. Although many experiments show alleviation of pain, the quality of the investigations, the number of subjects, and the varied techniques frequently preclude statistical verification. Currently, no universally accepted theory has explained the mechanism of either "laser analgesia" or "laser biostimulation." Modification of current lasers and innovative advances with biomedical laser instrumentation may eventually allow the physician to match optimally the laser and the treatment procedure with the lesion.[3] Low level laser therapy is still very controversial, and there are still studies that present conflicting results. However, as knowledge and techniques improve, a scientific explanation may provide an understanding of the cellular and molecular effects of LLLT.[19]

REFERENCES

1. Yuji, Y. & Y. Kunihiko. 2001. Cutaneous wound healing: an update. J. Dermatol. **28:** 521–534.
2. Nelson, J.S. 1993. Lasers: state of the art in dermatology. Dermatol. Clin. **11:** 15-26.
3. Walsh, L.J. 1997. The current status of low level laser therapy in dentistry. Part 1. Soft tissue applications. Aus. Dent. J. **42:** 247–254.
4. Cabrero, M.V., J.M.G. Failde & O.M. Mayordomo. 1985. Laser therapy as a regenerator for healing wound tisues. Int. Congr. Laser Med. Surg. June 27-28, 187–192.
5. Nussbaum, E.L., L. Lilge & T. Mazzulli. 2002. Effects of 630-, 660-, 810-, and 905-nm laser irradiation delivering radiant exposure of 1-50 J/cm2 on three species of bacteria *in vitro*. J. Clin. Laser Med. Surg. **20:** 325–333.
6. Schindl, A., M. Schindl, H. Pernerstorfer-Schon & L. Schindl. 2000. Low intensity laser therapy: a review. J. Invest. Med. **48:** 312–326.
7. Schindl A., M. Schindl, H. Pernerstorfer-Schon, *et al.* 1999. Diabetic neuropathic foot ulcer: successful treatment by low intensity laser therapy. Dermatology **198:** 314–317.
8. Nelson, J.S. 1993. Lasers: state of the art in dermatology. Dermatol Clin. **11:** 15-26.
9. Matic, M. *et al.* 2003. Low level laser irradiation and its effect on repair processes in the skin. Med. Pregl. **56:** 137–141.
10. Bosatra, M., A. Jucci & P. Olliano. 1984. *In vitro* fibroblast and dermis fibroblast activation by laser irradiation at low energy. Dermatologica **168:** 157–162.
11. Karu, T. 1989. Photobiological fundamentals of low power laser therapy. *Journal of* Quant. Electr. **23:** 1703–1717.
12. Dyson, M. 1991. Cellular and sub-cellular aspects of low level laser therapy (LLLT). *In* Progress in Laser Therapy: Selected papers from the October 1990 ILTA Congress. :221,222. Wiley & Sons, Inc. New York and Brisbane.
13. Enwemeka, C.S. 1988. Laser biostimulation of healing wounds: specific effects and mechanisms of action. J. Orthopaed. Sports Physiother. **9:** 333–338.
14. Van Breugel, H.H. & P.R. Bar. 1992. Power density and exposure time of he-Ne laser irradiation are more important than total energy dose in photo-biomodulation of human fibroblasts *in vitro*. Lasers Surg. Med. **12:** 528–537.
15. Basford, J.R. 1995. Low intensity laser therapy: still not an established clinical tool. Lasers Surg. Med. **16:** 331–342.
16. Pontinen, P.J. 1992. Guidelines for LLLT. Low Level Laser Therapy as a Medical Treatment Modality: A Manual for Physicians, Dentists, Physiotherapists and Veterinary Surgeons. P.J. Pontinen, Ed. :148. Art Urpo Ltd. Tampere, Finland.
17. Pessoa, E.S *et el.* 2004. A histologic assessment of the influence of low intensity laser therapy on wound healing in steroid treated animals. Photomed. Laser Surg. **22:** 199–204.

18. MANUSKIATTI, W. & R.E. FITZPATRICK. 2002. Treatment response of keloidal and hypertrophic sternotomy scars: comparison among intralesional corticosteroid, 5–fluorouracil, and 585nm flashlamp pumped pulsed dye laser treatments. Arch. Dermatol. **138:** 1149–1155.
19. TUNER, J. & L. HODE. 2002. Laser Therapy: Clinical Practice and Scientific Background. Prima Books AD. Grangesberg, Sweden. ISBN 91-631-1344-9.
20. KLEINMAN, K., S. SIMMER & Y. BRAKSMA. 1996. Low power laser therapy in patients with diabetic foot ulcers: early and long term outcome. Laser Ther. **8:** 205–208.
21. KARU, T.I. 1988. Molecular mechanisms of the therapeutic effects of low intensity laser radiation. Lasers Life Sci. **2:** 53–74.
22. STADLER, I., R.J. LANZAFAME, R. EVANS, *et al.* 2001. 830-nm irradiation increases the wound tensile strength in a diabetic murine model. Lasers Surg. Med. **28:** 220-226.
23. YU, W., J.O. NAIM & R.J. LANZAFAME. 1997. Effects of photostimulation on wound healing in diabetic mice. Laser Surg. Med. **20:** 56–63.
24. ENWEMEKA, C.S. 1988. Laser biostimulation of healing wounds: specific effects and mechanisms of action. J. Orthopaed. Sports Physiother. **9:** 333–338.
25. PINHEIRO, A.L.B., S.C. NASCIMENTO, A.L. VIEIRA, *et al.* 2002. Effects of Low level laser therapy on malignant cells: *in vitro* study. J. Clin. Laser Med. Surg. **20:** 23–26.
26. SHUTTLEWORTH, E. & K. BANFIELD. 1997. Wound care. light relief, low-power laser therapy. Nursing Times **93:** 74–78.
27. LANDAU, Z. 1998. Topical hyperbaric oxygen and low energy laser for the treatment of diabetic foot ulcers. Arch. Orthopaed. Trauma Surg. **117:** 156–158.
28. GUPTA, A.K., N. FILONENKO & N. SALANSKY. 1998. The use of low energy photon therapy (LEPT) in venous leg ulcers: a double-blind, placebo-controlled study. Derm. Surg. **24:** 1383–1386.

Closing Remarks

DR. MANTO TSHABALALA-MSIMANG[a]
Honorable Minister of Health of South Africa

Chairperson Prof. Girish Kotwal, local and international delegates, ladies and gentlemen:

It gives me great pleasure to address you at the end of what has been the most remarkable and unique 3-day conference here in Cape Town. It is unfortunate that I was not able to join you at the beginning of this conference. My duties as a member of the Inter-Ministerial Committee, tasked with the responsibility to coordinate the response to the Tsunami Disaster, that required that I attend a meeting on this matter in Geneva earlier this week.

It is fitting that this conference should be launched in the Wolfson Pavilion, the home of the Institute for Infectious Disease and Molecular Medicine, Cape Town, South Africa, which is to be inaugurated during the course of this year. This is an institute that seeks to use the techniques of molecular medicine to discover solutions to the diseases of the poor such as HIV and AIDS, TB, and malaria.

Judging by the program, the list of delegates, and the reports that I have received, it is clear that this has truly been a landmark conference in the field of natural products and molecular therapy. The conference has brought together a high caliber of international scientists from Japan in the East to Brazil in the Western hemisphere, from Iceland in the North to South Africa in the southern tip of Africa. The breadth of scientific disciplines represented has also been great, ranging from presentations by practitioners of alternative and complementary medicine to structural biologists studying computer-generated images of the molecular structure of natural medicines.

The determination by the organizing committee to bring together the disciplines of natural medicine and molecular medicine in pursuit of a greater understanding of natural products was truly inspirational. It is a strategy that has proven to be very successful.

We feel honored that South Africa was chosen to host this historic meeting, and we hope that this will not be the last interaction of this kind. Our country contains 25,000 species of plants, one-tenth of the world's known floral biodiversity, as well as diverse insects and animals and marine biodiversity. Our soils contain all kinds of minerals and trace elements, which provide the milieu for the growth of plants, and our habitat includes all the different climates of the world. Our peoples include racial groups from all over the world, and our San people have the oldest genes in the world. Therefore, this country is indeed an ideal place for further research of the issues discussed here.

[a]Delivered by Patricia Lambert.

Ann. N.Y. Acad. Sci. 1056: 494–496 (2005).
doi: 10.1196/annals.1352.043

The South African government has been at the forefront of reestablishing the traditional knowledge systems in the quest for human health. The Department of Health and other government structures have provided support for research into traditional medicine research as well as the promotion of other fields of inquiry in indigenous knowledge systems.

For us as South Africans, the study of indigenous knowledge systems is not simply a scientific endeavor. It provides an opportunity to reclaim our scientific and sociocultural heritage. This forum should expose the false dichotomy that had arisen between natural medicine and allopathic medicine, a division fostered by the need to make money from patented drugs by discrediting the use of natural products.

In order to make information on traditional medicines more widely known, we launched the National Reference Centre for African Traditional Medicines during the first African Traditional Medicine Day on August 31, 2003. This ceremony was graced by the presence of the Director-General of WHO, Dr. J. W. Lee, who recounted his own personal childhood experiences with the use of Korean traditional medicines.

We have also developed a legislative framework to encourage the development of natural products for human health as well as a support mechanism for African traditional medicine research has included a R6 million grant, channeled through the Medical Research Council of South Africa, for research into the safety, efficacy, and quality of traditional medicines used as immune boosters by persons living with HIV and AIDS. This field of research has shown promising results with completion late last year of Phase I trials in healthy human volunteers of one of the five candidate preparations.

We are also supporting research in our universities and science councils into the efficacy of many traditional medicines used for conditions such as tuberculosis, malaria, asthma, cancer, diabetes, anxiety and stress, and musculoskeletal disorders. In some instances, our scientists have extracted and characterized the active chemical moieties for possible development as novel drugs. In other cases, the approach has been to use the natural product in its native state and to study its safety and efficacy.

Our government also supports research into the role of nutrition in human health including the use of different foods, micronutrients, and vitamins in conditions such as HIV and AIDS, diabetes, and osteoporosis. Indeed, nutritional supplementation and traditional medicines are important components of our Comprehensive Plan for Management, Care and Treatment of HIV and AIDS. This approach has been vindicated by recent studies such as the Tanzania/Harvard study, which showed a 30% reduction in mortality of HIV-positive women in Tanzania with the use of multivitamins, and the Zambia/British MRC trial, which indicated a 50% reduction in mortality in children with AIDS when given cotrimoxazole.

Earlier, reports (Reuters) of a study conducted by Northwestern University in Chicago identified oleic acid, the main component of olive oil, as the main factor in protecting women who consume olive oil from developing breast cancer.

This conference has also unraveled the molecular mechanisms that explain the effectiveness of a whole host of everyday fruits, vegetables, and spices in promoting health and reducing the impact of diseases. This conference has highlighted the importance of the Mediterranean diet in heart health, particularly the centrality of monounsaturated fats such as olive oil. It has helped explain the observation that such a diet containing olive oil has been shown to reduce mortality after a heart attack by 70% in the Lyon Heart Study, a figure twice that achieved by statin drugs. It reduced death from heart disease by 30% in initially healthy Greek men consuming a Mediterranean diet.

The conference has also explored novel methods to prevent HIV transmission such as the use of *Punica granatum* (pomegranate) juice as an HIV-1 entry inhibitor and a candidate microbicide.

We look forward to the exciting developments ahead in this quest to use the methodologies of molecular medicine in the study of natural products. We are eagerly awaiting the outcomes of your work, which should assist us in better understanding the functioning of our own bodies, minds, and spirits and increasingly in understanding how we interact with our environment. The application of such knowledge should result in the promotion of health, prevention of sickness, and better management and treatment of many of the diseases that affect humankind.

Index of Contributors